Bleher's Discus

Volume 1

HEIKO BLEHER

*This book is dedicated
to the memory of three people
who contributed significantly to the spread
of our wonderful educational hobby
and to the worldwide interest
in discus:*

*Amanda Flora Hilda Bleher, née Kiel
Ferdinand Cochu
and
Hans Willy Schwartz*

Cover photo:
Symphysodon haraldi in an unusual (rocky) biotope,
in this case in the Caburí (Nhamundá region) during the high-water period.
Front endpaper:
Symphysodon discus in its biotope in the Igarapé Irauaú (Rio Negro region) at rising water.
Rear endpaper:
Symphysodon aequifasciatus in its biotope in the Lago Tefé during the low-water period.
Drawings: Andrea Maturi and Natasha Khardina

First published 2006

Bleher's Discus
Heiko Bleher
ISBN 88-901816-1-3
Printed on paper produced using chlorine-free bleach

Volume I
Discus in world history and in the history of Amazonia
The first discoveries of discus in the 19th and 20th centuries
Taxonomy old and new
The natural distribution of the genus *Symphysodon,* its species and variants
Discus habitats, water types, chemical parameters, and temperatures
The natural diet, collecting, discus communities and enemies

Copyright declaration:
All rights reserved. No part of this book may be reproduced, stored in any data retrieval system, photocopied or copied by any other method (electronic or otherwise), used for broadcasting or public performance, without the express permission of the publisher.

Disclaimer of liability:
While every effort has been made to ensure that the information in this book is accurate, no responsibility can be accepted by the author and publisher for any loss, injury, or any problem whatsoever experienced by any person using this book. Acceptance of this disclaimer of liability is implicit in the purchase of this book.

Aquapress Publishers
Via G. Falcone, 11
27010 Miradolo Terme (Pavia), Italy
E-mail: aquapress@pmp.it
www.aquageo.com – www.aquapress-bleher.it
Layout: Rossella Bulla – Aquapress Publishers – Printed in Italy
Publisher: Aquapress Publishers, Italy ISBN 88-901816-1-3

Bleher's Discus

Volume 1

HEIKO BLEHER

with photos by the author and others

Aquapress
Italy

Contents

Foreword by Heiko Bleher & Hans J. Mayland 8
Acknowledgements 12
Discus – an introduction 14
How to use this book 22

Chapter 1: The History of Discus 26
   First discovery 28
   The second discus 41
   Discoveries in the 20th century 48

Chapter 2: The Taxonomy of Discus 108
   Heckel's work 108
   Pellegrin's description 114
   Schultz's revision 118
   Kullander's synonyms 130
   Taxonomy comments by Jacques Géry & Heiko Bleher 134

Chapter 3: Distribution 140
   The distribution of discus 142
   Distribution of the discus variants 144
   Maps of the discus regions (maps 1-8) 146

Chapter 4: Discus Variants in Nature 162
   The Heckel discus (Symphysodon discus) 164
   The green discus (Symphysodon aequifasciatus) 178
   The brown discus & the blue discus (Symphysodon haraldi) 190

Chapter 5: Natural Habitats of Discus & Collecting 214
Habitats, including different water types, chemical parameters, and temperatures 216

1. Belém and surrounding area; Rio Tocantins; island of Marajó and Jari; Rio Xingú 216

2. Rio Paru to Santarém; Santarém-Tapajós region; Alenquer region; Óbidos and surrounding area 242

3. Oriximiná and Trombetas region; Rio Nhamundá region; Parintins region; Rio Uatumã; Rio Urubu 288

4. Maués region and Rio Abacaxís; Rio Madeira region; Manacapuru region 332

5. Rio Negro region; Rio Branco region 354

6. Purus region; Tapauá region 395

7. Coari region; Tefé region; Rio Japurá; Rio Juruá region 432

8. Rio Jutaí; Tonantins and Rio Içá; São Paulo de Olicença; Tabatinga-Benjamin Constant region; Leticia-Putumayo (Colombia and Peru); Nanay region (Peru) 468

Water types, chemical parameters, and temperatures – a summary 504

Discus nutrition in the wild 510
  – *Detritus* 512
  – *Vegetable material (flowers, fruits, seeds, leaves)* 514
  – *Algae and micro-algae* 540
  – *Aquatic invertebrates* 560
  – *Terrestrial and arboreal arthropods* 578
  – *A summary of the discus diet in the wild* 590

Discus communities – sympatric species & predators 596

Collecting 624

Glossary 640
References 647
Index 656
Photo credits 670
About the author 671

Foreword
by Heiko Bleher & Hans J. Mayland

Heiko Bleher: When this book was first conceived and planned, the intention was to produce a single volume covering all aspects of discus – history, taxonomy, biology and biogeography, nutrition, aquarium care, and breeding. The planned publication was announced to discus aficionados at selected events to gauge response, and the idea was met with great enthusiasm. "When will it be ready, Heiko?", everyone wanted to know.

However, the process of first writing a book; then getting the original German text translated into multiple language editions for simultaneous publication, so everyone could enjoy it instead of waiting months or years for it to appear in their language; checking and editing all those different versions; sourcing and selecting photos and writing captions (again to be translated); and finally, putting the jigsaw together – fitting all the different language texts into the same layout of text and illustrations (ie all the different languages have to be the same length, for every single page) – all this takes a very long time indeed. Add in other commitments – ongoing research and field trips, lecture tours, other publications, and simply earning a living – and nearly three years slipped past. "When IS your book on discus going to appear, Heiko?", was now the question on everyone's lips.

For this reason – and also because by autumn 2005 the book had grown to twice the size initially planned – it was decided that the best course was to publish the work in two volumes instead of the one originally planned. And here is the first of them. "When will we see volume two, Heiko?", you will probably now ask. Well, it would be foolish to even try and put a firm date on that, but we are working on it already. In the meantime, I hope you will find plenty to keep you occupied in the 670 pages that follow here!

In the current, first volume 1 will first of all relate the history of discus – which I have researched at length – precisely as it happened, no more and no less. If this runs contrary to what is generally assumed, that is because some people who tell the story are simply repeating what they have heard or read elsewhere, and which doesn't always match the actual facts.

As will be seen from the text itself, I have put a lot of work into drawing all the facts together to produce a comprehensive whole. I have taken the trouble to hunt through all of (world) history – in museums in three continents – to bring you the true story of discus in a degree of detail never before seen. Hopefully this will provide a firm basis which present and future generations, whether of laymen or scientists, can use to obtain detailed information on the history of arguably the most fascinating aquarium fish of them all.

In addition this volume covers the natural forms, their taxonomy, their distribution and their biology, down to practically the finest detail (details of the cultivated forms, care and maintenance, the first breeders, breeding today, championships and classification methods worldwide, and much more, follow in volume 2).

With the aid of leading scientists and systematists, and in close collaboration with molecular biologists and breeders worldwide, I have succeeded in producing a complete revision of the discus species, which is presented for the first time in the current volume. The new taxonomy is logical and represents the current state of evolution. In order to achieve it no effort was spared to enable the necessary DNA study to be performed. Together with Natasha Khardina I collected specimens from numerous discus biotopes, and these were evaluated at the University of Konstanz, under the supervision of Professor Axel Meyer. Thus the 1960s division into two species and three subspecies, as well as the subspecies of *S. discus* subsequently described by Burgess (1981) and Kullander's (1986) synonymy now become obsolete. The genus *Symphysodon* is now divided into three species:

S. discus Heckel, 1840 – the Heckel discus,
S. aequifasciatus Pellegrin, 1904 – the green discus
S. haraldi Schultz, 1960 – the blue or brown discus.

A scientific paper to this effect is currently in press, in parallel to this book.

But it is not only the new systematic arrangement that is presented in this book for the first time. The "history of discovery" in Chapter 1 offers an interesting, previously unpublished history of discus. The detailed distribution maps in the third chapter demonstrate for the first time how this element further supports the species division, showing the precise and limited occurrence of the three species (and their variants), down to the smallest *igarapé* and *lago*. As well as the types of water in which the species are "at home" (also covered in detail in Chapter 5 under *Habitats*). The recorded water parameters again represent totally new information. For the first time it can be demonstrated that each of the three species occurs only where these parameters are to its liking. This revolutionary discovery points to the conclusion that not only have species of the genus *Symphysodon* adapted to their specific water types in the course of evolution, but also that other fish species can now survive and breed only where the water parameters are appropriate. And that other fish species can likewise be differentiated (ie classified to species level) on the basis of water parameters. A whole new aspect of taxonomy.

The second part of Chapter 5, on discus nutrition in the wild, with extensive discussion of almost all the components that I have to date been able to record as being eaten by discus, also provides totally new information. This opens up new fish-food vistas, not only for researchers in the field of (Amazonian) fish nutrition, breeders, and hobbyists, but for the entire commercial fish-food industry. (And, along with the water parameters, indicates where each discus species feels "at home".)

But don't worry, at the same time there is something for almost everyone in this work, be it the beginner, the seasoned discus enthusiast (and expert), or the layman. As this book certainly contains more information on the history of Amazonia and other parts of South America, and on explorations and research expeditions there, than most other publications. You can read all about (mainly Amazonian) indigenous tribes – living and long disappeared – and their traditions and cultures, and about the life of the *caboclos ribeirinhos*.

But I don't want to give everything away here! Instead I will make way for a (discus) friend of many years standing, and the most successful author of discus books of all time: Hans J. Mayland. Sadly Hans died quite unexpectedly on 27th October 2004 in Rothenburg an der Fulda and the fish and discus worlds are still mourning the German author who wrote more books on discus and other fishes than practically any other aquarium-hobby writer in the world.

Hans always took the view, "that is an author's prerogative", and that "all that glisters is not gold", but there is no doubt that he himself shone out brightly in the (mainly German) discus literary scene, and published more facts and concrete data than any other discus author. I can say this not only because he accompanied me on various expeditions and I am familiar with his more than 50 books on fishes, but because I knew him for over 35 years and we were practically neighbours for around 30 years (Hans in Oberursel near Frankfurt am Main and I in nearby Zeppelinheim).

When, shortly before his death, I asked Hans to write a foreword for me, he said, "Heiko, are you sure? Who am I to contribute a foreword to your work? I'm really not that famous." He was a modest man, and a travelling companion who never caused any problems, unlike most of the others that I took along. Hans wrote the foreword overleaf only shortly before he died, and I am proud to be able to publish it. It is just a shame that he didn't live to see his words travel in more than eight languages around the (discus) world, the world he loved so much.

There is no doubt that Hans J. Mayland will live on in the memory of discus enthusiasts all over the world and that his learned, well-researched aquarium-fish books will bring pleasure to countless generations of aquarists to come, as they have to his contemporaries and myself. I certainly hope so!

Heiko Bleher
Italy, April 2006

Foreword by Hans J. Mayland

"Go travelling with Heiko?" Oh yes, that is great fun! He not only speaks all the commonly used languages and can communicate with people in every corner of the globe, but he also (almost) always achieves whatever he has undertaken. I have travelled many parts of South America as well as Central America and Australia with him: his stories about his experiences are totally credible, and if anyone is still in doubt, in the final analysis he can prove that the salient facts are true.

Discus have a broad distribution, encompassing almost all the area within the boundaries of Brazil, but in some places – particularly in the north-west – extending beyond its borders. Because an infrastructure is practically or totally non-existent in the vast jungle regions, unconventional methods of transport are a necessity, usually involving boats. However, the latter are more often than not equipped with only a low-powered diesel engine, so you either have to reconcile yourself to a trip that takes a long time and taking along all the necessary fuel and provisions, or looking around for a better, albeit usually more expensive method of transport. We have used most of them in our time.

Collecting discus suitable for export, and not only treating them properly but keeping them alive until they reach a holding facility – that means a huge amount of time and trouble must go into the preparations. Anyone who doesn't plan properly or leaves just small items at home, mustn't expect to be able make good these deficiencies at little shops in the "back of beyond". Of course, it is often the airline that gets the blame for a pack of nets being overlooked during loading before take-off, but pointing the finger of blame is absolutely no use at all when the missing equipment is urgently required! So it is sometimes very handy to have a polyglot companion by your side, who very often (but unfortunately not always) discovers some deficiency or other and can be relied on to improvise the required item indian fashion. On the other hand, I have never learned fear of aquatic creatures in Heiko's company, be they piranhas or alligators or whatever. Respect, maybe! If, as sometimes happened, the natives tried to warn us, Heiko had his own opinions regarding the dangerousness of aquatic denizens. So I always kept my spirits up and trusted in Heiko's opinion.

Nowadays travelling in Brazil is not a problem, provided you have the necessary booking confirmations, as it often happens that all the flights for the next few days are cancelled. It isn't possible to achieve much in the Amazonian border regions using tourist English, and even if you slip a high-value note in the appropriate currency into your passport, this can nevertheless sometimes lead to misunderstandings. In the local language, however, coupled with a "certain look", it's quite another matter!

In Germany, as well as in neighbouring European countries, the specialist trade in discus offers numerous opportunities to purchase the finest wild-caught specimens, and for this reason there isn't much point in aquarists travelling to Brazil, and enduring all the associated bureaucratic problems, on the basis that they can do better than the professionals in this or any other respect, and in a foreign country too. The country has so many sights to see, and offers us so many interesting insights into its natural and other history, that the sole object of such travels should be to broaden our personal horizons as well as receiving improved ichthyological insights.

As matters stand at present, it is totally inadvisable to catch fishes of any species without the appropriate permit from the correct (!) source and to thus try and get around the laws currently in force. The important point here is that you will be a foreigner in Brazil, and you should behave in accordance with its laws. If you do otherwise, and end up in court and then prison, then don't be surprised, and don't expect the staff at your embassy in the capital, Brasilia, to travel several times to visit you, spend the night, intercede with the law on behalf of their fellow-countrymen, and sort everything out free of charge. Such expectations are at best justified when the whole matter has repercussions "at a higher level". In recent years we have seen a whole series of examples that justify this warning, as after many months had passed the local lawyers still had a lot to do, and they don't do it for nothing!

Naturally discus will continue to be caught in the future, as the aquarium hobby will continue to exist and we won't want to see it diminished in any way. The following story is for those who regard foolishness as entertainment.

I had already prepared, but for certain reasons not yet published, an article. It was about three aquarists who arrived at the airport in Manaus with six styrofoam boxes full of fishes of various species, intending to fly thence back to Germany via Rio de Janeiro. Or so they thought. When they were asked for their collecting and export permits, they had no acceptable answer to give and were referred to the relevant department of IBAMA, an environmental protection organisation that in recent years has achieved a degree of notoriety for its crude methods. While two of these people were busy with their consignment at the IBAMA offices, the third went to the toilet. On his return he realised problems were looming, as the other two were in the process of being arrested. He made his escape and arrived back in Germany without further incident. The other two were initially thrown into jail at the end of February 2003. Repeatedly interrogated and hauled before the judges. The Foreign Ministry in Berlin, to whom I had written following the events in Manaus, kindly answered me as follows on 20.05.2003:

"The Foreign Ministry, the German Embassy in Brasilia, and the German Honorary Consul in Manaus have been quick to intervene in the affair. The German Embassy and the Honorary Consul in Manaus have conducted intensive consular activities on behalf of the two Germans: thus, for example, the Honorary Consul has visited them numerous times in prison. The head of the Legal and Consular Department at the Embassy in Brasilia has himself travelled from Brasilia for four court hearings (judicial hearing on 21.03, main proceedings on 28.03, 04.04. and 25.04.). Unfortunately further details cannot be given for reasons of the personal security of those involved".

Later on the two were released from prison, but their passports were taken away so that they couldn't leave the city or the country. They were allowed to stay at a hotel. A lot of money was paid, and after almost exactly six months the two returned to Germany. I'm not allowed to divulge the details of what happened. They have to be kept secret for good reasons.

Heiko has reported his experiences of this type in his magazine *aqua geōgraphia* (no. 21/2001b, page 88, in the commentary "Brazil – The world's greatest biodiversity in danger!"). This book, from the pen of Heiko Bleher alone, will no doubt contain up-to-date data on the biodiversity and additional information on the current status of the Amazonian environment.

Discus collecting sessions, as I have repeatedly described in my books and magazine articles, are without doubt among the most interesting and gripping experiences that an aquarist can share. The excitement begins as soon as you start to travel along a body of water to the chosen site to check whether the necessary preparations for collecting with the huge, heavy net have been worthwhile. It is with a similar adrenalin rush that I anticipate Heiko's work will undoubtedly be the most comprehensive ever to be published in the discus world so far.

Hans J. Mayland
Germany, June 2003

This photo shows, from right to left: Hans J. Mayland holding two plastic bags containing green discus and angelfishes, caught by the author in the Rio Tefé. Then Cristina Taras, a Chilean lawyer, followed by Jürgen Inge, a German pimelodid specialist, and Brazilian Segderu Esashika, son of the first discus collector in the Tefé region. The photo was taken in 1985, following a successful expedition to the Rio Tefé. Hans was obviously very happy...

Acknowledgements

My thanks go first of all to the indigenous peoples of Amazonia, and to the *caboclos* who have often indefatigably accompanied and assisted me – day and night – on my research trips; not least Manuel Torres, the "oldest" of the discus collectors.

Next I would like to thank the now deceased pioneer of discus exportation, Hans Willy Schwartz of Aquario Rio Negro, his family, and the current owner of the business, Asher Benzaken, who has expanded it tremendously and changed its name to Turkys Aquario. Asher and his family have always enthusiastically taken me under their wing, in every respect. Like Willy, they have welcomed me with food and lodging during my annual visits (often several times a year), organised *lanchas* and *caboclos* or *disqueiros*, helped me to hire aquaplanes and other small aircraft and helicopters, and placed their fish-holding facilities at my disposal.

I thank the staff of the Brazilian airline VARIG S.A. in Milan for their helpfulness as regards flights to and within Brazil. Nobody knows the country as well as VARIG.

I am grateful to Hans Petersmann and his wife, who have shared with me their experiences with discus collected by me, and taken splendid photos of their fishes and their discus aquarium, glorious orchids, and much more.

My heartiest thanks are due to Aquarium Dietzenbach and its enthusiastic owner Herbert Nigl, who has spared neither time nor effort in housing and looking after the numerous fishes I collected (after I closed Aquarium Rio); as well as to Valverde Aquarium and Paola Pierucci, who also looked after some of my discus for a long time.

I thank all my colleagues at Aquarium Rio – in Rio de Janeiro and Frankfurt – who in the course of more than 30 years tirelessly maintained and cared for the fishes I collected. They are too many to mention by name, apart from Mohamed Elkadir, who not only cared for my discus but also went collecting with me. Likewise his brother, Hamed Elkadir. Both helped me at Aquarium Rio for more than two decades, often working day and night, seven days a week. And dear Johanna, my long-term secretary in Germany, also deserves a thank-you.

Naturally I would also like to thank all the discus breeders who have provided information and in some cases received fishes from me, for example Willy Brockskothen, Willy Mikschofsky, Manfred Göbel, Peter Thode, Jürgen Weissflog, and Kitti Pananhiti, to mention but a few; they are too many to list in full, as I have visited hundreds in the course of the past 40 years and stayed in touch with many of them.

I am grateful to T.F.H. Publications, Inc. and Glen Axelrod for making available old photos from their American magazine. I would also like posthumously to thank Hans J. Mayland for his cooperation regarding text and photos, and for being a pleasant, easy-going comrade during our travels together. The same applies to my other travelling companions, again too numerous to list here. I would nevertheless like to give a special mention to the ladies who have accompanied me, including Petra Brönner, Christina Taras, Monique Nicolaï, Paola Pierucci, and my partner Natasha Khardina, who have all truly "stood by their man" and followed me "through thick and thin". More than can be said for any man!

Thanks are also due to Axel Mewes, the best cameraman I know, who has shot excellent films on discus for me and who lost all his film equipment during our last expedition (see Chapter 1: *Discoveries in the 20th century).*

I would like to offer heartfelt thanks to the heads of the various museums who have unhesitatingly made available type material and/or photos, and supplied essential information and copies; they include Christian Meunier, Jeff Williams, Heraldo Britski, Sven Kullander, Ernst Mikschi, and Marc Sabaj, as well as members of staff at Harvard University, the Natural History Museum (London), Claude Weber and Sonia Fisch-Müller in Genebra and many more.

I would also like to mention, and thank posthumously, a number of personalities who have contributed not only to the world of discus, but also to the aquarium hobby as a whole, helping make the care and maintenance of an aquarium, its fishes, and its plants, into the finest and most widespread hobby on Earth, but who are very rarely remembered today although we owe it almost all to them.

First of all there is my grandfather, Adolf Kiel, who more than 100 years ago became one of the pioneers of the modern aquarium hobby and brought into being the underwater garden full of aquatic plants. Walter and Karl Griem, pioneers of collecting and importing fishes, chiefly from the New World. Fred Cochu, who for more than 50 years populated aquaria worldwide, directly or indirectly; likewise Teo Way Yong, who, as long ago as 1920, was the first to send Singapore aquarium fishes to Europe and America and whose descendants are still active today. And naturally my mother, Amanda Flora Hilda Bleher, also belongs to this group of people whose activities led to the hobby growing and becoming what it is, but never received the recognition they deserved. Likewise Raffael Wandurraga, an indian who, starting immediately after the Second World War, for 30 years collected the finest Amazonian fishes, and later the first green discus, for Fred Cochu. And Old Cesar too. I am lucky to have had the opportunity to get to know and treasure all these truly unique people. No-one can take that away from me. And they are irreplaceable. Their kind is long extinct, but I nevertheless thank them here for the sacrifices they made for the benefit of the aquarium world. And I hope that they will never be forgotten as long as there are people to remember them.

I would also like to mention my closest and dearest living friends, who have continually supported my work with good advice, who have often helped me get over an unsuccessful expedition or an attack of malaria: Peter Frech of Memmingen, Germany, not only a very dear friend but also one of the best breeders (and collectors) that I know. And Jacques Géry of Sarlat in France, in my view the finest ichthyologist and without doubt the greatest expert on characins who has ever lived. Thank you, Peter and Lore Frech, for your untiring help with accommodation over the decades – for me, my companions, and my fishes. Thank you Jacques for your patience with me, for your perseverance, and for your knowledge.

I mustn't forget a man who deserves a special "thank you", Ross Socolof, the living fish-legend of America. Ross and I have been bound together by a very special friendship for more than 40 years. I did my first commercial fish breeding with him in Florida in 1963 and learnt a vast amount. He too has shared his discus experiences with me, as well as giving me the tapes he himself recorded with Fred Cochu. Thank you, father Ross.

I would also like to thank a few other friends and acquaintances who have supported me in every way, for their continuing help:

Adolf Schwartz, the son of Willy Schwartz, whom I dragged off at the age of 10 to the football World Cup and to go collecting fishes in Mexico, and who only recently provided me with a car which I practically wore out during my last expedition (December 2005). I would like to thank Axel Meyer, in my view the greatest of all molecular biologists, and his colleagues Walter Salzburger and Kai Nikolas Stölting for their painstaking research work, on discus DNA. Axel and I first met around 20 years ago when we were both invited to lecture on cichlids in New York. And to the indestructible Rolf Geisler, surely the only limnologist to have spent his childhood and grown old with aquarium fishes, especially the very small ones, I say: *"Muito obrigado meu amigo por sua ajuda."*

The staff of INPA in Manaus, in particular Efrem Ferreira and Geraldo Mendes dos Santos, have earned my gratitude, as have the staff and President of IBAMA in Manaus, for granting me permission to collect fishes.

And last but not least I would like to thank my friends and colleagues who have by now been diligently working for almost three years on the translation and editing of this work. First of all John Williams a top-notch copy editor, and for years correcting the Oxford English for Aquapress Publishers, but who has also provided help and advice (sometimes at great length....). Next Marcel Dielen, the best rainbowfish breeder in Europe, as well as an expert *par excellence* on fishes and natural history. *Merci* Marcel for your patience in translating into French, thank you for your hospitality and my very best wishes for your dream garden and your fantastic rainbowfish aquariums.

I would like to thank Igor Sheremetyev, from the Ukraine, one of the most prolific authors of aquarium books in the Russian language, for his perseverance in converting complex *caboclo* texts into Cyrillic – *sapsibo* Igor. I am also obliged to Spaniards Fernando and Roberto and their splendidly organised "Discusland" – *muchas gracias;* and to dear Raffaella Raganella for an excellent translation into Italian – *grazie*. That leaves only Marcel Notare and his colleague Ótavio, in Brazil, who have put a lot of effort into the Portuguese edition – *obrigado* Marcel. And Frantisek, who at the time of writing is still working hard on the Czech edition

And finally my deepest thanks are due to the two women in my life: Rossella Bulla, who has faithfully worked diligently for eight years at her computer designing layouts, and without whom this book would never have come to fruition. *Grazie* Rossella, *sei un angelo*. So is Natasha. Whether it be final corrections in the Russian or advice on the German, sitting working day and night over endless drawings or scanning and cataloguing photos – whatever needed doing, she has done it all, and been an unbelievable help to me. *Spasibo* Natasha.

Discus
– an introduction

"Brown Scalare", "Blue Scalare", "The Aristocrat of the Aquarium", "The Crested Cichlid", "Pompadour Fish", "King of Fishes", "King of the Amazon", "King of the Aquarium" and many more such names have been given to this, probably the most highly priced tropical freshwater fish of all time.

There is no other fish on Earth about which so much has been written and published, or which has so often been filmed, painted, or photographed, as the discus. A quite inconceivable flood of quarterly so-called "Discus Newsletters", "Discus Newspapers", and "Discus Journals", plus half-yearly volumes, often in book form, and annuals in more than ten languages. A good dozen new discus books appear every year. New videos and CDs – and now DVDs as well – are practically part of the daily round. Even the television has jumped on the bandwagon. In 1964, during my ichthyological studies and work at Gulf Fish Farm in Florida, I was invited to appear on one of the first professional aquarium fish and Discus TV shows. In 1978 I was offered the first major German production on the "King of the Amazon". "Expeditionsziel Aquarienfische" ("Expedition Aquarium Fishes") with the "King" as star, was translated into more than 10 languages and distributed worldwide. But that was just the beginning. Subsequently Americans, Japanese, Chinese, Brazilians, and many others produced and distributed discus TV films. During the last 30 years I have been invited to discus symposia in five continents – and I am asked to lecture on discus more than on any other fish, even though I have been deeply involved with all fresh- and brackish-water fishes for as long as I can remember. Innumerable magazine articles on discus have been published, in more than 40 languages. In Korea, not only were my discus lectures televised, but a so-called "Royal Green Discus", discovered by me, appeared, along with my humble self, on the front page of the most popular daily paper – in colour (see page 19). Then the Saint Petersburg TV channel recorded an interview with me on collecting discus and other aquarium fishes, and this was broadcast all over the former Soviet Union over Christmas 1990.

There are by now more clubs and associations for discus than for any other group of fishes. And if you log on to the Internet then at the time of writing you will find more than 11,2000,000 websites under "Discus" – each with at least 10 web pages.

So what does the discus have going for it that has enabled it to "oust" old friends such as the Nishikigoi (known to us as the Koi or coloured carp), highly prized for almost 200 years?! Not only in its homeland of Japan, but also in other Asian countries. And how come the discus has moved up to first place in just a few years in China, with its carp culture dating back for more than 2000 years? It has also supplanted the guppy, the "millions fish", popular for around a century. Indeed, it has even pushed the previous "King of Aquarium Fishes", the angelfish *(Pterophyllum scalare)*, down to second or third place in just a few decades. Without question the discus is today number one on the desirability scale (not, of course, in terms of numbers actually sold, where the cardinal tetra, guppies, and other livebearers are generally in front).

The discus was crowned the new "King" at the latest shortly after the Second World War, when my mother organised the first aquarium exhibition in the ruins of Frankfurt Zoo; and by the time it was finally bred successfully, and imported in large numbers towards the mid 1960s, it was so popular that I and my former import and export company put four different colour varieties on public display for the first time in Wiesbaden in 1968. Numerous further, albeit smaller, exhibitions followed. Often at clubs, during the "green week" in Berlin, and at the Interzoo, in the U.S.A., and in Canada.

I myself was responsible for bringing about the first international discus exhibition. This was in 1986 in Japan, where in the course of two days some 44,000 people marvelled at the more than 400 discus aquaria, filled mainly with fishes I had collected and those Dr. Schmidt-Focke had bred. This was followed in 1989 by the first Aquarama in Singapore (for which I was asked to arrange the judging, ie to invite suitable judges;

likewise for the two following Aquaramas in 1991 and 1993), and today almost every country in Asia stages an annual exhibition, including judging of the exhibits. Since 1996 what is now the world's largest discus championships have been held in Germany every two years (again one of my suggestions, which the energetic Herr Nobert Zajac enthusiastically turned into reality), and there are one or two exhibitions each year in neighbouring countries, as well as in the U.S.A. and elsewhere.

Now it is none too easy to provide concrete reasons for this almost incredible mania for discus. But the unusual form of the fish is undoubtedly at its root. Discus, as almost everyone knows, is the word (of Greek/Latin origin) for a "throwing disc" such as was already in use in the athletic games of ancient Greece. And because the fish is likewise flat and normally round, the Mannheim naturalist Johann Jakob Heckel, working on its description at the Vienna Museum in 1840, was reminded of a disc *(Subdisciformis. "Die Gestalt ist beinahe scheibenförmig,...")*. He thus gave the new species the name *discus* and placed it in a new monotypic genus, *Symphysodon*. No other fish, whether in fresh or salt water, has this sort of disc shape.

Secondly, "His Majesty" the discus fascinated fish enthusiasts right from the start with the way that it moved. Virtually no other aquarium inhabitant is as elegant and graceful. Then there were its brilliant colours, immediately catching the eye, of course, because of the large lateral surface. Even so, Man has not been satisfied with the splendid natural colours and has for years been breeding an ever-increasing spectrum of colour strains, today barely surpassed by that of the coloured carp! The popularity of the discus has increased even more because of these extreme colour varieties, especially in Asia, although these extreme colour forms are not to the liking of everyone…

In addition right at the beginning it was the high price that made the discus so desirable, but then the discus increasingly "gained ground" as its compatibility with other fishes became evident. And the "aristocrat" truly behaved with nobility. It could easily be kept with other fishes, and would do no harm to any tankmate. Aquarists had found a peaceful and extraordinary cichlid.

Of course there was yet another decisive factor: biology. For the first time we had an aquarium fish – a cichlid – which not only practiced brood care but also produced a special skin secretion that provided food for its fry to thrive and grow.

Today the "King of the Amazon" is known in the furthest corners of the globe, whether it be in Bhutan, New Zealand, the Philippines, Saudi Arabia or the Cape Verde Islands. There have been times when fish have changed hands for $10,000 or more. Some people have even sold their house to buy a particular discus – losing wife and children in the process. Nowadays, however, discus, including unusual cultivated forms, can be obtained a lot more cheaply, and it is not always necessary to lose one's house (wives are still, perhaps, another matter!).

Many citizens of our planet regard the discus as a status symbol. Although in the pre-war years and for some time thereafter the "Browns" were the only discus known – everywhere erroneously labelled as *Symphysodon discus* – during the 1950s they were the most sought-after of all fishes. Anyone who actually owned a "King" could easily obtain thousands of dollars for it! The discus was in such demand that my mother disregarded the dangers involved and in 1953 started making two-year expeditions "in search of the discus".

When, during the 1960s, I was able to bring the first "Royal Blue" individuals back to Germany from the Manacapuru region (see page 19), these were immediately an international hit although the retail price was still about a thousand marks (US$ 500). Later, during the 1970s, the most sought-after discus was the true "Rio Içá" (see page 19) – I was able to bring back only a single genuine Red specimen, which Dr. Eduard Schmidt-Focke was the first to breed. I compared this with the "dream machine" of every man, the Testa Rossa Ferrari. It was one of the most sought-after forms of all (just like Enzo Ferrari's masterpiece of the time). There were no "Red" discus then (and very few of the "Reds" from Modena). But it was also the decade when the solid-coloured cultivated forms became popular.

By the 1980s the "Cobalt Blue" discus had been produced from the "solids", and was the "latest rage", the new status symbol. Even so, at this time the "Green" discus with all-over lentil-sized red dots, brought home by me from the Coari region, were leading their first fry in Dr. Schmidt-Focke's tanks. These went first of all, via my company Aquarium Rio, to Hong Kong, and this was the start of a triumphal progress without equal in the history of discus. Today the "Red-Spotted" is the winner in its class, and the Grand Champion, at almost every championships.

The beginning of the 1990s saw the start of the Asian invasion, firstly with the "Pigeon Blood" discus, followed by cultivated forms such as the "Ghost", "Snake-skin", and "Red-White". The discus colour spectrum now knew no bounds.

Today there is no longer any "discus status symbol" in the true sense, although a few snow-white individuals with red spots have arrived on the scene. Fortunately people are by and large turning their backs on the unnatural forms. A new status symbol has been bred in Asia. They have their new "fish god". At the time of writing the so-called "Flower-Horn cichlid", a hybrid between two Central American cichlids, rules the scene. But for how long? (Hopefully for a long time, as the unnatural cultivated discus forms almost all originate from the area.) Among discus cognoscenti it is still the "Red-Spotted" or the "Red-Spotted Green" that holds sway, as well as selected strains of wild-caught varieties.

Fishes from the natural habitat – ie wild-caught – are nowadays available in much smaller numbers than in the past. Less than one per cent of the discus sold worldwide now originate from the Amazonas. It is no longer worthwhile for the *caboclo* – and even less for the indians – to continue collecting discus (see Chapter 5: *collecting*). Transportation costs have risen enormously and it takes thousands of litres of diesel or gasoline to bring back often only a few specimens. In addition, the customer may want only selected specimens; if you are lucky maybe one in 500 wild fish, if that. Then there are the unpredictable weather conditions – which have changed in the Amazonas, just as elsewhere. As has always

been the case, forecasts are unreliable. In two out of three cases you arrive at some far-flung collecting site only to find that the water level is far too high – even though it is supposedly the dry season – and the whole journey has been "money down the drain". Another point, which few people realise: for about nine months each year discus cannot be found, let alone captured. Hence the *caboclo* can feed his family by collecting discus for only three months of the year.

But captive breeding has largely compensated for the shortage, and today no fewer than 1.5 million discus are bred every month world-wide. In southern China alone the figure is more than 500,000 per month.

Without doubt Heckel can never have dreamt, back then in Vienna, what a brouhaha the name he invented – and, of course, Natterer's original discovery – would cause, or that the discus would rise to the status of best-known and most sought-after of all aquarium fishes. Or that over the course of the years hundreds of colour forms would be discovered in the wild; discoveries in which I have played no small part over the past 40 years. Or that so many of the best scientists would end up throwing in the towel as regards what constitutes a good species (see Chapter 2: *comments on taxonomy*).

Nowadays the discus adorns telephone directories, telephone cards (left), CDs, cups, mugs, and plates, watches, stamps, and calendars. There are key-rings, brooches, bracelets, and necklaces with the "King" in tin-plate, bronze, silver, gold, and even platinum. Wood carvings and paintings – which now fetch astronomical prices; and the discus can be seen on thousands of T-shirts. You can even have a discus tattoo (page 18). I doubt if any shark, or even a whale, can hold a candle to it. Not to mention the panda. There is, by the way, even a "Panda" discus – a cultivated form produced by my friend, the well-known breeder Jack Wattley.

The discus cult knows no bounds, and this fish will undoubtedly always be number one as long as there are aquarists and fish enthusiasts. But you can only truly appreciate the fascination of this elegant – truly majestic – fish if you have observed it in the wild. I first had the opportunity to do so as a child, and the thrill will never pall. It is something undescribable, but I will nevertheless make the attempt in this book, using numerous reminiscences and detailed descriptions of natural biotopes (and the paintings on the endpapers). I have also endeavoured to convey this to the public using authentic biotope aquaria at exhibitions over the past few years (see also Chapter 9, *wild discus vs. captive-bred varieties*). As Dr. Eduard Schmidt-Focke said back in 1956, "*Symphysodon discus* unites all the qualities desirable in an aquarium fish in a rare harmony of form and colour". Then there is the introduction to a later article, "Diskusfieber" (discus fever) (also the title used subsequently by my late friend Hans J. Mayland, the well-known author, for a book of the same name, possibly one of the most-sold discus books of all time), which really says it all: "Incredible that this is no fantastic figment of some aquarist's fertile imagination, but an actual reality". Eduard was quoting the words of the publisher of an aquarium magazine of the time. It was spring 1960, and they were standing in front of a display tank in the Rainbow Aquarium establishment in Chicago, which contained two displaying majestic green discus males, totally unaffected by the crowd around their new home, the noise, or the hundreds of flashes from the press photographers. They faced one another, their gills and undersides yellow. A delicate green shimmered on their backs and flanks, broken only by black-brown vertical bars; red-brown lentil-sized spots adorned the anal fin in the one, tiny stripes of the same colour in the other. Apart from the pectorals and the tail, the fins were all edged with red...

I am also reminded of the words of the famous ethnologist, Harald Schultz, who collected discus during the 1950s and early 1960s, and his "flowery" description of the "Blue" discus: "The body is a symphony of green, green-yellow, yellow, orange, red, blue, and brown." Or, before Schultz, the words of Dr. W. Ladiges: "The neon tetra is merely a starlet compared to the discus, which is an aristocrat."

The discus is one of the most beautiful and interesting creatures that the Amazon region – by no means poor in unusual

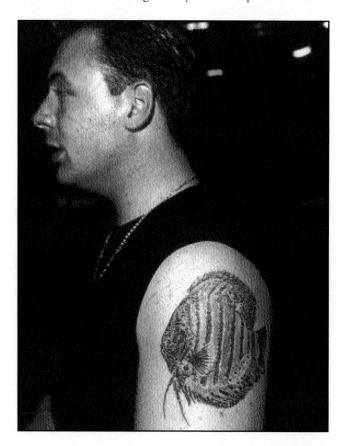

Nowadays there are almost countless magazines on discus, but *The Discus Digest*, founded in 1988, was the very first. The "king" features on the front pages of the major daily papers (left, in Seoul, Korea), clocks, mugs, T-shirts, matchboxes, stamps (see page 20), tattoos, and beautiful women (see page 18) – at least since my discovery of the "Royal Blue" in 1965 *(top left)* and the "Red Içá" in 1977 *(top right)*.

life forms – has brought forth. It is and will remain the "King of the Amazon", or, better still, the "King of Aquarium Fishes". And it can be found imortalised on paintings world-wide (like this one, with the city of Barcelos and cardinals), on T-shirts, machtboxes, stamps, mugs, keyrings, Bohemian glas sculptures, or covering the state of Texas. The "King" graced even the cover of the first *T.F.H.-* magazine (but as *Pterophyllum discus*)...

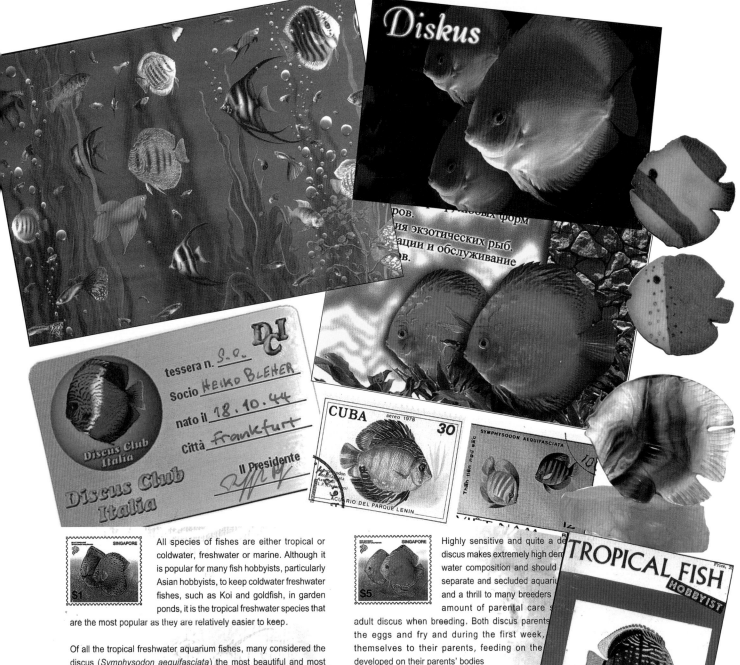

All species of fishes are either tropical or coldwater, freshwater or marine. Although it is popular for many fish hobbyists, particularly Asian hobbyists, to keep coldwater freshwater fishes, such as Koi and goldfish, in garden ponds, it is the tropical freshwater species that are the most popular as they are relatively easier to keep.

Of all the tropical freshwater aquarium fishes, many considered the discus (*Symphysodon aequifasciata*) the most beautiful and most graceful of all. Many dedicated hobbyists also regarded the discus as "the fish" as keeping and breeding of this fish is a challenge for experienced aquarists.

The highly prized discus is a flat, round fish and as the name suggests, disc-like. Belonging to the family of Cichlidae, discuses are found in the warm, soft waters of South America, particularly in the Amazon rainforests.

Highly sensitive and quite a de discus makes extremely high dem water composition and should separate and secluded aquari and a thrill to many breeders amount of parental care s adult discus when breeding. Both discus parents the eggs and fry and during the first week, themselves to their parents, feeding on the developed on their parents' bodies

Although the wild type variety of colours and pa also variants, the fish c (referred to as hybrid colour, shape or size the three. In this stamp of these regal fishes – the Blue Turquoise Red Alenquer ($5) and Red Turquoise ($10)

DISCUS – AN INTRODUCTION

Above: Discus transportation in 1960s was in hand-woven baskets lined with plastic.
Below: And in the 1970s in plastic containers designed by myself (along with a German company): these can be stacked inside or on top of one another.

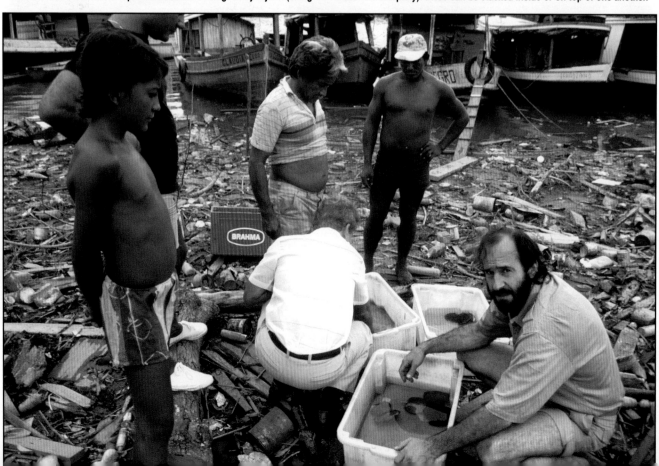

How to use this book

This book is laid out as follows:

The first chapter deals with the history of the discus, and I expect you will think "But I have heard and read all that time and again already...". I can well understand that you know something on the subject, but I can assure you that the most part will be new to you. Like all the other chapters it is the result of years of research – about 50 of them, in fact! – during which I have discovered all sorts of interesting facts, not otherwise generally known and in the main not previously published.

Thus it is well known that the Austrian explorer Natterer brought "home" the first discus (albeit dead). This was followed by the discoveries by the Thayer Expedition and those of the Frenchman Jobert (but they both brought only specimens in alcohol back to the USA and Paris, respectively) and subsequent collections (mostly live) in the 20th century, about which generally more is known. But not so much is known about the collectors before the Second World War, or about my mother, and Harald Schultz – whom I met – and his colleagues. Or about my own annual expeditions and discoveries from 1964-65 on – details of which remain largely unpublished.

In the second chapter you will read about the often-cited work of the naturalist Heckel – and without doubt you will learn something new. Followed by the describer of the second species, the celebrated Frenchman Pellegrin, with much information not hitherto mentioned elsewhere. Finally you will find extracts from the well-known revision by the American ichthyologist Schultz, whose subspecies were eventually placed in synonymy by the Swedish ichthyologist Kullander. And in conclusion a commentary on the taxonomy – an overview of why there are three good *Symphysodon* species. And comments on the taxonomy with the help of Jacques Géry.

The third chapter deals with the distribution of the discus. Here you will find a summary of my knowledge accumulated during more than 300 sampling and collecting trips in the Amazonas region over the last 40 years. I will show you first of all the overall natural distribution of discus cichlids in the Amazonas basin – more precisely than any previously published details – and then their unnatural distribution (where they have been introduced). You will then be able to see the currently (end of 2005) known distribution of the different colour variants (covered in greater detail in the following chapter). And, in conclusion, the distributions of the individual colour forms are presented again, but now in detail, on eight maps. This also clearly shows that there must be three good species with numerous colour variants.

The fourth chapter presents many of the individual colour variants found in nature – never before published in any book, except for my loose-leaf *DISCUS*. In each case starting with the four known colour forms of the three species, and also showing the "overlapping variants", especially the Blues and Browns (Blue and Brown = *Symphysodon haraldi*), ie in what areas the colour variants intersect. Each fish is labelled with its precise collecting locality – not, as in most publications, just a place name the writer or importer obtained from the exporter. (Hardly any exporter has to date visited a discus collecting site, hardly any importer, dealer, or writer was ever on the spot, with the exception of, for example Harald Schultz or H. J. Mayland.)

Chapter five discusses each of the eight regions (shown in Chapter three) in turn, with information on some of the history of (the discovery of) these discus regions; their drainage systems (rivers, lakes, *lagoas*, and dams); the discus habitats; the various water types; plus lists of chemical parameters and temperatures measured on the spot in each locality. One section is devoted to the nutrition of discus in nature, another to the fish species found with discus, as well as their enemies. And, in the final section, how discus were captured in the past and how they are collected today (insofar as they are still collected at all).

The sixth chapter (in Volume 2) deals with breeding. It tells about the first breeders before the Second World War: their successes – or, more to the point, their failures. Then the further progress after the War. About Dr. Eduard Schmidt-Focke, the pioneer of modern breeding – with extensive reference to his (largely unpublished) records, plus biographical material. And finally, at the end of the chapter, recent breeders from four continents. The best-known in various countries (some of their discus are shown in Chapter 7, in many cases detailing their success and how it was achieved – the methods and equipment they use for breeding, rearing youngsters, and feeding their fishes.

Chapter seven deals with discus – especially cultivated forms – around the world, covering four continents in turn (unfortunately there are as good as no professional breeders in Africa, just discus enthusiasts; and in the Antarctic it is a bit too cold).

Then finally the latest achievements on the world-wide discus scene, hopefully right up-to-date – at least at the time of printing.

The eighth chapter deals first of all with the history of classification and judging and the internationally recognised classification of the natural forms the author established in 1983, including that of the cultivated forms and hybrids, which today constitute the majority of the individuals submitted for judging. Followed by the evolution of discus exhibitions and championships, right from the very beginning, and including the most important events of recent years. And finally, the various judging methods used in different countries.

The ninth chapter will, hopefully, provide new information for anyone who keeps – or wishes to keep – discus. Naturally an incredible amount has already been written on the subject (some of it excellent, but equally much of dubious merit), but I am sure there will be something new to help both the beginner and the fully-fledged discus expert. The years of breeding experience – mainly commercial – in my youth, and the knowledge derived from my almost daily collaboration with Dr. Schmidt-Focke, in my view still the greatest discus breeder of all time, over three decades (he lived in Bad Homburg and I was in nearby Frankfurt). I will also detail my experiences with hundreds of thousands of discus in the course of more than 30 years running my former import/export company, Aquarium Rio, based originally in Brazil and later in Germany. But not only all that, but also information provided by breeders over the decades (and still today – those whom I still supply with new discus every year). On the basis of this information and my experiences in the wild, I venture to make a number of suggestions that differ from those generally current until now.

There is advice on tank size plus setting-up and optimal maintenance. Learn from my experience with filters and filter media and note my comments on heating and lighting. A special section of this chapter is devoted to the subject of quarantine, and another to the relative requirements of wild-caught and cultivated forms. These topics are generally overlooked and often (or almost always) lead to a "breakdown" in discus maintenance/acclimatisation. My commercial methods of combating pathogens, as well as disease prevention – based on three decades of experience, day-in, day-out. In conclusion, a selection of hints and tips based on practical experience.

The tenth and last chapter discusses the future of discus. Here you can read what is happening in the natural habitat and what (negative) implications this may or may not have for the discus. Unique photographic material is used to illustrate the relentless destruction taking place in the Amazonas region and the lunacy of the destroyer – Man.

Chapter 10 is followed by an **Appendix** providing a short list of discus clubs and associations worldwide. And, of course, the book would not be complete without an extensive summary of discus on the Internet and an overview of discus products (anyone with a further interest in these should contact Aquapress at the address given).

I each volume I have also added a **Glossary** – in particular, in the first, for the numerous words from Brazilian Portuguese and Indian languages. And, of course, a comprehensive **Bibliography** for those who (I hope) will want to learn more about discus from the best of the relevant literature, plus an **Index** for easy reference to specific topics.

And finally, an overview of the author's life and work.

In conclusion, I would like to say a few words about the **photos and other illustrations**. It should be noted – particularly as regards the photos of wild-caught fishes – that it was, of course, not always possible to bring back to Europe all the fishes captured in the course of my many sampling and collecting trips. Moreover it has occasionally happened that the discus collected were lost (consignment not kept cool enough, lost, or delayed). In other words, it was not always possible to photograph the fishes in the aquarium. For this reason – and in order to provide a truly comprehensive and previously unsurpassed overview – I have also used photos of some variants in the hand. The reader will, I hope, understand and excuse this. Almost all the photos are my own (the work of other photographers is indicated under photo credits) and selected from some 250,000 available. I have been at pains to select only the best, and where one or another is less than perfect, then it has been included only because none better was available to illustrate the form, race or species in question. None of the photos has been colour-corrected. The colours are all the originals! Also in the old photographs.

The drawings and paintings are from all over the world, and in many cases reproduced by special permission. Many originate from the Aquapress archives. The two splendid paintings on the endpapers are true to nature, and were painted over a period of months by the Italian natural history artist Andrea Maturi and Natasha Khardina often in my presence. They performed a miracle in putting on paper what I have seen time and again during my explorations. Unfortunately I have never been able to obtain a photo of equivalent quality because of the murkiness of discus waters and/or the speed with which the group of discus beat a retreat. This **is** how discus live in nature. And Natasha Khardina drew all the excellent maps.

The Quality Policy

The Quality Policy of Turkys Aquarium Ltd. is based on supplying Tropical Fish, according to the quality specifications, and also Keeping the respect with our collaborators, which allows the Company to atend the expectations and necessities of our customer.

"Our policy is quality", that is the motto of Ashers Turkys Aquarium in Manaus and will also be the motto of this book. Asher Benzaken is the successor to the pioneer of ornamental fish export from Manaus, Hans Willy Schwartz, and is married to Schwartz's daughter Adele. She spent part of her childhood with me and I took her to school. That was the start of a friendship which lasts into these days...

A *Symphysodon discus* variant from the Rio Nhamundá. Caught at the end of 1998.

CHAPTER

1

The History of Discus

FIRST DISCOVERY; THE SECOND DISCUS;
DISCOVERIES IN THE 20TH CENTURY

First Discovery

It was an extraordinary day in the history of Austria – in two respects – when the State Chancellor and Foreign Minister, Prince Clemens Wenzel von Metternich, who in his day wielded considerable influence at the imperial court and on Kaiser Franz I (II), sampled the very first *Sacher-Torte* in Vienna. It was later to become the most famous cake in the world, and the hand-written recipe has remained a closely-guarded secret right up to the present day. Round about the same time (1832), the assistant supervisor of the imperial natural history collection, and leader of the Austrian expedition to Brazil that had started back in 1817, became the first white man ever to handle a discus: the Austrian Johann Baptist Natterer. His fish was later (1840) described as *Symphysodon discus* by Jakob Heckel, along with many other species of fish discovered by Natterer (see below). During his 18 year expedition to Brazil, this Austrian collected a total of 1,671 fishes, 1,678 reptiles and amphibians, 12,293 birds, 1,146 mammals, 32,825 insects, 1,729 vials of intestinal worms, 1,024 mussels, 125 eggs of various species, 430 mineral samples, 192 skulls, 42 anatomical specimens, 242 seed samples, 216 coins, and 1,492 ethnographic items. But just as neither von Metternich nor the 16-year-old baker's apprentice, Franz Sacher, who created the *Sacher-Torte*, could have foreseen the subsequent world-wide success of that confection, likewise Natterer had no idea that his discovery would one day become the most popular of all aquarium fishes, the "King of the Amazon".

So what does this so influential State Chancellor von Metternich have to do with the very first discovery of the discus, and who was he anyway? Well, let me tell you.

The future Chancellor of the Austrian Empire was born in 1773, in the German village of Metternich, on the left bank of the Mosel, just a few kilometres from Koblenz, and was baptised Clemens Wenzeslas Lothar von Metternich-Winneburg. At that time Louis XV still ruled in France, Maria Theresa in Austria, Katharina II in Russia, and Friedrich II in Prussia; Napoleon Bonaparte was only four years old, just like Arthur Wellesley, Duke of Wellington, his eventual conqueror.

When Metternich died 86 years later, the rulers who survived him were Victoria I, Franz-Josef I, Alexander II, and a boy had been born into the Prussian royal family whose destiny was to bring about the demise of the old, monarchistic Europe which the aged servant of the Austrian Empire had supported against all change.

Between these two epochs there was hardly an event of any importance in which Metternich was not involved, be it as witness or active participant; there was no notable personality during his time that he didn't know personally. Thus it was he that arranged the marriages of the Emperor's daughters Marie-Louise and Leopoldine. The latter is of the most interest to us, in connection with the first discus, although in my view it was Napoleon who was to "blame" (or, at least, chiefly to blame) for Natterer being sent on his travels. But judge for yourself.

Metternich was already well aware, when he began to draw the threads of the marriage of Princess Leopoldine's darling older sister Marie-Louise together, that marriage was a favourite instrument of Habsburg politics. *"Bella gerant alii, tu felix Austria nube!"* (Let others wage war; you, lucky man, marry Austria). And marriage between Marie-Louise and Napoleon would mean that the Austrian princess, once established in Paris, could moderate the demands of the conqueror, lay his suspicions to rest.

It is important to realise that in September 1808 Napoleon had held discussions in Erfurt with Tsar Alexander I. Pawlowitsch, his "dear brother and ally" as he called him, and Metternich had been present. And that following 1805, the Corsican tyrant had occupied Austria for a second time in 1809. That the existence of the Austrian Empire did not accord with Napoleon's concept of world domination; and that the Emperor's declarations of goodwill towards the French had met with other than belief...

Metternich was also aware that by 1808 Napoleon was already

contemplating securing his succession via a new marriage – preferably marriage into an imperial family – and that he had repeatedly flirted with the dynasty of the Russian Tsars. This view was reinforced by the meeting in Erfurt.

However, the Tsar was apparently of a different opinion. When, in 1808, he learned of Napoleon's intentions, he immediately married his elder sister Katherina to the Duke of Holstein-Oldenburg. But there was still his sister Anna. She was then only 13 years old, and there would have to be a delay, but Napoleon wasted no time. In 1809 he despatched his emissaries to St. Petersburg with an official proposal. Meanwhile (on 16th December 1809) the Senate had pronounced the divorce of Napoleon and Joséphine.

The wily Metternich was thus fully aware of the danger hanging over Austria in the event of a marital union between France and Russia, which would threaten not only her own destruction but also the carving up of Europe between these two great powers. The fraudulently obtained peace treaty (October 1809) between France and Austria had already divided the country, requiring large areas to be forfeited, and access to the sea to be relinquished. She had also been compelled to give up all trade with England. But at least this undesirable treaty had allowed the Austrian Emperor to return to Vienna on the 26th November, and Metternich to take office as Minister of State and do what he had been planning: "...from the day of the signing of the peace treaty our strategy must be solely to resist, connive, evade, and cajole. Only thus will we succeed in surviving until the day of the very probably universal liberation."

On the 28th November – ie two days after his arrival at the Ballhausplatz – he summoned Count Alexandre de Laborde to his office. The latter was the son of a great financier, a member of the national government in Vienna, and acting as commissioner for the regulation of certain financial interests dependent on the recently implemented peace treaty. Metternich put forward the possibility of a marriage between the French Emperor and an Archduchess: "The idea is my own, I have not yet sought the views of the Emperor on this matter, but I am almost certain that he will look favourably on the scheme", he said. When Metternich then learnt from de Laborde that Napoleon would consent provided acceptance of his proposal was guaranteed (clearly the two Russian "flops" had not gone down well with the French ruler), and that the way was clear in Paris, he wrote to his master on the 7th February 1810, "The marriage scheme will undoubtedly develop to our benefit." Ultimately his "idea" not only had the desired effect, but also proved popular with the entire land-owning class when the news spread like wildfire. It even saw Austrian government securities rise by 30% on the stock exchange.

The 1st April 1810 was no matter for joking, but the day of the civil ceremony, with – who would have thought it – the religious service in the Louvre the next day. In the Louvre because the Pope had not given Napoleon his consent and for this reason he wished to avoid a wedding in Notre Dame. The elders of the Roman Catholic religion opposed a church wedding between the Archduchess and a divorced man.

Just as in the case of Marie-Louise (who was called Marie Ludovica before her marriage), the Emperor Franz I (II) did not dare oppose the all-powerful Chancellor when it came to Leopoldine. In 1816 Metternich suggested the marriage of his daughter to the heir to the throne of Portugal, Brazil, and Algarde – and, of course, as with Marie-Louise this was for political as well as personal considerations. The following is the background to the situation.

In 1807 the Portuguese court had decamped in its entirety to Brazil. Napoleon occupied Portugal, but himself abandoned it later. The neighbouring English took advantage of the situation, seducing the Portuguese people with their liberal policies. For this reason the Portuguese king, João VI, was interested in a closer liaison with the houses of Habsburg and Braganza, which promised greater security against Great Britain.

Following the agreement made at the Vienna Congress (1815), Portugal became part of Metternich's system of alliances, as he himself perceived that Austria would become stronger through her influence on Portugal and the New World. (Brazil was elevated to the status of kingdom because of the liaison with Portugal, and thus became the only monarchy in South America.) And Metternich knew that his Emperor, Franz I (II), desired the restoration of the monarchy in Portugal, which would effectively put an end to the liberalisation. And the instrument for that could only be Leopoldine.

As well as representing the political and economic interests of Austria, the State Chancellor had, of course, not forgotten to think a little further ahead. If only because his Emperor was a dedicated gardener – for which reason he became known to posterity as the *Blumenkaiser* (Emperor of flowers). Franz I (II) had greenhouses erected and parks created. His children were also enchanted with nature, and he commissioned the creation of a garden for them in Schönbrunn, a garden which they had to look after themselves as part of their education, and which served to instruct them in botany. Leopoldine herself loved country life and nature. She maintained her own orchard in Laxenburg – the summer residence of the imperial household – where she herself cultivated various berry fruits, as well as keeping white foxes, a parrot, and bantams from Angola, and breeding hares.

So what could be more appropriate than for her Highness and Austria to also derive some scientific benefit? He therefore suggested that a mission for the benefit of science and culture should take place in conjunction with the marriage of her Highness the Austrian princess. Of course permission was given, and the planning for this expedition was already under way in 1816. Metternich was in overall charge, and von Schreibers, the director of the *Naturalien-Cabinet* in Vienna, supervised the scientific side.

On the 29th November 1816 the betrothal of Archduchess Leopoldine and the son of King João VI, Dom Pedro, was sealed. Emperor Franz I (II) had not given his consent until the return of the Portuguese royal house to Lisbon was imminent. He did not find it an easy decision, but it would be the first time in the history of the world that an emperor's daughter had crossed an ocean to a virtually unexplored land.

Leopoldine began to study all the contemporary books on Brazil, as well as maps pertaining to South America. She learnt the Portuguese language.

The wedding took place on the 13th May 1817 in the church of St Augustine in Vienna, albeit in the absence of the bridegroom, who was represented *per procurationem* by the Archduke Karl. The 13th May because this was the birthday of Dom João. Leopoldine had sought in vain to make it a different date. She was superstitious, and moreover, her mother had died on a 13th, her darling sister Marie Louise had taken her leave of the imperial family on a 13th, it was on a 13th that Austria had lost a battle against France, and many other instances. The Portuguese ambassador from Paris, the Marquês de Marialva, who (armed with a healthy bank balance, diamonds and other precious stones, and jewelry) had previously officially asked for the hand of the Archduchess in the name of the son of Dom João and thereby promulgated the fiction of Brazil as a land of unsurpassed wealth, was originally supposed to represent the son at the wedding, but assigned his authority to the Archduke Karl. Nevertheless he arranged, from France, for festivities to take place over several days, an event that long remained in the memories of the Viennese.

Because Metternich also wanted international recognition for the expedition, rather than it being merely a collecting trip for the imperial natural history collection, he approached various scientists, including Alexander von Humboldt, a visitor to his salon. The latter presented him with a long "wish list".

Eventually Metternich had assembled a team totalling 14 academics, researchers, doctors, and painters. At the request of the Bavarian king, Max Joseph I, the expedition was to be accompanied for a while by the botanist Philipp Friedrich von Martius (1794-1868) and the zoologist Johann B. von Spix (1781-1826), a member of the Academy and Conservator of the zoological collection in Munich. The Grand Duke Ferdinand von Toscana even delegated the naturalist Joseph Raddi (1770-1829), at that time working at the Natural History Museum in Florence, to go on the trip.

Together with von Schreibers, Metternich suggested Johann Natterer as the scientific leader in the field.

Since 1806 Natterer had been working in the *Naturalien-Cabinet*, initially as a visiting researcher (until 1808), and thereafter as an unpaid member of staff. From the end of 1809 he received 300 florins per year. He had already won recognition from Director von Schreibers for his work capturing marsh and aquatic birds from the Neusiedlersee and the Plattensee for the imperial collection. He had collected in Croatia, Hungary, and Styria, and along the Adriatic coast. In 1808 he was commissioned to take charge of a consignment of natural and archaeological items that had arrived in Trieste from Egypt, and accompany it to Vienna. On his own initiative – and mainly in his spare time – he collected fishes and intestinal worms, the latter for his director, who was a learned doctor who had very early in his career established a collection relating to the worm diseases common in those days. Moreover the Emperor himself had personally awarded Natterer a certificate of commendation for this work, and granted him permission to work in the *Naturalien-Cabinet* without remuneration (this was, of course, a signal honour...).

As well as making several trips to Italy – to Calabria – for the Emperor, usually to bring back creatures arrived from overseas or to make collections himself, he gave private tuition in English, Italian, and French. He was also involved in the original repatriation of valuable items belonging to the *Naturalien-Cabinet* from Ofen (Budapest) in 1806. When Napoleon marched in during 1805 everything had been evacuated as it was well known that the French plundered collections. Napoleon always ensured that scientists accompanied the rearguard of his army, and during the Egyptian campaign this led to important discoveries such as the famous Rosetta Stone. There were two evacuations and repatriations, in 1809 and 1813, in which Natterer played a significant part. In 1815, on the order of his Emperor, he travelled with von Schreibers to Paris in order to organise the return to the Vienna *Münz- und Antikencabinet* of the art treasures, libraries, and other items that had been spirited away from Austria by Napoleon in 1809.

Natterer's involvement with the collections during the "French years" and repatriations was one reason why in 1816 he was promoted to "supervisory assistant" in the *Naturalien-Cabinet*, and ultimately was appointed scientific leader of the Brazil expedition in the field. Even so, shortly before the departure of the expedition there was a dispute, as there were moves to replace him with a certain naturalist from Prague, Doctor Johann Christian Mikan k.k. (of the Cabinet of Natural Objects), a professor of botany, a suggestion he opposed vigorously. In the end the expedition party was split into two groups, with each of them leading one part. Both groups had to follow the direct instructions of their respective leader, and in every case consult him with regard to suggestions for excursions and collections to be made.

A set of "Instructions for service for the naturalists Doctor Johann Christian Mikan k.k., Professor of botany from Prague, and Mr. Johann Natterer k.k., *Naturalien-Cabinet* assistant from Vienna, appointed to the expedition to Brazil" was given to them, and was binding on all the participants. *Inter alia*, Rio de Janeiro was to be the starting point for all excursions. Travel plans must be made in advance, and, indeed, include details of accommodation, routes, hazards, duration, details of the return journey, etc. It would appear that no-one at the court of Vienna had very much knowledge about Brazil and the jungle! Let alone the conditions in that vast country. Only consider, Natterer needed more than a year just to get from the coast to the Mato Grosso. (Something my mother, along with four children, managed in a month 125 years later, and which today takes only two hours by plane.)

I find the section of these instructions, "Notes and comments for the expedition to Brazil", particularly interesting: for example, they were particularly to "search for half savage aboriginals, long-bearded apes, gold-panning opportunities, etc." in Cantagalo, rather nearer the coast. Cantagalo is in the state of Rio de Janeiro and was already than a well known *Município*...

When the frigates of the Austrian navy, *Austria* and *Augusta*, left Trieste on the 9th of April 1817, it was the first time in the history of Austria that ships had ventured overseas. And there were still serious problems, such as a lack of navigational equipment. The only chronometer in the entire k.k. marine arsenal of Venice was defective, and not until Gibraltar did they obtain a functional one. There was likewise no sextant until the marine commandant fetched one of his own.

Aboard the *Austria* were Mikan and his wife, Spix, Martius (both latter pictured right), and others, while Natterer and his assistant hunter/conservator Sochor shipped on the *Augusta*. Both frigates were wrecked off the Adriatic coast in a storm only three days later. The *Augusta* lost all her masts and had to lay up in Pola (todays Croazia) for a long time; the *Austria*, on the other hand, was soon under way again and first reached Rio de Janeiro on the 14th June 1817. After her repairs the *Augusta* joined the Portuguese fleet of two ships in Gibraltar and all three ships put into Guanabara Bay off Rio on the 4th November 1817, where they were welcomed with cannon, bells, and fireworks.

Metternich, who had accompanied the Archduchess Leopoldine to Livorno, where she embarked, wrote letters to his family describing interesting events and detours during his journey across Italy. And I cannot resist including a few extracts here, in part because I myself have now lived in this beautiful country for several years. He enthuses about Padua, Ferrara, and Bologna, and then writes: *Florence, 14th June 1817:*

"We have been here since 11 o'clock yesterday morning... Everything I have seen so far exceeds my expectations. Good God! What fellows those men of the past were!.. The country is glorious... the climate is heavenly."

And because the arrival of the "cursed squadron" – as he termed the Portuguese fleet – continued to be delayed: *26th July 1817:*

"The squadron is, happily, now at anchor in Livorno... I am off on my travels again, to the Baths of Lucca. I plan to begin my treatment tomorrow..." (He had eye problems and even had an eye specialist with him.)

Baths of Lucca, 28th July 1817:

"I am staying in the house that Elisa (Napoleon's eldest sister) had built for herself, or rather, had converted for herself (todays Vila Reale outside of Lucca). That should tell you that it is comfortable and well located... News from Livorno... that the squadron will weigh anchor before 15th August."

Livorno, 10th August 1817:

"I arrived here at eight in the evening. I found the entire court here, along with 4,000 foreigners, I saw my princess and went to an evening entertainment."

11th August 1817:

"This morning I was aboard the Portuguese warships... You cannot imagine how many people there are on such a warship... As well as the Austrian ladies the entire Portuguese court is there... The number of officers of all ranks has been tripled. In addition there are a considerable number of cows, calves, pigs, lambs, 4,000 chickens, several hundred ducks, and 4,500 canaries, as well as large and small birds from Brazil; so you can imagine that Noah's Ark was just a child's toy compared with the *Johann VI*." (He was referring to the *Dom João VI*, one of the two warships transporting the bride. There were a total of 1,300 men on the *São Sebastião* and the *Dom João VI*. The numerous ornamental birds were for the entertainment of her Highness.)

13th August 1817:

"Today around four I escorted her Ladyship the Archduchess on board her ship... her apartment pleased her very well... it would have been difficult to have decked it out more elegantly."

And then from the Baths of Lucca on the

16th August 1817:

"I have bade farewell to my Archduchess. The squadron set sail yesterday morning around half past six."

Leopoldine had also said farewell to her much-loved sister Marie-Louise, by now elevated to Duchess of Parma, Piacenza, and Guastalla as a result of the Vienna Congress. At that time neither of them as yet had any idea that this was the last time they would see each other.

Metternich had completed his work, although he repeatedly had to intervene when there was trouble in Brazil – if disputes or financial problems arose. In fact this was the case only with Mikan, and after barely a year the latter returned prematurely to Europe, along with other members of the expedition, some of whom were seriously ill. In 1820 he also published a splendid volume, *Delectus faunae et florae brasiliensis* (The enjoyment of the flora and fauna of Brazil), notable not only for the high quality of its illustrations but also for a whole series of new descriptions – including a *Metternichia*, a tree-like shrub (family Solanaceae).

Leopoldine, who was deeply impressed with the rich nature of Brazil, wrote to her father on the 26th January 1818 that every

day she made new discoveries in the plant, animal, and mineral kingdoms. And that every morning at eight she rode out hunting with Dom Pedro. She sent back live plants and animals, hides, stuffed birds, minerals, and butterflies – mainly to Marie-Louise and her father. But the scientists didn't do badly, either.

Donna Leopoldina, as the Portuguese called her, was very active. She had a Viennese *Naturalien-Cabinet* established at the fortress of São Cristovão – this later developed into the Museo Nacional (the Brazilian National Museum). She was involved in the founding of the splendid Jardim Botanico in Rio, as well as the animal park at Santa Cruz. But Leopoldine was not only a blessing to science and nature – plant genera and species were named in her honour – but also an energetic woman who knew how to get her way, something that had already caused Metternich some concern. She took a lively part in politics and Dom Pedro listened to and asked for her opinion on difficult questions, as his role as Emperor had not yet gone to his head. She played an important role in the elevation of his rank, the liberation of Brazil from the Portuguese motherland, and the coronation of her husband as Emperor in 1822. The design of the Brazilian flag was her concept (with the southern cross in it, only the text: *Ordem e Progresso* – Order and Progress – was added later). She also earned long-term credit for her sorties into art and culture. Her charitable acts, her social work, her trips to the quarter of the poor (where she endeavoured to help in person) are still remembered today. Many Brazilian *ruas* (streets), *praças* (squares), *bairros* (parts of cities), *cidades* (entire cities), and *provincias* (provinces) bear her name. Sadly she didn't even attain her 30th year. On the 11th December 1826 the talented Empress of Brazil died – apparently of a broken heart. Dom Pedro had forced her to acknowledge his mistress, Donna Domintila, whom he had ennobled, as the first lady of the court and allow her to sit on the throne next to him.

The two Austrian frigates *Austria* (left) and *Augusta* by Thomas Ender (Kuperstichkabinett der Akademie der Bildenden Künsten, Vienna).

Meanwhile Natterer had many expeditions behind him and had despatched large collections from the region around Rio, São Paulo, Goiás, Minas Gerais, and Rio Grande do Sul back to Austria.

Originally he wanted to be in Mato Grosso by this time, but the Brazilian government refused him access, so Martius and Spix travelled before him into the paradise of birds, as they termed it. In 1829, along with Dr Jean Louis Rodolphe Agassiz, discoverer of the 2nd Discus much later *(page 42)*, a Swiss ichthyologist who subsequently emigrated to America, they published in Munich the book *Selecta Genera et Species Piscium quos in Itinere per Brasiliam Annis MDCCCXVII – MDCCCXX Jussu et Auspiciis Maximiliani Joseph I. Bavaria Regis Augustissimi* (Selected genera and species of fishes) with numerous drawings, two of which (of cichlids) are presented here *(right-hand page)*. But they didn't find any discus.

It is also worth mentioning that this work, which appeared in Latin, listed Spix as the author and describer of new genera and species (eg the characiform genus *Leporinus* and the cichlid species *Cichla monoculus)*, although he had already been dead for three years.

As luck would have it, Natterer also made collections in the Rio Tietê in São Paulo, in which all life forms except bacteria have now been extinguished by catastrophic environmental pollution – São Paulo is today in all probability the most heavily populated city on Earth, with probably more than 30 million inhabitants.

His son-in-law Julius von Schröckinger-Neudenburg later divided Natterer's travels into 10 sections *(see map, page 37)*. By the time of the third (to the Tietê and Curitiba) orders had arrived from Vienna for him to return to Europe (this subsequently happened twice more). He wrote to his brother that he must get to the Mato Grosso (like my mother later) and that he would from then on continue at his own expense. He took out a loan and thereafter sent his collections to England for sale. Whereupon he received permission to continue, along with

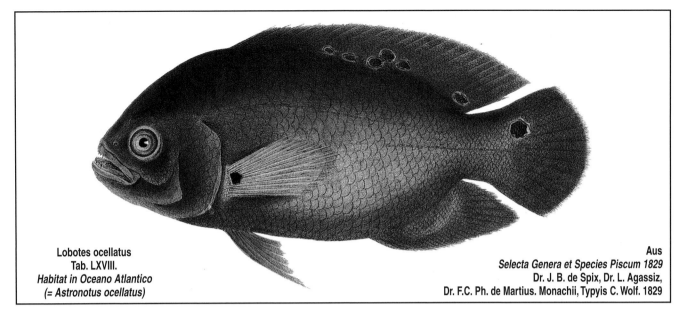

Lobotes ocellatus
Tab. LXVIII.
Habitat in Oceano Atlantico
(= *Astronotus ocellatus*)

Aus
Selecta Genera et Species Piscum 1829
Dr. J. B. de Spix, Dr. L. Agassiz,
Dr. F.C. Ph. de Martius. Monachii, Typyis C. Wolf. 1829

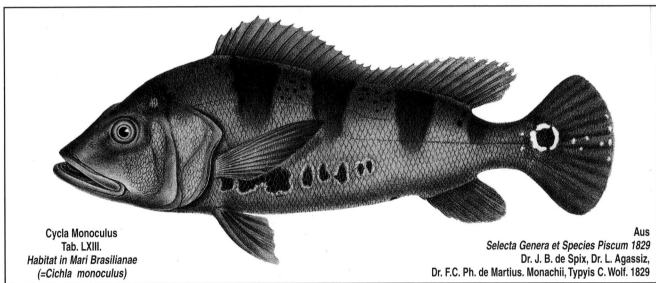

Cycla Monoculus
Tab. LXIII.
Habitat in Mari Brasilianae
(=*Cichla monoculus*)

Aus
Selecta Genera et Species Piscum 1829
Dr. J. B. de Spix, Dr. L. Agassiz,
Dr. F.C. Ph. de Martius. Monachii, Typyis C. Wolf. 1829

adequate financing, from Vienna. Metternich had without doubt realised the possible consequences otherwise.

Natterer planned to make his way from Curitiba to the Mato Grosso with 23 mules, five horses, and four dogs. He had purchased two slaves and rented another two. Special ox-hide sacks were made which the mules could carry on one side, with a cask of brandy on the other, for preserving the animals, amphibians, and fishes that were collected. He set off in October 1822 and, travelling via Goiás, reached Cuiabá, today's capital of Mato Grosso state, in December 1823. The trek was very arduous and lasted more than a year (it took my mother over four weeks by car through the "green hell").

The hardships were so great that he fell sick with an acute infection of the liver and had to remain there until 1825. During this journey he encountered indians for the first time and acquired large quantities of artefacts (like us). The Mundurukú indians made their dance head-dresses, arm-bands, aprons, and many other items from the feathers of parrots and bare-faced curassaw *(Crax fasciolata,* family Cracidae, localy known as *mutum – see also page 280).* The indians generally kept these birds specially for the purpose, repeatedly plucking out their feathers.

Natterer's sixth journey (from January 1825 to July 1829) likewise took place under an ill star. In Arraial de São Vicente his faithful companion Sochor fell ill with a very bad fever (undoubtedly malaria) and died. In this village of 600 souls there was no doctor, let alone medications. Then Natterer himself suffered attacks of fever – the malaria had him in its grasp as well. He survived only thanks to the intervention of a miller's wife named Gertrud, who took him to Vila Bela de Santissima de Trinidade, on the Rio Guaporé, where he was cured.

Natterer stayed in the Mato Grosso region and by the fabulous Guaporé until July 1829, collecting and making the bulk of his fish drawings *(two of which can be seen right),* including that of the piranha species later named after him *(Serrasalmus nattereri).* Although it is repeatedly stated that there are discus in the Rio Guaporé, this is not the case. Natterer found no *Symphysodon* there, nor have my mother or I during our numerous collecting trips years later. Discus are nowhere in their range found in the vicinity of rapids or waterfalls, let alone upstream of these obstacles. And the Guaporé-Marmoré (the lower Rio Guaporé is sometimes called the Marmoré after it is joined by the left-bank affluent of that name) has more than 20 north of Guajará-Mirim, before, together with the Rio Beni and Rio Abuná, it becomes the Rio Madeira.

Natterer followed the course of the Rio Guaporé, the centre of which forms the border between Brazil and Bolivia for more than a thousand kilometres in this region, where it is known as the Iténez. He successfully negotiated the rapids and the perilous cataracts at Teotonio on the Rio Madeira, and reached the village of Borba November 1829.

On his eighth journey (June 1830 to the beginning of 1831) he followed down the Rio Madeira passing to the south of the Island Tupinambarana, up the Amazon and the Rio Negro upstream to São José de Marabitanas near to the border with Venezuela. He surmounted the waterfalls at Uaupés (nowadays São Gabriel da Cachoeira) and on his journey upriver went some distance up three Rio Negro affluents, the Rio Içana, the Rio Xié, and the Rio Uaupés. (Only the Englishman Richard Spruce (1817-1893) managed to travel further up the Uaupés, 20 years later – see also Römer *et al., 1995, aqua geōgraphia* Vol. 11.)

Natterer's drawing of the piranha caught in the Rio Guaporé in 1829. Rudolf Kner described this fish as *Serrasalmus nattereri* in 1860.

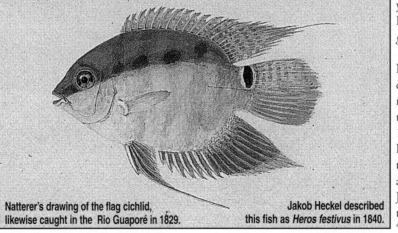

Natterer's drawing of the flag cichlid, likewise caught in the Rio Guaporé in 1829. Jakob Heckel described this fish as *Heros festivus* in 1840.

On the way back Natterer stopped at Barcelos, and from this base made the collections of the 9th "section" between 1831 and 1834, in the Rio Negro basin and in the Rio Branco upstream as far as Forte São Joaquim. It was during this period that the discus "found its way into his net" – in actual fact an indian caught it for him near the mission Moreré (later called Moreira), but not, I think, with bow and arrow as they usually do (see photos and drawing *on page 39).*

We cannot imagine what Natterer must have thought when he held this unique discus-shaped fish in his hands. Moreover he never made any drawing or wrote anything about it (or perhaps any relevant papers were stolen in Belém (see below) or went up in smoke in the subsequent fire in Vienna – nobody knows).

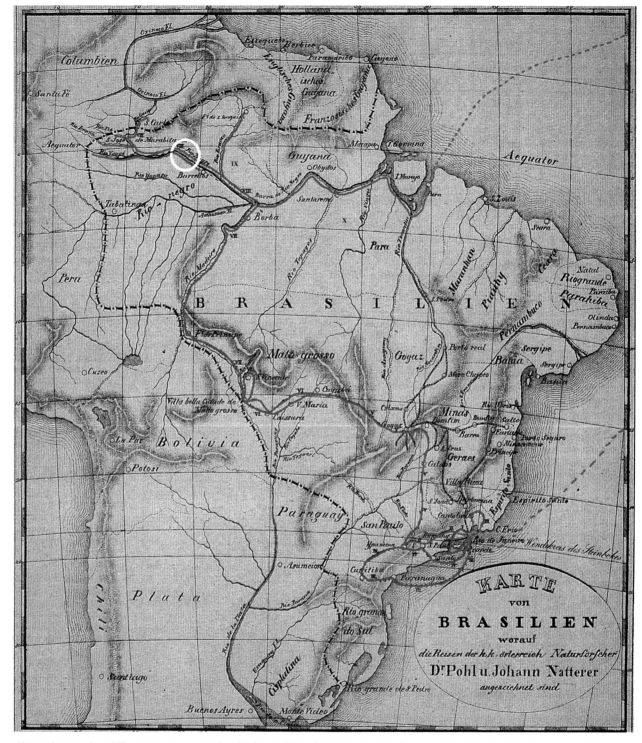

Natterer's route - in red. The maps were prepared at the time and also show the route taken by the botanist Pohl (in black). It is also possible to see (on closer examination) Natterer's 10 journeys (I, II, III, IV, V, VI, VII, VIII, IX & X). The white circle is where Natterer collected the first discus *(Symphysodon discus)*. The black dotted line shows the border of Brazil at that time. Pará at the mouth of the Amazon = Belém.
With permission of the Kupferstichkabinett der Akademie der Bildenden Künste, Vienna.

Be that as it may, one thing is sure – today the "prince of collectors" is at least as well known in scientific and ornamental fish circles worldwide as Herr Sacher among gourmets around the world for his cake. (With the slight difference that the young Sacher didn't invest as much time and didn't have to undergo any hardships to achieve his fame.) And by the way, Sacher's cake is today sent all over the world in boxes, just like live discus!

Natterer had married a Brazilian woman, Maria de Rêgo, in Barcelos (1831) and she bore him three daughters. On his tenth and last expedition (1834-35), accompanied by his family, he lost almost all the material he had collected. Civil war had broken out in Pará. The bloody Cabanagem popular uprising was raging in Belém and other parts of Pará. The province had declared independence from the Portuguese crown pending the proclamation of the majority of Pedro d`Alcântara (Pedro I had abdicated in favour of his first-born, five-year-old son Pedro d`Alcântara, and a council of regents had taken control while the successor to the throne remained a minor). In the city of Belém, at the mouth of the Amazon, from where Natterer planned to embark for home, he was robbed and lost the bulk of his material. Even his extensive collection of live animals, destined for the imperial menagerie, fell victim to the plunderers. They devoured the priceless tapir and other animals on the spot!

Natterer wrote, "...I had to leave all my things to their fate on land with three blacks, I went ashore with the Englishman and to my house, where I found everything in the greatest state of confusion; almost all my chests had been broken open and the contents scattered; everything of value – my clothes, three airguns, eight firearms, pistols, 600 florins, and many other items stolen. Almost all the menagerie had been killed apart from a few monkeys, parrots, and parakeets, including the tapir, all the turtles, everything that was edible. My blacks had several times narrowly escaped death. At great risk... every day I returned to the house to get the rest of my things together and pack, and at night I slept on the corvette. Because of lack of time and porters I had to leave some of my things behind, including a large chest with two 12-foot sawfishes and other large fishes that I had collected along the sea coast during the months of February and March..."

Luckily for us, the discus was among the items rescued and thus put to sea aboard a ship of the English navy along with Natterer and his family on 15th September 1835.

After 18 years in Brazil he arrived, along with his wife and three daughters, in Vienna on the 13th August 1836 (Leopoldine's unlucky number). But the uprooting was too much for his Brazilian wife. Frau Natterer and two of their daughters died. Only Gertrude – the third daughter, named after her father's "saviour" from the Mato Grosso – survived.

The first discus was described by Heckel four years later in his work *Johann Natterer's neue Flussfische Brasilien's nach den Beobachtungen und Mittheilungen des Entdeckers beschrieben. Erste Abteilung, die Labroiden.* (Johann Natterer's new river fishes from Brazil, described from the observations and specimens of the discoverer. Part 1, the Labroids.) (See also Chapter 2: *Heckel's work).*

Unfortunately the prince of collectors didn't live long after the publication of this work on the cichlids, but died of a pulmonary embolism on 17th June 1843 – very likely the price of

On the opposite page (above) is the most famous Brazilian ichthyologist, Prof. Heraldo Britski (left in the photo, with H. Bleher (with a broken arm) right, visiting the *Museo de Zoologia* in São Paulo after an expedition to the Amazonas), holding a *Symphysydon discus* that was shot with an indian arrow in the Rio Trombetas. (A detail is shown on this page, below.) But under normal circumstances the indians never kill discus, as they regard them as sacred and a symbol of fertility (see Bleher & Linke, 1991a: video *The World of Discus I),* although they have been known to shoot them with bow and arrow for the white man, as can be seen in the engraving from the book *Selecta Genera et Species Piscium* (1829). (How else were the naturalists of those times to obtain their material?). On page 38 (below) we see the holotype of *Symphysodon discus* Heckel, 1840. Natterer's specimen which was used for the description of the genus and species that remain undisputed to the present day. Unlike so many species, the Heckel discus has not undergone any change in its scientific name, which has remained unchallenged for more than 160 years.

the almost unimaginable hardships he had undergone in the Brazilian jungle – without having published his main work on the ornithology of the region. And, of course, with no idea that he would still be honoured, albeit posthumously, in his native land today.

His fellow-countryman, Rudolf Kner (1810-1869), was the first to immortalise Natterer. This son of a high-ranking government official from Linz had planned to become a doctor, but his talents as a naturalist were recognised while he was still at boarding school, at the age of just 15. He nevertheless acceded to the wishes of his father and studied medicine, graduating as a general practitioner and surgeon in 1835. During his studies in Vienna Kner had, however, had regular contact with the *k.k. Naturalien-Cabinet,* and in particular with Johann Jakob Heckel, helping – unpaid, of course – with the collections. And during this time the graduate in agriculture (against his will) and the doctor (against his will) became close friends (See Chapter 2: *Heckel's work).*

In the same year that Natterer returned from Brazil, Kner even took up a position as assistant to Heckel at the royal museum, on a salary that was, as he said, *"...zum Leben zu wenig, zum Sterben zu viel"* (too little to live on, but too much to die on). In 1840, when Heckel's work was published, they made a collecting trip to Dalmatia to investigate the riverine fish fauna. Hardly had they returned when Kner was invited to take up the post of Professor of Natural History and Agriculture at the University of Lemberg. But his new position made no difference to his attachment to ichthyology and his friendship with his mentor, Heckel. Far from it. Kner did the first good drawing *(lower)* of

Rudolf Kner, the well-known Austrian ichthyologist *(top)*, painted the first Heckel discus *(S. discus) (above)* in 1842. Meanwhile in 1865, during the Thayer Expedition, J. Burkhardt painted the first green discus (plus a juvenile) (right-hand page), although he didn't know it was a Green at the time.

Natterer's remarkable discus in 1842 – Heckel had provided only a rough sketch with his description (page 113). Kner's drawing was the first reasonably accurate illustration of the species *Symphysodon discus.*

Eight years later Kner returned to Vienna and in 1849 took up the newly created chair of zoology at Vienna University. From then on he again had access to the ichthyological collections and began to work on the parts with which he was most familiar and the *"Schätze von Johann Natterer"* (treasures of Johann Natterer) which had thus far remained unstudied (apart from Heckel's work).

First of all he published *Die Panzerwelse des k.k. Hof-Naturalien-Cabinetes zu Wien* (1853), which was followed by *Die Hypostomiden – Zweite Hauptgruppe der Panzerfische* (1854), which contained numerous new descriptions and remains the standard work on these catfishes to the present day. After several publications on South American catfishes, following Heckel's death in 1857 Kner worked on Natterer's characins, immortalizing the "prince" for the first time with the piranha species most commonly kept today, *Pygocentrus nattereri* Kner 1858. He described two new genera *(Rhytiodus* and *Bryconops)* and a total of 36 new species, including *Chalceus opalinus,* which was subsequently recategorised as *Brycon nattereri* Günther 1864 (*ex* Kner). Kner then joined the ranks of the great ichthyologists through his work on the monumental amount of material that the frigate *Novara* had brought back in 1859 from her three-year circumnavigation of the world. To assist him in this, the professor of zoology engaged a young law student who had changed tack to enter the field of natural history (apparently the fate of all Austrian ichthyologists!). His name was Franz Steindachner (1834-1919), and he was destined to become one of the best known of all ichthyologists internationally. They studied 1,600 specimens from 550 species, their work subsequently appearing in a major publication, *Reise der österreichischen Fregatte* Novara *um die Erde*, in 1868, one year before Kner's death.

As a result of his outstanding work, Steindachner was appointed to the post of head of the fish collections, which had remained vacant since Heckel's death. In the period from 1859 to 1868 he published no less than 55 ichthyological works (about 900 pages), including one on *Sternarchogiton nattereri* (Steindachner, 1868). This was followed by further species named in honour of their discoverer, which still remain valid today. In 1876, *Corydoras nattereri, Leporinus nattereri, Copella nattereri, Achiropsis nattereri,* and *Thalassophryne nattereri*. And in the years that followed, *Anchoviella nattereri* (Steindachner, 1879), *Trachydoras nattereri* (Steindachner, 1881), *Aphyocharax nattereri* (Steindachner, 1882), and *Farlowella nattereri* Steindachner, 1910.

Interestingly this extraordinary man made collections in South America for the second time in 1903, at the age of 69 (his first visit was during the Hassler Expedition, which covered almost all of the New World, in the company of the world-famous ichthyologist Jean Louis Rodophe Agassiz (1807-1873), Swiss by birth but living in America). He didn't bring any discus back from this trip – or, if he did, I can find no information on it. What I have discovered, however, is that he was possibly distantly involved with the second discus discovery – earlier than Jobert (see below). But whether or not Steindachner, who at Agassiz's invitation travelled to America in 1868 to study the collections of the Thayer-Expedition, saw any discus there, remains unknown.

THE SECOND DISCUS

Credit for the discovery of the second discus has usually been given to a Frenchman, Clément Jobert, who fancied himself as an architect, although he was really a naturalist and doctor of physiology. But it now looks as if he wasn't after all the "second discoverer" of the discus as previously assumed.

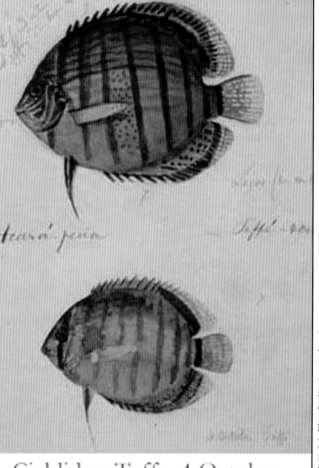

Cichlidae: Teffe, 4 October

The famous Thayer Expedition (New York-Brazil-New York, 1865-66), financed by the Boston merchant Nathaniel Thayer and led by Louis Agassiz, returned from Brazil with some 34,000 fish specimens. The expedition had collected at 156 sites including around Eda (= Teffé; both place-names are cited, but they are one and the same – Teffé (nowadays Tefé) was formerly called Eda and Ega), where Agassiz and his colleagues apparently netted green discus at the site THAYER138 (the label reads: "LAGO TEFFÉ; EDA, collectors L. Agassiz *et al.*, collection date 14 IX-22 X 1865"). Agassiz's companions included Jacques Burkhardt (1808-1867) (the man with the white beard, *see page 42, centre*), his "personal artist", who on the 4th October 1865 painted an adult green discus *(left)*, along with a younger specimen which does not yet show the typical red dots, but whose anal fin clearly demonstrates that it is likewise a Green.

These are the first known colour illustrations of a discus. Thus Agassiz and his companions *(page 42, centre)* must have had these fishes in their hands in 1865, at the place where the Rio Tefé enters the lake of the same name, without realising that they were a second species.

Burkhardt painted some 2,000 fishes in watercolours. These detailed paintings remained unpublished and after many years were passed to the ichthyologist George S. Myers around 1940, with the words "...maybe you can find something interesting among these paintings, otherwise I am sure you have a fire once in a while in California..." And it was not until a few years ago that they resurfaced in Cambridge, Massachusets (in the Museum of Comparative Zoology at Harvard University, where most of the fishes from the Thayer Expedition are stored). His discus painting can now be displayed for the first time in this book.

Jobert, meanwhile, netted a further two green discus some 13 years later, in the same place (or, at least, somewhere around Lago Tefé), and also collected additional specimens elsewhere (the precise number is in question – see Chapter 2: *Pellegrin's description*).

Unfortunately not very much is known about Dr. Clément Jobert. Even in his birthplace of Lyon there are today no records to be found of his birth or death. There is no biography, and the Internet has no entries under his name (except under *curare*, see below), although he was a well-known physiologist. All that remains is a number of his works, published between 1870-1881, in museum libraries.

Clément Jobert studied the sensory organs of various animals (mainly those of mammals, including humans) from 1870 to 1876 and the respiratory organs of terrestrial crabs (1875). He published material on the optical organs of cirrhipeds (with Georges Pochet in 1876); and, after (or during?) his time in Brazil, on the *curare* poison (1878) and diseases affecting Brazilian coffee pickers (1878). He also researched the evolution of insect genera (1881). But he wrote only three papers on fishes: in 1870, on the evolution of their sensory organs; in 1877 on the evolution of their respiratory systems; and in 1878 on the anatomical and physiological factors that led to the evolution of the

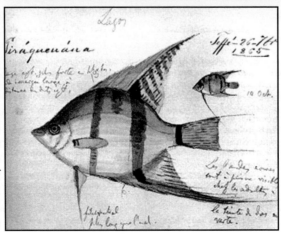

respiratory organs in fishes. He undertook one or more (?) research expeditions to Brazil in the 1870s and made collections in cooperation with the Brazilian government. He travelled to Rio Grande do Sul and the mysterious Serra do Estrello (= Mountains of the Stars, not found on any map, but possibly a contemporary name for the mountains around Petropolis), bringing back *inter alia* a new species of characin which his fellow-countryman Pellegrin described in 1909, naming a new subgenus after its discoverer – *Characidium (Jobertina) interruptum*. But more of that later.

However, he spent the longest time in the Amazon region, from which he brought back in 1878 what M. León Vaillant described in 1880 as a *magnifique collection* (magnificent collection). And this included three or four specimens of discus which Pellegrin described in 1904 as *Symphysodon discus* Heckel var. *aequifasciata var. nov.* (see Chapter 2: *Pellegrin's Description*).

Jobert collected mainly (or exclusively) in seven places:

in Pará (= Belém) and on the island of Marajó; in the mouth region of the Xingu; around Santarém; Manaos and Barra do Rio Negro (= Manaus); around Teffé (= Tefé); Tonantins (on the Solimões); and in Calderón (= Calderão or Tabatinga). The four discus subsequently studied by Pellegrin were collected at Santarém (1), Teffé (2), and Calderón (1). A further specimen collected at Barra do Rio Negro (= *S. discus*), was not included in Pellegrin's description.

However, during my researches into Jobert (with the aid of the world-famous characin expert Jacques Géry) I came across a number of very interesting stories that I would like to relate briefly here.

Dr. Clément Jobert was interested in architecture (as a hobby?), and decided to build a miniature version of the Chateau de Versailles at Tefé, on the edge of Lago Tefé, right in the middle of what was then the deepest Brazilian jungle. Apparently the mind-blowing heat and almost 100% humidity proved too much for him (and perhaps led to his early death (all traces of him disappear after 1881), the fate of so many explorers before and after him).

Be that as it may, he did in fact start to translate his dream into reality, as evidenced by the presence on the lake shore even today of frescos and palatial steps leading up to the Prefecture *(right)*. And the original structure of the *chateau* itself – or, rather, its beginnings – is still standing, and has been occupied by Jesuits for more than 100 years. It was the first massive (stone) building by the lake, and the stone must have been brought from far away by *lancha* - there is (was) nothing but sand and jungle in the vicinity of Tefé. Traces of the fabulous castle can still be seen in the sculptures hewn from stone *(photo on top)*.

The "architect" – as I will call him – was, however, also interested in something quite different: the production of the *curare* poison of the Tikuna indians (also written as Ticuna, Tecuna, and also Tukuna). He visited this tribe in the vicinity of Tabatinga, where he also found a discus – a Blue? – which Pellegrin mentions 25 years later as when he described var. *aequifasciata,* as mentioned above – but this specimen cannot be found anywhere (see Chapter 2: *Pellegrin's description*).

Apropos of which, it is interesting to note that Harald Schultz is also said to have caught the so-called "Blue discus" at Tabatinga – albeit some 80 years later. Moreover Schultz too sought out this indian tribe, only somewhat further to the north-east, and made a film about it (see also *discoveries in the 20th century*, below).

Back in Belém de Pará following his researches in Tabatinga, Jobert wrote as follows to French toxicologists *Sur la préparation du curare* (on the preparation of *curare*), describing in detail how the Tecuna indians (as he calls them) produce the *curare* poison:

"*Le Dr Jobert a pu faire préparer devant lui l'un des meilleurs curares américains, celui des Indiens Tecunas, au Calderão (Brésil), non loin de la frontière péruvienne. C'est un poison purement végétal.*

Les éléments principaux de la préparation sont:

1° L'Urari uva, plante grimpante, du type des Strychnées (peut-être le Strychnos castalnae de M. de Weddell);

Jobert, the "second" discoverer of the discus – the "architect" – started building a residence resembling the palace of Versailles on the shore of Lago Tefé, a structure full of wonderful, lavish, and artistic ideas. Today only the walls and various sculptures remain *(top, right, & centre)*. The palace built by the "Sun King" Louis XIV in 1678 was the pattern for many others, for example King Ludovic II's Linderhof, Frederick the Great of Prussia's Sans Souci in Potsdam, and Prince Miklós Esterházy's "Hungarian Versailles", Fertőd. Their names became immortalised – but Jobert is remembered only in the names of a few fishes such as *Jobertina (above)*.

2° L'*Eko* ou *Pani du Maharão*, plante grimpante offrant les caractéres des Menispermacées (peut-être le *Cocculus toniferus* de M. de Weddell).

Les éléments acessoires sont:

3° Une Aroïdée, le *Taja*;

4° L'*Eoné* ou *Mucura-ea-ha* (*Didelphys cancrivora?*), qui a le port d'une Amarantacée);

5° Trois Pipéracées (du genre *Artanthe?*)

6° Le *Tau-ma-gere* ou *Langue de Toucan*.

Ces plants ont été photographiées par M. Jobert, qui en rapportera des échantillons en Europe et pourra en donner une détermination plus exacte.

Voici comment les Indiens procédèrent à la préparation du poison:

Ils râclèrent la première pour écorce, fort mince, des rameaux les plus développés de l'*Urari* et de l'*Eko*, et mélangèrent ces râpures dans la proportion de 4 parties de la première pour 1 partie de la seconde.

Ce mélange, pétri la main, placé ensuite dans un entonnoir en feuille de palmier, fut épuisé à l'eau froide, qu'on reversa sept ou huit fois. Le liquide prit alors une teinte rouge. L'Indien le fit bouillir avec des fragments de tige de *Taja* et de *Mucura*, pendant environ six heures, jusqu'a l'amener à une consistance épaisse. On ajouta à ce liquide la râpure des ... ?

Sir Walter Ralegh (1552-1618)

("Dr Jobert was able to observe the preparation of several American *curares*, those of the Tecuna indians at Calderão (Brazil), not far from the frontier with Colombia. *Curare* is a strictly vegetable poison.

The main ingredients in its preparation are as follows:

1. The *Urari uva*, a climbing plant of the nightshade type (perhaps the *Strychnos castalnae* of M. de Weddell);

2. The *Eko* or *Pani du Maharão*, a climbing plant exhibiting the characteristics of the Menispermaceae (perhaps the *Cocculus toniferus* of M. de Weddell).

The secondary ingredients are as follows:

3. An aroid, the *Taja*;

Charles Marie de la Condamine (1701-1774)

4. The *Eoné* or *Mucura-ea-ha* (*Didelphys cancrivora?*), which has the habit of the Amarantaceae);

5. Three Piperaceae (genus *Artanthe?*)

6. The *Tau-ma-gere* or *Langue de Toucan*.

These plants were photographed by M. Jobert, who brought samples back to Europe and was thus able to provide a more precise determination.

Here is the way the indians prepare the poison:

First of all they finely scrape the most developed stems of the *Urari* and the *Eko* in order to obtain their bark, which is then mixed in the ratio of four parts of the former to one part of the latter.

This mixture is kneaded by hand and then placed in a funnel made from a palm leaf and squeezed out into cold water, which is passed through seven or eight times. The liquid thereby takes on a red colour. The indian boils pieces of *Taja* and *Mucura* stem in this liquid for about six hours until a thick consistency is achieved. The next ingredient to be added to the liquid is the bark of the..." (The text ends at this point.)

Jobert was in all probability the first to reveal the millennia-old secret of the indians – the composition of the poison with many names – different tribes or dialects variously pronounce it *woorari, woorara, curari, cururu, ourari, wourali,* or similar.

And this brings me to the European discoverer of this indian arrow-poison, Sir Walter Ralegh (sometimes spelt wrongly, as Raleigh) (1552-1618). This courtier, poet, explorer, and adventurer travelled the Orinoco on two occasions, in search of *"El Dorado"*. As early as 1595 he brought the poison (but not Eldorado!) back from his first expedition. This remarkable "all-rounder", a knight of noble blood and long-time favourite of Queen Elizabeth I as well as Captain of her Royal Guard (and lover?), was in 1592 imprisoned by Her Majesty in the Tower of London when she learnt of his secret marriage to Elizabeth Throckmorton, a maid of honour at the court. However, he was soon released again to resolve a dispute over a captured *carraque* (a Portuguese galleon) filled with treasure – only he could deal with the matter, as the expedition had been planned by him. And in 1595 he began his quest for *El Dorado*. The following year, after returning, without having found his goal, with just a few pieces of gold and the arrow poison, he published his first book, *Discovery of Guiana*, in which he wrote as follows on the subject:

"The fourth are called Aroras, and are as black as negroes, but have smooth hair; and these are very valiant, or rather desperate, people, and have the most strong poison on their arrows, and most dangerous, of all nations, of which I will speak somewhat, being a digression not unnecessary.

There was nothing whereof I was more curious than to find out the true remedies of these poisoned arrows. For besides the mortality of the wound they make, the party shot endureth the most insufferable torment in the world, and abideth a most ugly and lamentable death, sometimes dying stark mad, sometimes their bowels breaking out of their bellies; which are presently discoloured as black as pitch, and so unsavory as no man can endure to cure or to attend them.

And it is more strange to know that in all this time there was never a Spaniard, either by gift or torment, that could attain to the true knowledge of the cure, although they have martyred and put to invented torture I know not how many of them. But everyone of these Indians know it not, no, not one among thousands, but their soothsayers and priests, who do conceal it, and only teach it but from the father to the son."

Sir Walter never found out either. Around 1600 he was proclaimed Governor of Jersey, but his luck was running out. Because of political unrest over Essex's purported treason and execution, and because his enemies claimed he was against the accession of James I (in 1603), he fell from favour. He was removed from office, stripped of his property, and accused – without any actual proof – of plotting treason with Spain. In addition he had supposedly plotted to kill the king and enthrone Arabella Stuart in his place. But he was reprieved from the scaffold and instead sent back to the Tower. Here he devoted himself to science and literature, and began his never completed *History of the World*.

Ralegh was released in 1616 and shortly thereafter headed off to the Orinoco again, in search of gold. He was warned to leave the Spanish colonists in peace. This expedition was blessed with neither gold nor discus (of course, there are no *Symphysodon* in the Orinoco system – see Chapter 3: *Distribution*), and hence a failure. But his companion, Laurence Kemys, captured a Spanish settlement, and after Ralegh returned to his native

Top: a woodcut dating from 1848 showing Ticuna indian women, in typical costume, dancing. Jobert visited the Ticuna at Calderão – which was also where he collected a discus and studied the curare poison in which the indians dipped their blow-pipe darts in order to kill their prey (above). Jobert established that they derived the poison from *Strychnos castalnae, Cocculus toxiferus,* and *Didelphys cancrivora,* a plant of the arum family they called *taja*, and three species of Piperaceae (*Artanthe* spp.). They extracted the sap from these plants (without the bark) and boiled it in water (sometimes mixed with poison from snakes, ants, and frogs) for 2 days, and then let it dry out.

land the Spanish ambadassor asked the English crown to punish him (for something that he hadn't done).

Before he was finally executed at the Tower of London in 1618, Ralegh fingered the headsman's axe and declared, "This is a sharp medicine, but it is a physician for all diseases." It appears his last thoughts were of *curare*. Perhaps he was thinking about the contrast of *curare* being a slow painful death...

Unusually for those days, his wife was allowed to claim his head, which she had embalmed and kept constantly with her for 29 years, until her own death.

Before returning to Jobert, the second discoverer, I would like to make another digression, this time on a very well-known Frenchman, de la Condamine, an officer with a passion for the lathe. He was a bit touched – as Jobert with his Versailles – constantly preoccupied with the possibility of controlling and thus automating the turning process by fitting templates to the lathe, for example to produce portrait medallions. Tradition has it that he invented an apparatus for mass-producing cameos using portrait templates. He also developed, *inter alia*, a technique for the automated engraving of patterns on flat surfaces. His lectures on his work to the French Academy of Sciences are preserved in the annals of the Academy.

Charles Marie de la Condamine (1701-1774), was, however, primarily a mathematician, physicist, explorer, and geographer. La Condamine was sent to Ecuador in 1735 to measure the Earth at the Equator. He was the first European to make a scientific study of the Amazon region (he even collected about 30 fish species in Lago Tefé – but no discus) and he mapped the Amazon by following it by raft from the Andes to its mouth. La Condamine had already made adventurous expeditions to Algeria, Alexandria, Palestine, Cyprus, and Constantinople.

In Europe at that time learned people were still debating whether the polar circumference of the Earth was greater than the equatorial. The King of France and the French Royal Academy of Sciences had commissioned two expeditions in order to answer this question. One was shipped to Lapland (under the leadership of the Swedish physicist Anders Celsius) and the second to Ecuador. La Condamine initially set off with the second group, along with Louis Godin and the mathematician Pierre Bouguer.

When, in 1735, they landed in Colombia they had to cross the Isthmus of Panama on foot in order to sail on to Ecuador. La Condamine marched through the rainforest with Pedro Vincente Maldonado, the local Governor and a mathematician. They sailed up the river Esmeraldas and climbed over the Andes, reaching Quito, in Ecuador, on the 4th June, 1736, and completing all their measurements by 1739. When the news came from Lapland that the polar survey was finished and had proved that the Earth was flattened at the poles, La Condamine decided to remain in South America. For an additional four years he explored, performed scientific research, and mapped part of the Andes and the Amazon, finally returning to France in 1745.

His 10 year adventure was documented in his *Journal du Voyage fait par l'ordre du Roi á l'equateur*, published in 1751, in which he also mentions his experiences with *curare*. La Condamine also originated the idea of vaccination against smallpox (later developed by Edward Jenner), which he had suffered as a child.

I am telling the story of these adventurous pioneers here not just because they by and large achieved something unique, but also because, like so many, they failed to receive the laurels they had earned – far from it. And because they performed a certain amount of scientific research, far in advance of everyone else, even though neither of them brought back discus (although La Condamine did follow the "discus route" for many years). Or maybe they did and we simply don't know it...

We do know, however, that Natterer's was the first recorded discovery – albeit just one individual. And almost 30 years later Agassiz and his companions were second. But Jobert's three or four specimens were the ones described as the second (sub)species, although he was the third one to collect discus.

In fact the "architect" did not receive the honour he deserved until after his death (it is well known that people often do not become famous until they are dead). In 1880 M. Léon Vaillant published on the freshwater rays that Jobert had collected in Calderón; on species of "Siluridae" – including a species that he named *Otocinclus joberti* after its collector (the species was later transferred to the genus *Hypoptopoma*); and many more – but I do not know whether Jobert ever knew of this. Then, in 1902, Pellegrin wrote a lengthy paper on the cichlids from Jobert's collection (*Cichlidés du Bresil rapportés par M. Jobert*), in which he mentioned the *Symphysodon discus* Heckel it included, without going into further detail. This was, however, rectified two years later when, as already mentioned, he described these fishes as a new variety, *aequifasciata*. And in 1909 he published a work on a characin of the family Characidae from the Serra do Estrello, which he described as *Characidium (Jobertina) interruptum*.

The crowning accolade came in 1977 when Jacques Géry elevated the subgenus to generic status. Though, of course, Jobert was no longer around to know of this, or indeed that in 1993 and later the name was disputed and synonymised by some authors. But that is science (or rather, scientists).

Dr. Clément Jobert's discus *(all 3 shown, left)* have now gone to their eternal home in the vaults beneath the Musée d´Histoire Naturelle de Paris, in France, where they rest alongside numerous other cichlids, including *Pterophyllum altum*, likewise described by Pellegrin in 1904 – which, however, has never been found syntopic with discus in nature.

The three discus shown here are the original specimens that Jobert collected in 1878 (there was supposedly also a fourth specimen from Calderón). The fish on the left-hand page has the locality given as Santarém (Nr. 02-130[1]) – it is quite clearly a blue discus! The specimens above came from Teffé (02-134-135[2]) and are definitely greens. The adult fish (top) exhibits the typical pattern of markings in the anal fin, and the appreciably smaller specimen from Teffé *(above)* as yet has no markings, as is typical in Tefé wild-caughts; they first appear at an age of 8-12 months – often even later.

Discoveries in the 20th Century

No further discus collecting expeditions, let alone new discoveries, are known, or worth mentioning, from the 19th century after the collection by the Frenchman Jobert.

The beginning of the 20th century saw the beginning of the "run" of the first "King of aquarium fishes", *Pterophyllum scalare* (above in the photo to the right, and discussed in *Brehms Tierleben, Fische*, 1910). In Germany it was known as the *Blattfisch* (= leaf fish), a modified translation for the generic name of the Greek words *ptero* = wing, feather (= fin in fish taxonomy) and *phyllon* = leaf. Later on it was renamed the *Skalar* or *Segelflosser* (= sail-fin), and the Americans made it world-famous as the angelfish. This species is now known almost universally as the angelfish in English-speaking countries.

After several adventurous collecting trips in the Amazon region, the German Bruno Sagratzki first brought live specimens back from the Rio Negro near Manaus in 1909. Thereafter this cichlid was also known as the "King of the Amazon", and in 1913 J. Cvancar of Hamburg achieved the first spawning. The angelfishes spawned on a plant leaf. Nevertheless, up to the end of the First World War there had still been no significant breakthrough in breeding, and further expeditions followed in search of what was in its day the most desirable of aquarium fishes.

From 1920 Aquarium Hamburg, and its energetic founder, Walter Griem, began to stock fishes from many parts of the world in the Hanseatic city, in competition with C. Siggeklow (the company that had first imported angels before the war) and W. Eimecke, both likewise in Hamburg. Eimecke, after whom an angelfish species, *P. eimekei*, was named (in 1928, but later placed in synonymy), was a show-boxer in market places. He also dealt in elephants, crocodiles, and all sorts of other animals as well as live fishes. Then there was the company Scholze & Plötzsche, founded in Berlin in 1905, as well as Hermann Härtel in Dresden, since 1898. And my grandfather Adolf Kiel, who even earlier, around 1887, was one of the first to introduce the modern aquarium hobby, although he was predominantly an aquatic plant specialist (he gave the plants that he grew in the aquarium and sold the name *Wasserpflanzen* (= water plants), still used in Germany today). He had little to do with discus, although he did sell angelfishes.

The hunt for the then "King" ran almost like a gangster movie. Despite the post-war depression in Germany and later in America, and the resulting shortage of money, people in the USA nevertheless paid 200 dollars per pair. (An average month's pay was about 40 dollars.) The "dash for the Amazon" was unbelievable.

But already in 1921, when a sailor arrived in Hamburg with a consignment of fish cans from Pará (= Belém) for Eimecke, and a discus was discovered among the valuable angels, was there any inkling that angels weren't the be-all and end-all.

Even so, these different-looking, brownish, disc-shaped fishes with 8-9 bars (instead of the normal 5-6 of angels) were initially seen as no more than a variant of the angelfish, and hence they were called "brown angels". It can be seen from the first sketch of a discus by Werner Ladiges *(right-hand page)*, which appeared in the German *Wochenschrift* on the 23rd August 1932, that the artist was still thinking in terms of an angelfish. Almost all discus have 9 bars of similar width (the exceptions are the Heckel and a few variants) and Ladiges shows very variable, sometimes very angel-like bars (similar to

Until the beginning of the 1930s the angelfish *(Pterophyllum scalare)* (above) was the "King of aquarium fishes", but was dethroned as soon as the first live discus (labelled *Symphysodon discus* – a "Brown" discus, below) appeared on the scene. Since then the discus has been the undisputed monarch worldwide.

those seen in individuals assigned at that time to *Pt. eimekei* – below).

After that first discus, the angel continued to dominate the scene as the "King" of aquarium fishes for another seven years, although it was starting to be bred more frequently, albeit still not in large numbers.

Then, in October 1928 the second turning-point took place. Wilhelm Praetorius, a well-known fish collector at that time, was making frequent arduous trips by ocean steamer in order to bring back tropical "jewels" to Germany, at that time the world centre for tropical fishes. While searching for the much-prized angelfish towards the end of 1928 he had an almost unbelievable stroke of luck. During a collecting trip to the Tapajós, he found five discus in his net while collecting angels. I would emphasise the word "luck" here, as discus normally never find their way into a "normal" net of the sort used to catch angels (see also Chapter 5: *collecting*).

Of course he immediately realised that these splendid disc-shaped fishes were not angelfishes and that they must be valuable. He did everything he could to bring them back safely, along with the other fishes he had collected, for the company Scholze & Pötzschke in Berlin. He even set up aquaria in a cabin on the steamer – in addition to his normal fish cans – in order better to take care of the valuable goods. But, at that time it was the practice to collect anything aquatic – not just tropical fishes – that appeared interesting and saleable and turned up in the net. Hence Praetorius was also transporting some big Arrau turtles (*Podocnemis expansa*), one of the largest and most powerful fresh-

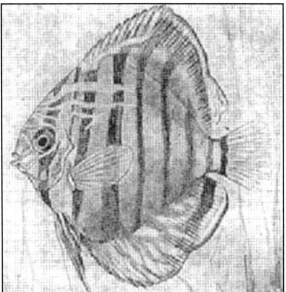

Werner Ladiges drew the first discus 1932 *(top)* **surely thinking of the close relationship to the angelfish described as** *Pterophyllum eimekei* **by E. Ahl in 1928** *(above)*. **The top drawing from Ladiges appeared in the** *Wochenschrift* **of 23rd August in 1932. The drawing above is by W. Schreitmüller, from** *Zierfische,* **1934.**

water turtles. Shortly before docking in Le Havre they apparently decided to "disembark" and in the process tore many of the airlines with their large claws. They were probably upset by bad weather in the Channel – or maybe they got a whiff of delicious French croissant... The five discus were literally left gasping and died. The turtles survived...

Apparently Praetorius, who, of course, was reluctant to give up (as I can very well imagine), subsequently had better luck at the end of 1931 – this time deliberately searching for discus with long, large nets. Some fifteen kilometres from Santarém – almost certainly again in the Tapajós drainage – he caught 51 specimens in the space of two days and another 26 later. In a subsequent article he mentioned that discus are found in waters where there are rocks and fallen trees in deeper water (Praetorius was thus the first to realise that discus live at greater depths).

He also wrote that they fled at the least disturbance and so everyone can imagine how difficult this makes them to catch (he was slightly out here, as by day discus almost always remain among deep-water cover for protection against predatory fishes). Sadly, Praetorius' attempts at discus importation seem to have taken place under an evil star, and this batch too failed to reach Germany successfully. Thereafter he seems almost to have lost his passion for discus, instead concentrating on collecting neon tetras in the years that followed.

In 1940, after feverish searches for the new King of aquarium fishes during the 1930s had generally resulted in consignments arriving dead or sick, Härtel senior wrote that the cause of these frequent mishaps probably lay in deficiencies in the holding accommodation in Pará (Belém). And he was not far from the mark. The problem was not restricted to those times but persists today. The majority of discus imported via Belém regularly "succumb". A terrible bacterial infection literally dissolves the protective mucus coating and the healing process, if feasible at all, is long-drawn-out and laborious. And often unsuccessful. And back then they had only a fraction of our modern treatment methods, let alone the medications (more on this in Chapter 9: *diseases*).

After these failures on the part of Praetorius (who, by the way, subsequently vanished without trace on the Rio Negro, along with a lot of money belonging to Aquarium Hamburg – it was said he was searching for a second neon tetra) however, Aquarium Hamburg wanted to know more.

Walter Griem made a collecting trip to the Amazon region together with his brother Karl, who was already living in Rio de Janeiro and constantly on the look-out for supplies of new tropical fishes. They learnt a huge amount during this expedition, even though they had much else to contend with, including the mosquitos and the ubiquitous *piúm* (tiny little biting midges). Karl in fact came away with a malaria infection which he never managed to shake off thereafter (just as happened to me later).

Their main objective was to search for discus, but they also collected many other splendid fishes, which in some cases now found their way into the aquarium hobby for the first time. These included the spotted headstander *(Chilodus punctatus)*, Schreitmüller's *Metynnis (Metynnis schreitmülleri)*, leaf fishes *(Monocirrhus polyacanthus)*, various pencilfishes *(Nannostomus trifasciatus, N. eques, N. marginatus)*, silver hatchetfishes *(Gasteropelecus levis)*, *Abramites* headstanders *(Abramites microcephalus)*, Ternetz's headstander *(Anostomus ternetzi* – at that time identified as *A. anostomus)*, banded *Leporinus (Leporinus fasciatus)*, graceful *Hemiodus (Hemiodus gracilis)*, shadowy *Prochilodus (Prochilodus insignis)*, pink-tailed *Chalceus (Chalceus macrolepidotus)*, numerous mailed catfishes (including, for example, *Corydoras julii)*, as well as large suckermouth cats *(Hypostomus* species – known at that time as *Plecostomus)*, and many more. Of course, they were eventually successful in landing discus as well as the second

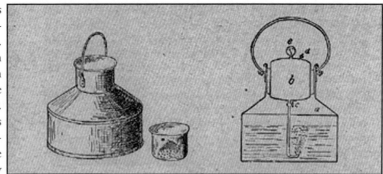

Top: Sketch of a fish (discus) transport can. Fish cans were used worldwide until the 1950s for the transport of ornamental fishes, until replaced by plastic bags. *Above:* a glimpse into a ship's cabin full of fish cans of all sorts, being aerated. This is how the first live discus – each in its own can – reached New York in 1932. The cans were insulated to keep them warm, using wood-fibre, straw, and jute sacking.

most sought-after ornamental fish at that time, the Oscar *(Astronotus ocellatus)*, albeit only 18 specimens (12 according to some reports, see below). They did everything possible (including adding salt) to ensure the valuable goods survived. Then the brother who lived in Brazil took the discus, each individually packed in a well-insulated fish can with about 20 litres of water, on the next ship to New York. Along with the Oscars, which he had promised to his fellow-countryman Mertens, who lived in the USA. Meanwhile Walter Griem took the steamer to Hamburg with the rest of the fishes. The Griems had undertaken the journey with the additional purpose of establishing permanent supplies to the United States and Germany. Karl landed in New York at the end of 1932. Tropical fishes had by then become all the rage in the land of unlimited possibilities as well, and the Americans were at that time getting all their stocks (apart from those they bred themselves or obtained locally) from Hamburg. Griem supplied the few importers – there were three or four in total – dealing in aquarium fishes, and Aquarium Hamburg even planned an offshoot there. It has not been possible to ascertain whether Karl Griem went to New York for this reason as well, or just because of the discus.

SYMPHYSODON DISCUS (HECKEL)
The Aristocrat of the Aquarium

BY *William T. Innes*

ALTHOUGH not new to science at the time, *Symphysodon discus* (pronounced sim-fy'so-don disk'us) was first introduced to aquarists in 1933 by Carl Griem, who carefully brought them from Brazil, each in its own container. This precaution was not taken on account of any fighting qualities of the fish, but in order to give each one plenty of room and individual attention. Accompanied by Mr. Griem as personal attendant, they landed in New York without loss. As the writer remembers it, there were about 18 of them, 12 of the lot being later taken to the great German breeder, Härtel. Those remaining here were valued at $75 each, and were disposed of at approximately that price.

it is reasonably probable that this is not their natural food. These worms come from foul conditions, and we know that the fish inhabits clean water. Furthermore, the worms here do not do well in warm situations, such as we would find in the Amazon and most of its tributaries. As Tubifex worms constitute the principal commercial live food in and about New York City, the Symphysodons were given this luxury upon their arrival. Thereafter most of them refused anything else. As we shall later see, this trouble can be overcome in properly educating the next generation.

Much study was given the original stock (and later importations, of which there have been several) in order to

The brother of the owner of Aquarium Hamburg, Karl Griem, brought the first live discus to the USA in 1932. He sold some 6 individuals for $75 each and the remainder went to Germany to the breeder Härtel (or the other way round – see text). William T. Innes, after whom the neon was subsequently named, wrote about this first importation in the American magazine The Aquarium, of which he was editor. Interestingly, in those days – and until as late as 1960 – almost all the discus collected were Browns. The first true *Symphysodon discus,* the so-called Heckel discus, arrived in the 1960s. It was apparently also Innes who first invented the name The Aristrocrat for the "King of aquarium fishes". The title "King" came a lot later.

One of Griem's American customers was the company Empire Tropical Fish, headed by two Germans, Richard Büttner and Carl Mertens. Mertens, a true Nazi, had close connections with Härtel, and that enthusiastic breeder from Dresden repeatedly made the long journey to the States, at that time

very time-consuming and expensive, in order to bring back fishes (and is it not still the case today that breeders do not mind long journeys, inconvenience, and expense as long as they can obtain the fishes they desire?). Härtel was in fact after the only recently discovered (by Praetorius, in 1929) Oscar *(Astronotus ocellatus)*. Because competition among fishkeepers never sleeps (still true today) Härtel wanted the fishes for breeding and knew from Mertens what of interest was available or going to arrive in the USA. So he thought nothing of crossing the ocean to buy a few fishes. (Nowadays people fly to the "Big Apple" just to bring back Christmas presents or to see a Broadway show, but hardly on account of a fish!)

Now as luck would have it, at this moment in time Karl Griem arrived with the first discus and went to Mertens and Büttner (by chance or by design?). They took 12 specimens and Härtel was left with six (although opinions differ as to the exact numbers – how many there were, and who had how many). The German literature states that Härtel took three discus as well as the Oscars. (The German discus breeder,

photographer, and well-known author Hans-Georg Petersmann spoke personally to Härtel junior in 1972, and the latter told him it was three discus. But, even so, that doesn't add up. (See also Chapter 6: *The early breeders.*) In April 1933 Innes wrote in his magazine *The Aquarium*, "The fish has been

Above left: The first discus photo (W. T. Innes). *Above:* Fred Cochu with A. Rabaut, the discoverer of the neon. *Below:* Fred Cochu in 1937. *Right:* Highlights of Fred Cochu's Paramount Aquarium.

PARAMOUNT HIGHLIGHTS

1929 - 1939: over 250 shipments from Germany
1934: Paramount Aquarium, Inc. is founded
1936: Fred Cochu flies first shipment of neon tetras to Lake Hurst, N.J. on the Hindenburg. The lone surviving tetra of the trip became famous and was named "Lonely Lindy" by an adoring public.
1937: The first of 204 expeditions up the Amazon by Fred Cochu for Paramount.
1939 - 41: Fred travels the Far East, supplies rare animals for Frank Buck, and brings shipment of fish from Singapore.
1939 - 47: Paramount worked for the government trapping electric eels
Stocked the Pittsburgh Aquarium Society's new aquarium with fresh water dolphins and other rare fish.
1946 - Fred Cochu explores Nigeria and introduced for the first time the elephantnose, the butterfly, the upsidedown cat, the rope fish, and many others.
1955: Fred Cochu and Paramount discovered and introduced the cardinal tetra
1961: Paramount leads expedition to King Salmon, Alaska for Beluga whales for New York Zoological Society.
1966: Expedition for Niagra Falls Aquarium to Rio Negro for fresh-water porpoises.
1969: Paramount delivered 21 porpoises to Sweden.
1970's: Fred Cochu made several trips up the Amazon to improve and modernize Paramount's collecting stations.

listed as an aquarium fish for some time, but has been regarded as one of these hopeless things that we never would get. This shipment of twelve magnificent specimens has just been personally conducted and landed in New York by Mr. Griem, each in its own separate can. The Empire Tropical Fish Import Company of New York (which was than owned by Büttner with Mertens being a partner) has the majority of them, the balance going to Germany". He added that he was grateful to have been permitted to take the first photo *(page 52)*. Later, in an article in October 1935 (an extract from which appears *on page 51)* in the same magazine, he wrote that the first importation consisted of a total of 18 specimens and that Härtel took 12 of them. Ie only six remained at Empire. But, whatever the answer, there is no dispute that they all arrived in New York in good condition and were truly the first living wild-caughts the world had seen. (What subsequently became of these discus is detailed in Chapter 6: *The early breeders).*

The third demonstrable turning-point involved Fred Cochu. Likewise originally from Hamburg, he was in and out of Aquarium Hamburg as a little boy of 10 (and his sister worked there as a secretary). He emigrated to the States in 1929, at the age of 19, arriving in New York where he worked in the hotel trade. Fred earnt his daily bread as assistant to a head waiter in the ballroom of the newly opened (1930) New Yorker Hotel, where his job was to arrange the disappearance of whisky bottles which the guests brought with them and hid beneath their tables (it was the era of prohibition). Then, when the guest and his lady returned from the dance floor, they would find, instead of the bottle, the head waiter with his hand

The first live (surviving) discus arrived in Germany via the company Härtel at the turn of 1932/33 (see Chapter 6: *The early breeders*). My grandfather, Adolf Kiel (here his advert in *Wochenschrift* from 1907, on page 312), was one of the pioneers of the modern aquarium hobby, and Scholze & Pötzschke was the second ornamental fish wholesaler in Germany.

out. Fred received 10% of the nightly takings and earned up to 100 dollars per week – an almost unbelievable sum in those days.

Ferdinand Cochu was born in 1910 in Hamburg. His mother came from Denmark and his father had French parents – his grandfather was a famous doctor in France (you will find a bit about this on the Internet). From the age of 15 he worked some of the time as an assistant at Aquarium Hamburg, where in 1926 he met William A. Sternke, formerly of Berlin, who was to influence his future. Bill, as the Americans called him, had lived in the States since 1908 and came on a visit to Germany to bring Walter Griem the first sailfin mollies (*Poecilia velifera*) from Florida (at that time five pairs paid for his passage). But Fred had little interest in Bill's mollies – he wanted to know all about the New World. Sternke was the very first to build a fish hatchery in Florida (in 1925), and, at the beginning of the 1950s, after more than 20 years in New York, Fred likewise went to Florida. They were business partners and close friends up until Bill's death (1976).

But before Fred got his dream job at the New Yorker at the age of 20, he worked as a dish-washer and did a lot other menial work in order to survive. When, in 1932, prohibition was lifted he went back to Germany on a visit and Griem offered him a place in the fish business. But he returned across the Atlantic before deciding, in 1933 – after six months

In 1933-34 Carl Mertens and the brother of the owner of Aquarium Hamburg, Karl Griem, who took the first live discus to New York, founded the import company Amazonica, Inc. and discus were first mentioned in their 1934 advert.

training at Aquarium Hamburg – to start his own company, Paramount Aquarium, and distribute fishes from the Hamburg firm. Prior to the Second World War he made the crossing more than a hundred times to select ornamental fishes personally and transport them back from Hamburg. His fish business soon expanded and he also bought in American tank-breds from Bill Sternke, Albert Greenberg, Herb Woolf, and other breeders. Fred was the first to import the neon tetra to America (in 1935; this fish has been discovered by the Frenchman A. Rabaut and in 1936 was named after the editor W. T. Innes), and also to bring the cardinal tetra to the USA (in 1955; later, at the beginning of 1956, a certain H. R. Axelrod bought this species from a petshop and had it described as his own discovery). When he started collaborating with Aquarium Hamburg, he of course also received discus via Walter Griem's brother Karl. However, the supply was too limited, and so he decided to build his own collecting station in Brazil.

At the beginning of 1937 Fred travelled up the Amazon for the first time, as far as Iquitos. *Inter alia*, he established a neon tetra monopoly, a trust, in Leticia, Peru, as well as at Tabatinga and Benjamin Constant – on the Brazilian side. For more than 20

The Goeldi Museum in Belém is named after the Swiss Emil August Goeldi (1859-1917). From 1894 to 1907 he turned the Museo do Pará, founded in 1866, into a worldwide centre for the study of the natural history of Amazonia, its rainforest, and its mammals. Unfortunately the institute was subsequently almost forgotten and much of it fell into ruin *(above left)*. Fred Cochu supported it financially between 1946 and 1954 and had a receiving station for his fishes there. Today it contains more than 2000 plant and 600 animal species in an area of 52,000 m², as well as a miniature Amazonian rainforest. And even the aquarium, founded in 1911, is up-and-running again *(above right)* with brown discus from the Xingú.

years Fred Cochu, along with Aquarium Hamburg, paid the Brazilian and Peruvian governments for sole collecting and export rights – it cost them a lot, but was profitable. Thus it is easy to prove where all the discus (neons, and many other Amazonian fishes) came from in the years that followed. But as early as 1932 Walter and Karl Griem had started a company in Rio de Janeiro, which was later (in 1934) reborn in a different form in New York as Amazonica Inc. Karl Griem went into partnership with Carl Mertens, who was to handle stock and sales in the United States (although Mertens was involved with Büttner in Empire Tropical Fish). But right at the beginning the company suffered a grievous blow. Karl Griem, who usually caught the fishes himself, or organised the collections, and always brought them to New York singlehanded, fell ill with a dreadful fever (as already mentioned, hardly any collector escaped malaria). Consignments failed to arrive, or attempts to collect fishes in the Amazon region during the rainy season proved a hopeless task (just like today). They were like a blind dog in a slaughterhouse, who can smell meat but not get any of it.

During a trip to the Amazon (1937) Fred found Karl Griem near to death in Belém and saved his life, arranging for him to return to Germany to recuperate. After that Karl sold Amazonica, Inc. to Fred, who from then on was the only person still receiving discus (and all other Amazonian fishes from Brazil, and even the first specimens from Guyana and Trinidad).

Nevertheless, before the Second World War discus only ever arrived in small quantities in America and Europe. And the newly-hired collector for the Aquarium Hamburg–Paramount Aquarium consortium, Hans Pietsch, lost more than a hundred individuals he had collected at the beginning of 1937 at Leticia, supposedly through spawnbinding. But Pietsch deserves a chapter to himself. On one occasion he brought to Hamburg more than 12,000 neon tetras, which he himself had caught during months of searching in the vicinity of Leticia. In Belém his competitors played a dirty trick on him – while he was ashore looking for fish food, somebody emptied hundreds of his fish cans and he was able to save only 1800 fishes. Later (at the turn of 1937/38), when he was commissioned to collect fishes in West Africa (which he had already done earlier), he turned his back on the consortium, planning to break the neon monopoly single-handed. Like so many others, he had fallen under the spell of the "red gold". He set off up-river from Belém by motorboat and was never seen again. Vanished while searching for fishes, like many before and after him.

Thus Karl Griem brought out the first live (brown) discus, and it was Fred who, before and after the Second World War, indefatigably supplied the global market with (brown) discus, repeatedly making collecting trips to Brazil and almost all of South America. Fred Cochu did more than any other to provide the aquarium hobby with tropical fishes. For almost 40 years he was the dominant force in this region, obtaining the relevant permits from governments and other authorities. Unmatched in his endeavours, he set up operations in Guyana, Venezuela, Colombia, Peru and Brazil. He used his planes – at times he had three or four in service (I flew with him twice myself) – to supply the entire American and European markets. His operation was almost unbelievable for those days. Just consider, at that time there were initially no planes at all, then only propeller-engined machines. There were no roads in the area. Packing materials – apart from German fish cans – had not yet been invented, and terms like styrofoam, plastic bags, fill-

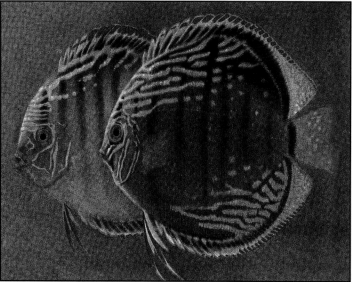

Top: The black and white drawing was made by Arthur Rachow in 1928. It is the first (?) discus drawing of the 20th century, and clearly shows that the live discus was not yet well-known. *Above:* The first (?) 20th century discus drawing in colour, by William T. Innes in Exotic Aquarium Fishes, which was (re)printed 10 times between 1935 and 1948. He wrote: "Our illustration presents the fish in breeding colors, although sometimes the blue color pattern extends more over the body, as can be seen in the following plate (= page 57) showing the male fanning the eggs... Occasionally the female is more brilliant golden yellow ..."

ing with oxygen, oxygen tablets, and insulated boxes were still unknown.

Fred died in 1994 at the age of 84 and worked with fishes right to the end – for more than 60 years, longer than anyone else in the field (apart from two, see below). Sometimes he was on the breadline, and often, when all the bills had been paid, had only five dollars in his pocket. But there were also times when his earnings were good, or even very good, but he always ploughed it all back into the company. Fred left behind only a few fishes bearing his name, eg the blue Peruvian tetra *(Microbrycon fredcochui)* and Cochu's mailed catfish *(Corydoras cochui)* (he couldn't tell me where that came from). He was happy right to the end and didn't regret a single minute of his life, as he stated on a tape recording made in 1993 by one of my best friends, Ross Socolof, who made it available to me. And before I continue the story of discoveries in the 20th century, I would like to reproduce here (translated from the original German) two post-War episodes recorded faithfully by the German ichthyologist Werner Ladiges. These are just brief moments from the life of a man who, in my view, dedicated his life to the aquarium hobby, and for whom no task was too difficult. The episodes are from the years 1945 and 1948.

Anno 1945

"In the first light of the rising sun, in the sunny land of Florida, a silver plane rolled along the runway at Miami. *Princess* was painted on its body *(see below)*. Lifting off, it spiralled upwards in the golden flood of light before heading south.

The ornamental fish breeder, who had just made his first tour of the pond complex not far from the airport, stood shading his eyes and following the silver bird with his gaze: 'Devil of a fellow, that Fred! Off to get neon tetras again!' *(for more on neon tetras, their history and discovery, look at Nutrafin Aquatic News No. 1 – or on the Internet at www.hagen.com)*, he commented to the little Japanese who was going round feeding the shoals of colourful cyprinids in their sun-warmed ponds. The plane floated away over the shimmering cactus hills of Cuba, across the improbably blue expanse of the Caribbean, dotted gold with floating seaweed, over the final outliers of the *cordilleras* of Colombia, stopping off to refuel in Bogota; then the shadow of the great humming bird passed over the endless dark green jungle carpet and light green *llanos* of Venezuela, until the silver network of the mighty river system of the Rio Negro, and ultimately that of the Amazon, gleamed below. The bird now followed the vast river, flying over tributaries and streams, small settlements, and again forest, green, green forest! As the sun

was about to set in the west for the third time, the end of the river was at last visible, a mass of channels and huge bays, and beyond the wide ocean. The coloured lights of the airport at Belém were already shining to the right. Journey's end once again.

After the plane had landed, the New York ornamental fish dealer, Ferdinand C., his pilot, and a mechanic disembarked following a flight lasting 60 hours from La Guardia-New York. In Belém all the concrete containers teemed with darting, snapping fishes, accumulated during weeks of repeated collecting and brought here from upstream. Can upon can totalling: 12,000 neons, 15,000 neons, 20,000 neons, plus 10,000 other fishes: hatchets, dwarf cichlids, pencilfishes, catfishes, and anything else available to fill up the consignment. *(Ladiges doesn't list discus here, but the first brown post-War individuals were by then included.)* Museum Goeldi received its share *(after the war Fred had leased the then run down Goeldi Museum and Zoological Garden in Belém as his collecting station and thereby supported the museum – see page 55),* his agent *(Old Cesar, who later worked with me and caught discus)* received his instructions regarding the next consignment, and then the engine started to thunder again, and the *Princess* flew away, this time heading north.

They had already made this flight some 20 times, and things didn't always go smoothly. Bad weather (storms, downpours, fog) necessitated detours, while all sorts of unforeseen incidents (revolution at the refuelling stop-over, engine failure, and the like) caused delays, but the shoals of neon tetras, tiny, gleaming and glittering jewels, always arrived safe and sound at their destination." *(Ladiges' article was about the neon.)*

Ferdinand Cochu 1992 *(top)* may not have discovered any discus (at least that I know of), but he certainly imported the majority (if not all) to the USA and Europe between 1934 and the mid 1950s. He was also the first (1955) to bring the cardinal tetra into the States (but someone else took the credit).

Anno 1948

From the *Richmond Times* of Thursday, Feb. 5. 1948:

"An importer of tropical fishes has landed here with rare species. Yesterday Ferdinand C. arrived out of the clouds with 350 cans of beautiful fishes, but he didn't have any kind words to say about the weather in Richmond. Hatchetfishes, pencilfishes, leaffishes, pirayas – every conceivable kind of tropical fish, more than a hundred species in total *(discus were also included),* spent the night swimming in a warm room in a Byrd Airport hangar.

Mr. C. was not very communicative, exhausted after a flight of 3000 miles without sleep. They had started out from the shore of the Amazon, where Mr. C., the head of a New York company importing live tropical fishes, had loaded his consignment of thousands of small inhabitants of that great river onto a plane headed for New York. He and Doc. M., the 'skipper' of the chartered plane, made the journey in the best of spirits until they reached Raleigh, where the second pilot slipped on the ice and broke a leg. As if this were not ill omen enough, they found themselves in thick fog and had to come down at Byrd Airport *(in those days they flew by sight).* When they learned of this unforeseen incident in New York, one of Mr. C.'s colleagues *(that was Hugo Schnelle)* came down right away to help him with the water changes essential for the fishes' survival, they need a temperature of more than 22 degrees Centigrade, to feed them, and to watch over them like they would small children.

'In the past year Mr. C. has made fourteen trips to South America for the same purpose, and one to Africa', explained the man from New York. 'If the trip can be made without interruption then the fishes arrive in good condition. There is a heater in the plane and the cans are carefully packed and encased in woollen covers. At the same time he has hired two young people at the airport and entrusted them with the care of his valuable cargo. They are using a hot-air machine and a wood-burning stove

These three photos by Amanda Bleher from the 1950s show indians fishing and a Xingú indian *(below)* with his catch in front of a typical Xingú men's house on the island of Bananal in Brazil. The reason I include these photos here is that after the Second World War photos of discus collecting were frequently published – also involving indians, as here. But all such photos are misleading, in particular as regards their captions, eg "Discus collecting in small *igarapés"*, as discus are never found in such shallow waters. Not in any igarapé (stream), nor in standing water, nor in white water. Discus spend the daytime in deep, black or clear, water (at least 2-4 metres deep). Only fishes such as the Xingú indian *(below)* is holding are found in *igarapés* – but no discus!

for the overnight accommodation, heating pans of hot water for the water changes, and doing everything to show southern hospitality to the unexpected guests.'"

Fred thus brought most, if not all, of the wild-caught discus onto the market in the post-war period – even though they were demonstrably exclusively browns. There are photos (by Innes) of one or two blues, apparently taken at the beginning of the 1950s (one magazine claims in 1934, which can hardly be right). But where they came from remains unknown. As mentioned above, Fred, who collected discus, worked almost entirely out of Belém. And by far the majority of his discus originated from the region around Santarém (Fred stated this in his interview, and it is also apparent from the photos that exist from that time.

The fourth discus turning-point occurred with Amanda Flora Hilda Bleher, who was undoubtedly the only commercial female fish-collector in history. (Even if she never actually caught discus herself – but I have made up for that...).

Frau Bleher, still Kiel at that time, began her adventurous travels at about the same time as Fred, having taken it into her head to search the Amazon region for discus and swordplants. My mother was born in the same year (1910) as Fred Cochu. She had a clear memory of her first "fish experience", when she sat, barely five years old, in the carriage of her father, Adolf Kiel, on the way to communion. He had a planted glass jar with fishes clasped in his hands, to deliver to a customer en route. And while Adolf raced across the bumpy Frankfurt cobbles in the open horse-drawn carriage, the water kept slopping over little Amanda's new, snow-white clothes and making them look grubby.

Unlike other little girls, instead of dolls Amanda was given aquatic plants, fishes, reptiles, and amphibians to play with. For as long as she could remember she had worked, or helped, in my grandfather's business – who already in 1900 owned the largest aquatic plant nursery and fish hatchery in the world. Little Amanda looked after the plant and fish ponds and was the only one of three children who grew up involved in the business of the "father of aquatic plants". Plants, fishes, and the daily involvement with nature weren't work in her eyes, but a vocation, or rather even a religion – or so I believe. She

Amanda Bleher, after her adventurous expedition with her four children to hunt for the discus *(page 62)*. After her return she became famous in the world press in 1955 and 1956, sometimes on the front page *(top)*. The adventurous family 20 years later *(above)* with her brother-in-law *(far right)* and two grandchildren.

also spent her last 30 years in the wild, albeit only 40 km from the gates of Rio de Janeiro, but in the deepest jungle nonetheless. Surrounded by thousands of orchids, bromeliads, heliconias, and countless other plants (including, of course, aquatic plants), plus fishes, monkeys, iguanas, and parrots. Far from people, whom she avoided as much as possible – apart from her children, whom she loved above all else and for whom she continuously made sacrifices. Even going as far as getting divorced after the war on account of the children; and refusing to sign any contract unless it included her children; and not making any journey – no matter how far – without at least one child going along; and taking all four children with her in 1953 to hunt for discus in the "Green Hell" of Amazonia and exploring South America for almost two years in search of the sought-after discus and other fishes, plants, and other creatures. And it was during this expedition (which will be described in a forthcoming book on the subject, entitled *Iténez – River of Hope*) that, at the age of only eight years, I too began to learn a lot about the secrets of the splendid discus (this appears to have been my vocation).

But in spite of this incredible undertaking and the deprivations involved, some of the time following the route taken by Natterer, my mother failed to catch any discus during this journey (it wasn't possible to get through to the Amazon). Not even at Cuiabá, cited as a locality where a *Symphysodon* was found in some literature. There have quite certainly never been discus there. But she did find almost innumerable new fish species, not one of which was named after her (the credit went to other people). Today only a few of the many bromeliads, orchids, and aquarium plants discovered by her bear the names *bleheri*, *bleherae*, *amandae*, or *osiris* (Lotus Osiris was the name of our company in Brazil at that time).

My mother was, however, undoubtedly the first woman to put discus, other fishes, and aquatic plants on display at the Frankfurter Zoological Gardens, after the war in 1948. (The brown discus, labelled as *Symphysodon discus* Heckel, came from Aquarium Hamburg/Fred Cochu.) This exhibition amid the post-war ruins of the once world-famous zoo was an enormous achievement. Amanda Bleher erected almost all of it herself, and it was a huge success (from which others profited...). Now and then at the beginning of the 1950s discus were also available at her Goki-Zooversandhaus (the first wild animal supplier in the world – *see left*), founded in 1948. They cost 1,000 Marks each, and Dr. Schmidt (see Chapter 6: *Dr. Schmidt-Focke's work*) was one of the first to buy them.

The *tolle* (mad) Amanda, as she was known in Germany (or the woman with the heart of Tarzan, beyond its borders), helped an awful lot of people, smoothing their path, taking them into the business, or sharing with them the knowledge acquired from her experiences and discoveries. To mention just a few: Emil Haas and his business Zoo-Haas, for more than 50 years the best-known in Frankfurt am Main; Hans Schmidt, whose wholesale operation Tropicarium Frankfurt became one of the largest in the world in the mid 1950s; Carlos Stegemann, originally a baker in São Paulo, but who, thanks to mother, rose to be the best-known fish dealer in Brazil at the time; and Harald Schultz, the ethnologist, who, building on her knowledge and discoveries, was acknowledged as the discus collector of the 1950s (and tracked down other fishes and plants which Amanda Bleher bequeathed him). And he marked the fifth turning-point (see below).

Amanda Bleher died after a long illness on the 25th May 1991, in the jungle home we children had built with our own hands. She went all alone and pennyless from this Earth (and, as she said, its ungrateful people). The daughter of the father of aquatic plants may not have collected as many fishes as Fred Cochu, and no discus, but, as already mentioned, she did blaze a trail for many others, as a result of which huge numbers of new species reached the finest hobby on Earth. And she did leave behind a wealth of plants and knowledge, unmatched in aquarium history. And in that she even outdid her father. In my view three people are largely responsible for the goldfish bowl developing into an incomparable hobby: in the course of well over 60 years Adolf Kiel, Ferdinand Cochu, and Amanda Bleher certainly made a longer contribution than anyone else (though I hope to emulate them) as regards turning the aquarium hobby into a multi-billion-dollar, worldwide industry so that today hundreds of millions of people can enjoy an underwater garden. And extending over almost a century... but on to the 5th turning-point in the discovery of discus.

Harald Schultz was of German ancestry, but born in Brazil. He began his career as an ethnologist under the renowned Marshall Rondon (1865-1958), greatly honoured in Brazil.

Candido Mariano da Silva Rondon came into the world in the poorest possible circumstances among the indians of the Mato Grosso, but was destined to go far. This pioneer of telegraphy had more 2,500 km wire laid across the deepest jungle of Brazil, founded most of the towns today found in the states of Mato Grosso, Acre, and Rondonia (named after him). He changed the map of Brazil, discovered rivers, established correct boundaries and geographical co-ordinates. And as long ago as 1910 he brought the *Serviço de Proteção ao Índio* (indian protection) into being. He visited countless tribes and impressed his personal motto on them: *Morrer, se necessário for, matar, nunca* (Die when necessary, but never kill). He was the greatest of men. He received titles such as *Civilisador dos Sertões* and *Marechal da Paz* (marshal of peace), and in 1913 acted as guide to the American President Theodore Roosevelt during his Amazon expedition. In 1914 he not only brought peace to the Xokleng in Santa Catarina but was also awarded the Livingstone Prize by the New York Geographical Society. And much more.

Thus at an early stage Harald had the pleasure of

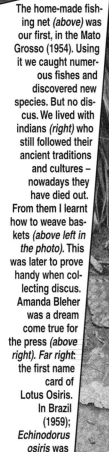

The home-made fishing net *(above)* was our first, in the Mato Grosso (1954). Using it we caught numerous fishes and discovered new species. But no discus. We lived with indians *(right)* who still followed their ancient traditions and cultures – nowadays they have died out. From them I learnt how to weave baskets *(above left in the photo)*. This was later to prove handy when collecting discus. Amanda Bleher was a dream come true for the press *(above right)*. Far right: the first name card of Lotus Osiris. In Brazil (1959); *Echinodorus osiris* was named after our company Lotus Osiris, which exported aquatic plants and fishes worldwide. The design was mine... aged 15.

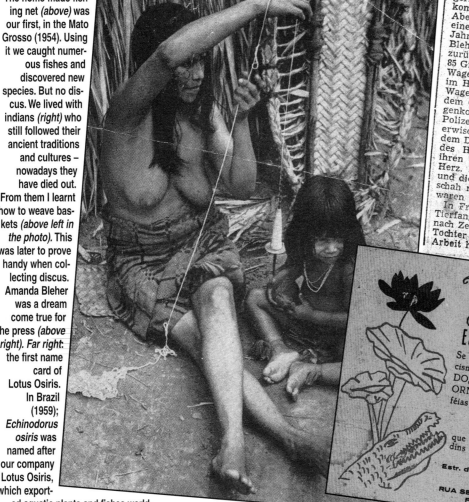

Die Frau mit dem Tarzanherzen
Die Frankfurterin Amanda Bleher ist Deutschlands einzige Tierfängerin

Aus Oubangien in Französisch-Äquatorial-Afrika brachte die 43jährige Amanda Bleher kürzlich in einem Spezialkoffer eine fünf Meter lange Pythonschlange nach Berlin. Das Tier war für eine Berliner Giftschlangenfarm bestimmt. Frau Bleher ist an seltsame Reisebegleiter gewöhnt. Sie ist Deutschlands einzige Tierfängerin und handelt mit Riesenschlangen, Alligatoren, Leoparden, Tigern und Löwen, die sie selber fängt.

Frau Bleher ist noch nicht lange dabei. Nach dem Kriege machte sie eine Handlung mit botanischen und zoologischen Raritäten auf. Das Geschäft ging gut, nur störte es die tüchtige Frau, daß sie so manchen Kunden, der nach Schlangen oder wilden Tieren fragte, nicht bedienen konnte. So kam sie eines Tages auf die Idee, „wilde Viecher" zu importieren.

„Selbst ist der Mann", ist ein Motto, das sich schon oft bewährt hat. Frau Bleher ist Mutter von vier Kindern. Sie hat zwei Jungen und zwei Mädel im Alter von acht bis vierzehn Jahren.

Ihr „Schoßhündchen" ist der Kaiman Purzel, ein Kleinkrokodil, das sie selbst aufgezogen hat. Fragt jemand die Tierfängerin nach aufregenden Erlebnissen, so bekommt er zu hören, daß sie ihr wildestes Abenteuer nicht im Urwald, sondern in einer deutschen Großstadt hatte. Es war im Jahre 1950, und noch heute denkt Frau Bleher mit Schaudern an die Begebenheit zurück. Mit zwei Spezialkoffern, worin 85 Giftschlangen steckten, war sie mit ihrem Wagen in München gelandet. Während sie im Hotel frühstückte, wurde ihr parkender Wagen bestohlen. Die Diebe türmten mit dem Gepäck und nahmen auch die Schlangenkoffer mit. Frau Bleher alarmierte die Polizei, aber die Spitzbuben wurden nicht erwischt. Vierundzwanzig Stunden nach dem Diebstahl sah Frau Bleher in der Nähe des Hauptbahnhofs ein paar Kinder mit ihren Schlangen spielen. Ihr stockte das Herz. Ein falscher Griff der Kinderhände, und die Reptilien bissen zu. Zum Glück geschah nichts dergleichen. Die Giftschlangen waren in der Novemberkälte erstarrt.

In Frankfurt bereitet sie jetzt eine neue Tierfang-Expedition vor. Diesmal soll es nach Zentral-Afrika gehen. Und ihre älteste Tochter wird sie begleiten und ihr bei der Arbeit helfen.

visiting and researching various indigenous tribes with this extraordinary peacemaker of indian origin. By 1944 he had started recording the traditions and cultures of the various tribes on 16 mm film – by the beginning of the 1960s he had made 60 films. A treasure beyond compare, as many of these ethnic tribes and almost all their traditions have by now long since become history.

In 1946 the Ethnologia Brasileira department was founded at the Museo Paulista do Ipiranga, and Harald was employed under contract, working there until his death in 1967. And it was there in São Paulo that the ethnologist met my mother, at Carlos Stegemann's. At that time everyone knew every German or other foreigner in Brazil and people were always eager to hear the news from recent arrivals. There was no Internet, hardly any telephones or trans-Atlantic flights. In addition, mother had an unbelievable wealth of knowledge – not just about fishes and plants – and was very well known. There had never been such an adventurous woman. Filled with enthusiasm for her searches for fishes and plants – and, of course, the hunt for the precious discus – he couldn't learn enough from her. I remember very well how he questioned her for hours on end. And I know that, after we had left the Mato Grosso, and in 1955 article after article about our expeditions had appeared in the world press *(page 60)*, as well as the two-part report *Jagd auf den Diskusfisch*

Harald Schultz, of German ancestry, here seen looking at some fishes in the home of Dr. Schmidt-Focke. Harald was regarded as the discus discoverer of the 1950s (see text).

(Hunt for the discus) in 1956 in the magazine *Rasselbande* of the time *(page 62)*, Harald and Carlos set out to follow in my mother's footsteps.

Thereafter he not only established contact with the National Geographical Society (who published five of his contributions on indians between 1959 and 1966) but also offered articles to a new fish magazine in America, *Tropical Fish Hobbyist*, and received a positive reply from the publisher as far as fishes were concerned. This spurred him to concentrate more on aquatic vertebrates as well during his ethnological travels. In the second half of the 1950s a number of articles appeared under the then publisher, H. R. Axelrod. Including some on catching discus, although all the discus were actually caught by indians or *caboclos*. Apart from a few small fishes, Harald hardly ever fished himself.

Harald Schultz wrote in 1959 that the first time he ever heard anything about discus during his ethnological work was at the end of 1955, from *caboclos* in the Tefé region. They actually told him about the *braruá (= Uaru amphiacanthoides),* the *acará roxo (= Heros cf. severus,* the red-spotted form), and many others. Only when he met someone among the indians who spoke the *lingua géral* (a widespread indian dialect) and told him about a *paneliquara* (which means the bottom of a pot) did he think that the discus must be meant. But at first he wondered if it might not instead be the *peléco?* This onomatopaeic term refers to the sound that fishes make at the height of the dry season (or during the so-called *friagem,* sudden drops in temperature), when they hang at the surface and snap up air, making a "peleck, peleck" noise. (Schultz didn't know at the time that this phenomenon applies to the Oscar *(Astronotus ocellatus)* rather than discus.) Only when he was brought a *paneliquara*, captured at night by bow and arrow in the Lago Jurity, did he understand. He described the fish thus: "The entire body was covered in gleaming emerald green dots and blue markings on a green background. Although it wasn't the breeding season the fish was in full colour. Undoubtedly an adult male." The ethnologist thought at the time that it was the blue variant previously pictured by Innes, but he suggested that the fish would be better described as green, "...because green is its most intense color..."

It was the wrong time of the year, it is rarely possible to catch discus so late in December. The rainy season had long since begun. Geronico the fisherman (whom I later met myself, and went discus collecting with him and his son) told Harald that this fish could possibly be caught in the May-June period,

Three photos by Harald Schultz from the time (1959) of his visit to the Ticuna and the capture of the first (?) blue discus. Today the Ticuna live in 114 communities in the Brazilian Amazon on the border with Peru, in Peru, and in Colombia. The numbers of this, the largest surviving indian tribe, have fluctuated between 18,624 and 32,613 (in 1998). Even today at least one Ticuna is shot each year by white settlers (gold miners or loggers). On the 28th March 1988 12 loggers and their companions killed 14 indian men, women, and children and wounded many more. The motive for the massacre was anger against the government policy of granting the Ticuna land rights. The Ticuna have lived in contact with the white man for more than 300 years. Arrows poisoned with *curare (above right)* are now used mainly to kill wild animals *(far left:* Ticuna with blow-pipe). They often keep "pets" *(left)* and survive by virtue of the rich supply of fish – but never kill discus, which they hold in reverence. They hunt, farm, and trade. The women sell their splendid *tucúm* bags, which they make from the fronds of a palm species, as in New Guinea.

after the first *vazamento* (the first short drop in the water level) or, for sure, in the middle of the summer (from August to October in the Amazon).

But not until the end of 1958 was he again in the region for ethnological work, and once again it was December. The water was rising by the hour and it rained every day. Nevertheless he visited the clearwater lake with Geronico. Harald wrote of this typical *várzea* region, "Certain species of trees and bushes are submerged beneath them to thirty feet of water for three quarters of the year, the only month except are August to October. Here in this thicket of millions of roots and branches, we find a paradise for the Blue discus and many other fishes. Here most species spawn and according to the fishermen also the discus...". Geronico told him that at best it might perhaps be possible to catch discus in a more distant *lagoa*. They paddled along various *igapós* until they could go no further. Then they trekked overland for three hours, sometimes passing gigantic Brazil-nut trees more than three metres in diameter. Once they had arrived they actually saw a large group of 50-70 discus, and Schultz wrote that "They resemble nothing more than a herd of stampeding cattle..." (apparently this was the first time that a white man realised that the discus is a shoaling fish). But first they couldn't capture any fishes with their 20-metre net. He wrote that the water in the lake was damned cold (something the breeders won't like to hear!), and only after they had tried for hours to drive the fishes into the net did they succeed in catching two specimens. They repeated the process all afternoon and eventually had 86 discus – but only one of them had red dots all over its body (= a dominant fish).

With 17 discus per (10-litre) can on their backs they battled their way back for hours through the knee-deep mud, reaching the landing place at dusk. More than half the discus were already dead, and despite efforts to save the rest (holding nets in the *igapó*) all but ten perished. They ate a few of the dead discus (normally nobody does this, neither the *caboclos* nor the *indios* – for reasons mentioned elsewhere), and Schultz preserved a few specimens. Only two of the others survived and eventually reached the publisher in New York, but died after their arrival (after they had been photographed).

Now this was not in fact a discovery, as without exception these were a green variant *(right)*, although they were repeatedly labelled as blue discus in the publications of the aforementioned publisher and elsewhere. When and where Schultz obtained the true blues *(centre)* that he took to Schmidt-Focke in November 1959 is difficult to establish. One thing is sure, he did take them to Europe and hand them over to Schmidt-Focke (see Chapter 6: *Dr. Schmidt-Focke's work*).

In June 1960 the well-known revision by L. P. Schultz appeared (see Chapter 2: *Schultz's revision)*, in which new discoveries were described, none of which was actually new. And it is a fact that Harald Schultz obtained the first live greens. And Swegles was apparently the second to do so (see Chapter 6: *Schmidt-Focke's work*).

Thus there were no (new) discoveries to speak of until the beginning of the 1960s. And so we come to my own small contribution, which is also one of the reasons for writing this book – to put my discus collecting and discoveries over the last 40 years in the right perspective.

At the age of eight to ten I spent almost two years hunting the "King" and also collecting fishes with my mother, but my real discus-collecting period didn't begin until 1964-65. Since then I have been indefatigable in the search for new discus (and all sorts of other fishes) during my annual exploration and collecting trips in the Amazon region. (At the time of writing – August 2003 – I have well over 300 trips to this region under my belt. Whether or not this is a record, I have no idea. Between 1937 and the 1970s – ie an equivalent period of time

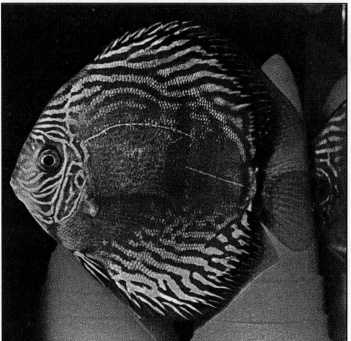

One of the three blue discus that Harald Schultz sent to the Doctor in November 1959. Their origin is unknown but they look like individuals from the Purus region.

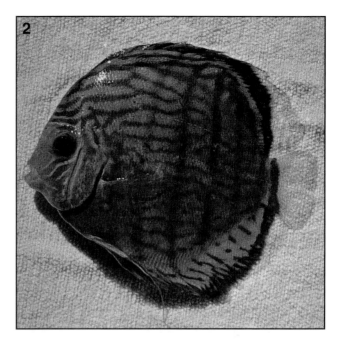
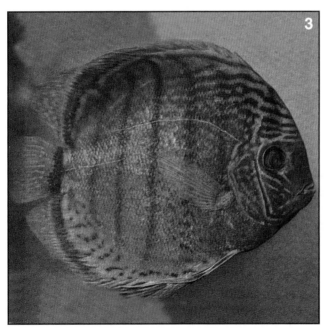

Above: Four green discus variants which Harald Schultz (according to my researches) collected. 1. One of the two individuals from the Lago Jurity that arrived alive in New York (at the beginning of 1959), a green that was termed blue at the time. The photo appeared in Schultz's 1959 article; then in the 1960 revision by L. P. Schultz, with the comment: "Lago Tefé"; and in 1970, in the Axelrod book *All About Discus*, as *S. ae. haraldi* (= blue discus) with the locality cited as Benjamin Constant. 2. Axelrod claims – in a recent publication – to have caught this fish in the Rio Tefé (*cf* page 83), but it has been pictured previously elsewhere. 3. This green specimen was once (1976) captioned: "...collected in various Brazilian Rivers... variation of *S. a. haraldi*' (= blue) and on another occasion (1984): "... near the city of Tefé ... another form of *S. a. aequifasciatus*...". 4. This green is likewise supposedly from the Rio Tefé (but elsewhere given as Lago Tefé) and caught in 1974.

– Fred Cochu made 204 collecting trips in the Amazon region. I know of no-one else who can claim to have researched discus in the wild for a similar length of time.) Below I have permitted myself to reproduce four extracts from my notes and some published work on new discoveries regarding discus – just as they occurred – over the past decades. Parts of it has previously appeared in American and Japanese publications as well as one in *aqua geõgraphia* in different language issues and deals with Royal Blues, the unique discus from the Trombetas, and the solid-coloured green forms from the Rio Tefé, as well as recent Nhamundá discoveries.

My (first) Royal Blue discovery

My first "discovery" – and this is quite certain – was the "Royal Blue" discus. Hans Willy Schwartz (page 70), who in 1959 founded his company Aquario Rio Negro at 106 Miranda Leão in Manaus, giving up his work collecting wild animals for zoos, Walt Disney, and others, helped me at the beginning. Immediately after my return from Florida I founded my company Aquarium Rio, in Rio de Janeiro – the loveliest city in the world (hence its name – *Cidade Maravilhosa*), installing 400 aquaria of 120 litres each and simultaneously starting to build my collecting stations (just as Fred had done) at various places in Brazil: on the Araguaia in the state of Goiás, in Mato Grosso (on the Rio Paraguay and on the Guaporé – of course), in Belém (with Old Cesar), and in Manaus with the Austrian Hans Willy Schwartz, who was the only in the Amazon region at the time, except for the Italian Erio Peretti who had just started in the early 1960s.

With Willy (as everyone called him, mostly prefixed by *Senhor*, or just *o alemão* (the German) – as to the Brazilians he

The fishes pictured here are all from Lago Manacapuru or Lago Grande de Manacapuru. Without exception they are termed Manacapuru discus in the hobby and among breeders. The following distinctions must be made, as I realised at the time of the discovery, as well as later: 1. Royal Blue are dominant individuals, so-called alpha individuals, and that seen here was such a fish – but still in the making! It was an adult that didn't

develop its full, all-over striping until a few months later. From the Lago Grande. 2. This is a juvenile Royal Blue from Lago Manacapuru. It will undoubtedly colour up further but not become a true, all-over striped, Royal Blue. This colour variant is often encountered. 3. & 4. These are both typical blue discus from Lago Manacapuru, where they are common. Neither is a dominant or alpha individual – just ordinary blues from Manacapuru.

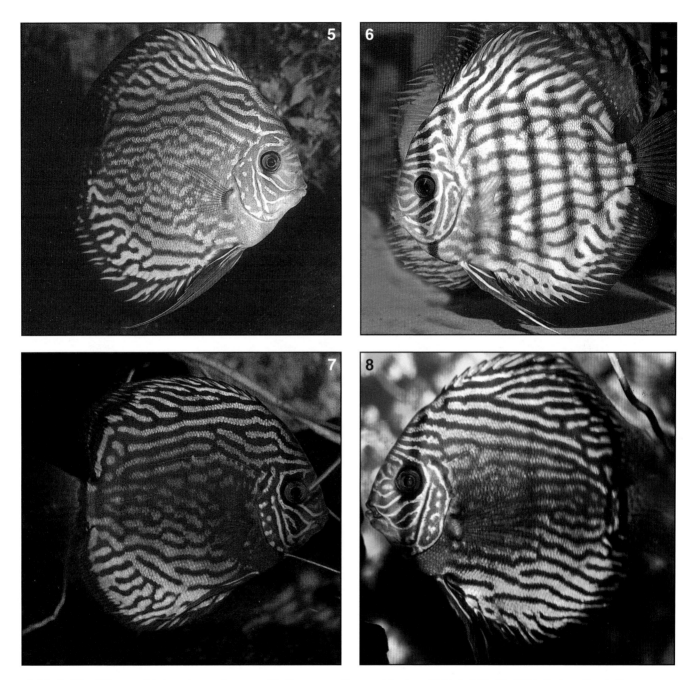

5. A typical Royal Blue – as it is termed everywhere except in German-speaking countries. I caught this individual in 1971 in the Lago Grande Manacapuru. H. J. Mayland photographed it for me. It is an adult alpha individual with the typical fire-red pectoral fins seen in almost all Manacapuru discus. 6. This is an individual from Lago Grande and identical to the specimens caught in 1965 (no photos available). It is an all-over striped Royal Blue *par excellence*, an individual that provided Schmidt-Focke with the basis for breeding his fantastic Red Turquoise. 7. I landed this Royal Blue from the Lago Manacapuru in 1986. Its base colour was extremely red, the pattern colour a brilliant turquoise, but not all over. 8. And this individual from Lago Grande may (in my view) be a separate species (see Chapter 2: *comments on taxonomy*). It is a variant in which the stripes invariably run straight back from the head to the extreme end of the anal fin. No other form exhibits this feature. And it is found nowhere else.

was a German) you could survive through thick and thin, although from the mid 1960s on he himself rarely went into the bush. But he always provided me with *lanchas* and a few *caboclos* to help me. And Manacapuru was one of my first goals.

When I got back, emaciated after more than two weeks of difficult travel by boat, very few of the fishes collected were still alive in the dreadful 20-litre kerosene cans, where they shot around and dashed themselves to pieces. But I had had a few indian baskets woven and lined with plastic, and most of the fish survived in these. Including the Royal Blue discus (The idea of the baskets came from my childhood in Mato Grosso, where mother had baskets of this type made for her plants – *see page 63.*)

At that time the town of Manacapuru was a tiny place with a few dozen wooden houses up on the *terra firme*, and there were as good as no sources of supplies. Nowadays it is the third largest Amazonian city with about 100,000 inhabitants (officially only 74,000), and there is even an 84 km metalled road to the Rio Negro, from where one can take a ferry across the Rio Negro to Manaus. Along with the *caboclos* I fished all day in the lower lake (called Lago Manacapuru or Lago Caballana), which at high water forms a continuous unit with the Solimões. But here I found only common blue discus. In the "upper" lake (as I term it – though in Brazil it is known as Lago Grande de Manacapuru), I bagged the first "Royal discus" (page 19).

These photos from the 1960s illustrate the early days of my discoveries:
1. At that time it was usual to employ a very long net and encircle the biotope where discus might be hiding along a bank like this one. The forest was cleared (a day's work) so that the net could be pulled out onto the bank (see Chapter 5: *collecting*).
2. I had indian baskets woven and lined them with plastic, as containers for my discus.
3. After the return to Manaus the discus had to be brought ashore from the lanchas in (on average) 200 litre containers and carried up the steep bank.
4. Then the discus were returned to my baskets and transported to Willy's establishment.
5. At 106 Miranda Leão Willy had aquaria built to my design for the King. Here, at the age of 19, I accommodate my first discovery.
6. Willy and his wife Robine often visited me in Germany, after I relocated Aquarium Rio to Frankfurt in 1967.

The Japanese were enchanted by the Royal Blue from Manacapuru right from the start (beginning of the 1970s). I was able to collect on average 2000 superb individuals from this region every year between 1967 and the mid 1980s, and the majority of them went to the Land of the Rising Sun. They were such a gigantic success that the first International Discus Show came into being in 1986 in Tokyo; I played a leading part in its organisation and caught almost all the fishes for it (see Chapter 8: *International Championships & Classification*). The fish shown above is an F_1 tank-bred from a Manacapuru Royal Blue discus that was exhibited in Japan and labelled as Manacapuru Red Turquoise. The owner of the excellent *Aqua Magazine*, Mr Matzusaka, kindly provided the photo.

My (second) discovery, in the Trombetas region

At the end of 1965 I first had the opportunity to visit this tributary of the Amazon, which rises in the mountainous north, on the Guyana Plateau. I made what was at that time an almost incredible journey, starting from Belém, and initially travelling by *lancha* to Oriximiná, which lies less than 50 km up the Rio Trombetas from the mainstream Amazon.

A *lancha* is made of wood and basically has one deck *(above)*. There is usually a small cabin towards the stern, used for cooking or where the helmsman sleeps, often with his wife and child(ren) living there too. In the middle, but set lower down, is the big diesel engine. And towards the bow another small cabin with the helm. The whole is some 12 metres long and three metres wide. Nowadays there are much larger *lanchas,* often with two or three decks, and known as *recreios*. But almost 40 years ago the basic *lancha* described above was the only way of travelling up the Amazon and its tributaries. And the *motorista* (driver of the boat) had to be ready for anything!

The boat was en route for a week, with 78 passengers aboard, plus the family of the *motorista* João (he had four children). Picture it thus: there were 78 hammocks hanging in a space somewhat more that 3 metres long and about 2.8 metres wide. When they were occupied each body was pressed tight against the next. When it rained – as was often the case during the trip – the rainwater swept in clean on one side and came out on the other pitch black. Each time the hammocks and their contents, including clothes, were washed clean. So I didn't bother showering in the morning or washing my clothes, as it rained almost every night and by day the wind created by our passage rapidly dried everything again. Unless a hurricane was forecast – nothing unusual in this part of the Amazon. More than 1,000 km from its mouth, being in the middle of the river can be reminiscent of the stormy ocean. Waves as high as houses breaking over the puttering *lancha*, with a lot being washed overboard. Now and then a passenger or two, but nobody noticed any difference in the degree of crowding. Often, soaked to the skin and kept warm only by the bodies above, below, and next to me, I prayed that the boat wouldn't founder.

A week later, on arrival in

The *lancha* – house-boat or Amazon bus – on the Trombetas.

78 hammocks were suspended on the deck of the *lancha*, next to loads of boxes.

Bleher with a successful catch being unloaded back on the shore.

This discovery from the region of the Trombetas (Lago Salgado) is unquestionably another new discus variant. I first managed to catch this discus at the start of the rainy season (1965). It had a very extreme pattern of straight parallel lines, almost invariably equidistant from one another. This type of pattern is rarely encountered in other discus variants. The only similar variant (again discovered by me) is one from the Manacapuru region *(page 69)*, but the stripes and intervals are narrower in the Trombetas form than in the Manacapuru. Could this perhaps be a distinct species? Unfortunately I was unable to photograph the original specimen, so my friend Mayland photographed it on a subsequent collecting trip.

Oriximiná, I was about three kilos lighter, my clothes hung on me like a sack, and my tennis shoes were floating somewhere in the direction of the mouth of the Amazon. Luckily I located a new hammock and sandals at a *boutequím* (stall); the latter were made from old rubber tyres (although at that time there were no cars along the Amazon, let alone the Trombetas).

I first of all filled my empty belly with *aipím, bananão* and *tracajá* (manioc, cooked bananas, and aquatic turtle), and then travelled on by dug-out to the Lago de Sapucuá, where there was said to be a very beautiful discus variant (apart from the Heckel discus).

Note that nowadays the "last" aquatic turtles are officially protected, especially in the Trombetas. During the dry season the Arrau turtles *(Podocnemis expansa)* clamber up from the riverbed onto the kilometres-long sand-banks that constitute their largest egg-laying grounds in South America. Even so their days are numbered. These giant creatures can attain a carapace length of 90 cm and weigh more than 90 kg. But even at the time of Humboldt's journey (1800) the colonists devoured more than 33 million turtle eggs per year. And when the English naturalist Henry W. Bates visited this region years later he wrote in 1863 in his book *The Naturalist on the River Amazon,* that people along the Amazon harvest about 45 million turtles eggs each year to manufacture oil from it.

The indians have lived for millennia in a healthy balance with nature, but since the colonisation everything has changed

Photos from the 1960s, again showing the early days of discus hunting:
1. Typical Trombetas discus biotope.
2. A village of the Wai-Wai indians (children, *right-hand page*) on the Mapueira (a Trombetas tributary).
3. A receiving station for discus outside Manaus: ponds in the ground, reinforced with planks.
4. A discus collecting boat, a *lancha* in which I travelled for weeks.
5. My homemade indian baskets lined with plastic bags were ideal for my discus – and also for quick and easy water changes.
6. Despite the night photography, the Heckel discus, here from the Rio Negro, can clearly be seen at the bottom of the basket. I caught discus also after midnight in the Trombetas, once I had discovered that it is much easier, and better for the discus, to catch them by night (see Chapter 5: *collecting*).
Above right: Wai-Wai indian children on the Mapuera.

throughout the Amazon area – and not only as regards the King of the Amazon, the discus...

Paddling to the great *lagoas* west of Oriximiná meant another week of suffering. But this was rewarded with fantastic fish species and an unusual discus, a new variant *(see page 73)* for a Trombetas region. A year later (1966) a Brazilian scientific expedition collected a Heckel discus variant here by bow and arrow. The specimen is stored in the Museo de Zoologia in São Paulo *(page 39),* and labelled as *Symphysodon discus...*

However, 40 years ago I was (unfortunately – or luckily?) not yet as interested in the determination and scientific study of new species. I took the fishes back to Rio de Janeiro (to Aquarium Rio), where I got good money for them – albeit not the fabulous prices of today. And thus this "discovery", like many thereafter, went unrecorded. (Luckily my friend, the author of many discus books, the late Hans J. Mayland had an opportunity years later to photograph the form for me.)

The Trombetas lured me back repeatedly, and especially the discus found in the Lago Salgado, a glorious colour form with a wine-red base colour and, rather like the Urubu individuals, narrow, often regular, longitudinal stripes running backwards. But I never saw a specimen with a bright stripe along the base of the dorsal fin as the Urubu variety. A unique pattern.

The discovery of the solid green discus

Tefé has long been a magical name for discus enthusiasts and scientists – and at least since the 1960s a synonym for "green discus" among all discus fans. The place is as significant to them as Mecca to the Moslems (except that Tefé is more difficult to get to). And if you surf the Internet under Tefé – eg at www.google.com – then you will find about 70,700 pages to click on, most of them mentioning the green discus (or, better, "Tefe discus").

Today this city of 100,000 inhabitants (about 70,000 according to the 1998 statistics) has grown into yet another metropolis on the largest river on Earth – and in a very short time. When Sir Walter Ralegh explored South America, Tefé did not yet exist. La Condamine likewise found no village here. A settlement was first established in 1759, set somewhat back from the Rio Solimões, right at the mouth of Lago Tefé, 663 kilometres by river upstream of Manaus (at that

Podocnemis expans

Turtle eggs

time called Barra do Rio Negro), and initially called Ega (Eda). When the German naturalists Martius and Spix, who between 1817 and 1820 travelled Brazil in the service of the King of Bavaria (see above: *First discovery*), passed through here in December 1819 it was still Ega. Note that Spix travelled the upper Solimões from here only as far as the border with Peru, while Martius explored the Yapura (= Japurá) upstream as far as the waterfalls of Arara-Coara on the border with the then vice-regency of New Granada (= Colombia). At the beginning of February 1820 Spix was back in Ega, and because Martius had not yet returned from his Yapura trip, he used the time to explore the Rio Negro until the end of the month (without finding any Heckel discus – Natterer was the first to do that). On the 11th March 1820 they met up again in Barra do Rio Negro, and continued travelling the Amazon until, on the 16th April 1820, they arrived in Belém, by which time Spix was gravely ill. After embarking their collections and live animals they sailed on the *Nova Amazonia* to Lisbon, where they arrived on the 23rd August 1820. From there they travelled across Spain and France to Munich, which they reached on the 10th December 1820. As a result of this dreadfully arduous journey Spix died only a few years later.

When, in September-October 1865, Agassiz and his companions collected the first green here (Thayer Expedition), the name had changed, but they mentioned both in their locality descriptions: Teffé in some cases and Eda in others. The name change was not yet in universal use.

Anton Ludwig von Hoonholtz (1837-1935), of German ancestry, was honoured with the title Barão de Tefé. He was a diplomat and naval officer from Rio de Janeiro, and contributed significantly to the Brazilian victory in the war with Paraguay (1865-1870). He negotiated the border treaty with Peru and was the founder of the Brazilian Hydrographical waterways.

The main means of travel to and fro is still the *recreio* (as there are no roads leading to Tefé). It was then, and remains so today. (Only then the *lanchas* were much smaller.) A ticket from Manaus, for example, costs (end 2003) 16 Reais = 6.00 dollars and the journey lasts some three days and two nights, with an average of 100 people on board.

When I went there for the first time – exactly 100 years after Agassiz & co – it was even possible to fly by DC3, once a week. Nowadays jets fly at least twice daily from Manaus. But the flights are almost always fully booked. Even after 11th September 2001.

But going back to the time of my first trips there, in the 1960s and 70s – it was still always an adventure, albeit not comparable to that of my predecessors in the 19th century. Even so, I experienced similar difficulties to Harald Schultz.

Now it wasn't me that discovered the green discus, nor Schultz, but I did collect them shortly after him, and at the

The new airport at Tefé – the former Ega.

Tefé as it looked at the time of one of the first trip in search of discus.

Most of the village consisted of wooden huts built on stilts.

end of the 1960s made them available to the discus world in larger numbers for the first time, via Aquarium Rio. Public aquaria from San Francisco to Sydney and Tokyo profited thereby, as did wholesalers all around the globe (in 1969 I was the first to bring Greens and Royal Blues to Japan), as well as breeders. And, of course, the best alpha specimens that I collected ended up with Schmidt-Focke in Bad Homburg.

But why am I telling you all this? In order to tell you about my initial experiences of collecting Greens? No, the point here is the discovery of the solid green discus. And this is how it came about: after Schmidt-Focke line-bred the first solid individuals from my wild-caughts in Bad Homburg at the beginning of 1970, I started to wonder where this inherited characteristic came from "It's all in the genes", said Eddy (Schmidt-Focke) repeatedly, and I said to myself deep down, "Somewhere in the wild there must be solids and I am going to find them" (see also *Discoveries in the Nhamundá*, below).

There followed annual expeditions (often three to five or more per year) to the furthest corners of the Amazon region, and I discovered a number of things (see also page 101-2). *Inter alia*, for the first time in 1976, the splendid "Royal Green" discus, as I christened them. They came from the vicinity of Lago Tefé and from the Japurá area. Schmidt-Focke received these jewels and bred them just a year later (since then there have been Royal Greens worldwide and the name has become established). The collecting as well as the transport was a breath-taking experience (an extensive report appeared in *Tropical Fish Hobbyist* magazine (April 1984)).

I transported these fantastic discus personally for the entire journey, from dug-out to *lancha*, to the DC 3-plane, by car and bus, until they were safely on the plane to Frankfurt. But in the cabin their Highnesses began to thrash around so violently that they tore almost all the plastic bags, despite being triple-bagged and padded with newspaper, and only two survived. Water continuously trickled from my hold-alls and transformed the Boeing 747 into a miniature Amazon. The stewardesses wondered where all the water was coming from and had to keep mopping it up.

But it was not until a collecting trip in October 1985 that I found the answer to my years-old question. Here is my report, part of which has previously appeared elsewhere:

I stood in the foyer of the fantastic Tropical Hotel on the Rio Negro in Manaus with the Chilean photographer and lawyer Cristina Taras, the German author Hans J. Mayland, and the catfish expert Jürgen Inge, also German. We had spent a pleasant evening with Senhor Efrem Ferreira of INPA (Instituto Nacional da Pesquisa da Amazonia), who had, *inter alia*, worked for years on stomach-contents investigation of Amazonian fishes. Our stomachs filled with tasty *calderada de tampaquí* (a type of Amazonian *bouillabaisse*), we discussed our plans for the following day and our forthcoming flight to Tefé. Then I fell fast asleep in a large, snow-white hammock on the veranda of my comfortable room next to the hotel's menagerie.

As the sun arose I was awakened by a woolly monkey who clambered into the hammock with me and lay on its back on my stomach so I could tickle its belly. At 5 o'clock sharp a breakfast fit for the gods was served – a choice of nine different local fruits: *papaya, cupuaçú, graviola, guyaba, melão, açaí, fruta de conde, banana-maca,* and *jabutí*. Plus seven different

My first flight to Tefé (1965) with the then Cruzeiro do Sul (DC 3).

The Green from the Japurá – one of my "discoveries" in the Tefé region.

juices, eight types of bread, cakes, and tarts, flaked oats and muesli, yogurt, sweet rice, caramel pudding, *tapioca* (a delicacy made from manioc, prepared according to an indian recipe), maize (ground and boiled, then mixed with manioc meal and wrapped in maize leaves – served warm), sausages, ham, cheese, marmalade, biscuits, five hot dishes, hot milk and chocolate, and, of course, tea and coffee. Where else could one get a breakfast like that?

At 6.30 we arrived at the airport, but the plane was full – even though I had booked and paid two months previously. Just five minutes before take-off, after much argument with the manager and with a few dollars less in my pocket, he took four passengers off the plane (I never did find out how he managed that) and we climbed aboard.

It was still early morning when the plane touched down at the pretty little airfield at Tefé under a clear daylight sky. We were the only passengers to alight here (nowadays it is rather different). Eight taxis were waiting there – all unregistered VWs and the drivers undoubtedly had no licenses. We couldn't cram our almost 500 kilos of equipment into a single taxi, so a second was summoned. Half an hour later we reached the town, travelling at 10 kph as the old rattletraps couldn't go any faster on the horrendous roads (all now metalled).

We were straight away the main attraction in the town – people stared at us as if we were from another planet and showed us to the only hotel (and the only three-storey building), where we had to drag all our luggage up the almost vertical stairs to the third floor. (This reminded me of the Maya temple steps in Tikal, Guatemala, where the sacrificial victims were so exhausted by the time they got to the top that they hardly noticed their hearts being cut out.) We also banged our heads on every landing (so we too lost some blood!), as the ceilings were very low, designed for the small people of the Amazon. When we reached the "heights", the pleasant receptionist said *"Bom dia",* and her smile, plus the view over Lago Tefé, the great square, and the church (the hotel was right in front of it), did a lot to make us forget our wounds. But the rooms and beds were a further shock – they too were designed for Amazonian "dwarfs" and our feet hung way over the ends of the beds. The air-conditioning unit, apparently fitted into the wall using brute force and ignorance, worked only when there was electricity, which wasn't often. But despite all this, and if you ignored the giant American cockroaches, the accommodation was top class for the time and place.

The hotel owner, Senhor Deostede, who owned half the town as well as this, its tallest building, tried to find a certain discus catcher for us, the *japonés* (Japanese) as they called him. But he learnt from the *colonia dos pescadores* (colony of fisher-

1. Since the 1970s the Hotel Tropical (owned by Varig Airlines) in Manaus has been the leading hotel in Amazonia. This fantastic establishment, built in the colonial style, stands right next to the Rio Negro and includes a private zoo. Here (1989) with the Japanese Kamihatasan, Monique Nicolaï from the USA, and Bleher, prior to departure on a new expedition. 2. Arrival in Tefé (1985) for the hunt for the solid green discus. Hans. J. Mayland, Cristina Taras, and Bleher (from right to left). 3. Aerial view of Tefé (1997). Lago Tefé with the town *(right in the photo)* and the Rio Solimões in the distant background. Nowadays Tefé is an Amazonian metropolis. 4. Tefé at the time of our arrival in 1985.

men) that the man was on his way to Manaus in his *lancha* with 4,000 discus, with a stop-over at Coari. But his son, Segderu Esashika, an agronomy student who was out performing field studies but supposedly due back that afternoon, would be able to tell us more. Meanhile we discovered that there was no *lancha* available – although I had reserved one by letter a month previously.

The tropical rain started to fall and within a few minutes everything was flooded, half the town was under water, but this didn't in any way spoil our meal at Donna Maria's romantic restaurant, complete with lake view. She cooked my favourite dish, *pirarucú cocido no molho de legumes* (*Arapaima gigas* – the largest freshwater fish on Earth – in a vegetable stew). Half an hour later – when the sun was once more shining over the huge lake and the sky was again cloudless – we were enjoying this delicacy. A few rays of sunshine fell on the sculptures on the steps at the entrance to the prefecture, a legacy of the collector of the first green discus, Jobert (page 42).

The entire scene was perfection and the blue and yellow aras flying above us were the icing on the cake. And as we sat over a *cafezinho* I thought of how the Brazilians rarely suffer from stress or heart attacks (and the *cablocos* certainly don't), as everything is accepted *com calma* (calmly). Then Segderu appeared. Together we strolled through the little (in those days) town, past the only barber's shop, to the fishermen and boat-owners. As dusk drew in we found ourselves at the home of the second *japones*, who owned the other half of Tefé. He was fast asleep, despite it being so late in the afternoon. A long siesta, I thought, and had him awakened. Over our second *cafezinho* I arranged the hire of one of his big *lanchas*, plus an aluminium boat with outboard, for 100 dollars per day.

The *Grande Barão* (after the Baron who gave Tefé its name) was a typical 14 metre long Amazon houseboat, but luxuriously equipped with a kitchen and toilet, a cabin for the helmsman, and a 245 horsepower Volvo diesel engine.

Loading the *Grande Barão* on the lake to go discus collecting in Rio Tefé.

After we had loaded everything the next day and strung our hammocks on deck, we initially headed in the direction of the Lagoa de Caiambé, where we arrived by night. I wanted to look for greens here as well, and had brought my 1,000 candle-power underwater lamp (which was a huge weight on land). I dived deep in order to take a leisurely look around through my mask, very cautiously moving the beam among the branches.

The water was not very clear, but I could see fishes some 2-3 metres away. I pushed on through the tangle of branches and frightened some *Mesonauta* sp., *Cichla ocellaris*, and a number of extremely red-spotted *Heros cf. severus*. Naturally there were no aquatic plants, as the water rose too high here for them.

I made a further three dives, and then I began to freeze wretchedly, as the temperature had dropped to 24.9 °C at the surface although it was not yet midnight, and was even colder at the depth (some 2-3 metres) to which I was diving. Just as I was about to give up, I saw a few discus huddled together motionless, like leaves on the tree-trunks. No nocturnal predator would recognise them as prey. Despite it being night, their coloration was fascinating: light bluish with an intense red base colour, a broad black band running across all their fins (except pectorals and

Segderu with his provisions for the trip – *cascudos* (= *Liposarcus* sp.).

ventral fins) red fin edgings, and a pitch black band through the eyes.

It was impossible to catch them. I could find nowhere to stand, no bottom – it must have been 4-6 metres deep – nor any footing on the steep bank. All my attempts were in vain. Either there were too many branches, or the fishes jumped out of the hand-net however quickly I raised it (discus are damned fast!).

The next morning our *motorista* headed back to Lago Tefé and we crossed the lake, taking almost four hours to traverse the 70 km. During the voyage Segderu, who had arrived with his friend Raimundo (a commercial fisherman), told us about his catch a few days ago. They had captured more than 77,000 *jaraquí (Semaprochilodus insignis)* with one haul of their net. Of course, the latter was 400 metres long and 6-8 metres deep. But they had not previously caught so many fishes at one time. In his opinion the *jaraquís*, which weighed two kilos on average *(below right)*, were undoubtedly migrating to spawn.

And he continued, "In the course of a year about 10-12 tonnes of fish are caught every day in Lago Tefé. And it is all for the personal use of the local people. Every morning the fishermen arrive with their wares at the market near the shore or offer them for sale while unloading. Mostly *tucunaré (Cichla ocellaris)*, *pacu (Colossoma macropomum)*, *aruaná (Osteoglossum bicirrhosum)*, *Pseudodoras niger*, and large *cascudos (Hypostomus* species). Less frequently – mainly during spawning migrations – *matrinchã (Brycon melanurus)* and *jaraquís (Semaprochilodus insignis).*" (Note that the name *jaraquí* is applied frequently elsewhere to other Amazon species of the genera *Semaprochilodus, Prochilodus* and *Curimata*)...

I have rarely seen such a quantity of freshwater fishes in one place, but must add that there were never any discus among them, not in hundreds of visits to markets here or elsewhere on the Amazon. Neither the *caboclos* nor the indians eat the *acará-buceta* as it is known locally (which I won't translate, it is a Brazilian translation from the Yanomami language. The indians have used this name for the discus for thousands of years, and it is a symbol of fertility in their mythology. The story goes that a woman who had borne no children was put in the water with various fishes, but only the discus did the trick. Since then it has been treasured and revered by the indians.)

I asked our boatman to stop in the vicinity of a typical discus biotope. Full of dead branches and trees trailing in the water, sometimes to quite a depth, and hence impenetrable and almost impossible for diving. In such places discus spend the day in 2-4 metres depth. Quite different to Heckel discus in the Rio Negro, which are almost invariably found beneath live trees lying in the water, or that have recently fallen in, as well as overhanging bushes.

The water in Lago Tefé was rather murky (although it is a clearwater lake), with a pH of 5.9 to 6.0 and a conductivity of 15 μS/cm (and to think there are people who think salt is needed for keeping discus). The general hardness was thus 0.1 dGH° and the carbonate hardness 2.0 dKH°. The temperature at the surface measured 29.9 °C at nine am. But the discus were at depths where the thermometer showed below 27 °C and there was water movement – currents that can lead to considerable fluctuations in temperature.

I dived and discovered a small group in some 1.5 metres of depth, but they immediately fled to deeper water.

1. In my hammock on the *Grande Barão* on the Rio Tefé, making notes. 2. The daily catch of food-fishes in Lago Tefé is enormous. Back on shore, each fisherman immediately deals with the fish he has caught. 3. 77,000 of these *jaraquís (Semaprochilodus insignis)* were caught in the lake in a single haul of the net.

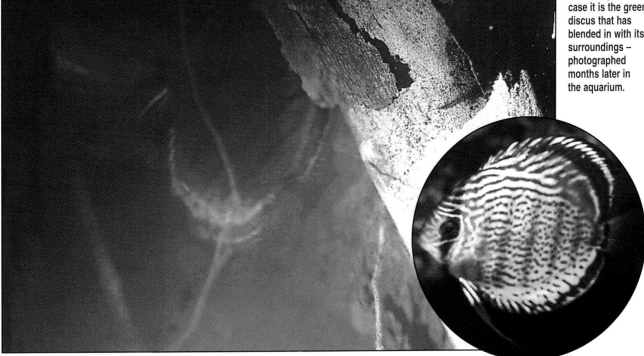

Left: Bleher during a typical night fishing session – or rather, researching the discus by night. He almost always searches for the "King" after midnight (and on moonless nights). Since the mid 1960s he has found that it is easier to locate the fishes at night and they can then be caught without any damage to the fish. Although a lot of patience is required. For at night – in contrast to during the day – discus are almost always found singly and, as can be seen here *(below left)*, hiding close to, or among, branches in clear or brown water. Perfect camouflage. In this case it is the green discus that has blended in with its surroundings – photographed months later in the aquarium.

Segderu later told me that the water level in the affluent river hadn't fallen in years. This was attributed to environmental destruction (logging and clear-felling, plus burning). In the inundation zones the water level no longer subsides and as a result millions of trees are dying, as their roots can no longer obtain oxygen *(see below)*. And it was true that a large part of the area looked like a tree cemetery. And the discus had moved away too. There was no longer any shelter (shadows) for them, and no food, as the latter consists mainly of the flowers and fruits, the seeds and leaves, of the trees.

It is a fact that catches of green discus in the lake have declined drastically over recent decades (in contrast to what is published in discus magazines). The few occasional collectors that remain today have gone off to Coari, the Japurá, and elsewhere. Segderu told me that catching discus was no longer profitable, and the high cost of petrol and diesel (almost as high in Brazil as in Europe) was also a contributory factor. Segderu had long experience with discus, as his father was the first commercial discus collector (said to have started as long ago as 1962).

Raimundo, Segderu and I dived together repeatedly at various spots in the Rio Tefé as we puttered our way up-river. We ventured into the inundated areas where there really should have been discus, but there was nothing to be seen. I came face to face with a *sucurí* (the green anaconda, *Eunectes murinus* – the largest snake on Earth), but it fled away in another direction. Numerous *Cichla*, *Cichlasoma*, *Colossoma*, and *Serrasalmus* swam in front of my mask, but no discus.

I enjoyed that single night in my hammock on board the *Grande Barão*. It was so gloriously peaceful and utterly remote from civilisation. I gazed at the star-filled sky, which looked close enough to touch. Not a cloud in sight, just millions of stars, each one gleaming and twinkling. And the whole accompanied by an incomparable jungle concert. With no mosquitos to bother me, I slept like the proverbial log, forever dreaming of green discus with dream colours. When the concert abruptly stopped in the morning, I knew it would be only minutes – or seconds – before the sun came up (in Equatorial regions this happens faster than the speed of thought) and it would be broad daylight. Then eight aras arrived to herald in the morning, flying squawking over our boat towards the west, followed by two toucans.

Along a large part of the Rio Tefé the water level no longer subsides.

A typical discus biotope in Lago Tefé – at a spot where no people live.

With my toothbrush between my teeth I watched as a *tucunaré* squabbled with a metre-long *aruaná* over breakfast. The bony-tongue made a mighty leap out of the water to snap up a beetle, a feat that the *Cichla* could only dream of and look on silently. But the brilliant yellow cichlid with its pitch black spots grabbed a few small characins and seemed happy with that. Was I in paradise, I wondered, when Cristina, who was also already awake, handed me a bowl of muesli with fresh fruit.

The goal of our journey was a *lagoa* (lagoon) on the central Rio Tefé. Years ago some *caboclos* in Tefé had told me about it, but apparently none them had fished, or even been there. It is not marked on any map. It would take us another six to eight hours or so at full speed to reach the branch upstream that guaranteed access at high water (and it was high, despite being the middle of the summer).

The sun burnt down mercilessly. Inge, who was stretched out on the roof of the *Grande Barão*, was as red as a lobster. "Ouch", he moaned, looking around him. Poor soul, it was the first time in his life he had been in the jungle and he wanted to see everything. And the view from the cabin roof was, of course, the best.

Left: The *lagoa* (lagoon) on the middle Rio Tefé where I disovered the solid green discus.
Below right & left: a detail of the solid colored discus discovered in 1985). He had heard about this spot years previously, but nobody seemed to have explored this far. In addition no *caboclos* or indians lived here – the area appeared to have always been uninhabited. The water in the *lagoa* was still, with very little movement, and a very slight layer of algae covered the water's surface.

Even though a lot of the time we were travelling past dead forest where no tree remained alive and hardly any other life was to be seen – after all, what was there for any creature to live on? No forest, no life. That is as certain as night following day.

Moreover no indian tribe lived there any more. Only a couple of *botos (Ina geoffrensis* – pink dolphins) who accompanied us and apparently lived there. And they put on a show for us, as if to say, "We are intelligent and know a few things."

It was already almost dark as we turned into the *lagoa*. Nature appeared totally unspoiled here. The water was tea-coloured, like that of the Rio Negro. (And yet some people persist in maintaining that the southern tributaries of the Amazon contain only white or clear water. I have by now learnt that this is not true at all.) But we couldn't get any further, even though our *lancha* was specially designed for shallow water. Segderu, Raimundo, and we four climbed into the aluminium boat and hurried away before it became completely dark. For more than an hour, with our electric lamps illuminating the pitch-dark jungle night, we felt our way forwards. Raimundo repeatedly lifting the outboard out of the water lightning-fast at Segderu's warning shout, so that the propeller wouldn't become fouled. And then finally we reached the interior of the great lagoon. Raimundo turned off the motor and unbelievable peace reigned.

I just had to get into the water. Nothing could stop me from diving in right away. And I had been swimming for only about five minutes among the branches before, in a metre of depth, I saw something flash. A green pattern such as I had never seen before. I was so excited that I shot high out of the water, shattering the deathly silence with a loud cry of "Green discus, a fantastic colour form, brilliant solid green". The five in the boat were stunned. They thought something had attacked me. (Raimundo had told us about the 40-metre anaconda that was supposed to live here and swallowed everything that ventured into the lagoon where it held sway.)

It was pointless to try and spread the net in this biotope. I suggested that we should come back after midnight and use my tried-and-tested night-fishing technique. Back on the *lancha* we ate *pacu* with *farinha*, rice, and black beans. Then Segderu, Raimundo, Cristina and I once again left our houseboat in the *voadeira*, as aluminium boats of this type are called, en route for the *lagoa*. By now it was 23.00.

This time large *Curimata, Cichla,* and *Prochilodus* kept leaping into our *voadeira,* at full speed, apparently frightened from their nocturnal rest. The air was cold (23-24 °C) but the open water was warmer (29 °C). Cristina was frozen, especially by the wind created by our movement. She pulled on a warm jumper, while Segderu and Raimundo warmed themselves up with a bottle of *cachaça* (sugar-cane brandy).

Arrived in the lagoon, Segderu and I, with no concern for the giant anaconda, slid simultaneously into the dark water which the beam of my light barely penetrated. I first swam to the exact spot where I had discovered the discus, as fishes usually return to their (original) places. When I arrived I thought they had gone, but then I discovered them a little way above me! They had migrated closer to the warmer surface – the later in the night, the higher they move, as the water cools from the bottom up. They looked just like leaves protruding from the branches. A clever piece of camouflage. But the closer I got – always with the light full on the discus, to dazzle them – the further they gently moved away along the branches. Perpetually out of the reach of my metre-long hand-net. I drank in their fascinating coloration with my eyes and was able to shoot a few photos with the underwater camera, but that was all I got. It was gone midnight when we returned to the *lancha*, Cristina was fealing terrible cold.

There followed a long, hard day of work at the *lagoa*. We went back to the old method and encircled part of the lagoon with the 60 metres long and eight metres deep net, fastening both ends to trees on the bank. Then we hacked away for some six hours at the dead and rotten branches to clear a circular area of about 50 metres radius in the bank region, and threw the rubbish ashore. And when, full of anticipation, we finally pulled the net to land in the afternoon, there were actually two fantastic solid green discus in it *(see page 83),* as well as *Leporinus, Cichla, Hypostomus,* and three different *Heros cf. severus* colour forms (3 different species?).

So now I had my discus and had confirmed my theory. The rain was by now bucketing down but it didn't bother me (I simply jumped back into the warm water of the lagoon – which had a pH of 5.2, a general hardness that was barely measurable (as was the carbonate hardness), and a conductivity of less than 25 μS/cm).

1) A typical green discus from Lago Tefé (caught there, *biotope page 82).* 2. This *Heros cf. severus* variant is often found with the green discus in the lake (and likewise caught there). 3. We netted the *tucunaré (Cichla cf. monoculus)* in the Rio Tefé along with the solid green discus *(page 83).* 4. This characin, *Curimata vittata,* also turned up in the discus net. 5-9. A selection of our provisions for the Rio Tefé expedition: *farinha* (5), milled from the *aipím* root (7) and roasted; *suco de açaí* (6), prepared from a small coconut; *rabo de jacaré* (crocodile tail) (8); and *tambaquí (Colossoma macropomum)* (9).

Discoveries in the Nhamundá

It is impossible to include all my discoveries of different discus variants in this book, but I will, however, describe one of my recent collecting trips – which has previously appeared in part elsewhere – with appropriate comments thereon.

The Nhamundá region is an almost unique melting-pot for discus in the Amazon region. More variants occur there than in any other place I know. And they include forms never seen before. My third trip to this indian region proved to be quite a costly adventure. Here is the story, just as it happened:

"*Temos que seguir as instruções da Funai!*" said the *tuchaua* (chief) of the Hixkaryana indians.

The only indian of the tribe who knew a few words of Portuguese had reported our landing on the jungle airstrip by radio to the regional head of the indian administration (Funai) in Parintins. And from there had come the order, "*Confisca a filmadora deles*" (confiscate their cameras), when the indian explained to him that we – that is, Axel Mewes, the pilot Paulo, and I – had landed without permission.

We had flown there to film the Rio Nhamundá, and in particular the rapids upstream of which no discus can penetrate, from the air. Paulo, who had amassed more than 15 years experience in the Amazonas as a bush pilot, and knew almost every airstrip as well as the administrative regulations, had said to me, "*Aqui podemos aterisar*" (here we can land) as we flew over the indian village, and that was the only reason I had agreed. But hardly had we touched the ground – with the propeller still running – before the indians came rushing up, hauled us out of the little Cessna, and dragged us off to the village.

There they first of all convened a meeting of the village elders, which we attended without being able to understand a single word. An hour later the translator explained briefly that he would have to inform Funai of our landing, and when the reply came, in Portuguese, the chief made his decision: eight muscular indians ripped Axel's professional film equipment, including his special lenses, from his hands. Axel yelled (in German) "You can't do that, that is all I own, my work, the tools of my trade.", but there was no reprieve. Everything I said to the *Tuchaua* (via the translator) was in vain. They dragged us out of the hut and back to the plane, and closed the doors behind us. The translator shouted, "If you don't take off immediately we will set your plane on fire." Paulo was practically shitting himself (there really was a smell, perhaps the sweat of fear) and started up the engine...

The Hixkaryána indian village on the upper Rio Nhamundá.

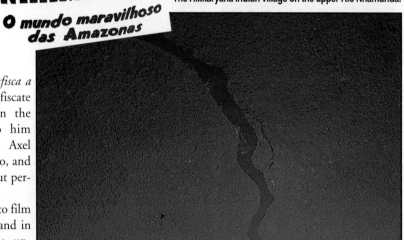
The Nhamundá is a blackwater river and still largely unspoiled.

Thus the equivalent of about 50,000 dollars has by now been lying for more than three years in an indian hut in the deepest rainforest of the northern Amazon region, close to the border with Guyana, and probably mouldering away completely in the high humidity.

Our attempts at rescue – notably by Asher Benzaken – likewise proved a total waste of time. The indians had become cunning, and tried to turn the mistake by Funai to their own advantage. They demanded two large aluminium boats with

40 hp outboard motors, worth almost as much as the film equipment. Of course Funai didn't have the money – despite international economic aid – and the regional official responsible for starting the entire fiasco by failing to look into our case more closely, let alone hear our side of it, was never brought to book. Likewise my subsequent indefatigable trips and representations to the President of Funai in Brasilia (in the meantime changed for the sixth time – because of stress?), and to the German embassy and other authorities, proved in vain.

Now, I am repeatedly asked what impels me to do such things, to take such risks. But I grew up with it, and remember how at the age of only five I lived an unusual life with my mother in the deepest jungle among the aboriginals of central Africa. And later, aged seven, among cannibals in South America – but no harm ever came to us. So why now, for the very first time, after so many expeditions? Has the attitude – or the self-view – of indigenous peoples changed? It is certainly the case that the attitude of the indians of Brazil has altered, and they are no longer the friendly natives that I learnt to know over a period of almost half a century.

There are now barely 100,000 indians in Brazil – out of more than six million formerly – but nowadays they know what they want, especially since the advent of Funai...

Family Benzaken (from right to left): Asher, his wife Adele (daughter of the export pioneer Willy Schwartz), Bleher, and the family's 3 children Nina, Zev and Tamara.

The *Eldo* II, the recreio to Nhamundá, with three decks, carved wooden doors, etc.

Asher, the owner of Turkys Aquarium in Manaus, is undoubtedly the only exporter in South America who acclimatises discus properly using immersion heaters.

It all began when I invited Axel, my friend of many years, a professional cameraman but also a passionate aquarist with many breeding successes, to join my third Nhamundá expedition, to film the splendid fishes long the unexplored Rio Nhamundá and in the lake, as well as capturing the discus varieties pictorially for the first time. The region is a meltingpot for four forms and thus perhaps unique in the realm of the discus.

After months of preparation we arrived, laden with heavy equipment, in Manaus at the end of November. Asher, the owner of Turkys Aquarium and successor to Hans Willy Schwartz, the pioneer exporter, had arranged a *recreio* (houseboat) for us to make the three-day trip down the Amazon and up the Nhamundá to the village of the same name, which lies on an island in the middle of the Lago Nhamundá, sometimes also called Faro after the only other village on the lake.

A *recreio* is the local equivalent of a bus and remains the chief mode of transport in the vast Amazon region, where roads are few and far between. They putter out of the floating harbour in Manaus (below) to head in every conceivable direction. Some 40 or so of these *lanchas* lay before us when we arrived at 6 am, with superstructures consisting of one to four decks. Normally the accommodation consists of hammocks strung on the decks – up to 100 at a time. Our *recreio* seemed very comfortable when I thought back to a trip I made years previously from Belém to Oriximiná (see *Trombetas discovery*). This time, however, it was almost like being on a passenger steamer. The boat was called *Eldo II* and had three decks with two between-decks for *redes* (hammocks) – the bottom deck was reserved for pigs and fishes. Asher had chosen the boat of Frankie Azevedo Cruz because it had four *camarotes* (cabins) – an unimaginable luxury – that were normally reserved for special passengers such as mayors and *deputados* (deputies). We had two of them. There was just one problem: the Amazon people are mostly very small and hence the cabins and the couches were only 170 centimetres long. Neither Axel nor I (1.78 metres) could fit in. Only after I had bloodied the back of my head half a dozen times did I learn to adjust to the length. But at least it was a safe place for all our copious equipment.

When we first looked around we were not a little surprised to find a gigantic parabolic antenna on the upper deck, TV and disco speakers, chairs and tables, as well as a small bar. Down below other passengers were streaming aboard, heavily laden with everything from gas refrigerators to bicycles, guitars to billy-goats, hens and cockerels, even motorbikes, ladders, tables, chairs, and masses of provisions. Sacks full of the staple food of the *caboclos*: *arroz e fejão* (rice and black beans), and cases of the national drink, *guaraná* (made from a wild fruit and drunk by the indians for millennia, as well as being sold at a high price in France as an aphrodisiac for about 150 years), plus Coca Cola and *cerveja* (beer).

Aboardáa recreio on the Amazon, en route to the third Nhamundá expedition.

The world-famous *encontro das águas* where the Rio Negro enters the Amazon.

Heavily laden with almost a double load, the *Eldo II* left the *porto flutuante* (floating harbour), and shortly before we reached the *encontro das aguas* (= meeting of waters) I looked back one last time at the town, nowadays grown to metropolis size but which I once knew when it had only about 20,000 inhabitants. Today Manaus has passed the three million mark and is barely recognisable. Oil refineries, timber works and sawmills, cement factories, dockyards, and much more line the shore down to where the Rio Negro enters the Amazon, and beyond.

The meeting of the waters of the "black river" with those of the sediment-rich Amazon (the *encontro das águas – centre*) is always a magnificent sight to see, and one I didn't want to miss, even for the umpteenth time. Various fish species from both the black- and the white-water are found together here. And, of course, the spot is a permanent stronghold for cormorants. They repeatedly dive into the mingled flood, returning to the surface with their latest prey. They also roost on the treetrunks and roots that come down the mighty stream, or on huge bushes washed out from the banks – the constant erosion caused by the relentless activity of the *fazendeiros*. It is totally catastrophic. Nowadays cattle are even grazed on the fertile *varzéa* regions! That is the worst abuse the Amazonas has had to endure. A good ten years ago this was not yet the case, and at that time I thought, "Even if they fell, burn down, or otherwise eliminate the entire rainforest, they cannot destroy the *varzéa.*". That is now just more "snows of yesteryear". During the dry season the areas above water are clear-felled and planted with grass, and grass grows quickly on this rich soil. Zebu cattle are then transported there on huge rafts for six months, and by the time the rainy season arrives the cattle will have eaten themselves fat and be ready for slaughter, and are transported away again.

Man knows no limits, and there will probably never be enough meat to satisfy his appetite. When I asked Frankie who would eat all that meat, he just looked at me and laughed (see also Chapter 10).

But the catastrophic consequence is not just the disappearance of countless life forms – a biodiversity such as is no longer to be found elsewhere on Planet Earth – but in addition the rains have failed to come (at least at the time of writing) as a result of the relentless deforestation. It is the middle of the rainy season, but not a drop of rain has fallen in Manaus for three months.

Frankie steered the boat safely down the Amazon past ocean-going giants and container vessels, and after a few days turned off into an incredible maze of waterways: the Calderão, a whitewater river. He told me that there had been no rain for a long time, otherwise he could have taken a short-cut through an *igapó*. Along both banks of this river too the jungle had completely disappeared. Nothing to be seen but pasture for cattle. Then a herd of water buffalo crossed the Calderão. Frankie slowed right down until the party of some 30 individuals had reached the other bank. At the next curve I saw a 1.5 metre *jacaré* (crocodile) slip into the water and a throng of white herons fishing along the bank. Shortly thereafter we turned into the Nhamundá. The water was now clear, and in front of our bow *tucuxi* (dolphins, *Sotalia fluviatilis*) leapt out of the water. Some of them twisted right over in mid-air, performing repeated somersaults even though this was the wild rather than a dol-

Deforestation has led to tremendous erosion.

Faro, also located at Lago Nhamundá.

The island of Nhamundá.

phinarium. I asked whether there were any *botos (Inia geoffrensis,* the pink porpoise) and he replied, *"Aqueles que o Jacques Cousteau filmo e chamou cor de rosa?"* (You mean those which *Jacques Cousteau* filmed and termed pink?). And then an individual about two metres long surfaced to our left and shot a huge spout of water into the air. There is, in fact, more than one species of *Inia*, and only some are pink in colour. The one we saw here was grey.

Shortly after 5 pm on the third day we put into Terra Santa (the name of the village means "the holy land") *(centre)*. Right next to the landing-place, where another *recreio* was already anchored, there was a large notice, *"Mais Aqua, Mais Saude e mais Desenvolvimento"*. The sense being that more (clean) water brings health and progress.

Abigaiú, our lady cook, served us yet another meal of *galinha*, *arroz*, and *fejão* (chicken with rice and black beans) – for the fourth time in succession – before we finally arrived at Nhamundá at 8 pm, by which time it was already getting dark. The village was a sleepy little place, with only a few streets and houses painted in light colours, among which the light blue Prefeitura Municipal de Nhamundá (the town hall) was conspicuous. There were no cars, just bicycles and a couple of mopeds. The sandy beach – which is exposed only during the dry season – was almost as white as snow *(left)*. The lake water was clear and mainly pure. The aquatic scene was dominated by brisk boat-traffic.

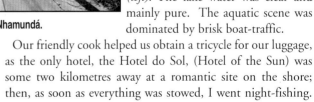

Our friendly cook helped us obtain a tricycle for our luggage, as the only hotel, the Hotel do Sol, (Hotel of the Sun) was some two kilometres away at a romantic site on the shore; then, as soon as everything was stowed, I went night-fishing.

And success justified my efforts, as after a few kilometres hike along the bank I came to a rocky outcrop, and the water there held a number of loricariids, an unknown species of the genus *Peckoltia*. Splendid black-and-white fishes, almost like the zebra catfish *(Hypancistrus zebra)*, but with appreciably finer markings. "Another candidate for the hundreds of L-number catfishes", thought I.

The next morning we met Manuel Torres and his beautiful young wife Efanny *(right)*. Asher had called and asked him to help me with my collecting, and to place his *lancha* at my disposal. Manuel *(far right)* has been catching discus for 25 years – currently for Turkys and previously for Aquario Rio Negro. He had now settled here after discovering that there were various discus variants in the lake. He has little interest in other fishes, except to eat! However, on this trip I wanted to investigate not just the discus, but also all the other fishes in the unexplored Nhamundá. Manuel's *lancha (centre)* was ideal for the local conditions. We could sling our hammocks, stow our equipment safely, and house any fishes we collected optimally. Efanny even had a little galley, and there was a toilet. The river was our bath-tub...

First of all we made floating containers – so-called *viveiros* – to hold the fishes *(right)*, while Efanny prepared a delicious lunch: large characins *(Laemolyta striata)* with *farinha, arroz*, and *fejão* as well as a tomato sauce. In the afternoon we puttered north up the lake as far as the Serra do Guariba, where in 1997 I had managed to catch a spotted discus, although it was the only one out of 600 collected. The others were all typical brown discus, *disco comun* (common discus) as the *caboclos* call them. No doubt it was an alpha individual – but spotted? Green discus have never been found here, not by me or anyone else. Moreover no greens have ever been discovered east of Manaus!

The lakes to the south of Nhamundá, Lago Terra Santa and Lagoa do Jacaré, likewise contain only *disco comun*, but there too

I have caught very attractive blue-spotted discus. On the shore of the Serra I caught two different *Geophagus*, a *Pimelodus* species, *Rhineloricaria cf. castroi, Triportheus cf. angulatus*, a *Fluvifilax* species, and numerous tricomycterids. The place was a dream: snow-white sand, complete with an empty palm-hut. A nice place to live!

Later on we passed the Vermelho, a red cliff some 80 metres high. The surrounding area had been totally deforested so that at high water cattle could be brought there from the *várzea* regions.

Further along the right-hand bank of the lake we came to the Praia do Matapi, whose point extends well out into the lake but was by then almost submerged. The water level was already rising. Towards evening we came to the northern end of the Lago Nhamundá, where half a dozen huts stood on a long tongue of land, while behind them grew innumerable *castanha do Pará* (Brazil nut) trees – hence the spot is sometimes known as Castanhal. Here two rivers enter the lake – close by, from the northwest, the blackwater Rio Nhamundá (the outflow of the lake bears the same name), and further to the left, flowing from the west, the clearwater Rio Paracatú. Interestingly, to date no discus have been found in the Paracatú, while the Rio Nhamundá harbours the Heckel discus *(Symphysodon discus)* and variants, which, however, differ considerably from the type material.

As already mentioned, the Nhamundá region is a melting-pot of discus forms. I was able to find four variants in the lake as well as additional previously undiscovered colour variants (see also the video: *Abenteuer Nhamundá* (Nhamundá Adventure)). In fact *Symphysodon discus* occurs only in the river (where the pH is considerably lower than in the lake). Only 1.5 kilometres to the south, at the Serra do Espelho, I found a colour variant very similar in appearance to the Heckel, though at first glance one would think it a cross between *Symphysodon haraldi* and *S. discus (see page 92)*.

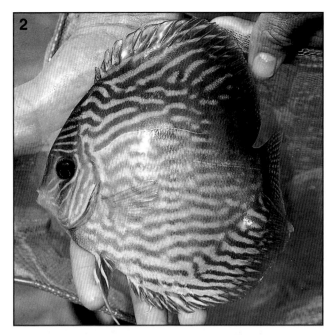
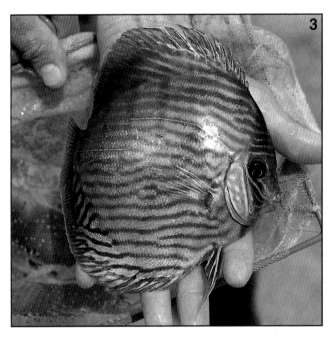

Lago Nhamundá, also known as Lago de Faro, is a discus melting-pot. I was able to demonstrate this for the first time in 1997. No other region I visited has so many variants. Perhaps the Nhamundá drainage provides additional proof that the brown and blue discus are a single species (see chapter 2). In the lake there are Browns with hardly any markings (1), only vestiges on the head and anal fin regions. (This form has entered the trade as Nhamundá Rose.) In addition there are various blue variants: one similar to the Marancapuru discus fishes (2); one that resembles a Manacapuru alpha individual *(page 69, photo 8)*, only more boldly marked, with an almost completely regular, horizontal, blue-turquoise pattern (3), a dream discus; and a Royal Blue (4) – this form may also originate from the Purus region. But all have bands 5-7 noticeably prominent!

I had asked Manuel to take his *lancha* right up the Rio Nhamundá so that I could collect along the way, but after just a few metres this plan seemed doomed to failure. At the river mouth the water was so shallow – barely a metre deep – that the large *lancha* grounded. To our right the bank was lined with hundreds of *guiaranas* (the indian name for trees that die off if the water level remains constantly high – the influence of *El Nino?*). We pushed the *lancha* over the obstructing sandbank and Manuel was then able to motor a little further, but soon we went aground again. So he decided that we would have to continue from there on in the *voadeira* (aluminium boat). That was the plan for the next day. He now steered into a *paraná* (sidearm) and anchored. I took the opportunity to do some more night fishing, and came back with countless colourful *Biotodoma wavrini (below)*, a pair of *Rhineloricaria* – the female with spawn clinging to the underside of her body – plus innumerable tiny fry, doradids, a pimelodid, small light-blue characins, *cachorros* of the genus *Acestrorhynchus*, *pescadas* (immigrant marine fish), *apapá-branco (Pellona flavipinnis, below right)*, a flatfish (Soleidae), and delicate tetras of the *Hemigrammus marginatus* group. Shortly before midnight I collapsed, dead tired, in my hammock. Axel and the others were already snoring away.

The next morning Manuel scoured the area in the *voadeira* for another *paraná* where he might be able to get further upstream. Meanwhile I investigated a large bay full of *Eleocharis* plants and thought I had netted my first

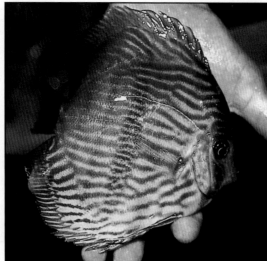

Biotodoma wavrini

Apistogramma, but it turned out to be a *Taeniacara* – a colour variant differing somewhat from *T. candidi*. Could this be a second species of this monotypic genus? I also caught dwarf pike cichlids (*Crenicichla* sp.), *Bryconops* sp., *Moenkhausia* sp. (another beautiful new species), *Geophagus* (two species), *Hoplias malabaricus*, a widely distributed predator (always lying in wait for prey in the shallow water), and many more. Meanwhile Manuel had found a way through, and now steered his *lancha* along a winding course through various natural channels, where we came across three *caboclos* in a smaller *lancha* laden with a wealth of fish they had caught. They were all two *Cichla* species: *C. temensis*, known locally as *pinima* but elsewhere as *tucunaré pintado*; and the *açú* or *tucunaré preto*, which has pitch-black spots, larger than those in *Cichla ocellaris*. This genus is problematical, as there are far more species than are scientifically recognised to date. The fishermen told us that the *açú* is found only in northern tributaries of the Amazon, while the *pinima* is found on both sides. In the Solimões there are the so-called *potoca* or *tucunaré verde* and the *toa*, also known as *tucunaré vermelho*. The latter is a strictly whitewater fish. They had several hundred large specimens stored on ice, destined for the *mercado* (market) in Manaus, more than 1,200 kilometres away

Manuel battled on through the *paranás*, eventually coming back to the 100-200 metres wide Nhamundá, but it was not long before he was stuck again, and this time very badly. This was quite close to the very last settlement on the Nhamundá, Portuguesa.

Efanny prepared a lunch of baked bananas in the middle of the river. Then we went to visit "Capello", which means hat – the nickname of a *caboclo* called João because he always wore one. His wife welcomed us with sugary-sweet *cafezinho* (strong Brazilian coffee) on our arrival.

João and his family were the last non-aboriginal people living here – thereafter there was nothing but jungle and indians – and had a few cattle, mango and *cajú* trees, and by now nine children. The eldest daughter, at 15, already had two of her own. Her father was the first white man to visit the Rio Nhamundá 21 years previously, and had settled here north of the river mouth. He first came here as a logger for big business and had organised the clear-felling. Later, when all the valuable hardwoods had been transported away, he started to breed cattle. In the meantime two more *fazendeiros* had settled some 30 minutes away by *voadeira*.

After we had agreed a price for an eight-day expedition up the Nhamundá, we first of all had to obtain fuel. Capello declared himself willing to go by *voadeira* to Nhamundá (a 12-hour trip) and buy it for us at the floating filling-station. Towards five the next morning he returned fully laden through the densest of fog. To this day I have no idea how he managed it. In the damp morning mist visibility was less than 50 centimetres – you could hardly see your outstretched hand. Despite the fog and bitter cold we reached the elevated *fazenda* of Senhor Mirinda about seven. We were permitted to pick as much fruit as we liked from a huge *cajú* tree (my favourite) while he told us how he and his family of 10 made their living from the *fazenda*. He had 17 zebu cattle, and additional income from renting out his *terra firme*, a clear-felled piece of land where the cattle-breeders could put their zebus at high water before transporting them back to the *várzea* regions when the flooding abated.

After this breakfast, João steered onward undaunted through the dense fog, though his eyes were by now very red. Above our heads, blue-yellow *aras* flew cackling in the direction of the rising sun as the mists slowly lifted. Soon afterwards we came to a splendid *lagoa* on the left bank (below). Capello said it was called Jacitara. At its mouth stood a palm-thatch, and here I found a lovely *Myriophyllum* species, a fine-leaved *Eleocharis* species, and numerous *Nymphaea*. The wealth of fishes was also overwhelming, and I even netted a large colourful *Crenicichla* species, undoubtedly a member of the *marmorata* group *(see photo below, left)* after a long chase through the crystal-clear water. Later, towards nine, when the sun was burning quite brightly, we encountered a canoe there in the lonely wilderness. The two occupants had paddled all the way up here from Faro some days previously in order to fish. They had specimens of two *Semaprochilodus* species, two different piranha species, a *Curimata* and

In the mouth region of the Rio Nhamundá, where black meets clear water, there are unusual discus variants *(left-hand page):* naturally occurring hybrids? The Nhamundá area is home to Wa-Wai and Hixkaryana indians *(above),* splendid *Crenicichla* species *(left),* and the captive fish otter *(right)* in the house of one caboclo.

a *Myleus* species, as well as a splendid *Geophagus* with extremely long fins, and a *Pseudoplatystoma* species (a new species?). All aquarium fishes!

The days flew by. João steered the boat safely through the deathly quiet of the Nhamundá. For two whole days we encountered not a soul, but we did repeatedly collect more new fishes. On the third day we reached a settlement of Wai-Wai indians (see page 93) who had migrated here from the upper Mapueira (see *aqua geõgraphia* no. 4).

But I had almost forgotten: on the evening of the first day we arrived at another *caboclo* hut. One of Capello's daughters lived here, with her husband, Biseo. She already had four children and a fifth on the way, although she was not yet 19 years old. Biseo collected *copaiba* oil (obtained by boring into the *copaiba* palm) here and sold it in Nhamundá. This was not enough to survive on, so he had single-handedly cleared a piece of forest and planted manioc. But that didn't bring in much either, and he told us he was now planning to burn down more woodland, that was quicker – and besides, he no longer had an axe, let alone a saw.

They had a baby fish otter captive (page 93) – they had eaten the mother and sold her pelt. This South American dwarf otter species has now become almost extinct – but only after it was listed in Appendix I of the Washington Convention for Species Conservation. The *lontra*, as I call it, immediately awakened sad memories. At the age of eight, when I lived with my mother among the Cabixí indians, I had a *lontra* which slept with me in my hammock every night, caught piranhas for lunch, taught me to swim – something no swimming teacher had been able to achieve previously. He also guarded everything in our hut – better than any guard dog. For six months this animal was dear to my heart, and it was the most intelligent creature that ever crossed my path. Then an indian boy poisoned it, because he was caught stealing and bitten by the *lontra*. So now I sat there and took the hungry animal in my arms and stroked it, and it immediately settled down on my lap just like back then. Someone had tied a string round its neck, almost throttling it – my *lontra* was allowed to live free, and he never ran away. The animal made me sad, and I couldn't hold back my tears. I would dearly have liked to take it away with me, but the species is "protected" and may not be kept in captivity. It has thus been condemned to extinction...

Axel helped me catch a few fishes which the otter immediately devoured hungrily. Then, despite the darkness, I insisted we continue our journey – I wanted to get away...

The next morning fantastic rock formations appeared on the left bank of the river – rather like those in Bryce Canyon in the USA, only smaller. The Rio Nhamundá was so calm that the silhouette of the shoreline was reflected in its surface. But that wasn't all. The rock formations along the bank, and their pastel colours ranging from pale pink to silver-grey, were also mirrored in the water. Cormorants dived and dolphins leapt in the peaceful river. Fishes turned somersaults (for joy?) on the surface and a caiman, almost four metres long, slid lazily into the water. *Tracajá* (aquatic turtles) sunned themselves on a protruding tree-trunk. Two toucans flew cackling away from us, while three black ibises glided in the opposite direction. A natural spectacle beyond compare. "Is this true, or am I dreaming?" I asked myself, as I could hardly believe my eyes. Here everything was pure wilderness, no oh so clever white men far and

"The Lord is my shepherd, therefore shall I lack nothing" was written on the *recreio*, which transported me and my discus.

wide. They had yet to discover this paradise. Long may it so remain...

But back to the Wai-Wai. They were very friendly, immediately giving us bananas and *caju*. The chief called himself Antonio Viana and he spoke a few words of Portuguese. When I asked him his real name, he replied "Kanahtxe". He had come here with ten members of his family. At the age of 39 he had eight children, the youngest three and the eldest 20. They were called Telma, Samuel, Sandra, Celma, Lazaro, Marinete, Salomão, and Rachel. His wife, Maria Isabel, was busy preparing *farinha* for sale (a *caboclo* cannot survive unless he gets his daily *farinha*). They cultivated manioc, bananas, *cajú, cana* (sugarcane), *batata* (similar to potatoes), *milho* (maize), *melancia* (water melon), and *fejão*. The lived on what they could produce. Now and then they went hunting, mostly for *porco do mato* (wild pig), *anta* (tapir), *javali* and *paca* (both rodents).

We pitched our tent and went night-fishing. There definitely ought to be discus here. I searched for the *maracaraná* brush known as *jaranduba* here on the Nhamundá. It also grows along the Rio Negro and is the place where Heckel discus spend the night. This brush is spiny and has long roots – an ideal place to hide and shelter from predatory fishes. Moreover, in the course of numerous fishing and collecting trips, I had learned where various fish species occur in the wild, ie their favourite haunts. For example – to stay with discus – the *aracá* tree is the place where

Only *S. discus* live in the Rio Nhamundá, but there are different colour variants there: an almost solid alpha individual *(top)* (a new discovery), individuals with a blood-red base colour *(centre)*, or emerald green *(above)*.

brown and blue discus lurk. They are widespread right up the Amazon basin to the tributaries of the Rio Purus. But never in blackwater rivers. And not in the Nhamundá.

This was really to be my night! After a lot of effort I managed to catch some fantastic Heckel discus, with colours such as the world had never before seen, one by one from among the undergrowth. Some were completely solid blue, others the entire anterior part. There were also individuals with a dark red base colour and brilliant turquoise dots on the operculum. Simply a discus sensation!

The trip was a success apart from what happened later, and is described at the beginning of this extract. Ie that I hadn't reckoned on Axel losing his camera equipment before we could film the Nhamundá. It does happen that expensive cameras fall in the water – this happened to me again recently in Mozambique – or are stolen. But I never dreamt that a Funai official would "simply" order indians to relieve us of our equipment without even hearing what had happened!

I got to know Brazil and its indians long before Funai existed, and remember well the very friendly natives with whom I spent a lot of time. Wonderful times that I would not have wanted to miss. It is a pity that nowadays everything on Planet Earth is money-oriented and that as a result everything is getting worse. This is bad for Nature – which nobody actually protects (except on paper) and likewise for aboriginal peoples. They were peaceful people and anyone who says or has said otherwise is lying. But "savages" have been mercilessly destroyed simply because they were "savage". Whether in the name of God, through land seizure, by gold miners, or simply by contact – they had no immunity to the white man's diseases. And now they are disappearing completely where they are "protected", or turning the tables as happened with us and many others.

"Wild animals" have suffered a similar fate. First of all the white man slaughtered them for food. Then they were collected for zoological gardens and private collections (a good thing, as many species – including fishes – survive today only because they are maintained and bred in "captivity"), or killed for their pelts, feathers, or aphrodisiac properties. But despite all this the majority of species survived – until "protection" dealt the death blow. We would do well to learn by experience – including what happens, ever since Adam and Eve, if something is forbidden…

I can count myself lucky to have seen Spix's aras in their thousands – since they became "protected" they have become totally extinct (if Spix had lived to see this he would undoubtedly have died of grief), just like many other creatures. Anything that is "protected" disappears in the twinkling of an eye. Including fishes and plants. But I will carry on trying to save some of those.

My urge to collect and study the majority of the discus races/forms/populations will remain with me for as long as I live. I am convinced that there are still fantastic colour forms (or species?) to be discovered. And there are a number which I will write about in future articles and books. It is also worth looking through my book *DISCUS,* as it is the only one on the market that illustrates and lists the precise locality for every race/form I know of, and is kept constantly up-to-date via its ring-binder system. At present it contains more than 1,000 different wild-caught and cultivated forms of discus.

Enjoy these miracles of Nature, the Kings of the Amazon. Fishes that deservedly bear that name, the most beautiful crea-

tures in the Mar Dulce. And please endeavour – this is aimed mainly at breeders – to breed only pure strains and not the frightful forms that for years have flooded the market from Asia. Dr. Eduard Schmidt-Focke, who specialised in breeding wild-type strains throughout his "discus life" and did more good for the discus community worldwide – and, of course, the fishes too – than anyone else, should still be regarded as the criterion, the good example.

I conclude this chapter with 10 extra double pages of photos, as follows: firstly, of past trips in search of discus; then a number of photos of additional discus discoveries; a double page on the city of Rio de Janeiro, as that is where it all began – for Natterer, Jobert, Amanda Bleher, and my humble self (and many others); two pages of photos of Manaus, the capital of Amazonia – the starting point (or main stopping-off place) for discus collectors and the number one export base – including pictures from past and present; and finally a brief glimpse of the indian tribes visited, in the past and during recent trips. After all, **they** were the real first discoverers of the discus (only that is generally forgotten). All the other discoverers mentioned here have merely contributed to scientific determination, collecting, transport (often under almost impossible circumstances), export and import, as well as the "coronation" of the "King of the Amazon" and helped to establish its continuing popularity among lovers of nature, plants, and fishes. But the indigenous people were the true first discoverers…

The Rio Nhamundá is a dream river – in the truest sense of the word. I had the good fortune to be the first white man to explore it ichthyologically. Accompanied only by the *caboclo* João, his son, and Axel Mewes (João´s son and Axel *left-hand page).* Hardly any people live there apart from a few settler families in the mouth region plus the above-mentioned *caboclo* family. Along the middle and upper river there are only indians. By far the majority of the area is completely unspoiled, pure Nature. Primary forest as far as the eye can see, and fabulous black water. Often like a mill pond (3-6), so that it is difficult to tell where air ends and water begins. But in the morning the amount of moisture in the air is incredible, because of the high humidity and the temperature fluctuations resulting from nocturnal cooling (1-2). Early in the day – often until as late as 9 am – you can hardly see your hand in front of you (1).

Heiko Bleher in search of discus — Heiko Bleher in search of discus

Heiko Bleher in search of discus — Heiko Bleher in search of discus

During my discus field trips in the course of the last 40 years I have come into contact with just about every possible type of local people, be they indians or *caboclos*, *serigueiros* (rubber collectors), *mestizen* or mulattos, black or white Brazilians, Peruvians, or Colombians. But, no matter what their race or the colour of their skins, they have all helped me and always been friendly. So it is thanks to them as well that I have managed to achieve so much. The expeditions have not always been of the easiest – a degree of suffering must always be entered into the equation, often lasting for days or even weeks. And these hardships have not always been rewarded with success. These 2 pages show just a few souvenir photos from my expeditions – from the very beginning and during my decades of collecting: 1. 1975, as the first white persons along the just openend Transamazonian Highway in search for discus and other fishes.

Heiko Bleher in search of discus — Heiko Bleher in search of discus

Heiko Bleher in search of discus — Heiko Bleher in search of discus

2. 1997, with Wai-Wai indians in the central Rio Nhamundá. 3. 1983, measuring the water parameters in Lago de Erepecuru. 4. 1989, among a typical Trombetas discus biotope. 5. 1965, in the Breves region on the island of Marajó (showing the extent of the influence of the tides in the lower Amazon region). 6. 1966, collecting in the Abaetetuba region (no discus found). 7. 1968, loading fishes onto a chartered twin-engine plane in Porto Trombetas. 8. 1986, exploring the northernmost distribution of *S. discus* in the São Gabriel region using a chartered helicopter (the helicopter doors were left behind). 9. 1982, checking whether this stick holds up the TABA plane in Barcelos (Rio Negro). 10. 1983, en route by hydroplane to search for discus biotopes in Lago Canacari. 11. 1982, en route from Manaus to Porto Velho, with an anaconda that (obviously!) didn't attack but instead fled away. 12. 1989, departing to explore the Rio Branco.

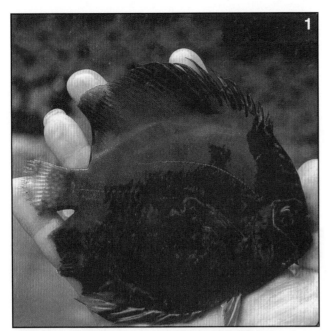

New discus discoveries in the 20th century

I am showing only a few my discoveries in the last century (see also pages 68-97). The question still remains, what constitutes a new discovery? A new colour form, a new variant, or only a new species or subspecies? From a taxonomic viewpoint, no new discus species was discovered in the 20th century – not even by me (or the description has not yet been published). During the last 50 years I have certainly succeeded in catching a huge number and variety of forms, perhaps more than anyone else, and have made most of them available to the aquarium hobby as well as making sure that they are bred worldwide. For example:
1. The only black discus – to date. I caught it in 1983 in the lower Rio Uatumá (20 years later I found *S. discus* (unique colour forms) for the first time in the central region of this blackwater river – see Chapter 4: *The Heckel discus*). A number of people have tried to find this variant again. 2. The same applies to the so

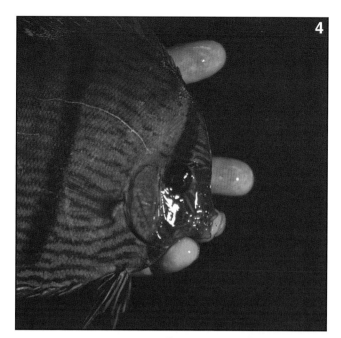

New discus discoveries in the 20th century

called "Coari". I discovered this (alpha) individual in the end os 1983 deep in an affluent (Rio Urucu) of the gigantic Lago de Coari. Subsequent collectors have fished in the Rio Coari and, finding only Blues, assumed there were no Greens there. 3. I was also the first to fish in the Caiambé lagoon (1982). Schmidt-Focke crossed a green with the the Içá brown. 4. I caught blue-Headed Heckel discus at the beginning of the 1970s in the Rio Xeruiní, and have often returned there since. 5. This blue-Headed Heckel discus originated from the Nhamundá (1998). 6. Browns from the Alenquer region were exported for the first time at the end of 1968, almost 20 years before the name became a household word. 7. And the red from the Rio Içá region remains a thorn in the flesh for many people, as no-one else has caught them and they provided the basis for the majority of reds worldwide.

Indigenous Peoples of Amazonia in Bygone Days

Indigenous Peoples of Amazonia in Bygone Days

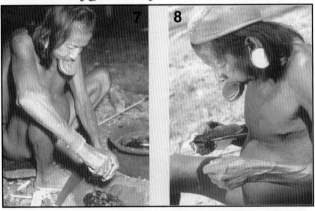

Indian tribes during my early years of discus exploration and from the present day. All the photos on this page are from the 1950s: 1. 1954, Suyá chief with a *tucunaré* (*Cichla temensis*) captured by bow and arrow, Mato Grosso. 2. 1954 Suyá indians preparing to go fishing fish using the toxic *timbó* tree root. The roots are pounded to free their sap and then they are placed in the (standing, or barely flowing) water, where it numbs and usually kills the fishes (the poisoned fishes are nevertheless safe for humans to eat). Nowadays *timbó* is used by ichthyologists worldwide for fish censuses. 3. 1954, Pauserna indians from the Iténez. 4. 1955, Waurá indians from the upper Xingú, obtaining mineral salts from *Eichhornia azurea*. During the dry season huge quantities of this floating plant are collected, dried on the bank, and then burnt. The plant ash is lixiviated with water and the solution evaporated. What remains is a light grey salt

Indigenous Peoples of Amazonia Today

Indigenous Peoples of Amazonia in Today

containing potash (consisting largely of potassium chloride). 5+6. 1954, Cabixi indians (formerly cannibals). 7+8. 1955 Suya indians preparing the *curare* poison (7) and smearing it on the tips of their arrows (8). The photos on this page are from the last two decades: 9. 1998, Kanamari indian girl from the Rio Jutaí, Amazonas (south of Benjamin Constant). 10. 1998, Kulina indians hunting – still by bow and arrow, but wearing T-shirts – in the Rio Japurá drainage, Amazon. 11. 1998, a Kulina indian boy with blonde hair! 12. 1999 Kulina indian girl in primary forest by an unspoiled tributary of the Japurá. 13. 1989, Wai-Wai indian women and children by the Mapuera, the largest tributary of the Rio Trombetas in northern Amazonia. 14. 1989, Wai-Wai indian woman suckling her baby. Rio Mapuera.

Manaus : Capital of Amazonia, and still the main centre for discus

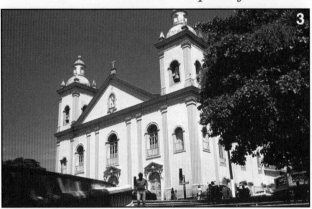

Manaus : Capital of Amazonia, and still the main centre for discus

Manaus, the capital of the largest state in Brazil and the starting point for almost all Amazon expeditions. Including collecting discus. All the early discoverers passed through here, and some even discovered *S. discus* in the Rio Negro, at whose confluence with the Amazon Manaus lies. Barra, Barra de Manaos, or Forte, as it was formerly known, was originally a small stronghold and grew hardly at all over the course of the centuries until the rubber boom began around 1870. Even so, after 1915 what had been the richest and most splendid city on Earth (around 1900) was almost forgotten. Only animal collectors, pelt-hunters, and adventurers (including Willy Schwartz, the pioneer of discus export) continued to make the town their base. During the 1930 to1950s , when barely 20,000 people lived here, it was a way-station for neon tetras. Evidence of its former wealth includes the opera house (1), which took 15 years to build (all the materials came from Europe – even the timber was

Manaus : Capital of Amazonia, and still the main centre for discus

Manaus : Capital of Amazonia, and still the main centre for discus

sent there for preparation) and was ceremonially opened in 1896. (José Carreras came in 1996 for the centennial celebrations). The name Amazon originates from Greek mythology (2), and because of a wild fantasy on the part of one of Orellana's companions has been applied not only to the largest river on Earth and the state, but also the opera house (Teatro Amazonas) – but there were never Amazons here. In Manaus there are still churches from the colonial period (3) and other fine buildings from the time of the rubber boom (11+12), as well as the market (10) designed by no less a person than Eiffel, where the fishing boats tie up every morning (9). Nowadays skyscrapers (4) dominate the scene in this metropolis of more than 3 million people – and the construction boom continues unabated (6). Since the creation of the free trade zone (1967) the city (8) has experienced an incredible resurgence (photo 8 shows the same place as photo 7 – but 50 years later...)

São Sebastião do Rio de Janeiro – the starting point for expeditions

São Sebastião do Rio de Janeiro – the starting point for expeditions

Its full name is São Sebastião do Rio de Janeiro (St Sebastian of the River of January), as when the Portuguese first anchored in Guanabara Bay on a January day in the 16th century they thought they were at the mouth of a river. Since then this fabulous place has been the starting point for almost all expeditions to Brazil (including that of the discus discoverer, Johann Natterer). Rio's geographical and geological situation is surely unique on Earth; not for nothing was it soon known as the *cidade maravilhosa* (marvellous city), and in the eyes of the Brazilians it is the loveliest in the world. Around 1950 barely 50,000 people lived here, but in the meantime Rio has grown to a veritable Moloch of more than 15 million inhabitants, but it is still marvellous and the starting point for most travellers in Brazil. On these two pages a few photos: 1+2. One of the best-known beaches on Earth: Copacabana. View from the Sugarloaf (1) and a glimpse

São Sebastião do Rio de Janeiro – the starting point for expeditions

São Sebastião do Rio de Janeiro – the starting point for expeditions

of life on the beach on a summer working day. 3+4. Rio is synonymous with *carneval* (or vice versa). The two-day parade of the samba schools (3) – often 10-14 of them, each with 6,000-12,000 people in fantastic costumes (costing up to 250,000 dollars each) on a specific theme – is quite unique. 5+6. The best-known monuments include the fallen soldiers on the Flamengo beach and the unique statue of Christ on Rio's highest mountain peak, the Corcovado. The statue is more than 37 m high and 34 m wide and was carved from a single block of stone by the disabled Alejadinho. 7+8. Little remains from the colonial period: Rio's unique tramway and curved bridge (7), and the old market, built by Eiffel, which now houses a restaurant. 9+10. Brazil is also synonymous with football (five times world champions) and Pele (8). 9+10. Rio has the longest bridge on Earth (9) and the loveliest bays (10).

CHAPTER

2

The Taxonomy of Discus

Heckel's works; Pellegrin's description; Schultz's revision; Kullander's synonyms; Comment on Taxonomy

Heckel's Work

The name Heckel has long been synonymous with the King of the aquarium fishes – at least among discus enthusiasts worldwide. If only because the fish he described in 1840, *Symphysodon discus*, was the first discus to receive a scientific name – which remains valid to the present day. And the name Heckel discus is used much more frequently than the scientific one.

So who was this Johann Jakob Heckel, who in his day rose to the position of curator of the *k.k. Hof-Naturalien-Cabinet,* and became a member of the kaiserlichen Aklademie der Wissenschaften (Imperial Academy of Sciences) and much more?

Heckel was born on 23rd January 1790 in Mannheim (Germany) and took ichthyology (the study of fishes) in Austria to its first peak; from 1820 onward at the *k.k. Hof-Naturalien-Cabinet* in Vienna, whose fish collection took a significant upturn thanks to him.

Like Natterer, the first discoverer of the discus, he was no "learned zoologist". At the wish of his father he had studied agriculture, and in 1811 – after the death of his father – took over the family smallholding at Gumpoldskirchen (near Vienna). His real interest, however, was the natural sciences, in particular botany and ornithology. He was, moreover, a very good artist and a skilled dissector, whose friends declared he had two right hands. In the course of identifying a few rare specimens in his noteworthy collection of bird skins he came into contact with the then curator of the *k.k. Hof-Naturalien-Cabinet* in Vienna, Josef Natterer, the elder brother of Johann, at that time away in Brazil. This encounter was critical in his decision to devote himself exclusively to the natural sciences. Because fishes were extraordinarily poorly represented in the collection of the *k.k. Hof-Naturalien-Cabinet* at that time, Heckel decided to devote himself to this particular field. In 1820 he took a job as dissector and began to work hard to acquire all the relevant professional training he lacked. The word quickly spread internationally that there was a new ichthyological broom in Vienna, and Heckel maintained close contact with the great ichthyologists of his time, such as Cuvier, Valenciennes, Bonaparte, Müller and Troschel. Heckel's own collecting and travelling activities were limited, he worked mainly on material that other collectors and explorers brought back to Vienna. "His" collectors included Baron von Hügel, Russegger, and Kotschy, whose activities in Kashmir, Syria, Persia, and Egypt significantly enlarged the Vienna collection. As well as concentrating on the ichthyofaunas of Asia Minor and north Africa, Heckel dealt with, *inter alia*, the cichlids from Natterer's Brazilian collections (see below), and ultimately turned to fossil fishes as well. From a systematic and taxonomic viewpoint his work on cyprinids (carp-like fishes) was of particular note. His *Neue Klassifikation und Charakteristik sämtlicher Gattungen der Cypriniden* (New classification and characterisation of all the genera of the Cyprinidae), in which he dealt extensively with the significance of the pharyngeal dentition as a systematic character, is – like many others of his over 60 scientific publications – a standard work in the ichthyological literature. Heckel didn't live to see the appearance of the work on which he had laboured for 24 years and which he regarded as his crowning achievement, *Die Süßwasserfische der Österreichischen Monarchie Mit Rücksicht auf die Angränzenden Länder* (The freshwater fishes of the kingdom of Austria with reference to the adjacent countries). He died on the 1st March 1857 at the age of 67 years as the result of an infection which he probably picked up while removing the skeleton from the corpse of a sperm whale, stranded on the coast of Istria, for the museum. The publication of the work was taken over by his former student, assistant, and friend, Rudolf Kner, a year after Heckel's death. Kner had helped earlier – without payment, of course – with the work on the collections and it was during this work that the graduate in agriculture and the medical doctor had become friends.

In his work published in 1840 Heckel described as *Symphysodon discus* a specimen that Natterer had collected during his ninth expedition (1831-34) (see original description below). The work formed part of the Annals of the Vienna Museum (Vol. 2). Along with the discus genus and species, other cichlid genera and species from Natterer's collection were described by Heckel in the publication *Johann Natterer's neue Flussfische Brasilien's* (Johann Natterer's new fluviatile fishes from Brazil): *Acara* (with 15 species), *Batrachops* (2 species), *Chaetobranchus* (2), *Cichla* (3), *Crenicichla* (8), *Geophagus* (7), *Pterophyllum* (1), *Heros* (13), and a *Uaru* species. (The characins, catfishes, and other fish groups were not determined until many years later, and then by various authors – for the most part only after other specimens of the same species had been collected and Natterer's discoveries had at long last been described).

Because Heckel devoted himself mainly to the study of Natterer's cichlids it is hardly surprising that the majority of the names for which he is responsible relate to this zoological group.

Beneath Natterer's excellent illustrations in the recently printed book *Johann Natterer und die Österreichische Brasilienexpedition* (Johann Natterer and the Austrian expedition to Brazil) we find some of the names, ie *Acara crassipinnis, Acara nassa, Acara margarita, Acara viridis, Chaetobranchus flavescens, Crenicichla adspersa, Crenicichla johanna, Crenicichla lepidota, Geophagus megasema, Geophagus pappaterra, Heros modestus* and *Heros festivus*, which Heckel published in 1840 (*teste* Heckel, 1840).

Of *Symphysodon discus* Heckel writes (translated): "In life this species, which may likewise be described as rare, and is found at Barra do Rio-negro, in the river itself, has a very attractive coloration. The base colour of the entire fish is violet-grey, the longitudinal stripes ochre-brown, the stripes on the forehead and cheek, leading to the mouth, turquoise blue; the pectoral fins are transparent yellowish, the ventrals brown-red, and their spines, and the following soft ray, as well as the spines of the anal fin, are gold-green with a beautiful turquoise blue stripe along their length; the dorsal fin is dark grey, the upper edge of its soft part bordered in red, the caudal fin greyish with blackish dots. The base colour of the iris is black-brown, with a narrow yellow band around the pupil, then a black ring, and a cloudy brown-red spot to the rear. Length of the described specimen 5 inches."

Symphysodon discus Heckel, 1840, Holotype, NMW 35612, 98.6 mm SL. The original specimen collected by J. Natterer at Moreré (= Moreira), northwest of Barcelos.

The meaning of the genus and species names is as follows:

Symphysodon is derived from the anatomical term symphysis (from the Greek συμ (or συν) = sym (or syn) = together and φυσειν = physein = grow), plus οδων = *odon* = tooth. Heckel used this name to refer to a diagnostic feature of the genus, the small number of teeth that are found only on the symphysis of the lower jaw, the area where the two branches of the lower jaw are joined ("grown together") anteriorly *("ad symphysin utriusque maxillae plagam parvam occupantes; reliqua maxilla edentata"* = occupying a small area at the symphysis, on either side of the maxilla; rest of maxilla without teeth); the name translates as "symphysis tooth". The specific name discus is the Latin word for a disc or discus (originally derived, by the Romans, from the Greek δισκος = *diskos* = disc, discus). The name refers to the body shape. We do not know whether Heckel was referring simply to the disc shape of the fish, or to a fancied resemblance to the discus, the throwing disc used in classical and modern athletics. The latter is often assumed.

THE ORIGINAL DESCRIPTION OF THE FIRST DISCUS BY J. HECKEL 1840

SYMPHYSODON nob.

Character generis.

Corpus *valde compressum, elevatum, Chaetodonti simile.*

Dentes *velutini, uncinati, fortiores, ad symphysin utriusque maxillae plagam parvam occupantes; reliqua maxilla edentata.*

Ossa pharyngea *parva, inferiora planum triangulare aequilaterale formantia,* dentibus brevibus velutinis, uncinatis, munitum. (Tab. XXX, Fig. **21** et **22**).

Arcus branchialis externus *in latere concavo papillis obsoletis, arcus reliqui aculeis minimis.* (Tab. XXX, Fig. **23** et **24**.)

Radii branchiostegi 5.

Partes operculi *leves.*

Nares *geminatae ori aproximatae.*

Os *parvum.*

Apertura analis *sub pinnis pectoralibus.*

Pinna dorsalis et analis *basi elongata, squamata, radii osseis validis, sensim sensimque longioribus.*

Pinnae ventrales *ante pinnas pectorales inseretae.*

Squamae *minutae.*

Linea lateralis *interrupta.*

Der Körper dieser höchst merkwürdigen Gattung hat viel Chaetodon-artiges, indem er ebenso wie an diesen gegen die verticalen Flossen zugeschärft ist; die kleine Gruppe von Kadenzähnen auf der Symphyse der übrigens zahnlosen Kiefer, zeichnet sie vorzüglich aus. Die Flossen sind nicht zugespitzt; die Schuppen bedecken die Basis des weichstrahligen Theiles der Rücken- und Analflosse dergestalt, dass diese sich nicht niederlegen können; in der Analflosse befinden sich mehr Stachelstrahlen als in der Rückenflosse. Stirne, Suborbitalknochen, Vordeckel und Unterkiefer sind nicht beschuppt.

SYMPHYSODON DISCUS nob.

Moreré am Rio-negro. Natterer.

Subdisciformis. Fasciis tribus verticalibus, striis 17—18 longitudinalibus; pinna caudali seriebus punctorum 3—4 ad basim.

caudalis.	pin. caudalis.	Pinna dorsalis.												
	$6\frac{1}{4}$	$6\frac{7}{8}$	$12\frac{1}{2}$	$18\frac{3}{4}$	$22\frac{1}{2}$	21	18	$11\frac{1}{2}$	—	$11\frac{1}{2}$	$9\frac{1}{4}$	$7\frac{1}{4}$	5	$1\frac{3}{4}$	$\frac{1}{2}$	''
70	55	$52\frac{1}{2}$	52	46	36	28	$19\frac{1}{2}$	$13\frac{1}{2}$	13	10	8	6	4	2	1	0
Apex pin.	Basis	$6\frac{1}{4}$	6	$11\frac{1}{2}$	$18\frac{1}{2}$	$21\frac{1}{2}$	$19\frac{1}{2}$	$16\frac{1}{8}$	—	$12\frac{1}{2}$	$10\frac{1}{2}$	$8\frac{7}{8}$	$7\frac{1}{4}$	5	$3\frac{1}{2}$	3	1
		Pinna analis.							P. ventr.		Symph. clavic.					Os.

P. $\frac{\frac{2}{8}}{3}$ V. 1/5.   D. 9/31.   A. 10/24.   C. $\frac{\frac{3}{14}}{2}$ Squamae 46.

Die Gestalt ist beinahe scheibenförmig, gegen den Mund zu einen sehr stumpfen Winkel bildend, der Schwanz so kurz, dass seine Flosse aus dem durch Rücken- und Analflosse beschriebenen Halbzirkel kaum hervor tritt. Die grösste Dicke des Rumpfes über den Brustflossen übertrifft nur wenig jene des Kopfes, die $1^{3}/_{4}$mal in dessen Länge enthalten ist, diese Kopflänge macht zugleich den vierten Theil der Gesammtlänge des Fisches und den dritten Theil von dessen grösster Höhe aus. Beyde Kiefer sind gleich lang, die Mundspalte beginnt ganz vorne, einen Grad unter der Achse, ist sehr kurz (einem halben Augendurchmesser gleich), denn der hintere Rand des Oberkiefers ($2^{1}/_{2}$ Lg. $2^{1}/_{4}$ Tf.) erreicht nicht die Mitte zwischen Auge und Mundspitze, und liegt unter dem Suborbitalknochen verborgen; die kleinen Lippen sind sehr fleischig. Die kleinen runden Nasenlöcher stehen vertical über dem hinteren Oberkieferrand, etwas entfernt über einander (das untere $2^{1}/_{3}$ Lg. $^{3}/_{4}$ Hh.), das obere beynahe am Rande des Profils ($2^{1}/_{2}$ Lg. $1^{3}/_{4}$ Hh.). Das Auge, dessen Diameter (5 Länge-Grade) $^{2}/_{7}$ der Kopflänge ausmacht, liegt (der Mittelpunkt $8^{1}/_{2}$ Lg. $1^{1}/_{4}$ Hh.) in der Mitte der oberen Kopfhälfte, mit seinem unteren Rande um $^{1}/_{2}$ Grad unter der Achse, und mit seinem hinteren Rande um einen Längegrad von der oberen Einlenkung des Vordeckels entfernt; dieser letztere zieht sich von 12 Lg. 1 Hh. schief vorwärts bis auf $8^{1}/_{2}$ Lg. $4^{1}/_{4}$ Tf. herab, wo er eine kleine Rundung bildet, und in 5 Lg. $4^{3}/_{4}$ Tf. sich der Einlenkung des Unterkiefers nähert. Der Deckel endet nach hinten ($17^{1}/_{2}$ Lg. $^{3}/_{4}$ Tf.) in einen sehr stumpfen Winkel. Die Stirne ist über den Augen sehr hoch; der grosse Suborbitalknochen misst einen Augendiameter; an jeder Seite des Unterkiefers sind zwey Poren.

Die Brustflossen beginnen senkrecht unter der Deckelspitze ($17^{1}/_{2}$ Lg. $3^{3}/_{4}$ Tf. der oberste Strahl) sind so lang wie der Kopf und schief zugerundet, die 2 oberen und 3 unteren Strahlen sind ungetheilt. Die Bauchflossen, welche um $4^{1}/_{2}$ Länge-Grade vor den Brustflossen entspringen, sind beynahe eben so lang als diese, aber zugespitzt, ihr Stachelstrahl ist mässig stark und erreicht die halbe Länge der darauf folgenden längsten getheilten Strahlen. Die Rückenflosse beginnt beynahe vertical über den Bauchflossen, ihre ganze Basis, von welcher die Stachelstrahlen $^{2}/_{3}$ einnehmen, ist mit der beschuppten Haut des Körpers überzogen, wodurch jedoch diese Stachelstrahlen, deren letzter 6mal länger als der erste, 11 Längegraden gleicht, am Niederlegen nicht gehindert werden; die weichen Strahlen sind nur einmal getheilt, können sich nicht niederlegen, schliessen sich gleichmässig an den letzten Stachelstrahl, dessen Länge sie wenig übertreffen, an, und vereinigen sich zuletzt beinahe mit der Schwanzflosse. Die Wimpel der Stachelstrahl-Membrane sind sehr zugespitzt. Die Analflosse ist wie die Rückenflosse gestaltet, nur nehmen die robusteren Stachelstrahlen $^{2}/_{5}$ der Flossen-Basis ein. Die Schwanzflosse, deren zwei mitteren Strahlen 2mal, die übrigen, mit Ausnahme der Seitenstrahlen, nur einmal getheilt sind, ist gerade abgestutzt. Der Anus liegt $2^{1}/_{2}$ Längegrade vor der Analflosse.

Die Schuppen sind in der Mitte des Körpers $^{2}/_{5}$ des Augendiameters gleich, werden aber nach oben und unten zu, kleiner; 7—8 schiefe Reihen kleiner Schuppen, die jenen am Kopfe und an der Brust gleichen, sitzen auf jeder Backe; beiläufig 46 horizontale Schuppenreihen befinden sich zwischen Rücken- und Bauchflosse, und eben so viele einzelne Schuppen in einer geraden Linie zwischen Kopf- und Schwanzflosse; der obere Theil der Seitenlinie aus 16 Röhrchen-Schuppen, folgt Anfangs der Wölbung des Rückens, und hat 17—18 Schuppenreihen über sich; der untere gerade Theil (an gegenwärtigem Exemplar nur auf einer Seite vorhanden) zählt 11 Röhrchen-Schuppen. Die Gestalt der Schuppen aus der Mitte des Rumpfes ist beinahe rund, an dem freien Rande mit einer kurzstachligen Binde, die gegen die Mitte zu breiter ist, versehen; gleich hinter dieser Binde liegt im ersten Viertheile der Schuppe ihr Strahlenpunkt, von den feinen concentrischen Ringen rein und deutlich umgeben; 13—14 Radien durchziehen sie vorwärts auf der bedeckten Fläche, und machen am Rande eben so viele leichte Einkerbungen.

Die Hauptfarbe ist gegenwärtig hellbraun auf der obern Hälfte und graubraun nach unten zu. Drei dunkelbraune vertikale Binden von der Breite eines Augendiameters umgeben den Körper, die erste geht durch das Auge selbst, die zweite um die Mitte des Körpers von der Basis der ersten weichen Rückenflossen-Strahlen herab, und die letzte umgibt die Schwanzflossen-Basis; 18 etwas wellenförmige, schmale, rostbraune Längsstreifen, welche eben so schmale Zwischenräume haben, durchziehen den Rumpf in paralleler Richtung; nur der 2., 3. und 4. Streif vom Rücken herab fliessen, indem sie die mittere Vertikalbinde durchkreuzen, in einen einzigen Streif zusammen. Auf dem Kopfe liegen zwei weissliche Querstreifen vor dem Anfang der Schuppen, von einem Auge bis zum andern; zwei ähnlich gefärbte Längsstriche ziehen sich von jeder Seite der Stirne zum Munde, und abermals zwei am hinteren Rande des grossen Suborbitalknochens vom Auge bis zum Mundwinkel hinab. Brust- und Schwanzflosse sind gelblich, letztere hat 3—4 Querreihen schwarzer Punkte gegen die Basis; alle übrigen Flossen sind einfärbig schwarz.

Im Leben bietet diese eben so ausgezeichnete als seltene, bei *Barra do Rio-negro* im Flusse selbst vorkommende Art, ein sehr reizendes Farbenkleid dar. Die Grundfarbe des ganzen Fisches ist violetgrau, die Längsstreifen ocherbraun, die zum Munde führenden Streifen der Stirne und Wange Türkiss-blau; die Brustflossen sind durchsichtig gelblich, die Bauchflossen braunroth, ihr Stachelstrahl, der darauf folgende weiche Strahl, so wie auch der Stachelstrahl in der Analflosse, hat auf goldgrünem Grunde der Länge nach, einen schönen Türkiss-blauen Streif; die Rückenflosse ist dunkelgrau, ihr weicher Theil am obern Rande röthlich gesäumt, die Schwanzflosse graulich mit schwärzlichen Punkten. Die Grundfarbe der Iris ist schwarzbraun, um die Pupille ein schmaler gelber, dann ein schwarzer Ring, nach hinten zu ein wolkigt braunrother Fleck.

Länge des beschriebenen Exemplars 5 Zoll.

In 1862 Kner published *(Sitzber. in Akad. Wiss. Vienna,* **vol. 46, P. 299, Table 2) a work with the title** *Pterophyllum* **etc. in which this sketch, apparently made by Heckel, is reproduced as "Fig. 2** *Symphysodon discus"*. **It must be the very first drawing of a discus, but the striking central bar is missing.**

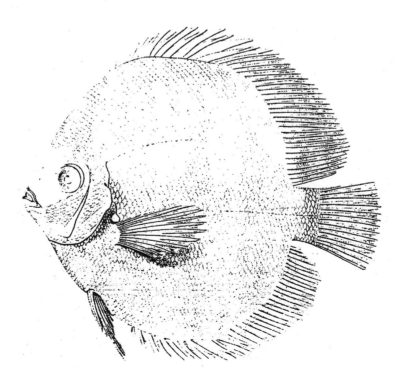

PELLEGRIN'S DESCRIPTION

Jacques Pellegrin (1873-1944) was undoubtedly one of the most extraordinary ichthyologists that ever lived. Hardly any other published as much as he did: 548 scientific works between 1898 and 1936 alone, 447 of them on fishes (almost all new taxa). In other words, every month he produced one or more new pieces of work. (Nowadays a taxonomist is happy if he manages 1-2 scientific publications per year.)

However, the Frenchman was not only a diligent and acknowledged scientific author but also received high honours such as *Chevalier de la Légion d'honneur au titre de la guerre* (after his war service 1914-1919), *Officier du Méritee agricole, Officier de l'Instruction publique*, and many more. He was the chairman of a number of conferences; vice-president of the committee for international exhibitions of live insects, fishes, birds, and small mammals, and of the international commission of applied zoology, etc; president of several societies and conferences, for example the aquarium and terrarium section of the Société d'Acclimatation from 1920 onward, and the colonial fishes section of the 12th Paris congress of fisheries and maritime industry in Paris (1931), and others. In addition he was a member of a whole series of scientific organisations such as the International Commission for Zoological Nomenclature, the Zoological Society of London, the Madagascan Academy, the Portuguese Association for the Natural Sciences and that of Morocco, to name but a few.

Pellegrin's thesis for his doctorate was on the cichlids (1903) – as well as his 1899 graduate medical thesis on *Les Poissons vénéneux* (the venomous fishes). The former was a complete monograph on the anatomy, biology, and taxonomy of one of the largest families in what he termed the Acanthopterygii (nowadays the subclass Actinopterygii), with a catalogue of all the specimens in the museum. Thus he was destined to describe the second discus. Apparently, however, Pellegrin never met Clément Jobert, the collector of the second(?) discus (see Chapter 1: *The second discus)*, as he first began to publish 20 years after his compatriot returned from Brazil with the specimens (in 1878, when Pellegrin was just a child of hardly five).

Pellegrin published his doctorate thesis as *Contribution à l'étude anatomique, biologique et taxonomique des Poissons de la famille des Cichlidés* in the *Mémoires de la Société zoologique de France*, 1903, pp. 41-399, which, however, wasn't published until 1904.

In this work Pellegrin described not only the new *Symphysodon discus* var. *aequifasciata* **var. nov.** (see original publication) on page 214, but in the chapter *Étude des Poissons de la Famille des Cichlidés* he also described a new *Astronotus* var. *zebra* **var. nov.** from six specimens also collected by Jobert (Teffé (3), Manaos (2), Calderón (1)) and many other species of the family Cichlidae. He also mentioned the altum angelfish, *Pterophyllum altum,* described by him in 1903, giving the type locality as "Atabapo, Venezuela" (where discus have never been found).

The Frenchman had already published (1902) a summary of the cichlids collected by Jobert *(Cichlidés du Brésil Rapportés par M. Jobert. Par M. le Dr Jacques Pellegrin)* in which he mentioned the following cichlids:

CHÆTOBRANCHUS FLAVESCENS Heckel. — Marajo, Santarem, Manaos, Teffé, Tonnantins.
— SEMIFASCIATUS Steindachner. — Manaos, Teffé.
CHÆTOBRANCHOPSIS ORBICULARIS Steindachner. — Marajo, Tonnantins.
CICHLA OCELLARIS Bloch et Schneider. — Para, Manaos, Teffé, Tonnantins.
— var. *argus*, Valenciennes. — Manaos.
— TEMENSIS Humboldt. — Santarem, Manaos.
ACARA TETRAMERUS Heckel. — Marajo, Santarem, Manaos, Teffé, Tabatinga.
— THAYERI Steindachner. — Teffé.
— VITTATA Heckel. — Tonnantins.
— (HYGROGONUS) OCELLATUS Agassiz. — Marajo, Santarem, Manaos, Teffé.
— (ACAROPSIS) Heckel. — Santarem, Manaos, Teffé, Tonnantins.
HEROS (CICHLASOMA) BIMACULATA Linné. — Marajo, Manaos, Teffé, Tonnantins, Tabatinga.
— — CRASSA Heckel. — Manaos, Teffé, Tabatinga.
— — CORYPHÆNOIDES Heckel. — Manaos.
HEROS SPURIUS Heckel. — Santarem, Manaos, Teffé, Tonnantins, Tabatinga.
— AUTOCHTHON Günther. — Rio-Grande, Rio-de-Janeiro.
MESONAUTA INSIGNIS Heckel. — Manaos, Teffé, Tonnantins, Tabatinga.
UARU AMPHIACANTHOIDES Heckel. — Santarem, Manaos, Teffé.
PETENIA SPECTABILIS Steindachner. — Para.
CRENICICHLA BRASILIENSIS Bloch et Schneider, var. *lenticulata* Heckel. — Manaos, Tonnantins.
— — var. *lugubris* Heckel. — Manaos, Tonnantins.
— — var. *johanna* Heckel. — Marajo, Para, Teffé, Tonnantins.
— MACROPHTHALMA Heckel. — Manaos.
— SAXATILIS Linné. — Santarem, Manaos, Tonnantins, Tabatinga.
GEOPHAGUS (MESOPS) CUPIDO Heckel. — Marajo, Santarem, Manaos, Teffé, Tonnantins, Tabatinga.
GEOPHAGUS (MESOPS) THAYERI Steindachner. — Santarem, Manaos, Teffé, Tonnantins, Tabatinga.
— — TÆNIATUS Günther. — Teffé, Tonnantins, Tabatinga.
— (SATANOPERCA) ACUTICEPS Heckel. — Para, Santarem, Manaos, Teffé, Tonnantins.
— — JURUPARI Heckel. — Marajo, Santarem, Manaos, Teffé, Tonnantins, Tabatinga.
— — BRASILIENSIS Quoy et Gaimard. — Rio-Grande, Rio-de-Janeiro.
— SURINAMENSIS Bloch. — Marajo, Para, Santarem, Manaos, Tonnantins, Tabatinga.
SYMPHYSODON DISCUS Heckel. — Santarem, Manaos, Teffé.
PTEROPHYLLUM SCALARE Cuvier et Valenciennes. — Marajo, Teffé, Tonnantins, Tabatinga.

Thus he had already listed the *S. discus* collected by Jobert before they were first described, but not the specimen from Calderón! Or the locality. In this work he mentions only the towns of Santarém, Manaos, and Teffé. That is interesting because the addition of the fourth discus specimen from Calderón, from the Colombian-Brazilian border, was by hand, in ink, after the printing of the original description. And later Pellegrin mentions that Jobert had collected in seven places: Pará (= Belém) and on the island of Marajó; in the Xingu mouth region; around Santarém; Manaos and Barra do Rio Negro (= Manaus); around Teffé (= Tefé); at Tonantins (on the Solimões); and at Calderón (= Calderão or Tabatinga) – but again the last-named locality is everywhere noted only by hand.

Thus Pellegrin had gathered together the three (or four?) discus collected by Jobert (which can be seen on the right) and described them as a new variant in 1903 (officially published and/or printed in 1904): *Symphysodon discus* var. *aequifasciata* var. nov.

He noted under "Description" that Jobert's discus comprised two different types, the first (from Manaos) was *Symphysodon discus* but the specimens from Santarém (1), Teffé (2) and Calderón (1) were a variant, which he described as new. In summary: the first type had three prominent bands and a large number of parallel longitudinal lines, alternating clear and dark in colour (= Heckel discus). The other (var. *aequifasciata*) had nine bands at uniform intervals and no parallel longitudinal lines. And it was not a question of the age of the specimens, as they were of comparable size. (That is not quite correct – see commentary on taxonomy, below.)

Pellegrin may – as some believe – not have been the best taxonomist, but his achievement was immense and should not be underestimated. Moreover he was honoured for it, and not just with decorations. Some 30 fish species bear the specific name *pellegrini* and there is a genus *Pellegrinina* (for the characin species *heterolepis*); cichlid fans will be most familiar with *Geophagus pellegrini*, and there is no doubt that *Symphysodon aequifasciatus* Pellegrin, 1904 (after it was subsequently raised to specific status) will forever be synonymous with the green discus among all discus experts and fans.

For many people (cichlid fans and scientists) a number of his works will remain as unforgettable as his person, for example:

Pellegrin, J. 1900. Cichlidés nouveaux de l'Afrique équatoriale. *Bull. Mus. Natl. Hist. Nat.*, vol. 6 (no. 6): 275-278.

Pellegrin, J. 1902. Cichlidé nouveau de la Guyane française. *Bull. Mus. Natl. Hist. Nat.*, vol. 8 (no. 6): 417-418.

Pellegrin, J. 1902. Cichlidé nouveau du Congo Français. *Bull. Mus. Natl. Hist. Nat.*, vol. 8 (no. 6): 419-420.

Pellegrin, J. 1903. Description de Cichlidés nouveax de la collection du Muséum. *Bull. Mus. Natl. Hist. Nat.*, vol. 9 (no. 3): 120-125.

Pellegrin, J. 1903. Cichlidé nouveau de l'Oubanghi appartenant au genre *Lamprologus*. *Bull. Mus. Natl. Hist. Nat.*, vol. 9 (no. 5): 220-221.

Pellegrin, J. 1905. Sur deux poissons du genre *Crenicichla* de la collection du muséum de Paris. *Bull. Soc. Zool. Fr.*, vol. 30: 167-169.

Pellegrin, J. 1913. Diagnoses préliminaires de Cichlidés nouveaux du lac Victoria recueillis par MM. Alluaud et Jeannel. *Bull. Soc. Zool. Fr.*, vol. 37: 311-314.

Pellegrin, J. 1919. Sur un Cichlidé nouveau de l'Ogôoué appartenant au genre *Pelmatochromis*. *Bull. Soc. Zool. Fr.*, vol. 44: 102-105.

Pellegrin, J. 1929. Cichlidés de Madagascar recueillis par M. Georges Petit. Description d'une espèce nouvelle. *Bull. Soc. Zool. Fr.*, vol. 54: 252-255.

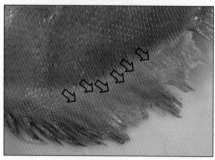

Top: The 2 jars in the MHNN containing the types of Pellegrin's var. *aequifasciata*. He didn't consider (or see) the striking differences in colour pattern (still visible in alcohol). The specimen from Santarém (2nd to 4th photos from top) clearly shows the pattern of what is now *S. haraldi* (= blue/brown discus, arrowed). *Right-hand page:* Only the 2 Tefé specimens exhibit the very typical pattern of dots seen in the green discus (= *aequifasciatus*).

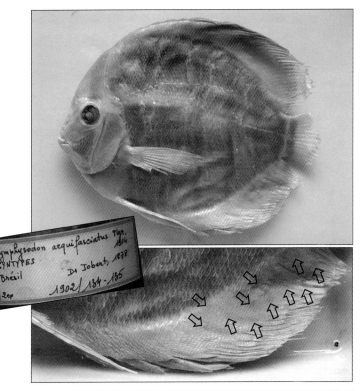

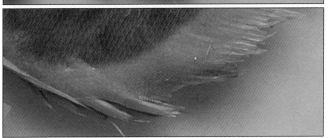

THE ORIGINAL DESCRIPTION OF THE SECOND DISCUS
BY J. PELLEGRIN IN 1904

19. Symphysodon.

HECKEL, 1840 Ann. Wien. Mus., II, p. 332.

Corps très court, élevé, comprimé. Dents coniques, très peu nombreuses, occupant seulement la symphyse. Maxillaire non visible. Branchiospines rudimentaires (5). Ecailles cténoïdes, moyennes (46-56). 2 lignes latérales perçant des écailles plus grandes. Rayons mous de la dorsale (28-31), beaucoup plus nombreux que les épines (8-9). 7 à 9 épines à l'anale.

1 espèce. Brésil.

1. SYMPHYSODON DISCUS Heckel.

Symphysodon discus HECKEL, 1840, Ann. Wien. Mus. II, p. 333 ; GÜNTHER, 1862, Cat. IV, p. 316 ; STEINDACHNER, 1875, Sitz. Ak. Wissen. Wien., 71, p. 106 ; EIGENMANN et BRAY, 1894, Ann. N.-York. Ac. Sci., VII, p. 623.

H. 1-1 1/6 ; T. 3 1/3-3 2/3 ; Œ. 3-3 3/4 ; D. VIII-IX 28-31 ; A. VII-IX 26-31 ; P. 12 ; S. Den. $\frac{1\,(2)}{1\,(2)}$; Ec. 15-16 | 46-56 | 29-32 ; L. lat. $\frac{21\text{-}23}{10\text{-}12}$; Ec. J. 7-8 ; Br. 5.

Lèvre inférieure continue. Bouche petite, oblique. Dents au nombre d'une dizaine, placées sur 2 rangées, à chaque mâchoire. Maxillaire étendu un peu au-delà de la narine. Ecailles moyennes sur l'opercule. Branchiospines vraiment rudimentaires. Ecailles cténoïdes, celles des lignes latérales plus grandes que les autres. Pectorale arrondie aussi longue que la tête. Ventrales prolongées en filament. Epines dorsales et anales croissantes. Dorsale et anale recouvertes en grande partie par de petites écailles. Pédicule caudal presqu'absent, extrêmement court. Caudale arrondie.

Brun-olivâtre avec 9 barres foncées transversales, la première traversant l'œil, la 5e divisant le corps en deux et la 9e en travers de l'origine de la caudale parfois plus larges et plus foncées que les autres, d'autres fois les 9 lignes semblables entre elles (var. *æquifasciata*).

Brésil.

02-131 à 133 [3] Manaos (Brésil) : Dr JOBERT.

var. *æquifasciata* var. nov.

02-134-135 [2] Teffé (Brésil) : Dr JOBERT.
02-130 [1] Santarem (Brésil) : Dr JOBERT.
04-270 (1) Calderon "   :   "

La coloration de ces spécimens peut-être ramenée à 2 types. Dans l'un, la 1re barre qui traverse l'œil est très accentuée, la 2e, la 3e, la 4e disparaissent presque complètement, la 5e qui divise le corps en deux est très large et très marquée, la 6e, la 7e, la 8e sont assez nettes, la dernière à la base de la caudale est aussi très prononcée ; en résumé il y a 3 barres prédominantes. De plus sur tout le corps s'étendent un grand nombre de lignes longitudinales parallèles, alternativement claires et foncées.

Dans l'autre (var. *æquifasciata*), les 9 barres transversales sont minces et égales entre elles, aucune n'est prédominante, il n'y a pas de lignes longitudinales parallèles. Ce n'est pas une question d'âge car les séries de spécimens sont de tailles équivalentes.

SCHULTZ'S REVISION

Leonard P. Schultz (1901-1986) was born in Albion, Michigan, and even as a child developed a passion for biology that was to lead to his graduating in that subject from Albion College in 1924. Two years later he obtained his Master's degree under the well-known ichthyologist Carl Hubbs at the University of Michigan. Thereafter (until 1936) he taught ichthyology at the University of Washington, where he also obtained his doctorate in 1932. Between 1930 and 1936 he participated in several ichthyological collecting trips in the north-western Pacific and described numerous new taxa. The university has Schultz to thank for a quarter of its museum collection.

In 1936 he started work as Assistant Curator at the National Museum of Natural History in Washington, D.C. and took charge of the Division of Fishes. Just two years later he assumed the position of Curator-in-Charge, which he retained until 1965 when he was promoted to Senior Scientist. In 1968 L. P. Schultz retired and wanted nothing more to do with fishes. He had published some 214 scientific works, the majority of them on fishes, the first appearing in 1927 and the last in 1969.

During his time in Washington, D.C., Schultz made a number of collecting trips, including to Phoenix (Arizona), Samoa, the Marshall Islands, Venezuela, and the western USA. Particularly noteworthy among his books are *The Ways of Fishes* (with E. Stern), his ground-breaking *magnum opus, Fishes of the Marshall and Marianas Islands*, as well as his compilation *Shark Attacks of the World*. He worked in close association with the publisher T.F.H. Publications, Inc. in New Jersey and published a series of articles in their magazine *Tropical Fish Hobbyist* right from its first issue (1952). He also rose to the position of their Advisory Editor and, after the publication of the *Handbook of Aquarium Fishes* (co-authored by the owner of the publishing house, H. R. Axelrod) in 1956, a closer relationship developed between author and publisher. This also led to the at the time highly contentious scientific description of *Cheirodon axelrodi* Schultz, 1956 (now *Paracheirodon axelrodi*). A year previously two other well-known ichthyologists had prepared an original description of the cardinal tetra, discovered by Professor Sioli (1952), first imported to the USA by Fred Cochu (1955), and given to the ichthyologist Myers for study, and its name should have been *Hyphessobrycon cardinalis*. Axelrod had "discovered" this fish in February 1956 in a New Jork shop and (in great haste) got Schultz to describe it and name it after him. Schultz (or better Axelrod) rushed into print before the other name could be published. As the first published is valid.

The association between the two of them grew closer when the publisher began to collect violins, as Leonard had played the violin since his youth. And was undoubtedly further reinforced by the rapidly prepared and published description of the discus subspecies.

After Axelrod received the first *Symphysodon* from Harald Schultz in Brazil and himself returned from Brazil at the end of 1959 with additional specimens, he would easily have been in a position to persuade Schultz to undertake an immediate revision of the genus *Symphysodon*. In order to print it right away in his magazine, *Tropical Fish Hobbyist*. (Barely two months elapsed between the handing over of the discus material and publication.) The chronology was as follows:

After Harald Schultz (see Chapter 1: *Discoveries in the 20th century*) sent the first discus to Axelrod in New York at the beginning of 1959, and the specimens had died shortly thereafter, the publisher promised to go to Brazil to bring back "his own" discus. In the autumn (in the northern hemisphere) of 1959 Axelrod set off on his second trip to Brazil, his first to the Amazon region. (A year previously he had flown to São Paulo with Rosario Lacorte after killifishes, and had told Rosario that this was his first visit to South America.) On this new (1959) trip the two men, H. R. Axelrod and H. Schultz, met in person for the first time. Whether they actually caught discus together at that time cannot be ascertained, as no precise details are available. I know only from my own, subsequent experiences (expeditions that I organised and led) that Axelrod caught hardly any fishes. He always had his hired *caboclos* with him, or I did the collecting when we were together (or when I was by myself). But that is by the by. The demonstrable facts are as follows.

Right at the beginning of the text of L. P. Schultz's revision it is clear that the publisher had told the author an untruth, as three years previously Axelrod had never caught a discus (see above). The specimens that Schultz obtained from Axelrod at the end of 1959/beginning of 1960, and which led to the *Symphysodon* revision, are as follows (with the relevant data as recorded in the National Museum for Natural History in Washington, D.C., which, however, often differ from those cited in the revision – much is contradictory):

1. *Symphysodon discus* Heckel, 1 specimen (USNM 00179828), 76 mm long (SL), Locality Manaus, Brazil, no date. Collector: Axelrod, H. R. (see page 124).

Schultz began his revision with this specimen – which undoubtedly originated from Willy Schwartz's business, Aquario Rio Negro. He didn't examine Natterer's specimen, the holotype (through lack of time?).

2. *Symphysodon aequifasciata aequifasciata* Pellegrin, 104

specimens (USNM 00179611) with a length of 69 to 148 mm (SL), which originated from Lago Teffé and have the collector given as Harald Schultz (in the NMNH Axelrod, H. R. is given as the collector for this catalogue number) (see page 125).

From these he described this subspecies with the previously published type localities of Lago Teffé and Santarem, Brazil (of course the Frenchman Pellegrin had already published on this subject, but Schultz didn't examine his types).

He then described the following subspecies as new:

3. *Symphysodon aequifasciata haraldi* new species (although he described a subspecies), 1 specimen (USNM 00179829), holotype, 117 mm (SL).

At the same time he synonymised as a *nomen nudum* the subspecies *S. discus tarzoo* described by Lyons a few months previously (on the 28th November 1959) with the type locality Leticia (Kullander has subsequently refuted this to some extent – see *Kullander's synonyms*, below); and likewise placed in the synonymy of this new subspecies the drawings (and name) of *S. discus* by Meinken *in* Holly, Meinken and Rachow, from the book *Aquarienfische in Wort und Bild* (see page 127). Supposedly Schultz's holotype originated from Benjamin Constant and the collector was Axelrod, H. R. No date was given and, moreover, Axelrod was not there in 1959, just in Manaus and the surrounding area (according to Willy Schwartz). Benjamin Constant lies almost 2000 kilometres from Manaus (by river). Perhaps, however, the fish originated from Harald Schultz, who was there with the Ticunas at the beginning of 1960, making a film, or – to the best of my knowledge like all Axelrod's specimens – they were from the business of Hans Willy Schwartz – as Willy himself often told me.

But that is still not all as regards the mystery of the locality and collector details: in the NMNH Fish Collection there are additional specimens of this subspecies with the annotation, Collector: Axelrod, H. R., viz:

1 specimen (USNM 00224864) that he supposedly collected on 25th August 1976 in the Rio Ipixuna (see page 129).

Which is particularly interesting, as at that time Axelrod was with me (along with Jacques Géry, Adolf Schwartz, and others). And never anywhere near the Ipixuna, but at Willy Schwartz's establishment. (He also gave it to be understood that the fish was caught in six metres of depth.)

And further:

3 additional discus specimens (USNM 00326731) that he supposedly caught in November 1963 in the Rio Purus, "Taparaua 380 km by water from Manaus in small stream" (see page 129).

Now, he was actually in a tiny village named Tapauá, which lies on the Purus at the mouth of the Rio Itaparaná (see Chapter 4: *the blue discus*). Along with the German Fritz Terofal and Harald Schultz he visited this village in November 1963, but they didn't catch any discus (or *Paracheirodon innesi* as claimed). I witnessed this myself on the spot, and have had it confirmed by people still living. They brought not a single discus or neon back from Tapauá. Thus this datum too is totally false.

Moreover, it is even more interesting to note that in the part of the revision covering the new subspecies *S. aequifasciata haraldi* three photos of live fishes were published (the colour description was apparently based on these and hence he designated these specimens "blue"). One of these photos is the often published picture (see page 122) that crops up throughout the publisher's publications with a variety of locality data ("photo Axelrod"), while the other two are the well-known (two) fishes from Harald Schultz (see pages 137 & 139), which Schultz sent to Dr. Schmidt-Focke in 1959 ("photos Schmidt-Focke"). Thus none of the fishes illustrated was deposited in any collection.

The culmination of all this is to be seen in the caption on page 13 of the revision: "This blue discus, *Symphysodon aequifasciata haraldi,* differs considerably in its shade of colour from the individual shown top on page 8. But the only difference is that the fish on page 8 is in an aquarium with black gravel, while this one was kept in an aquarium with a white substrate." (see page 126). **But** two totally different specimens (subspecies, as he described them) are involved. The first (Axelrod) photo (page 8 of the revision) is quite definitely a fish from the lower Rio Urubu (and must thus be *S. a. haraldi*, though the deposited material from the Urubu is annotated as *S. a. axelrodi*). That with the caption cited is by Schmidt-Focke, and is one of the specimens (ex Harald Schultz) he bred. There is no photo of the holotype, alive or dead.

He described the next subspecies as:

4. *Symphysodon aequifasciata axelrodi* new species (again actually a subspecies), 1 specimen (USNM 00179831), holotype, 105 mm (SL), with no synonyms.

The holotype was supposedly from Belém, Brazil, and the collector Axelrod, H. R. (see page 121). Here too a date is lacking. Now, practically everyone with any knowledge of discus knows that no discus are found in the vicinity of Belém. The nearest proven locality for discus is in the region of Cametá on the Tocantins or near Breves (on the island of Marajó). Dias Lopes, a collector and exporter, was in Belém at that time and undoubtedly gave Axelrod the specimen (or was it Schwartz? As he was importing brown discus via Belém at that time and

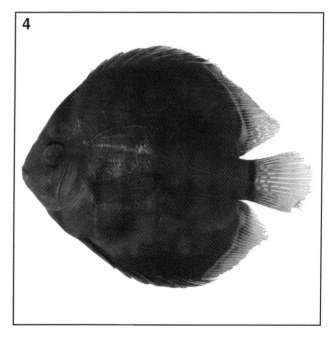

1-4. Type material from Schultz's revision in the NMNH: 1. *S. aequifasciata axelrodi,* 105 mm (SL), holotype (USNM 00179831) supposedly from Belém and collected by Axelrod, H. R. (no date). 2. *S. a. haraldi,* holotype (USNM 00179829), supposedly from Benjamin Constant and collected by Axelrod, H. R. (no date). 3. There are 48 specimens of S. a. haraldi measuring 75-139 mm SL (USNM 00179610) – one specimen shown here – all with the collector given as Axelrod and the Rio Urubu as locality (date given as 1958). 4. The data cited for this *S. aequifasciatus* (USNM 00191597) – locality Rio Araguaia near Aruaná, Brazil, and, again, collector Axelrod – are again incorrect. Discus have never been found upstream of so many rapids, and never so far south.

selling them to Louis Chung in Georgetown). The collecting locality is unknown and Axelrod was not the collector. The fish looks like those from Breves (see Chapter 4: *The blue and brown discus*).

9 specimens (USNM 00179609), paratypes, with a length of 105-122 mm (SL), are additionally cited by L. P. Schultz, and were supposedly caught in the Rio Urubu, Brazil near Itacoatiara, with the only date 1958 and collector Axelrod, H. R. (However, the latter was not on the Amazon in 1958, though in 1959, along with Harald Schultz, he was at the Rio Preto da Eva, which lies far from the Urubu and never contains discus. Undoubtedly these specimens, and those that follow, were also from Aquario Rio Negro.)

48 specimens (USNM 00179610), with a length of 75-139 mm (SL), are further listed by the scientist, and likewise deposited in the NMNH with the date given as 1958, the collector as Axelrod, and the locality as Rio Urubu.

The museum also contains additional lots under the name *S. a. axelrodi*:

103 individuals (USNM 00179611) with the same locality, the Rio Urubu (although the label says Rio Uruba, but also "near Itacoatiara"), collector Axelrod H. R., and date 1958. These are designated topotypes.

7 specimens (USNM 00331512) with the locality Rio Manacapuru, Brazil, no date, collector Axelrod.

7 specimens (USNM 00191597) – but labelled only as *S. aequifasciatus* – with the locality Rio Araguaia, near Aruaná, Brazil, and Axelrod again listed as collector.

Now this is a complete fabrication as *Symphysodon* have nowhere ever been found upstream of rapids or waterfalls (and there are an awful lot of those in the almost 2000 km long course of the river between Aruaná and its confluence with the Rio Tocantins). And, moreover, never so far south. In addition, I myself (and many well-known Brazilian ichthyologists) have undertaken 14 collecting trips in the Aruaná region on the Araguaia (my brother and my friend Heiner Nolden have made even more). There are no discus there. Plus a fisherman who worked daily along the Rio Araguaia for 30 years confirmed this and many other fishermen based in Aruaná supported this view.

The pictorial material which L. P. Schultz used as the basis for the description of the coloration of *S. aequifasciatus axelrodi* consists of two photos: one is by Axelrod with the remark: "the discus comes from Belem" (but it is a typical specimen of the Breves form mentioned above), and the second is a dead discus – a so-called blue discus from the Rio Urubu (in this case the locality does ring true, as blues occur the lower Urubu). And the point of the story is this, *axelrodi* is not the blue discus. Even assuming that the division into subspecies is valid at all, then all forms from the Rio Urubu should be *S. aequifasciatus haraldi* and not *axelrodi*! (cf Chapter 4: *the brown discus*, and also *ibidem, the blue discus*).

This is one of the three **S. a. haraldi** photos which Schultz included in his revision without specifying a locality. In Tropical Fish Hobbyist magazine (4, 1984) it is stated that Axelrod caught and photographed the fish at Benjamin Constant, in a western branch of the Purus.

Thus, leaving aside the numerous contradictions and false data (and Kullander's very correct analysis – *qv*, below) the work of L. P. Schultz does not stand up. Except that, in my opinion, Schultz was not the main culprit. He undoubtedly acted in good faith and merely cited the details he had been given. Nevertheless he did make a positive contribution as far as aquarists are concerned: the concept, still valid today, of categorising discus by colour (even though opinions vary over the details): green, blue, and brown (or common).

Finally, it is praiseworthy that – even though L. P. Schultz did not himself benefit from it – the publishing house converted its existing T.F.H. Fund into the Leonard P. Schultz Fund, in order to retrospectively honour and remember the scientist.

The original description of the Discus Subspecies by L. P. Schultz in 1960

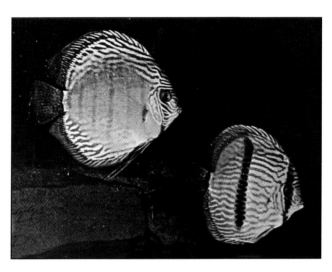

Fig. 1. *Symphysodon aequifasciatus haraldi* on the left, with *Symphysodon discus* on the right. Photo by Dr. Eduard Schmidt.

A Review of the Pompadour or Discus Fishes, Genus *Symphysodon* of South America

By Dr. Leonard P. Schultz
United States National Museum
(From the June 1960 issue of *Tropical Fish Hobbyist*)

Nearly three years ago Herbert R. Axelrod informed me that he had specimens of at least four different color varieties of *Symphysodon* and asked if I would like to study them. Naturally, my curiosity was aroused and I wrote him that I would need specimens of each variety, so Mr. Axelrod made a trip to South America to collect specimens with the assistance of Harald Schultz. These specimens were donated to the United States National Museum and form the basis of this paper. Much credit should go to these two men for their great effort in obtaining specimens of the discus fishes.

For several months I have been counting and measuring various characters on specimens of *Symphysodon*. In the meantime, Mr. Axelrod has furnished several excellent color transparencies of the living specimens, that were later preserved and sent to me.

This review of the genus *Symphysodon* leaves certain questions unanswered and it must not be considered the last word on this group. We do not know the extent of the area occupied by each species or subspecies nor do we know the amount of integration for the color pattern between the subspecies. It may be possible for aquarists to explore some of these problems and thus advance our knowledge.

The purposes of this paper are to make known to aquarists and to scientists that at least four different kinds of discus, *Symphysodon*, do occur in South America and that a great amount of study is still needed to learn about their life history and ecology.

Genus *Symphysodon*

Symphysodon Heckel, Ann. Wiener Mus., vol. 2, p. 332, 1840 (type species, *Symphysodon discus* Heckel); Eigenmann and Bray, Ann. New York Acad. Sci., vol. 7, p. 623, 1894 (Revision); Regan, Ann. Mag. Nat. Hist. Ser. 7, vol. 16, p. 440, 1905 (Revision).

Characters in common for all species: A single pair of nasal openings; ctenoid scales rather small, from 44 to 61 vertical rows from rear edge of head to base of caudal fin rays; scales occur about halfway to tips of dorsal and anal fins; basal third of caudal rays scaled; pectoral and pelvic fins naked; cheek below eye scaled; opercle, subopercle and interopercle scaled; gill rakers absent on upper part of first gill arch with about five short ones on lower part of arch; lips thick and fleshy; teeth small, conical, in a single row; branchiostegal membranes joined across isthmus and free from it; nine vertical dark bars on head and body.
Dorsal rays: IX or X, 30 to 33; anal: VII to IX, 26 to 33; pectoral rays: ii, 7 to 9, iii or iv; pelvic rays: I, 5; branched caudal rays: usually 7 + 7, sometimes 7 + 6.

Key to the species of *Symphysodon*

1a. Vertical scale rows about 44 to 48 from rear of head in a straight line to midbase of caudal fin rays; three of the nine vertical dark bars, numbers one, five and nine notably much darker than the others; background coloration of alternating blue and reddish-brown lengthwise streaks; eye blue; no partch of isolated scales behind eye. *discus* Heckel

1b. Vertical scale rows 50 to 61; all of the vertical dark bars of approximate intensity except first and last may be a little darker, but the middle or fifth bar not darker than adjoining bars.

3a. Lengthwise streaks dark brown on a dark green background; eye reddish brown.
aequifasciata aequifasiata Pellegrin

3b. Lengthwise streaks bright blue on a light brown background, eye bright red.
aequifasciata haraldi, new subspecies

2b. No lengthwise streaks present on body or fins, although a few blue streaks may occur on forehead, vertical dark bars purple on an olive colored background; eye red, no isolated patch of scales opposite dorsoposterior part of eye a little above the upper end of the preopercular groove.
aequifasciata axelrodi, new subspecies

Symphysodon discus Heckel

Symphysodon discus Heckel, Ann. Wiener Mus., vol. 2, pp. 332-333, pl. 30, figs. 21-24b, 1840 (type locality Rio Negro); Kner, Sitzber. Akad. Wiss. Wien, vol. 46, p. 299, pl. 2, fig. 2, 1862; Steindachner (in part), Sitzber. Akad. Wiss. Wien, vol. 71, p. 46, 1875 (Amazon at Teffé, Rio Xingu at Poroto do Moz; Rio Madeira at Maues; Rio Negro).

Since I have not been able to examine the specimens on which the following references were based, they may refer to species of *Symphysodon* other than *S. discus* Heckel: Günther, Catalog of the Fishes in the British Museum, vol. 4, p. 316, 1862 (Rio Cupai, Brazil); Eigenmann and Eigenmann, Proc. U.S. Nat. Mus., vol. 14, p. 71, 1891 (Amazon); Eigenmann and Bray, Ann. New York Acad. Sci., vol. 7, p. 624, 1894; Pellegrin, Mem. Soc. Zool. France vol. 16 p. 230, 1903 (Manaus); Regan, Ann. Mag. Nat. Hist. Ser. 7, vol. 16 p. 440 1905 (Rio Cupai; Rio Negro; Teffé; Manaus; Ihering, Revista Mus. Paulista, vol. 7, p. 336, 1907 (Amazonas; Rios Negro, Madeira, Xingu); Eigenmann, Repts. Princeton Univ. Exped. Patagonia, 1896-1899, vol. 3 Zool. Pt. 4, p. 479, 1910 (Amazon and tributaries); Haseman Ann. Carnegie Mus., vol. 7, nos. 3, 4, p. 372, 1911 (Manaus, Santarem); Hegener, Blät. Aquar. Terr. Vol. 47, no. 11, p. 241, 1937.

In addition to the references cited here Meinken *in* Holly, Meinken and Rachow (Die Aquarienfische in Wort und Bild, Lieferung 75-76, pp. 769-773 1943) give numerous references for *Symphysodon*.

Specimens examined: USNM 179828, a single specimen, 76 mm. From tip of snout to midbase of caudal fin rays, collected at Manaus, Brazil in the Amazon River.

Description: Counts are recorded in tables 1 and 2 for this species. Although no significant difference in number of fin rays found between *S. discus* and the other species, *S. discus* definitely has a fewer number of vertical scale rows from rear of head to caudal base, 44 instead of 50 to 61. The scales on the head above the eye reach to the edge of the bony orbit dorsally but not on the fleshy rim; the patch of free scales behind posterior edge of eye are lacking; scales on dorsal part of head do not reach quite to a line between the pairs of supraorbital pores; posterior preopercular row of scales reaches a little above the dorsal end of preopercular groove.

Fig. 2. *Symphysodon discus*, the real discus! This fish comes from the Rio Negro near Manaus, Brazil. Photo by Harald Schultz.
(Another of the eight photos accompanying the Schultz's revision, a Heckel discus from the Rio Negro near Manaus, Brazil).

Color pattern: Background coloration of body composed of alternating horizontal light reddish-brown and bluish streaks, numbering about 15 to 18; these streaks beginning behind operculum and on forehead at mid anterodorsal line, continue a somewhat wavy course posteriorly, thence disappearing at base of median fins, the usual 9 vertical bars are present, three of which are dark blue, the others scarcely distinct; the first is a broad one through the eye to instmus, next three very light tan, the fifth is a broad dark blue one, attaining its greatest intensity on lower midside of body but not reaching to base of anal fin, the next three are very light tan; the last is a broad bar across base of caudal fin sharply contrasting with general background coloration as do numbers one and five; the usual dark band on bases of dorsal and anal fins indistinct, distal parts of median fins light blue with scattered light blue spots; pectoral fin light blue; oblique light streaks on cheek, and two or three vertical ones on opercle; outer pelvic ray blue; other rays yellowish.

Remarks: Steindachner (Sitz Akad. Wiss. Wien vol. 71, p. 46-47, 1875) had two species of *Symphysodon* because he mentions that in small specimens the vertical scale rows were 46 to 48, whereas in large specimens the scales numbered 52 to 56. This is exactly what I have found for *S. discus* and *S. aequifasciata* Pellegrin, the latter having the greater number of scales.

All of the species and subspecies, including the two new subspecies, are distinguished and diagnosed in the accompanying key.

Symphysodon aequifasciata aequifasciata Pellegrin

Symphysodon discus aequifasciatus Pellegrin Mem. Soc. Zool. France, vol. 16, p. 230, 1903 (type locality, Lago Teffé and Santarem, Brazil).

Symphysodon discus Steindachner (in part) (Sitz. Akad. Wiss. Wien, vol. 71, p. 46-47, 1875 (Lago Teffé); Eigenmann, Repts. Princeton Univ. Exped. Patagonia 1896-1899, vol. 3, Zool. Pt. 4, p. 479, 1910 (Santarem; Teffé); Ribeiro, Fauna Brasiliense, Arch. Mus. Nac. Rio de Janeiro, vol. 17, p. 69, fig., 1915 (Amazon and tributaries); Flower, Mus. Hist. Nat. Javier Prado, Univ. Nac. Mayor San Marcos, Lima, p. 252, fig. 87, 1945 (Peruvian Amazon); Innes, Exotic Aquarium Fishes, color plate, p. 439, 1950 (Amazon); Axelrod and Schultz, Handbook of Tropical Aquarium Fishes, p. 655, color plate, 1955 (Amazon).

Specimens examined: USNM 179611, 104 specimens, 69 to 148 mm. Collected in Lago Teffé, Brazil, by Harald Schultz.

Description: Certain counts and measurements are recorded in the tables. Scales on head above the eye reach to bony edge of orbit but not on fleshy margin of eye; a patch of two or three isolated scales behind eye dorsally, posterior preopercular row of scales reaches above dorsal end of preopercular groove and is almost continuous with the patch of isolated scales.

Color pattern: Background coloration dark brownish green with nine dark brown vertical bars all of about the same intensity except last is darkest; these dark bars have the same positions as the homologous dark bars in the other species of *Symphysodon*. Basal three-fourths of dorsal and anal fins blackish, distally these fins are light olive green with scattered light spots basally; caudal fin translucent with scattered light spots; horizontal blackish streaks on head, dorsally on body and on dorsal fin, mostly absent on midsides, but distinct ventrally on anal fin. Alternating light blue and dark oblique streaks on cheek; three vertical, dark, bluish streaks on opercular; iris reddish-brown, pelvic

Fig. 3. The Green Discus from Lago Teffé, Brazil, is *Symphysodon aequifasciatus aequifasciatus*. Photo by H. Schultz. (This is the only accompanying photo of a Green Discus and is the individual from Lago Jurity – see p. 67).

green, outer ray blue, and tips of pelvic rays dark brown; pectoral fin translucent, its base dark green.

Symphysodon aequifasciata haraldi, new species

Symphysodon discus tarzoo Lyons, Tropicals Magazine, Holiday issue 1960, vol. 4, no. 3, pp. 6-10, 4 fighs., Nov. 28 1959 (Amazon in vicinity of Leticia (nomen nudum).

Symphysodon discus, Meinken in Holly, Meinken and Rachow, Die Aquarienfische in Wort und Bild, Lieferung 75-76; pp. 769-773; 41, 6, photographs, color drawing, 1943 (Biology of discus); Fowler, Mus. Hist. Nat. Univ. Nac. Mayor San Marcos, Lima p. 253, fig 87, 1945 (Peruvian Amazon).

When an unknown or new aquarium fish is introduced to aquarium hobbyists and a name is printed such as the one Mr. Earl Lyons introduced for the blue discus, confusion and disagreement among those interested in scientific zoological nomenclature is likely to occur unless the rules of zoological nomenclature are followed. I have shown Dr. Curt Sabrosky the article by Mr. Lyons and he agrees that the name *tarzoo* does not have nomenclatorial standing because the following two rules were not fulfilled as defined by the International Rules of Zoological Nomenclature and adopted at the 1927 Budapest Congress.

(1) After January 1, 1931, the specific name must have been published with a statement in which the author attempted to indicate differentiating characters or with a summary of characters which distinguish the species from other species, (2) the publication of a figure of the species with a scientific name does not meet these requirements.

Therefore, I must conclude that Mr. Lyons' article does not establish any scientific name.

To avoid confusion in the future in our Zoological Nomenclature it seems wise to describe as a new subspecies the blue discus according to the Rules of Zoological Nomenclature.

Holotype: USNM 179829, a specimen 117 mm. in standard length, collected at Benjamin Constant, Brazil, in the Amazon, by H. R. Axelrod and Harald Schultz.

Description: Certain counts and measurements are recorded in tables 1 and 2. Scales on head above the eye reach to pores above eye but not quite to edge of bony

Figs. 4-5. Caption to the photo by Dr. E. Schmidt shown top in the original description (4). This blue discus, *Symphysodon aequifasciata haraldi*, differs considerably in its shade of colour from the individual shown top on page 8 of the same work (5). But the only difference is that the fish on page 8 is in an aquarium with black gravel, while this one was kept in an aquarium with a white substrate. The caption to the photo (5) shown top on page 8 reads simply *"Symphysodon aequifasciata haraldi"*. Photo by Herbert R. Axelrod. Fig. 6. This photo is from Aquarienfische in Wort und Bild: *Symphysodon discus;* Schultz regarded this discus (on the basis of the photo) as synonymous with *S. a. haraldi*.

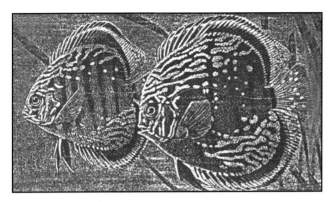

orbit; a patch of two or three isolated scales behind eye dorsally; posterior preopercular row of scales reaches above dorsal end of preopercular groove and almost continuous with patch of scales; scales on dorsal part of head do not reach to the supraorbital pores.

Color pattern: Background coloration of body brownish anteriorly, darker brown posteriorly; head purplish; head and body crossed with nine dark blue vertical bars, first and last darkest, first bar through eye across cheek to isthmus; second from spiny dorsal origin to upper edge of opercle; third across base of pectoral fin to baseof pelvic fin; next four across body; next to last from rear edge of dorsal fin base across caudal peduncle to rear edge of base of anal fin; last dark bar almost black across base of caudal fin rays; basal three-fourths of dorsal and anal fins dark balckish-purple that blends in with the background coloration of body; distal areas of soft dorsal and anal fins light yellowish with scattered light spots mostly basally; pelvic fin dark brown, outer ray blue, distally the tips of rays are pinkish; pectoral fin purple, its base brown; horizontal wavy blue streaks irregularly interrupted, cover the entire body, except breast and behind head, and distal parts of median fins; forehead with four or five blue horizontal streaks, two oblique blue streaks below eye and two vertical blue bars on operculum.

In alcohol the purple bars preserve as dark brown and the blue streaks almost completely disappear.

Remarks: This new subspecies was named *haraldi* in honor of Mr. Harald Schultz, São Paulo, Brazil, who has collected numerous new and rare South American fishes.

Symphysodon aequifasciata axelrodi, new species

Holotype: USNM 179831, a specimen 105 mm. in standard length collected at Belém, Brazil, Amazon River by Herbert Axelrod.

Paratypes: USNM 179609, Rio Urubu, nine specimens, 105 to 122 mm. in standard length, collected by Herbert R. Axelrod and Harald Schultz.

USNM 179610, Rio Urubu, 48 specimens, 75 to 139 mm. in standard length, collected by Herbert R. Axelrod and Harald Schultz.

Description: Certain counts and measurements are recorded in tables 1 and 2. Scales on head above the eye reach to edge of bony orbit but not to fleshy margin of eye; no isolated patch of scales behind eye; posterior preopercular row of scales does not reach to dorsal end of preopercular groove; scales on dorsal part of head do not reach to the supraorbital pores.

Color pattern: Background coloration light yellowish-brown to dark brown, overlaid with nine vertical bars, the first through eye is dark purplish-brown to blackish and the last through base of caudal fin rays blackish, these two (first and last) notably darker than the other seven, all of which are light purplish to purplish-brown; the second at origin of spiny dorsal fin ends at upper edge of opercular opening; the third passes through base of pectoral fin thence ending at base of pelvic fin; the next four across body; the next to last vertical bar occurs on caudal peduncle and extends from rear edge of base of dorsal fin to that of the anal fin. Soft dorsal fin and soft anal fin basally with a broad dark purplish band that is continuous with the dark crossbar at base of caudal peduncle and constrasts notably with the light brown background color of body; soft dorsal and anal fins distally light olive, with scattered lighter spots mostly basally; pelvic fin reddish brown with distal tips of rays yellow or orange; anal spines blue, membranes brownish; pectoral fin light blue to light purplish; bases of pectoral rays purple; operculum dark purple; iris red; forehead with about four horizontal light blue streaks, anal fin with blue streaks basally.

In alcohol the vertical purple bars preserve as dark bars against a lighter brownish background.

Named in honor of Herbert R. Axelrod who collected most of the specimens used in this study. This new subspecies is diagnosed in the accompanying key.

Table 1. Counts recorded for the species and subspecies of *Symphysodon*.

NUMBER OF FIN RAYS

Species and subspecies	Dorsal								Anal									Pectoral					
	VIII	IX	X	29	30	31	32	33	34	VII	VIII	IX	26	27	28	29	30	31	32	ii	7	8	9	iii	iv
discus aequifasciata	—	1	—	—	1	—	—	—	1	—	—	—	—	1	—	—	—	2	2	—	—	2
aequifasciata	1	19	3	1	3	7	7	4	1	—	20	2	—	2	1	1	10	5	3	19	—	12	7	10	9
haraldi	—	1	—	—	—	1	—	—	—	—	1	—	—	—	—	1	—	—	—	2	—	2	—	—	2
axelrodi	1	20	7	—	6	14	7	—	—	2	21	5	2	1	7	3	9	4	2	29	—	20	9	15	14

NUMBER OF PORES IN LATERAL LINE

	Anterior lateral line						Caudal lateral line					Total pores								
	18	19	20	21	22	23	10	11	12	13	14	28	29	30	31	32	33	34	35	36
discus aequifasciata	—	1	—	—	—	—	—	—	1	—	—	—	—	—	1	—	—	—	—	—
aequifasciata	1	4	4	8	4	2	1	4	7	9	2	—	—	1	3	6	4	4	4	1
haraldi	—	—	1	—	—	—	—	—	—	1	—	—	—	—	—	—	1	—	—	—
axelrodi	2	5	8	5	7	2	4	8	9	8	—	1	3	3	10	7	4	1	—	—

Table 2. Counts recorded for the species and subspecies of *Symphysodon*.

Species and subspecies	Number of vertical scale rows from upper edge of opercular opening in a straight line to base of caudal fin.																		NUMBER OF VERTEBRAE								
																			Abdominal			Caudal				Total			
	44	45	46	47	48	49	50	51	52	53	54	55	56	57	58	59	60	61	12	13	14	17	18	19	20	30	31	32	33
discus																													
Steindachner	1	—	—	—	—	—	—	—	—	—	—	—	—	—	—	—	—	—	—	1	—	1	—	—	—	1	—	—	—
(1875)[1]	—	—	x	x	x	—	—	—	—	—	—	—	—	—	—	—	—	—	—	—	—	—	—	—	—	—	—	—	—
aequifasciata																													
aequifasciata	—	—	—	—	—	—	2	1	5	4	3	4	1	1	—	1			1	13	—	—	9	2	1	—	10	3	1
haraldi	—	—	—	—	—	—	—	—	1	—	—	—	—	—	—	—	—	—	—	1	—	—	1	—	—	—	1	—	—
axelrodi	—	—	—	—	—	2	2	4	2	4	4	4	2	1	2	—	—		1	14	2	1	15	1	—	1	14	2	—

Counts from literature

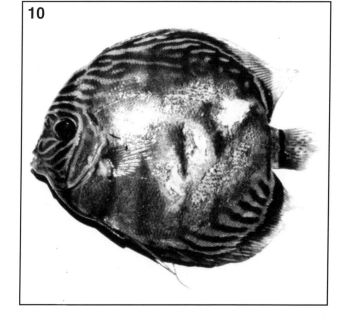

Fig. 7-8. These 2 photos are also from Schultz's revison, and were captioned : *"S. aequifasciata axelrodi,* the common discus. The top fish comes from Belém, Brazil. Photo by H. R. Axelrod. The lower fish, also *S. aequifasciata axelrodi,* was collected at the Rio Urubu, Brazil. Photo by H. Schultz." (The latter is as paratype and a blue discus – so should be *S. a. haraldi* and not axelrodi... Com-pare also in Chapter 4, the forms from the Belém Region and the Urubu, respectively.)

Fig. 9-10: The two photos above are not from Schultz's revision, but from "Axelrod's collection": 9. One of the three specimens in the NMNH [USNM 00326731] which Axelrod purportedly collected in the Rio Purus (no discus occur in the Purus) in November 1963. 10. This specimen is also stored in the NMNH [USNM 00224864], and was supposedly caught on the 25th August 1976 in the Ipixuna, but on that day he was travelling on the Transamazonica with me...and we didn't catch any discus.

KULLANDER'S SYNONYMS

The Swede Sven O. Kullander is Senior Curator at the Naturhistoriska riksmuseet in Stockholm and head of the departments of Ichthyology and Herpetology as well as the Vertebrate Collection of the museum. He has travelled indefatigably, formerly often alone, in recent years with his wife Fang Fang and colleagues, in South America and more recently also in Myanmar, Indochina, and China (he has just revised (with Ralf Britz) the Asian family Badidae, described seven new *Garra* species together with Fang Fang, and much more).

Sven is without doubt the scientist who has published the most work on the family Cichlidae and its taxonomy over the last two decades. And especially the cichlids of South America, which he certainly knows better than any other ichthyologist. He has been publishing work on cichlids (and other subjects) since 1981 and has brought out one revision after another. His 431-page book, *Cichlid fishes of the Amazon River drainage of Peru* (published in 1986 by the Swedish Museum of Natural History, Stockholm – see page 133), effected rather dramatic changes in the world of South American cichlids and also in the taxonomy of the genus *Symphysodon*.

In this book Sven addresses the problem of the species of *Symphysodon* and places all the subspecies described for the two species, *S. discus* and *S. aequifasciatus,* in synonymy, as follows:

1. *Symphysodon discus* Heckel, 1840
    Synonym:
       Symphysodon discus willischwartzi Burgess, 1981
       Trop. Fish Hobby. 29 (7): 32-42.
2. *Symphysodon aequifasciatus* Pellegrin, 1904
    Synonyms:
       Symphysodon discus var. *aequifasciata* Pellegrin, 1904.
       Mém. Soc. zool. Fr. 16, p. 250 (Teffé (Brésil); Santarem (Brésil)).
       Symphysodon discus tarzoo Lyons, 1959.
       Tropicals Mag., p. 6, fig. p. 7 (Leticia, Colombia).
       Symphysodon aequifasciata axelrodi Schultz, 1960.
       Trop. Fish Hobby. 8 (10), p. 14, fig. p. 9 (Belem, Brazil, Amazon River).
       Symphysodon aequifasciata haraldi Schultz, 1960.
       Trop. Fish Hobby. 8 (10), p. 11, fig. p. 8 (Benjamin Constant, Brazil in the Amazon).

Unfortunately in the years since its appearance this work by Kullander has been almost completely overlooked or disregarded in the aquarium literature, and the subspecies continue to be mentioned.

Later Kullander published an article, "Eine weitere Übersicht der Diskusfische, genus *Symphysodon* Heckel", in a special publication by *DATZ (Deutsche Aquarien und Terrarien Zeitschrift)* in October 1996 *(Diskus*: 10-16), where he again addresses the problems of the descriptions – including the conflicting scale counts – and confirms his researches prior to 1986.

There is also a very good page on the South American Cichlidae (Guide to the South American Cichlidae) on the website of the Swedish Museum of Natural History (Naturhistoriska riksmuseet) in Stockholm, maintained by Sven; just click on
www.nrm.se/ve/pisces/acara/cichalfa.shtml

Here one can find practically all the cichlid genera and species of the continent and learn (almost) everything about them. And, of course, also his views on discus, under *Symphysodon* Heckel, 1840.

However, his map of the distribution for the genus and species is very limited (as in the publications mentioned above), reflecting the fact that he had only a small amount of museum material available (there are not that many voucher specimens of *Symphysodon* worldwide – this fish has always been very difficult to catch and cannot even be obtained by scientists using fish poisons such as *timbó* or rotenone).

Kullander summarises the distribution thus in his website:

"*Symphysodon aequifasciatus* is collected along the mainstream Rio Solimões and Rio Amazonas in Brazil, and also in the Río Putumayo-Içá in Peru and Brazil (and probably Colombia). A record from the Rio Muiuçu (Rio Tapajós above Jacareacanga) is doubtful. There are many more localities reported in aquarium literature, but within the same general area of distribution. *Symphysodon discus* on the other hand is restricted to the lowermost parts of the Rio Negro, Abacaxís and Trombetas" (Kullander, 2003).

His statement that the localities for the genus reported in the aquarium liturature fall within the same general area of distribution is definitely not correct (see Chapter 3). He also omits the distribution for *Symphysodon discus* north of the Amazon in the Nhamundá, Jatapu, middle Urubu, and the middle Rio Negro region and its affluents, and to the south of it in the Marimari. And he seem to be unaware that the green discus has a very limited range – certainly the smallest distribution of them all (see comments on taxonomy, below).

During his study of the meristic characters Kullander discovered that *S. discus* have on average lower counts than *S. aequifasciatus,* but that *S. aequifasciatus* from western Amazonia (= Greens) have lower counts than those from central Amazonia and *S. aequifasciatus* from the eastern Amazonas area, and these meristics are similar to those of *S. discus* (this reflects the fact that *S. discus* occurs closer to – often almost overlapping – the blue and brown forms than to the green – cf also Chapter 3: *Distribution*). In addition, the description of the differences between the two species recognised is restricted to the dead material available to him, and he writes that the pattern of markings (pattern colour) is practically the only way to distinguish them (which is true): *"Symphysodon discus* has a distinct pattern of dark undulating lines covering the sides and a dark vertical bar across the middle of the side much more intense than the other vertical bars. *Symphysodon aequifasciatus* has all vertical bars similar in width and intensity; horizontal stripes do not show in preserved specimens except occasionally on nape and on anal fin base." (Kullander, 1996)

Here I must point out, that there is no Royal Blue or fully-striped specimens in the collections (alpha individuals are always sold immediately for good money and never end up in museum collections). He also writes (Kullander, 1996) that a form from Maués, unknown from museum material but mentioned in the aquarium literature, may be distinct. It is said to be similar to *S. aequifasciatus* except that the central bar (as in the so called "Rio Içá" discus) is especially prominent but no wider than the other bars. But such individuals are found widely distributed (see chapter 5).

The most interesting point is that he notes (Kullander, 1996) the following (colour) details for *Symphysodon aequifasciatus* from Tefé – although he nevertheless does not regard it as distinct (except from *S. discus*):

"*Symphysodon aequifasciatus* from Tefé can be recognized by yellow anterior sides and red spots over much of the sides. Life colours vary considerably in *S. aequifasciatus* but it remains to be investigated how the variation correlates with distribution, environmental conditions, sex and ontogeny."

Kullander's diagnosis *(DATZ* 1996, here translated into English), is as follows:

" The only known characters that clearly distinguish *S. discus* from *S. aequifasciatus* are provided by the pattern of markings. The entire body and the bases of the dorsal and anal fins exhibit wavy, mainly horizontal blue stripes with red or brown intervals; the dark bars through the eye, on the centre of the body, and at the base of the caudal fin are deep black, and the central bar is wider than the rest. By contrast *S. aequifasciatus* rarely exhibits blue stripes on the flanks, but does so frequently at the base of the anal fin and across the nape; although the vertical bars through the eye and across the base of the caudal fin are more distinct than the rest, all the bars on the flanks are the same width and of the same intensity.

Symphysodon discus has on average fewer scales in a longitudinal series (48 to 62, usually 53 to 57, vs 53 to 60, usually 55 to 59), fewer dorsal fin rays (D usually IX.30 vs IX.31 or X.30) and fewer abdominal vertebra than *S. aequifasciatus* (13 + 18 vs 13 + 19), but there are no known morphometric or meristic characters that separate the species unequivocally."

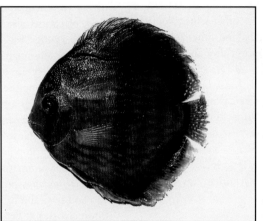

Holotype of *Symphysodon discus willischwartzi* Burgess, 1981

And his concluding remarks:

"Our knowledge off the phylogenetic relationships and the primary taxonomy of the genus *Symphysodon* remain only superficial. The results of further research may completely support, correct, or overturn what is stated here. It would be sensible in the first instance to investigate the relationships between the heroine cichlids in greater detail, and such research should make reference to anatomical characters, which to date have been virtually unstudied. The investigation of the geographical variation of *S. aequifasciatus* should be conducted without reference to the commercial aquarium hobby, which is potentially a source of 'manipulated' specimens, incorrect locality data, and other nonsense; it should, however, take more account of coloration and markings; such studies have provided useful distinguishing characters in the case of other cichlids from the same region. Meristic characters exhibit significant variation, but must be augmented using additional specimens from additional localities. The morphometrics have again hardly been studied hitherto, but appear to be less promising."

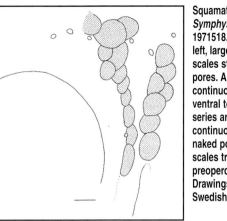

Squamation of temporal region of head of *Symphysodon aequifasciatus,* NRM THO/1971518.4071, 73.6 mm SL; anterior to the left, large semicircle skin border of orbit, scales stippled, small circles lateralis pores. Anterior vertical scale series continuous with squamation dorsal and ventral to that figured, posterior vertical series are anteriormost scales of continuous head and flank squamation; naked portion between vertical series of scales traces dorsalmost portion of preoperculum. Scale 1 mm.
Drawings: Courtesy of Kullander 1986/ Swedish Museum of Natural History.

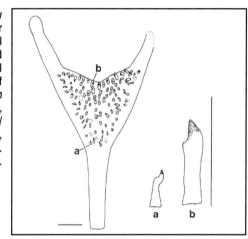

Occlusal view of lower pharyngeal tooth-plate and teeth in medial aspect of *Symphysodon aequifasciatus,* NRM SOK/1981325.3322, 110.7 mm SL. Scales 1 mm.

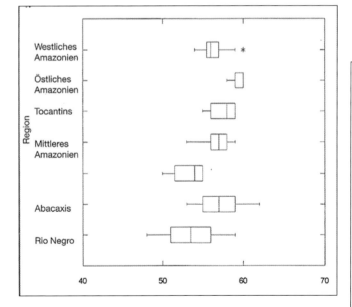

Symphysodon, scale counts of the geographical sub-groups of *S. aequifasciatus* (western, central, and eastern, Amazonia, Tocantins and Belém) and *S. discus* (Trombetas, Abacaxís, Rio Negro).

Symphysodon, geographical distribution based on museum collections.
Drawings: S. O. Kullander, 1986.

Comments on Taxonomy I
Remarks on the type localities of the species and subspecies of the genus *Symphysodon* Heckel

During a historical survey of the genus *Symphysodon*, the need for precise type locality information for some of its different taxa became evident (see also discussion in Kullander, 1986, 1996, and Klausewitz in Schmidt, 1961). It became apparent that the review by Schultz (1960) was based on too few specimens of uncertain or concealed origin (owing to the commercial value of these fishes), most of them probably coming from the tanks of an exporter in Manaus, and on identifications made without study of the types of the two species.

(1) The type locality of the type species of the genus, *S. discus*, known as the Heckel discus among aquarists, is given as "Barra do Rio-negro" (= Manaus) in the final paragraph of the original description (Heckel, 1840), although the species does not now occur at Manaus or close to it. Interestingly, it appears at first glance that the type locality is designated by the line of text following the species heading, "Moreré am Rio-Negro. Natterer." According to ancient maps, Moreré lay to the north of Barcelos on the right bank of the Rio Negro between the mouths of the Rio Cuiuini and the Rio Ararirá, which would have been on Natterer's route. However, it was Heckel's practice to give, here at the beginning of each of his descriptions, the common name of the species used in the locality where it was found. Thus Heckel's text signifies, "According to Natterer the fish is called Moreré along the Rio Negro". This is confirmed by Heckel's table (*op. cit.*, p.442) and by Fowler's Os peixes de agua doce do Brasil (4a *entraga*, *Archivas de Zoologia* vol. IX: 318, 1954), and by H. Bleher's findings on the spot (2004). Bleher found in the library at Barcelos that *moreré* was the original name for the *acará-disco*. And also that in the past (and sometimes still today) a particular animal was named after the place or region where it was discovered or occurred. Now Moreré, although no longer extant today, did exist in Natterer's day and is still shown, as Moreira, on modern maps (IBGE 1998). It follows that the type locality must be Moreré (= Moreira), on the middle Rio Negro, at 63°30'W and 0°35'S.

(2) Pellegrin (1902) examined three lots of *Symphysodon* collected by Dr. Jobert, from "Santarem, Manaus and Teffé". In 1904, he published the description of a new variety, *S. discus* var. *aequifasciata*, from specimens from Teffé and Santarem, without designating a holotype.

In 1960, L. P. Schultz, acting as first reviser, raised *S. aequifasciata* to specific rank, and selected the locality Teffé as its type locality; this last action was not performed formally, because he had not seen the types, but by indication: he founded his description on 104 specimens from Lago Teffé (now spelt Tefé), leaving aside the second locality mentioned by Pellegrin (1904).

The type material deposited in the Museum National d'Histoire Naturelle (MNHN) consists of three specimens: 2, No. 221-68-2-2, 90.5 and 122.5 mm SL, marked "Brésil", are those mentioned by Pellegrin as collected in Teffé, according to their catalogue numbers. The third specimen, No. 221-68-2-1, adult, but without size, also labelled "Brésil", is from Santarem. It is imperative (in view of the principle of taxonomic stability and the importance of the form among aquarists, where it is called the "green discus") that the action of L. P. Schultz (followed by Burgess) be complemented by the the designation of Pellegrin's two Tefé specimens as lectotype and paralectotype, and that the type locality be confirmed as Lago Tefé, on the right bank of the Rio Solimões at approximately 64°50' W and 3°20'S.

(3) In the same work, L. P. Schultz described a new subspecies of *S. aequifasciata*, wrongly termed "new species", which he named *S. aequifasciata haraldi*, based on a single specimen said to have been collected at Benjamin Constant by H. R. Axelrod and Harald Schultz. This form is known among aquarists as the "blue discus". The locality given, Benjamin Constant on a tributary of the Rio Javari, Upper Solimões basin in the region where Colombia, Peru and Brazil meet, opposite Leticia and Tabatinga, is the subject of controversy. Despite the fact that the blue discus was signalled by aquarists as originating from Leticia (under the invented name "discus tarzoo"), it appears that only the green discus is collected in its natural habitat in the upper Amazon basin (Lago Tefé, and upstream) south of the Solimões; the similarity *in vivo* of the green forms from Tefé and those from "Colombia" (Leticia, etc.) was pointed out as early as 1961 by E. Schmidt (*Tropische Fische*, Nov. 1961, pp. 516-526), while the editor (W. Klausewitz) remarked at the end of the article that the occurrence of the blue discus in the upper Amazon was yet to be delimited: "*Unklar ist bisher nur noch die Frage, auf welchen Teil des Amazonas-Oberlaufes die blaue Unterart beschränkt ist*" ("It is as yet unclear to what part of the upper course of the Amazon the blue subspecies is restricted" – but this book gives an accurate answer). It is now believed that the green form was acclimated in the vicinity of Iquitos for commercial purposes, and was thus not "naturally" present. It is now known, from several collections, that the blue discus lives in clear water, with a distribution extending at least from the Rio Purus basin downstream to the Rio Nhamundá, in the Alenquer and Tapajós region. "Benjamin Constant" is thus apparently an incorrect indication or a *lapsus* (error) on the label; and, according to ethnological films, Harald Schultz collected the blue discus for the first time in the Lagoa Beruri, a side-arm of the lower Rio Purus.

This is likely to be the "true" type locality of *Symphysodon aequifasciatus haraldi*. Note that in a 1961 paper *(Tropische Fische,* July 1961, pp 404-310), Harald Schultz mentions only the discovery of the variety in brooks with clear water deep in the forest of the upper Amazon ("*...in Büchen weit in Innern unzugänglicher Wälder des oberen Amazonas"),* without naming the river basin, a secret of the trade at that time. (And the Javari and its tributaries are permanently muddy whitewater rivers – see also Chapter 5).

(4) A second new subspecies, again wrongly termed "new species", was described by L. P. Schultz as *S. aequifasciata axelrodi,* and is called the "common discus" or "brown discus" by aquarists. The locality given for the holotype ("Belém, Brazil, Amazon River") is obviously incorrect: discus cannot live in the white water of the delta of the Amazon, and the nearest places where the brown discus actually occurs are the Rios Tocantins, Xingú, and Tapajós. As, at the time of the original description, the centre for collecting expeditions for brown discus was Santarem, it is likely that the type locality of *S. aequifasciatus axelrodi* was actually somewhere in the clearwaters of the Rio Tapajós drainage around Santar8Em. The locality given for the 57 paratypes, said to come from the Rio Urubu, is likewise suspect. Despite H. Schultz's assumption *(Tropische Fische,* July 1961, pp 304-310) that the Rio Urubu hosts the *"rein braune Variante"* (pure brown variant), the fish illustrated by L. P. Schultz *(op.cit.,* p 9, bottom) as also being *S. aequifasciatus axelrodi* from the Rio Urubu is a blue discus. And the Rio Urubu, which has approximately the same water parameters as those of the Rio Negro, is in fact home to *S. discus* in its middle and upper course, and to the blue discus, ie the subspecies *haraldi,* near its confluence with the Amazon.

In consequence, it is possible that the type material of the subspecies *axelrodi* should be restricted to its holotype.

(6) On the other hand, the most recently described taxon, *S. discus willischwartzi* Burgess 1981, is correctly described as endemic to the Rio Abacaxís, an acid (pH 4.5), blackwater river situated south of the Amazon. This has been confirmed by recent collections. Note that this river, which belongs to the Rio Maués system (which joins the Amazon upstream of the Rio Tapajós), and is also connected to the Rio Madeira in the rainy season, lies outside the remaining distribution of *S. discus*. The other forms are found in the blackwaters (in the broad sense) of the northern tributaries of the central Amazon, from the Rio Negro downstream to the Rio Trombetas.

While the type locality of *willischwartzi* is indisputable, its taxonomic status is not. *S. discus*, particularly in the Rios Abacaxís, Nhamundá, and Jatapu, is a polychromatic form like many other cichlid taxa, and any typological approach to its taxonomy seems worthless. However, the ssp. *willischwartzi* is said to have distinctly more longitudinal scales than the forms of the northern tributaries, and the river Amazon may be an effective geographical barrier for speciation.

In conclusion, if the above distribution pattern is accurate, the following hypotheses can be put forward:

– *Symphysodon discus* appears isolated by its ecological requirements: for example, in the upper part of Lake Nhamundá, where the black water of the river enters the clear water of the lake, there may be a distinct reproductive barrier. (But for further details on this see H. Bleher's remarks and findings elsewhere in this work and the DNA analyses of Stölting *et al.* (2006) in Konstanz, Germany under Prof. Axel Meyer and his collegues.)

– The nominotypical form of *S. aequifasciatus* (the green discus) could be in the process of isolation. It is to some degree differentiated from the other forms by its red anal-fin pattern, formed by irregular rows of dots, as if the uninterrupted red lines or bands seen in the other forms had been fragmented during evolution; moreover these dots extend onto the body, in contrast to the other forms. It has a relatively restricted distribution limited to the upper part of the Amazon basin, with a population in the Rio Putumayo and another extending from the basins of the Rios Japura, Jurua, Jutaí, Tefé, down to the Coari drainage. No hybrids have yet been found or suspected in nature. In the aquarium, in the hands of the best breeders, it does not interbreed freely with the other forms; in the event of success, the hybrids degenerate after the F_4 generation (but this is also the case in some other inbred forms). According to E. Schmidt *(Tropical Fish Hobbyist,* August 1962, pp 42-49), the male has a characteristic behaviour, different to that of the other forms: this may explain the putative isolation.

– By contrast, the two forms currently known as the brown and blue varieties of *Symphysodon aequifasciatus* interbreed readily in nature where their distribution patterns overlap (for example in the affluents of the lower Rio Madeira and of the middle and lower Purus, as well as in Lake Nhamunda and many other locations – see chapter 3: *Distribution).* They also interbreed in the aquarium, and produce stable hybrids, such as the much-prized Turquoise. They could thus be considered geographical varieties of *S. aequifasciatus,* despite the fact that the brown discus, like the Heckel, is said to lack the patch of scales on the head behind the eye shared by the two other varieties, the blue and the green. This character needs to be studied in F_1 hybrids as well; it is probably no longer significant after the F_1. However, breeders have confused the situation to such an extent that arguments based on crossing are extremely feeble.

The three discus species and their diagnostic characters: 1. The Heckel discus, *S. discus*, is always recognisable by its invariably prominent broad central (fifth) band, which is always covered in vermiform or horizontal, 1-3 mm wide, grey to turquoise stripes (see photo, inset). 2. The green, *S. aequifasciatus*, can be recognised by its striking rust-brown to dark red, dotted spots, which may range from just a few to extending all over the body, and may create a broken or continuous pattern of rows of dots (as can be seen on the upper body of the specimen shown here). This striking pattern is present, without exception, in the anal-fin region of adults, and no other species exhibits this colour pattern. 3. The so-called browns and blues comprise a single species, *S. haraldi*, whose colour pattern is incredibly variable but never resembles that of the two species above. It can range from yellow through yellow-brown to dark blue or black, It is normal to start with the lightest (yellow) and progress to the darkest. Yellow-brown is normally darker than yellow, and from without stripe pattern to all-over striped; it can have 9 to 16 bars (the other two species always have only nine). Individuals of this species with a striking central (fifth) bar are not all-over striped.

But as the specific isolation of the green discus *S. aequifasciatus* is demonstrated in this work (and by molecular biology methods), then this putative "good" species constituted by the blue and brown varieties should be known as *Symphysodon haraldi* Schultz, 1960 (the blue discus), the first name proposed, in accordance with the principle of priority (even if of only four pages) and the biogeographical considerations discussed above.

Needless to say, these hypotheses, based on first and second-hand information and extensive research in the field, should be enough to confimation the actual status of 3 "good" *Symphysodon* species.

Despite thousands of publications, discus are very poorly known for several reasons. They are the stars of the freshwater aquarium and, as such, they suffer from journalistic exaggeration. They live in the richest (in fishes and other groups) region of the world, the central Amazon basin, which is far from completely sampled, and only in limited areas, in great shoals protected (during the day) by the depth of the water (minimum 1 metre) and by the roots and fallen trees along the steep banks of the rivers; and they have strict ecological requirements (water not turbid, extremely poor in electrolytes, with a high iron content; relatively high temperatures; specialised nutriment composed mainly of small, bottom-dwelling arthropods, detritus, fruits, seeds, etc.) They are thus difficult to observe, to acclimate, and to induce to reproduce.

The systematics of discus, a distinctive genus with comparatively few taxa, are based mainly on their colour *in vivo*, which depends on genetics as well as environment, in a complicated schema. It has never been tackled by a specialist in cichlids (except for a few pages in a book on upper Amazonian cichlids by Kullander (1986, *The Cichlids of the Peruvian Amazon*, pp 226-231) and his paper in 1996 *(Eine weitere Übersicht der Diskusfische, Gattung* Symphysodon *Heckel*, DATZ-Sonderheft, S. 10-19) while their ecology and biogeography have never been studied by a scientist, except by Geisler (see, for example, *Aquarama* No. 59, 3. 1981, pp 37-41), who was able to spend only a few weeks during each of four expeditions to some of their biotopes (a *paraná* in the lower Rio Purus, and tributaries of the lower to middle Rio Negro). Likewise, most of the findings in the aquarium (behaviour, courtship pattern, results of hybridisation, etc.) are usually reported in aquarium journals by breeders with a "green thumb" but not trained in scientific methodology. Some, for commercial or other reasons, keep their results to themselves. Very few of these reports can be taken into consideration and most of the data are second-hand. Finally, it is a general rule, chiefly in ecology and biogeography, that negative results (such as the absence of a taxon or of a hybrid in a biotope that has not been studied thoroughly over a long period) are not significant.

But surely this monograph, the result of decades of detailed study, will provide answers to many existing questions.

By Heiko Bleher with the assistance of Jacques Géry, 2006.

COMMENTS ON TAXONOMY II
Remarks on the species of the genus *Symphysodon* Heckel

As I sat in a crowded plane flying from Miami to Guayaquil – on my way to seek out an ichthyologically unexplored region in the extreme south of Ecuador – I couldn't stop thinking about the problem that had occupied me for decades: how many *Symphysodon* species are there in reality? I had just visited my friend Jack Wattley, the world-famous discus breeder, in Florida, and (like so many other breeders – see below) he had further reinforced my long-nurtured hypothesis.

Taking into account the contradictions in the descriptions of L. P. Schultz, the results of my research in Paris (Pellegrin's specimens), Kullander's arguments, my vast amount of experience with discus in the field, statements from numerous breeders well known to me, and the observations of Dr Schmidt-Focke during 30 years of breeding, I came to the conclusion that the genus *Symphysodon* must include three good species (which DNA research would subsequently confirm), as follows:

First species: the pompadour or Heckel discus

Symphysodon discus Heckel, 1840 is unquestionably a good species and always easily recognisable. So in this case there is no argument. Its characteristic markings are well known (see page 139). All attempts to date to cross the Heckel discus with other species have sooner or later resulted in failure (the offspring were sterile at the latest after the F_3 or F_4 generation (see also Chapter 6: *Schmidt-Focke's work* and *recent breeders)*). And the aberrant forms I have found in the Jatapu-Uatumã and Marimari-Canumã regions as well as in the mouth region of the Rio Nhamundá (in Lake Nhamundá) (see Chapter 4: *The Heckel discus* and Chapter 5: *Habitats)* also proved to be sterile natural hybrids.

In addition:

– They have nine bars; are on average the smallest of the three species, and their distribution is limited to blackwaters with pH values below 5.1 (mostly between 3.8 and 4.8 – see Chapter 5: *Habitats).*

Second species: the green discus

Symphysodon aequifasciatus Pellegrin, 1904 is likewise undoubtedly a good species. Albeit restricted to the colour form with which many are familiar, the so-called green discus from the Tefé region. The description of this variant by Pellegrin *(qv)* was, in my view, where the first error in the taxonomy crept in, as the specimens collected by Jobert include one (that from Santarém) which is not a green discus, and which I assign to the third species of *Symphysodon (see below,* under *S. haraldi).*

In addition:

Compared to the other two species, *S. aequifasciatus* occurs in a relatively small distribution region in the vast Amazon basin. And to the present day (2006) it has never been caught, or even seen, with any other species or form. In other words, it is not sympatric with any other species. Natural hybrids have not been found to date.

Moreover:

S. aequifasciatus also has a very distinctive pattern of markings, which, like that of *S. discus,* is recognisable even at a distance, viz:

– *S. aequifasciatus* almost always has a pattern of red dots on the anal-fin region, which may, however, in rare cases include short red dotted lines;
– adult individuals always have red dots, sometimes only a few (1-3 mm in diameter) on the body, but sometimes (often in alpha individuals) all over the body which again in some rare cases may include red dotted lines *(see page 139);* this pattern of markings has to date not been found in any colour variant of any other species;
– *S. aequifasciatus* never attains the length or circumference of the third species, *S. haraldi (see below)* – only *S. discus* remains on average even smaller than *S. aequifasciatus;*
– *S. aequifasciatus* cannot be crossed long-term with either of the other two species listed here;
– the species has nine bars; and
– they live almost exclusively in blackwater regions with pH values of below pH 6.0 (mostly between 5.0 and 6.0 – see Chapter 5: *Habitats).*

Third species: the so-called brown and blue discus

Symphysodon haraldi Schultz, 1960. As mentioned above, one of the three specimens collected by Jobert – that from Santarém – should be assigned to this species, along with the numerous different colour forms from the central and lower Amazon basin, from totally brown discus to the blue individuals with all-over parallel striping. (Note that Pellegrin mentions in his work that one specimen (this one) has nine bars at uniform intervals like the others, but no parallel longitudinal lines (= brown), and that this was not a question of the age of the individuals, as they were of comparable size.)

In 1959-60 Harald Schultz, after whom L. P. Schultz named this species (albeit as a subspecies, *S. a. haraldi),* visited an indian tribe in the vicinity of Tabatinga, where he purportedly discovered a discus. L. P. Schultz termed this species the blue discus.

Likewise the second subspecies described by L. P. Schultz in 1960, *S. a. axelrodi,* which Axelrod supposedly caught (but which probably originated from Schwartz's establishment or from his discus collectors), should be assigned to *S. haraldi.* Schultz termed this subspecies brown, and the type material supposedly came from the Rio Urubu and Belém. (Note that the paratypes from the Urubu *(see page 129)* are blues and thus these *axelrodi* paratypes should have been designated *haraldi* following L. P. Schultz.)

The blue and brown are one and the same species, as demonstrated by the following facts:

– they have a clear-cut distribution (see Chapter 3) and are nowhere found with either of the two other species – but brown and blue are very often found together throughout the entire lower and central Amazon basin (in fact almost everywhere);
– the colour variants can range from pure brown via yellow to blue to black, and usually have either no pattern colour – not a single parallel longitudinal stripe - or stripes all over the body; just fragments (= pearl pattern) of stripes; a very fine stripe pattern (= snakeskin); but not the striking colour pattern of the other two species (and if they do, they are occasional natural hybrids – see Chapter 4);
– they can differ in size, on average from 10 to 20 cm (body length) – larger specimens are extremely rare (but in any case *S. haraldi* grows on average larger than *S. discus* or *S. aequifasciatus);*
– all the colour variants known to date are relatively easy to breed; all the popular variants are easily crossed with one another and the offspring are always fertile. Almost all the cultivated and colour forms seen today originate from this species;
– they can have nine or more bars (in some cases up to 15 or 16); this has never been recorded in either of the other two species;
– all forms of this species occur only in clearwater habitats at pH 6.0 or higher (up to 7.8).

It should also be noted that the knowledge I have accumulated over the years, reinforced on every new collecting trip, confirms that the blue and brown are just a single species. And not only the DNA analyses recently conducted by Professor Axel Meyer and his assistants, but also the recorded water parameters (see Chapter 5: *Habitats),* as well as the researches of breeders worldwide, support my theory.

Heiko Bleher, 2006.

CHAPTER

3

Distribution

THE DISTRIBUTION OF DISCUS;
DISTRIBUTION OF THE DISCUS VARIANTS;
MAPS OF THE DISCUS REGIONS (MAPS 1-8)

THE DISTRIBUTION OF DISCUS

There are almost innumerable texts on the distribution of the discus, but hardly any of the authors has ever been to South America, let alone the Amazon region. And the few that have been there didn't catch any discus. They travelled at most once or twice (only a few more frequently) to Manaus, Tefé, Alenquer, Santarém, Belém or Iquitos, and thence with the *disqueiros* (discus collectors) to a discus region, and sometimes went along on a collecting trip as well. Thereafter they published extensive texts on distribution, writing "...I caught discus there". I even know of one who took a discus from an export station, held it in his hand, and had a photo taken of himself on the beach of the Tropical Hotel in Manaus with the Rio Negro in the background (at a place where discus are never found). And the caption read, "...with his newly-discovered discus in the natural biotope." Another stated that he had travelled in Amazonia for more than 20 years and caught discus, even though he had been to Manaus just once and then to visit exporters. Or a third, who had "...been catching discus since 1947" but actually made his first visit to the Amazon in 1959 and never caught a discus then or at any time (except in Schwartz's station). The same man also stated, *inter alia,* that he had found discus at Belém (where there aren't any – see page 145 and Chapter 5: *Natural Habitats of Discus).* In addition, it is a fact that the collectors *(caboclos* and indians) as good as never give away details of their collecting localities, and certainly not those for new variants. No exporter has ever caught discus and they can only repeat the localities given by the *caboclos.*

On the basis of these few examples we can see just how inexact published data can be. We must be (very) sceptical about distribution details published previously, especially given that 38 variants supposedly originate from a single river (Rio Madeira), where no discus has ever been found. Some of the correct type localities are given only in three, at most four scientific works (Chapter 2). It was for this reason that in the 1970s I started publishing my own distribution data – in 1984-1986 even in Japanese and Chinese – and in 1992 up to 2000 in the Ringbinder *DISCUS*. And now, my resume and collected data from 50 years of discus collecting trips. On the right the distribution of the genus, geologically and politically. From the geological map it can be seen why discus occur only in the central and lower Amazon basin: they stay clear of any natural barriers such as gradients, waterfalls and rapids. The political map shows how discus are (almost entirely) "Brazilians". There is just one natural population along the Peru-Colombia border (Putumayo), one questionable (circle), and one that has been introduced in the Nanay (spot). The species are discussed on the following pages.

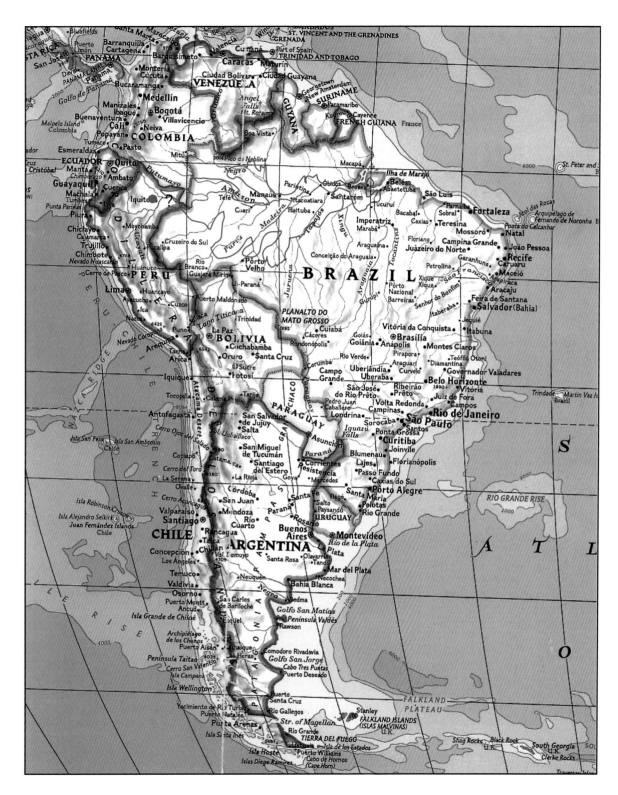

THE DISTRIBUTION OF DISCUS

BLEHER'S DISCUS

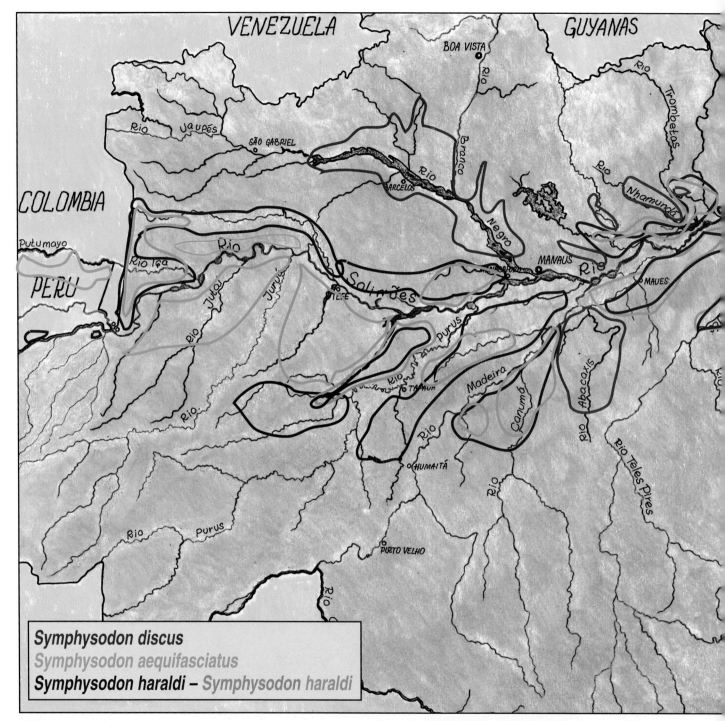

Symphysodon discus
Symphysodon aequifasciatus
Symphysodon haraldi – Symphysodon haraldi

On this map we see not only the distribution regions of the three *Symphysodon* species, *S. discus* (Heckel discus), *S. haraldi* (split into the colour forms known to aquarists as blue and brown discus), and *S. aequifasciatus* (green discus), but also the three water types: the light blue represents clearwater biotopes; the brown the muddy (sediment-rich) whitewaters (where discus are never found); and the aquatic biotopes marked in near-black are the so-called

DISTRIBUTION OF THE DISCUS VARIANTS

The map on the left shows the distribution regions, as recorded by me, of the four main, naturally-occurring discus variants known in the aquarium hobby: the Heckel discus (*S. discus*) in red, the Green (*S. aequifasciatus*) in green, and the colour variants of *S. haraldi*, the Blue and Brown, in blue and brown respectively. I have shown the latter two separately to demonstrate the extent to which the so-called Blues and Browns intersect in the wild (and to reinforce the point that division at species or subspecies level makes no sense). From the map it can clearly be seen that none of the three good species live together in the wild. (The detailed distribution and localities of the individual species/colour forms can be seen on the following eight pairs of pages.)

Heckel discus (*Symphysodon discus*). The distribution is restricted exclusively to blackwater biotopes in five regions: 1. The northern and southern affluents of the middle and lower Rio Negro. In part also in the Rio Negro, where it was first discovered – but not in the Rio Branco, a whitewater river. 2. In the Rio Abacaxís area and its blackwater affluents. 3. In the middle and lower Rio Jatapú region. 4. In the Rio Nhamundá. 5. In the the blackwater biotopes of the lower Rio Trombetas basin. (Only in the last few years have I been able to discover new localities, eg in the Jatapú, a Uatumá-tributary, in the Marimari and in the lower Nhamundá). Heckel discus are found only at pH values of less than 5.0 (usually 3.8-4.5), and only beneath bushes of the genus *Licana*. Note that *S. discus* was formerly found in the vicinity of Manaus, but there have not been any there for a long time. (The capital of Amazonia is now home to more than 3 million people.)

Green discus (*Symphysodon aequitasclatus*). The Green has the smallest distribution of the three species. To date it has been found only in the Coari region (Rio Curuá – the easternmost extent of its distribution), the Tefé region, the middle and lower Japurá regions, in the vicinity of the lower Juruá, as well as in the Jutaí and its affluents plus the Tonantins and Rio Amaturá. Its natural distribution is very clearly defined, the only exception is the population in the Rio Putumayo (Peru-Colombia). The Nanay population consists of translocated individuals from Lago Tefé.

Brown discus (*Symphysodon haraldi*). Brown discus – those with no or only a few pattern markings – are restricted to clearwater biotopes and are encountered throughout almost all of the middle and lower Amazon basin. So far I have been able to find variants with more pattern/stripes only upstream of the Alenquer region, but never in so-called white water (at most in "mixed water" during the rainy period when the Solimões rises). Numerous colour variants from the clear tributaries and *lagoas* of the Rio Madeira and Purus region, the Manacapuru, the Lago Nhamundá (also called Lago de Faro), to name but a few, are also classed as Browns. These are often (almost always) found in regions that are defined as part of the distribution of the Blue. I have found Brown variants in the Rio Içá region and even further west (see Chapters 4 & 5).

Blue discus (*Symphysodon haraldi*). These have the widest distribution of all the colour forms (or species), and are undoubtedly even more widely distributed than we know at present. They live in the same waters as Browns, often together with or overlapping the latter (demonstrably one and the same species). Their highest concentrations are encountered in seven regions: 1. Trombetas and Lago Nhamundá (only in clearwater biotopes). 2. Maués. 3. Madeira tributaries and *lagoas*. 4. In the clear and tea-brown affluents of the Rio Purus and its numerous *lagoas*. 5. From the Manacapuru and Anamá regions to the Coari area. 6. North of the Solimões to the Rio Içá and Tabatinga region (Brazil) and north to the Rio Apaporis.

blackwater rivers and lakes. Note that there are no discus in the whitewater regions even where these lie within the discus distribution regions shown on this map! For detailed distribution see the following eight maps.

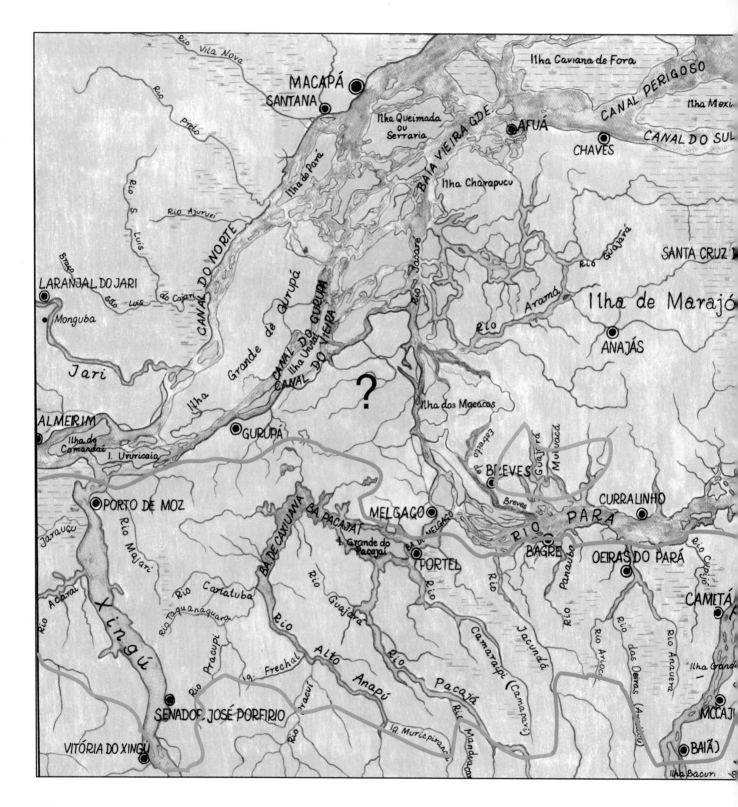

Map number 1 of the discus regions:
Belém and vicinity; Rio Tocantins; the island of Marajó and the Jarí; and the Rio Xingú.

There is no doubt that there have never been any discus in the vicinity of Belém. All the specimens deposited in museums and elsewhere with the locality given as "Belém" are inaccurately labelled, and just as incorrect as, for example, the locality "Atlantic Ocean" (with which a number of freshwater fish species have actually been deposited). I have travelled the region around Belém practically every year since 1965 – often several times per year – and tried to find discus, but the nearest place where *S. haraldi* (brown variants – almost without pattern colour) are to be found is in the lower Rio Tocantins.

Here in the clearwater river discus are found along the island region (eg Ilha Grande do Jutaí and Ilha do Bacuri – the latter is to date the most southerly known discus distribution in the region – it lies south of Baião), as well as in the still intact, often blackwater, branches and lagoons (eg the Rio Parurú or the Igarapé do Grito). However, my researches indicate that they are no longer collected for export (see Chapter 5) following the construction of the gigantic Tucuruí hydro-electric dam in 1984. The Tocantins region also constitutes the easternmost boundary of the distribution of the genus.

Very little is known of the Rios Cupijó, Anauera, and Araticú (also known as the Rio das Oeiras) further to the west, or of the Panaúba drainage region. I learned from the local *caboclos ribeirinhos* that they had seen but didn't catch *acará-disco,* and I personally have so far been unable to find them there. Though I have been successful (starting in the 1960s) in the region of Breves, at the western end of Marajó, the largest river island on Earth, where the "brownest" of all the *S. haraldi* variants are found. Plain solid brown without any pattern colour. They live in clear water (Igarapé Grande) which is, however, also termed *agua escura* (dark water) locally. I know of no other locality for them on the island. In the eastern steppe region of Marajo discus are absent – no habitat as well as very large tidal influences. Likewise they are not found in the northern Amazon tributaries in the state of Amapá, or in the Rio Jari, where typical biotopes are generally lacking (or have been destroyed – see Chapter 5), and in addition the numerous floods, rapids, cataracts, and waterfalls prevent their occurrence there.

In the gigantic aquatic region of Portel, where the great Rios Jacundá, Camaraipi (Camaraipe or Camapari), and Alto Anapú flow from the south into the Baía das Bocas, the Baía do Melgaço, and the vast Baía Caxiuana respectively, the genus is represented by the species *S. haraldi* (brown individuals), likewise in the Rio Pacajá. In the region south of Portel, in clearwater lagoons (Rio Camaraipi region) and in the Rios Cariatuba and Taquanaquara (both left-hand affluents of the Alto Anapú), which at high water are connected with the Rio Majari (or Matari), a right-hand arm of the Xingú, I found almost circular brown *S. haraldi.* Very similar to the variant that occurs in the lower Xingú.

But this region too is largely (ichthyologically) unexplored and I personally have been there only twice. The distribution of the discus is adversely affected by the muddy Amazon water – expecially during the rainy season, when the major part of the region is a whitewater zone and the discus are generally obliged to retire to the *igapós* and *lagoas*.

The genus *Symphysodon* (at the time thought to be *S. discus* but actually *S. haraldi*) was first discovered in the lower Rio Xingú, in the vicinity of of Porto de Moz, by the Thayer Expedition (1865-66). They are circular fishes, just like the discus of athletics, and brown with a lighter shade of colour (often verging on yellowish to quite yellow). During the 20th century not only myself, but also other researchers and collectors, repeatedly managed to find discus there, upstream as far as Victória do Xingú. The American Mike Goulding even found an individual in the "bend" of the Xingú at Belo Monte, downstream of the huge rapids – where a hydro-electric dam is now being built. There are no discus anywhere in the entire Xingú basin upstream of the falls. Nor in the actual mouth region (= white water), and there are likewise none known from the Almeirim region opposite.

Map number 2 of the discus regions:
Rio Paru to Santarém; Santarém-Tapajós region; Alenquer region; Óbidos and vicinity.

Although the Rio Paru de Este (there is also a Paru do Oeste to the west) is cited as a locality for discus in popular publications (and on the Internet), no *Symphysodon* have been found there to date, either by myself or by scientists or discus collectors (see also Chapter 5). There are virtually no typical biotopes but numerous rapids. Likewise no discus are known from the next left-hand Amazon tributaries, the Urumu and Jutaí, which lie partially isolated between spurs of the Guyana Shield. However, I have found browns with only a small amount of pattern in the lowlands opposite, in the Lagoa Urubuquá. There is no evidence of discus in the Rio Parauaquara (also known as the Curminaú), and the *lagos* along the middle Rio Jauari have so far not been ichthyologically investigated. In the Lago Maripá, west of Prainha, I found Browns with variable pattern colour. During a research trip in the 1970s I was unable to find any discus in the numerous lakes and *lagos* on the right-hand side of the Amazon and in the Pará do Uruará. But at the beginning of 2004 the local ribeirinhos told me that they saw *acará-disco* there during the dry season. And in the Rio Cucari, as well as the Lagos Ururu and Maraco. All contain typical discus habitats but are inundated with muddy Amazon water during the high-water period. The lower Curuá-Una was formerly home to more discus, although since the hydroelectric station was opened in 1977 there has been no further sign of them.

The Lago Grande de Monte Alegre (in the north), along with the Rio Maicuru and its numerous *lagos*, harbours an interestingly patterned *S. haraldi* variant, in which the majority of males exhibit considerably more pattern colour than the females. And from here on, heading west, colour forms of *S. haraldi* are widespread.

After my first research trip (1975) to this region, discus variants from numerous *lagos* and rivers appeared on the market as "Alenquer". These were caught (and to some extent still are) in the area around Alenquer, in the Lagos Aningau (splendid Reds with a red eye, but the lake is accessible only by land for practically the entire year), Araparí, Uruchi, Curumú, Capintuba (or Capituba), Paracarí, Samauma, do Jauari, and Cuipeuá (called Curipera or similar on the Internet and in discus literature). In the Rio Curuá (cited as Barra Mansa in the popular literature, but the name refers to a small village on the Curuá and no discus are found there) as well as in the northern part of the Lago Grande. I have visited all these sites during the dry and rainy seasons.

Very attractively striped discus are also found in the vicinity of the settlements of Corréa and Jaraquítuba, as well as at other places in the region.[1] Discus variants are encountered almost everywhere here during the dry season (end July to October – sometimes November), except in the huge Lagos dos Botos and Itandéua, as well as two large *lagos*, Itarim and Pacoval, in the Ilha Grande do Tapará, as these all contain muddy water for practically the entire year. It must be borne in mind, however, that when the water rises in this region many of the clearwater *lagos* (and the Curuá for many kilometres upstream) are inundated by the waters of the Amazon, and accordingly the discus retreat into the mixed-water zones or clearwater *igapós*.

There are similar colour variants in the Santarém region, although not with such an intense pattern colour as in some individuals from the Alenquer region. Data on the Internet or in the discus literature are misleading, as can be demonstrated by records of collections (since 1928 – Chapter 1) as well as specimens deposited in museums. The majority of discus come from the Lagos Verde and Maicá and additional lakes to the east. From the Rio Ituquí and the surrounding area, as well as the lower Rio Tapajós region – which contains clear water all year round (it is also termed *verde* (= green)). And from the Lago Grande de Santarém, which is so large that the shore cannot be seen from its centre. I personally have found lots in this area, even in the Arapiuns, a blackwater tributary, where I found a very large brown – almost grey – discus of more than 22 cm total length. The most southerly distribution limit lies in the Tapajós region, barely 200 km upstream, before Itaituba. No discus are known from the out-size lakes further west, the Lago do Poçâo Grande and Lago Grande do Curuaí (Luxemburg could be submerged in the latter!), which contain white water for the entire year. Their clearwater affluents have not been investigated. But in the vicinity of Óbidos, in the Mamaurú, called lago de agua preta but contains rather clear water, I discovered some very beautiful discus with pattern colour – similar to some Alenquer-region individuals. Otherwise there are no discus to be found in the vicinity of of Óbidos, except in the Lago Curumú. All whitewater.

---

[1] The localities cited by importers, eg Mata Limpa, Traira, Inanu, Mangal, Itelvina, Jaraki, Carananzaru, Curipera, or Lago Preto, must all be invented names. None of the locals has heard of them, nor has the IBGE (Instituto Brasileiro de Geographia). The exporter in question declines to comment, likewise the importers who offer discus with these locality names.

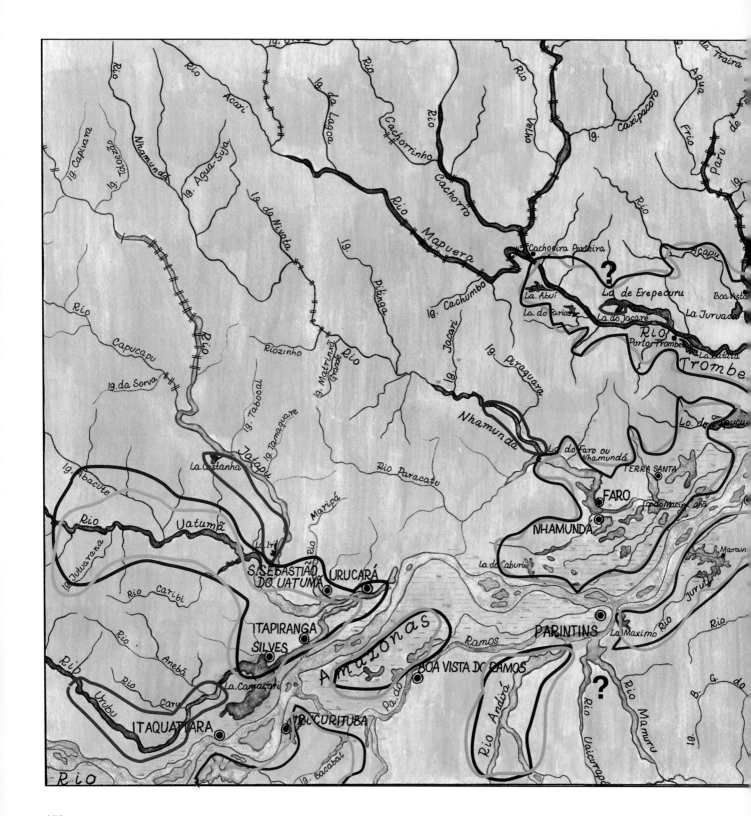

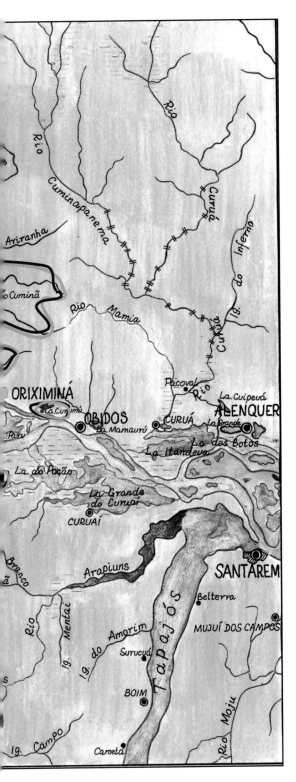

Map number 3 of the discus regions:
Oriximiná and Trombetas region; Rio Nhamundá region; Parintins region; Rio Uatumã; and Rio Urubu.

Unlike Óbidos, Oriximiná[1] lies on a clearwater river, the Rio Trombetas, although the local people call it *agua preta* (= black water – and it is shown as such). The truth is that the upper Trombetas, its main branch (the Mapueira) and the majority of its affluents, as well as the large Lagoa do Erepecuru, do in fact contain genuine black water. The Trombetas region has been known for discus since my discovery of them there in 1965 (Chapter 1), confirmed in the same and the following year by a scientific expedition. In fact the ichthyologists also investigated the Lago Jacaré (or Lagoa do Jacaré) at the western end of the Erepecuru lakes, where they collected three specimens of *S. discus*, the Heckel discus, preserved in the MZUSP. Interestingly, this is the first time (upstream from the mouth of the Amazon) that we encounter this, the first-described discus species. It is the easternmost known population of *S. discus*, virtually isolated (like that in the isolated Abacaxís region), and hence interesting as only *S. haraldi* is found throughout the entire Trombetas region from here downstream. I was able to find *S. haraldi* variants (brown, blue, and fully-striped) in the following *lagos*[2]:

1. On the right-hand side in the Lagos Batata (where there are definitely no *S. discus*, but fully-striped individuals of *S. haraldi*), Água Fria, Acarí, Samauma I & II, Apé, Jibóia, Arajasal, Achipicá, Camija (or Camicha), Jereuá, and Sacurí, and in the two Carimó lakes (I & II).

2. On the left-hand side in the *lagos* and *lagoas* Tarumá, Bacabaú I, Vajão, Bacabaú II, Cuminá, Agua Fria, Acapuzinho, Curupira, Castanha, Xiriri, Caipurú, Iripixi, Paraquí, and Parauacú. No discus are known from the lower Rio Paru do Oeste and its rich region of lakes, but they are found there, or so the *ribeirinhos* told me. At least from the *lugarejo* (tiny settlement) of Delfina downstream. The lake region south of Oriximiná is connected to the Amazon and contains white water all year, and thus offers no suitable habitats. But *S. haraldi* is found in the western branches of the gigantic Lago Sapucuá. Generally individuals with pattern colour, only rarely fully-striped.

On the right-hand side of the Amazon I have found dark brown discus (*S. haraldi*) in the blackwater biotope at the foot of the Serra Capiranga, east of Juruti. But discus also occur in the clearwater biotopes of the Rio Juruti and its lakes. Some of them with pattern colours.

The next large lacustrine distribution region of the genus to the west is the gigantic Apaes de Nhamundá region, unexplored until the mid 1990s, which contains a network of almost innumerable *lagos* (enormous, large, and small), endless channels, and *rios*[2]. An incredible aquatic labyrinth. Discus occur in almost all these bodies of clear water, except where the waters of the Amazon penetrate all year or there are cattle pastures. Only *S. haraldi* variants, ranging from colourless to light brown through pink to vivid red. Individuals ranging from those with turquoise spots or a few stripes to fully-striped and Royal Blue. Even so-called "snake-skin" (see Chapters 1 & 5). *S. discus* is found in the Rio Nhamundá, and only there. (But not together with *S. haraldi*). The Lake Nhamundá is a melting-pot for the genus. No other distribution region known to date has so many variants of *S. haraldi*, plus *S. discus* close by in its affluent. However, during the high-water period many of the numerous *lagos*, *paranás*, and *rios* of the region become turbid with the muddy waters of the rising Amazon, whose influence sometimes extends even to the Lake Nhamundá (once every 10 years in recent times).

The next known distribution is in the Lago Maximo (browns with beautiful pattern colour), east of Parintins. A lovely clearwater lake. But browns have also been found in the Lago Zé-Açu[2] The Rio Andirá harbours half-striped blues, sometimes very beautiful. This too is a clearwater river. To date no discus have been reported from the two *rios* south of Parintins, the Mamuru and the Uaicurapá, which flow from further south. However, there are undoubtedly discus there, because of veryfied habitats. Likewise in the large lake region north and west of Boa Visa dos Ramos, although this area is inundated with muddy Amazon water for most of the year, and contains numerous cattle pastures but not very many discus biotopes.

The Uatumã-Jatapu region is a further major centre for discus. So far I have records of Heckel discus at several sites (eg Lagos Iri & Castanha) in the Jatapu region. And innumerable brown *S. haraldi*, from partly to all-over striped, in the lower Uatumã. The Jatapu has Heckel discus upstream to the Lago Castanha. Blues are caught almost exclusively in the Uatumã – usually just off São Sebastião do Uatumã[3]. There is virtually no commercial collecting in the Rio Urubu, where I discovered very attractive, fully-striped discus, in the mouth region and in the Lago Camaçari. The Rio Urubu itself is home of Heckel-discus.

[1]There is no Rio Oriximiná, as often stated. [2]It is not possible to show all the tiny *lagos*, *lagoas*, and *igarapés* here – see also Chapter 5. [3] The discus (*S. haraldi*) offered on the Internet and by importers as Jatapu A, Jatapu B, etc, all come from the lower Uatumã region. Foreigners are taken to the Uatumã and told it is the Jatapu.

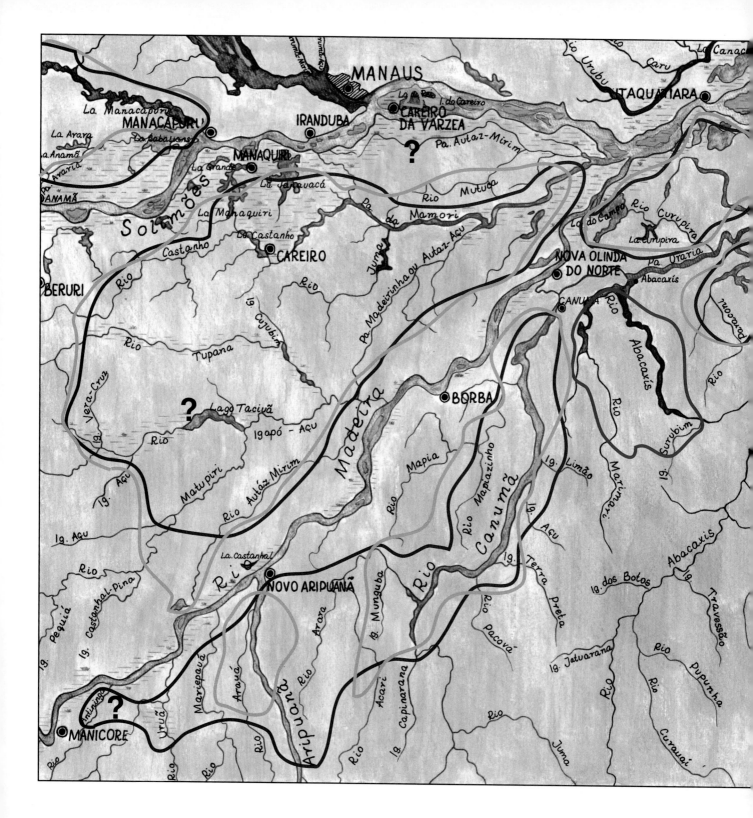

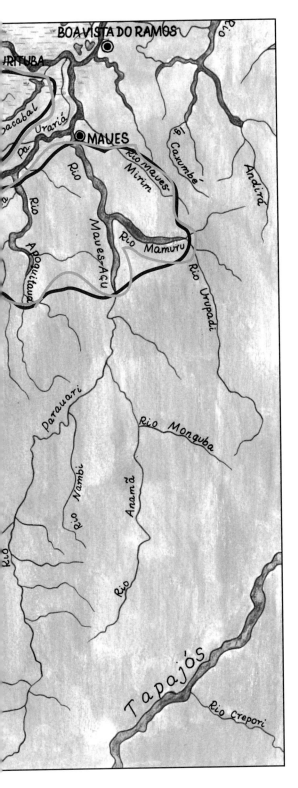

Map number 4 of the discus regions:
Maués region and Rio Abacaxis; Rio Madeira region; and Manacapuru region.

In the region of Maués, blue and brown discus *(S. haraldi)* are found in the Rios Maués-Mirim, Maués-Açu, Mamuru (also known as the Maraú), Parauarí (often with fully-striped individuals), Apoquitauá, and Paraconí (sometimes called Pacoval), as well as the Lago do Campo, Lago (and river) Curupira, Lago Grande and Lago Curuçá, and the smaller *lagos* to the left and right of the long whitewater Paraná Uraría (which links the Madeira with the Amazon all year), eg Maria Curupira, Maripauá, Juruparí, and Amaniú (and many more). From this large region come numerous striped individuals with attractive pattern colour, but also some that exhibit hardly any stripes. The rivers and lakes are usually termed *água preta* by the *ribeirinhos* (and also on many maps), but basically they are all clearwaters. Including the above-mentioned rivers flowing from the south.

Only the waters of the Rio Abacaxís and Rio Marimari constitute an apparent exception. Both are black (like the Rio Negro) and harbour *S. discus*, the Heckel discus. Just as in the Nhamundá and Urubu region (map 3), the two species almost overlap here. The effects of water chemistry can be really interesting: at one spot in the Rio Marimari, where I found Heckel discus on the right-hand shore, a *paraná* enters the river on the left-hand side, forming a navigable link with the Rio Canumá during the high water period. And this contained blue discus *(S. haraldi)* – only 400-500 m from the Heckel discus collecting locality. But the water parameters were totally different (more on this in Chapter 5).

Thus the Abacaxís and the Marimari are pure Heckel discus habitats and no other forms are found in them for a long way upstream. In the Rio Canumá, by contrast, there are lovely Blues, some with very striking pattern colour and broad stripes, for a long way upstream. This too is a clearwater river but is often shown or referred to as blackwater. The same applies the Rio Aripuanã further to the south – also termed *preto* but actually clear.

I have often found very light (yellowish) discus *(S. haraldi)* with little pattern colour in the Aripuanã, upstream as far as the Rio Jumá (bottom of map). I know of no *Symphysodon* further south, although the mayor of Manicoré told me that there are *acará disco escuro* (ie dark discus) in the adjoining region, that of the Rio Atininga. This may well be the most southerly distribution of the genus in the Madeira drainage region (there are none in the Madeira itself, which is a muddy white-water river).

In the gigantic lake and river region[1] between the Rio Madeira and the Rio Solimões (whose details are not included here) there are very numerous clearwater and *água escura* (dark water) biotopes where discus occur – but also permanent whitewater zones. I discovered an interesting reddish variant of *S. haraldi* with very little pattern colour in the large, branched Lago Juma, through which flows the river of the same name. Blues *(S. haraldi)* are known from the Lago Castanhal opposite Nova Aripuanã. *Symphysodon haraldi* have been recorded in the Lago Taciuã (whose eastern part is also known as Lago Grande and the western as Lago Manianrã), in the lake region of the Rio Igapó-Açú and Tupanas, and in the Lagos Castanho and Manaquiri – although museum deposits are labelled *S. aequifasciatus* or *S. discus*. The majority of these specimens from the region are brown individuals without any prominent patterning, but discus with a partial stripe pattern are also (or used to be) caught. And their distribution is, in my experience, almost everywhere in this regions clear waters with discus habitats.

A single specimen (of *S. haraldi?*) is known, from 1987, from the Ilha do Rei, the gigantic island which begins practically at the mouth of the Rio Negro where it joins the Solimões/Amazon, but I know nothing of its colour pattern. And the locality is problematical, as the gigantic Lago do Rei almost always contains white water.

The Manacapuru region has been world famous for discus since 1965 and my discovery there of the Royal Blue. (Only today hardly any fully-striped specimens are caught there). They are found in the *rio* itself and the innumerable *lagos*, in the Lago Grande and further upstream, and in the upper Rio Manacapuru. The colour spectrum ranges from light brown without stripes to all-over-striped, and even includes turquoise-spotted individuals.

---

[1] The region of *lagos, paranãs, rios,* and *igarapés* between the Madeira in the east, the Amazon in the north-east, the Solimões in the north-west, and the Purus in the west is gigantic and cannot be shown on a map because of the considerable changes that take place seasonally.

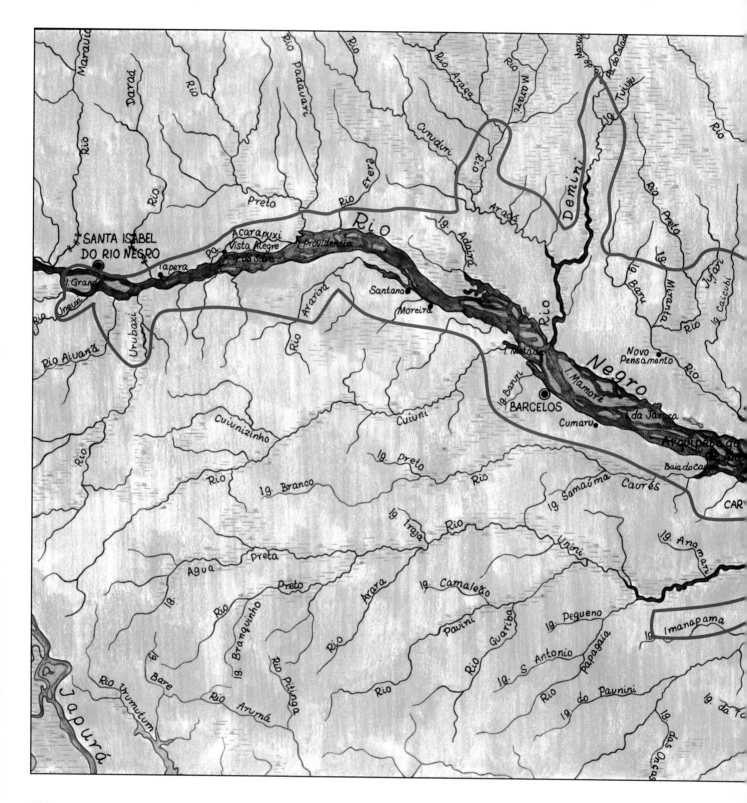

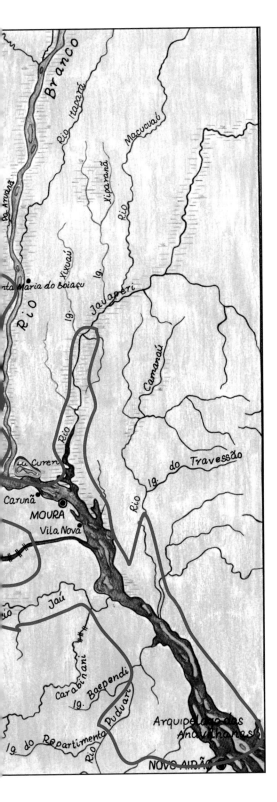

Map number 5 of the discus regions:
Rio Negro region; and Rio Branco region.

The Rio Negro and its drainage are reserved almost exclusively for the Heckel discus (*S. discus*), whose distribution lies in the black water of this gigantic river. I (and, of course, many others) have recorded it at the places listed below (although it is undoubtedly more widespread – I believe almost throughout the entire lower Rio Negro region):

Naturalists collected Heckel discus very early on (C. Jobert around 1878; D. Merlin *et al.* in 1923; and J. D. Haseman (precise date unknown)) in the mouth region of the Rio Negro (around Manaus, then known as Barra), and in the Tarumã system. Nowadays discus are no longer to be found in the vicinity of the Amazon metropolis – or at most odd individuals that may have escaped from an exporter. The habitats in the Tarumã have been destroyed, the river and its affluents polluted beyond redemption. The most southerly natural distribution today appears to lie at Novo Arião, in the Ariaú (which some people call the Ariuaú) and around the Archipelago das Anavilhanas.

They occur further upstream (left-hand Rio Negro affluents) in the system of the Rio Apuau, in the Rio Canamaú, and in the Jauaperí upstream to the Igarapé Xixuaú – in the indian reserve. In the last-named river I discovered a Heckel discus variant with a fine all-over stripe pattern, while almost everywhere in the other two the typical Rio Negro Heckel discus population, which we know from hundreds of publications, is found, although nowadays the majority are collected in the region around Boca do Rio Branco (the mouth region of the only large, left-hand, whitewater tributary of the Rio Negro). Here, mainly on the right-hand, Rio Branco side, we find a contrasting blackwater biotope only metres from the shore of the muddy Rio Branco. A natural spectacle, which puts the *encontros das águas,* the confluence of the Rio Negro with the Rio Solimões, (and that of the Tapajós with the Amazon) well and truly in the shade. And here, right next to "hostile" habitats, lives the true "King", the Heckel discus. (Because it was the first discus discovered, it deserves the title even though live individuals didn't reach the aquarium hobby until the beginning of the 1960s – long after the brown/blue and even later than the first greens.) Heckels are found in the majority of the blackwater *lagos* and *igarapés* on the right-hand side of the lower Rio Branco. I have found the blue-headed Heckel discus only in the glorious Rio Xeriuini (Xeruini), which runs almost parallel to the Branco, initially in the 1980s and again in the 1990s.

The next left-hand tributary is the Rio Jufari, in whose lower course and branches typical Rio Negro Heckel discus are found. A lovely Heckel discus variant with a wavy stripe pattern occurs in the next river, the Demini (also known as the Demene, Demine, Deme, and other names); I recorded this form in 1988 (as well as another with fewer "wavy lines" in the Aracá, a left-hand tributary), albeit only upstream as far as the Paraná do Marium/Paraná do Calado. This is apparently the overall northernmost limit of the distribution of the Heckel discus. To date none have been found further upstream. But they do occur opposite Barcelos, in the Igarapés Peixe-Boi, Ariaú, Tukano, Zamula, and Buibui, and in the gigantic mouth region of the Igarapé Andairá. Likewise in the lowest part of the Ereré, plus a very attractive light variant in the Rio Padauarí. By the Paraná Acarapuxí at Visa Alegre the species is already less frequent, and the Rio Negro distribution appears to end round about the island of Ilha Grande in the Rio Uneuxi, a right-hand Rio Negro tributary (= westernmost limit of the species' distribution). And on this side I have found Heckel discus in the lower Urubaxí; in the Rio Arariá; in the Cuiuni (also called the Quiuni) – splendid, majestic, light-blue, all-over-striped individuals; then a light, almost yellow form in the Baruri in the vicinity of Barcelos; in the Baía do Caurés; a very broad-striped form (that with the broadest stripes of all the forms known to date) in the lower Rio Unini – well below the first rapids; in the Jaú – the typical Rio Negro form; and in the lower Rio Puduari (not to be confused with the northern Padauari). As already mentioned, in recent years I have no longer been able to find any south of the Archipelago dos Anavilhanas. But the Rio Branco phenomenon still remains.

As frequently mentioned already, there are essentially no discus in whitewater rivers or lakes. I have been able to demonstrate this unequivocally (for the reasons see Chapter 5). They cannot survive there (immediately eaten by large predators). But, back at the beginning of the 1980s, I found a half-striped individual, not at all like a Heckel discus in appearance, in the Lago Cureru – in the immediate vicinity of the mouth of the Rio Branco and its inundation zones. Unfortunately since then I have not been back to this rather clear, isolated lake. Either an isolated population of *S. haraldi* lives there, or the specimen was a natural hybrid. (I hope to find out more soon...)

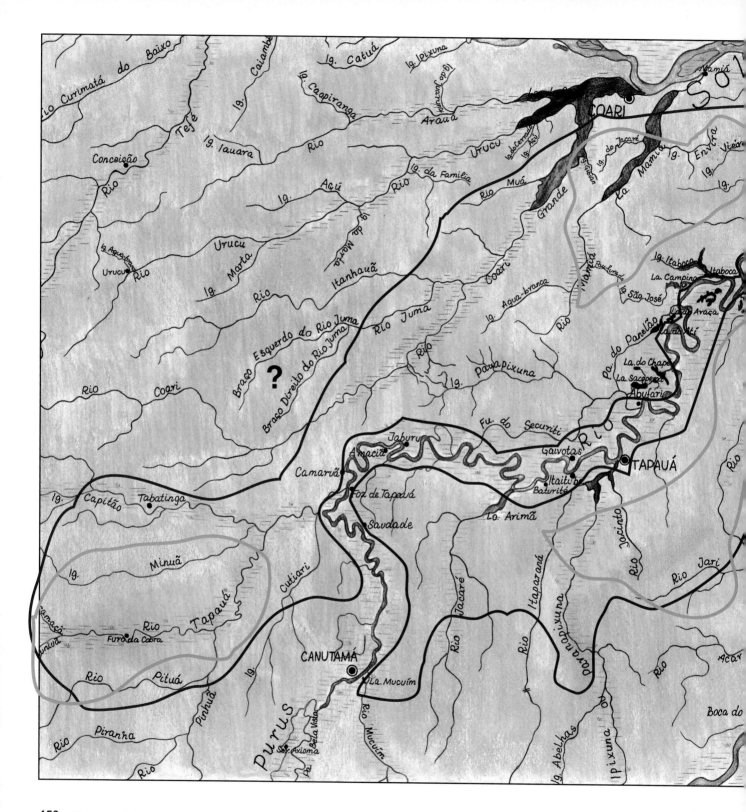

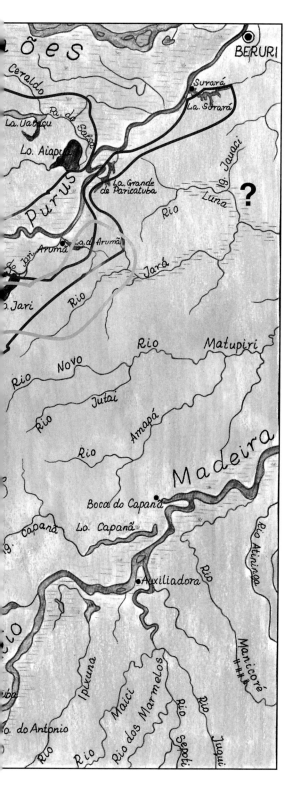

Map number 6 of the discus regions:
Purus region; and Tapauá region.

Since the end of the 1960s the Purus region has been renowned for blue discus. Although, as in the case of "Alenquer" and "Santarém" discus, the term "Purus discus" relates to the region and doesn't mean that discus live in the river or town in question. The Purus is a muddy river except during the dry season in places where it continues to flow, and downstream of the mouth of the Rio Jari where water draining from the numerous tributaries and blackwater lakes transforms it into a clearwater river.

The details are as follows – as far as I have ascertained to date – beginning with the left-hand bank, travelling upstream: There are *S. haraldi* – Blues, often even all-over striped individuals – right away, in the Lago Aiapuá, a blackwater lake. And there are also various Blues in the numerous medium-sized and small clearwater lakes to the north (not shown here) upstream to the Lago Uauaçu. Likewise in the Lagos Piraiauara, do Sacado, Caãpiranga, Comprido, Bacurí, Itaboca (and its *igarapé),* and Campainha – which has a *furo* (link) to the Rio Mamiá, via the Igarapés São José and do Pão Furado, where blue variants occur as well. Further along the left-hand side I have found equally attractive individuals in the virtually isolated lakes (accessible only by land during the dry season) Araçá, do Atí, and Panelão (along with the *igarapé* of the same name). But here they are all similar, with more or less broad stripes on the upper and lower body. Upstream of Abufari are the Lagos do Chapéu and Sacopema. These are fed by blackwater streams – the Igarapé Pauapixuna, the Furo do Securiti, and the Paraná do Abufari – and *S. haraldi* are to be found (almost) everywhere there, including Blues with more or less of a stripe pattern.

I haven't investigated the Lago da Ponta at Tapauá, but I have been to the Lago Solitário (only accessible by land and not shown on any maps), where a less strongly patterned variant occurs. Further upstream we find the Lagos Quatí, do Tambaquí, Mapixi, and do Cachimbo, among others, but none of them has been investigated. However, the fishermen in Tapauá say that they all harbour *ácara-disco*. In the Rio Tapauá, which contains clear water (though it is termed *água preta),* there are completely blue discus and Royal Blues. Likewise in the Capitão and its main affluent, the Igarapé Minuã. And, back in the 1980s, I found particularly numerous *S. haraldi,* some totally without pattern colour (= browns) in the main tributary of the Tapauá, the Rio Cunhuá (also called the Cuniuá). Also in the lower course of the Rio Camaçá (the most southerly distribution of the genus?). Nothing is known about the *lagos* further south. A few are extremely isolated, eg the horseshoe-shaped Lago Riberâo, the curved Lago Caratiá, the serpentine Lago do Ronca, and others; but they may harbour discus. Likewise nothing is known about the region around Canutamá.

Along the right-hand bank of the Purus, heading upstream from the mouth, we find Blues right away in the following *lagos:* Surará (south of Beruri); Ubim (with regular inundation); Ipiranguinha along with Ipiranga to the southwest; do Matias; and Cavania (which often forms a unit with the Purus, whereupon the discus disappear into the *igapós).* After walking cross-country I found attractive blues in the isolated Lago Água Fria (whose temperature never drops below 26.7 °C). *S. haraldi* also occurs in the huge clearwater Lago Grande de Paricatuba and the isolated Arapari; the Lago do Arumã; the Lago do Mira; and in several smaller, more isolated lakes around the mouth region of the Rio Jari.

In the Rio Jari itself, a splendid large blackwater river with a gigantic lake, there are very large numbers of discus, ranging from brown to blue, and from without pattern colour to almost fully-striped, lighter-coloured individuals. They are found far upstream and in the smaller *lagos* and *lagoas* (not shown), as well as in the larger *igarapés*. Between the Jari and the Purus there are almost innumerable *lagos,* in all of which *Symphysodon* ought to occur – at least the *caboclos* assert this is the case. I have recorded them in the Lagos Barbara, Jurará, and Jenipapo Segundo; in the large Lago Pereira; in the Lagos Baixo and Cabeçeira Grande, and in the isolated Tambaquizinho; and in Lago Pupunha and a number of nameless lakes. On this side of the Purus, around Tapauá, I was unable to find any in the Lago Ausencia, but I was successful further upstream in its blackwater tributaries, the Rio Jacinto and the Parapixuna (almost as far as the Labrea-Humaitá road); in the large Lago and the Rio itaparaná (crossing the lake), where numerous individuals exhibit a stripe pattern, but sometimes only fragments of one (such fishes are called *pintados).* Also in the smaller lakes such as Ancorí and Jatuarana; in the larger Lago Arimã; as well as in the blackwater Rio Jacaré.

Further to the south this side of the Purus has been little explored, but the *riberinhos* say that there are *ácara-disco* in the isolated *lagos* such as Jamarí, Sobral, Canariã, Tamanduá, and Jadibaru. There are attractively patterned discus (none fully-striped) in the Lago Mucuím as well as in the Rio Mucuím. According to the local *seringueiros* (from the Seringal Axioma) there are discus in the Paraná Bela Visa, which is connected to numerous *lagos*. So far none have been found further south.

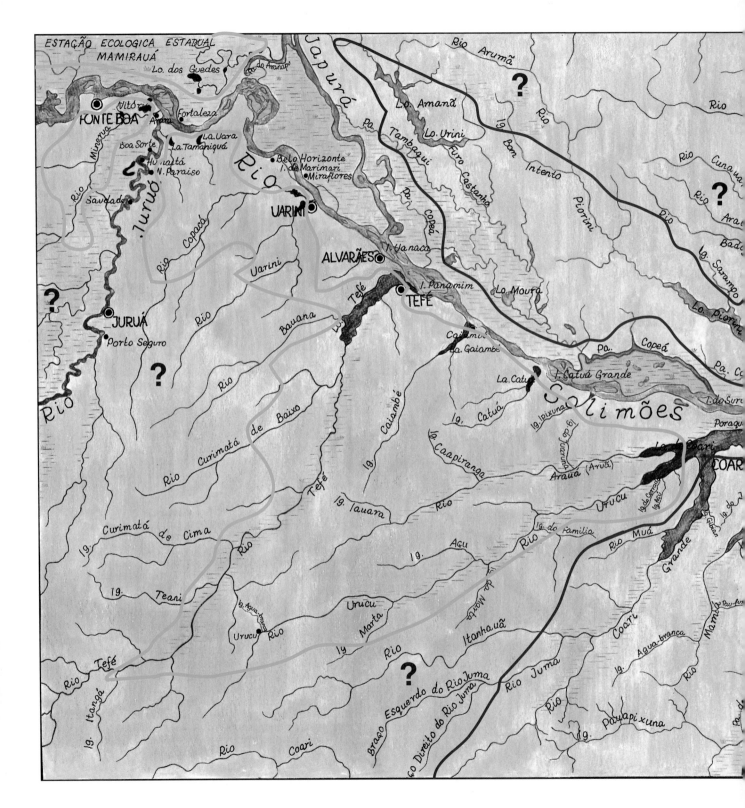

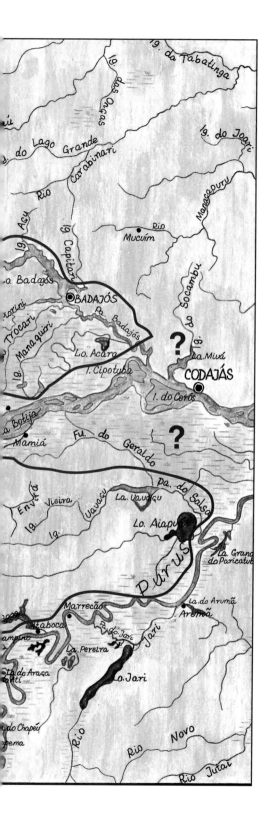

Map number 7 of the discus regions:
Coari Region; Tefé region; Rio Japurá; Rio Juruá region.

Blue discus are well known from the Lago Arara and the large Lago Anamã (both on map 4, and both regarded as blackwater lakes). No discus have yet been found (as far as I know) in the vicinity of Anori (not shown on the map) or in the Codajás region, including the Igarapé do Socambu (also called the Miuá), which feeds the gigantic Lago Miuá. These are clearwater regions which, however, like many others, are made turbid for months on end by the influence of floodwater from the muddy Amazon (Solimões). Further west lies the Lago Badajós, which extends for more than 70 km. Here there are blue discus – similar to the Lago Grande de Manacapuru variants. *S. haraldi* has also been found in the Igarapé Açú, the Rio Badajós, and the Cunauaru, as well as in the Igarapé Manaquari and its *lagos*. By contrast, the approximately 80 km long Lago Piorini and its affluents are so far virtually unexplored, but local *ribeirinhos* and *pescadores* (fishermen) confirm the presence of discus, including in the *lagos* further south.

Much has been written about the Coari region, but (almost) all of it incorrect. This vast region, like a number of other discus distribution areas, contains the dividing line between two species, in this case *S. haraldi* (blue/brown) and *S. aequifasciatus* (green). In the Lago (and river) Mamiá, as well as its tributaries and *lagoas*, there are blues of variable colour pattern. Its water, like that of the three rivers to the west (Coari Grande, Urucu, and Aruá), is described as *água preta* (= black water), but I have found typical blackwater only in the Urucu and Aruá (see Chapter 5). We also find variable Blues – often broad-striped to fully-striped – in the faintly tea-coloured Coari Grande. They occur quite a long way upstream, to (and in) the Rios Itanhauã and Juma, but are also found in the lower right-hand *igarapés* of the Coari Grande (eg the much-branched Gibian), at Boarí, and on the left bank at São Sebastão dos Flores and Samauma. A number of the right-hand *igarapés* are connected to the Mamiá at high water, eg the Igarapé Jacaré, which flows into the Mamiá.

The Rio Coari Grande forms the dividing line between blue and green. In its lower course, before it opens into the Lago Coari (into which the Urucu and Aruá flow as well), it is a gigantic river. With a length of more than 80 km and a width of up to 10 km, this section hardly equates with the term "river" (hence "Grande"). The easternmost distribution of the green lies in the two blackwater rivers to the west (which are darker than the Coari or Mamiá and have other water parameters). *S. aequifasciatus* occurs in the Urucu drainage. I found it here several times back in the 1980s in the lower river region (including large-spotted (alpha) individuals), as well as in the right-bank Igarapés Cajú, Cerrado, and Açú. (Because the collecting site was in the vicinity of of the Lago Coari, back then I named its green discus variant "Lago Coari"– but I have never mentioned the Rio Coari Grande as a locality.) In the drainage of the Rio Urucu (whose upper course is connected to the Rio Tefé) there are only Greens. And to the present day (the beginning of 2006) not a single Blue or Brown *(S. haraldi)* has been recorded in the distribution region of the green (see also map 8). Between 1997 and 2004 I also found greens upstream in the Igarapés Juaruná, da Familia, and da Maria, as well as in the region of the link to the Rio Tefé. (However, as is well known, a huge petroleum and gas source was discovered at the upper Urucu, near todays village Urucu, some years ago and pipelines were laid to Coari.)

I learnt from fishermen along the lower Urucu that *ácara-disco pintado* are are found in the blackwater Aruá, as well as in its numerous tributaries (not shown). In 1983 I recorded the first greens in the Lago Caiambé and its *igarapés* (which almost all contain black water), which likewise lie on the right-hand side of the Solimões. Also in the nearby Igarapés Ipixuna and Catuá which likewise empty into the Solimões via a *lago*. And since 1865 they have been known from the Lago Tefé, another blackwater habitat. But I have found green individuals, in most cases almost solid-coloured, not only in the approximately 60 km long Lago Tefé but also in the Rio Tefé, upstream to the Itangá. But so far not in the Rio Bauana. Further west they occur in the Rio Copacá and in the mouth region of the Juruá, in the blackwater *lagos* such as Uará and Tamaniquá, at Arara and Boa Sorte, but mostly only slightly spotted. In 1983 I found the first finely dotted Greens between the Solimões and the Japurá. In this enormous region they are widely distributed in blackwater habitats, in most of the Estação Ecologica Estadual Mamirauá, as well as in the Lago dos Guedes, and many other places. However, there are no records of Greens north of the left-hand bank of the Japurá – only Blues *(S. haraldi)* in the gigantic Lago Amaná. In addition, the Japurá is often described as a whitewater river, but I have always found it to be an "almost" clearwater river at various times of the year; I have never found green discus in the river or to the north of it.

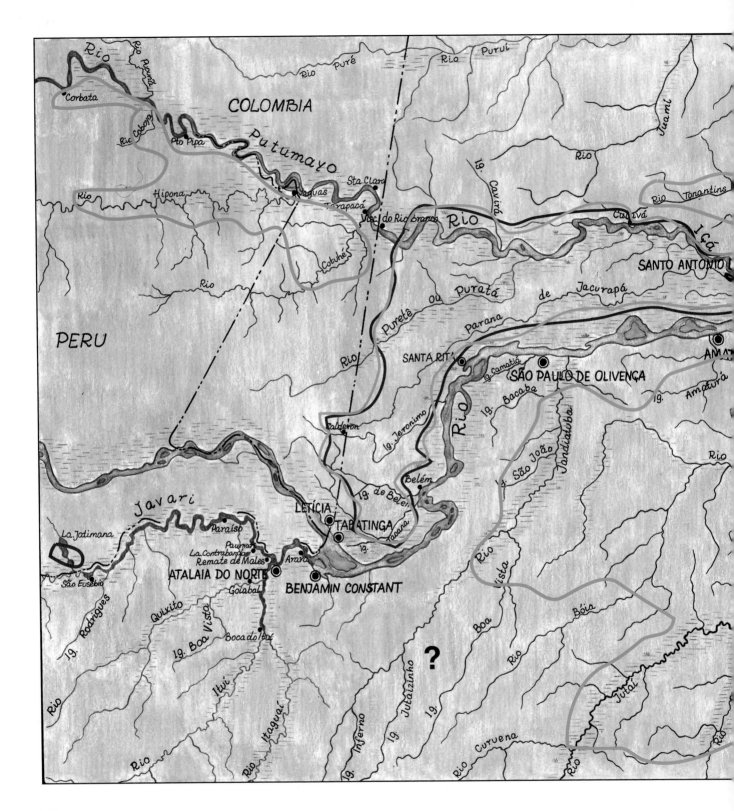

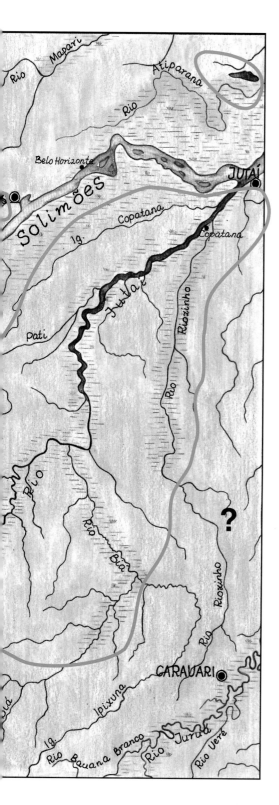

Map number 8 of the discus regions:
Rio Jutaí; Tonantins – Rio Içá; Putumayo; São Paulo de Olicença;
Tabatinga-Leticia-Benjamin Constant-Nanay region.

Greens *(S. aequifasciatus)* are widespread in the Rio Jutaí. A very attractive colour variant, sometimes with large red dots, as well almost solid-coloured individuals. They occur in the Rio Jutaí upstream to Curuena. I have also found greens – but no other species – in the lower courses of its main tributaries, eg the Igarapé Copatana, the Rios Riozinho, Pati, Bóia, and Mutum. Always in black water. No records are known from the Rio Altiparana, a northern Solimões tributary, with its mixed water. But greens occur in the blackwater Tonantins and its *lagoas*. These colour variants are very similar to those from the Rio Jutaí. The same applies to the blackwater Rio Amaturá and Rio Jandiatuba, both right-hand tributaries of the Solimões. And then there is what I find the most interesting point of all in the distribution of the discus species: the distribution of *S. aequifasciatus*, the green discus, south of the Solimões ends abruptly with the Jandiatuba, but extends north to well into the upper Tonantins, ending in the upper Rio Juami, with a second, limited distribution starting later in Colombia and extending along the Peru-Colombia border, west to Porto Pipa.

To the present day neither I, nor the local people and fishermen, nor the only other collector, the Swede Hongslo, have found green discus anywhere in the entire Rio Içá and its tributaries and lakes (see Chapter 5: *habitats*). In this region there are exclusively *S. haraldi*, individuals with a blackish, brownish, or reddish base colour, with no, few, or numerous stripes. In 1977 I found a very striking red-brown discus with a pronounced central bar in the Lago Jacurapá – but only a single (alpha) individual. This fish entered discus history as "Rio Içá" (and was bred with other browns). I have recorded *S. haraldi* variants in the Lago Jacurapa and tributaries of the Içá such as the Pureté (Puratã or Puriti) and Cuvirá (and Hongslo at Cuiavá – not Cuiabá, a transcription error in the original record of the locality and on numerous maps). And they are said to occur in the Içá drainage upstream to Visconte de Rio Branco on the Colombian border. And additional blues *(S. haraldi)*, but with a very interesting colour pattern, occur even further to the north as far as the lower course of the Rio Apaporis, in lagoons north of Vila Bittencourt, and in the border region. But, interestingly, practically exclusively on the Brazilian side. To date no discus have been found in the Río Caquetá itself. The northernmost distribution of the genus lies in this region.

Further west along left-hand side of the Solimões we come to the Igarapé São Gerônimo (known as the Calderón on the Colombian side), the Igarapé do Belém, and the Igarapé Tacana. (On some old maps this region is shown as the island of Calderão.) In all three of these blackwaters I have found blue- to brown-coloured individuals – mainly dark specimens with little pattern colour, and what there is in the form of dots or streaks. Chiefly in the *lagos* and bays. Their distribution in these *igarapés* extends into Colombia.

In none of the right-hand Solimões tributaries in this region (including on the Peruvian side where the river is once again called the Amazon) have discus ever been captured. Any remarks to the contrary in the scientific or popular literature refer to false data or misinformation. Almost all these *rios* and *igarapés* contain muddy water for the entire year. And there are no discus habitats. I have been able to confirm this on numerous research trips, and the natives (Ticuna indians) and *caboclos riberinhos* have also told me so. Most recently at the beginning of 2004. There are just two clearwater lagoons, or rather *lagos*, where, according to the Ticuna, discus (which they identified as Blues) occur. They call one of these lakes Jatimana; it is about 20 km long and empties from the north into the Javari (and must represent the westernmost distribution of the genus). The second is called Contrabando (= contraband) and lies at Paumari, south of the Javari, with which it is connected at high water.

I have found more Blues on the Colombian side north of Leticia, in the rivers and *igarapés* that flow towards Brazil. No natural discus distribution further west is known. Neither I nor anyone else has to date been able to find discus further into Peru or Colombia (any more so than in Venezuela). Apart from the deliberately introduced green discus (from Lago Tefé) in the Rio Nanay in Peru (see pages 138-139). But, as is by now well known, this has also been represented as an "accident" (specifically, that some individuals escaped at high water and established themselves in the lower Nanay). That was more than 20 years ago, and made life a lot easier for the fish exporters in Iquitos, as Tefé is more than 1000 km away by river...

CHAPTER

4

Discus Variants in Nature

The Heckel Discus;
The Green Discus;
The Brown Discus & The Blue Discus;

Discus Variants in Nature

The purpose of this chapter is to look in detail at the natural forms of the discus (genus *Symphysodon),* ie the three species and their (colour) variants. Each fish is labelled with a precise locality (place of capture) or the region where it occurs. The only forms discussed here are those where I have been able to confirm the locality personally, usually by catching them myself. I have ignored trade information published up to the end of 2005, as in most cases it is inaccurate (often deliberately so), and hence cannot be confirmed. I have likewise not taken the details of distributions in recent publications, as well as on the Internet, into account, as these too cannot be authenticated (as the suppliers of the information sometimes admit themselves). This photographic documentation – along with the work *DISCUS H. Bleher & M. Göbel* – is intended to provide a clear and unequivocal overview, just as I have tried to supply elsewhere in this book as well. The categories used are as follows:

Symphysodon discus Heckel, 1840 – the Heckel discus

Symphysodon aequifasciatus Pellegrin, 1904 – the green discus

Symphysodon haraldi Schultz, 1960 – the brown & blue discus

The Heckel Discus
Symphysodon discus Heckel, 1840

The distribution is limited, as can be seen in Chapter 3, to the region comprising the middle and lower Rio Negro and its blackwater affluents; the blackwater region of the middle Rio Urubu and Rio Jatapu (Uatumã region); the lower Rio Nhamundá (well below the first rapids); blackwater biotopes of the Rio Trombetas (well below the Cachoeira Porteira, in the Lago Jacaré three specimens have been collected in 1966); and the only *S. discus* variant known to date which occurs in a blackwater region to the south of the Amazon, in the middle and lower Rio Abacaxís and the Rio Marimari drainage. (I succeeded in finding not only the first Heckel discus in this region near to the Paraná-Urariá, but also in the Rio Nhamundá and in the Rio Jatapu – localities that were previously unknown.)

There are very few *S. discus* among the museum specimens, and Natterer collected only a single individual. For this reason the scientific works published to date are full of gaps as far as the species' distribution is concerned. Many exporters have generally cited (and still do) incorrect collecting localities – often already falsified by the *disqueiros* (discus collectors) – in order to avoid giving away the actual location. This has contributed to a continuing state of confusion regarding distribution, which it is intended to dispel here once and for all.

As already mentioned elsewhere, *S. discus* occurs exclusively in blackwater biotopes, and only those with specific parameters (see Chapter 5: *Natural Habitats).* A further peculiarity of the Heckel discus is its habitat. It (almost) always occurs only in blackwaters where the *acará-açú* tree grows. (This plant is also found in the distribution region of the Green – but less with browns or blues.) Its foliage not only provides shelter, but is also the basis of the species' diet, and hence its survival (see Chapter 5: *Natural Habitats ... Discus nutrition in the wild).*

S. discus at no place overlaps with *S. aequifasciatus,* but with *S. haraldi* in some locations. In regions such as that where the Rio Nhamundá enters the Lago Nhamundá (or Lago de Faro), as well as in the lower Rio Jatapu-Uatumã region, and likewise at Marimari – where there is a connection with the Rio Canumã, I have found variants that were naturally-occuring hybrids (see page 177). This somewhat typifies what also happens under aquarium maintenance: breeders have often managed to cross Heckel discus (usually males) with an *S. haraldi* (a blue or brown variant) (see Chapter 6) and produced a so-called Heckel discus cross (or Heckel cross). Dr. Schmidt-Focke was the first to do this. But neither he, nor any other breeder (eg Göbel, Homann, Wattley, Weißflog, Zell and many others), has succeeded in breeding further from such an F_1 generation. It has often proved possible to produce an F_2 or F_3 generation from these tank-breds, but hardly ever further. And the above-mentioned F_2, F_3 or beyond have without exception become progressively smaller. In every single case of which I have seen or heard. And the naturally-occuring hybrids found by me at such localities have (generally) been just as difficult to breed further as the offspring of crosses in the aquarium.

Worldwide breeding experiments involving crosses have shown that the majority of the offspring exhibit one character or another of a true Heckel discus. With a few exceptions, each juvenile exhibits a different colour pattern. Wild Heckel discus, by contrast, are almost always uniform in their coloration within a given biotope. These experiments, as well as the wild Heckel discus variants shown here, indicate that *S. discus* is a good, separate species and clearly distinct from the two others. (Moreover, to my knowlege to date the number of breeders worldwide who have successfully bred the pure Heckel discus can be counted on the fingers of one hand.)

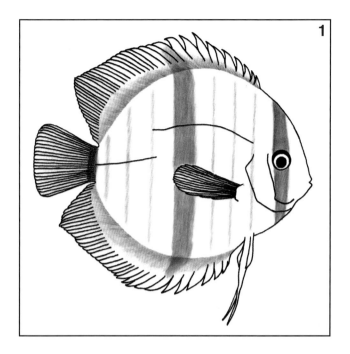

1. *S. discus* markings showing the defining characters for identification of the species: 9 bars – but 3 dominant – with the central (5th) always the most prominent (broadest). On the following pages I have in each case grouped together individuals collected in a single region. All the individuals illustrated are wild-caught from the precise location indicated. (Additional colour variants of this species can also be found on other additional pages in this work and in Bleher & Göbel, 1992). RIO NEGRO REGION: 2. Rio Negro. Adult individual after months in the aquarium (taken with powerful flash). It originates from the region where Natterer probably caught his specimen (Moreira). 3-4. Additional typical, semi-adult Rio Negro individuals in the aquarium. Collected south-east of Barcelos.

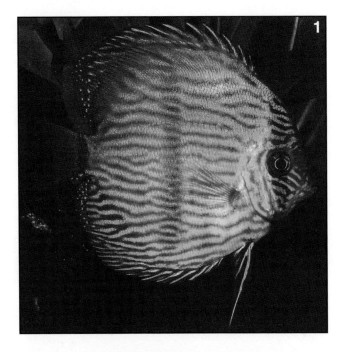

RIO NEGRO REGION:
1-2. Both *S. discus* are from the Rio Cuiuini (also written as Quiuini– see text and habitat, page 381), a right-hand tributary of the Rio Negro not all that far from Natterer's collecting site, north of Barcelos. The pattern of broad horizontal stripes running close together (pattern colour) is characteristic of the variant from the Cuiuini. 3-4. An individual from the Rio Novo Airão (3), (today) one of the most southerly localities for Heckel discus in the Rio Negro system (page 370). It is the typical Rio Negro colour variant. This specimen (4), caught in a *lago* not far away in the Airão mouth region, is very similar.

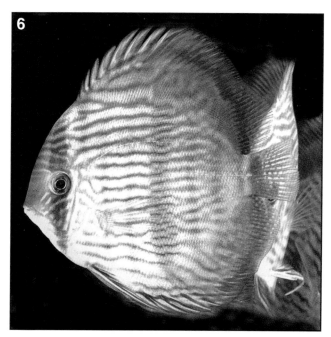
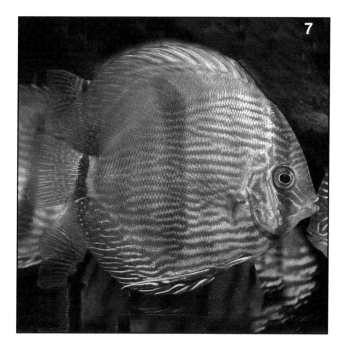
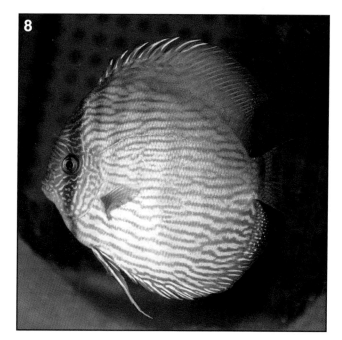

RIO NEGRO REGION:
5-8. All the Heckel discus shown here originate from the Rio Unini. Again a right-hand Rio Negro tributary whose mouth lies to the south of the Rio Branco. In its mouth region there are numerous lagoons and *igapós* – often difficult of access. In one of these (page 388) I discovered a variant with the broadest horizontal pattern colour (5) known to date. Closer to the mouth region there are forms similar to the typical Rio Negro variant (6), and further from it, narrower pattern colour in individuals from one of the isolated *lagos* in the Jaú National Park (7-8).

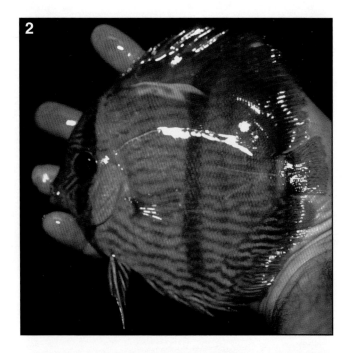

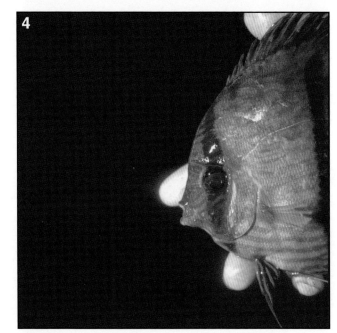

RIO NEGRO REGION:
1. I found this Heckel discus variant in the blackwater *lago* of the lower Rio Jauaperí (a clear left-hand Rio Negro tributary, which runs almost parallel to and east of the Rio Branco). It has a fine stripe pattern. 2-4. By contrast, in the blackwater lower Rio Xeruiní (also called Xeruiuní, or similar) drainage region I found the first Blue-headed Heckel discus at the beginning of the 1970s (page 101). Nowadays also known from other collecting sites (I have found individuals with splendid head colour in the Rio Abacaxís and the Rio Nhamundá, for example). Usually these are single alpha individuals.

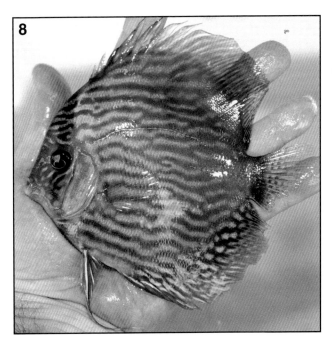

RIO NEGRO REGION:
5-8. These four variants are also similar to the typical Rio Negro Heckel discus. That is hardly surprising as one (5) is from the Igarapé Tucano, whose mouth lies on the left bank south of Barcelos. The others (6-8) are from the Igarapé Irauaú. Again a blackwater in the Rio Branco mouth region, where the majority of *S. discus* are caught today. The latter have a basically light blue (6+8) to turquoise (7) pattern, which may be narrow (8) to wide (6). Blue-headed Heckel discus are (7) only rarely found among them. (See also pages 378-379.) (DNA samples taken from 6-8.)

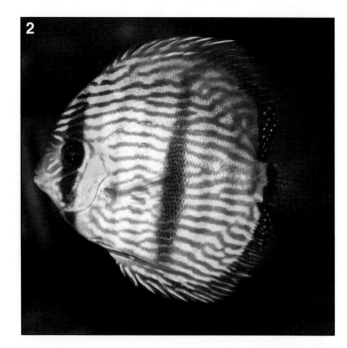
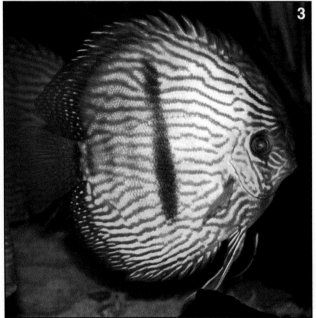

RIO JATAPU REGION:
1-4. Only Heckel discus live in the Rio Jatapu region; by contrast blue (so-called Royal Blue) and brown discus live in the Rio Uatumã, into which the Jatapu empties. This discus region is very similar to the Nhamundá region. Here too there is overlap in the mouth region – as at the mouth of the Rio Nhamundá – and hybrids are regularly encountered *(see right hand-page)*. Three of the individuals shown here (1-3) are from the Lago Iri– hence their similar colour pattern. This alpha individual (4) with a violet-blue anterior body originates from the Lago Castanha (further upstream) (see also pages 322-327).

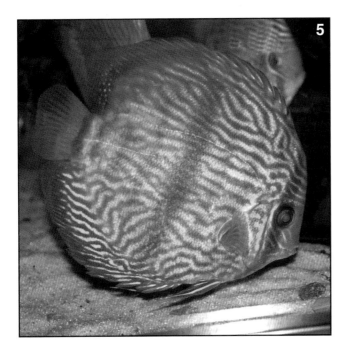 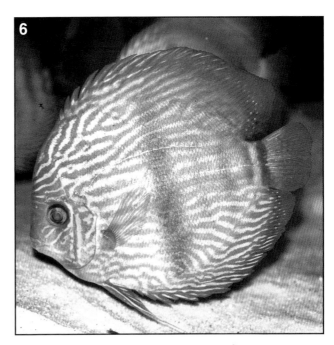

 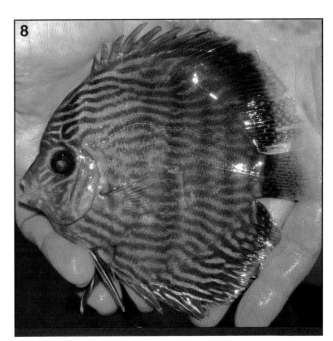

RIO JATAPU REGION:
5-6. Both photos show the same individual *(S. discus* variant) from one of the *lagos* in the Jatapu mouth region: one of its right-hand (5) and one of its left-hand (6) side, in order to illustrate the variable pattern colour. (Note how it is looking for food in the sand – in the wild just as in the aquarium – given the correct substrate.) 7-8. These two specimens are also from the lower Jatapu, close to the Uatumã drainage. It can be seen clearly how they deviate from the normal Heckel discus pattern. DNA samples were taken from both latter ones.

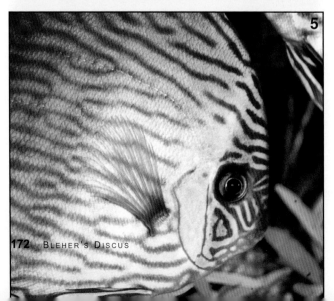

RIO NHAMUNDÁ REGION:

1-5. The Rio Nhamundá, where I found *S. discus* for the first time in 1997, harbours a number of unique – almost solid blue – Heckel discus variants. One specimen (4) was freshly caught (at night), and the others (1-3) are individuals that had already been acclimated to the aquarium for several months. But even in the Nhamundá river such individuals are rare (see also page 311, photo 4). It is more common to find normal Heckel discus (page 95), often with striking markings on the operculum (page 311, photo 3), or variants (5) with a wine-red base colour. (More text and photos on Nhamundá on pages 86-97 and 306-311; in Bleher, 2003b, "Nhamundá", and Petersmann, 2003 "Blue Heckel discus"; as well as in Bleher, Supplements 99/2000 loose-leaf *DISCUS*)

Note: I would like to add that in my view the Rio Nhamundá – and not, as generally assumed, the Rio Trombetas – represents the easternmost distribution of *S. discus* (see right-hand page and page 201).

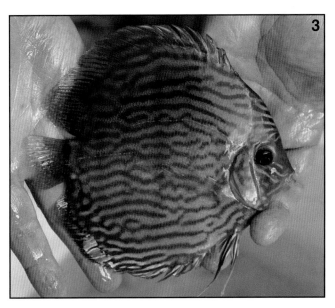

RIO NHAMUNDÁ REGION – RIO TROMBETAS REGION:
1-3. Individuals from the Rio Nhamundá, caught close to where it flows into the Lago Nhamundá (or Lago de Faro). In one (1) the three prominent Heckel discus bars are clearly visible – although the base colour isn't typical of *S. discus*. The second individual (2) shows only traces of bars and a lot of the typical blue discus colour pattern. In the third individual (3) at most only the middle bar is still discernible. DNA study (1b2) has shown that it is a natural hybrid *(S. haraldi x S. discus)*. 4. But whether this fish, shown as *S. discus* in Axelrod *et al.* 1985 page 372, is a natural hybrid or a true Heckel discus, can no longer be ascertained. It is supposedly from the Lago Batata (Trombetas), where I was unable to find a *S. discus* – only *S. haraldi* with a rather broad centre bar. 5. It is also questionable whether the specimens collected in 1966 in the Lago Jacaré and deposited in the MZUSP are *S. discus*. Again, I found none there. Just *S. haraldi* with a prominent centre bar, as in several other Rio Trombetas regions as well. (See also page 201.)

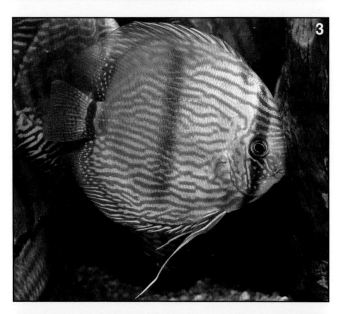

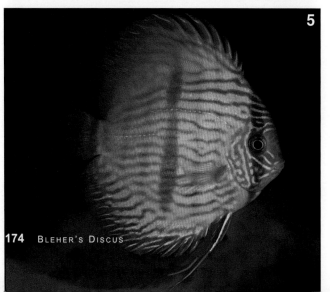

RIO ABACAXÌS REGION:

1-5. The subspecies *S. discus willischwartzi* was described from the Rio Abacaxís by Burgess in 1981 (page 132). This is the holotype (1), shortly after capture (but almost dead). The other specimens also originate from the Rio Abacaxís region. Two of them (2 and 3) from the lower Abacaxís, where they are often reddish. Because this collecting locality lies close to the confluence with the Paraná Uariá, they are often labelled with this (incorrect) location. The Uariá is a muddy whitewater *paraná*. The other two (4 and 5) were caught in the middle Abacaxís and have a similar pattern to the holotype. But all discus from the Abacaxís are incomparable *S. discus* and do not belong to a separate subspecies.

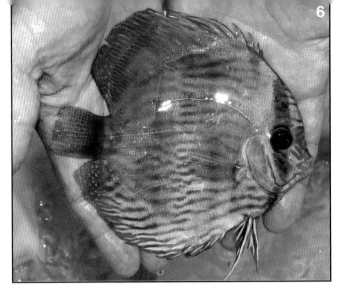

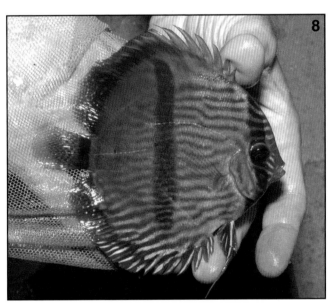
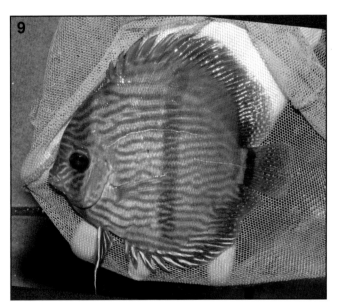

RIO ABACAXÍS REGION:
6-10. These five specimens are likewise from the Rio Abacaxís. Four of them (6, 8, 9, and 10) were taken immediately after capture and one (7) months later in the aquarium (the flash is in part responsible for the colour rendition). Interestingly almost all Heckel discus variants from the Abacaxís and Marimari regions (for the latter see pages 176-177) have a more or less distinct red fin edging (mainly in the dorsal fin) and often a red base colour (8-10). And it is here too that splendid, almost solid blue or turquoise, alpha individuals (6) are to be found – but not in the Marimari. (See also page 340). DNA samples were taken from specimens 8-10.

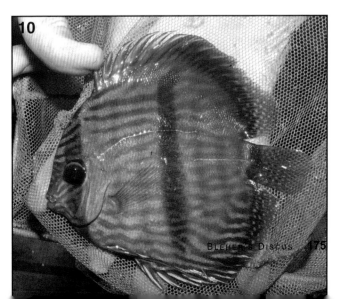

RIO MARIMARI REGION:

Until a few years ago it wasn't known that *S. discus* also occurs in the largest tributary of the Rio Abacaxís, the Rio Marimari (in its lower course). This is indian territory and it is undoubtedly for this reason that it had never been investigated for discus prior to my discovery. This splendid alpha individual (1) stood out in a collection made near the Mucajá settlement. In the Marimari the majority also have a reddish fin edging (3-4) and in other respects look very similar to the Heckel discus variants from the Abacaxís.

But in the Marimari (as well as in the Jatapu and Nhamundá) the two species, *S. discus* and *S. haraldi* come into contact at extremes of high water, as I have been able to demonstrate on the basis of a number of (natural) hybrids (2 and 5 here). (See also pages 340-342.)

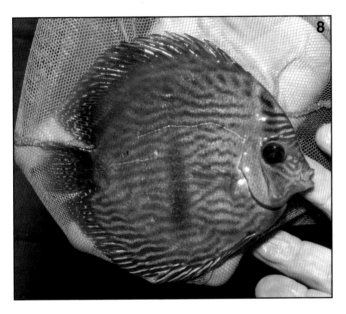
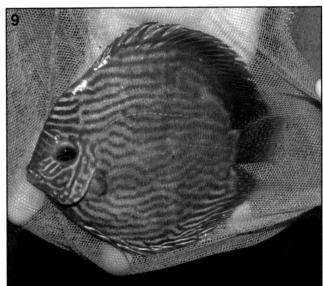

RIO MARIMARI REGION:
6. This individual is undoubtedly a hybrid beween the two species mentioned. I found it in the Igarapé Bem Assim, which at high water forms a direct link between the clear Rio Canumã *(S. haraldi* habitat) and the Paraná Urariá. The fish was infertile and stopped growing at 68 mm SL.
7-8. Two *S. discus* from the lower Marimari. In this case too an alpha individual (8) with an emerald green anterior body. 9-10. From these two, as the two previous specimens (7-8), DNA samples were taken at the University of Konstanz, Germany, and none of the four can be ascribed to *S. discus*.
(See also Stölting, Salzburger, Bleher &/ Meyer, 2006).
They came from the upper Marimari where I have never yet been able to find any true Heckel discus.

THE GREEN DISCUS
Symphysodon aequifasciatus Pellegrin, 1904

The green discus, like the Heckel, has a limited distribution, but differs in that it is not split between various distributional "islands" but is found in just one, almost continuous, region (see Chapter 3). Its most easterly distribution begins in the Coari region, in the Rio Urucu, and extends along the southern side of the Solimões to the Rio Jandiatuba, where it ends abruptly at the last blackwater biotope. On the northern side of the mighty whitewater river its distribution encompasses blackwater regions only, from the lower Rio Japurá region north to the Putumayo in Colombia and Peru, with just one interruption, where the uplands (mountainous regions) intervene.

There are no further natural distribution areas known to date (2006), and the green is, like the Heckel discus, found exclusively in black water. Though not all black water is the same and there is a considerable difference between the parameters of the Heckel discus habitats and those of the green discus (see Chapter 5: *Habitats*).

Unfortunately there are only a very small number of green discus among the museum specimens – even fewer than for *S. discus*. This is in part because, right from the start (beginning with the specimens collected by the Thayer Expedition in 1865/66) material was deposited under the name of the first described species *(S. discus)*. When Pellegrin (in 1904 – see Chapter 2) described the variant *aequifasciata* (though only two of the three or four specimens were greens – again see Chapter 2), he may have coined a new name, but nevertheless all the specimens were stored and catalogued as *S. discus*. (And until after 1960 aquarists likewise called all discus imported alive *S. discus*.) In fact there are a total of barely 20 specimens in five museum collections besides those mentioned in Chapter 2 with the locality given as a site where greens occur, such as Tefé or Putumayo.

The fact that so few specimens were deposited as type material is not just a function of the above plus the small distribution region, but also because discus are much more difficult to catch than most other Amazon fish species (for example, for more than 10 years representatives of INPA collected in the Rio Negro and were able to find almost 500 species, but not the *S. discus* that occur there). To summarise: by far the majority of the discus material deposited in institutional collections – worldwide – is *S. haraldi*, even if it is labelled as *S. discus* or *S. aequifasciatus, S. aequifasciatus haraldi,* or *S. aequifasciatus axelrodi*.

This is also one reason why details of the distribution of the green are more than fragmentary in the scientific works published to date. And the data provided by aquarium fish exporters have done little to help (and the bulk of the deposited material originates from such businesses). Apart from Tefé (lake and river), and the localities cited in the book *DISCUS by H. Bleher & M. Göbel,* to date there are hardly any other accurate data on their distribution (apart from occasional mentions of the Putumayo drainage and the Tefé individuals translocated to the Rio Nanay, in Peru). In this work the overall distribution region of the green, *S. aequifasciatus,* is shown for the very first time, as never before. And on the following pages you will find a whole series of local variants (most of them discovered by me personally, on the spot) from the majority of regions for this species.

On the basis of these numerous local variants it can be seen that – as in the Heckel discus, *S. discus* – we are dealing with a single well-defined species. And not, as often assumed (mainly earlier) to be regarded as a subspecies. Or as one and the same species as the brown or the bblue variants of the species *S. haraldi* – which at least academenics call *S. aequifasciatus,* since Kullander's work in 1986. This is indicated not only by the unambiguous striking coloration and the isolated distribution (allopatry), but also the following fact:

Since Dr. Schmidt-Focke succeeded in breeding the first live green individuals at the beginning of the 1960s (and from 1970 onwards almost continuously), breeders have tried to cross the true *Symphysodon haraldi* (then termed *S. aequifasciatus haraldi* or *S. ae. axelrodi*) with green discus (formerly *S. ae. aequifasciatus*) and produced them as infrequently as the true Heckel discus (*S. discus*) with a second species. Even if a first generation from such crosses is produced successfully, then at the latest after the F_2 generation the fry degenerate, and viability ceases completely after the F_3 or F_4 generation (if one reached that far…). There can be no better proof of the distinctness of a species.

I would like to point out once again that greens live in brownish, so-called blackwater regions. They are very rarely encountered in mixed water (sometimes during the rainy season) and never in fast-flowing waters. They prefer calm, very slightly flowing, or almost still waters. By day they typically remain among dead tree trunks, branches and roots (except during the rainy and spawning seasons), but also beneath the a*cará-açú* tree.

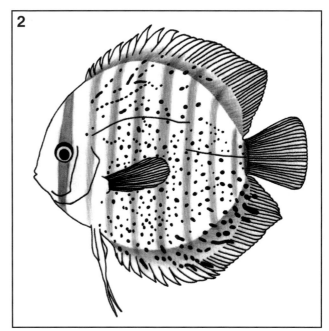
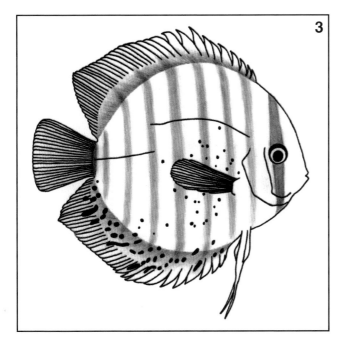

1-3. *S. aequifasciatus* markings showing the typical characters and range of variation – for identification of the species: The red spots can vary, for example, they may appear only on the anal fin (1), a few on the body (3), or all over the body (2); and sometimes verge on stripes (eg in the anal fin). All (more than 30,000 to date) the individuals studied have nine bars (also noted by Pellegrin), and essentially, that through the eye is the most prominent (like the middle bar in *S. discus*). The individuals on these pages are in each case from a single region. All the wild-caught specimens are from the stated location.
LAGO TEFÉ REGION: 4. A typical green, *S. aequifasciatus*, from the left-hand bank of the Lago Tefé, shortly after capture (DNA: 1A2).

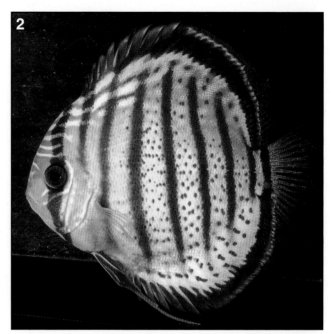

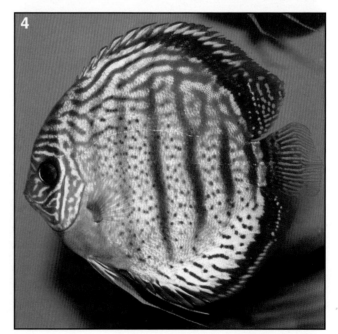

LAGO TEFÉ REGION:
1-4. Almost all the greens caught to date in the Lago Tefé have a similar coloration. It is true that the number of red spots in the anal fin and on the flanks may vary, from a few on the anterior part (1) or the posterior part (2) to odd ones in the middle (3) or – in alpha individuals – all over the side (4) – but the diameter of the spots varies little (1-2 mm). Nor do the eyes (almost always red) or the stripe pattern (uniform) – except when there is no striping (4). All have a fine red dorsal fin edging and almost always (in the wild and after acclimatisation, like these) a yellow-orange base colour.

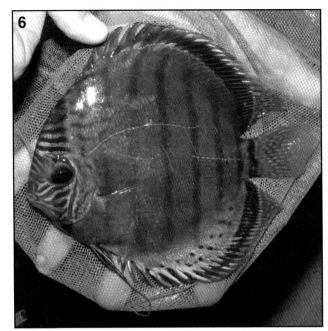

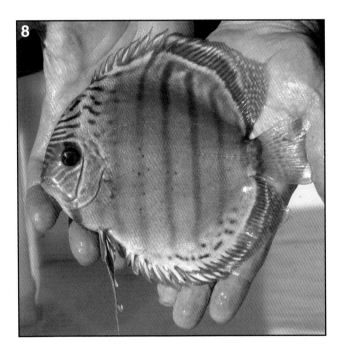

RIO TEFÉ REGION:

5-8. In the *lagos* of the middle Rio Tefé the colour pattern differs from that of individuals in the Lago Tefé. In 1985 I discovered the almost solid green discus that occur only here (see also page 83). It is notable that the majority of specimens collected to date in the Rio Tefé have no red spots on the body (5-6); or only a few (7-8); and that the all-embracing greens sometimes leads to the anal fin spots being barely visible (7). The solid colour is not apparent in fresh-caught specimens (7); it can be seen clearly only in acclimated individuals (7-8). Gold-yellow specimens like this (an alpha individual?) are rare in the Rio Tefé. (DNA 6-8).

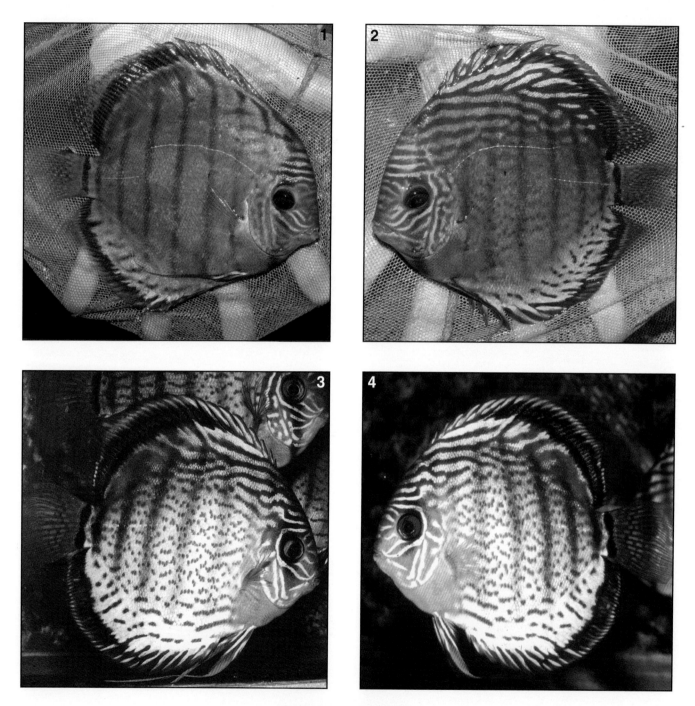

RIO URUCU (COARI) REGION:
As already mentioned (page 159) a number of "discus experts" have assigned the discus I discovered in the 1970s and 1980s and labelled "Lago Coari" to the Rio Coari. But there are only blues and browns in the latter. There are greens from the Rio Urucu on, and the Urucu, like the Rio Coari Grande, empties into the Lago Coari. *Caboclos* regard the lower courses of both as part of the Lago Coari (hence the name). My first specimen (3+4 show the right- and left-hand sides of the body) originated from the Urucu region (see pages 444-445). Urucu individuals usually have larger red spots (1.5-2.5 mm ø). They have the same markings further upstream (2), but fewer in the *furo* to the Rio Tefé (1).

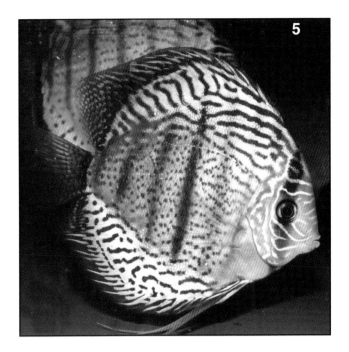

RIO URUCU (COARI) REGION – LAGO CATUÁ REGION – LAGO CAIAMBÉ REGION:
5-8. Another beautiful (alpha) individual from the Rio Urucu (5). These are among the most coveted in the hobby, but nobody collects them (any more). But they have been bred world-wide – from individuals which I collected and which Dr. Schmidt-Focke was the first to breed. These F_4 specimens (6) originate from his strain, which he always selected extremely rigorously – the best from the Coari (Urucu) region – which led to the Red-Spotted we see today. I have only once managed to find greens in the Lago Catuá region (7) – and this was the lightest green variant. This lovely variant (8) came from west of there, from the Lago Caiambé, just before Tefé, in 1983.

RIO JAPURÁ REGION:
1-4. Discus from the Japurá region. It is important to be aware that greens occur only south of the right-hand bank of the Japurá. The mainly clear Japurá forms the natural boundary between blue/brown and green. Discus from this region (Japurá-Solimões) often have fine red spots verging on dotted stripes (1, 2, and 4). Only in the Lago dos Guedes have I found Lago Tefé-like variants (3). When, in 1983, I managed to bring home a brilliant metallic green individual (4) from the Lago Jauarauá, it entered discus history as the Royal Green. Later, individuals from the Igarapé Pirini (1-2) as well. (All those shown are acclimated.)

RIO JURUÁ REGION:
1-4. The situation regarding the Rio Juruá is very similar to that of the Purus: greens have never been found in the Juruá – they couldn't survive a single hour in there. It too is a muddy, sediment-laden stream, large catfishes would swallow them within seconds. Moreover the parameters are wrong. Greens are found only in blackwater *lagos* such as Tamaniquá and Urará, from where these (1-4) originate. The greens in this region have a variable number of spots, which are never red but more of a faint rusty brown. The fifth to seventh and ninth bars are almost always prominent (4). All photographed on the spot.

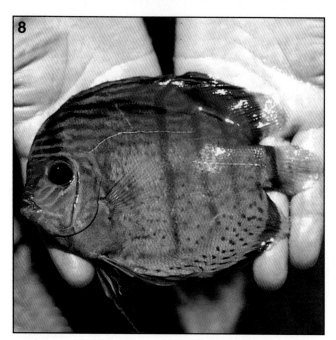

RIO JURUÁ REGION:
5-8. These four too are from the mouth region of the Rio Juruá, from the blackwater lakes there. These specimens likewise have spots, but more rusty brown than red (note on 5&8). They also have nine bars, but the fifth to seventh and ninth are almost always prominent (5,7& 8). Although there are also specimens that exhibit more pattern colour (stripe pattern) (6) and fewer spots (except in the anal fin). Greens from this region rarely have red eyes (see also pages 185 and 465) and their base colour can vary from yellow to olive green, light blue to rusty brown. Again, these specimens were photographed on the spot, immediately after capture.

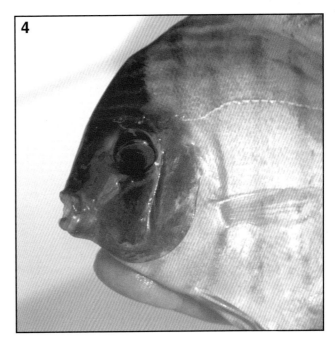

RIO MINERUÁ (FONTE BOA) REGION:

1-4. Greens *(S. aequifasciatus)* from the Rio Mineruá drainage, to the east of Fonte Boa, have to date only rarely been caught and exported. And when they have, then they have sometimes been sold as Fonte Boa discus – although Fonte Boa is a town on the muddy Solimões and there are certainly no discus there. In this region – unlike in other green distributions – I found a number of very large individuals (1-2), which essentially had very few spots on the body – and these were often generally small. Just in the anal fin (1-3). Among these wild-caught Mineruá specimens there was also one individual with a nervous system dysfunction in the head region (4). (See also page 465.) (See more of greens from the Mineruá at the end of this volume.)

RIO JUTAÍ REGION:
1-4. In the Jutaí region there are fabulously patterned greens. They are very rarely exported (too far from Manaus). Along with Urucu specimens they have the largest red spots – up to three mm in diameter (1-3). They are also a brilliant, almost emerald green colour (1&3). But some individuals are a solid green colour almost all over the body (4) – practically without spots, except in the anal fin. I have even found alpha individuals with this coloration (see also page 470). In addition there are individuals from this region that exhibit a yellow-brown base colour (2) and in many of them the fifth to seventh and ninth bars are more prominent.

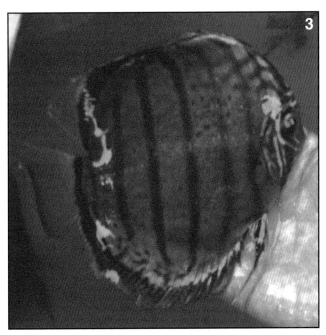
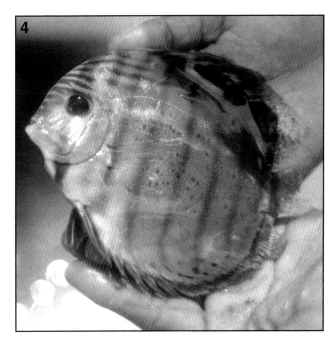

RIO PUTUMAYO REGION – RIO NANAY REGION:
1-2. The westernmost natural distribution of the discus lies in the Rio Putumayo region (see page 495) – Colombia/Peru – where only greens *(S. aequifasciatus)* occur. I found this variant (1) in the Rio Cotuhá region; its red spots are almost blurred and it has a light blue pattern colour. Even further west (in the Pto.-Pipa region) a very beautiful, very finely red-spotted variant (2). 3-4. Green discus (undoubtedly individuals from the Lago Tefé) were introduced into the Rio Nanay – Peru – more than two decades ago; they have bred well there and look like typical Tefé greens.

THE BROWN DISCUS & THE BLUE DISCUS
Symphysodon haraldi Schultz, 1960

"Why lump brown and blue together?" you will ask. Well, there is now no question but that the discus termed thus are just a single species (see also below). I use these popular names here only because they have become entrenched in the aquarium hobby and among discus enthusiasts – at least since the revision of the genus by Schultz (1960) (see Chapter 2).

In fact these popular names are not (now) strictly relevant as we find an almost incredible spectrum of colour variants among this species. The colour range of the base colour extends from light brown to brown, dark brown, or black; from yellowish to golden yellow to orange; from beige to reddish to red, or speckled red; and recently some whitish variants came from the Canumã region. There are Blues without pattern colour, but in addition there are individuals which have these base colours decorated with a pattern colour: with just a minimal stripe pattern on the anal fin; on the anal and dorsal fins; or both; on the head region; or all three; with a stripe pattern on the upper body; on the lower body; on the anterior part of the body; on the posterior half of the body; distributed over most of the body; or even striped all over (= Royal Blue). Again, these stripes can be very narrow (= Snakeskin), medium width, or broad. And they can be more or less close together or far apart. Running horizontally, diagonally, in a zigzag pattern, or even in part almost vertical. There are also local variants that exhibit only fragments of a stripe pattern (= *pintado*, or Pearl). In addition this pattern colour (stripe pattern) can itself vary in its shade: from beige to light blue, bright blue to dark blue; from yellow-green to turquoise (= Turquoise or Türkis).

As regards the vertical bars seen in every discus (all species), *S. haraldi* too normally has nine, dark to black (rarely light, or lighter) vertical bars of almost equal width. But there are also numerous individuals in which these vertical bars are irregular; run diagonally; or one or more is missing. And I have seen individuals with 10-16 vertical bars – albeit very rarely. (The cultivated forms such as the Snake-skin and similar originate from such individuals.) Then there are specimens with a pronounced fifth vertical bar (as in the Heckel discus – but not as broad), but these too are very rare and almost always alpha individuals (eg the Royal Blues and similarly heavily patterned individuals – more of this on the following pages and in Chapter 5: *Habitats*). And this species includes individuals where the fifth to the seventh (or eighth) vertical bars (or even the sixth to eighth vertical bars) are particularly prominent, in negative or positive colour.

Now, it is certainly not easy to define a species with such a wide range of (colour) variants – especially when the morphometrics barely differ from those of the other two species. But there are a number of points that do nevertheless distinguish this species clearly from the other two:

1. It has the largest – practically continuous – distribution region of the three species. It is found from the extreme east of the Amazon basin (in the Tocantins) to across the border with Colombia and Peru in the west (albeit only to a limited degree) – and with all these colour forms mixed at random. (And not as stated in many publications and on the Internet, eg "blues are found only there; browns only here") I have found so-called "browns" from the Tocantins through the lower and middle Amazon basin to north of Tabatinga and those listed as "blue" from the Lago Grande to Tabatinga and far upstream to the Apaporis (and almost everywhere in between). You can forget almost everything all previous publications (except *DISCUS* by H. Bleher & M. Göbel) say about the distribution of these colour variants.

2. Although in very many distributional locations browns through to fully patterned (*S. haraldi*) occur together, to date such they have never been caught or found together with either green (*S. aequifasciatus*) or Heckel discus (*S. discus*).

3. *S. haraldi* lives almost exclusively in clearwater zones (for exceptions see Chapter 5: *Habitats*) and tolerates mixed water better than the other two species because. The Amazonian mixed waters occur mostly in the rainy season and have a much higher pH value (up to neutral and more) and higher conductivity value (30 to 70 μS/cm and more). Quite different to the parameters for the two other species. (For details and the average parameter of each one of the species see Chapter 5: *Habitats*.)

4. Induvidduals of complete different appearance are breeding together and not only in nature (another reason why new variants are continually discovered – often in the same biotope), but also in the aquaria of breeding establishments worldwide – even with the hobby breeder. (Which cannot be said for the other two species – *qv.*) Practically all the cultivated forms that exist today originate from this species. (And for this reason it has always been relatively easy to linebreed the most unusual colour forms – whose numbers continue to increase almost every month.)

In the future the species may one day split into more species – particularly at the known locations (eg Lago Nhamundá, Uatumã region, etc.). But that can take a long time (perhaps 100, 1,000 or 10,000 years?). This is an evolutionary process that has begun with the first form of life on our planet – from the very beginning – and will continue possibly also for discus fishes if they and their habitats survive...

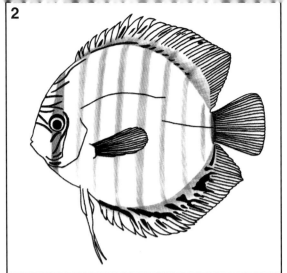
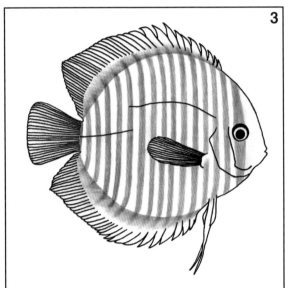
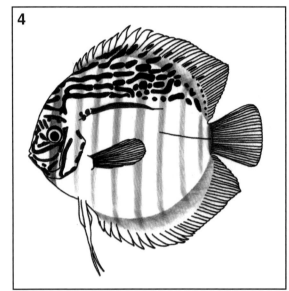
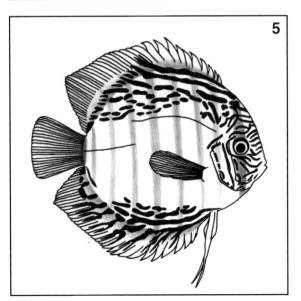
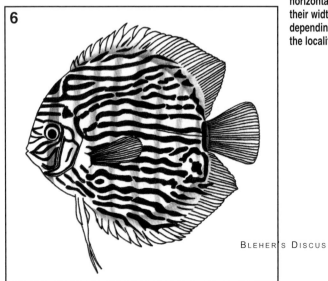

1-6. *S. haraldi* (brown and blue discus) markings, illustrating here its most prominent characters and the most important variations, for correct identification of the species: The species can occur totally without bars (or they may not be visible) and pattern colour (1); with nine bars and pattern colour at the base of the dorsal and anal fins (2); with bars ranging from nine to 16 (3) – with or without pattern colour (Snake-skin); with pattern colour only on the head and back (4); or also on the lower part of the body (5); or even all over the body (6) in wavy or totally horizontal lines, their width depending on the locality.

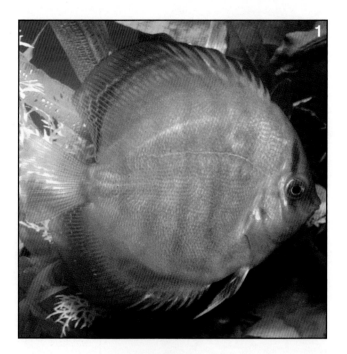

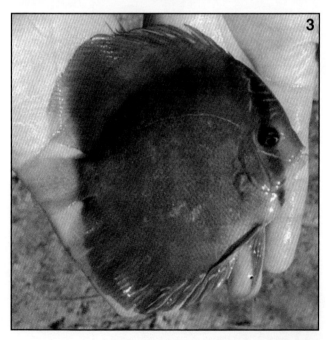

BREVES REGION – RIO TOCANTINS REGION:
1. An individual from the Breves region (the western part of the island of Marajó): completely without bars (or bars barely visible) and without pattern colour (but I have also found discus here with nine bars, although always with a pink-red base colour – see also Bleher & Göbel, 1992, *DISCUS*). 2-4. In the Rio Tocantins region too there are individuals almost without pattern colour (3) or just a minimal amount on the anal fin (2&4). Here too the bars may not be visible (3) or appear only "in negative" (2). The Tocantins discus are from the region in the vicinity of Cametá, where I also found discus with more pattern colour (see also page 223).

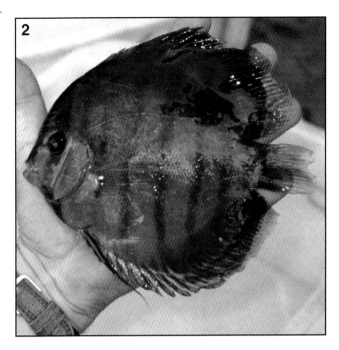
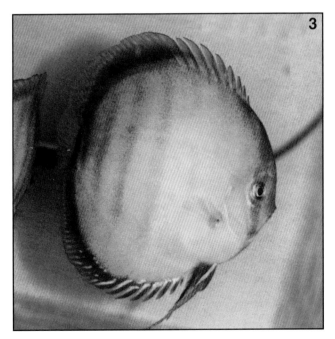

RIO XINGÚ REGION:
1-4. The lower Xingú (off Victoria) harbours discus variants whose coloration (practically without pattern colour but with nine bars) is sometimes very similar to that of those from the Tocantins (2). There are in fact also such browns (like those at Victoria) closer to the mouth, in the *lagos* around Porto de Moz (as the Thayer Expedition established back in 1865/66 – see page 238), but also a very yellow colour variant (the yellowest of them all). I established this in the mid 1970s (1). These too can completely lose their bars (4) or show them prominently (3). The last two specimens (3&4) are particularly round and were bred by Brian Middleton (UK).

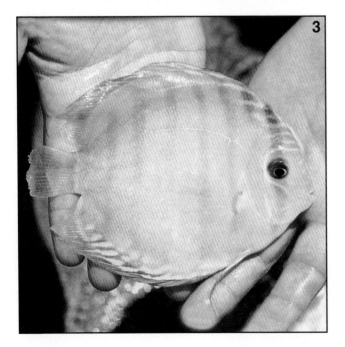

SANTARÉM REGION:
1-4. I have some reservations about assigning the Rio Maicá (to the east of Santarém – right-hand side of the Amazon) to the Santarém region (there is a Paraná de Maicá, a Furo de Maicá, and a Lago de Maicá). But there are interesting *S. haraldi* in the *paraná* and *lago* there. Large, almost round specimens with a brown (1), grey-yellow (2), almost completely yellow (3), or light brown (4) coloration, rarely exhibiting any pattern colour, but always with the fifth to seventh and ninth bars very prominent – sometimes "in negative" (3), but usually "positive" (1, 2, and 4). In addition, a typical Tapajós individual (Lago Verde) can be seen in the photo on page 265 (similar to photo 1 on this page).

SANTARÉM REGION:
1-2. There are also yellow discus in the Santarém region. But not quite as yellow as the *S. haraldi* variant from the lower Xingú (page 193); some exhibit nine bars (2), others none (1). They come from the Lago Verde region (1) and the Paraná do Ituquí (2), where the first discus were collected in the 1930s.
3-4. But there are also individuals with pattern colour (3) in the Santarém region, as in the Alenquer region. However there is no such place as Traira, to which such variants (3) are ascribed, nor Inanu (for all-over striped), or Trinidad (see text on page 264).
My first red (4) also came from the Tapajós drainage (in 1970).

SANTARÉM REGION:
1-4. The blackwater Rio Arapiuns is also to be regarded as part of the Santarém region; it empties into the Tapajós at the Lago Grande de Santarém (west of Santarém). All the specimens on this page originate from the lower Arapiuns. These discus are often extremely large – I have found specimens of 22 cm SL (Ø) (3). Usually with a greyish base colour (2&3), rarely yellowish (1) or light brown (4); often hardly any pattern colour, just in the anal fin and on the forehead (1-3); rarely with a partial stripe pattern (4); very rarely with stripes all over the body. The majority of those shown here (1, 2, and 4) are acclimated, the largest (3) freshly caught.

ALENQUER REGION:
"Alenquer" is a far-reaching term and nowadays no longer really tenable. When I brought back the first individuals from this region in the 1970s they were sold as browns. Not until the mid 1980s was the name changed, and Alenquer was suddenly on every (discus enthusiast's) lips. The Lago Grande (de Monte Alegre) is to be regarded as the easternmost part of the Alenquer region. It is home to brown (and reddish) females with little pattern colour (1+4); males often exhibit more red on the body and an irregularly striped (2) or pearly (3) pattern colour, but are never all-over striped. (See also page 249.)

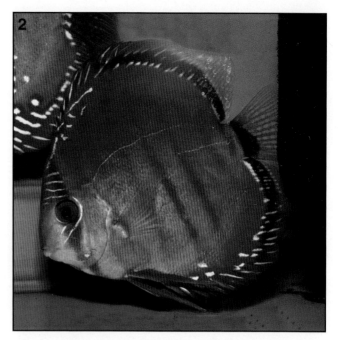

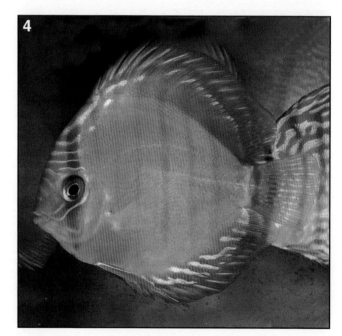

ALENQUER REGION:
1-3. The first Red, which I collected in the mid 1970s in the Lago Cuipeuá (1) and took to the Doctor (as a discus from Alenquer), was the basis – along with my later discus from the Içá – for the first Red discus. This red (2) with 10 bars (more frequent in *S. haraldi*) is another individual from the same lake, as is this one with a pronounced fifth bar (3) – incorrectly labelled "Rio Içá" (see page 279). Cuipeuá discus are also often sold with names like Mata Limpa or Curipera, *inter alia* – places that don't exist here. 4. I found another red – with red eyes but little pattern colour – in the Lago Aningau (accessible on foot).

ALENQUER REGION:

5-7. In the *lago* at Jaraquituba I found another interesting variant with red shades on the upper and lower parts of the body (5); males have fewer stripes, but females of the F_1 generation of this variant develop more (6) and the F_1 males more still (7). The Jaraquituba race usually has a yellow-gold base colour. All-over striped individuals have also repeatedly been found in the Alenquer region (see also page 201). I found a partially striped variant, albeit with a reddish base colour (except in the head region), close to the village of Corréa. But all-over striped are rare here.

ALENQUER REGION:

1-5. As already mentioned (page 198), in the Lago Cuipeuá I found individual colour variants with a prominent fifth bar, and this also applies to the Rio Curuá (at the Boca) (5). And at Barra Mansa – a locality for discus enthusiasts, although it is only a little *comunidade* on the right bank of the middle course of the river (page 276). There I found a similar variant (4); plus browns with little pattern colour; blues with more, but rarely all-over striped or red. Red only because there is a link between Curuá and Cuipeuá at extremes of high water. But while the majority of Alenquer individuals are ascribed to the Lago Grande in the discus world, they are actually from the Rio Curuá, the Boca region, or Cuipeuá. The same applies to the three other discus here (1-3), which some label as Heckel cross but are actually unique specimens of *S. haraldi*. One has the fifth to eighth bars coloured prominently "negative" (2); one faint (1); and one – Henry Bak´s pride – "positive" (3). But there isn't room to show all the Alenquer variants here. The lake region is around 250 km long, with four Lago Grandes alone, and at extreme high water virtually constitutes a single lake.

ALENQUER REGION – ÓBIDOS REGION – TROMBETAS REGION:
1. The *S. haraldi* called *pintado* by the natives; it has a pattern of red spots – usually clustered on one part of the body (beneath the pectoral fins) – and appears on the Internet with the locality given as Trairaha, which doesn't exist. This individual even had 11 bars, and originated from the Alenquer region (and dates from many years ago). 2-3. Two variants from the Lago Paracarí. One photo from the 1970s (2) and one from the 1980s (3). A mainly striped variant – with more (3) or fewer stripes (2). And sometimes the fifth to eighth bars are also prominently "positive" in colour (2). 4-5. From the Lago Curumú in the vicinity of Óbidos come these two *S. haraldi*, notable for their deep-bodied form and 10 bars. Some people also regard the lake as belonging to the Oriximiná region on the Trombetas, as at high water it can be reached from there by boat, but the people of Óbidos say it is their lake. These variants are typical for many Trombetas discus: the 10 bars; the pattern colour (except that more often they are all-over striped – page 73); and even more, the pronounced fifth bar. (Is that why science thought *S. discus* had been found there? – see page 172.)

NHAMUNDÁ REGION – MACURICANÃ REGION:

1-5. There is a link between the Lago Macuricanã (also known as Maracanhã) and the Trombetas, as well as with the northern Terra Santa lakes and the Lago Nhamundá. And, as can be seen, this gigantic, almost connected lake region is a discus melting-pot. This is also why the adjacent variants are very often similar. For example, as in the Trombetas region, here too there are individuals with 10 bars (2). As in the Rio Curuá (Alenquer region) discus with a pronounced fifth bar (3&5), or, as also seen in other Alenquer lakes, the fifth to seventh bars particularly prominently marked (2, 4&5, and page 310). And almost completely normal browns like those to be found in the Santarém region (1). Like myself, the Belgian Hustinxs, the owner of these five acclimated Macuricanã individuals, was able to demonstrate in the field that there are also red and almost completely striped individuals there. *Note:* Some individuals have an even greater red component – given appropriate feeding and/or lighting. (Warning: nowadays photos are often altered – with lots of magenta and yellow added).

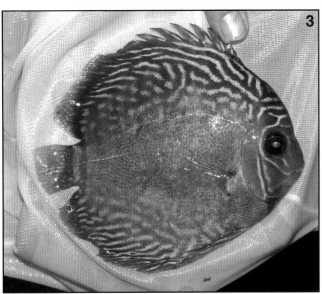

JURUTI REGION – RIO ANDIRÁ REGION:

1-3. In the Juruti region (often incorrectly written as Juriti) there are clearwater lakes behind the village on the right-hand bank of the Amazon (page 304), for example the Lagos Salé, Araçá, and Juruti-Miri. In the last of these there are two quite different variants, one of which is now sold as Juruti Red (page 302, photo 3) and a variant with very irregular pattern colour (3). In the Lago Araçá too there are individuals with partial striping (1&2), although the base colour here may also be faint greyish yellow (1) to light brown or even reddish (3). But commercial fishing is rare here. 4-5. Further west of here lies the Rio Andirá, which empties into the Parana do Ramos south of Parintins. Here there is a light brown variant, verging almost on yellowish, with very little or no pattern colour (almost like those in the Tocantins.). The fishes were caught here at the end of the dry season and were emaciated without exception. One had even been bitten in the anal fin region (5), but it was already healing.

THE BLUE & BROWN DISCUS

LAGO NHAMUNDÁ REGION:
1-4. South of the Lago Nhamundá there is an adjoining series of lakes with discus habitats. In the Lago Algodoa I found a very interesting variant with a pearl-like pattern (1). And in the Lago Xixiá-mirim a pointed-headed discus form (2-4). However, neither of the latter had an atractive base colour, mainly greyish or a dark brown. The collections were made at the end of a long dry period (December 1998) and the fishes were emaciated in the wild. Wimple piranhas *(Catoprion mento)* had eaten scales from some of the discus. In one individual (3) it can clearly be seen how the slanting mouth of a wimple piranha has torn off the scales literally in rows.

5-6. As already mentioned this is a gigantic lake system and here there are more discus variants in a single place than practically anywhere else. One finds browns, practically without pattern colour, individuals that today are often offered as Nhamundá Rose at horrendous prices (5-6). They can become very attractive (see also page 91, photo 1). Unfortunately the majority of discus sold as Nhamundá Rose are from totally different locations. None of these traders, breeders, or authors who call those names, were ever there.

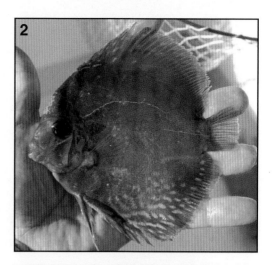

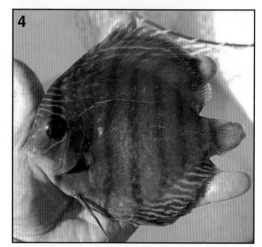
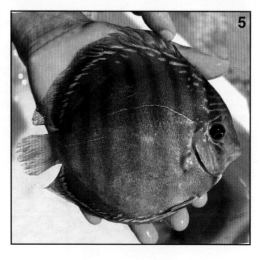
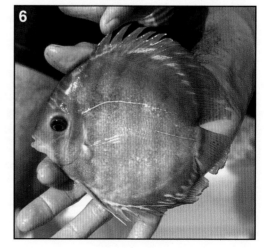

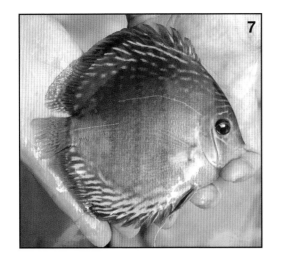

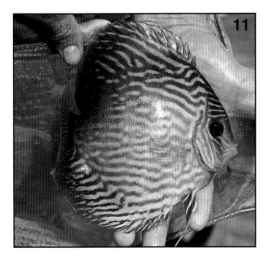
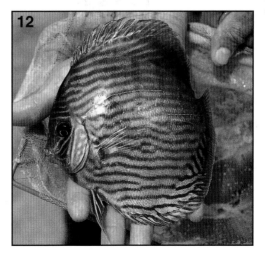

LAGO NHAMUNDÁ REGION:
7-12. The range of *S. haraldi* variants from the Lago Nhamundá is almost inexhaustible: base colour pale brown verging on reddish (7), or gold-yellow (9), pale yellowish with pink (10), golden with red (11), or a very red base colour (12). The pattern colour (striping) is likewise enormously variable, including slight (tiny) pearl-like markings (7), fine stripes on the head and anal region (8), the same but broader (9), fine stripes across the body (10), broad stripes all over the body – sometimes fragmented (11), or splendid specimens with very broad stripes all over (12) – but these are very rare (alpha individuals). And in the Lago Nhamundá too there are individuals which have the fifth to eighth bars more prominent than in other variants. (More on the Lago Nhamundá in Chapter 1, text on pages 86-97, and photos on page 91; photos 1-4 on page 308; photo 1 on page 311; page 206 *(top);* as well as in Bleher, 2003b, Nhamundá, *aqua geõgraphia* no. 24.).

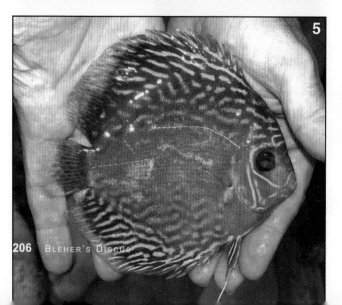

LAGO NHAMUNDÁ REGION – UATUMÃ REGION:

1-2. Only a few years ago Manuel Torres managed to find individuals in the Lago Nhamundá that had a very fine pattern on part of (1) or all over (2) the body. With nine (1) or 10 (2) bars. This was a sensation, as it was previously thought that the so-called Snake-skin *(S. haraldi* cultivated form) was purely an Asian "creation", with such individuals originating from the genes of a wild form, just like in the Pidgeon Blood (as shown in *Bleher, supplements 1993, DISCUS,* page 142b).

3-5. In the Uatumã region the situation is quite similar to that in the Lago Nhamundá region: *S. discus* occurs in the main affluent (in the case of Uatumã the Rio Jatapu and for the Lago Nhamundá the Rio Nhamundá). The *S. haraldi* variants in the Uatumã region are likewise enormously variable. Browns (often with greyisher or grey-brown base colour) with little or no pattern colour (3), all-over striped individuals, often with bite wounds (4), or with a pearl-like pattern.

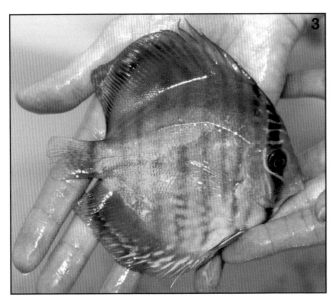

UATUMÁ REGION – RIO URUBU REGION:
1-3. Uatumã discus may also exhibit a red (1), grey-blue (2), or greenish (3) base colour. With (usually) nine bars, but repeatedly individuals are encountered with 10 or more bars (3). As already mentioned the pattern colour is also enormously variable. A pattern of broad, fragmented stripes (1); or a pattern only on the upper part of the upper and lower part of the lower halves of the body (2); or with very fine, fragmented stripes (3). Two individuals collected in the Lago de Jaquarequara (1+2), another (3) opposite the town of Uatumã. But the variation in the pattern is even greater (see also text, pages 322-330 and photos, page 322). But as regards the numerous all-over striped that supposedly occur in the Jatapu, there are definitely none there, they are all Heckel discus *(S. discus)*. 4-5. In the lower Rio Urubú and its mouth region, the Lago Canaçarí (or Camaçarí), I found blues with a broad stripe pattern (4) (see also page 329, photo 13), and on a single occasion a black discus (5) in the lower Uatumã region.

RIO CANUMÁ REGION;
LAGO JANAUARI REGION;
LAGO JUMA REGION;
LAGO MAMORI REGION:
1-2. The blue and brown discus (*S. haraldi*) from the Rio Canumá, its drainage region, and south to the Rio Manacori are often regarded in the hobby (and trade) as discus from the Madeira and publishers have also adopted the name Madeira Discus. Although not a single discus occurs there. Canumá individuals often have a lighter (pale yellowish or pink) base colour (1&2). Their pattern colour usually appears as an irregular stripe pattern – including sometimes vertical (1) – mainly in the anal fin region. If the stripes are halfway straight then they are on the upper part of the body (2). Exporters also label them "Gipsy". Individuals with a pronounced middle (fifth) bar also occur here. But are rare. They also occur in the Maués region, the Lago Campo Grande (in the two small photos – but see page 337), again Blues and Browns, but none with a special pattern colour.
My findings in the lake region south of Manaus also relate to browns with very little or no pattern colour, for example in the Lago Janauarí (3) or the Lago Janauacá. Not until still further south, in the Lago and Rio Juma, do we find very light, gold-yellow individuals with a lot of red in the breast region and fins (4). By contrast, in the Lago Mamori they are more light greyish (5) to pink (6) in base colour. But they too have little or no pattern colour.

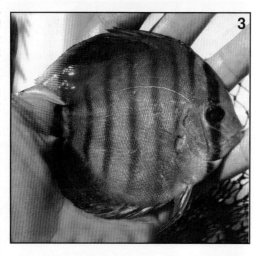

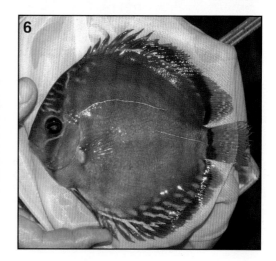

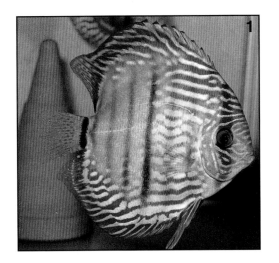

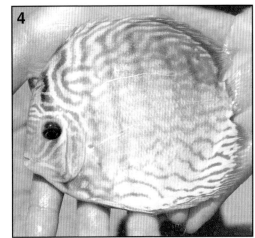
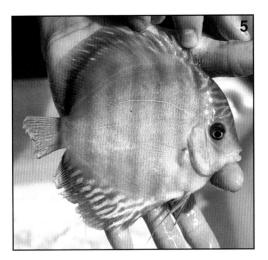
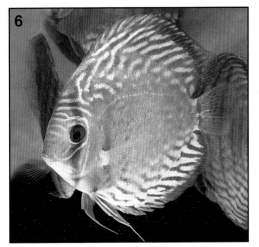

**RIO BRANCO REGION;
LAGO ANAMÃ REGION;
MANACAPURU REGION:**
1. The only evidence of *S. haraldi* from the Rio Negro region is from the Lago Cureru in the vicinity of the Rio Branco mouth region (1).
2. West of the Manacapuru region lies the Lago Anamã. In its drainage region, which has a connection at high water with that of the Manacapuru, there are blues, sometimes with a red-brown base colour, and also individuals (as in Manacapuru) with stripes all over the body (2). But alpha individuals are rare here too.
(See also page 398.)
3-6. I have written about the Manacapuru region itself extensively in Chapter 1 (where there are numerous photos of acclimated individuals on pages 68, 69, 71, and 352). And about the Rio Manacapuru and the two lakes (Lago Grande de Manacapuru and Lago Manacapuru) as well as the headwaters. Here just a few the variants discovered – there can be a lot of variation in appearance there (it is mainly a question of in which of the four habitats you look/collect). Here a few very differently coloured, freshly-caught specimens (3-5) and an acclimated individual (6). One, from the upper Rio, with unique pattern colour – almost Royal Blue (4). Two from the Rio Manacapuru (3&5) and one from the Lago Manacapuru (6).

RIO PURUS REGION:

I have written a lot more about the Purus region in Chapter 5 and illustrated a number of the *S. haraldi* variants (no other species occurs in the entire region) (page 410 – Jari region; page 418 – Itaparaná drainage; page 423 – Lago Tambaquí; page 428-429 – Rio Cuniuá). 1-5. Here a few more from the Lago Jari, where, as well as in all its tributaries and the smaller *lagos* of its drainage region, all individuals look very similar. The discus here can have a yellowish base colour (1&2); they may be greenish (page 410); or brown verging on reddish (3-5). The patterning is almost always only on the upper part of the upper and and lower part of the lower halves of the body, with a variable number of stripes. And they always extend to the extreme tip of the anal fin (typical for Purus region variants). All-over striped individuals are very rarely encountered in the Jari region. The nine bars are almost always equally prominent, it is rare for bars 5, 6, 7 and 8 to be prominent (1&5).

RIO PURUS REGION:

1-3. Further up the Rio Purus, in the Tapauá region (Rio Itaparaná, Rio Parapixuna, etc) there are very blue (really blue) discus. Their base colour is in fact yellowish-grey, but their overall appearance is almost always deep blue. Here too all-over striped are very rarely seen (if at all). As in large parts of the Purus region the pattern colour is restricted to a few stripes on the upper and lower parts of the body (1) or just fragments are visible (2&3), but all extending to the tip of the anal fin. And of their nine bars, the fifth to seventh and often the eighth are prominent (1-3). These specimens were photographed directly after capture; two are from the middle Itaparana (2+3) and one from the Parapixuna (1) (also written Paraipixuna or Paranapixuna).

4-5. These individuals were caught in the Rio Cunhuá (or Cuniuá) – where they exhibit a very light, often completely yellow (4) or yellow-grey (5) base colour. Here too they are almost without pattern colour but always have nine bars.

THE BLUE & BROWN DISCUS

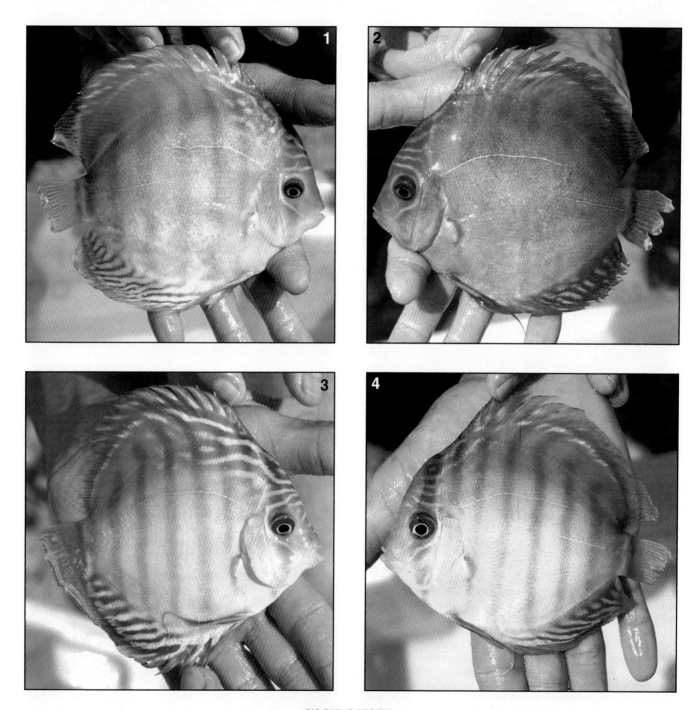

RIO PURUS REGION:
1. Discus from the Lago Tambaquí, where, as in the Cuniuá drainage (page 211), they are very light and exhibit hardly any pattern colour. 2-4. These three specimens – which have also appeared in print as Juruá specimens – likewise originate from the Cuniuá region. But they were (are) thus termed only because the Cuniuá flows through part of the middle Rio Juruá region (and there are even connections at extremely high water, but no discus can survive in the muddy Juruá) known as the province of Juruá. Their base colour is again golden (2), yellowish grey (3), or yellow (4), with little (3) or no (2&4) pattern colour, just in the anal fin.

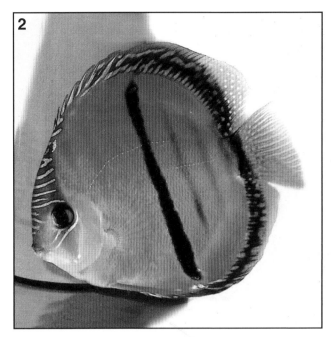
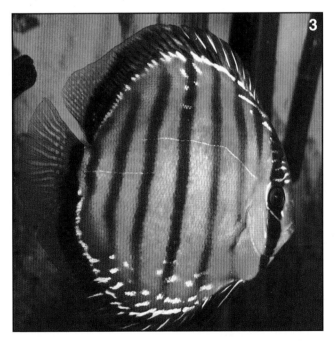
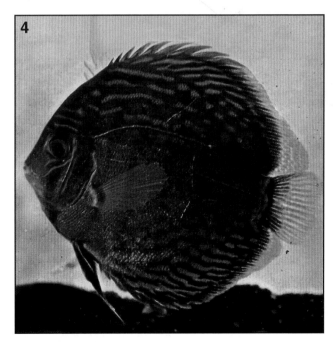

RIO COARI GRANDE – LAGO AMANÃ REGION – RIO IÇÁ REGION – TABATINGA REGION:
1. The Rio Coari Grande region is home to blues, very similar to the Purus region variants (see page 444), but including all-over striped individuals like this one from the Lago Amanã (1). However, the Amanã so far remains little explored. This alpha individual is one of the few collected. 2-3. Brown variants occur in the Rio Içá drainage (and I found a number of blues – see also text and photos, pages 477-478); those with a prominent middle (fifth) bar (2+3) are very rare. 4. Blue discus like this (4) occur in the Tabatinga region– to the north and northeast in the Igarapé Tacaná and the Igarapé do Belém. Likewise in the Apaporis.

CHAPTER

5

Natural Habitats of Discus & Collecting

**HABITATS, INCLUDING DIFFERENT WATERS TYPES,
CHEMICAL PARAMETERS, AND TEMPERATURES;
DISCUS NUTRITION IN THE WILD;
DISCUS COMMUNITIES – SYMPATRIC SPECIES & PREDATORS;
COLLECTING**

Habitats, including different water types, chemical parameters, and temperatures

In this chapter I will discuss the various habitats of the discus in nature, both those known previously and those I have explored myself. For each of the eight discus regions shown on the eight maps in Chapter 3 *(distribution of the discus variants)* we will look at the individual (discus) distribution zones, places where they do not occur, and areas that to date remain unexplored; sections of their history (and that of Amazonia); the main centres of population; the rivers, lakes, and drainages; the habitats of each discus biotope; water types; and lists of the chemical parameters and temperatures recorded. (All parameters in **bold** throughout this chapter).

As on the maps in Chapter 3, we will start at the mouth of the Amazon and work upstream, covering the following eight regions:

1. *Belém and surrounding area; Rio Tocantins; island of Marajó and Jari; Rio Xingú.*
2. *Rio Paru to Santarém; Santarém-Tapajós region; Alenquer region; Óbidos and surrounding area.*
3. *Oriximiná and Trombetas region; Rio Nhamundá region; Parintins region; Rio Uatumã; Rio Urubu.*
4. *Maués region and Rio Abacaxís; Rio Madeira region; Manacapuru region.*
5. *Rio Negro region; Rio Branco region.*
6. *Purus region; Tapauá region.*
7. *Coari region; Tefé region; Rio Japurá; Rio Juruá region.*
8. *Rio Jutaí; Tonantins and Rio Içá; São Paulo de Olicença; Tabatinga-Benjamin Constant region; Leticia-Putumayo (Colombia and Peru); Nanay-region (Peru).*

1. Belém and surrounding area; Rio Tocantins; island of Marajó and Jari; Rio Xingú.

Belém and surrounding area: The correct (full) name of Belém is actually Santa Maria de Belém do Grão-Pará. It is the capital of the state of Pará, the second largest Amazonian state (after Amazonas), and the second largest metropolis – after Manaus – on the largest river on Earth, with more than a million people. Translated from the Portuguese, the name means "Holy Maria of Bethlehem on the great river" (in fact the last word, *pará* (= river) originates from the Tupi-Guarani indian language). In the past the city was often called Pará as well as Belém, and shown thus on the maps. It lies about 120 km from the Atlantic Ocean on the right bank of the Baía de Guajará, which is fed by the clearwater rivers Moju, Guamá, and Acará and the sediment-rich whitewater river that forms the right-hand branch of the mouth of the Amazon, known locally as the Rio Pará. Belém stands on a site consisting of several islands, the *ilha das Onças* being the longest (19 km), sheltered from the ocean winds that can blow up to 1,200 km upstream.

The Spanish navigators Vicente Yañez-Pinzón and Diogo de Lepe (Amerigo Vespucci was supposedly there as well), the first Europeans to drop anchor in the gigantic Baía de Marajó, likewise found shelter here at the end of 1499. (The Baía de Marajó is the continuation of the Rio Pará.) But the Portuguese Pedro Álvares Cabral, who didn't arrive until later (1500), is credited with the discovery of Brazil, although he never ventured up the Rio Grande del Mar Dulce (= the great river of freshwater sea), as Pinzón had christened the Amazon, nor did he land here. (He landed on the coast of todays Bahia.) It is a fact that Francisco de Orellana was the first to stay for any time (6th-20th August 1542), resting here with his surviving men after an eight-month voyage by raft, down the Amazon from Ecuador to its mouth region. From there he started his journey back to Portugal on the 26th August. After his exploration the mother of all rivers was re-named again, as the Rio Francisco de Orellana, and thereafter there were repeated skirmishes over the vast river-mouth region. (And in 1555 the river was again renamed, this time as Rio de las Amazonas, as purportedly the Amazons lived there.) Everyone wanted a slice of the cake constituted by the former "freshwater sea" and the surrounding region (or all of it). The English, French, and Dutch all tried to muscle in, until in 1616 the Portuguese, who claimed to be the discoverers, naturally thought it was all theirs, decided they had had enough and despatched an *expedição com cerca de 200 homens e 3 embarcações: Santa Maria da Candelária, Santa Maria da Graça e Assunção* (an expedition of about 200 men and three ships: the *Santa Maria da Candelária*, the *Santa Maria da Graça*, and the *Assunção*). Under the command of Francisco Caldeira Castelo Branco they annexed the territory and laid the foundations of the *casa forte* (= strong building) which would later become the fortress *Forte do Castelo*, which still survives today. From then on the region was called Nossa Senhora de Belém (Our Lady of Bethlehem), as the Portuguese crown was strongly Catholic. In 1515 the architect D. Manuel built the world-famous Torre de Belém at the harbour in Lisbon, where the Emperor Don João II, along with Garcia de Resende, had planned a fortification to defend the harbour. There followed almost innumerable Portuguese roads, squares, buildings, villages, towns, and even regions under Portugal's rule, that were given the name Belém.

In April 1823 a terrible war of independence broke out. The province of Grão Pará wanted to be free from the empire (In 1822 Dom Pedro had had himself crowned emperor in Rio

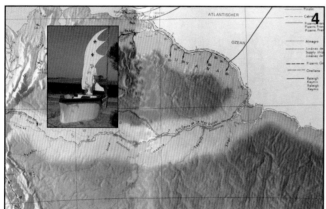
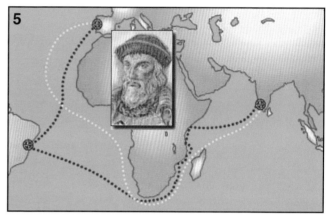
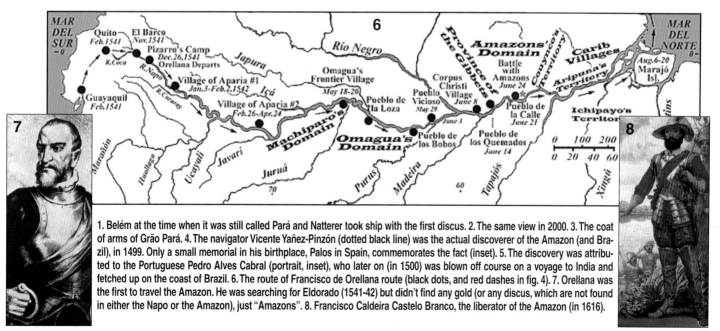

1. Belém at the time when it was still called Pará and Natterer took ship with the first discus. 2. The same view in 2000. 3. The coat of arms of Grão Pará. 4. The navigator Vicente Yañez-Pinzón (dotted black line) was the actual discoverer of the Amazon (and Brazil), in 1499. Only a small memorial in his birthplace, Palos in Spain, commemorates the fact (inset). 5. The discovery was attributed to the Portuguese Pedro Alves Cabral (portrait, inset), who later on (in 1500) was blown off course on a voyage to India and fetched up on the coast of Brazil. 6. The route of Francisco de Orellana route (black dots, and red dashes in fig. 4). 7. Orellana was the first to travel the Amazon. He was searching for Eldorado (1541-42) but didn't find any gold (or any discus, which are not found in either the Napo or the Amazon), just "Amazons". 8. Francisco Caldeira Castelo Branco, the liberator of the Amazon (in 1616).

and was compelled by his wife Leopoldine to support the move for independence from Portugal). The civil war in the region raged until 1840, when the *cabanos* (see *Glossar*) finally ceded their weapons or were killed – and in the mids of that war-period (1835) Natterer arrived in Belém with the first discus... (see Chapter 1: *First discovery*).

Around 1870 the Amazon was opened up to international shipping, and the rubber boom began. Even before the end of the 19th century this city was the greatest export and trade centre for the by now enormous production of raw rubber in Amazonia, which until 1910 led to unbelievable wealth for the entire region, but by 1915 was followed by total collapse and the end of the empire.

Just consider: in 1915 exports from the city totalled 323,003 tonnes – all of it rubber. About 80 years later the tonnage may have increased by a factor of two and a half, but was almost all wood, plus pepper and palm-heart *(palmito)*, fish, shrimps, Brazil nuts *(castanha-do-Pará)*, and manioc meal – but long since no rubber. Today the main markets are Japan, Europe, and the USA. But the export trade didn't begin again until the 1970s; prior to that there was very little apart from discus and other aquarium fishes. Belém was the place from (or rather via) which practically all Amazonian fishes came until the end of the 1950s (see Chapter 1: *Discoveries in the 20th century*).

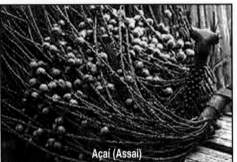
Açaí (Assai)

When, during the 1960s, I started to go there it was still a quite dreadful place. There was, as Fred Cochu so neatly put it, "only the Grand Hotel were giants rats run around and Madame Peréz's borthel – that was it!" And, that apart, it was impossible to lay hands on anything. I even had to take my own packaging materials and oxygen cylinders with me. There was only one exporter, Dias Lopez. *Disqueiros* (people with boats for catching discus – see p. 22) brought him discus. Apart from him there was just the old Japanese gardener Takase, who caught fishes together with his son Renato. Later, at the beginning of the 1970s, exporters sprouted like mushrooms, but most had fallen by the wayside by the beginning of the 1990s. Nowadays there are again a few, but mostly small, exporters. The vast majority of Amazonian fishes come from Manaus and Iquitos.

Where discus are still shipped from Belém they are just browns from the Tocantins and the region around Santarém, Alenquer, plus now and then from the lower Rio Xingú, although the last-mentioned fishes come mainly via Manaus.

There are no discus habitats in the area around Belém, or in the direction of the mouth of the Amazon. Not only is there no biotope in which discus could survive, but the essential building-blocks of their diet (certain trees and plants, detritus, etc. – see also *Discus nutrition in the wild*) are lacking and the relevant chemical parameters are wrong. The influx of salt from the ocean is too high at flood tide.

However, Belém is not just about discus, wood, and (formerly) rubber, but has much more to offer and is currently experiencing a whole new golden age. Everywhere there is repainting, building, renewal, and improvement. The wonderful Theatro da Paz (Theatro de Nossa Senhora da Paz, the Belém opera house) is experiencing a "rebirth" as well. This splendid building, in the middle of the city park, was ceremonially opened by the architect Vicente Sall on the 15th February 1878. The Brazilian aristocracy strutted noiselessly around on the rubber-based floor of a foyer crammed with Ming vases from China, Lalique sculptures from Paris, and Tiffany lamps. Not to mention the Venetian candelabras and the gilded doors and window-frames. The lavish granite masonry had been shipped all the way from Portugal. The salon upholstery was the finest imported velvet, as were the hand-made curtains, the seats, and the drapes for the balconies and boxes. The walls and ceilings had been painted by European artists, and much more of the same – a real rival for La Scala. But the opera house too fell into disrepair after the collapse.

Nevertheless, after 125 years this splendid building has been reborn and established as one of Brazil's historic monuments. Attempts are being made to attract international collaboration, as when the (rubber) money flowed only the cream of the cream appeared here. For example, the *Compagnia Lirica Italiana* of Tomas Passini, from Italy, who performed Verdi's *Ernani* here in 1880. Or the at that time world-renowned orchestra of maestro Enrico Bernardi. Even after the slump had begun (in 1918), the opera house was host to the unforgettable, world-famous Anna Pávlova and her ballet company, still remembered today.

Anyone who comes to Pará might possibly overlook the opera and the discus (which he would have trouble finding), but not the saying that has become synonymous with this region: "*Quem vai ao Pará, parou, bebeu açaí, por lá ficou*". Which translates as, "Anyone who goes to Pará stops, drinks *açaí*, and stays". *Açaí* is a small coconut *(centre)* from which a reddish

1. Belém lies on a peninsula between the mouths of the Rios Guamá and Acará. Much like in Manhatten, innumerable skyscrapers extend skywards in this city of about a million people. 2. In 1965 I established the first collecting station there, with new methods of collecting, holding, and packing. The first road to the Amazon was by then open and I fetched all the materials from Rio, more than 3,000 km away. I brought back discus from the Tocantins and Breves regions in woven baskets covered with banana leaves (protection against the sun). 3. And initially had to put up with losses. 4. I had aluminium containers constructed as holding tanks (there were no aquaria). 5. The Opera House, with its sculptures imported from France at the time of the rubber boom, is experiencing a revival, as is Belém. 6-8. Cathedral and palaces are being renovated, telephon booth molded of giant macaws have been put up (6), and relics of the wealthy period polished to a high sheen (7-8).

HABITATS ... BELÉM AND SURROUNDING AREA   BLEHER'S DISCUS   **219**

juice is pressed *(page 218)*. It is a must for every Paráense (as the inhabitants call themselves), just as the Italians must have their pasta every day; if he doesn't get his *açaí* then he eats nothing. He would rather go hungry all day. But *açaí* can be found practically everywhere here: frozen, as a drink, cooked, or as a fruit to be carried around in a pocket and nibbled at one's convenience. It has developed into a vast industry: the palm is planted and processed extensively in the Amazon region.

But *tacacá*, the pride of every Paráense, is likewise not to be missed. They maintain that their cuisine is the most authentic in Brazil, as they use only ingredients obtained from the Amazonian flora and fauna. And the majority of their recipes originate from the indigenous peoples, mingled with a black African influence to produce unique delicacies, incomparable flavours found nowhere else.

Tacacá, for example, is produced from sticky tapioca juice (manioc), the indian *tucupí* (a poisonous liquid obtained from the raw manioc root and requiring special preparation), *jambu* (a plant of the region, whose leaves, if chewed, numb the lips but produce a pleasant tingling sensation), and dried shrimps. *Tacacá* is traditionally eaten in the late afternoon, from hot calabashes or earthenware pots, and enriched with a piquant pepper sauce. In Belém there are numerous places where one can slurp or eat *tacacá*. On almost every corner there are so-called *tacacázeiras*, a typical feature of the local culture. They also sell other traditional foods such as *tapioquinha, mingau de milho, bolo de macaxeira, pupunha*, and the juice of local fruits.

1. Tacacá. 2. Regional foods from Pará. 3. Pato no Tucupí.

Just as trying these dishes is a pleasure not to be missed, another must is a visit to a centuries-old institution, the two markets in Belém: the meat market, imported from Scotland as a unit, and the fish market built in France. Both at the time when Belém looked more like Paris than the capital of Brazil, Rio de Janeiro. Note that the fish market is generally known by the name Ver o Peso (= see the weight).

Every morning an incredible spectacle takes place here. At some unearthly hour innumerable fishing boats arrive bringing huge quantities of fresh-, brackish-, and salt-water fish. Wooden boxes weighing up to 100 kilos are balanced on shoulders and carried across a narrow gangway to be weighed and sold by weight. Whoever offers the most gets the box. A sight that must be seen. It is a tradition like the handicrafts of the suburb of Icoaraci, where Marajoara pottery is made, a legacy of the Aruã indians of the island of Marajó (see *aqua geõgraphia* no. 15). A concentration of skilled hands can be found making clay pots in the *bairro do paracuri*. In addition one can enjoy fresh fish, crab, or sea-food there, and also watch wood-carvers at work producing their indigenous artefacts.

But our subject matter is the discus habitats, and there are effectively none here – at most the ponds of the exporters, or fishes that have escaped during the flooding that follows a cloudburst, and survived in the waste-water (but that is very doubtful).

No *disqueiro*, Renato Takase, or myself has been able to find any *Symphysodon* species at any time during the last 40 years in the nearby clearwater rivers – the Rio Guamá (which in recent years has often contained muddy-white water), the Rio Moju, and the Rio Acará (= cichlid river). (Neither could Fred Cochu's fishermen, nor Karl Griem or Wilhelm Praetorius.) And, of course, there are no discus in the whitewaters around Belém. Plus the the government-commissioned Geo-Imagen Consultoria & Projetos Ltd of Belém, which in 2000 conducted a population survey of the rivers in connection with the construction of new roads and bridges, was able to find several cichlid species, but no discus. It is quite evident that discus have never penetrated this far into the Amazon mouth region.

The most easterly distribution of the discus is found in the **Rio Tocantins region,** but nowadays the discus situation there is no longer so good. The influence of the Tucuruí hydro-electric dam (since 1984) has led to many changes. Before it was

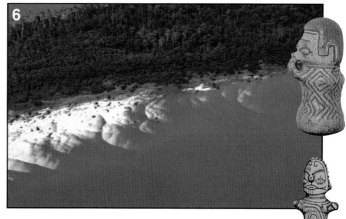

1. A glimpse of Belém's Ver o Peso market (the building with towers) from the fortress dating from 1616, where the fishing-boats put in early in the morning and the fish is sold by weight. 2. There are splendid basketwork and pottery goods there (more about the inserted Aruá indian ceramics on p. 229), plus fruit and vegetables and much more besides. More than 145 wild fruits from the Pará region are sold in the *mercado fluvial* – the floating market – since the 17th century *(left-hand page)*. In those days a tax, based on weight, had to be paid to the Portuguese crown (hence Ver o Peso). 3. Crates of fish weighing up to 100 kg are carried, balanced on the head, from the boat up to the weighing-house. 4. Back in the 1960s I searched the Belém region for discus without success. 5-6. The Baía de Guajará (5) off Belém always contains muddy water. There are no discus there, or in the rivers Moju and Acará which flow into it (6), or in the Rio Guamá.

brought into commission I was still able to land a few discus, but after 1985 I have been unsuccessful and no subsequent imports from the region are known.

It is also interesting to note that population surveys were performed by well-known Brazilian ichthyologists from INPA (Instituto Nacional de Pesquisa da Amazonia), both before and after the filling of the hydro-electric dam, and after 1984 not a single *Symphysodon* could be found in the more than 500 km of the middle and lower Rio Tocantins (and there were once many browns in the latter section).

The Tocantins rises at an altitude of some 1,100 m in the *planalto* of the state of Goiás and flows from south to north for 2,400 km to join the Amazon (the southern arm, here known as the Rio Pará). Its main tributary is the Rio Araguaia, another long and major river flowing from the south and which contains almost innumerable rapids, mainly in its lower course. Over its first 1,060 km the Tocantins has a fall of 925 m. In the next 980 km it drops about 149 m over numerous rapids. Only over the last 250 km, downstream of Lake Tucuruí (the second largest hydro-electric dam in Brazil), is there no longer any noteworthy fall, but there are still rapids, huge rock massifs, and sand-banks, with an average depth of 2.5 m – impossible, or at least difficult, for some ships to navigate. Only between Cametá and its mouth is it deeper, and navigable even for larger ships.

Tucuruí itself, at the northern end of the giant lake created by the dam, was built as a tourist centre. With selected fishes stocked for anglers, boats to hire on the lake, and artificial beaches to attract bathers. A Centro de Proteção Ambiental (centre for wildlife conservation) was created to vie with the ruins of the flooded villages as the main attraction. It lies 280 km from Belém as the crow flies, 400 km by river, and 385 km by road. Discus are not found in this area. The water is too shallow and full of rocks. And without doubt the hydro-electric dam did the rest. At any rate there is (no longer?) any discus habitat.

The huge Tucuruí hydro-electric dam has caused a lot of change. Since 1984 not only have characin species such as *Anodus elongatus* become extinct, but populations of *Curimata cyprinoides* have almost disappeared, and the lower Tocantins fishery has collapsed. The gigantic dam prevents fishes from swimming upstream to their spawning grounds, and the discus food supply has vanished almost without trace. The clear-felling and the almost lifeless water of the reservoir have (almost) wiped out the once rich planktonic biomass.

The region around **Cametá**, by contrast, is (or was?) made for discus, with its more than 90 offshore islands (very similar to the Anavilhanas, the gigantic archipelago in the lower Rio Negro). The name of the town (and the people who once lived here) originates from the Tupí indian language and means "staircase in the forest", as the aboriginals used tree-trunks to create stair-like steps in the ground in order to reach their settlement on the *terra firme*, 150 m above sea-level, and to go hunting from it.

With over 100,000 inhabitants (in 2000), Cametá is nowadays only nine hours from Belém by boat. (I remember the first time I went there, in 1965, it took me more than two days and nights.) And, as in the past, this region can be reached only by water from Belém. There are no roads thither. Officially Cametá was founded at the end of 1930, but the surrounding area was colonised as long ago as 1620 by the Capucin Frei Cristóvão de São Josétegrantes, who wished to convert the Camutá indians who lived there at that time. In 1713 it was already described as a *vila* (= village) and by 1848 even a *cidade* (city). Eventually it was designated a *município* (municipality) in 1930.

The numerous islands here in the Tocantins are an oasis for fishes. As in the Rio Negro (but note, the Tocantins is a clearwater river) discus have plenty of suitable hiding-places here, spending the day sheltering in several metres of depth and finding food among the

1. Cametá translates as "staircase in the forest". For decades it was the starting place for the collection of brown discus. 2. There are almost countless discus habitats around the numerous islands in the lower Tocantins. 3. And discus habitats exist in the Igarapé do Grito (stream of cries), as can be seen from this photo from the 1970s. 4. But even then the migrant *caboclos* from Ceará had started to clear the forest. 5. I have found a brown variant of *S. haraldi* with little pattern colour in the Igarapé do Grito. 6. Downstream of Cametá, the Rio Parurú is home to almost uniformly brown discus, as well as the unique tree-frog *(Hyla flavoguttata)* and the guaruba parrot *(Aratinga guarouba) (inset right and left, respectively).*

trees that are still present. The clearwater rivers such as the Rio Cagi and the Rio Tambaí are (or were?) also discus biotopes. But, as already mentioned, just browns, without any special pattern colour (mostly none at all).

The water parameters on 7.10.1999 at 1500 in the Parurú were as follows: pH 5.98; LW 24 μS/cm; daytime temp.: 33.5 °C; Surface area of water: 28.5 ° C and in 2 m: 27.9 °C. In the Rio Tocantins itself on the same day at 0800 at an Island-discus-habitat: pH 6.55; LW 43 μS/cm; daytime temp.: 27.5 °C; Surface area of water: 26.5 ° C and in 2 m: 25.9 °C. The surrounding area intact; water slighty flowing; no aquatic vegetation; river bottom sandy. For discus from that region see pages 192 and 223.

The measurements were taken at the (former) discus habitat on the date shown, as at the time of my first collecting trips I had no measuring equipment and had hardly ever taken a photo (nobody had the money in those days). But H. J. Mayland (in his book *Diskusfieber*) was able to take similar water parameters at this location in 1986.

Now, for the sake of completeness I have also performed a few researches in the area around **Abaetetuba**, the *Medellín brasileña* (= the Medellín of Brazil) as it has been dubbed for some time by the media in Bogotá and São Paulo. This Pará city, situated on a small right-hand arm of the mouth of the Tocantins about 60 km southwest of Belém, is supposedly the centre of the Brazilian cocaine trade (this has been known only since 1997). The cocaine arrives here (by ship and plane), far from any police intervention, and is then shipped on to Europe. Interestingly this was discovered by an Italian organisation, La Fondazione Giovanni e Francesca Falcone. After the dreadful assassination of Judge Falcone, his wife, and their escort in 1992 in Capaci, Italy, the Falcone Foundation was brought into being by Falcone's sister Maria, a teacher of law, who has made it her life's work to bring the "question of legality" to the young people of the country. Not only in the schools of Italy, but far beyond its borders, Maria Falcone gives lectures in which she reminds people of her brother's sacrifice. In addition the foundation endeavours to discover and expose the worldwide activities and connections of the Mafia, organises international congresses (including two in Brazil to date), and much more. In 1986, as Italy's leading judge, Falcone brought a mammoth court action in Palermo against 471 men and women of the Cosa Nostra. After an indictment running to 8,460 pages (and 660,000 pages of evidence) the court awarded 14 life sentences, 327 custodial sentences totalling 2,665 years of imprisonment, and 114 acquittals. In Rome, Judge Corrado Carnevale had to resign the presidency of the highest court of appeal (his nickname was *ammazzasentenze* (= judgement-killer), as he had overturned or annulled hundreds of judgements against *mafiosi*). For the first time in more than 60 years (during which a secret protection system had guaranteed *mafiosi* freedom from imprisonment) the judgements were enforced. And it is worth noting that after all the to-do and Falcone's murder an unwonted peace reigned in Sicily. The Mafia was no longer in the headlines. The bloody Mafia wars of the past seemed light-years distant.

Moreover, did you know that the Mafia is actually called Cosa Nostra (= Our Thing) and its members regard the word *mafioso* as an insult; that they describe themselves as *uomini d'onore* (= men of honour); and that the *onorata società* (= honorable society) has a code of honour that is absolutely binding? And the moral of the story: the parliamentary anti-Mafia commission wrote that the Cosa Nostra is a parallel state, with an estimated annual turnover of 310 billion Euros (almost 400 billion dollar), whose financial reserves and international connections are optimally suited to the age of globalisation.

And the moral of the story is: for some years now I live in a street named in honor to that brave judge and his escort *Giovanni Falcone ed alla Sua Scorta* (street sign below).

But to return to **Abaetetuba**: its former counterpart was Medellín in Colombia, the main centre for the cocaine trade in the in 1980s under the regime of the resident drug cartel and its leader, Pablo Escobar, *el dueño* (The Chief or the owner – of Colombia...). And as an aside: begin of the 1980s my fish supplier from Bogotá, Colombia arrived on my doorstep in Germany with Escobar (although I didn't learn until much later, from the TV, that he was the famous drug king). They tried to persuade me to import cocaine hidden in fish boxes destined for my company, Aquarium Rio. Nobody would notice and I would get 10,000 dollars per consignment. Without further ado I threw my supplier and his companion out of my house – but another Frankfurt importer agreed to the deal and ended up in prison a few years later (in 1988) ...

It is said that light aircraft carrying up to 1.5 tonnes of cocaine land on unpoliced airstrips in the jungle and some 50 drug ships travel the river, trans-shipping in Abaetetuba; but there are no discus. To date I have never been able to find any in the surrounding area, but I did discover an interesting piece of fish-related information. In the past the approximately

1-3. The main collecting stations for discus in the 1960s and 1970s were in the Tocantins region. The fishermen lived in houses on stilts and have build specially-designed fish-traps (2&4). 3. Once a month a *lancha* arrived from Belém to collect the discus from the *ribeirinhos* (river people). Initially in woven baskets lined with plastic *(see page 219)*, and later in specially designed plastic containers (visible on the deck of the *lancha*). 5-6. Typical discus habitats in the Tocantins region during rain. 7. The discus collected at night in that habitat here in a plastic container.

100,000 people here lived mainly from fishing and *palmito*, but for a while now also from aquaculture. Because, after the Tucuruí hydro-electric dam was commissioned, there were no longer any drinking water and fishes, here in the largest river system on Earth. The region's aquatic inhabitants found the essential migration routes to their spawning grounds cut, and so they could no longer breed.

When I asked a fisherman, Antonio, who was carrying a *tambaquí (Colossoma macropomum)* in his arms, whether there were any in the area, he replied, *"Não, bem pouco, alguns que escapulia, né? Por exemplo a nossa região aqui ela foi um pouco ou muito, prejudicada com o projeto da barragem de Tucuruí."* (No, only a few, or when they escape from the breeding facilities. We have been affected to some extent, sometimes very badly, by the Tucuruí hydro-electric dam.)

He then told me that breeding facilities had been set up, out of necessity, on the islands off Abaetetuba, in order to provide the local community with essential protein in form of *tambaquí, curimatã,* and *aracu* (the last two are *Prochilodus* and *Leporinus* species, respectively). But because the fishes do not spawn in the ponds, they required a laboratory here in which the fishes are stripped, the eggs hatched, and the fingerlings raised. I visited the facility and discovered that they had just two wild *tambaquí* females, named Ossinho and Cesinha, which they fussed over like a mother hen her chicks. Each produced over a million eggs per year.

In the flooded forest – only during the rainy season – *tambaquí* gorge themselves on seeds that fall from the trees; then, in the dry season, they undergo a long period of hunger. In the breeding facility it is quite different. Here they receive a *ração* (an enriched seed mixture), and are given hormone injections to bring them into spawning condition. This is how people have to alter nature in order to survive. Later technician Raul Ferreira told me that it was the idea of the local community to build a laboratory and they had all helped, even digging out the ponds with their bare hands. They received financial backing for the laboratory from various sources including POEMA (Programa Pobreza e Meio Ambiente da Amazônia = the Amazonian programme for poverty and the environment). Today they can pay the workers and technicians themselves as some 280,000 fingerlings of *tambaquí, curimatã, and aracu* are sold each year, at a price of 40 *reais* per thousand. There are 62 *comunidades* on the islands, with 300 *lagos* (= ponds) for growing on the young fishes. A number of them are by now privately owned.

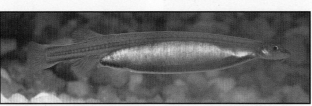

When I visited breeder Marcos Agostinho he told me, *"Isso aqui era um igarapé, nós represamos. Agora nós temos tilápia, temos curimatã e temos peixe da região também que se reproduz aqui sem que a gente coloque: aracu, vários tipos de acará. Tambaquí só se colocar, nós temos pouco aqui. Nós já tiramos agora, esse fim de ano, uns 100 quilos de peixe que foi vendido."* (I have dammed my stream and stocked it with tilapias, along with *curimatã* and other local fishes that breed there unassisted, eg *aracu* and various types of *acará* (mainly *Chaetobranchopsis orbicularis*). I stock with *tambaquí* only when there are very few. At the end of a year I have already harvested 100 kilos of fish to sell.) Antônio added that earlier they had suffered very badly from the disappearance of the fish, but now, with over 21 thousand square metres of ponds and 32 men working there, the picture was very different. The people again had food to eat, could even afford a beer and do not have to worry so much anymore...

Heading west from the Tocantins we come to **Oeiras do Pará**, a town of 23,255 souls (in 2000) that was founded in 1758 on the left bank of the large confluence of the turbid Rio das Oeiras and the clear Rio Araticu (also as Rio Anauera on IBGE maps). Oeiras do Pará, like so many towns in Brazil, was named by the Portuguese after a place in their own country, simply adding the name of the relevant state (or province). The colonisation was undertaken by the priest Antônio Vieira, although by 1653 there was already a large Jesuit mission further south, right on the Araticu. At first it was called Vila de Araticu, until in 1758 Mendonça Furtado gave all the settlements in Amazonia Portuguese names and christened it Oeiras. The name Araticu derives from plants found in this region, the genus *Rollinia* (family Annonaceae), some 65 species of which are known but only a few bear edible fruits,

Left: The *tambaquí (Colossoma macropomum),* also known as the pacu, is the main food-fish of Amazonia. In the past it was possible to find specimens up to 1.5 m long, but today individuals as large as 80 cm are rare. Like the discus and other fish species, the *tambaquí* is attacked by the notorious *candiru* (Trichomycteridae), which feeds on it parasitically. The Brazilian ichthyologist Ivan Sazima has been able to demonstrate unequivocally that one such parasitic species *(Vandellia chirrosa)* in the Amazon seeks out the gills of a *tambaquí* and "digs" itself into the lamellae, where it sucks the blood to fill its belly *(detail, left-hand page).* These unique underwater photos show traces of blood drifting in the current (and there are people who think that vampires are evil...)
Below: The farming of *tambaquís* (and other fishes) arose through necessity when the population of the Abaetetuba region (Pará) were left without protein after the commissioning of the Tucuruí hydro-electric dam. When they failed to breed naturally in ponds, a laboratory was built for stripping and artificial fertilisation, and two wild-caught females, Ossinho and Cesinha, are maintained for the purpose. The photos show Antonio with a *tambaquí* in his arms, Cesinha being examined to see if she was ripening with eggs, and a rearing pond at Abaetetuba.

for example the one well-known all over Brazil, the *biriba (R. mucosa),* also called *biriba de Pernambuco, fruta da condessa, jaca de pobre, araticu, araticum,* or *araticum pitaya.* Eaten raw it has an excellent taste, but it is also fermented into wine. The fruit *(centre)* is often used to make a lemonade-like drink or for its medicinal properties (analeptic and antiscorbutic). And the powdered seeds are supposedly a good treatment for enterocolitis. The wood of the tree is a yellow hardwood and is much used for making rudders, masts, chests of drawers, and cupboards.

Unfortunately very little is known about the fishes of this drainage. Discus may well occur there, but to the best of my knowledge this has never yet been investigated.

Bagre is the next largest town to the west, and lies on an (Amazon) island at the mouth of the Rio Panaúba. The main thing that astonished me about this *caboclo* city was its simplicity. Its few inhabitants live their lives as if this was the only place on Earth. Everyone knows everybody else and everything about his neighbour, from the day he was born. It is difficult to feel anything but good here. Bagre is surrounded by at least three rivers, the Rios Jacundá *(jacundá* is the local name for pike cichlids, *Crenicichla),* Panaúba, and Mocajatuba. The first contains water that is a real wow, so transparent that you can see every single fish from far off. Its banks are high and thus there is a fantastic view over this piece of Nature's miracle. It is true that anyone who visits the Alto Jacundá (upper river) will not find any discus there, but he will see an incredibly beautiful natural paradise full of birds and other creatures that are threatened with extinction in many places. But, as already mentioned, I have to date been unable to find *Symphysodon* in any of the rivers. There are no biotopes of the type essential for discus, which are as good as never encountered in crystal-clear water.

Island of Marajó and Jari: The island is divided into two basins, the Bacia do Marajó Ocidental and the Bacia do Marajó Oriental. By far the majority of this, the largest river island on Earth – Switzerland would be lost in it – (see also *aqua geõgraphia* no. 15, *Marajó*) is today farmland and swamp. The western part (Ocidental) includes regions that also contain clear water, and here lives the brownest of all discus forms. A colour variant which truly has no pattern colour at all.

The starting point for discus habitats is the **Breves region**. Millennia ago indians of the Bocas tribe migrated here. During the Portuguese colonisation the brothers Manuel Breves Fernandes and Ângelo Fernandes Breves became the first settlers. On the 19th November 1738 the supreme commandant of Pará, João de Abreu Castelo Branco, granted the Portuguese brothers a piece of land along the Rio Parauhaú, and not long afterwards this was named Engenho dos Breves after the first white settlers. Towards the end of 1850 the settlement was given the status of Freguesia (the Portuguese name for a kind of station were people could stop, stayover and/or get supplies – name for a place smaller than a village) and the following year granted the status of *vila*, with the official name of Município de Breves. Since 1909 Breves, which lies on the Rio Parauhaú, has been the capital of the *município*. And almost all the ships that travel the Amazon stop here. From Belém – via Porto Bom Jesus, Porto Líder, Porto Custódio, Porto Munducurus, or the Companhia de Docas do Pará – one can reach Breves within 12 hours, seven days a week. (When I came here for the first time, in 1965, the journey lasted three days and two nights.) Nowadays one can also fly here by light plane – there are three small airline companies: Aérea Meta, Puma, and Taca.

Breves itself now has a large industrial park for the timber trade. Timber exportation is the main activity of the surrounding area and the community has a project in progress whereby in future the wood will in part be made into furniture in the Marajó style. Not only are the valuable hardwoods felled, but the deforestation extends further in order to produce charcoal and to clear land for the *palmito* palm (the delicious palm-heart and the popular *açaí* – both are produced from this palm *Euterpe oleracea* – see also page 230) and agricultural products such as rice, corn, manioc, oranges, bananas, and lemons. As well as for the pig-, buffalo-, and cattle-breeding which by now has expanded to a vast extent at this end of the island.

The Estreito de Breves is truly the dividing line between island and mainland, ie the South American continent. In some places the *estreito* (= strait) is so narrow that it looks as if large ocean-going ships couldn't get through!

For discus habitat you must make your way to the Igarapé Grande, which is a clearwater river but looks like a blackwater (the Brazilians call it an *igarapé de coloração escura e transparente* = dark-coloured, transparent *igarapé*). It is bordered by numerous palm trees. And this is where the above-mentioned totally brown discus is found.

The first collectors here were the members of the Thayer-Expedition. From the 19th to 21st August 1865 they collected 20 species of "forest fishes" (no *Symphysodon)*, 15 of which were undescribed. However, I discovered discus in this region in 1965, when I was building my collecting station in Belém *(page 219).*

On the left is a view of the largest river island on Earth (approx. 49,000 km²). Unfortunately the Ilha de Marajó has largely been clear-felled and reduced over the centuries to pasture for farming cattle and water buffalo (as long ago as 1746 there were already about half a million). The eastern half is the worst affected. The smaller waterways have dried up. Annual burning of the savannah inexorably increases the drying out of the land. The water level of the enormous inland sea is likewise dropping and the Arari is now navigable only during the rainy season. There is in fact still a wealth of fishes – the survival specialist *Hoplosternum littorale* has undergone a population explosion thanks to the continual influx of cattle dung – but the former species diversity is diminishing noticeably. Only in the western and south-western parts of the island does any primary forest remain, and here there are still unspoiled *igarapés* and *lagos* – but for how much longer? For although the island has been declared a conservation area and ecotourism is being promoted, timber concessions are continually being granted and the palm kernel industry is on the increase, creating vast areas of monoculture. We can only hope that the rare scarlet ibis *(Eudocimus ruber)* plus the numerous herons (shown here) and other birds will survive. I have been able to find discus only in the Breves region (in the ó of the word Marajó) (see also in the text). The unique Aruá indian ceramic-ware also comes from Marajó (see page 221). (In *aqua geographia* no. 15, the history of this unusual island is covered in detail.)

The water parameters at noon time in September of 1965 were as follows: pH 6.30; LW: 55; Daytime temp. (Air): 32°C; Surface area of water: 27.6 °C; the water clear; sandy riverbottom; no aquatic vegetation; surrounding area primary rainforest; discus variants from the habitat on page 231.

The measurements were taken at the discus habitat (page 231).

Portel is the next region to the west. Here the clearwater rivers Anapú, Pacajá, Camaraipí (also spelt Camaparí) discharge their waters. The basin of the Rio Camaraipí ends at Portel, which lies south-east of Breves but on the mainland. It is nevertheless included in the *mesoregião* of Marajó and Portel. The town is situated at 01º 55' 45"S and 50º 49' 15"W. The surrounding area still consists largely of dense jungle, although a strip 1.5 km wide along the riverbank is just open mixed secondary forest, bush, or grassland. Primary forest is also lacking along the right bank of the lower Rio Anapú, where there are so-called *campos graminosos úmidos, em depressões assoreadas de sedimentos arenosos* (= damp grassy flats with unimportant secondary vegetation on sandy sediments). Thus here too there are no discus habitats to be found.

On my first visit in 1986 the unspoiled nature in this region had been reduced by only some 5% of its original extent, mainly along the rivers Pacajá, Anapú, and Camaraipí, as well as in the *baías* (= bays) of Melgaço, Portel, Pacajá and Caxiuanã. Hardly at all at the numerous *cachoeiras* (= rapids) such as Grande do Pacajaí, Pimenta, Piranha, Piracuquara, Pilão Grande do Tuerê, and Comprida. Since then the 2,000 km² Floresta nacional do Caxiuñã (Caxiuñã national forest reserve) has been created along the left-hand shore of the Baía de Caxiuanã and the Museum Emílio Goeldi now plans to build an ecological station there in order to rationalise the tree-felling, ie to keep it within justifiable limits.

The three large rivers mentioned run north-west from the south. The Anapú flows into in the Baía de Pracuí and thence into the Baía de Caxiuanã. Its main affluents are (right-bank) the Rio Marinaú, the Rio Tuerê, and the Igarapés Itatira, Muriapiranga, Janal Grande, Umarizal, Marapuá, Atuá, and Majuá, and (left-bank) the Rio Pracuruzinho, the Rio Curió, the Rio Pracupi, and the Igarapés Carumbé, Itatinguinho, Itatingão, Poção, Jacitara, Cocoajá, and Tapacú.

The Rio Pacajá discharges its waters into the Baía de Portel, off the city, and then unites with the Rio Camaraipí. Its main affluents are (left- bank) the Rios Urianã, Aratari, Manduacari, and Guajará, and the Igarapés Damiana, Capoeirão, Grande, Pajé, Limão, and (right-bank) the Rios Jacar-Parú Grande, Jacaré Paruzinho, and the Igarapés Vinte e Nove, Angelim, Do Ouro, Pereira, Ana, Tucumanpijó, Mineiro, Candirí, Maratuba, Cajú, and Araú.

The third of the group, the Rio Camaraipí, likewise opens into the Baía de Portel. Its main right-bank affluents are the Rio Banã, the Rio Pirico, and the Igarapés Esmeralda, Macaco, Açaituba, Meratuba, Grande, and Cariatuba; and on the left bank, the Rios Pitinga, Acangatá, Paca, and Ajará, and the Igarapés Taquera, Tamaquerinha, Tanquera, Arumã, and Otá.

Palmito palms and the valuable kernel

As one can see, it is an immense drainage system and little explored ichthyologically to the present day. Unfortunately, since my first discus trip the deforestation has increased dramatically, as has the population – by now, including Breves, over 300,000 people live here.

At the end of 2002 IBAMA (Instituto Brasileiro do Meio Ambiente e dos Recursos Naturais Renováveis) published a study entitled *Madeireiras Derrubaram Milhares de Árvores* (forestiers stealing thousands of trees). From the 30th September to the 31st October, along with the BPA (Batalhão de Policiamento Ambientai) they had conducted a blitz in Mosqueiro, Icoaraci, Outeiro, Barcarena, Acará, Abaetetuba, Igarapé Miri, Moju, Melgaço, Curralinho, São Sebastião da Boa Vista, as well as in Breves and Portel, during which they investigated 167 *embarcações* (cargos of timber), of which 69 were contraband and had a fine of R$ 1,726,355.60 (about US$ 850,000) imposed. During this blitz 1,121 terrestrial and aquatic animals were confiscated along the rivers Guamá, Barcarena, Acará, Abaetetuba, Moju, Pará, Pacajá, Aruanã, Anapú and Camaraipí, along with 411 turtle eggs and 1000 kg of game (pacas, armadillos, monkeys, and capybaras); 6,710,794 cubic metres of tree-trunks and 278,499 cubic metres of sawn wood (and a clandestine sawmill in Breves was closed); 2.5 tonnes of fish; 1,100 wild *palmitos*; 14 motor boats and 400 cubic metres of wood-chips. Conservationists calculate that in this region alone more than 100,000 hardwood trees and an incalculable number of *palmi-*

1. The mouth region of the Amazon is gigantic – up to 500 km wide, and contains the island of Marajó (the enormous savannahs can be seen on the satellite photo). 2. I found this brown variant with red eyes and no pattern colour in the Breves region – at the western end of the island. 3. Marajó is washed by the muddy Amazon water and at the south-western tip there is still a forested area. The region is constantly inundated by the tides and during the high-water period. 4. Discus can survive only in the small number of clearwater regions (dark-coloured inland waters).
5. The Igarapé Grande at Breves (and its discus habitat) has in part been converted into a *bânho* (bathing place).
6. And by now there are more than 200 sawmills in Breves.

to palms are illegally felled every month. The *caboclo*, battling for survival, is obliged to sell the *toras* (hardwood trunks) to the *madeireiros* (timber dealers) for only 8 *reais* (less than 4 dollars) each. This low price is forced on him. The *madeireiros* process the illegal timber, selling it to sawmills or sending it out of the country as a piece. Most of these smuggling businesses in Pará are in the *municípios* of Breves, Portel, Anajás, Curralinho, Pacajá, and Melgaço.

I encountered a typical *caboclo* in Eloi dos Santos Pantoja, who lives on the Igarapé Acagantá, an arm of the Rio Camaraipí where I was searching without success for discus. He told me that he had inherited 300 hectares of primary forest, rich in *massaranduba, angelim, jatobá, quaruba*, and many other valuable hardwoods – worth a small fortune. He and his family had to eat. Growing anything here was very difficult, but the *madeireiros* bought *toras* from him. Then IBAMA caught him illegally felling. The buyer escaped, leaving several hundred trunks behind. Pantoja was fined R$ 34,000,00 (some US$ 15,000), but when he provided IBAMA with an *atestado de pobreza* (evidence of poverty) he was let off the fine.

Transport on the rivers is by *jangadas* (raft-like ships), which convey the huge numbers of *toras* downstream. (Nowadays more than 90% of the gross income of the region comes from the sale of timber.) The spectacle that I saw on the Rios Pacajá, Aruanã, Anapú, and Camaraipí was almost incredible. Around Portel (and Breves) at full moon you might think the moon was made of wood, so many trunks are to be seen floating downstream on the *jangadas*. I was present at one "capture" on the Rio Pacajá where IBAMA seized 950 *toras* from the Marajó Islands Business. They

fined the company R$ 195,000 because their ATPF *(Autorização de Transporte de Produto Florestal* – freight permit), had expired. The *jangada* was carrying 1,950 cubic metres of hardwood trunks such as *quaruba* and *sumaúma, inter alia*. But this was less than 1% of what is estimated passes through the area every month.

The relentless ravishing of nature has taken an immense toll on the discus habitats in this region. Recently I was unable to find a fraction of those I saw on my first trip, and I asked myself what had become of the good spirits? (A Caruana legend tells of *espíritos do bem que habitam as águas e protegem as plantas os animais e os homens* (good spirits that live in the water and protect the plants, animals, and people).

Here too there are only Brown variants, but I found them only way below the last catarats in the Pacajá river. And they are quite similar in color pattern as the ones from the Rio Taquanaquara and Cariatuba.

I collected the following data end of July, 2002 around 1700 right at the discus habitat.
The water parameters were as follows: pH5.98; conductivity 65 μS/cm; daytime temp. (Air): 33 °C; surface area of water: 29.3 °C; in 2 m: 27.9 °C. The biotope was unharmed; besides some *Utricularia foliosa* no aquatic vegetation; rver bottom sandz and some stones (pebbles); surroundng area only partly deforested; discus variants almost same as those from the lower Xingú.

Further north-west, the last large left-hand tributary, the **Rio Jari**, flows into the Amazon, before its mouth. Although I found no discus there (or arrived too late), I would like to say a little about it.

The river rises to the north (on the Guyanan Plateau) and is often confused by

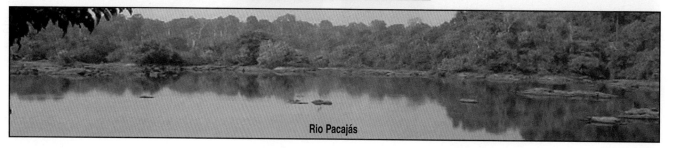
Rio Pacajás

Left-hand page: In spite of IBAMA and almost innumerable "rainforest conservation organisations" (or similar) worldwide – set up to protect the Amazon – the Amazon timber industry is growing faster than ever before. Concessions are continually being granted and valuable hardwoods are being felled like never before. Including in the Pacajá region, upstream of the the rapids *(lower photo)* and along its affluents. *This page:* A typical discus habitat that remains unspoiled and where I was able to find them.. But for how much longer? This ichthyologically little explored region around the three great rivers (Anapú, Pacajá, and Camaraipi) is about to be put up for sale. Above (with Brazilian names) are just 50 of the valuable hardwoods that occur there and which have become the number one export in Belém.

discus fans with the right-hand lower-Purus tributary of the same name, the home of splendid blues. (There are also a Ponto Jari and a third Rio Jari, the latter linking the Rio Tapajós with the Amazon – See map 1 on page 142.)

The Rio Jari is navigable only as far as the imposing Cachoeira de Santo Antônio (a waterfall – *page 235)*, some 110 km upstream. Its maximum depth is 4 m in the rainy season and drops to 2.4 m in the dry season. And it is that deep only because the river has been dredged from Monguba (Munguba) downstream so that ships can travel to Monte Dourado to the Jari project (upstream of Monguba), which already numbers some 16,000 inhabitants.

Until 1967 the Jari Indústria e Comercio S/A (a timber company) was sited here, and then the American multi-billionaire Daniel Keith Ludwig bought up the company and over 16,000,000,000 m² of jungle. Ludwig owned the largest private fleet of tankers and other ships on Earth – appreciably larger than those of Stavros Niarchos and Aristotle Onassis. He had a majority holding in a series of banks and hotel chains; owned office buildings, realestate worldwide; was the head of coal and iron mines in the USA as well as numerous other mining companies from Australia to Mexico; owned petroleum and petrochemical refineries in Florida and Panama, and much more. Ludwig wanted to "tame" the largest rainforest on Earth and to establish industry in Amazonia where there had never been any before.

After the finalisation of the purchase contract in the same year, and with the agreement of the Brazilian government, he founded the Jari Florestal e Agropecuária Ltda, which was to make history in the world of cellulose as "Projeto Jari".

The "invisible billionaire", as he was known, who was of German ancestry, foresaw that around 1985 the price of cellulose and paper would reach a peak (this actually occurred) and planned accordingly. He proposed to plant some 160,000 ha of land – some 10% of the total area – with trees. (Subsequently over 100,000 ha of fast-growing trees were planted between 1968 and 1982). The first cellulose factory was projected to come online in 1978. It was constructed in Japan and towed from Ishikawagima on two gigantic floating platforms, across the China Sea, over the Indian and Pacific Oceans, up the Atlantic, into the Amazon and up the Jari to Monguba, where the platforms still remain today. On one platform was the factory, which produces 220,000 tonnes of white cellulose fibre per year, and on the other a generator to deliver 55 megawatts of electricity – the power necessary for the factory. So that the toxic waste-water from the factory would not get into in the surrounding area and the Jari, an expensive waste-water regulation system was developed, including an 184 ha stabilisation pool. The industrial waste-water was filtered over a distance of 12 km before reaching the stabilisation pool, and thence passed, harmless, into the surrounding area. Parallel to all this he built a plant to produce high-grade kaolin for whitening the paper. A suitable source of kaolin existed a few kilometres away in the jungle. Ludwig also made provision for agriculture: along the lower Jari, in the inundation zone, he created huge rice fields, fully automated with the aid of the ebb and flow of the tides, which in this part of the Amazon and in the lower Jari follow their natural course. Completely modern pig- and cattle-breeding facilities, which developed genetically engineered crosses, including a method of breeding water buffalo, a model previously unknown but today imitated in temperate and tropical zones worldwide. (His buffalo reproduced around 98% faster than others.) Using simple methods Ludwig continuously expanded the Jari project without reference to anyone else. He took no notice of the negative opinions of the media, or of the envious. But by 1981 he had had enough. Because of the protracted Brazilian bureaucracy, which kept delaying or failed to grant permission for a power station (which is now being built) and other necessary structures, the goose that laid the golden eggs gave up. He had put more than two billion dollars into this unique project – and lost it. The Brazilian government took it all over and sold it to a consortium of 23 national business organisations for 290 million dollars, to be paid over 35 years. The entrepreneur Augusto Trajano de Azevedo Antunes was appointed general manager – the man who had already achieved success in the timber business in Amapá, after bankruptcies with the Madeira-Mamoré, the Fordlândia, and other enterprises. Augusto Antunes negotiated with Ludwig and the Brazilians and arrived at an agreement in 1982. The Companhia do Jari was formed, and his Caemi group took over its management. The kaolin works was thenceforth called Caulim da Amazônia,

This page (top): The Cachoeira de Santo Antônio in the Rio Jari forms a natural boundary for many fish species. However discus are found neither upstream nor downstream of the rapids.

This page (below): A photo, dating from 1980, of the gigantic Jari Project when it was in full swing at Monguba. It began in 1964 when the Brazilian military regime went seeking foreign capital from the multi-billionaire Ludwig, who had just started to invest in tropical trees in Panama. When he met with the then President Castello Branco, the latter couldn't believe that Ludwig planned to finance it all himself. Ludwig invested US$ 2 billions in the project, and when it was crowned with success the politicians found ways of depriving him of his due. They planned, *inter alia,* to keep nationalising his land until he gave up. And thus the dream came to an end.

Left-hand page: Ludwig required a vast quantity of wood in order to manufacture cellulose fibre. In the *Gmelina* from eastern India and Myanmar (formerly Burma) he found a tree that grows up to 30 cm a month, and immediately had 100,000 ha planted with it – supposedly the fastest-growing forest on Earth...

the cellulose operation became Cia. Florestal Monte Dourado, and the rice cultivation São Raimundo AgroIndustrial.

Although the new bosses had little idea, in 1986 the project was able to show a profit for its shareholders. And after 20 years the Cia. Florestal Monte Dourado has also purportedly shown positive results. Some 47,000 ha of land (less than a third of the original plan) are covered in what is said to be the fastest-growing forest on Earth, by now in its sixth rotation (2002). This means that the lowest cost factor per tonne of wood has been achieved. More than 100,000 people are by now directly or indirectly employed. Cellulose production has been increased from 220,000 to 340,000 tonnes annually. Unfortunately, after the death of its mentor, Antunes the Jari Project – like innumerable companies in the the Caemi group, as well as the group itself – was sold (2000) and none of those who took over has managed to come anywhere near the results achieved by the former greatest entrepreneur in Brazil, Augusto Trajano de Azevedo Antunes. Despite increased production the project has gone increasingly into the red. Economic experts say that instead of exporting the *cavacos* (wood) to customers 20,000 km away, which would increase profits several-fold by virtue of the highly-specialised industrial processes available there (production of special paper), they still "cook up" cellulose and kaolin at the Jari the way they did 20 years ago, and hence achieve only a small profit from the cellulose paste (cardboard) they produce.

Rio Jari

Be that as it may, the Jari region today looks more like a desert than a forest. The incomprehensible clear-felling and the afforestation with low-value timber such as Asian *Gmelina*, plus *Eucalyptus* and conifers, has transformed the area into an almost intolerably dry Amazonian region. The former, naturally homogeneous, primary forest and the ecosystem are now history. Unfortunately I first looked for discus here after the ecological disaster from 1980 on and thus was unable to find either typical habitats or any *Symphysodon*. Or any fishermen (almost all were recent immigrants or their descendants) who could tell me anything about the occurrence of discus. The only remaining indian tribes, the Waiapi and Wayana, live far above the waterfalls, where nobody goes.

Ludwig's legacy has, even so, brought the Brazilians some benefits: he constructed transport systems throughout the region, including three large harbour complexes, in Belém and Monguba and at the kaolin processing facility, where ships of up to 35,000 tonnes can unload. He organised personal transport to and from Belém via a fleet of *balsas* (rafts); these are towed and offer space for 292 passengers. He had an airport built where air traffic, ranging from twin-engine Islanders to FH Hirondelles with 44 passengers, can put down. And across the vast area of land he laid 60 km of asphalted roads 16 m wide, 1,500 km of dirt roads with a minimum breadth of 8 m, and 6,000 km of jungle roads 4 m wide. Plus a railway line for timber transportation, planned to be 220 km long, of which 68 km had been completed by his departure (whereupon the construction came to a full stop).

Say what you will, it was a Utopian project right from the start, that much is certain. It ripped an indescribable hole in the hemisphere, and whether or not it will ever return a profit is in the lap of the gods – as with the majority of the hydroelectric dams, one of which is currently being built here at Laranjal do Jari at a cost of 150 million dollars. But the invisible Ludwig, who practically never gave an interview throughout his almost 100 years of life and died childless in 1996, created jobs and ways to survive for hundreds of thousands of people. And I mustn't forget to mention, he bequeathed his fortune to the LICR (Ludwig Institute for Cancer Research), the world's largest academic cancer research institute, which still ranks above the Red Cross, the World Health Organisation, and the World Bank in the realm of cancer research. It is the only institute that sponsors its own clinical research and therapies itself and receives no support from pharmaceutical interests!

The last (discus) area in this region, which likewise belongs to the *Bacia do báixo Amazonas* (the lower Amazon basin), comprises the second largest right-hand Amazon tributary in this region, the Rio Xingú.

The **Xingú** was made world famous in 1967 by the three Villas Boas brothers, imvolved in ethnology since 1943, and the actual founders of the Parque Indígena do Xingú – the first large indian reserve (over 30,000 km² – about the size of Holland) in Brazil, to which they translocated 15 tribes who still live there today. (The then President of Brazil, Jânio Quadros, had ratified the laws accordingly for the park and the indians and appointed Orlando Villas Boas as Park Director.) It was

1. The habitats in the Rio Jari are, as can be seen here, not generally discus-friendly: fast-flowing and murky. In 1981 a *lancha* sank in the rapids at Ponta Alegre and 400 of her 600 passengers drowned. During my most recent visit to the region (February 2004) two *lanchas* collided at the same spot and seven people drowned. 2. The Cachoeira de Santo Antônio seen from downstream. The steep wall rises for about 30 m like an altar (hence "St Anthony"). I found two loricariids here: a *Peckoltia* species (note the variable pattern of markings in the five specimens) and a large *Pseudancistrus* species. 3. A mysterious German expedition was supposedly commissioned from 1935 to 1937 by the Third Reich to find a way to attack French Guiana from Brazil from high up the Jari (or so the story goes). 4. The members of the expedition travelled way up the Jari under the swastika flag. 5-6. They also tested planes here for the campaign in the tropics, but their leader, Otto Schulz-Kampfhenkel, established that the floats of the aquaplanes were inadequate. 7.& 9. One of the members of the expedition, Joseph Greiner, died of malaria on 2nd January 1936, and a year later the survivors gave up. A 3 metre high wooden cross with inscription (between smaller ones, and seen here below) on the bank of the Jari bears silent witness. (Note that there are no discus habitats here, either.) 8. The Jari Project attracted thousands of workers who founded the settlement of Beiradão, which was later renamed Laranjal do Jarí and has more than 50,000 inhabitants now, 70% of whom live in houses on stilts without any sanitation (it all goes straight into the river), while 6,000 of them commute daily to work on the opposite shore in Monte Dourado, which has more than 16,000 inhabitants.

also Claudio and Orlando (Leonard died early) that brought FUNAI (Fundação Nacional do Índio – the national indian authority) into being in 1967 and initially headed it. Orlando even held office as President of the administration on three occasions.

In 1967 the brothers received the gold medal of the Royal Geographic Society in England. At the conferment ceremony, Sir Gilbert stated that the the brothers Villas Boas had devoted themselves to the welfare of the indians in the Xingú region, arranged for the construction of roads in the area, ensured that no indigenous person came to harm, and in addition opened up a previously little known and dangerous region for benefit of the country of Brazil and of science. (In 1943 the three brothers had undertaken a trip lasting over 11 months on foot across this region of the upper Xingú in order to open it up for the government.) They were twice nominated for the Nobel Peace Prize, in 1971 and 1972. But, after 30 years of legendary daily toil on behalf of the indians of Brazil, they became tired of the battle with the authorities and retired in 1972.

"*Deixamos a vida de sertanista porque nos convencemos de que cada vez que contactamos com uma tribo estamos contribuindo para a destruição do que ela tem de mais puro.*" (Which roughly translates as, "We are leaving the stage as we are convinced that every time we make contact with a tribe we are contributing to the destruction of the possession that is their birthright, their freedom.")

They battled to save the indigenous cultures and peoples from being overwhelmed by the white man, but stated that they ultimately abandoned the jungle as they never received any thanks for their sacrifices and toil (or only from overseas). Orlando added that 9,000 km² of the Parque Nacional do Xingú had already been taken over again by agricultural interests; clear-felling for cattle-breeding was in full swing in the south of the park – in a region assigned to the indians during the rule of the previous government; while in the north of the park a vast chunk had been taken for the construction of the Xavantina-Cachimbo national highway (BR 080). And that despite the fact that FUNAI had assured the Villas Boas, when they retired, that the park would remain in its original form. Since then the park has been eroded to about half its former size, with no end in sight.

But the Xingú was in the news as long ago as the 19th century. In 1882 the German ethnologist Karl von den Steinen (1855-1929), a member of the first German South Polar expedition, parted from the group in Buenos Aires on his way back from the South Pole. From there, in 1884, he journeyed with his cousin, the painter Wilhelm von den Steinen, across Cuiabá in the source region of of the Batovi river. They followed the river to its confluence with the Xingú and continued thence to the Amazon. They were undoubtedly the first white men to travel down the more than 1800 km long Xingú and investigate it ethnologically. One can only imagine the dangers posed by the almost innumerable hazardous rapids, rock massifs, and jagged rock formations in the Xingú, as well as the months of deprivation. Nevertheless, by 1887 von den Steinen had returned to Brazil for a second Xingú expedition. He was the first to make contact with the previously unknown Bakairi and Kustenau indians (the latter have now died out, while 570 Bakairi still survive).

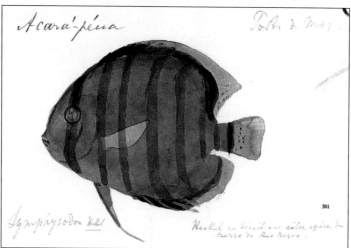

As regards discus, it was apparently the Thayer-Expedition (page 41-42), under the leadership of L. Agassiz, who on the 28th August 1865 collected the first two discus, with the comment: 1°45'S 52°10'W, Rio Xingu at Porto do Moz (= Porto de Moz). By the 23rd August Agassiz wrote in his log that they had caught 27 fish species at Gurupá and 57 at Porto do Moz; 84 in 12 hours, 51 of them new.

The Rio Xingú is a (very) clear river, like almost all of its headwaters. Only the Rio Jarauçu, a lower left-bank tributary, sometimes carries black water.

In addition I found the first live brown discus (often with a yellowish, or almost completely yellow base colour) here in 1975, to the south of Porto de Moz.

Porto de Moz itself today numbers more than 24,000 inhabitants, and lies about 500 km from Belém on the right bank of the Xingú some 60 km from its mouth. On the opposite bank, somewhat further downstream, the Paraná do Aquiqui runs in a north-westerly direction, linking the Xingú with the Amazon

Porto de Moz (whose coat of arms is shown *above right*) lies on the right-hand bank of the lower Xingú, and is the spot where the Thayer Expedition (1865-66) discovered their first discus and collected two specimens. The painter Jacques Burkhardt, who accompanied the expedition, made a colour illustration of one specimen *(left-hand page),* which is almost identical to the fishes that I found here around a century later (see page 241). But the lower Xingú region also harbours an almost quince-yellow variant *(above)*, which I have also found and which is today seen in the hobby. It is an almost perfectly circular discus. There are no discus in the Xingú's largest tributary, the Rio Iriri, but I have caught new fish species there, including the recently described *Moenkhausia heikoi (above right).*

at Almeirim, on the left bank of the latter. It is 66 km long. Porto do Moz, like the majority of towns in the state of Pará, depends mainly on the timber industry. Little is cultivated there and tourism is virtually unknown. The town has a relatively small built-up area, beaches, and a few open spaces, so-called *campos alagados* (wetland) with water buffalo, a number of *morros* (raised areas), *serrados*, (elevations) and *florestas densas* (dense jungle). There are even supposedly still lots of wild animals such as jaguars and pumas.

The town stands on land that once belonged to the Montorus indians, who have long since been wiped out and consigned to history. The tribe possessed a very advanced, for Amazonia, ceramic culture. Pieces of their handiwork can be seen in the Emílio Goeldi museum in Belém. (And possibly some can still be found at Porto de Moz.) Upstream, for a stretch of less than 300 km, discus habitats are to be found, in the lower course of the Jarauçu and in the Rio Acaraí. Further upstream there are rapids with a considerable drop, creating a natural barrier for all discus (and many other fish species).

The Xingú's headwaters rise at an altitude of some 600 m in the Serra do Roncador and the Serra Formosa. The total length of the system measures some 2,045 km, but only 1,815 km of these bear the name Xingú. The river drains a gigantic basin of about 531,250 km^2 with an average width of some 350 km and a length of 1,450 km. Despite containing this enormous mass of water, the Xingú is rarely navigable, or only to a limited extent in some places. Only the last 360 km before its mouth are open to shipping all year. Upstream of the first major rapids (a waterfall in the dry season) only small aluminium boats can navigate between the innumerable rapids – and then not in the dry season (many of those who have tried lost their lives).

The lower Xingú thus consists of the stretch from Belo Monte to its mouth, which drops very little, just like the Amazon itself. Here it is very broad, almost like a giant lake where the far shore cannot be seen. It narrows only at its confluence with the Amazon, where it is some 7 km wide. and the ebb and flow of the tides still have a considerable influence, clearly apparent. In its mouth region there is a series of islands, often submerged, but many of them now clear-felled for cattle pasture, covered in grass or used for agriculture during the dry season.

Of course I must also mention **Altamira**, which lies upstream of the waterfalls, but can be reached via the Transamazonica road network (unlike Porto de Moz which has no road links). There is also a road from Altamira to Victoria do Xingú where the so-called Porto de Altamira has been constructed so that ships' cargos can be transported onward by road from the lower to the upper Xingú (and vice versa). In addition, Altamira has daily air links.

The Xingú has been world-renowned among catfish enthusiasts for some 25 years. Everybody has known its name since at least 1975, when I saw the zebra catfish (subsequently described as *Hypancistrus zebra*) for the first time, and it subsequently appeared on the covers of aquarium magazines worldwide. (I mean the Rio Xingú, not *Hypancistrus zebra*, which is still the most popular of the loricariids in the hobby, but unfortunately since 2004, because of overfishing protected. No collecting or export is allowed until further notice.)

But this enormous river system harbours not only discus and loricariids. New fish species are forever discovered there, with no end in sight. As recently as 2002 Natasha Khardina and myself found several new characins and cichlids in its largest tributary, the Iriri, a to date ichthyologically unexplored river, just like large parts of the almost inaccessible middle and upper Xingú, which may have been investigated by my mother and others, but where so far only the surface has been scratched. (See page 239 for one of the new discovered species, and *aqua Journal of Ichthyology and Aquatic Biology* vol. 9(1).) Just recently a new serrasalmid genus was described from the Xingú, with more to follow.

Only in the case of the brown discus variants, which occur exclusively in the above-mentioned lower course, is there little likelihood of new discoveries. Experts are of the view that the Rio Xingú is poor in minerals and nutrients, but in that case what do its numerous fishes live on? Including the eight to nine species of freshwater rays?

Here are my data for the habitats I have taken – **the water parameters north of Viktoria do Xingú on 24.07.2002 at 1400 were as follows: pH 6.55; conductivity: 21 μS/cm; daytime temp. (Air): 36°C; surface area of water: 28.5 °C; in 2 m: 27.9 °C; no aquatic vegetation; water clear; sandy bottom, in some areas rocks; surrounding area: parts of the river banks inhabitant; some cattle farming; secundary forest. Discus variants brown to light brown and yellow near Porto de Moz (see pages 239 & 241).**

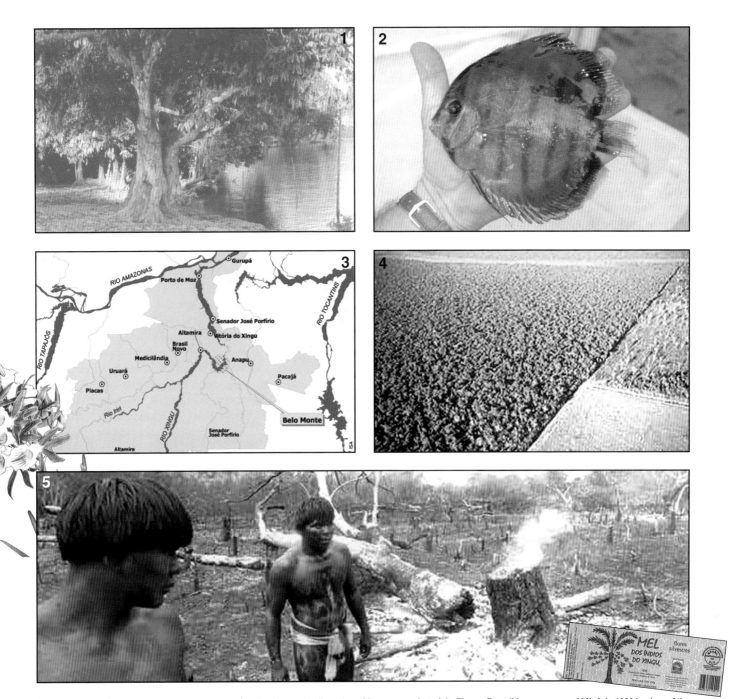

1. A discus habitat downstream of Viktoria do Xingú. 2. The brown that lives there (the same as that of the Thayer-Expedition – see page 238). 3. In 1988 leaders of the indigenous peoples(ie Paulo Paiakan, a Kayapó from the village of A'Ukre), as well as environmentalists worldwide, managed to put a stop to the construction of the power station at Kararaô (now Belo Monte). But since then the might of the electric interests has prevailed. The construction of the hydro-electric dam and the destruction of the jungle at Belo Monte *(map)* are in full swing. The Xingú is (was) one of the last large Amazon tributaries to remain almost unspoiled, surrounded by disappearing indian reservations. 4. The indians protect their forest (clearly visible in the aerial photo – the open areas are cattle ranches). 5. And it is not only the dam that is a blow to the forest people, but also the increasing burning in the south, where the largest national park in Brazil is relentlessly being reduced to cattle ranges.

2. Rio Paru to Santarém; Santarém-Tapajós region; Alenquer region; Obidos and surrounding area.

Rio Paru to Santarém: The Rio Paru is the first large left-hand tributary of the Amazon further upstream from the Xingú. The village of Almeirim lies on the left bank at its mouth *(map 1)*. The vegetation is mixed and unaltered, with dense primary forest on the mountain chains of the Serra Tumucumaque – an outlier of the Guyanan Shield – along the upper and middle Rio Paru, which is actually called Rio Paru de Este (of the east), as further to the west there is another Amazon tributary with the name Rio Paru de Oeste (of the west). The so-called *mata ciliar*, a terrace-like forest, extends from the middle Paru, immediately south of the Equator, to near the mouth region, and it is only to the north of the Serra Pataquara and on the Planalto Maracanaquara that the shore is covered in open woodland, known as the *floresta aberta latifoliada* (open broad-leaved forest) or *cipoal*. The mouth region itself consists of *campos* (open plains) and primary inundation-zone forest.

The composition of the rock formations is complex, just as along the Jarí, and is typical of the Guianan Shield. There are rocks from the Pre-Cambrian including those of the Guiana complex (granite, migmatite, granulate etc); the Vila Nova group (schists, serpentinite, quartzite, itabirite, and many more); the Uatumã group with Mapuera granite, Serra do Mel granodiorite, and the alkaline Iricomé rock. In the middle Parú region the Palaeozoic sediments of the Amazon basin appear, running in a SW-NE direction, and including the Trombetas (Silurian), Curuá (Upper Devonian), and Itaituba formations from the middle Carboniferous. In the mouth region there are Cenozoic (Tertiary) sediments of the Barreira group and recent rock types.

The course of the Paru is for the most part set with waterfalls and rocky rapids *(right & page 244)*, making navigation virtually impossible. Only in the lower reaches are there large accumulations of sand. Its most important affluents are the right-hand tributaries, the Rios Citaré, Itapecurú, Tucuranã, Paicuru and Urucurituba. The Paru flows parallel to the Jari in a NW-SE direction and its waters are sometimes crystal clear. The Waiana (or Wayana) indians inhabited its shores when it was first discovered in the 17th century; they lived in the middle Paru region and upstream to its headwater, the Rio Citaré, as well as along the upper Rio Jari and the rivers Litani and Paloemeu. Nowadays their territory is restricted to the border zone with French Guiana, Surinam, the Maroni, Paloemeu, and Litani, and a second tribe, the Apalaí, who formerly inhabited the Amazon area to the south, the lower Curuá, Maicuru, and Jari, live along the middle and upper Paru. Both tribes languages have their origins in the Carib language. For more than 100 years they have often shared the same villages and also intermarried, and as a result they are by now thoroughly mixed. Since 1997 they have been allocated an area of 3,071,067 ha in Brazil, the Parque Indígena Tumucumaque, which includes the northern parts of the communities of Almeirim, Óbidos, Alenquer, and Oriximiná. In addition there are indian villages which in 1997 were proclaimed Terra Indígena Rio Paru de Leste. In 1998 the total Brazilian population of the two tribes was estimated at 415 living in 13 villages along the middle Paru de Leste. (In 1996 there were still 711 in 16 villages in French Guiana, and some 450 in 12 settlements in Surinam in 1998.)

At the beginning of 1998 I again visited their region on the Brazilian side, in order, as is so often the case, to map the distribution of the discus. I found none. But in the nearby mountain caves I discovered so far undated, unearthly-looking paintings that have been little studied and are worth a visit. And that the Waiana remain true to their traditions and culture. They still believe firmly in their cosmic origins, and this is reflected in their rites and ceremonial objects. Their almost innumerable masks likewise reflect this. A number have antenna-like structures like those of the Hopi indians in the USA. During one dance the men cover themselves in straw so as to recall their ancestors *(right)* and the so-called *marake* ceremony is the most important in the life of a Waiana. This ritual can be undergone a maximum of eight times, and involves the use of the mystical *orok* masks, depicting a the mysterious bird that represents the creation. This ceremony ends with the application of the *kunana*, three mystical figures filled with *tucandeira* ants which are placed on the chest of the youth or maiden. The bite of this ant species is extremely painful (I've had my share...) and, in order to be accepted as adults of the tribe and granted the responsibilities that entails, the youngsters must remain silent while about 50 of these little monsters bite deep into the flesh of their chests.

The stories of cannibalism among the Waiana are untrue and were originated by the *cabanas* (see Chapter 1: *First discovery*). These outlaws spread such rumours in order to justify their own frightful massacre. Accompanied by savage dogs, they

Old photos of the Waiana (Apalai) in the Parú region: 1. Indians with a *Myleus* sp. (discus do not occur in the Rio Paru de Este). 2. The Paru is full of rapids. 3. The lives of the Waiana are (were) enriched through decoration, a social behaviour in which they transform their physical bodies into cultural objects. And not only with masks, feathers, and painting, but also through the ritual of blister decoration, produced via slits in the skin. As well as the piercing of the ear-lobes in the people along the Rio Citaré. The basis of their art and for their daily lives is the local ichthyofauna, from which they derive inspiration and their aesthetic sense. For example, the colour models provided by the *pirarara* (= the red-tailed catfish, *Phractocephalus hemiliopterus* – feather decoration, *left-hand page*) or the common *piaba* (= tetras). 4. The *marake* ceremony with *olok* bird mask. Inset: modern handiwork.

penetrated the region in canoes when the Waiana men were out hunting. They tied the women and children to the huts and sliced off pieces of flesh from their legs, arms, and bodies, straightaway eating the meat raw, with a little salt added. They mixed the blood flowing from the still living, terribly suffering people with *farinha* and consumed it as a side dish. What remained was packed in boxes and sold in the market. They were accursed along the Paru for their repressive behaviour. Not until around 1840 were the *cabanas* finally wiped out. It is a sad fact that, throughout the history of colonisation worldwide, aboriginal peoples have always been grievously treated by so-called civilised people – killed, massacred, wiped out, flayed, and dismembered. The aboriginals, by contrast, have hardly ever been guilty of such atrocities, though repeatedly accused of them.

When I visited this river region for the first time in 1986 the forested area had been reduced by about 5% (according to LANDSAT-TM only about 2.85%). The Parú de Este with its numerous rapids was navigable only with a small *lancha*. But neither I (both then and again in 1998) nor any scientific expedition to date has been able to find any *Symphysodon* species there. Only in the popular literature and on the Internet are discus found there. But if we consider that the river drops more than 600 m in the stretch of less than 600 km from the border zone to its mouth, the absence of the King of the Amazon is hardly surprising. They don't ascend rivers of this type. But I was able to find a few loricariids in the rapids below the Cachoeira de Barreiras *(centre)*, for example a *Baryancistrus*, a *Guianancistrus*, a *Panaqolus*, a *Pseudancistrus*, a *Lithoxus*, and two *Hypostomus* species (one of which is currently sold as *Chaetostoma* sp. "Rio Parú" in the hobby), a beautiful little *Teleocichla* species, and a *Krobia* species. The four vertical bars behind the lateral spot and the interorbital stripe are diagnostic characters of the last-named cichlid genus. *Krobia* are found mainly in the north of South America, specifically Surinam, French Guiana, and in the northern tributaries of the Amazon such as the Rio Jarí and the Parú. I have also found *Krobia* in the Rio Araguaia and in the Mato Grosso.

The Parú de Este is, like the Jarí upstream of the waterfall, a glorious river and still largely unspoiled. Only around Almeirim and to an altitude of 235 m in the hinterland has everything been clear-felled, and a catastrophe threatens in the near future as a result of the continuing erosion. Even so there is a glimmer of hope to the north. In 2002 the superlative Tumucumaque National Park *(right)* was brought into being; at 3,870,000 ha, the largest rainforest national park on Earth to date, larger than the two US states of Massachusetts and Connecticut put together. Not before time, as this region is so unique that, when you visit it, you would think yourself transported thousands of years into the past. But the region is, as already mentioned, almost impossible to reach because of the numerous rapids. There are no roads, and (luckily) no landing place in the entire protected area.

Cachoeira – Rio Parú, and anaconda in the Paru region

The park occupies 27% of the state of Amapá and 53% of Pará State. It was brought into being by Conservation International (CI) in collaboration with the Brazilian government, in order to conserve the biodiversity of the region. Although to date it is virtually unexplored, scientists estimate that more than eight monkey, 350 bird, 37 lizard, and numerous other animal species that have become extinct or are endangered in other parts of Amazonia occur here. In other words, 42% of all Amazonian lizards, almost half the bird species, and about 10% of the the Brazilian primates live here. "Since Tumucumaque is one of the greatest unexplored places on Earth," says CI President Russ Mittermeier, "we can only imagine what undiscovered mysteries will one day be found in the park." Interestingly, I have heard that the indian tribes there are safe from AIDS (HIV) as there are no roads. Moreover the construction of the planned Perimetral Norte has been halted, and it is in the lap of the gods if and when this northern Amazon axial route will ever be built.

Further up the Amazon, and again on the left bank, we pass the Rio Urumu and the Rio Jutaí (the latter not to be confused with the river of the same name near the Peruvian border), along with smaller tributaries and *lagoas;* and on the right-hand side the Rio Guajará (not the Guajará near Belém, also in Pará State). So far discus have not been found in any of these, or in the lagoons, and it seems unlikely that any will be. There are no typical *Symphysodon* habitats to be seen.

1. The Rio Parú de Este (as well as the Rio Jari, *inter alia*) rises in the Serra do Tumucumaque. A region which extends along the northern Brazilian border of the state of Amapá to the state of Pará. It is a gigantic region that has recently been declared a national park, the Parque Nacional das Montanhas de Tumucumaque. It harbours a rich — in part endemic — flora and fauna, which to date remains little studied scientifically. Species of animals threatened with extinction in large parts of Amazonia are still found here, for example: 2. The toco toucan *(Ramphastos toco).* 3. The Guianian cock-of-the-rock *(Rupicola rupicola).* 4. The harpy eagle *(Harpia harpyja).* 5. The jaguar *(Panthera onca),* and 6. the green Boa canine *(Corallus caninus).*

By now we have reached the *município* of **Prainha**, also on the left bank. The original settlement area was along the shore of the Rio Urubuquara (also known as Furo do Outeiro), at the village of Outeiro. But the access (by water) was very difficult and so everyone moved to the mouth region, next to the Amazon. In 1758 the settlement was raised to the status of *freguesia* (like a small Municipality) by the Portuguese Francisco Xavier de Mendonça Furtado, and in 1881 to *município* by the Brazilian government. In 1988 the region was split into in two *municípios*, Uruará and Medicilândia, and the villages Prainha, along with Pacoval, were granted an additional *distrito* (district) status.

Shortly before Prainha and after the the Rio Jutaí, the Rios Parauaquara and Jauari enter the Amazon, but they too originate in the mountainous Guianan Shield, are full of rocks and rapids, and offer no discus biotopes. Except for the areas of lakes in the lower Jauari region, like the Lago Ipaiva, which has not been explored, and Lago Maripá at the end of the Furo de Outeiro, were recently I was able to discover Browns *(S. haraldi)*.

On the right-hand side of the Amazon at this point we find two more tributaries belonging to the so-called Baixo Amazon (lower Amazon), the Pará do Uruará and the Cucari. Here too there is no evidence of discus, although both have *lagoas* with typical discus habitats in their mouth regions. (I fished the Rio Uruará several times in 1975 and 1976.) The Uruará is connected to the Amazon mainly via the numerous *lagoas*, especially at high water. Unfortunately, according to LANDSAT-TM, by around 1986 about 11% of the primary forest had vanished. But the real damage started after that. The *terra firme* region and in particular the *várzea* islands, as well as their entire ecosystems, have been senselessly ruined. The large islands are now just pastures for zebu cattle. *Acará-açú* (the plants discus dwell under, protect and nurish them), Itanduba, Muiratuba and many others are history...

Monte Alegre is the next stop. It lies on the *terra firme* some distance from the Amazon, and there is actually a road, albeit a bad one negotiable only by four-wheel drive during the dry season, leading from the village to the Serra da Lua 35 km to the north. There may be no discus there, but there are incredibly beautiful caves with paintings said to be 9,000-12,000 years old. There are also rock formations sculpted by wind and weather, and the mountains Paituna and Ererê are well worth seeing *(right-hand page)*.

But there are Browns (one of the so-called Alenquer variants) in the Lago Grande, also known as the Lago Grande de Monte Alegre. It is also interesting to note that it is here that the first (relative to the mouth of the Amazon) discus *(S. haraldi)* with patterning on the body are encountered. In addition, in this habitat the discus exhibit sexual dichromatism. In other words the males are essentially far more colourful (more pattern colour and often a greater amount of red) than the females *(right-hand page)*.

The water parameters in the Lago Grande on 16.02.2004 at 05:30 hours were as follows: pH 6.57; conductivity 22 μS/cm; daytime temperature (air): 28.5 °C; at the water's surface: 27.2 °C; and in 2 m of depth: 26.4 °C. The surrounding area was natural, pristine; *acara-açú* scrub adjoining the habitat, partly submerged; water hardly flowing; floating aquatic vegetation *(Utricularia foliosa and Pistia stratoites);* **lake bottom sandy.**

For discus from that region see pages 247-249.

From here on, heading west, the regions are pervaded by even more water. Innumerable *paranãs, lagoas, rios, igapós,* and in particular gigantic lakes, criss-crossing the Amazon basin to the Peruvian border and beyond. But first let us take a look at the Rio Curuá-Una, which joins the Amazon immediately to the south of the Lago Grande de Monte Alegre. It flows from the SSW and is fed by two large tributaries, the Curuá do Sul (formed by the Tutuí and the Uruará and also known as the Curuá-tinga), and the Rio Moju (with its tributary, the Rio Mujuí). It is a clearwater stream (at least it was until the completion of the hydro-electric dam of the same name). The hydro-electric plant 80 km upstream was commissioned in 1977 to supply power to Santarém. Unfortunately for some time it has been producing less and less energy so that the people of the third largest city on the Amazon experience continual power outages, at least during my last two visits in 2003 and 2004.

Its largest affluents, the Moju and Mujuí, both left-bank tributaries, join it in the the reservoir area. And this left-hand bank of the reservoir is characterised by deforestation, settlement, erosion, silting, and poor water quality. The right-hand bank is very different. The dense tropical forest is still intact there,

Lago Grande – Discus Habitat

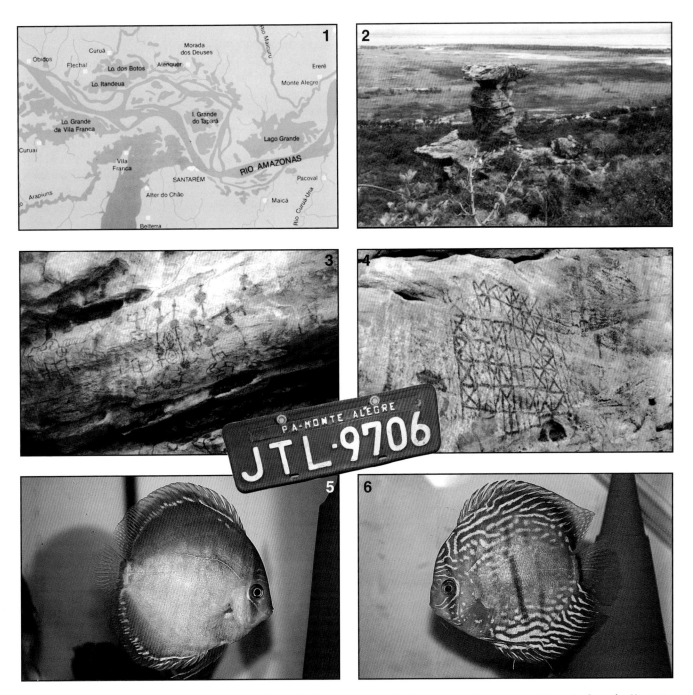

1. On this map we can see the gigantic size of the Lago Grande (de Monte Alegre – right in the photo) and the waterways of the entire Santarém-Alenquer region. 2. A view of the Lago Grande from the Serra da Lua with its unique rock formations. 3-4. The cave paintings in the Serra da Lua are said to be 9,000 – 12,000 years old, though in the course of time various cultures have frequently painted over them and the top layer of the approximately 170 m² painted surface is said to date from the 16th century. It is possible to make out humans, animals, and geometric patterns. Wallace discovered them back in 1853. Since then similar art has been found at Balbina, Alenquer, and Óbidos 5. Female discus from the Lago Grande. 6. Male discus from the Lago Grande.

the area is only sparsely inhabited, and there is virtually no pollution.

My first journey here in 1976 resulted in the discovery of brown discus some 30 km downstream of the by then already completed dam. They didn't have any special pattern colour.

The water parameters in the typical discus habitat in the Curuá-Una in August (unfortunately the day is washed away from my notes) of 1976 around mid-day were as follows: pH 5.5; conductivity 25 μS/cm; daytime temperature (air): 27.5 °C; at the water's surface: 29.3 °C; in 1.9 m of depth: 28.4 °C; oxygen saturation 75%, with a concentration of 5.8 mg/l. The surrounding area was intact; the habitat full of roots and fallen in trees; water slighty flowing; no aquatic vegetation; river bottom sandy. Discus only Browns.

During my second visit, in the 1980s, I saw thousands of *jaraquís (Semaprochilodus brama)* swimming to and fro beneath the tall hydroelectric dam, trying to make their way upstream to their habitual spawning sites. And I was told that various species upstream of the dam had died out, for example, the *pirapitinga (Piaractus brachypomus)* and *jatuarana (Brycon amazonicus)*. Though a *fazendeiro* told me: *"Tem muito tucunaré e charuto na represa"*. Which means that *Cichla ocellaris* and *Boulengerella maculata*, two predators, have multiplied in large numbers, and I know from the ichthyologist Efrem Ferreira, who works at the INPA in Manaus, that in the first two decades after the dam was built also

a piranha species (*Serrasalmus rhombeus*) had bred in greater numbers than ever before known.

Now, despite assertions to the contrary, there is no doubt that this dam was ill-conceived. This is confirmed by the following facts. 1. A statement by the *engenheiro* Antônio Ramalheiro, coordinator of the Usina Hidroelétrica de Curuá-Una, published on the 4th February 2003, declares: *"A empresa Rede Celpa precisou reduzir a potência da usina para um terço da capacidade em função do nível das águas do rio, que está bem abaixo do necessário para movimentar as três turbinas. Apenas uma delas está funcionando, com capacidade de geração de 10 mil kilowatts."* In other words, the water level has dropped so far that only one turbine is still working. 2. When the planning began in the 1960s four turbines were proposed, but the fourth was never commissioned. Of the three that were installed, only one is still working and so the plant produces only a quarter of the energy originally envisaged. Meanwhile Santarém experiences frequent black-outs. 3. The Museu Paraense Emílio Goeldi has demonstrated that the shortage of the water required by the turbines is due to clearfelling along the river banks. The lack of vegetation has had a detrimental effect on the soil, and this has led to erosion on a gigantic scale, so that the river bed is becoming continually shallower. Today the water level is only half what it was originally when the dam was built. 4. Following the damming of the waters, eutrophication (an increase in undesirable nutrients leading to the growth of useless plants (algae)) took place, and persists unabated. As in the past, the lake contains an excess of nutrients and diminishing oxygen concentrations in the somewhat deeper layers of the water. In addition there is an excess of phosphorus (from the *fazendas* along the Transamazonica highway to the south), and this is a limiting factor for biological productivity. 5. The eutrophication was accompanied by an excess of iron which has long since caused considerable corrosion damage to the turbines, which inhibits the outflow severely. 6. After some three decades there are still thousands of dead trees protruding from the lake, impeding the boat traffic and the anglers. Their decay has been limited by the absence of fly larvae, which have never been unable to gain a "foothold" due to the dramatic fluctuations in the water level.

And now, after 30 years, they will take an inventory of the construction site and the turbines of the Curuá-Una dam – including for safety reasons...

None of the Amazon hydro-electric dams has so far resulted in a positive balance. Not, historically speaking, in terms of supply of energy, and most certainly not as regards nature and the environment. The results are obvious for all to see, and confirmed by the Curuá-Una hydro-electric plant.

As regards the world of discus, neither the dam nor the numerous *comunidades* (communities) along the banks of the middle and lower Curuá-Una, for example, Nova Canaã, Bananeira, São José do Aru, Santa Maria do Aru, Xavier, Tambor, Castanheira, Porto Alegre, and Porto Novo e Corta Corda, will have done anything to ensure the continuing existence of discus habitats below the dam. The days of the "King" are numbered here just as in the Tocantins, or they may already be history.

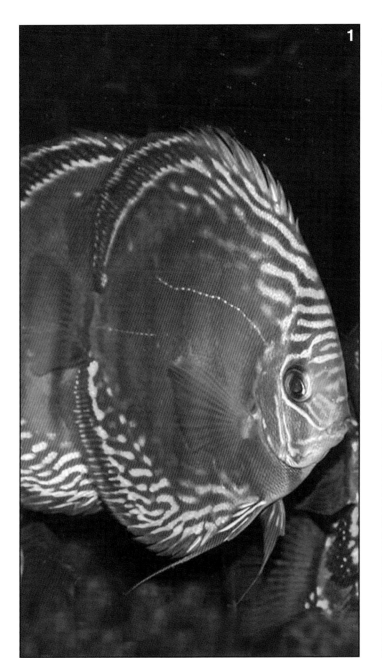

1. As already mentioned, the majority of individuals from the Lago Grande exhibit sexual dimorphism *(see also left-hand page:* male in the background, female in front). The male shown here is one of the variants that entered the hobby prior to 1975 as so-called "Alenquer" discus (see the book *DISCUS* by Bleher H. & M. Göbel). The town of Alenquer itself lies not far away, but somewhat further to the west of the Lago Grande, and there is also a lake of the same name there (see also main text). 2. Some individuals from this region, for example this male from the Lago Grande, have a considerable amount of pattern colour and a lot of red in their base colour. (The red initially made discus from the Alenquer region very popular with breeders.) 3. In some juveniles from the Lago Grande a red component can clearly be seen at an age of only 7-9 months.

By good chance, in 1976 – shortly before the commissioning of the *hidroeletrica* – Jacques Géry, along with the INPA institute, performed a census of the fishes and so we at least know what species used to be there. In addition, an English team researched the reptiles at that time and found 19 different lizard species alone that occupy the tiniest of specific microhabitats and are found nowhere else – only a few are predatory and go hunting. Interestingly, a short time ago the UFJF-ICB (Departamento de Zoologia – Instituto de Ciências Biologicas – Universidade Ferderal de Juiz de Fora) and INPA performed a new census and established that a number of species have died out, although numbers of predatory fish have increased greatly and dominate many areas, mainly upstream of the dam. But among the 3,052 fishes caught over a six month period there was only a single brown discus, from site VI, 20 km below the dam. And that was it.

Santarém-Tapajós region: Santarém was founded on the 22nd June 1661, on the site of a Tupaiús indian settlement, by the Jesuit Jōao Felipe Bettendorf, who came from Luxemburg and had studied art and civil law in Italy before entering the Jesuit order at the age of 22. The region had by then already been inhabited for a long time. Recent finds of ceramics indicate that people were already living here 7,000-8,000 years ago. They had an important and complex culture, composed of tens of thousands of communities, which was apparently established here long before the Mayas, Aztecs, and Incas (for more details see Bleher H. *et. al* 2003a: *Pre-Columbian*). They possessed a social structure comprising several classes, and were already masters of ceramic techniques to a degree unknown from any other contemporary culture. We do not know what they were called, but we do know that at some stage they were overthrown by the warlike Tupaiús (also written Tapaiús or Tapaju) from the Tapuiuçu region. The Tupaiús were of larger and heavier build by comparison, but they didn't have any cannibalistic tendencies and allowed themselves to be taught the art of ceramics and how to burn it.

The aboriginals called their river the Paraná-pixuna, or just the Ipixuna, which approximates to "black river". The invaders re-named it Tupai-paraná – the river of the Tupaiús. When Orellana and his men passed through in 1542, one of his comrades, the Dominican monk Frei Gaspar de Carvajal, noted Tapaiós (or Topaiós) in the records, and this eventually became Tapajós. He also wrote about the indians' population density, their ruler Nurandaluguaburabara *(centre),* and that the Europeans were shot by these indians with poisoned arrows while plundering the Tapajós maize fields. (Orellana was supposedly the first to bring maize to Europe.) Thus from the time of the discovery of the Amazon the Europeans called the tribe and their village Tapajós, and their river still bears the name today. But it isn't black, it just looks black where it mingles with the whitewater Amazon at their confluence (see page 262).

The settlement was re-named Santarém in 1758 by the then *governador* of the province of Grão Pará, Francisco Xavier Mendonça Furtado, who planned to turn Amazônia into a second Portugal and gave the native villages (which he called barbarian settlements) the names of Portuguese towns. And I can't resist relating the very interesting history of the "mother city" on the Iberian peninsula. It is my belief that almost every Portuguese is proud of this lovely town on the Tejo (see pages 252-253).

The story goes that in 1215 BC, Prince Gergoris (or Gorgoris) Melícola ruled in the kingdom of Lusitânia. He bore the name Melicola (Lat. *mel* = honey, *cola* = one who cultivates) as it was he who discovered how to extract honey from the comb and thus brought prosperity to the land. One day Odysseus (Homer's famous hero) landed at the mouth of the Tejo. He found there a wonderfully peaceful oasis where he rested, and he planned to remain in the country where the honey flowed until the opportunity arose to sail back to his home in Greece. Gergoris had a lovely daughter of enchanting beauty who bore the name Calypso, and Odysseus immediately fell in love with her. And the result of this love was a prince whom they named Ábidis. When Gergoris learned of this he was seized with rage and sent his soldiers to arrest Odysseus, but the latter immediately hoisted his sail and departed, regretfully leaving Calypso

Top: A copper engraving from the expedition of Martius and Spix to Brazil, during their explorations in the region of Santarém, in canoes with Tupajú indians (the name of the erstwhile tribe can still be seen on the boat inset left). This what it looked like then (around 1819). *Above:* And this is what it looks like at Santarém today (inset: a typical car number-plate). The city is now the third largest Brazilian Amazon metropolis.

and Ábidis behind. To avoid news of Calypso's transgression getting out, the ruler ordered the child to be placed in a basket on the Tejo to drift out to sea. But instead the river carried the basket upstream and it ended up on the bank close to a herd of deer. A hind took the baby as her own and suckled him, and he grew up in the forest, living on fruits and plants. Not until he was 20 was he discovered living there in Gergoris's kingdom. Calypso immediately recognised him as her own offspring. The ruler made peace with Ábidis and named him as his successor. The young prince is said to have been the best and and fairest ruler of all time. And, to thank the wild animals, he founded what is now Santarém in the place where he had grown up, only then it was called Esca-Ábidis (= the place Ábidis came from), and turned it into a paradise on Earth (it is still known as "Paraiso do Portugal" today). The name Santarém didn't come into use until much later (see page 253), and is derived from the name of Saint Irena (Eirena or Herena), ie *Sancta Irena* became Santarém.

The Brazilian Município de Santarém occupies an area of 26,058 km² and the city of Santarém lies on the *terra firme* only 29 metres above sea level, on the shore of the crystal-clear green-blue Rio Tapajós, right where it joins the muddy waters of the Amazon. The average temperature is around 30 °C, and lower than 22.6 °C has never been recorded there. By 2003 the population numbered some 300,000 people, but when von Martius passed through on the 18th September 1819 and almost foundered in the ocean-high waves of the Amazon off Santarém during a storm, he noted in his records that barely 2,000 souls, mainly Tapajós indians, lived there.

The inhabitants, the so called *santarémense* live mainly by agriculture (including cattle- and pig-breeding), fishing, and commerce. Naturally, the timber industry accounts for the bulk of exports.

Tourism is as yet hardly worth mentioning, although there are very interesting places to visit both in and around Santarém. For example, the nearby Alter do Chão is being developed as a weekend and holiday resort on the Tapajós, right on the splendid white sandy beach. And ceramic remains from the distant past can be found not only in the vicinity of Santarém, but also in Alter do Chão, where more than 2,500 pieces from various cultures and tribes can be seen in the Centro de Preservação da Cultura.

But the region that is today the Município de Santarém has not only known discus discoverers such as Agassiz *et al.*, Jobert, and many more, but was also the place from which the first live specimens were exported for the aquarium hobby (see Chapter 1: d*iscoveries in the 20th century)*. It is thus world-famous for discus as well as for its beautiful "mother city" in Portugal, but perhaps even more renowned for the part it played in the history of the harvesting of natural rubber. A story that changed the world and has been told in many versions, but which apparently had its beginnings on the Tapajós.

La Condamine, who passed through here in 1743 and was the first European to perform scientific research in the Amazon region (see also Chapter 1, pages 44-45), brought a substance which he called *India caoutchouc* to the Institute de France in Paris. In his publication *Relation Abrégéé d`un Voyage fait dans l`Intérieur de l`Amérique Méridionale*, which appeared in 1745, he devoted a section to *caoutchouc*, a name from which the term for rubber derives in many languiages today).

From a chemical viewpoint, rubber is a complex-molecule hydrocarbon contained in the milky sap of some 500 plant species (but the sap of *Hevea brasiliensis* is the best). It is a *cis*-1,4-polyisoprene with the gross formula $(C_5H_8)n$. The milky sap (technically known as latex) is contained in a system of capillary vessels at a pressure of 1.2 to 1.5 mPa in the bark portion of the trunk and branches (not in the wood). Tapping severs the capillaries and leads to a drop in pressure to about 0.2 mPa.

Condamine had already written (in 1736) that the South American indians made the sap into bouncing balls to play with, and

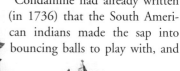

Santarém in Portugal *(left)* is the "mother town" of the Amazon city of the same name, from which came the first live discus. Santarém has a fascinating history (see page 250). Ábidis, the then ruler of the region (Lusitânia), was chosen to create a "Paradise on Earth" here. It was called Esca-Ábidis until Julius Caesar and his army annexed the Iberian Peninsula after the defeat of Hannibal (see Bleher, H. 1995: *Tunesia*) and christened it *Proesidium Julyum*. But in 653 AD, when the Swedes ruled the town, calling it Escalabicastro there occurred the miracle of the maiden Irena, the Santissimo Milagre (memorial, *left)*, and from then on it was called Sancta Irena until the Moors occupied it in 715 and called it Chantarin, Chantirein, and also Xantarin. Not until after the liberation by the Spaniards did this metamorphose into Santarém. It was the *cidade mais bem defendida de toda a Espanha* (the best-defended citadel in all Spain), as the citadel *(left-hand page)* was never captured in its almost 3,000 years of history. The cathedrals *(left-hand page)* are:
1. Drawing of that of Santarém in Brazil, and
2. photo of that of Santarém in Portugal.

HABITATS ... SANTARÉM-TAPAJÓS REGION

as a glue for assembling their feathered decorations. In the Ecuadorian provinces of Quito and Esmeralda he had had the opportunity to observe the natives harvesting rubber. In his report he remarks that the trees were called *he ve* (from which the Latin name was later derived) and produced a white milky fluid that turned brown or black as soon as it was exposed to the air. This was the first mention in a scientific work of the milky sap that would (much) later revolutionise the world. A fluid which the Spaniards soon afterwards named *látex*.

In his report from Tapajós (Santarém) he wrote that the natives there called it *caaochú* or *Caaochó* (*caa* = wood; *ochu* = tears, hence "weeping tree") and used it to make shoes and flasks. They prepared an earthenware mould and covered it with the (latex) mass, then left it to dry. They repeated this procedure until they had achieved the desired thickness. The mould was then smashed, the pieces removed, and there remained a perfect shoe or flask. In similar fashion they impregnated cloth to make it waterproof. The missionaries turned this to their profit, getting the Tapajós to produce water-tight shoes and starting a trade in rubber goods even back then. As well as shoes, they traded balls plus flasks and rubber containers (bags) for transporting water or wine (the Portuguese called these bags *borracha*, which remains the Brazilian word for rubber even today).

The Iberian conqueror initially called the trees *Syrinx*. Greek mythology was in vogue at the time, and, just as the name of the Amazons had been adopted for the river, an analogy was at hand for the "weeping" tree. The story goes that the dryad (wood nymph) Syrinx, patroness of the law, of equilibrium and harmony, of fairness, and of respect for natural law, spurned the advances of Pan. Fleeing, she found her way barred by the river Ladon. She begged her sisters, the river nymphs, to save her, and they turned her into a handful of reeds, whereupon the trees wept at their loss. The wind blowing through the reeds made music which enchanted Pan, so he bound some of them together with wax to create his famous panpipes, on which he thereafter played his tunes, calling them *syrinx* (the Greek word for panpipes) in memory of the nymph. These pipes were used for theatre music in Graeco-Roman times and subsequently entered into European folk music via the stage, *cf* Papageno in Mozart's *Magic Flute*. In Brazil *Syrinx* became modified into *sering*, hence a number of Brazilian terms still in use today: *seringa* for the trees (or *pão de seringa* = *Syrinx* wood) and *seringueiros* for the rubber collectors. (Chico Mendes, who was murdered in 1988, posthumously became the most famous of all the *seringueiros*.)

All the rubber products manufactured by the indians were now exchanged for hocus-pocus. It went so far that by 1755 rubber shoes had become the height of fashion in Pará and Lisbon, and King Joseph even sent his riding boots to Brazil to be waterproofed with latex, so he could go hunting in Portugal without worrying about getting his feet wet...

In La Condamine's description we find two rubber processing methods still in use today: dipping objects in latex and impregnating textiles with it. La Condamine is known as the father of the latex industry, but it was his colleague François Ferneau (1703-1770) who first (in 1762) scientifically analysed the chemical reactions of the natural product in the test tube. Then, in 1770, the self-taught chemist Sir Joseph Priestley (1733-1804, a theologist by profession, with no formal scientific training) discovered something revolutionary. Using one of La Condamine's pieces of rubber he established that the material was ideal for removing pencil and charcoal marks from paper (he actually used the term "rub off", which led to the English name India rubber; elsewhere in Europe it was called *peaux de nègres*, = negro skin). But he also established that it didn't last long and was susceptible to extremes of temperature. (Note also that Priestley was, *inter alia*, responsible for the invention of soda water; discovered that graphite is an electrical conductor; was the first to isolate carbon dioxide and describe its properties; discovered nitrogen and oxygen; and was the first to discover that photosynthesis takes place in green plants, and is the assimilation of carbon dioxide with the aid of sunlight to form carbohydrates, in accordance with the equation $6CO_2 + 6H_2O = C_6H_{12}O_6 + 6O_2$. And that photosynthesis is the most important biochemical process on Earth.) Moreover it is said that at about the same time (or earlier) the English engineer, Edward Naime, accidentally picked up a piece of La Condamine's material instead of breadcrumbs to erase something he had written. After this discovery he produced pieces of rubber commercially and sold them as erasers.

Thirteen years later, on the 27th August 1783, the people of Paris were witnesses to a spectacular event. More than 50,000 people were astounded and amazed by the ascent of the brothers Joseph-Michel and Jacques-Étienne Montgolfier's first

Sir Joseph Priestley

Above: The Englishman Joseph Priestley was the first chemist to study gases: he was responsible for the invention of soda water *(above left),* paving the way for the drinks industry worldwide; and established a record that remains unbroken to the present day by discovering eight different gases (1772-74): NO_2, N_2O, HCl, CO, NH_3, O_2, SO_2, and SiF_4 *(above). Left:* After the name Amazon had been borrowed from Greek mythology, it was appropriate to do likewise with the "weeping" tree *(Hevea) (left),* which was named *seringa* (= rubber) after the nymph Syrinx. Fleeing to escape Pan's advances, she was turned into reeds and her name became the word for the Pan-pipes he made from them *(far left).*

hydrogen-filled balloon, constructed by a physics professor, Jacques Alexandre César Charles. This technological achievement was made possible by the milky fluid, which he used to build up the layers of the balloon.

In Europe there was increasing speculation with rubber, catheters were manufactured in large numbers, and towards the end of the 18th century more products as a result. But then, because the natural rubber was susceptible to temperature changes, things came to a halt for a long time, with no further breakthroughs until 1839.

Charles Goodyear (1800-1860) had already been playing around with latex in the USA for many years when he carelessly dropped some latex mixed with sulphur on the hotplate of the stove in his miserable kitchen *(right)*, which doubled as his experimental laboratory. And lo and behold, instead of melting the mass was "cured" (= hardened = vulcanised). It was still flexible the next day, and thereafter the process of heating a latex-sulphur mixture was called vulcanisation (after the Roman god of fire, Vulcan). When vulcanised the rubber lost its susceptibility to temperature changes.

This discovery opened the way for hundreds of practical applications for rubber. In June 1844 Goodyear patented his "accident" and experimented further – but this required expensive investment. When he died, in 1860, he owed 200,000 dollars, and his last words were, "I die happy, others can get rich." Moreover, as a result of Goodyear's discovery breadcrumbs were finally replaced by rubber erasers, and to the present day this type of "rubber" is used to erase just about everything (though there are others). And in 1858 the Philadelphian Hyman Lipman patented a pencil fitted with a rubber eraser, and these too still exist today.

But someone else had already been experimenting with rubber before Goodyear – Charles Macintosh (1766-1843) of Glasgow, Scotland, whose father had since 1777 been running a factory where violet-red dye (powder) was produced from lichens. And Charles began to show an interest in chemistry at an early age. In 1818, while researching how to make further worthwhile use of the by-products of coal-gas production, he stumbled upon coal-tar-naphtha, which liquified india rubber. He took a piece of woollen material, coated one side of it with the diluted india rubber, and laid a second piece of the same material on top. When it had dried he established that it let no water through. He then worked for a long time, together with a friend, the chemist Thomas Hancock, until in 1824 this led to the development of reliable, rain-proof tarpaulins and coats. The latter became known as mackintoshes. And the mackintosh too still exists today, although the copyrights are now owned by the Dunlop Rubber Company.

Demand increased, and the raw latex balls, as well as the utensils made from them, could still be obtained only from the natives in South America, the Tapajós. (The indians got hardly anything out of it, the missionaries a lot, and the Europeans the most). But, by and large, it wasn't enough.

In 1842, following Goodyear's discovery, some of his vulcanised material came into Hancock's hands. Goodyear wanted the English patent (and not only the English) and had sent samples from the USA. The piece which Hancock obtained exhibited slight traces of yellow which he was immediately able to interpret as sulphur. And thus in 1843, a few weeks before Goodyear, Hancock secured the patent for himself in England.

Hancock refined the process and made his fortune from it. When, in 1845, R. W. Thomson invented the tyre, plus rubber hose, and even an iron wheel, the bicycle was born. By 1850 almost innumerable rubber goods were being produced – toys, balls, dolls, grips for tennis rackets and golf clubs, and many more. The *vélocípède* (first bicycle) was invented (by Michaux in 1869), and eventually the car tyre. And with the advent of Hancock's "pickle", which masticated rubber, or rubber scrap, into tiny pieces ready for processing (a process he had invented, using a home-made machine – the masticator or "pickle", as it was also known), the manufacture of inflatable cushions and mattresses, hoses, pipes, solid tyres, shoes, packaging materials, and springs began. Hancock became the largest manufacturer of rubber items in the world and planned his own rubber tree plantation. In 1853 he suggested to the Royal Botanical Garden in London that they themselves should cultivate rubber plants. As demand increased attempts were made to produce artificial rubber as well, but this didn't become technologically possible until around 1910, and then in the 1930s Buna, neoprene and latex products became widespread. (Buna was the patend and trademark name for the first synthetic rubber produced by the German I.G.-Farbenindustrie. At the peak they had 500 partner companies around the

The Para rubber tree *(Hevea brasiliensis)* is the most important rubber-producing plant, representing more than 90 % of the world harvest. (Rubber is found not only in higher but also in lower plants such as fungi of the genera *Lactarius* and *Peziza*. And not only in the milky sap of the bark, but also in leaves, stems, twigs, roots, and fruits. In the case of a dandelion species, *Taraxacum* sp., cultivated in the former USSR, the rubber is extracted from the pulverised roots.) The natural distribution of the up to 30 m high tree with a diameter of more than 1 m is Amazonia (see page 261), but nowadays it is found in Papua New Guinea (1-2) as well as circumtropically elsewhere (3). There are magazines about rubber (4), and innumerable books (5).

world, plus 400 in Germany and 200 large plants. With the increased demand for synthetic fuel and rubber during world war II a 1 billion Reichsmark-factory was built in Auschwitz in 1941, were 25,000 prisoners died during its construction. After the war it became Polish and is today the largest employer of the Auschwitz area...)

Thus Santarém became famous, thanks to La Condamine's bits of rubber, that, at least after Goodyear's "accident" the whole world wanted *he ve*. People had in fact repeatedly experimented with extracting rubber from more than 100 different plants from Africa, Central and South America, and Asia, but, quite simply, the *Hevea* species produced the best stuff.

The interests of the chief world power of the time, the Commonwealth, were, of course, at the forefront, spurred on by Hancock. And Sir Joseph Dalton Hooker (1817-1911), the most famous botanist of the 19th century and subsequently director of the Royal Botanical Gardens at Kew, didn't need to be told that twice. He was already Assistant Director, and had learned a lot from the Director, his father, William J. Hooker, during the translocation of the *chinchona* or Peruvian bark tree *(Chinchona officinalis,* said to be named after the Countess of Chinchon) from whose bark is obtained the alkaloid quinine (whose name derives from the indian term *quinaquina);* a medication which the South American indians had long used successfully against the dreaded malaria, from which millions of people have died and are still dying today (more than from any other infectious disease). It had been successfully transplanted to India where it was significantly cheaper to exploit than the wild plants.

He not only headed the Royal Botanic Garden – at that time the largest in the world with 4,500 live species and 150,000 herbarium specimens – from 1865 to 1885 as his father's successor, but from 1844 on was a well-known botanical taxonomist and Darwin's closest friend. With Darwin coming genuinely to rely on Hooker. This had started by 1844, when, on the 14th January, Darwin wrote a letter to Hooker, in which, he said, *inter alia,* "I am almost convinced (quite contrary to the opinion I start-

Sir William Jackson Hooker

ed with) that species are not (it is like confessing a murder) immutable", and "I think I have found out (here's presumption!) the simple way by which species become exquisitely adapted to various ends". The "simple way" was, of course, via natural selection, and Hooker was the first person in the world to learn of Darwin's secret.

On the 4th June 1873 James Collins, Curator of the Museum of the Pharmaceutical Society in London, who had also recognised the importance of *Hevea* for the plantation industry, had donated a few hundred seeds to the Garden, but barely a dozen plants germinated, and these had been sent to India, to Calcutta, where they died due to the climate. Hence, after these first experiments with *Hevea* seeds from Brazil had failed, and having got the green light for his scheme from the British government, Hooker commissioned his fellow-countryman Henry Alexander Wickham (1846-1928) to bring viable seeds to Europe at the beginning of 1876.

After an adventurous trip via Trinidad, up the Orinoco and its waterfalls, down the Rio Negro and its rapids, and then down the Amazon to Santarém, Wickham smuggled 70,000 seeds and out of Brazil via Pará. The majority (or all) of them were collected from the Tapajós, and on the 14th June 1876 he brought them safely to London. But at Kew the germination rate was only 4%, so that only about 2,800 seedlings were available. On the 9th August 1,919 young plants were successfully freighted to Ceylon (now Sri Lanka) in Wardian cases (forerunners of terraria and aquaria – see Bleher, H. 2004b; *Aquarium History part 4),* and three days later 50 to Singapore. But the freight costs hadn't been paid on time and the valuable cargo rotted. Hooker then sent the last 22 seedlings to Singapore on the 11th June 1877, and it was supposedly these 22 that wrote a new chapter in the history of rubber and changed the world forever.

And these 22 plants made it possible that 30 years later, the first 6,000 tonnes of plantation rubber arrived on the world market, and before long it was millions. And the major part of this revolutionary development was due to the Director of the

1. *Hevea brasiliensis* seeds photographed in the Congo Republic, where the Belgians established enormous plantations.
2. Henry Ridley *(left)* was a pioneer of the rubber industry. As Director of the Botanic Gardens in Singapore (1888-1911) he was involved in Malaysia becoming the largest rubber supplier in the world. He discovered the herringbone-cut; that when the latex was being harvested, the removal of a thin layer of bark at the end of the cut would allow the sap to flow better and the bark to regenerate more rapidly; the optimal method of planting and starting to harvest, etc.
3. It was thanks to Henry Wickham that *Hevea* seeds reached Singapore from Brazil, thereby disenthroning Brazil.
4. And Italian monks still process *Hevea*-seeds into *Amaro Tonico,* an antique medicine...

Botanic Garden in Singapore, Henry Nicholas Ridley (1855-1956), who had learnt his trade as a gardener at Kew. He also discovered the herringbone-cut-method, which is still used by rubber-tappers worldwide. Nowadays, as in the past, the largest plantations are in Malaysia and Indonesia, but the rubber tree is also widely cultivated in Sri Lanka, Thailand, Vietnam, India, Myanmar (formerly Burma), and New Guinea, and in tropical Africa and America.

Now, after 1913 the commercially-oriented Asian rubber cultivation weakened Brazil's economy tremendously Brazil, but provided work for millions elsewhere. (There are now more *Hevea* trees than people and at least three million rubber-tappers worldwide, with a turnover of more than 6 billion dollars by 1970.) But the Brazilians shouldn't be so upset about the "bio-piracy" that everyone talks about nowadays, as they had coffee smuggled in via Cayenne by Francisco de Melo Palheta (after whom the largest coffee brand in Brazil is named) long before their *Hevea* seeds left the country, and as a result Brazil became the largest coffee-exporter on Earth while Ethiopia went down the tubes. Much the same happened with the soya bean (for which they are now world leaders), mango trees, sugar cane, and many others.

I just wanted to mention that, and likewise that it was La Condamine who started the worldwide economic miracle from Santarém and that the Tapajós played a part in it. However, La Condamine had something else, subsequently of considerable use to ichthyologists, "in his bag of tricks": fish poison. It has many names, such as *menico* among the Huaorani indians; *timin* among the Shuar; *barbasco* among the Quichua, and *timbó* in the *lingua geral*. The Brazilians also call the "poison suppliers" (ichthyotoxic plants) *oreja de negro, pacará, timbó colorado, timbó cedro*, or *tamboril*. And the majority of these plants belong to the family Fabaceae. All species, and especially the frequently utilised *barbasco (Lonchocarpus nicou)*, also known as rotenone *(below centre)*, are widespread in Amazonia and the Suhar in Ecuador even cultivate them. The roots and leaves contain rotenone, which is a readily degradable insecticide also used in agriculture. The indigenous people have been using this nerve-paralysing poison for fishing for millennia *(right-hand page, below)*. Roots or twigs are pounded, or shredded with a knife then pressed, in order to obtain the poisonous milky sap.

The ichthyotoxin is best used only in still waters with a low water level – and not during the rainy season. The poison is added to the river at suitable spots, where it spreads out and clouds the water. After a few minutes the fishes, paralysed and unable to breathe, tumble helpless to the surface and are collected by hand or with nets. Because the poison is highly toxic to cold-blooded creatures, but virtually harmless to warm-blooded, the fishes caught in this way can be safely eaten by people.

Its use is now banned in Brazil and other countries, but this method of fishing continues to be practised, often using other poisons or explosives instead. Interestingly, discus have never been caught with rotenone. Not by the natives, nor by scientists or representatives of institutions.

Kayapó

Lonchocarpus nicou

But before I bring my history of rubber to an end, I would just like to mention Henry Ford (whom everyone has heard of), who founded Fordlandia and Belterra upstream on the Rio Tapajós (see page 263). In typical American style he had a gigantic rubber plantation established there at the end of the 1920s, in order to break the Indochina monopoly. He had 70 million *seringueiras* plant out a million hectares of land. At that time Ford controlled 50% of the automobile market worldwide and had the small change needed to re-awaken the Amazon region from the slumber into which it has fallen since the demand for rubber stagnated.

He proposed to produce 300,000 tonnes annually – half of the world production back then (today it is more than six million tonnes). But he hadn't reckoned with the tropical rainforest of Amazonia, and in 1946 he gave up. Today both Belterra and Fordlandia are tourist attractions and worth a visit – like the Jari project, symbols of American gigantism. Adventurers, explorers, naturalists, botanists, gold miners, collectors, or fishermen, none have reckoned with the largest tropical rainforest on Earth. To the present day few have understood it.

Today, unlike in the past, Santarém is easy to reach – by ship (daily *recreios* from Manaus or Belém), by plane (several flights per day with Tam, Varig, Penta, Tavaj, Meta, or Rico), and via the Transamazonica or the Cuiabá-Santarém (although these earth roads still represent something of an adventure). The city lies at 02° 25'30"S and 54° 42'50"W and is now the third largest on the Brazilian Amazon. And, as already mentioned,

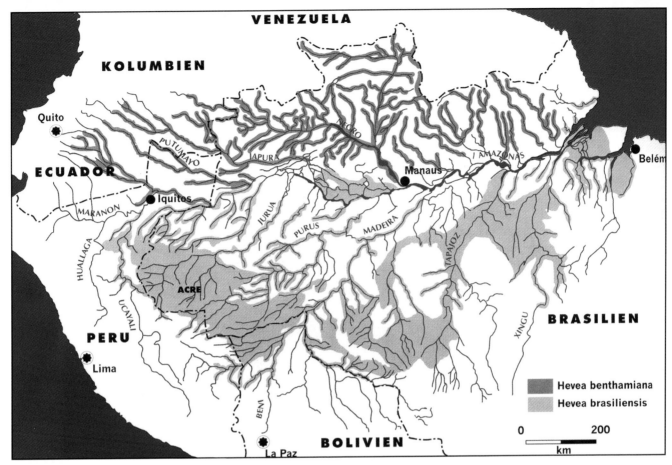

Above: The natural range of the *Hevea brasiliensis* rubber plant, a member of the family Euphorbiaceae, is predominantly the southern Amazon region. (Two varieties of this species are recognised: on the one hand *'Latifolia'* with broad leaves, white bark, and a little-branched trunk, and on the other *'Angustifolia'* with narrower leaves and a thick, soft, dark-coloured bark. The latter variety produces the better quality rubber.) The second species, *H. benthamiana*, whose latex has never achieved the popularity of that of *H. brasiliensis*, is found in the more northerly parts of South America.
Left: For thousands of years the Kayapó indians have used plant toxins for fishing. They beat the roots and leaves on the surface of still water so that the poison dissolves in it and stuns and kills the fishes, which they can then gather up with ease. However, not a single discus has ever been caught this way (see text).

the *fazendeiros* are expanding. Cattle pastures extend south for many miles around, over almost all of the offshore islands, and upstream along both sides of the Tapajós. There is no end to it in sight. In 1986 the region had lost only 11.2% of its primary forest, while by 2003 it was already more than 25%.

The *encontro das águas*, where the Tapajós with its green-blue water – which looks almost black here *(right)* – mingles with the muddy yellow Amazon water, is a natural spectacle, though here too it is impossible to overlook the increase in environmental pollution. Not from the vast harbour (500,000 m²); instead the bulk of the pollution originates in the thousand or more Tapajós tributaries and *igarapés* where, since the 1970s, *garimpeiros* (gold prospectors) have turned most of the formerly crystal-clear watercourses into muddy, mercury-tainted drains.

The Tapajós is the fifth largest tributary of the Amazon, and its basin drains 489,000 km² (almost the area of Spain). It is formed by two large rivers, the Juruena and the São Manuel (also known as the Teles Pires), which meet in the border region of the states of Amazonas, Pará, and Mato Grosso (the last two share 95% of the basin). Its source lies in the vicinity of the town of Cuiabá, some 200 km to the north of the Pantanal, and by the time it joins the Amazon it has covered about 2,700 km; and its greenish waters remain visible for about 130 km further downstream before they are lost in the muddy Amazon soup. Its basin lies for the most part in the Brazilian shield, but the first major rapids and waterfalls don't begin until about 200 km upriver, at Itaituba *(see below)*. Discus don't occur that far upstream – I have checked on several occasions.

For its last 150 km the Tapajós forms a lake some 20 km wide, with one bank barely visible from the other. There are no discus in the lake itself – there are no suitable habitats – only in the surrounding area. During the dry season sandy beaches more than 100 km long are exposed along both shores.

I mustn't omit to mention that Santarém is also famous for its annual *Festa do Çairé*, which always takes place in September Groups interested in folklore dress up as the *boto tucuxi* (a true dolphin – *Sotalia fluviatilis*) and the *boto cor de rosa* (the pink Amazon dolphin – *Inia geoffrensis)*, dancing, engaging in mock battles, and holding a parade – rather like the Amazon festivals in Parintins, where two groups, the *caprichoso* (the moody) and the *garantido* (the certain) dress up as bulls and dance all day long; or in Barcelos, where it is instead the *acará-disco* (discus) and the *cardenal* (cardinal tetra) (see pages 384-390).

There is a rather nice story about the *boto*. It is said to change itself into a man by magic in the evening, then flirt with the young women by the river and get them pregnant *(left)*. This too is enacted in the *Festa do Çairé*. Elegant, clad all in white, with a hat covering the its spiracle to reduce the smell of fish, the *boto* mingles with the pretty girls and tries to seduce them. Always an effective tactic. First he dances with them, and then he invites them to go for a walk to a suitable spot. The magic ends with the first light of dawn when the *boto* leaps back into the water. In some parts of Amazonia it is firmly believed that the *boto* is responsible for children whose father is unknown (or has vanished).

I never cease to be enchanted and fascinated by the stories of the natives and *caboclos*. But I will now try to enchant you with tales of the discus habitats that actually begin here, south of Santarém, in the Lago Verde. It must also have been here where Jobert collected his specimen around 1878 (see page 43), that Ternetz sampled three fishes in 1924, Praetorius caught the first live individual in 1928, Harald Schultz a single specimen in March 1958, and I myself my first in 1966. In 1971 people from INPA caught a discus in the Rio Maicá, and Kilian and Seidel a semi-adult specimen in October 1991, in the system of lakes some 10 km to the south. All these discus are typical brown individuals such as those Fred Cochu transported for decades to the USA and Europe.

The lake contains clear water *(centre)*, although it is described locally as *verde* (= green). It can be seen from the underwater photos *(page 265)* that it is fairly clear. But it is practically impossible to photograph discus in their habitat.

The water parameters on 16.02.2004 at 10:00 were as follows: pH 6.38; conductivity 17 μS/cm; daytime temperature (air): 38 °C; at the water's surface: 29.5 °C, in 2 m of depth: 28.9 °C. The surounding area was still largely intact. Biotope: overhanging branches, tree-trunks in the water;

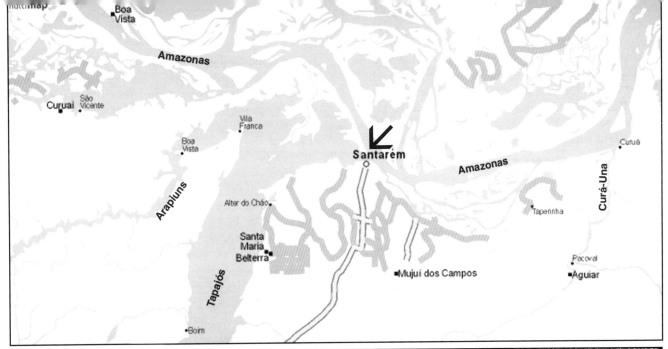

Map, top (dry season): Note how the Tapajós enters the Amazon (arrowed in black on the satellite photo) just off Santarém. The yellow road is the BR165 from Cuiabá, Mato Grosso to Santarém. (On both, the map on top and satelitte photo above, the extent of the enormous lower Tapajós can be seen (the black area on the left in the satellite photo)). *Satellite photo above* (rainy season): The white patches are clouds (mainly in the right-hand part of the photo); the completely dark red areas indicate the remaining primary forest (2003); all the light red areas are burnt or clear-felled regions; the pink areas are cattle farms (note that almost all the *várzea* regions are cattle pastures); whitish-green areas are settlements (eg Santarém, arrowed); light green areas are whitewater regions (eg the Amazon); light blue is clear water (eg Lago Grande); and pitch black is black water.

water slighty flowing; no aquatic vegetation; river bottom sandy. For discus from this region see page 265.

In 1976 I also found brown discus further east of Santarém, in the Paraná do Ituquí (Bom Jardim). Years later, at the end of September 1994, T. B. Andersen collected a specimen barely 9 cm (SL) long there. This is relatively small, and demonstrates something also known to the *disqueiros* (discus collectors):

In the Santarém region (south of the Amazon) the rainy season doesn't usually begin until late in January. Praetorius was undoubtedly the first to discover that the spawning time of the discus in this region begins in January/February (at the start of the local rainy season), and I was able to confirm this. It was Praetorius too who established that there are no underwater plants in discus habitats – at low water there are just masses of dead terrestrial plants to be seen, along with stands of *Caladium* spp. in marshy spots. He did, however, find floating plants such as *Eichhornia* and *Azolla* species, as well as *Pistia stratoites* in the lagoons and lakes. And it is just the same today. Depending on the time of year, they can be found in the Curuaí at Vila Franca (in the north) or in the Verde and the Muiraquitãs at Alter-do-Chão. Praetorius also noted that the water temperature fluctuated between 18 and 40 °C – but these are not the relevant figures for discus, as the latter swim in the temperature zones that suit them.

Tapajós

And the smallest number of fish species, and most definitely no discus, are found in the open water, where the sun has the greatest warming effect.

Following the Paraná in an easterly direction we come to another large area of lakes. The Furo de Maicá leads to the Lago de Maicá, a typical discus habitat. Here too they are Browns (see page 194).

The water parameters on 17.2.2004 at 10.30 hours were as follows: pH 6.25; conductivity 22 μS/cm; daytime temperature (air): 36.5 °C; at the water's surface: 28.9 °C; in 2 m of depth: 28.2 °C. The surrounding area was no longer as good - *comunidades* (small settlements) and *fazendas* with cattle pastures almost everywhere. **Biotope: tree-trunks in the water; water almost stagnent; floating aquatic vegetation (*Pistia stratiotes, Salvinia auriculata, Eichhornia crassipes*); river bottom sandy. For discus from this region see page 265.**

During the dry season this particularly rich world of birds, crocodiles, iguanas, and freshwater dolphins can be reached only via the labyrinthine várzea regions, which now consist mainly of cattle pastures. Chiefly via the above-mentioned Paraná Ituquí, which passes the Sítio de Taperinha, which belonged to the Baron of Santarém, Cel. Miguel Antônio Pinto Guimarães, in the 19th century and includes a mansion built during the slavery period. It was his summer residence, where, along with his partner Romulus J. Rhome, he laid out vast sugarcane plantions for the production of *cachaça,* the popular sugarcane brandy. A few years after the death of the Baron in 1882, Taperinha was sold to the zoologist Dr. Godofredo Hagmann of the Emilío Goeldi Museum in Belém, and it remains in the ownership of his family to the present day. Here too ceramic fragments have been found, dating back to 4000-5000 BC.

But let us leave the eastern system of lakes and visit the western area – the gigantic Tapajós mouth region, which is also known as the Lago Grande de Santarém and contains numerous islands. There are hardly discus habitats here; the first typical biotope isn't to be found until further west, "diagonally opposite" Santarém, in the Tapajós blackwater affluent, the Rio Arapiuns. (And there are to be earth roads and jungle tracks to get to the *lagoas*, which in recent years have become known as Mata Limpa and Traira (Thraia, Trahira or Tariraha). Destinations which for some time are bookable by discus fans via the Internet. But my research proofed that such locations do not exist here. Neither the Brazilian geographic institute (IBGE), nor in the City Hall of Santarém, farmers or *caboclos,* have ever heard of such names – the same applies to the location Santarém-Inanu. The exporter who gave those names, when asked, kept silent.)

The Arapiuns is a large left-bank tributary with numerous affluents such as the Rio Aruã and the local Rio Branco, plus

Tapajós

Tapajós

the Igarapés Braço Grande do Arapiuns, Curí, Caranã, and others on its left-hand side. On the right bank just one large *rio*, the Mentaí, flows into the Arapiuns. There are no further discus habitats known upstream, just continuous rapids and before long a waterfall, the 20 metre high Cachoeira de Aruã, where the river Aruã enters the Arapiuns.

Unfortunately there is increasing deforestation taking place in the Arapiuns area, formerly an incredibly beautiful region. It is generally the *madeireiros* (loggers) that are transporting the tree-trunks down-river and recently this has been the cause of a considerable amount of debate here (see Chapter 10: *...In nature*). But before I come to the unique flora and fauna of this region, a little more about the Arapiuns itself: I was able to find a 22 cm SL greyish discus, probably the largest in nature so far (page 196). I discovered the first orange-red headed *Geophagus*, possibly a *G. proximus*-variety. (At the international exhibition *Zierfische & Aquarium 2003* in Duisburg, Germany, I displayed some of those in the Nutrafin biotope stand, just as they live in nature). I also found beautiful loricariids in the area of the Cachoeira de Aruã *(centre)*, but no discus. The further you travel upstream towards the Cachoeira the rarer their habitats become.

My measurements, at a place where I found brown discus with a little pattern colour and the water really was black, gave a pH of 4.5 in 1-2 m of depth at night, but near the mouth of the Tapajós this rose to 6.5. For discus from this region see page 196.

But I would like to mention – as recently this has again become a subject for heated discussion – a number of the so-called L-number catfishes which I found at the Aruã waterfall area (and mentioned above), don't appear in the latest (2004) publications, therfor I will show a few of those here *(right-hand page)*. The Arapiuns is also rich in freshwater rays – I was able to find at least three or four species. (I exhibited a few specimens also on the Nutrafin biotope stand at the *Zierfische & Aquarium 2003*, just as they live in this river.) Often actually in the discus biotope, deep down and buried in the fine white sandy bottom.

The Tapajós itself forms a boundary for many lifeforms in the Amazon region – including, in part, the discus (see also Chapters 3 and 4). The biodiversity of the region includes 161 mammal and 556 bird species. For example the white-headed Capucin monkey *(Cebus albifrons)* and the saki *(Pithecia hirsutus)* occur only on the western side and not to the east, where instead we find the bearded saki *(Chiropotes albinasus)*, the titi ape *(Callicebus moloch)*, the grey-necked night monkey *(Aotus infulatus)*, and the spider monkey *(Ateles marginatus)*. The river itself is home to the spectacled caiman *(Caiman crocodilus)*, black caiman *(Melanosuchus niger)*, yellow-spotted sideneck turtle *(Podocnemis unifilis)*, Amazon or river manatee *(Trichechus inunguis)*, and the dolphins mentioned earlier *(Ina geoffrensis, Sotalia fluviatilis)*. Large white-lipped peccaries *(Tayassus pecari)*, collared peccaries *(T. tajacu)*, pumas *(Puma concolor)*, jaguars *(Panthera onca)*, tapirs *(Tapirus terrestris)*, and brocket deer *(Mazama* spp.), plus many others. I will mention just a few of the interesting birds, such as ospreys *(Pandion haliaetus)*, harpy eagles *(Harpia harpyia)*, toucans *(Ramphastos vitellinus)*, little chachalacas *(Ortalis motmot)*, nine tinamou species *(Crypturellus* spp., *Tinamus* spp.), and seven different aras *(Ara* spp.) including the hyacinth ara *(Anodorhynchus hyacinthinus)*. Numerous parakeets still live here, species of the genera *Paratinga*, *Pyrrhura*, and *Brotogeris*, along with Amazon *Pionus* parrots plus hoatzins *(Opisthocomus hoazin)*. And the outlook isn't good for these and all the other endemic species.

The Transamazonica and the Cuiabá-Santarém highways have carved considerable "holes" through the region. But it is

Cachoeira do Aruã – Rio Arapiuns

Podocnemis unifilis

Opisthocomus hoazin

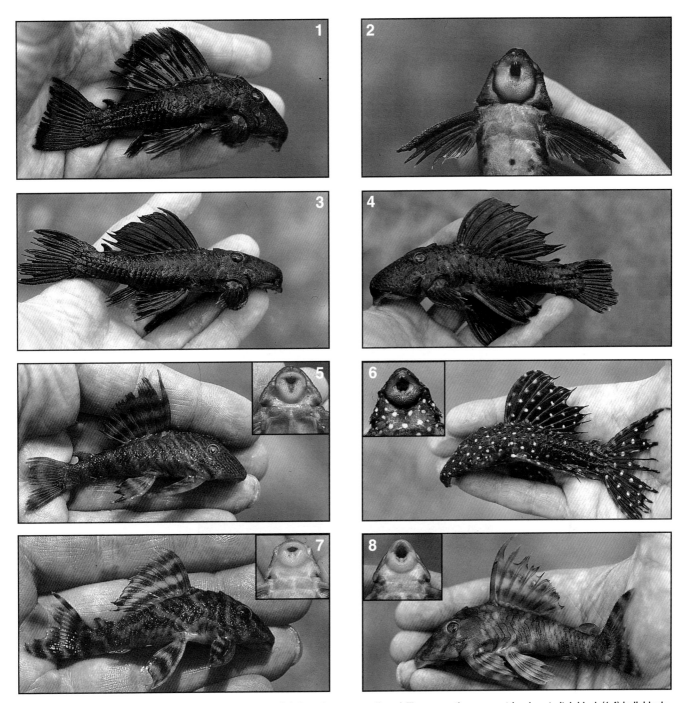

Loricariids that I found in the Rio Arapiuns (Cachoeira de Aruã) (where there are no discus): There were three sympatric, almost pitch black (1-4) individuals, the first of which (1 and 2 – an extremely spiny fellow) may be a *Leporacanthicus* species. The second (3) may be a new *Pseudacanthicus* species (there is no previous evidence of specimens of this genus from the Arapiuns). The third (4) is undoubtedly an interesting new *Parancistrus* species. The black catfish with enormous fins and ivory-coloured dots (6) may be a new *Leporacanthicus*, and the splendidly patterned small ones (5 and 7) may be two *Panaqolus* species (they have different mouths and teeth). The last (8) is undoubtedly a *Peckoltia* species.

what came with them – colonisation – that spells death for the primary forest and its innumerable lifeforms, bringing clear-felling and continual, usually uncontrolled, burning for the endless cattle ranches. Some areas already resemble desert landscapes. The region from Santarém to Aveiro is now a single gigantic cattle range and the little *igarapés* of the area are history. Everywhere water quality is increasingly deteriorating. There are hardly any protected regions here. To the south, at Itaituba, there is officially the Parque Nacional da Amazonia (9,935 km²) but it is rarely patrolled, just like the Floresta Nacional at Aveiro and even the Gleba Arapiuns. The timber exports from Santarém are also a contributory factor. Apart from a few discus and L-number catfishes (mainly from the Itaituba region), wood is practically the only export, and more every year: official records show 46,551 tonnes in 1993, but by 1999 this had risen to 74,881 t (only in the year 1997 it was even more: 95,500 t). But nobody knows the true figures.

Before I move on to the Alenquer region to the north, I would like say a little about **Itaituba**, further upstream. In the language of the indians Itaituba means roughly "little stone", and those are found here in large quantities – but no discus (they don't come that far upstream). It is a town of records, but not all are reasons for celebration.

Until 1993 Itaituba was the largest *município* in the country (165,578 km²), larger than some European countries. (It was subsequently divided into four, and today is 62,565 km² with 160,000 inhabitants). During the gold-prospecting boom (1970s to 1990s) the town recorded the highest *per capita* income; the busiest airport; and the largest deposits of gold in the world. *"Na época do garimpo o dinheiro era transportado em sacos na traseira das D20. A ostentação era tanta que havia disputa entre os garimpeiros para ver quem gastava mais"*, I was told by the military commandant Antônio Airton Celestino Vieira in 1999. In other words, the gold was dragged though the region by the sackful and the gold prospectors vied with one another to see who could spend the most.

Captain Pedro Teixeira, who in 1616 came from Portugal to Grão Pará, as part of Francisco Caldeira Castelo Branco's "purge", to drive out the Dutch, French, and English, can have had no inkling of this when, in 1626, he first travelled the Rio Tapajós, nor when he returned in 1639 with the Jesuíts. Nor could his fellow-countryman, Francisco da Costa Falcão, who in 1697 built the fortress at the mouth of the Tapajós (not a single stone of which remains to be seen today). And still less Miguel João de Castro, who in 1812 was the first to mention Itaituba in writing, when it was still just an indian settlement.

Now before we head north, a final tip for visitors to Santarém who may perhaps fancy a little gourmandising during their search for discus: you must sample their famous *piracai*, a local delicacy consisting of the fishes *curimatã (Prochilodus* sp.), *tambaquí (Colossoma macropomum), acari-bodó (Hypostomus* sp.), *jaraquí (Semaprochilodus* sp.), *tucunaré (Cichla* sp.). For starters there are *bolinho de piracuí* (fish cakes) and *charutinho frito* (fried strips of fish, mostly *Boulengerella* sp. or *Ctenolucius* sp.), and *piracaia* (fish spit) as main course.

Top: The Tapajós lake (with the Arapiuns in the background) was full of algae when I flew over it in February 2004. It looks like the gold prospectors are still hard at work. *Centre:* The Tapajós and its (black?) water. *Below:* Arapiuns discus habitat after heavy rain (biotope of the discus shown on p. 269).

A discus variant from the Rio Arapiuns (habitat on the left-hand page below). 2. The almost grey discus variant, mentioned in the text, which I likewise found in the Arapiuns (with individuals up to 22 cm in diameter). 3. A *Geophagus proximus* variant (or another species?) which occurs together with discus in the black water of the Arapiuns. The upper part of the head is brilliant orange-red (see 5, far left). 4. In the Tapajós, by contrast, a *G. proximus* (?) variant occurs with the discus, and is only slightly pink-coloured on the head (see 5, far right). 6. This *Geophagus* sp. is also found with discus in the Tapajós.
7. I found *Potamotrygon leopoldi* and two other rays in the white sand in the Arapiuns. 8. The "black water" of the Rio Tapajós. (Tapajós in the background.)

Alenquer region: The best way to get to Alenquer from Santarém is via the daily ferry, the *jato*, which takes 2.5 hours (and costs R$ 18 = US$ 6), or the much cheaper *recreio,* which takes 4-5 hours.

Alenquer lies on the left bank of the Amazon, on a branch which is also known as the Rio Surubiú (see below), at 01° 56' 56" south and 54° 45' 38" west. The *município* extends north to Almeirim, south to Santarém, west to Monte Alegre, and east to Óbidos. The inhabitants of the town, which lies barely 36 metres above sea level, call themselves *Alenquerense,* but are also known as *Ximango* (after the *gavião,* a local bird species). Alenquer itself they term the *cidade de amor* (city of love). And the city of love has been internationally famous since the 1970s, when I exported the first Alenquer discus via Belém. Nowadays the name is a household word to enthusiasts and almost anyone familiar with discus. (Unfortunately since then numerous variants have been labelled "Alenquer" even though they actually originate from other regions.)

I should also mention the following: in the past (until the 1990s), when a discus variant came from such a large region of lakes, lagoons, and rivers as the Alenquer region (where there are hundreds of such bodies of water, each with its own name, but no discus known – see also Chapter 3), people (often including myself) would name it after the largest settlement in the region. Discus from various collecting sites in the Xingú region were called "Xingú discus"; those from the Tapajós or Santarém region likewise; and specimens from the gigantic Manacapuru lake and river system received only a town or village name, just like "Nhamundá" or "Coari" – although two *Symphysodon* species occur in each of the last-named two regions, often very close together, practically "next door" to one another, although hardly ever together (see also below). Excluded from those has to be the so called "Madeira discus", as never a single discus came from there. Some exporter or importer invented that name recently...

The Barés (or Abarés) indians lived for millennia in the vicinity of the modern Alenquer. When Capucin monks came to the region in 1729 they landed at the mouth of the Rio Curuá, where they encountered these indians. Together with them they built a village on the right bank of the river (where Curuá stands today). But because of transport problems during the dry season (because the *furos* and *paranás* for the most part dried up) and a developing malaria epidemic, they abandoned the village and moved to the Rio Surubiú (today usually known as the Paraná Mirim or Igarapé de Alenquer). With the aid of *indios* from the Trombetas they built a new settlement on the left bank at the mouth of the Rio Itacarará. This site also appeared better suited strategically and could be reached by water all year round.

Thereafter the settlement was called Surubiú.

Then, in 1756, the *Governador,* Francisco Xavier de Mendonça Furtado, changed the name of the village to Alenquer (after his home town) as part of his well-known "Portugalisation" of Amazonia. And in 1881 it was raised to the status of town.

I can in fact heartily recommend the fascinating little Portuguese "mother town" – and especially its wine: the red is a balanced young wine with a fresh bouquet, while the white is full-bodied and very aromatic. Good places to sample them include Casa Santos Lima, Cerca do Rei, Espiga, Gorjão, Palha Canas, Quinta de Abrigada, Quinta de Don Carlos, and, of course, others.

Now, geographically speaking, Alenquer lies on the left bank of the Amazon, on the arm known as the Igarapé de Alenquer, but the network of waters here is gigantic and has many names. It extends almost continuously along the northern side of the Amazon from Óbidos in the west to Prainha in the east *(see right-hand page).* Europs largest lake could be contained in it several times. It is fed by the already-mentioned *igarapés* and *rios* in the east, and in the Alenquer region by the Igarapés Curuçá and Mamauru (at Óbidos), the Piquiá and Cabeleira (Rio Mamiá tributaries), the Igarapé do Lago (which also feeds the enormous Lago de Itandeua), and the Igarapé do Inferno (a left-hand arm of the Rio Curuá).

The Amazon is first and foremost among the *rios* that partially flood this region during the rainy season. It also feeds the Arapiri *varzéa* island, the lake of the same name, and the Paraná Mirim de Alenquer (= Igarapé Alenquer). Then there is the Cuminapanema (a right-hand tributary of the Curuá, around 300 km long) and its main branch, the Rio Capitari. The Curuá itself (also known as Curuá do Norte, although that name applies only to the northern part) has numerous discus habitats in its lower course (their distribution extends from the town of Curuá upstream to Pacoval) and in the north is composed of two waterways: the Cuminá de Leste flowing from the north-west and the Curuá-Panema from the northeast. Both are virtually unnavigable owing to their numerous rapids and waterfalls, and are little more than channels during the dry season (but nevertheless harbour a fantastic flora, *inter alia* thousands of Brazil nut, *copaíba,* and *salsa* trees.) The Curuá opens into the Lago dos Botos, which contains clear water during the dry season and mainly mixed water in the rainy season, and in the west mingles with the Lago de Itandeua. And the latter feeds the Lago Grande and the Paraná-Mirim de Alenquer (which also receives water from the

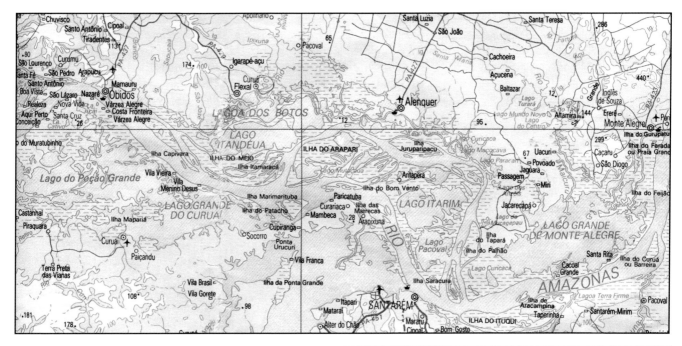

Top: The IBGE (Instituto Brasileiro de Geographia) map shows how gigantic the network of water in the Alenquer region is. The scale is 1: 1,000,000. The red lines are the roads in the region around 1998, but by now there are far more, especially to the *fazendas*. *Above:* The Lago Grande at Alenquer (= the "small" lake on the map, to the left near Alenquer), which we crossed at this point along with our equipment and fuel reserve. The photo shows clearly the size of one of the "smaller" lagos in the Alenquer region. None of its shores can be seen from the centre (just the link to the Lago dos Botos on the horizon) and it takes more than an hour to cross (at around 60 kph). The Lago dos Botos and Lago Itandeua, among others, are even larger still.

Rio Itacarará) via the Igarapé do Lago. The mouth region of the Rio Curuá is also known as the Boca do Curuá (see below).

The town of Curuá lies a few kilometres upstream (the old Curuá on the right and the new on the left bank). Further upstream, the Rio Mamiá, a whitewater river, flows from the west to join the Curuá, but is navigable only by *voadeiras* and canoes. Here I was present during a huge spawning migration of *Semaprochilodus* (two species – one with small and one with large scales), *Curimata* (one species) and *Laemolyta* (ditto). Here too I was able to catch some very beautiful angelfishes *(Pterophyllum)* (see page 281), but no discus. No habitat.

The Rio Curuá (by some also known as Curuá-Panema, its northern branch name) is some 520 km long and nobody knows precisely where it rises, only that its source lies some 400 m above sea level. There are also very large numbers of smaller rivers in the region, but none relevant to discus. But it is worth mentioning the *furos* (connecting channels) such as the Antônio Pedro, which links the Lago Grande do Jauari (also called the Janari) with the Lago dos Patos and the Boca do Curuá. (And when the water level rises the latter forms a direct connection between the Amazon and the Igarapé de Alenquer, but rarely during the dry season.)

There are huge numbers of *lagos*, such as the Amari (which is connected to the Rio Curuá), the Curumú (some 3 km long and some 1.5 km wide – see below), and Lago Vitória, close to the right-hand bank of the Rio Curuá.

The Lago dos Botos (in the northern part of which I have found discus habitats during the dry season), and the Lago Itandeua, a gigantic whitewater lake with no discus habitats. Then there is the more than 800 km² Lago Grande de Vila Franca at Alenquer (also known as just Lago Grande). And additional *lagos* found to contain discus variants, for example Lago Grande do Jauari, Lago dos Patos, and Lago Aningau (which can be reached only by land and harbours a splendid red variant of *S. haraldi* in which all individuals have red eyes). And the Lagos Arapaí (Araparí), Uruchi, Curumu, Capintuba (also known as Capituba), Paracarí, Samauma, and Cuipeuá (erroneously called Cuipiera, Curipera, or similar on the Internet, among discus specialists, and elsewhere), from which I brought the first red "Alenquer" to Europe. And I have found striped variants in the Corréa and Jaraquituba.

The sea of islands is almost unimaginable and the majority of them are submerged at high water, and then countless islands of grass are formed (but these – or parts of them – float away and eventually end up in the Atlantic Ocean). The waterfalls and rapids are likewise almost innumerable. The best known large falls are in the Curuá drainage: the Berimbau in the Curuá, 15 km from its confluence with the Rio Cuminapanema and the Boca closer its confluence; the Brigadeiro in the Curuá do Norte; and smaller ones in the Curuá, for example the Cachoeirinha close to the Cachoeira da Lontra and the Cajueiro. In the Rio Curuá do Norte, some 10 km upstream of Vila Curuá, there are the Cajuti and the Mundurucu falls, about 24 km upstream of the mouth of the Cuminapanema. (Note that the Frenchwoman Madame Coudreau reports in her book *Voyage au Rio Curuá* that in 1900/1901 black slaves, the *quilombos do Pacoval*, were already living here.)

But let me tell you more about Alenquer *(left and right)*. After it was raised to the status of a town (1881), a second story about its origins started to do the rounds. A long time ago a fisherman from Arcozelos (nowadays Curuá) had supposedly discovered a portrait on a tree on the river-bank. Puzzled, he carried the picture to the chapel of Arcozelos where the missionaries told him it was Santo Antonio, who bears a child in his arms. The next day, when the fisherman returned to the chapel, the picture had vanished. Weeks later – when he had almost forgotten his discovery – he passed the same spot again and found the same picture. But this time he didn't take it to Arcozelos again, but with much love and the help of the natives he built a small *palhoça* (palm-thatched hut) on the bank of the Rio Surubiú (where Alenquer stands today). And made his home there. The inhabitants of Arcozelos, confronted by constant transport problems during the dry season (when the Lago Grande almost dries up) and with a high mortality rate through malaria, heard about the fisherman and his *palhoça* and moved there. And, where once stood the palm-thatched hut, the Igreja Matriz e San-

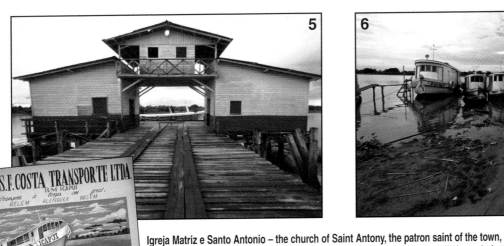

Igreja Matriz e Santo Antonio – the church of Saint Antony, the patron saint of the town, in Alenquer *(also left-hand page,* with monument). 2. Compre-se Peixes (fish vendor) on the bank of the Igarapé de Alenquer (formerly the Rio Surubiú). 3. The Hotel da Bettinha. 4. The government (mayor) of Alenquer has done little to restore the town, which has a gloomy aspect. 5. The former landing-place at Alenquer. 6. Today the *lanchas* and our *voadeira (centre)* tie up by the muddy bank. But there is plenty of transport to and from Alenquer, and nowadays also lots of cars.

HABITATS ... ALENQUER REGION

to Antonio *(centre)* was subsequently erected. Saint Anthony became the patron saint of the town and his festival is celebrated for several days every year. So perhaps the Capucins weren't the founders after all.

At the end of 1996 Alenquer numbered exactly 43,394 souls – today there are far more. The very run-down town (the new mayor, who has money, is, following his election in 2004, doing some good – or so it is said) lives by exporting the *cumarú* fruit *(Dipterix odorata),* which grows in the region. A valuable perfume is made from its hard shell and shipped via Belém, 1012 km away by river. *Cumarú* wood *(centre)* is exported as well. It is graded 1 (unlimited) for resistance, in other words it can be used for any purpose without impregnation. Jute is planted as well and exported, as are rice, black beans, and maize. There is cattle- and water buffalo-breeding, and enormous soya plantations in the hinterland.

When I was last in Alenquer, in February 2004, I stayed at the Hotel da Bettinha, which is situated right on the river-bank promenade. It has been open for three years and radiates cleanliness. The entire interior covered by beautiful tiles, with plants and flowers everywhere (Bettinha has green fingers). One evening, sitting in a rocking-chair outside the inn, I asked her husband, Senhor Halim, about the numerous *fazendas* I had seen on the way here along the road from Óbidos, and he said, "You would never believe how many there are. The *gauchos* come from the south of Brazil and buy their way in here. The land is cheap. Only recently an area 50 x 50 km was sold, and is to be clear-felled all the way to the border with Surinam for cattle-breeding. Roads are hacked relentlessly through the jungle so that every last corner or land can be turned into cattle pasture. And the big money comes from leasing – 6-7 reais per head of cattle per month. But the owner does well out of it too. A cow puts on about 100 kg per year, which is immediately worth 300 reais (= about 125 dollars). If we deduct around 72 reais for the lease that still leaves a considerable profit. For six months the cattle fill their bellies on the grass that grew on the *terra firme* during the rainy season, and during the dry season they are taken by raft to the lush (clear-felled) *varzéa* regions to grow even fatter."

When I asked him about soya (Brazil is by now the largest soya producer on Earth), he continued, "The first gigantic plantations were established last year. It is decided at government level (IBAMA and other official organisations) whether an area of jungle is worth anything or not. The 'worthless' areas are consigned to the flames, either for *fazendas*, or, as recently, for soya plantations." (Nobody worries about the flora and fauna, above and below water – only if one wants to take them out of the country to save them from the inevitable death in the Amazon, than these live forms are worth something to IBAMA and the people who carry them are considered *bio-piratas*...)

When I told him about my mapping of the *acará-disco* biotopes, my decades of work cataloguing the freshwater fishes on Earth, my journals and my books, he said, "In this region fishes are becoming ever fewer. Everyone catches what they want, and there are no regulations. Round here people even eat *acará-disco*. And the worst time for the fishes is during their spawning migrations. The rivers are fished clean."

He recalled a poem by the Rio Surubiú poet Aldo Arrais about the destruction of an *igarapé* and its fishes (which I reproduce here in the original Portuguese):
Arrastões, malhas finas como pingentes Depredação total nos rios pequenos Não mate, pescador imprevidentes
Tantos peixes, com bombas e venenos
Malhando peixes magros e carnudos Confinados nos lagos e nos rios Apanham peixes grandes e miúdos Pingando da folhagem verdejantes O sereno das noites de verão Molha os lábios do rio agonizante Riozionho Condenando, que tristeza Tuas águas fazem estranha procissão No lento funeral da natureza.

The Wood *(Dipterix ordorata)* of the Cumarú-fruit

(Which translates roughly as: Large nets with very small openings to fish the little rivers empty, Fisherman, do not kill so many fishes, slender and plump, with explosives and poison, The little river weeps in sorrow in the summer nights. A slow procession of death into the grave of Nature.)

Senhor Halim is an *alenquerense* of advanced years (69), and he too has a *fazenda*, but with 2,000 water buffalo, which bring 20 centavos more per kilo. He told me a lot more that evening, but when I started to tell him about Alenquer in Portugal and the origin of the name he was not a little surprised. "Nobody here knows anything about that", he said.

It is known that Surubiú comes from the Barés language and is derived from *surubi* or *surubim* (the large catfishes mostly of the genus *Pseudoplatystoma)*. Also that it was originally pronounced with a "y" or "yu" and *y* means water. Thus among the Báres *Surubiy* means "*surubim* in the water". But Alenquer?

The Portuguese town is supposedly mentioned as long ago as the sixth century BC. First the Celts, then the Carthaginians, the Lusitanians, and the Turdus lived in the region. From the second century BC to 418 AD it belonged to the Roman Empire and was called Arabriga, Gerabriga, or Iera-briga. From 418 until 714 it is recorded as Alanos, and was controlled by the Alencar family. The Alanos were barbarians from the Sea of Azov

1. *Caboclo* family fishing. At the beginning of the rainy season there are masses of young fishes among the temporary grass islands of *Echinochloa polystachya*. 2. In the Alenquer region almost all the *várzea* is occupied by cattle farms. 3. Discus habitats can be seen on the horizon in the Lago Grande; the *Echinochloa polystachya* here was still under water. 4. Inundated discus habitat. 5-6. The town of Curuá. No roads lead there. 7. The "filling station" for our boat in Curuá. 8. Fisherman on the Rio Curuá. 9. Brazilian duck, *Amazonetta brasiliensis*. 10. Spawning grounds for discus in the Curuá (at high water).

(Iranian in origin?). In 374 AD they were conquered by the Huns and fled to Europe, where, like other groups of barbarians living at the edge of Europe, they had to regroup, and eventually reached Romania, where they united with the Vandals and a group of Suevi. In 406 they crossed the frozen Rhine and together conquered many parts of Europe, plundering and destroying everything in their path. In 409 they settled in Gallaecia (Callaecia – approx. the current Galicia of Spain and the north of Portugal). The inhabitants welcomed the barbarians as they were tired of their Roman rulers. The Suevi (Seubi) integrated completely into this region. The Vandals dominated another region which they called Vandaluzia (later Andalusia). The Alanos called the region where they settled Alenen-Kerk *(kerk* = church), Alano Kerk, or Alankerk (= church of the Alanos). When, after 714, the Moors invaded and thereafter assumed rule, they called it Alain-Keir (= blessed spring) and El-Haquem (= governor). After the 24th June 1148, when the town was taken by the the Portuguese, it was called Alãoquer, Alunquer, Alonquer, Alanquer, and Alemquer, and then finally Alenquer. But the origin of the name is to be sought in the family name Alencar, which has been written with a "qu" since the Middle Ages. And many famous people have borne the name, for example, Pedro de Alenquer, leading helmsman for the fleets of Bartholomew Diaz (1487), of Vasco da Gama (1487), and even of Pedro Alvares Cabral, the "discover" of Brazil. Alenquer was also the birthplace of, *inter alia*, Queen (later Saint) Isabel and Portugal's best known writer and poet, Luís de Camões.

But let me also say a little about the habitats of the Alenquer discus. The parameters measured are reproduced below.

Boca do Curuá – Rio Curuá mouth region (see page 275):

1. Claudio (left in the photo), is a *caboclo* who has been collecting discus commercially since the end of the 1990s, along with 19 other collectors. Exclusively at night, in the region from the lower Curuá (about an hour's journey by *voadeira* upstream of the mouth, as this area consists solely of cattle farms both left and right of the shore) upstream to Pacoval. 2. This palm-thatched house stands on the *terra firme* downstream of the settlement of Barra Mansa on the Curuá (the fish containers double as wash-basins). 3. Discus also occur along the bank region at Barra Mansa. 4. Barely 80 people live, virtually autonomously, in the tiny *comunidade* of Barra Mansa. 5. Like most *caboclo* families in Amazonia, they live mainly by producing *farinha*. And discus containers double as storage for the *farinha*.

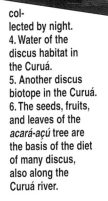

1. Runaway black slaves, the so-called *quilombos,* settled in Pacoval – as well as further up the Rio Curuá and its tributaries – more than 100 years ago.
2. There are no discus habitats in the Curuá on the Pacoval side. But on the opposite shore of the river, up- and downstream of the village, there are thousands of *acara-açú* trees.
3. A typical discus habitat, where numerous *S. haraldi* of various colour patterns were collected by night.
4. Water of the discus habitat in the Curuá.
5. Another discus biotope in the Curuá.
6. The seeds, fruits, and leaves of the *acará-açú* tree are the basis of the diet of many discus, also along the Curuá river.

The water parameters on 14.02.2004 at 14.00 hours were as follows: pH 6.39; conductivity 25 µS/cm; daytime temperature (air): 32 °C; at the water's surface: 31.9 °C; in 2 m of depth: 32.4 °C. The surrounding area (Jungle) largely clear-felled (cattle farms), still intact in only few places. Biotope: overhanging branches, tree-trunks in the water; some floating gras but no aquatic vegetation; water slightly clouded at the beginning of high water, otherwise clear; hardly any current. For discus from that region see page 200.

Rio Curuá – at the settlement of Barra Mansa (see page 276).

The water parameters on 15.02.2004 at 11:00 hours were as follows: pH 6.81; conductivity 26 µS/cm; daytime temperature (air): 38 °C; at the water's surface: 30.7 °C; in 2 m of depth: 30.2 °C. The surrounding area jungle largely intact. Biotope: overhanging branches, tree roots in the water, *acará-açú* brush; water quite clear, slow-flowing; no aquatic vegetation. For discus from that region see page 200.

Rio Curuá – at the settlement of Pacoval (see page 277), northernmost distribution of discus in the Curuá system. The water parameters on 15.02.2004 at 15:30 hours were as follows: pH 6.57; in 2 m of depth: 6.37; conductivity 26 µS/cm; daytime temperature (air): 37 °C; at the water's surface: 30.4 °C; in 2 of m depth: 30.3 °C.

The surrounding area intact – there was only a single *fazenda* on the terra firma at Pacoval (the same as at Barra Mansa). Biotope: overhanging branches – but chiefly *acará-açú* scrub along the ed-

This page: In the upper two photos we are battling along the clearwater Furo Pedro Antônio to the Lago Jauari, but enormous floating grass islands, which break loose during the rainy season, blocked the *furo*. Only by pushing and shoving did we make any progress. Then finally the Lago Jauari, which at the time contained mixed water (at other times it is clear). *Right-hand page:* The upper photo shows a discus habitat in the Lago Cuipeuá (discus, below left), where I also found the snake (inset). The author, with *disqueiros*, in Cuipeuá. The lower photo shows the totally different discus habitat in the Cuipeuá! A few years ago they were also found in the brush here (see *Collecting*). The Brown with stripe is also from Cuipeuá.

ges of the habitats; water clear (see page 277), slow-flowing; no aquatic vegetation; river bottom sandy. For discus from this region see page 200.

Note: the discus collected commercially in this region over the past five years can be found as "Barra Mansa" on the Internet and in the trade. Previously as "Alenquer".

In the Lago Grande the water parameters on 15.02.2004 at 6:30 hours were as follows: pH 6.53; conductivity 19 µS/cm; temperature (air): 28.9 °C; at the water's surface: 26.9 °C; in 2 m of depth: 26.3 °C. The surrounding area partilly intact – nowadays there are numerous *fazendas* in the area and the *várzea* regions. Biotope: overhanging branches, brush, numerous floating plants and temporary grass islands; water clear, but murky in the southern part – during the high water period most of the lakes are white water; almost stagnant (see page 271, below). For discus from this region see page 200.

In the discus biotopes *(right-hand page)* at Cuipeuá the water parameters on 15.02.2004 in the evening (18.00 hours) were as follows: pH 6.43; conductivity 74 µS/cm; temperature (air): 30 °C; at the water's surface: 32 °C; in 2 m of depth: 30.2 °C. The surrouning area was intact. Biotope: roots, tree-trunks; few floating plants; very little forest, only in a few places; water calm, almost still, and clear (rather murky at high water). In the nearby area, where discus also occurred and the Cuipeuá collectors normally work, the biotope looked quite different *(page 279, below):* brush, and temporary grass islands; numerous

HABITATS OF AVENTUEIR REGION: CUIPEUÁ   BLEHER'S DISCUS **279**

floating plants, *inter alia, Utricularia foliosa, Ceratopteris cornuta,* two *Salvinia* species, *Ludwigia helminthorrhiza* (with aerial roots), and even waterlilies. **For discus from this area see page 198.** *Note:* At this site I also found, *inter alia*, brown individuals with a central black band.

Óbidos and surrounding area: Our next stop is the former Villa Pauxí, Villa Bella, and Obydos, right on the left bank of the Amazon. Nowadays Óbidos, again named after its Portuguese mother town, is also known as *Cidade Presépio* (= crib town) *(bottom right)*. It is dominated by the cathedral of Nossa Senhora Sant'Ana on the heights, and, somewhat lower down, by the barracks of the Fourth Coastal Artillery Regiment, the Fortaleza Gurjão (named after the governor of Pará in the middle of the 17th century), on the Serra da Escama. This was originally built in 1906, during the rubber boom, to quarter a regiment for the defence of the region and to man Forte Pauxís. (The latter has been standing since 1697 and has four cannon with a long range – *bottom).*

It should be noted that off Óbidos the mighty Amazon is relatively narrow, only 1892 m wide at its narrowest point. Elsewhere it can be 50 km or more wider from one bank to the other. And the muddy water flows through this "strait" at a speed of some 8 km per hour – whereas elsewhere in its middle and lower course it achieves barely 2.5 km per hour. This narrowing is apparently also the reason why at precisely this spot the river attains its greatest depth: 82-93 m – appreciably more than its normal depth of 30 to 60 m. In addition, measurements taken in 1963 by the United States Geological Survey, the Universidade do Brasil, and the navy indicate a flow of 216,342 m³ per second off Óbidos. Twelve times more than the Mississippi and more than 20 times as much as the Nile. And just consider, after Óbidos it is additionally fed by further large tributaries – the Tapajós and Xingú on the right bank, and the Maicuru, Parú, and Jari on the left.

The river narrows were the reason the Portuguese built a fort here. No unauthorised passage was to escape their cannon (although it is said that no shot was ever fired). Little now remains of the Forte de Óbidos, as the fortress is also known – just remnants of the walls. But a yellowed sign *(below)* hangs at the police station, which stands on the site of the erstwhile fort.

A decisive role in the founding of this historic Amazon town was played by Capitão Pedro Teixeira, who discovered the narrows during a voyage upstream and suggested building a fortress there. That was in 1637, but it was to be 60 years before the construction actually took place. Only when the *Governador* of the province, Capitão General Antonio Albuquerque Coêlho de Carvalho, travelled many years later to the Forte de São José do Rio Negro (nowadays Manaus) – the westernmost fortress built by the Portuguese in Amazonia at that time – did the idea come to fruition. He instructed Manoel da Motta e Siqueira to abandon the already-started construction of Forte Itaqui at Santarém and instead to build a fortress here on the eminence. But what was it to be called? "Forte da Angustura" (fortress of the narrows)? "Forte do Trombetas" (as the mouth of that river is not far away)? Eventually they decided on Forte dos Pauxís to honour the aboriginal people of the region, who had also performed the construction work under Portuguese supervision. At the same time Motta e Siqueira mobilised the Capucin monks, who since 1663 had been performing their missionary work among the Pauxys (= Pauxís), half an hour upstream of the fort. At the Lago do Arapucú, or so the story goes. They had brought the indians there from the Trombetas to build a village which they called Aldeia Pauxys,

1. A pool on the *terra firme* at Barra Mansa. Although it dries up completely every year, it contained two *Pyrrhulina* species (an Ommochrome form and a *P.* sp. *aff. brevis*), as well as a brightly coloured *Laetacara* species (3), an *Apistogramma* species (5), a "red-top" *A. agassizii* (?) male with female, as well as a tropical relative of the European great diving beetle, *Dytiscus marginalis* (7), and quantities of *Acaronia nassa* cichlids (8), which the *caboclo* youths catch. 2. The Rio Mamiá (a Curuá affluent), a whitewater river with no discus, but home to *Pterophyllum scalare*. And it was here I once saw a gigantic spawning migration of *Prochilodus nigricans*, *Curimata* and *Schizodon* species (6) (latter not shown).

but which became known as Aldeinha. And thus the modern Óbidos came into being under the sign of the cross and of the sword, even though it was originally called Pauxys.

But whence "Pauxys"? The name is said to originate from the *lingua geral* (= common language) term *espaua-chuy* or *espauchy*, and *espaua* means *lago*, hence "the people who live by the lake". But I discovered another version in the archives in Lisbon, originating from the ethnologist João Barbosa de Faria. He states that "Pauxí" is a bird species of the family Cracidae, generally known as the *mutum* in Brazil. Although there are in fact at least three of these beautiful birds with a gifted vocal cord in Brazil, and they taste delicious (as a child I enjoyed eating *mutums* with the indians) – the *mutum-cavalho (Mitu mitu)*, the *mutum-poranga (Crax fasciolata fasciolata)*, and the *mutum-pinima (C. f. pinima)* (1, 2, 3 respectively on page 280, only males). João also wrote that during a visit to the region his *indio* guide Asarubi explained that "Pauxy" came from his ancestors the Caxuíana, who named these gifted birds *pauicé* and the white men subsequently corrupted the word. In addition, that the Trombetas indians were the Caxuíana, and that it was only because their name was impossible to pronounce that they were called Pauxís, after the bird that lived there.

Apropos of which, it is worth mentioning that Linnaeus had described two Cracide species by 1766. One was *Mitu mitu*, whose distribution is, however, restricted to Alagoas, Pernambuco, and Bahia, and which has supposedly become extinct. The other is *Pauxis pauxis*, which, however, occurs only in Colombia and Venezuela (perhaps Linnaeus was unaware of this at that time). In addition, it was Pelzeln who in 1870 described *Crax f. pinima* and posthumously honoured his Austrian colleague by giving the bird the popular name "Natterer's Hokkohuhn". It would appear that our discus discoverer had also collected the *pauicé* in this region. Unfortunately it has now died out, as have the Pauxí indians.

They were exploited and enslaved. The colonists, and the missionaries as well, exploited them, and by 1727 the so-called Pauxys had already practically disappeared. Not until 1754, when the priests Frei Domingos da Ribeira de Niña and Frei José de Vila Nova arrived in Pauxís, were the surviving Pauxys able to flee with their aid. Their leader, the *cacique* (chief) Manocassary, led them to Mocambos, on the upper Rio Trombetas. But what are, or is, Mocambo(s)? Not long ago I visited the bay and village of Mocambo in Mozambique on a field trip. From where, I found out, the Portuguese had brought slaves to Brazil back in the 16th century (mainly from their colonies). And *quilombos* is the name given in Brazil to negros who escaped from slavery early on. There are more than 700 *quilombo* communities in Brazil. They lie mainly in remote regions (to which the slaves fled in order to avoid being found) such as along the – very difficult of access – middle and upper Curuá and Trombetas.

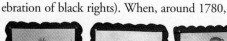
As long ago as 1604, slaves from Mocambo, Mozambique ran away from the *fazendas* and became the *quilombos*.

The *quilombo* movement began back in 1604 at Palmares, on the Brazilian coast. Forty slaves, originally from Mocambo, had run away and founded the first settlement of their own. Palmares was attacked for more than 100 years by Dutch and Portuguese troops, trying to get the slaves back and deter others from following. The penultimate king of the slaves was Ganga Zumbi, who was killed by his nephew because he signed a treaty with the Portuguese whereby the *quilombos* were to no longer give refuge to slaves belonging to the *fazendeiros*. The nephew fought on heroically, but died in battle on the 25th November 1695. A day that was to become the Dia Nacional da Conciência Negra (the national day of celebration of black rights). When, around 1780,

This page, left: The Amazon off Óbidos. Here the otherwise lethargically-flowing Amazon is at its wildest – at least as regards the current. *Top:* The *recreios* which ply almost daily between Óbidos, Parintins, Santarém and even Manaus and Belém. *Above:* The blackwater Curuçamba, which feeds the Lago Mamaurú, close to its mouth. *S. haraldi* occurs here, in the lake Mamaurú.
Left-hand page, below: Far left three gourds that originated from Óbidos on the Amazon but were hand-painted in Portugal. *Centre:* A *quilombo* girl; and right: three paintings showing the three types of people at Óbidos: a Pauxí, a *caboclo*, and a *quilombo*.

cocoa plantations and cattle-breeding began in the region of Óbidos (and Santarém), African slaves were brought there as well. And they brought their own *quilombo* movement into being. The settlements on the Curuá to which they fled (such as Inferno and Cipotema) were repeatedly burnt down. But they refused to be subdued. A number even travelled to the Rio Erepecuru and Cuminã. A hard road, full of rocks and enormous waterfalls such as the Barracão de Pedra and many more.

Around 1821 there were about 2,000 *quilombos* in this region, trading with the inhabitants of Óbidos, chiefly in their home-grown tobacco – but the persecution continued nonetheless. At the end of October 1870 the Governor of Pará, Manuel José de Siqueira Mendes, even issued a mandate for the elimination of all *quilombos*. Once again their settlements on the Curuá were destroyed, until 1888 when, on the 13th May, the slave trade was banned in Brazil as well. Ever since then the *quilombos*, like the indians, have been demanding that their title to their land should be officially recognised. One *quilombo* leader, Silvano Silva from Oriximiná, declared, "*Aquela história de abolição, pra nós, não é bem válida, não. Porque s ofremos pressão de mineradoras, madeireiras e até do IBAMA. E continuamos a ser escravos na nossa própria terra.*" (The end of slavery has no meaning for us. We are oppressed by the mining and timber companies, and even by IBAMA itself. We remain slaves on our own land.) But after more than 300 years there are signs of a change of fortune for the former slaves: in 1995, in Boa Vista (Pará), a *quilombo* community was for the first time granted title to the land on which they were living. (Boa Vista is one of the 21 *quilombo comunidades* in Amazonia.) Since then the *quilombos* of Pacoval (see page 277) and Água Fria (see Trombetas region) have been promised their land, and the new President Lula has, following his inauguration in 2003, signed a decree that all the *quilombos* should be given title to their land. At present 18 (out of 724) communities have such title.

But I would like to say something about the mother town in Portugal. Scholars are in disagreement regarding the age of this lovely town (which I can warmly recommend to everyone as worth a visit), likewise over the origin of the name.

Excavated remains of walls indicate that in 308 BC the Celts erected a defensive town here (=*Oppidum*). But Roman remains have also been found beneath the castle *(centre)*, which the Arabs erected as a fortress and the kings of the first dynasty (12th century) then repaired and extended. Meanwhile a fabulous inn, the Pousada do Castelo, was constructed inside, its walls reinforcing the lofty, 1,565 m long town wall, which remains completely intact and has stood for almost a thousand years. Within lies the fantastically restored old town *(right)*.

The fortress of Óbidos in Portugal – and below the yellow wave that signifies water (origin of the name?). But Óbidos is also the *ginjá* = the Portuguese for cherry laurel (*Prunus lusitanica*) from which the secret of the kings was brewed, a very tasty liquer.

There are two hypotheses about the name. Firstly, that it is derived from the Latin word *oppidum*, which means town. The second hypothesis says that the words *ob-id-os* signify, roughly, water movement (marine inlet, underground river), because such existed there (as is also reflected in the view – *centre*).

But we want to learn something about the discus of Óbidos in Brazil as well. In this respect the great discoverers such as J. Natterer (1835), H. W. Bates (1849 and 1859), and A. R. Wallace (1850) had no luck. Nor did Agassiz, Bentos, James, and Hunnewell Burkhardt (the members of the Thayer Expedition mentioned earlier, whose sole purpose was ichthyological) find discus at their stations in Villa Bella (and Obydos), THAYER 75, THAYER 76, and THAYER 77 between the 27th and 30th August 1865. However, I was amused by the comment of Agassiz's indian companion Laudigári, which reminds me of many of my own experiences with the natives, "Why are you attached to the fishes, especially the little ones, which appear to me only fit to throw away."

In the final analysis it is no wonder, as none of these gentlemen visited the Lago Mamaurú (there were no roads), which

The more than 2000-years-old Portuguese Óbidos (1,2,5,7), and its coat of arms (2). A fabulous city, now completely restored and surrounded by a still intact, thousand-year-old defensive wall (1). Óbidos on the Amazon (3,4,6,8) and its coat of arms (4); this city, by contrast, is run down after only about 200 years years of history. But it has a good (fish) restaurant (8); an interesting museum (6) which is also an example of Portuguese architecture (cf 5); and the Germanic cathedral of Nossa Sra. Sant'Ana on the highest point, with splendid gardens and a view of the Amazon (3). When Bates collected here (1850) the settlement was home to barely 1200 people – today there are more than 50,000. The rubber and jute booms of the past are (almost) finished. A certain amount of cocoa, Brazil nuts, and fish is still shipped, and soon there will be more (see text). But the city celebrates at least six large festivals every year, and a *carneval (right);* the Festival do Jaraquí *(left)* (the *Semaprochilodus* fish festival); Boi-Bumbá; Gracinha; Círo; and Nossa Sra. Sant'Ana.

lies some 12 km north-east of Óbidos and is fed by three *igarapés*. Two contain clear to tea-coloured water and mixed water when it rains *(right-hand page)*, the third (the Rio Branco) is described as a whitewater, but to my eye is clear, based on observations at various times of the year. The Lago Mamaurú is described as *água preta* but is in fact clear (and mixed at high water).

In the Lago Mamaurú the water parameters on 13.02.2004 at 08.30 hours were as follows (unfortunately I was unable to take any measurements there previously): pH 6.10; conductivity 27 µS/cm; daytime temperature (air): 28.5 °C; at the water's surface: 27.9 °C; in 2 m of depth: 25.4 °C; oxygen 1.17 mg/l. The surrounding area was only partially still intact, practically nothing but *fazendas* all around. Biotope: overhanging branches, brush, numerous floating plants; the water was almost motionless; lake bottom sandy. For discus from this region see page 201.

The only other habitat at Óbidos where discus still occur is in the Lago Curumú to the north-west (not on the map). It can be reached only via a 28 km dirt road and then 1-2 hours by rough tracks across the *fazenda* of Senhor Coreolano Sarrasin da Silva. At high water it can also be reached by canoe via the *furos* of the Trombetas (ie the Trombetas drainage region). It contains clear water and harbours *S. haraldi* – sometimes individuals with pattern colour. (For parameters see under Oriximiná-Trombetas, below.)

On the opposite side of the Amazon, in the Lago Grande do Curuaí, a large permanent whitewater lake (larger than Luxemburg), there are neither discus nor suitable habitats. Far from it. A number of years ago the large-scale commmercial cultivation of *curauá (Ananas lucidus),* a plant of the family Bromeliacae from the lower Amazon, was begun here, because it had been found to have more potential uses than even sisal *(Agave sisalana).* For a long time the natives used its fibre to make cord, hammocks, baskets, and other useful items. But in 1996 the fibre was researched in detail under the auspices of Emater (Empresa técnica de assistencia e extenção Rural), and since then no expense has been spared in developing plantations. Millions have been invested. It all started with an area of just 400 ha, and the plan is to increase this to three million ha, creating 2,000 jobs. More than 400 families cultivate *curauá* beside the Lago Grande. They supply the market with around six million plants. A single hectare of lime-enriched, fertilised land holds 20,000 plants. Each plant produces up to 98 leaves during its lifespan of around five years, and 12 leaves yield 1 kg of fibre. Ie about 15 tonnes per hectare, every September. Large buildings have been erected for extracting the fibre and further processing. By now even the Brazilian Volkswagen factory is using the fibre for the luggage space in its cars – it is an ideal substitute for fibreglass. More than 200 *Polos* come off the production line daily, and require six tonnes of fibre monthly (in 2004). Mexican, Spanish, and German car factories are queueing up for this much-prized natural fibre. And, of course, the families around the lake (and elsewhere) cannot produce enough. So it is now planned to exploit the enormous jungle areas between Santarém and Parintins to meet the demand worldwide.

Above: The green and red variants of the Amazonian *curauá* plant. Its fibre *(below)* has almost unlimited uses, a substitute for the best cellulose *(below right).* The jungle must make way ...

"Se gosta da sua vida, preserve a naturesa" (If you value your life, then preserve Nature) reads the sign in the lovely Igarapé Curuçambá, 8 km from Óbidos. But every day hundreds of cattle are driven through (2). They destroy, inter alia, the plant *Pontederia diversifolia* (3), and the habitat of the last natural (true) *Hyphessobrycon bentosi* (wild-caught ♂ (4) and ♀ (5)); two *Bryconops* spp. (on photo 7 shows 2 species living sympatric; photo 6 = detail of upper = *B.* sp.1; photo 8 of lower = *B.* sp. 2, possible mimicry); and characins that are undoubtedly still undescribed (photo 9 = sp. 3; photo 10 = sp. 4). But I cannot show all the endangered species of the biotope here, it is not a discus habitat, too shallow and too fast-flowing. But it is a major stream which feeds the Lago Mamaurú, which does contain discus (2).

3. Oriximiná and Trombetas region; Rio Nhamundá region; Parintins region; Rio Uatumã; Rio Urubu.

Oriximiná and Trombetas: This region is known as the Mesorregião do Baixo Amazonas (Microrregião de Óbidos), and it is the westernmost Amazonian region in the state of Pará. It includes Terra Santa, Faro, Juruti, and Óbidos. The Rio Nhamundá forms the border with the state of Amazonas, and to the north it is bounded by Surinam, Guyana, and the state of Roraima.

Some 60,000 people live in the Oriximiná region, which can be reached almost daily via *recreio*. The nearest airport is Porto Trombetas (with flights from Belém or Manaus), from where one can travel down the Trombetas to Oriximiná. The speedboat *(jato)* from Manaus makes interim stops at Óbidos three times a week en route to Santarém, and from Óbidos the dirt road already mentioned runs to Oriximiná.

Oriximiná has a number of places to stay and restaurants (in February 2004 I paid just 1 R$ = US$ 0.36 for a lunch, including fruit juice) and there are festivals here as well. At least nine take place each year, each lasting for two weeks: in January (or July) the Festival de Música de Oriximiná (a music festival); in February the Carnaval (carnival); in April the Semana Santa Cultural (Easter culture week); in May the Festival do Theatro (a theatre festival); in June the Festival da Castanha (the Brazil nut festival); in August the Festividades de Santo Antônio – Círio de Santo Antônio (where, as in Alenquer, Saint Anthony is celebrated – only here illuminated boats sail to and fro on the Trombetas at night); in September the Festival do Tucunaré *(Cichla* fishing festival); in November the Festival de Cultura (culture festival); and in December the Aniversário de Oriximiná (anniversary of the founding of Oriximiná).

The town was founded by the priest José Nicolino de Souza, who arrived from Faro on the 13th June 1877 and laid the foundation stone, although he named the settlement Santo Antônio do Uruá-Tapera. It wasn't until much later that it was again given its original indian name. The people make a living from (how else?) valuable hardwoods, jute, fish, Brazil nuts, and nowadays mainly from bauxite. Almost every day the ocean-going steamers of the MRN (Mineração Rio do Norte) travel down the Trombetas fully laden. Since 1979 bauxite has been transported to the refineries and exported as well. It comes from the Lago Batatas region, on the right-hand bank of the Trombetas in the vicinity of Porto Trombetas.

When I last visited the region in 2004 the situation was thus: the MRN were celebrating their success as the largest bauxite company on Earth following the enlargement of the mines and increased production. The President of the company, José Carlos Soares, was proudly announcing targets achieved. With 230 million dollars recently invested production could be increased by around 50% to 16.3 million tonnes per year (although with the worldwide recession in 2004 only 14.5 million tonnes had been mined). Aluminium is manufactured from the refined bauxite – drink manufacturers require millions of aluminium cans every second. Which cause ever-increasing damage to the environment. (The environmental damage from the processing of the material in space shuttles is much less ...)

The lion's share of the bauxite is processed in Brazil – nowadays the South American giant is the sixth largest aluminium producer on Earth with a turnover of more than US$ 6 billion (3.4% of of Brazil's national product). The Lago Batata bauxite reserves run to more than 800 million tonnes – some 50 years of production at current rates.

The lake belongs to Oriximiná, Pará (and lies at 1° 30' S, and 56° 20' W), and there were splendid discus biotopes there when I visited the region for the first time (1965).

It was a fabulous, isolated, unspoiled piece of nature. No-one knew about the bauxite and little felling of hardwoods was taking place. Another reason why I would like to say a few words about the current situation: Between 1979 and 1989 around 24 million tonnes of waste water were released into the Lago Batata. The ecosystem was almost completely destroyed. The bauxite effluent contained small particles of argil with a very high content of silica, aluminium, and iron, which formed a layer up to 4.5 m thick on top of the natural sediment. More than 30% of the total area of the lake was affected. An algae explosion took place, and fishes disappeared along with a large part of the vegetation. At the end of 1988 attempts were begun to "rescue" the lake, and measures for waste-water control were introduced. Recent studies commissioned by MRN report that since then there have been positive changes in the lake and its environment, and that the fishes have returned. But in February 2004 I could find no discus in the lake, although formerly there were three very different variants (all *S. haraldi):* a dark type with pearly pattern colour (as in individuals at Terra Santa, which is

1. Oriximiná in February 2004 (the beginning of the rainy season here); the normally clear Trombetas water is inundated by the Amazon. 2. Oriximiná with clear Trombetas water (dry season). 3. Lago Batata at the home of the *pescador*, Joaquím the *caboclo* (the water is almost clear) 4. His family. 5. Discus habitat in the Lago Batata with water test (colour), February (2004). 6. Ocean-going steamers which travel down the Trombetas daily, laden with bauxite. The bauxite is transported from the Lago Batata via the 30 km railway to Porto Trombetas where the steamers are loaded with an enormous conveyance belts before departing for Belém and the world.) 7. A bauxite monument, a new landmark in Oriximiná.

HABITATS ... ÒRIXIMINÁ AND TROMBETAS   BLEHER'S DISCUS   **289**

not far away, and where bauxite has now been discovered as well), a brown to reddish form, and a red variant (the last two with little pattern colour). But I did catch *Geophagus* and *Bryconops* species in the habitat.

Nowadays around 50 *caboclos* live around the Lago Batata, which has a circumference of some 250 km and contains several islands. I spoke to Joaquim, a *pescador* on the lake. He told me that since the beginning of the bauxite extraction the fish population has decreased by more than 50% and there has been no improvement. And that in the dry season the *lago* is a red soup. (At this time the harmful deposits can clearly be seen, although MRN asserts otherwise. Interestingly MRN has never been brought to book for this environmental catastrophe. Or for the clear-felling of far more than 4,000 ha of primary forest. Bauxite is found here only at a depth of 8-12 m and the whole forest has to go. Though by now 2074.2 ha have been re-planted with 5,185,500 trees. We can only hope this works.)

When I asked whether he still caught *barucas* (the indian name for discus in this region), he replied, *"Si, e a gente come"* (yes, and we eat them). And when I commented that a *baruca* has hardly any meat on it, he replied, *"Mais são gostoso, e si não tem outro peixe a gente come o que tiver"* (but they taste alright, and beggars can't be choosers).

Formerly this was unthinkable. None of the *caboclos* used to eat discus (and the indians still don't). There were always enough larger fishes. But the continual destruction of the Amazon region over the last 50 years has seen the loss of many vital habitats (through clear-felling and burning) and spawning grounds (through dams). The environment has suffered increasing damage (even the smallest Amazon settlements have a sewage outfall – *rios* and *igarapés* have become sewers). And the population explosion (from a few hundred thousand to more than 20 million white people) has resulted in per-

Two typical discus habitats on the Trombetas:
Top: habitat of *S. haraldi* with pattern colour; and *above: S. discus?* habitat, in the Lago Jacaré.

manent over-fishing. Everywhere you go you hear that edible fishes are becoming ever scarcer (and are smaller).

In the Lago Batata *(see page 289)* (in the part of the lake where I formerly caught discus – and where Joaquím said they were still supposedly found) the water parameters on 12.02.2004 at 13.00 hours were as follows: pH 5.83; in 2.5 m of depth: 5.95; conductivity 14 µS/cm; daytime temperature (air): 34 °C; at the water's surface: 30.5 °C; in 2.5 m of depth: 30.4 °C; oxygen 1.17 mg/l.

Biotope: trees, roots, leaves, overhanging branches; no floating plants; jungle only partially still intact. The bauxite mining and cattle-breeding in the surrounding area have taken their toll. The water was relatively clear *(see page 289)* and almost motionless. For discus from this place *see page 173*.

Now although ongoing studies claim that the lake is slowly recovering; that there is less clouding; that the concentration of organic material in the sediment is increasing; that the plants in the *igapó* are returning; and that there are fishes once again; the local people say quite the opposite. It is also doubtful whether the existing clear-felled area of 11,470 ha has been completely re-afforested (the fauna certainly hasn't been restored).

But let me describe the Trombetas region in greater detail. The Rio Trombetas rises in the Guianan Shield, at an altitude of some 800 m in the Acaraí Mountains on the border between Guyana and Surinam, and flows into the Amazon after some 750 km, some 2 km downstream and to the south of Oriximiná, where it joins an arm of the great river. The Trombetas basin is the second smallest Amazon tributary drainage (the Nanay is the smallest). Its most important right-hand tributaries are the Rios Turuna, Cachorro (or Inambú, Imabú), and the long Mapuera. On its left-hand side it is joined by its largest affluent, the Rio Parú do Oeste (also called the Rio

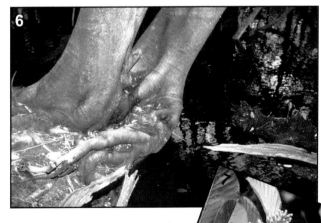

1. Typical discus habitat in the Lagoa Abuí, the northernmost distribution in the region. 2. Parts of the rocky Guianan Shield can be seen along many parts of the Trombetas shore (discus habitat only left in the photo). 3. During my Trombetas explorations I came across indian tribes such as these Katuena by a blackwater Mapuera tributary. 4. Discus habitat in the Lago do Jacaré. 5. Detail. 6. In the Lago do Jacaré the water is black. Discus from here have been catalogued as Heckel discus *(see text)*. *Left and right:* seeds and flowers which I have found in the stomachs of discus.

Cuminá or Erepecuru). In total its drainage encompasses around 134,000 km².

The local people call it a *rio de água preta*, but it is more of a clearwater river, at least in the parts (in the lower Trombetas) where the majority of discus occur. However, almost all of its tributaries contain true tea-coloured water. And from March (sometimes from February onwards) until July (or August), the lower Trombetas region, including parts of the enormous Lago Sapucuá, is imundated with muddy Amazon water.

The Trombetas is normally (but not always) navigable as far as Cachoeira Porteira, where the major rapids begin around 260 km from the river mouth. Large (ocean-going) steamers and *recreios* can travel up to 200 km upstream all year round. (The depth during the dry season is only 2.1 m at Oriximiná and 1.5 m at Cachoeira.) But not up the tributaries, as they are full of rapids and some have waterfalls up to 30 m high. As far as its larger tributaries are concerned, I have found discus only in the lower Paru to upstream of Porto Trombetas, although they used to be in the majority of the *lagos* and in the Trombetas itself. (That was from 1965 to 1989. By now there are numerous cattle farms in the lower part and in the *várzea* and *igapó* regions from Porto Trombetas to the mouth). But here is the list, starting with the northernmost distribution and habitats.

I explored the region around Cachoeira Porteira several times up to 1989 and was unable to find any trace of discus either upstream or downstream of the village on account of the rapids. Nor in the Mapuera *(right-hand page)*, which flows into the Trombetas at Cachoeira. The entire region is discus-hostile, with rocks and stones everwhere, and often metre-high rock walls of volcanic origin or granite. There are no discus habitats; the latter don't begin until much further downstream, specifically in the Lagoa Abuí *(see page 291)*. At least according to the local *quilombos* at Abuí, where 50 families live by agriculture, hunting, and fishing. There are also said to be discus in the Paraná do Abuí, where there is another *quilombo* settlement of around 150 former slaves. All *S. haraldi* – discus with a stripe pattern, often very prominent, or so it is said. And on the other (left-hand) side of the river almost opposite, where the Lagoa Grande lies a few kilometres inland and is connected with the Trombetas only at high water, there are said to be the same sort of colour variants.

Further downstream along the right-hand bank lies the large Lagoa de Farias *(see page 295)*. The water is tea-coloured and discus have not yet been collected or even found here. But it does have suitable habitats. Likewise, to the best of my knowledge nobody has to date looked for discus in the region of the *comunidade* of Jacaré, on the left bank of the river. (Jacaré is not to be confused with the Lagoa Jacaré – see below.) On the other side lies the Igarapé de Tapagem, which flows through a lake system before entering the Trombetas on its right-hand side. The local people, again *quilombo* fishermen and farmers, told me that there were *barucas*, but that they were seen only during the dry season and were *pintados* (painted).

Somewhat further south, again on the right-hand side, lies the Água Fria region, an *igarapé* that empties into a small *lagoa* before joining the Trombetas. When I last (2004) visited this habitat, the water was rising and had brought in enormous grass islands *(see page 297)*. Água Fria is also the name of the *quilombo comunidade* with its 14 families, who live by fishing, hunting, and agriculture.

The water parameters in the discus habitat on 12.02.2004 at 09.00 hours (unfortunately I was unable to take any measurements there previously) were as follows: pH 5.5; in 2.5 m depth: 5.81; conductivity 20 μS/cm; daytime temperature (air): 30 °C; at the water's surface: 29.9 °C; in 2 m of depth: 29 °C; oxygen 0.00 mg/l (not measurable).

Biotope: washed-up grass mixed with lots of *Utricularia foliosa* and with dead branches in amongst it (lots of tree-trunks in the deep water), and the dominant tree (bush) was the *acará-açú*; jungle in the main intact, there was only the *quilombo* settlement in the immediate vicinity. The water was almost motionless and blackish.

Água Fria lies so-to-speak on the border of the gigantic bauxite-mining region (wherein lies the likewise right-bank Lago Batata mentioned above). From here downstream – to shortly before Oriximiná – we encounter an almost endless succession of islands, which often divide the Trombetas into several channels. And many of the islands harbour *lagoas* and *igapós*.

After the Igarapés Pedra and Arajá, the Lagoa Palhar Grande, and the Igarapé Moura, which are all little explored ichthyologically but lie right in the clear-felled and bauxite mining region, we come to Porto Trombetas, the port of embarkation for the bauxite harvest (see photo, *centre)*. Here alone live more than 6,000 MRN mine-workers and their families. There are

1. The rapids in the Trombetas, right next to Cachoeira Porteira. 2. Cachoeira Porteira, the northernmost settlement on the Trombetas (apart from *quilombos* and indians). 3. A *quilombo,* from one of the 27 *comunidades* in the Trombetas region. 4. The Mapuera is like a mill-pond upstream of the first rapids. 5. In the Trombetas, immediately downstream of Cachoeira, I found no discus but I did find, inter alia, dwarf *Crenicichla* (7) and crab (8). 6. Upstream I didn't find any discus either, but there was this catfish *(Pinirampus pirinampu),* among others (see page 296).

Above: a section of the Rio Mapuera region (1) that I explored. Only Wai-Wai indians (see *aqua geōgraphia no. 4*) live there. It is a reserve of more than 10,000 km² and extends north to Guyana, ends to the west in the state of Amazonas (at the Nhamundá reserve), and its eastern part lies in the state of Pará. It is completely isolated. The Wai-Wai village (3) can be reached from Cachoeira only by surmounting more than 43 rapids and waterfalls. In discus-type biotopes (2) and stony habitats (4) I found interesting fishes *(see also page 296)*, including a *Satanoperca* (5), possibly a new species, but no discus. Not until the Lagoa Abuí, or so I was told by the local *quilombos* whose protein source is fish, which they catch daily (6). Some of them are blonde and evidently have indian blood. The daily "bread" of the Wai-Wai is prepared from manioc (7 and 9). 8. Arrau turtle nests in the Trombetas conservation region.

10. Lago Curupira. The old fisherman Ricardo with his grandchildren on the lake. 11. San Lazaro is a *caboclo comunidade,* on somewhat higher ground and with a small church. 12. *S. haraldi* (with little pattern) habitat in the Lago Curupira (water rising). 13. Discus habitat, Lo. de Farias. 14. *Fluviphylax* sp. (new?), from a discus habitat (13), also 15: *Spatuloricaria* sp. with an extreme mouth. 16. Discus in the Trombetas feed on these fruits as well. 17. And on these. 18. Discus habitat in the Lago Acarí (during rain).

completely modern hospitals, restaurants, hotels, and the above-mentioned airport, from which there are almost daily flights to one of the larger Amazon cities.

Although the entire region is designated a conservation area *(see page 301)*, the bauxite, gold, and diamond mines it contains are continually being expanded and new prospecting rights constantly granted. The mining in one region has been halted for a short time because archaeological finds, ceramics from the Kondouri culture from the period 1,000 years BC, came to light – but for how long? By 1988 51 such sites had been discovered (but how many have been buried?).

But I will not say any more about the mines, the destruction, and the environmental damage, but instead discuss the habitats found in the region. Efrem Ferreira, the well-known Brazilian ichthyologist, together with his colleagues from the INPA Institute, performed a census of the fishes of the Trombetas, but they found no discus, as they collected only with rotenone (fish poison) and commercial nets. (On this page you can see some of the species I found in the upper Trombetas.) And the Thayer Expedition (1865-66) didn't find any discus, either. However, three *Symphysodon discus* are noted both on the list for the area and elsewhere. They were collected in the Lago Jacaré by an expedition made by the Goeldi Museum between the 20th September and 13th October 1965, all with bow and arrow *(see page 173)*. They were classified as *S. discus,* ie Heckel discus, but so far I have been unable to confirm this identification. Likewise Axelrod's Heckel discus, which purportedly originated from the Lago Batata *(see page 173)*. My collecting and studies of discus in the Trombetas region have turned up *S. discus*-like variants, but no true Heckel discus *(cf Chapter 4, page 201)*. I have collected discus in the Lagoa Jacaré *(see page 291)*, as the *lago* at the western end of the Erepecuru lake is also known, but no true Heckel discus as I understand that species. Should it turn out, however that these were not naturally occurring hybrids (as I believe), then this must be the easternmost distribution of *S. discus*. And this would be an isolated region for the species, similar to the Rios Nhamundá, Jatapu, Abacaxís, and Marimari *(qv)*.

The Erepecuru region also lies in the Reserva Biológica and is full of rocks and stones at its north-western end. The Igarapé Candeeiro, flowing from the north-east, and

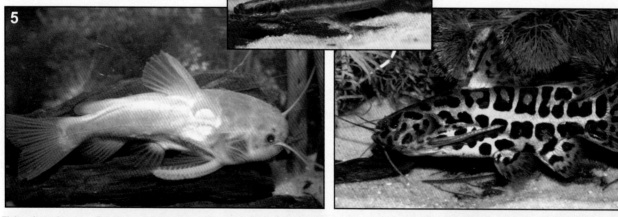

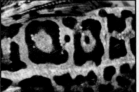
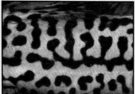

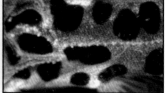

Fishes from the upper Trombetas (where there are no discus): 1. *Corydoras* **sp. 1. 2.** *Corydoras* **sp. 2. 3.** *Apistogramma* **sp. 1. 4.** *Anostomus anostomus* **– first record from Amazon-drainage. 5.** *Liosomadoras oncinus* **- albino. 6.** *Liosomadoras oncinus.* **Photos below:** *Liosomadoras oncinus* **variants which I found there.**

1-2. Água Fria, with another *quilombo* settlement (1). These two shots were taken when the water was rising and hence the *lago* was full of grass islands (2), which disappear later but make forward progress very difficult at such times (February). 3. A typical discus habitat beneath an *acará-açú* tree, full of leaves, branches, roots, and floating *Utricularia foliosa* as well as grass (the last of these only when the water is rising). 4. The black water here looks like this. 5. There are also tributaries that are actually full of aquatic vegetation (but no discus habitats). 6. Upstream to Porto Trombetas the discus habitats are for the most part swamped by grass islands when the water is rising. As in the *lagoa* seen here.

other tributaries are full of rocky massifs, stones, and rapids and contain no discus habitats. The water is almost black everywhere and harbours a wealth of fishes, for the most part unstudied. Piranhas (at least three species) abound here, and there are numerous spectacled caimans *(Caiman crocodylus)* in the dry season.

The lower Rio Parú do Oeste (called Erepecuru or Cuminá by many people) region with its wealth of lakes is little known. But discus do not occur there. The *caboclos-ribeirinhos* and *quilombos* told me that they lived downstream of the *lugarejo* (tiny settlement) of Delfina and Boa Vista, and I found them further south.

When I arrived in this region for the first time (1980s) I also heard, from a *quilombo* in Boa Vista, that there were still indians, deep in the jungle to the north-east, who wore unique lip discs and to date had had no contact with the outside world. Unfortunately I had little time and no opportunity of venturing into this inaccessible region. But on the 5th November 1987 they left their hidden retreat and were seen by the public for the first time at Esperança. A group of these so-called Zo'é can be seen below. It was a tense situation for the inhabitants of Esperança. People tried to "convert" them with gifts and got arrows in exchange, albeit with the tips broken off. More and more arrived within a few days, erecting palm-thatched huts on a height and even laying out fields.

According to FUNAI missionaries had previously attempted to make contact with the isolated tribe, but the indians wouldn't allow this, and then came to Esperança out of curiosity? In Esperança the Missão Novas Tribos then attempted to proceed with conversion in three stages: first learning their language (nobody understood them); teaching them the alphabet using the Bible; and bringing them the word of God.

They lived peacefully alongside the whites. But, four years later, after 45 had died of flu and malaria, the FUNAI evacuated the whites and denied the missionaries any further contact. There were only 133 Zo'é still left. (However, since then they have supposedly recovered to 152 or more.)

The Zo'é *índios* live between the Rio Cuminapanema (which flows in a valley almost parallel to the upper Rio Parú and subsequently joins the Curuá), in an enormous mountainous region which is designated as a *reserva indígena*. The region is criss-crossed by countless small *igarapés* and is full of Brazil nut trees – their main food after manioc. And they use the bark and bast to make all sorts of goods and materials, including for sitting and sleeping mats and hammocks.

"Zo'é" means "us" in their language (which belongs to the Tupí group). And they recognise no ethnic differences apart from Zo'é (themselves), *kirahi* (whites), *apam* or *tapy'yi* (enemies). The latter are chiefly other indian tribes, also called

1. The Zo'é were initially called Poturu, but that is actually the name of the wood which the Zo'é insert in the lower lip of girls as well as boys at the age of six. At first a small piece, and then increasingly large pieces, up to 20 cm long, as they grow older. They don't know the origin of this custom, but it is an ancient tribal tradition, and distinguishes them from all others.
2. The women also adhere to the tradition of birth control. They have a baby only every three years and achieve this using plants unfamiliar to any doctor. Deformed babies are buried before they can utter a single cry, they are thought to bring bad luck and endanger

further conception.
3. They are semi-nomadic and live chiefly by hunting. They also cultivate manioc, bananas, and sweet potatoes, as well as their *urucú* (or *urucum* -see page 360). From which, by crushing the fruit, they extract a red juice with which they paint their bodies.
4. Monkeys, which are still abundant in remote regions, are their favourite food, although they never kill young specimens, but take them with them and care for and nurture them as if they were their own children. They also catch fishes to eat, but chiefly using *timbó* – fish poison. There are no discus in their region, and they are unfamiliar with them.

kirahi, and by the same token white men such as missionaries are known as *kirahi ete* and other white men as *kirahi amō*.

They have ancient traditions *(see page 298),* for example that of the fire that is handed down from generation to generation and never extinguished. Or that they allow their children to grow up independently – from the moment that they start to walk they do as they please and fend for themselves in this Eden. They are punished only if they do something really bad; in such cases wounds are inflicted on them with sharp fish teeth (those of piranhas, or *Acestrorhynchus* sp.) to let out the bad blood.

They occupy an enormous region and recognise neither boundaries nor the concept of ownership. The land, they say, belongs to no-one. Nominally this land is protected, but back in 1997 the Canadian company Seahawk Minerals Ltd was able to sign a contract with the Brazilian mining company Pará Metais Nobre to give the former all of the mining rights in the 140,000 ha Erepecuru region (=Zo'é). The deal was completed after 120 days and three million dollars. Since then gold has been discovered. But no Zo'é can comprehend why the *kirahi amō* go digging for the yellow metal.

Interestingly there are almost 30 *quilombo* villages, with around 6,500 descendants of slaves, adjoining (and in some cases within) the region of the Zo'é at its southern edge. By the Parú, the Apacu, and the Trombetas. The largest settlement on the Parú is Boa Vista. This *comunidade* of 112 families was the first to be granted land (1125 ha) here in 1998.

I know of no evidence of discus from this region, but it does contain suitable habitats and, as already mentioned, according to the *quilombos* they occur as far as Delfina. I did, however, find interesting stone-carvings further upstream, mainly in the vicinity of rapids or waterfalls. Many of these have not yet been studied. They exhibit four different anthropomorphic patterns, and those belonging to the first of these, which are found north of Delfina, are the simplest. Further upstream there are figures with facial details, and even traces of body painting are discernible. At the waterfalls of the upper Parú (Erepecuru) there are faces with (over-) large eyes, and recognisable body and head decoration. Rarely zoomorphic figures. Only monkeys, birds, and quadrupeds.

As well as the immense Brazil nut woodlands there are fantastic tree ferns (chiefly the species *Zamia lecointei)* and much more to be seen in this still unspoiled area. But the region is relevant to aquarists only as regards loricariid catfishes.

There are supposedly discus *(S. haraldi)* in the Lagoa Encantada (on the other side of Boa Vista) and in the Lagoa Grande further inland, as well as on the large Ilha Grande de Cuminá, washed by the Parú do Oest. Further south, in the Lagoas Arúba and Salgado (again on the left-hand side of the Parú) all-

1. The Rio Parú do Oeste also harbours discus, or so the *quilombos* say. Here a *quilombo* youth with a pimelodid catfish netted in the vicinity of the discus habitat (background). They term the clear water *agua preta*.
2. The *quilombos* also make a living by collecting Brazil nuts, often walking all day through the Brazil nut woodland and filling hand-woven baskets with around 60 kg of nuts before carting them for up to six hours to the boat.
3. They then transport the nuts to Boa Vista by boat.
4. They erect palm-thatches in the jungle where they collect their Brazil nuts, as shelter from the almost daily rain (which can last for weeks). The use of the environment of the descendants of the *quilombo* slaves is inseparably bound up with their history and ethnicity, and in conflict with that of the mining and bauxite extraction and IBAMA.
5. The nuts "ripen" in their hard shells surrounded by an even thicker outer rind.
6. The "unit measure" for selling Brazil nuts.

over striped discus were found by H. J. Mayland (Lagoa Salgado) and myself (in both lakes) *(see page 73)*. But they undoubtedly occur everywhere in the vast labyrinthine middle and lower Rio Parú do Oeste region.

Along the right-hand bank, downstream of Boa Vista, we come to the mouth of the Rio Apacu. Its name derives from the *apacu* tree, which grows in this region and is used for house- and boat-building, parquet flooring, and many other purposes.

To get from the Paru to the Apacu you have to cross the great Lagoa de Juruacá (Jaranaca or Jarauacá), where I have found part-striped as well as all-over striped individuals. Travelling up the Apacu *(centre)* is an adventure in itself. A fascinating blackwater river, almost untouched and practically uninhabited apart from a few *quilombo* settlements. Discus habitats a-plenty, including in the lower courses of its right-bank blackwater tributaries such as the Igarapé Jutaí, among others.

That leaves the Trombetas region downstream of the afflicted Lago Batata. Over the years I have checked out almost all the *lagoas* here. And found splendid discus habitats (and only *S. haraldi* – some with a prominent centre bar) in the following:

Along the right-hand bank of the Trombetas in the Acarí, in the Lagos Samauma I and II, in the Lagos Apé, Jibóia, Arajasal, Achipicá, Camija (or Camicha), Jereuá, Sacurí and in the two Carimó lakes (again called I and II). Then on the left-hand side in the *lagos* and *lagoas* Tarumá, Bacabaú I, Vajão, Bacabaú II, Cuminá (and the river), Água Fria (this name crops up hundreds of times in Brazil), Acapuzinho, Curupira (where the Thayer Expedition collected, only no discus (see page 295)), Castanha, Xiriri, Caipurú, Iripixi, Paraquí, and Parauacú.

The lake region to the south of Oriximiná, including the gigantic Lagoa Parú, is permanently linked with the Amazon, contains white water all year, and has no habitats. There are *S. haraldi* in the western part of the great Lago (or Lagoa de) Sapucuá *(right-hand page)* and in the lower course of its northern affluents, such as the Igarapés Araticum and Saracuá, as well as in the *lago* off Almeida. Mainly individuals with pattern colour, rarely all-over striped.

As regards the Trombetas, it remains to say that hopefully the planned Cachoeira Porteira hydro-electric dam will never be built. For a number of new fish species have been described from just that one site, for example *Crenicichla regani,* Ploeg 1989 and others described by INPA workers. The majority are, however, undescribed (see also page 296). Apropos of this plan, I approve of the action of the local mayor in suggesting this lovely spot should instead be turned into a tourist attraction. It not only has the Reserva Biológica (385,000 ha, founded in 1979) out back, but fantastic rapids and waterfalls (and the above-mentioned stone-carvings) and the fascinating Mapuera on the doorstep. A unique natural paradise.

During my last-but-one trip to the region (2001) I was also particularly impressed that the so-called Pé de Pincha

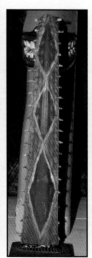
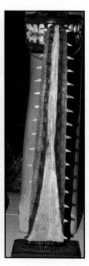

In the Rio Trombetas, in the lower Paru, around Oriximiná, and even in the vicinity of Óbidos fishermen formerly often netted sawfishes, but there have been none for years now. In the 1980s I found a *Pristis* species I didn't recognise *(centre)* on seven occasions. It always had a slightly conical rostrum and all individuals had 19 pairs of teeth in the average. For 30 years I have endeavoured to study these unusual fishes (the salt-water species for less). In the museum at Óbidos I found additional (painted) rostra. The two in the middle are from the same species. The two outer rostrums appear to belong to *Pristis perotteti*.

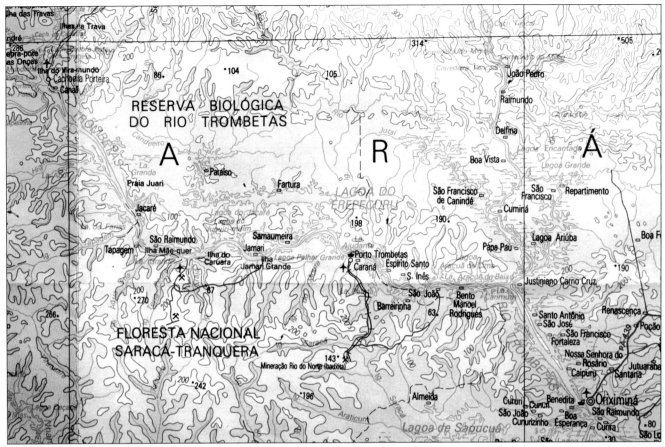

The IBGE map shows how the lower Trombetas – almost down to its mouth – flows through the outliers of the Guianan Shield. In the valley can be seen a large part of the immense lake system where discus may wander (or used to). As mountains, rapids, and rocks are natural barriers (see also map 3, pp. 146-147). The red lines denote dirt roads, and at Porto Trombetas, where there is the only airport in the region, the 30 km bauxite railway past the Lago Batata. (At Oriximiná there is only an airstrip for small private planes.) It also shows the two large conservation zones, the Reserva Biológica do Rio Trombetas (where there are still plans to construct a hydro-electric dam at Cachoeira Porteira) and the Floresta Nacional Saracá-Tranquera (which contains the largest bauxite mine on Earth – see text). Left: two photos of the Lagoa de Sapucuá with discus habitats (distant and close to) and detail of the water, which in the dry season is clear rather than tea-coloured.

Project had been initiated to conserve the endangered aquatic turtles in the Lago Sapucuá. The MRN had wanted to start digging for bauxite there (on the Almeida plateau) as well, but the people of Almeida *(see map, page 301)* sucessfully vetoed the idea. They didn't want to see a new Lago Batata catastrophe in the Sapucuá. In retaliation, so to speak, this project had been brought into being in Oriximiná and Terra Santa, to save the *tracajá* and *pitiú* turtles. 600 adults and children took part. It was carried out in three stages. Firstly, between October and November people trained by IBAMA collected eggs from along the sandbanks at around 5 am. The eggs were then put in incubators. After 45 days the young that hatched were transferred to a rearing tank for 90 days. And since then I have seen how the project has taken root. When I was last there 40,000 juveniles had been released into the wild.

Before we move on to the next discus-stop in Juruti *(below and right)*, a few parameters which I measured for discus habitats in the Trombetas region:

1. **Lago do Farias** *(see page 295)* on 15.06.1989 at 11.00 hours: pH 5.9; conductivity 18 µS/cm; daytime temperature (air): 33 °C; near the water's surface (35 cm depth): 27.5 °C.

Biotope: tree-trunks in the water, roots and branches, overhanging branches and foliage everywhere; floating plants – just some *Eichhornia crassipes*; jungle almost completely intact. The water was tea-coloured, moving only slightly, level dropping.

2. **Lago Jacaré** *(see pages 290-291)* on 19.06.1989 at 09.00 hours: pH 5.65 in 1 m of depth; conductivity 16 µS/cm; daytime temperature (air): 31 °C, in 35 cm of depth: 26.9 °C .

Biotope: lots of tree-trunks, roots, branches and trees festooned with flowers; no floating plants; jungle intact. The

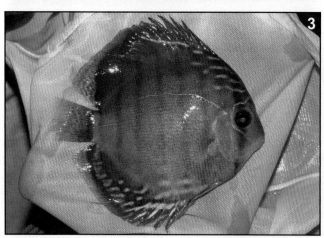
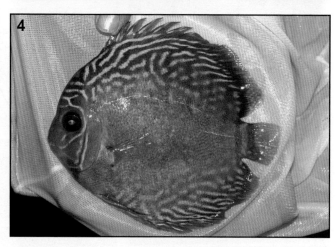

1. The water of the Trombetas tributaries is almost always black (tea-coloured). Along with a *quilombo* we emptied the boat to go fishing for discus in the Paru. (The boats not in use are filled up with water and sunk to prevent them being carried away by the current or stolen.) 2. I have often explored regions using chartered aircraft. Such small planes can carry only very limited cargo. 3-4. Two discus from the Lake Juruti Velho (caught together), left the so-called "Juruti Red".

Top: The Amazon just before Juruti – gigantic with gigantic islands. *Above:* Juruti and its floating harbour. Here sacks of rice were waiting to be loaded, edibles were sold to the *recreio* passengers, and vehicles waited for incoming goods. Juruti lies on the right bank of the Amazon and has no road connections (just a dirt track to the Igarapé do Retiro). But these days it is known for its fascinating annual (since 1955) spectacle involving the indian tribes, Mundurukús v. Muirapinima, which more than 18,000 people attend in the last week of July in the stadium. It is the largest, but not the only festival in Juruti...

water was black and almost motionless, but the level dropping.

Discus variants: *see page 173*.

3. Lago Acarí *(see page 295)* on 12.02.2004 at 14.15 hours: pH 5.7, in 3 m of depth: 5.52; conductivity 12 µS/cm; daytime temperature (air, during rainfall): 25 °C; at the water's surface: 30.4 °C; in 2.5 m of depth: 30.0 °C; oxygen 0.87 mg/l.

Biotope: trees, roots, leaves, overhanging branches; no floating plants; jungle for the most part intact, only a few *fazendas*. The water was tea-coloured and almost motionless, but had risen 2 metres.

4. Lago Curupira *(see page 295)* on 12.02.2004 at 17.00 hours: pH 6.05. in 2.5 m depth: 5.9; conductivity 23 µS/cm; daytime temperature (air): 30 °C; near the water's surface: 30.5 °C; in 2.5 m of depth: 30.0 °C; oxygen 0.60 mg/l.

Biotope: trees, roots, leaves, branches. No floating plants; jungle on the southern side of the lake still intact, on the northern side and around S. Lazaro only cattle farms and agriculture. The water was relatively still and clear (slightly tea-coloured), but rising.

5. Rio Trombetas (upstream of Oriximiná) on 12.02.2004 at 18.00 hours: pH 6.36; in 2.5 m of depth: 6.30; conductivity 26 µS/cm; evening temperature (air): 30 °C; at the water's surface: 30 °C; in 2.5 m of depth: 30 °C; oxygen 2.58 mg/l.

Biotope: trees, roots, overhanging branches; no floating plants; jungle only partially still intact, *fazendas* almost everywhere and hardly any primary forest remaining. The water was relatively clear, flowing slightly (it had just rained). Discus variants: *page 201*.

6. Lago Sapucurú on 3.10.1998 at mid-day: pH 6.2; conductivity 18 µS/cm; daytime temperature (air): 34 °C; near the water's surface (30 cm of depth): 30 °C; oxygen 4.9 mg/l.

Biotope: trees, lots of roots in the water, and grass as well in a few places, but no floating plants; jungle still for the most part intact. A few cattle farms in the surrounding area. Water very clear *(see page 301)* and almost motionless (murky and moving at high water).

Juruti, on the right bank of the Amazon with the large island of Maracá offshore, was formerly a small settlement of indians of the Mundurukú tribe A missionary was first stationed here in 1818. It has long since grown to a town of 31,210 inhabitants (2000), and its *município* borders the state of Amazonas and Faro to the west, Terra Santa and Oriximiná to the north, Óbidos – Santarém to the est, and Aveiro to the south. Its co-ordinates are 02° 09'08"S and 56°05'32"W. It lies at 36 m above sea level and belongs to the Mesorregião do Baixo Amazonas (Microrregião de Óbidos). It has a number of passable hotels, eg the Beira Rio (right on the waterfront – two rooms with a view of the Amazon), but the Santel, Juriti, and Thais hotels lie also above the floating harbour and have one or two rooms apiece with river views. The environment is largely intact and contains numerous *lagos* that are full of fishes but little explored ichthyologically. For example the Lagos: das Piranhas, Jará, Juruti Velho, and Preto, and the great Curumucuri, which at high water is also connected to the permanently whitewater Lago do Porção Grande and then contains partially mixed water – but is otherwise, like the rest of the list, clear. Also the Lagos Salé, Araçá, Juruti Miri, and Santana.

I have found discus *(S. haraldi)* in the different Juruti lakes *(see page 302, for a collecting site)*. Often different variants at a single site, red (termed "Juruti Red") and sometimes also partially striped.

Then there is the Cachoeira do Rio Branco (a whitewater waterfall –

The sign is the only thing left standing in a clear-felled region, and reads: "Conservation Zone". 2. Here in the Pa. do Nhamundá at the mouth of the Lo. Caburi there are nothing but *fazendas*, with no tree left standing. 3. During the rainy season the Amazon water penetrates to the Lo. do Panauarí discus habitat.

Leptotila verrauxi

Top: A small igarapé flows into the Caburi here. In the deeper zones (trees) there are discus, in the shallower bank zones I found an incredible wealth of species (February, 2004). *Above:* Lava rock discus habitat in the Caburi at high water (murky, otherwise clear). Some fishes from the *igarapé*: 1.*Pamphorichthys* cf. *scalpridens* 2. *Rivulus* sp.. 3. *Fluviphylax* cf. *simplex* (*simplex* = from Parintins). 4. *Ctenobrycon hauxwellianus.* 5. *Odontostilbe?* sp.. 6. *Prionobrama* sp.. 7. *Raphiodon* sp., juv.. 8. *Parotocinclus* sp., juv.? 9. *Acarichthys heckelii*, juv.. 10. *Mesonauta* sp., juv.. 11. *Heros* sp., juv.. 12. *Cichlasoma* sp.. 13. *Brachyhypopomus?* sp.. 14. *Eigenmannia* sp.. 15. *Mylossoma* sp., juv.. 16. *Colossoma macropomum*, juv. (mimicry?).

without discus habitats) and the Igarapé Maracanã, *de água fria* (cold water) as they say hereabouts (but to the best of my knowledge it never drops below 25 °C and also has no discus habitats), which is clear and full of rocks and sandbanks. Important *igarapés* include the Arauá and the da Sabina, which, however, empty into the Rio Mamuru.

The only significant high ground is the Serra de Santa Júlia and the Serra Capiranga, at 90-150 m, which contain blackwater *igarapés*. Olimpio Viera da Silva, a *caboclo*, has often caught dark *acará-disco* with little pattern colour in his *malhadeiras* (gill-nets) in the Capiranga drainage.

On the other side of the *serra* (south of it), lies the Andirá-Maraú indian reserve (4,658.68 km²), but more of that later.

Unfortunately during my visits I had no measuring equipment with me and hence have no parameters from this region, but I can tell you a bit about the name Juruti and the indians: Juruti comes from the Tupí-Guarani (indian) language and is a mysterious pigeon-like bird (as a matter of a fact one species of pigeon *(Leptotila verrauxi)* has even been given the popular name *juruti* or *juriti, see page 304)*. In the Tupí legends a young man is said to have forsaken the daughter of a chief for another, whereupon she died of a broken heart. Her father transformed his dead daughter into the *juruti-pepena (pepena* = paralysed/crippled) to thenceforth punish all those that were unfaithful. And the indians are still afraid today when they hear a *juruti* sing but can't see it. It might, after all, be the *juruti-pepena* come to punish (cripple) some one unfaithful...

The gigantic (indian) festival that has taken place annually since 1994 is (more than) worth a trip. For at least three days the Mundurukú "battle" against the Muirapinima for first prize, every time with a new story, costumes, spectacular decorated vehicles, singing and dancing. Interestingly the story goes that once upon a time (in the days of the Mundurukú settlement) a child with lighter skin colour and reddish hair came into the world in Juruti. The chief of the tribes and his underlings wanted to cast out the child, so the family, along with a few other Mundurukús, fled to the bank of the Lago do Juruti-velho where forests of the much-prized *muriapinima* wood grow. (The Portuguese later used it to manufacture their expensive colonial furniture.) In honour of the trees they called themselves Muirapinima, made red their special colour, and thereafter were hostile towards their former Mundurukú brethren (whose colour is yellow). And this is still reflected today in the annual festival.

Rio Nhamundá region: We will now cross the Amazon and explore the region to the north. But first we must traverse a truly incredible large lake region and an almost inconceivable area of *paranás*, *várzea*, and *igapós* (which are shown differently on every map of the Amazon – and not only their names).

I have already *(see pages 86-97, and elsewhere)* described the Nhamundá region – the *lago*, the *rio*, the habitats, and much more – and my research trips there from beginning to end, and so won't go into them again here. But I would like to say something about the history of, and the habitats in, the adjacent regions, and about the Amazons (more of which below).

Officially the Nhamundá and its drainage are part of the Trombetas basin (perhaps the reason *S. discus* have their easternmost distribution there is that in the distant past it was all one?). The river itself actually ends in the Lago Nhamundá (Lago do Faro). At least as far as I am concerned. Thereafter there is a gigantic network of waters which, depending on the time of year, flows in quite different directions down to the Amazon, or stagnates. In addition, in

Top: Manuel Torres with his wife and myself in front of their house in Nhamundá (2004) when I visited them six years after my previous trip there *(p. 90)*. And took the opportunity to give him and his wife, who in the meantime had borne two children, T-shirts showing the *S. discus* variants discovered in the Rio Nhamundá. **Above:** In the nine months when discus cannot be collected Manuel builds accurate replicas of *recreios*.

this region only the Rio Nhamundá contains true black water, as can be seen from the extreme differences in parameters *(see page 504)* and its colour *(see page 309)*. How you get to Nhamundá – which *paranás*, which *lagoas* or *igapós* you use to reach the village – depends on the water level. A *pratico* (experienced) *caboclo* is indispensable for the purpose. (Although you can instead charter a light plane in Parintins or go by *recreio.)* On one occasion Natasha and I were up the Caburi and found an *igapó* that went through to the Paraná do Nhamundá, as the water had already risen 2 m (see pages 304-305). And the Caburi is very interesting, discus-wise. It is true that in recent years *fazendas* have arrived by the **Lago Caburi** (there is also a small settlement of the same name, often spelt Cabori or Cabury), but it still has discus habitats – which are quite different to normal ones. Only here and in the Lago Nhamundá region I found discus among lava formations (see page 305).

The water parameters on 11.02.2004 at 08.30 hours were as follows: pH 6.36; conductivity 26 µS/cm; daytime temp. (air, during rain): 25 °C; at the water's surface: 30°C; oxygen 2.58 mg/l.

Biotope: lava rock, roots, overhanging trees; no floating plants; jungle partially intact, a few *fazendas*. The water was murky (rainy season), showed some movement, and had risen by some 1.5-2 m.

The discus *(S. haraldi)* that occur here are often dark brown and have little in the way of pattern, very like the individuals at Terra Santa, and from the Lago Algodoa and the Xixiá-mirim *(see page 204)*. They often have a pointed head *(like on page 310 – photo 2)*.

The entire system of *lagos*, *igapós*, and *paranás* (which contain clear water – apart from the Paraná do Nhamundá and Calderão) contain discus habitats; the latter have disappeared only where there are now *fazendas* and where clear-

felling (or universal burning) has taken place. But in the Lagos Macuricanã (I and II), Turéré, and Pirapuacá – to mention but a few – there are still numerous habitats. They all contain clear water, like the Lago Nhamundá; at most once or twice every 10 years the Amazon water rises as far as the Lago Nhamundá (in the more southerly *lagos* this happens frequently).

The island of Nhamundá occupies an area of only 2.2 km^2 and first became known (internationally) because of the discus in the lake. The *lago* is, however, often called "Rio", although it is around 60 km long and in a few places almost 5 km wide. The Rio Nhamundá, by contrast, is at most 200 m wide. Only since I discovered it has this paradise become all the rage. In recent years not only its splendid beaches but also the history of the region have become tourist attractions, and festivals have been celebrated, eg that of the angler who catches the largest *Cichla* (Festa da Pesca do Maior Tucunaré), and the Miss *Cichla* election (Garota Tucunaré), and many more.

But the most fascinating tale is that of the *muiraquitã (murakitá* = stone). It involves amulet-like sculptures discovered near the upper Rio Nhamundá *(centre),* which have presented archaeologists as well as historians with unsolved problems. (Similar stones were also found later at Óbidos and near the Tapajós.) They are said to actually originate from the Amazons, the so-called Icamiabas indian women, and are made of the jadeíte (best variety of Jade), nephrite (greenstone or kidney-stone), diorite (black granite), sedimentary rock, or crystal. Many represent frogs, but there are also fishes and other creatures. The legend goes that the Icamiabas lived without men, but were exceptionally skilled with arrow and bow. And Iací (the moon) was their protector. Once a year they took Guacaris warriors to them *(nha=* fetch, *munda =* men), as if they were their

Top: The *muiraquitã* amulets are supposed to have originated with the Amazons. Excavations at the Espelho da Lua (the lake where discus occur in the Nhamundá) brought these ancient sculptures to light. **Above:** poster for the annual Festa da Pesca ao Tucunaré *(Cichla* fishing festival) and Miss election.

husbands. Any boys born subsequently they handed over to the men, and the girls they kept. Before their tryst with the men they bathed at midnight, at full moon, in their sacred lake of the moon (Espelho da Lua), which at such times was smooth as a millpond (as is the Rio Nhamundá) so that Icaí could be seen reflected in it, and which they enriched beforehand with pleasantly-scented essences. From its depths they fetched up stones (or clay) which they formed or worked into animal figures, and they then gave these *muiraquitãs*, on cords of their own hair, to the chosen Guacaris warriors to hang around their necks. (And it is said still to work today.)

But I would like to mention my recent encounter *(see page 306)* with Manuel, who has been catching discus here since 1997. Last year he caught a single discus with a prominent (fifth) bar in the Lago Nhamundá. (He knew that in 1977 I found a fish like this in the Rio Içá and that discus from the Alenquer region likewise include one such specimen in approximately 20,000.) Manuel caught it in the part of the Nhamundá known as the Serra Guariba – which is also the only other known place where discus live alongside lava rock, besides the one in Caburi.

Further habitats which he now fishes during the three months of discus collecting are: Tigre or Boca (mainly Browns with little pattern colour – *see below*); Arijú in the vicinity of Faro (half striped and to date a single "snake-skin" *see page 311)*; Aibi (Browns and "semi-royals", as he calls the half striped); then Serra Guariba (Browns, half- and completely striped); Matapí (as at Guariba); Espelho da Lua – the lake where the first amulets were found – which lies at the mouth region of the Rio Nhamundá and where I had previously found so-called Heckel discus natural hybrids *(see below)*; Castanhal (where I likewise discovered unusual variants) and, of course, Heckel discus in the Rio Nhamundá *(right-hand page)*.

Although it was raining hard Natasha and I measured the water parameters of the Tigre discus habitat in the **Lago Nhamundá as the water level was rising (and I caught a splendid *Apistogramma*, *right-hand page)*: these parameters, on 11.2.2004 at 13.20 hours, were as follows: pH 5.75; in 3 m depth 5.6; conductivity 17 µS/cm; daytime temperature (air): 32 °C; near the water's surface: 28.8 °C; in 3 m of depth: 28.6 °C; oxygen 2.72 mg/l.

Biotope: trees, lots of roots in the water, and branches; no floating plants; jungle still for the most part intact; a cattle farm on the northern side; water slightly clouded (see page 309) with little movement.

Faro, which lies on the left bank and belongs to the state of Pará (the border runs through the centre of the lake), is the northernmost village on Lago Nhamundá. At the last count (1996) it had more than 6,000 inhabitants and there are even

1-2. Two quite different variants from the Lago Nhamundá (but there are lots more – *see pages 204-206)*. The first individual was caught in 1998 in the locality of Aibi. It is termed semi-Royal, as it is not fully striped. The other (2) was caught in 2003, in the region of Tigre (also called Boca – *right-hand page)*, likewise in the Lago Nhamundá. It too is a *S. haraldi*, but an almost brown individual. It is worth noting that in this variant bars 5-8 are almost always darker than the rest *(see also page 311, photo 1)*. 3&4. Both photos are of the same fish – one of each side. It is undoubtedly an alpha individual. But interestingly the collecting site lies at the southern end of the Rio Nhamundá, in the area where it enters the Lago Nhamundá *(photos 5-6, right-hand page*, note that the water is no longer black and the parameters are different). And the individuals found there now and then include what I believe to be natural hybrids. (DNA studies have been performed and proofed to be a hybrid *S. discus* x *S. haraldi*, but it is still a splendid individual with an extremely large amount of red.)

1. The Rio Nhamundá in its middle course. *S. discus* occur to around here. The blackwater river here is uninhabited, there are neither natives nor white people in this area. 2. The same Rio Nhamundá, but much further upstream, with no discus habitats and full of rocky massifs, rapids, and waterfalls (not shown). The territory of the Hixkaryana indians. 3. The part of the Lago Nhamundá known as Tigre, at high water (February). The otherwise clear lake is murky. 4. South-east of the Lago Nhamundá there are endless *fazendas* with cattle-embarkation places (for the *várzea* regions) along the *paranás* and the numerous *lagos* (eg Macuricanã). 5-6. Here I found, *inter alia,* a fully striped alpha individual *(page 308)* at night beneath the *acará-açú* brush. *Inset: above left and right:* plants of the region, *and centre:* a lovely *Apistogramma* from Tigre – *A. agassizii* group.

some foreigners, including Dutchmen. The former bauxite pioneers (from Alcoa) have long since retired, but keep coming back to visit. Only one has stayed – "Seu Alberto" (Albert Garrits). He has built himself a home on the splendid sandy beach and is even in the process of fulfilling his dream of bringing eco-tourism to the area. And is having a fantastic lodge built for the purpose in the jungle at Céu Estrelado, 40 minutes by boat up the Nhamundá.

Uabuís, Cunuris, and Guaicaris indians were living here long before it achieved the status of *vila* in 1758. Faro is a regular destination for *recreios* (just like Nhamundá), but the only road link in the region runs from **Terra Santa** to Porto Trombetas. Terra Santa (= holy land) is another clean and very pleasant community with around 14,000 people. It lies to the east of Lago Nhamundá on the enormous Lago Algodoal which is likewise ringed with glorious beaches during the dry season. And out on the *lago* you can often imagine yourself on the open sea. Metres-high waves beat in your face during the crossing. The reason for this – or so they say – lies in the huge rocks in the *lago* (which are, however, slowly vanishing as *recreios* travel there frequently and are supposedly causing them to crumble away). It lies strategically positioned right in front of the town and is the subject of an interesting legend. Where Terra Santa stands today there was formerly an indian settlement, long before the colonists arrived here. The rocks were repeatedly circled by the indians and were a sacred place for the women. One day the indian women were stricken by a terrible flu epidemic (against which they had no immunity and which had fatal consequences). One night, as they were dying one after another, the shaman had a vision: the women would be cured if they bathed on the rocks in the *lago* – and so it turned out. And ever since then it has been the "holy land".

In addition, you can try a splendid *piaracaia* (fresh fish baked in the sand) on the nearby beach, a real delicacy. But you have to take your own drink.

We now leave the "freshwater sea" full of discus – the only place I have never found any was in the clear Rio Paracatú and, of course, there are none in the whitewater *paranás* – to discuss Parintins and its surrounding area.

Who hasn't heard of **Parintins**? It has by now achieved international recognition on account of having the largest Amazon festival, which puts the majority of *carneval* festivals in Brazil in the shade. Every January the city´s black cow called *Caprichoso* (= the moody) – and the white cow called *Garantido* (= guarantor) groups is already seething with activity relating to the forthcoming largest folklore festival of Brazil, even though it doesn't actually take place until the last week in June. From the 12th June

1-2. *S. haraldi* from the Terra Santa region. Caught in November 1998 (the ones in the bowl, *far right*). They are mainly dark (as in Caburi as well). One is from the northern part of the Lago Algodoa (1). The other (2) from the adjoining Lago Xixiá-mirim. Those from this habitat are mostly pointed, quite unique; so far I have found variants with a similar body shape only in the Rio Jari and Caburi. 3. This specimen originates from the Lago Macuricanã (also known as Maracanhã) and was photographed at Aquaristik Hustinx, in Belgium. Individuals (*S. haraldi*) from that lake are often red and have prominent bars (5th to 8th) as in the Lago Nhamundá. 4. An interesting feature of (discus) aquarists is that they often know everything (usually better than the collector himself), especially as regards collecting localities: this specimen originates from the Alenquer region and, as mentioned elsewhere, there is one such (with the striking central bar) in 10-20,000 indviduals, throughout most of *S. haraldi´s* distribution.

 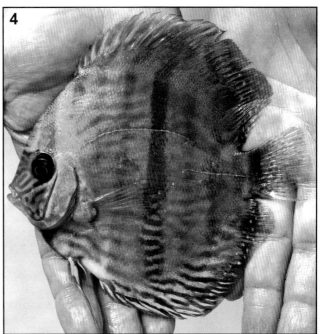

1. This splendid fish is from the Lago Nhamundá, at Arijú. 2. Likewise this uniquely patterned variant. A unique discovery by Manuel Torres (January 2003). The fish photo went aound the (discus) world. Apparently proof that the cultivated "snake-skin" forms must have a wild-caught *S. haraldi* (=snake-skin) in their genes. 3. *S. discus* from the lower Rio Nhamundá. Immediately after capture during the dry season (October 1997). An alpha individual, badly injured by piranhas, it survived and its fins are regenerating. 4. Another *S. discus* from the Rio Nhamundá, with almost solid blue coloration.

onwards there are displays in the Bumbódromo (stadium), which accommodates more than 35,000 onlookers. The origins of the festival lie with immigrants from the north-east of Brazil, who arrived in the region more than 100 years ago along with their families, baggage and cattle.

The last days of June are – as the entire city – reserved for *Caprichoso* and *Garantido*, the *boi-bumbá* rivals. The ritual lasts for almost three nights. There is dancing and singing in a masquerade that goes beyond all imaginable bounds. Gigantic trucks and other vehicles, decorated in accordance with a theme, are paraded; time and again new songs are composed on the theme of the presentation – there are no repeats. More than 50,000 visitors assemble from all over Brazil and the world to participate in the ritual of Pai Francisco, Mãe Catirina, Tuxauas, Cunhã Poranga, Pajé, and the innumerable indian tribes with their endless legends and ceremonies. Some of the time people dance in circles holding hands to the rhythm of *cateretê* (indigenous music), *carimbó* (*caboclo* music), and marches. The festivities begin with a firework display such as is usually seen only at New Year.

The apparently untiring "battle" for first place even has its own language: *arraial* (typical foods and activities at the festival); *Contrário* (the name of the opposition's cow – everybody takes sides); *Bicho folharal, Dona Aurora, Neguinho do Campo Grande*, etc (names of the individual characters in the "battle"); *galera* (rivalry); *cunhã poranga* (pretty girl); *marujada de guerra* (*batucada* = the orchestra of the moody); *palminha* (two pieces of wood that provide the rhythm); *QG* (the place where the costumes are sewn); *toada* (music); *tripa* (the person in the cow costume) and lots more.

What more can I say, there is hardly anything like it anywhere. You need to have seen it.

But now to Parintins itself. The city was originally founded in 1793 on an island which by now, with more than 100,000 people, it has spread to cover. The island lies on the right-hand side of the Amazon, encompasses 7,069 km², and was formerly (and is sometimes still today) known by the name of Ilha Tupinambarana. The Tupinambás, Maués, and Sapupés lived in this region long before the first whites.

It is surrounded by waterways and relatively large amounts of forest, *lagos, várzea* islands, *igapós,* and even a small *serra* (the Serra de Parintins), right on the border with Pará, as Parintins belongs to the state of Amazonas, lying where the Igarapé do Valério empties into the Amazon. To the south are the Andirá, an already well-known discus river *(see page 319),* and the Uaicupará. Parintins is 420 km by river from Manaus. There are daily flights and *recreios,* and even a *jato* three times a week. Nowadays it is an important centre for traffic in Amazonia.

The showpiece of the city is the cathedral of Nossa Senhora do Carmo. (Or perhaps the house of Dona Maria Ângela, who has covered her entire house with red objects and red paint in honour of the *Garantido*. As each group has its own colour. That of the *Caprichoso* and their founder, the folklorist Simão Pessoa, is blue.)

The most important *rios* are the Paraná do Ramos, which permanently contains white water (and hence no discus habitats), and the equally murky Paraná do Espírito Santo and Paraná do Limão. The Rios Uiacurapá and Mamurú, however, contain clear water. The *lagos,* such as Macuricanã on the opposite shore, as well as Aninga, Paranemá, Macurani, and the Lagoa da Francesa, which provide the town with water, are almostly muddy at high water *(right-hand page).* I could find discus habitats in the Lago Maximo *(see page 317)* and the Rio Andirá *(see page 319).* But they are undoubtedly more widespread in the clearwater biotopes.

But before we move on to that I would like to say a little about the Amazons, about whom endless amounts have been written (and filmed), although the interpretation often deviates from that in Orellanas logbook. And because it demonstrably occurred off Parintins, at the *Foz do Nhamundá,* I have referred back again to the entries and will cite text from them (translated into English):

"On 24th June 1542 our Spanish boats were attacked by indians, accompanied (or led?) by 10 or 12 women. They shot

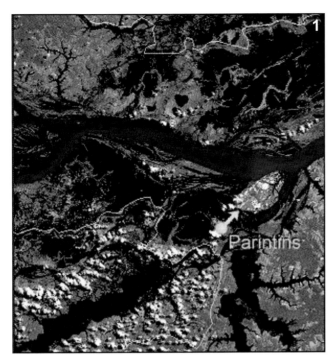

1. The island of Parintins (arrowed) can be seen in this satellite photo, and the bodies of water in the region can also be readily recognised and differentiated (light blue = white water, dark blue = clear). Part of the Nhamundá region with its *lagos* can be seen in the upper part of the photo. 2. Plan of Parintins. 3. At high water the lagos around Parintins are white. 4. *Recreios* in the harbour. 5. Ocean-going steamers (for tourists) tie up in Parintins. 7. The fish market. 8. Even the telephone boxes are dedicated to the *Garantido* and the *Caprichoso*.

so many arrows that many a brigantine resembled a porcupine. One arrow took out Carvajal's eye. The women among them were couragous, demonstrated strong leadership, and fought like 10 men. Eventually we managed to kill seven or eight of them, whereupon the rest withdrew. But we knew that this was likely to be only a short respite, as they outnumbered us. The Spaniards fled from the mainland to an island in the middle of the river. After this Orellana avoided approaching that shore despite having many wounded and practically no reserves of food and munitions. Not until the end of June/beginning of July did he halt, at a larger river island that appeared to be uninhabited. But the bank opposite was obviously inhabited by warlike savages."

Two weeks before the above events, Orellana had several times attacked and plundered indian villages on his way downstream. In one case, where the natives took refuge in their houses, he had even burnt everything down. (Centuries later the village was still called Pueblo de los Queimados (village of the burned) and probably corresponds to the modern Itacoatiara). Two days before the battle with the Amazons he passed a village, on the left bank, which purportedly had gleaming white houses, then went ashore later at a large settlement with a road which they called Pueblo de la Calle (= village of the road).

The following comment originates from Frei Gaspar de Carvajal, who accompanied them: "...they are very slim and tall with

Top: The painting shows a little of what people have been able to experience for more than 90 years at the end of June, all day and all night, in Parintins.
Above: A picture of *caboclos* (their island life); flora; fauna and shaman.

long matted hair on their heads. Full-bosomed (ie with both breasts) and running around naked with only their pubic hair covered, with bows and arrows in their hands and fighting like 10 indians".

Carvajal also wrote that they killed a number of the indians led by the women but nevertheless had to flee, though they did take one indian prisoner. Who later declared (I do wonder how the Spaniards knew the language of this unknown tribe?) that he was a member of a *tribu* whose chief ruled the entire (Nhamundá) region, but was subject to the women, who lived in the interior. As subjects the men followed and paid tribute to the warlike women, who were often in the company of his chief, the *conhori*. And the captive replied to further questions from the commander that the women were not married and that there were 70 *aldeias* (settlements) of them. That their houses had massive doors of stone and the *aldeias* were heavily guarded. That they brought children into the world without being married. When they felt the urge to do so they forcibly abducted men of neighbouring tribes and kept them with them until they fell pregnant. They killed any boy-children or gave them to the father, but raised the girls according to their traditions. He also told of their wealth, the gold and silver in their possession.

Now most people know the legend about the Amazons in Greek mythology, which dates from long before Christ. Also that they were so-called because they lacked one breast (in Greek *a* means "no" or "without" and *mazós* "breast"), the right one, which they burned or cut off when young in order to be better able to shoot with arrow and bow. A story often cited in the Middle Ages in Europe and which has inspired many writers and painters *(see centre)*. And they lived in Cappadocia, in Asia Minor. But what does that have to do with the indians who attacked Francisco de Orellana and his companions? Even if there were women among them? Carvajal is certainly to blame for incorrectly interpreting them and the situation.

I have lived among indians and visited many tribes. Almost all of them have a very clear distribution of labour. The men are responsible for hunting and often away for days (sometimes weeks) in the forest. They seldom have other duties. For this reason there are no men, or only a few, in the villages by day. I have also learned that the indians have very good memories and can recall the smallest event decades later; that they are highly intelligent and that their "jungle drums" really work. Thus even before Orellana landed the Icamiabas women knew exactly how he had burned down the settlements of other tribes. And thus it is no wonder that they didn't welcome him in friendly fashion (as the natives almost always do). And because, for the reasons mentioned above, there were only a few men in the village, the women who today are still often stronger than the men, simply because of their hard work in the fields and the dragging home of firewood and manioc – grabbed bows and arrows to chase away the villains. The Icamiabas had been forewarned by the Pueblo dos Queimados.

That may be what happened. Although an unusual experience I had in the 1960s in southern Pará, among the Kubenkräkein-Kayapó, throws the matter open to question nonetheless: accompanied by the *benadiôro* (chief) Tikiri, I was allowed to attend a women's ritual that took place in the centre of the *aldeia*. The men sat around the women, looked after their children, and refrained from entering the ring of celebrating women.

The ritual is called *mebiôk* and is a revolt by the women *(top)*. For a week long they are masters of the *aldeia*. They leave their huts and live in the *ngóbe* – the men's house, normally taboo for women. In the week of the ritual the men do the (house-) work and look after the children. In the evening they have to ignore the provocative calls of the women and demonstrate their steadfastness. On the final night the sexual act takes place in the totally dark *ngóbe*. The men come in from one side and the women from the other. At the signal of the *pajé* (shaman) each woman grabs the nearest man and sleeps with him. Until just before first light when the *pajé* announces the new day and the women leave the *ngóbe*, bathe in the river, and go back to their normal work.

This ritual (I don't know if it still takes place the traditions and cultures of indigenous peoples

Top: Kubenkräkein-Kayapó indian women during their *mebiôk* ritual (photo from the beginning of 1960). *Centre* and *above:* R. W. Vermehren Stevenson, the priced painter. The son of a German and an American, who travelled the Amazon in 1964 and lives since 1973 in Manaus, has immortalized the Amazons often. (Only as models he used beautiful Brazilian girls...)

are disappearing more rapidly than people realise) has two purposes: to remind the men that if the women aren't treated properly, they will go back into the forest and make war (= Amazons?). The second is insemination (which sometimes doesn't always take place successfully with the husband). And if a new-born child is fathered by another man during the *ngóbe*-night, it is nevertheless accepted into the family without any falling-out.

It is an ancient ritual. Nobody knows where and when it originated. Nobody can say now whether it has anything to do with warlike women from earlier times or if there really were "Amazons" (but with two breasts) in South America. It is more likely that an inter-continental exchange of traditions and cultures took place thousands of years before Columbus (see Bleher *et al*, 2003a: *Pre-Columbian*) and that perhaps the *muiraquitã* stones – which were often sculpted from jade, a stone that doesn't occur in America – came from Asia. Long before our time.

Back during the 1950s to 1980s a Dane, Peter Hilbert, found evidence of the prehistoric settlement of the Amazon – and in particular this region. In 2001 alone 6,578 ceramic and 58 worked stone fragments were found here. Cylindrical, anthropomorphic, and zoomorphic artefacts. Among which the *muiraquitã* amulet of green jade in the form of a frog is outstanding. The first were discovered in the 17th and 18th centuries in the Nhamundá river and are today in museums (and private collections) worldwide. In Brazil, in the museum at Santarém. And nowadays there are huge numbers of replicas, mostly original sculptures by the local sculptor Laurimar Leal *(see page 307)*.

From Parintins we make a detour to the **Lago Maximo** *(below and pages 317-318)*. It lies to the east of the island of Parintins and to reach it you have to go some way up the Paraná do Ramos (also written "dos") and then left into the Igarapé Maximo. This is the only approach to the lake. The last time (2004) the waterways were partly clogged, firstly with floating plants and tree-trunks in the Amazon and also to some extent in the *paraná*, and then grass everywhere in the *igarapé (see below)*. It was a pig of a job and took hours to get through the drifting grass. But at the end of February/beginning of March the way is largely clear again and readily navigable until December or the beginning of January.

The Lago Maximo is a small lake, "forgotten" as regards ichthyological study. The only person I know of who has visited the Lago Maximo for the same reason as myself, was Agassiz, during the Thayer Expedition. A few species from his Maximo-064 collection were described as new (most of them by Steindachner), but there were no discus among them. It is worth noting that the

1. The mighty Amazon at Parintins, at the beginning of the high water period. Thousands of tree-trunks, roots, and branches, millions of floating plants and grasses torn free by the masses of water. Small boats, like our *voadeira* here, often have to travel in zigzags to avoid damage. 2. Villa Amazonica at the mouth of the Paraná dos Ramos is actually the spot where Parintins originally stood. 3. The entrance to the Lago Maximo. The photo was taken in February, when the water was rising and grass completely barred the way. This happens every year. And if you want to get there at this time then you have to battle through it (there is no road). 4. Close-up of discus habitat in the Lago Maximo, when the water was rising. Grass has been washed onto the *acará-açú* plant, plus *Utricularia* and *Najas* (visible in the centre).

1. The Lago Maximo, at the spot where you have to climb 107 steps to the village. This is where people fetch water, wash, and fish. 2. View from above (the lake is not that large!). 3. A bay where I found lots of fishes (inter alia, photos 7-12) beneath floating plants *(Salvinia* species, *Pistia stratiotes,* and *Ludwigia helminthoirrhiza).* 4. The church, and 5. the Bar da Amizade (Friendship Bar) in Maximo. 6. *Caboclo* children. 7. The first live photo of *Megalamphodus (=Hyphessobrycon) eques.* 8-9. *Colomesus asellus* – distinguishable from *C. psittacus* only by the black spot (arrowed). 10. A perhaps new *Leporinus* (with 15 bars). 11. A possible new *Aphyocharax* species. 12. An unknown *Hemiodus* species (the only one so far with two spots).

HABITATS ... PARINTINS REGION: LAGO MAXIMO

Austrian described in 1875 the cichlid *Geophagus thayeri* from here, later placed by Kullander in the synonymy of *Acarichthys heckelii* Müller & Troschel, 1849. Now I have not found any *A. heckelii* in the lake, but I did catch next to some interesting fishes, ie two beautiful geophagine species in the discus habitat *(below)*. One of them could well be Heckel's *Geophagus acuticeps*, and the other *G. leucostictus* (placed today in *Satanoperca*, and *acuticeps* as well). Now, as already mentioned I was possibly only the second person to explore here, but the first to find discus *(S. haraldi)*. They have a predominantly brown base colour and little pattern.

During my last trip the mayor of the village Maximo, where 47 families live, known as Presidente Wilson, was very helpful to me. And he confirmed that my discovery of discus was no accident. *"Acará-buceta tem muito aqui, na epoca da seca, demais!"*, he said. (There are lots of *acará-buceta* in the dry season.)

The water parameters in the discus habitat on 9.02.2004 at 16.15 hours were as follows: pH 6.27; in 2.5 m of depth 6.59; conductivity 31 µS/cm; daytime temp. (air): 30 °C; near the water's surface: 29.5 °C; in 2.5 m of depth 29.4 °C; oxygen 2.09 mg/l.

Biotope: overhanging trees, roots, branches, leaves, and algae; floating plants: *Pistia stratiotes, Utricularia foliosa*. and floating gras (but only this time of the year). Jungle in part intact, a number of *fazendas* around the lake, but only on the *terra firme*. **The water was almost clear *(see below)*. The bottom of the discus habitat was sandy.**

As regards the vicinity of Parintins, it remains to mention that there is an interesting *caboclo* settlement in the area, on the Paraná do Limão. Here vegetables have been cultivated for generations in a type of aquatic garden in the *várzea* region (see page 319). The vegetable beds are laid out on so-called *jirau* (wooden tables standing in the water). When the *caboclo* Marchão came here around 100 years ago he can have had no inkling that his descendants – by now 40 families (he had 22 children by two wives) – would perpetuate his idea for cultivating *couve, alface, cebolinha, cheiro-verde, mixixe, berinjela,* and *fejãozinho*. Today the *comunidade* of Limão supplies the market in Parintins daily with cabbage lettuce, cabbage, onions, small beans and several native vegetables, transported by *rabeta* (an extremely long outboard motor fixed to the canoe, which can travel across very shallow water). Edmilson (49), one of Marchão's descendants, told me that at high water he has to travel to and fro between the *jiraus* in a *casquinho* (little canoe) in order to tend the vegetables and water them – often three times a day. It was very tiring, he would prefer to walk through the water,

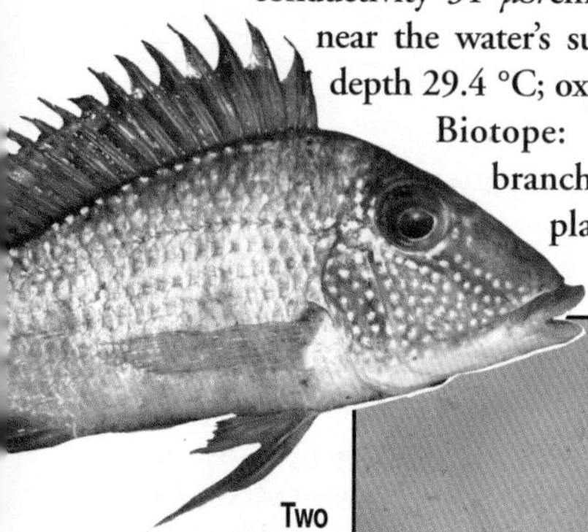

Two geophagine species: one *(right)* may be *Satanoperca acuticeps* and the other *S. leucosticta (above)*. *Below:* A *Crenicichla* cf. *marmorata*, likewise from the discus habitat.
Above: A discus habitat in the Lago Maximo, locality for the community species pictured here. *Below:* The water of the Lago Maximo, in the discus habitat (February).

but the *arraias* (stingrays) *"...são a maior praga"* (are a major plague). He continued, *"Agora tem um pouco menos, mais já teve ano de a gente matar 12 por dia"* (there aren't as many now, but there have been years when we have killed 12 a day), and showed me the numerous scars on his feet.

I enjoy listening to the *caboclos*' stories, and still remember how Wilmar told me about a six-metre crocodile. *"Meu filho matou, no Paraná do Ramos, um jacaré de seis metros. O bicho estava boiando, na frente da canoa. Ele atirou e, mesmo ferido, mordeu e furou a proa do bote de alumínio, rapaz!"* (It swam in front of the boat in the Paraná do Ramos, and although my son shoot it, the beast still bit a hole in the bow of the aluminium boat.) But you have to be wary with the bulk of the *caboclos*. I also have my doubts about his anaconda story. The *sucuriju*, as the snake is called locally, was supposedly eight metres long, and had got through the fence and into the barn in search of chickens a few years previously. There it had devoured 29 of the 30 birds and was then so bloated that it got stuck in the fence on the way out. The next day, when Wilmarsão (as he was called because of his boasting – "são" translates as "big", and his name is normally Wilmar) discovered the snake, he chopped off its enormous head, half a metre in diameter.

To the south of Parintins two large rivers, the Rio Mamurú and the Rio Uaicupará, enter the Paraná do Ramos (which from Parintins to opposite Silves practically constitutes a branch of the Amazon). But before that they flow into an enormous *lagoa* full of islands. Augmented by numerous *igarapés* such the Jacu, Itatuba, Sabina, Aduacá and many more. (No wonder that this region is also known as Planeta Água (Planet Water).) These are clearwater biotopes and the majority of them are still intact, although here too *fazendeiros* are encroaching incessantly and gigantic areas of *terra firme* are securely in their possession.

There are discus habitats in the lower Mamuru and on several occasions specimens have been collected and exported. So far I know nothing as regards the Uaicupará, but discus probably occur there as well as in the **Rio Andirá** whose mouth lies further west. From August to February the banks of the Uaicu-

1. The region around Parintins changes in June (at the time of the festival) to a "planet of water", and in the Paraná do Limão the *caboclos* lay out their vegetable beds on *jiraus* (wooden tables) in the water. 2. Despite it being high water the water is clear in the numerous *lagos* and *igapós* of the hinterland of the "enchanted forest" (Parintins is called *noçoken* (= enchanted forest) in the Maué language), and there are copious discus habitats. 3. In the dry season sandbanks are visible. 4. Discus from the Andirá.

pará are out of the water and have splendid long white sandy beaches (which are discus-hostile). And during those months there are frequently people on the river islands of Pacoval, das Onças, and Guaribas, enjoying the fabulous beaches and the slightly tea-coloured water there.

The Rio Andirá is likewise a clearwater river which flows into the whitewater Paraná do Ramos, and here we find ourselves in the Município de Barreirinha, which is part of the state of Amazonas. *Furos* without end form connections between rivers and lakes, *paranás*, and *lagoas*, right up to the Amazon, as well as to the west, east and also to the south. Some of the best-known *Furos* are the Uaicupará and the Furo das Colheiras, then we have the Paraná do Limão, the Paraná do Limãozinho, the Paraná Urucurituba; and the great Lago Arapapá. The region here is so beautiful that many years ago the best-known architect in Brazil (and the world?), Lucio Costa, built himself a wooden house (like a *caboclo*'s), with an area of 120 m², in Barreirinha (on the Paraná do Ramos), in order to get away from big-city stress (he lived in Rio) and find peace here in the jungle.

In case the reader doesn't know, he was not only responsible for the construction of the most modern city on Earth (Brasilia), but also much more. Lucio Marçal Ferreira Ribeiro de Lima e Costa was born in Toulon, France in 1902, of Brazilian parents (his ancestors on one side were descended from the penultimate Tsar of Russia). (Toulon was also the birthplace of the French discus breeder, Jean-Claude Nourissat, who died of a terrible malarial attack in 2003, aged 62; see Chapter 6: *recent breeders.*) Costa studied at the Royal Grammar School in Newcastle, and at the Collége National in Montreux. Learned painting and architecture at the Escola Nacional de Belas Artes in Rio. In 1936 he persuaded Le Corbusier to come to Brazil to appraise his plans for the Ministério da Educação (Ministry of Education) skyscraper in Rio. In 1938, together with Oscar Niemeyer, he was responsible for the Brazilian Pavillion at the World Fair in New York, and in 1957 he won the competition for the building of Brasilia (still the most modern city on Earth, a

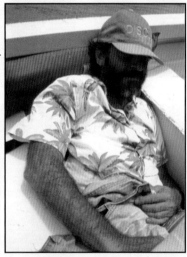

Travelling in the Amazon region is often exhausting. The distances are unimaginable (Europe could be lost in the Amazon region), and the only transport is *recreios* or *lanchas* with hammocks (often for weeks), then onward by canoe or *voadeira* to the habitat.

monument which, like Venice and the Great Wall of China, has been declared a world heritage site by UNESCO).

The British urbanist and president of the panel of judges at the time, William Holford, who supported Costa's project, said, "His project is a work of genius and one of the greatest contributions to contemporary urbanism."

In 1960 Costa was awarded the title of Professor *honoris causa* by Harvard University in the USA, and four years later called in to supervise the restoration of Florence after it was partially destroyed by flooding. In 1976, along with the Italian firms of Nervi and Lotti from Rome, he evolved the plans for Nigeria's new capital, Abuja, and at 96 he passed away in his house in Rio, on the beach at Leblon, poor as a church mouse.

The man who will go down in history as the greatest architect of the 20th century had retired to the solitude of the Andirá, as did the famous Brazilian poet Thiago de Mello. Remarkably the approximately 20,000 people of Barreirinha, who live by fishing and agriculture, didn't know whom they had as a neighbour. Except insofar as *na casa do Lúcio Costa* (the house of Lúcio Costa) was always open for *xarope de ervas* (Brazilian drink) or for just good advice.

Thiago also lived in Costa's house, but preferred an even more remote solitude for writing his books – the Freguesia do Andirá on the Andirá. The journey thither can be long or short depending on the season of the year. Between May and October there is normally a short-cut open from Barreirinha, via a *furo* from the Paraná do Ramos to the Igarapé do Pucu, and with a good *voadeira* you can reach the Boca do Andirá in 15 minutes. At low water you have to go along the Paraná do Ramos, which takes more than three hours using the same transport. The world-famous poet also built himself a wooden house at the Ponta da Sfadeza.

As far as discus habitats are concerned, there are very few worth mentioning in the adjacent region to the west, the almost

1. The Furo do Jabutí forms a link with the Uatumã (2). Here there were once huge numbers of discus habitats, but now the area is full of *fazendas*. But the *caboclos* say that *acará-disco* can still be found in the dry season. 2. The Uatumã is a mighty river. São Sebastião do Uatumã lies on the left bank in the centre of the photo (where the river narrows). 3-4. Discus habitats opposite the town. 5-6. Details of the discus habitats in the Rio Uatumã.

unimaginable "planet of water" (still regarded locally as part of the Ilha Tubinambarana). It is largely inundated by the Amazon, *várzea* without end, and the whole washed by the Paraná do Ramos (to the south) and the Amazon (to the north). Of course discus may be there depending on the season of the year. But so far there is no proof. Not until the hinterland of Urucurituba (there are two places with the same name in the region, I am referring to the one on the Amazon) where there are permanent clearwater lakes and I found half striped specimens many years ago. And some have also been collected now and then (as far as I know, not recently).

Rio Uatumã region: Immediately to the north, on the left-hand side of the Amazon, there are outliers of the Guianan Shield, and it is not until Urucará, capital of a *municípo* with the same name which has around 17,000 inhabitants (a few are of indian descent or have some *caboclo* blood), that we find discus habitats. Heading west and north-west from here, up the Uatumã and Jatapu, there are almost innumerable discus biotopes, and they have been commercially exploited in recent years. I will go into these in greater detail.

The town Urucará, with approximately 6,000 inhabitants, can be reached only by *recreio* (or chartered light plane). There are no roads, the few they have end behind the city after a few kilometres at Vila Nova, Vila Amanay, and Colonia Marajazinho.

However, the majority of discus habitats that are fished today lie immediately off São Sebastião do Uatumã, on the opposite, right-hand bank of the Uatumã *(see page 323),* further to the west. In October 2003, when I was staying in Urucará, the mayor was convicted of taking public money to pay for his Viagra...

The Rio Uatumã enters the Amazon via a labyrinth of waterways north-east of Urucará, but forms a gigantic lake, similar to the Lago Nhamundá, almost 60 km long and up to 5 km wide, to the west of São Sebastião do Uatumã. It is like an inland sea. Its water is clear, although the natives often term it *áqua preta*. Only upstream of the mouth of the Rio Jatapu was the water once black like the Rio Negro, at least until the building of the Represa Balbina, a huge hydro-electric dam in the north which was supposed to solve the water and electricity supply problems of Manaus (but which, like most of the other hydro-electric dams, has proved a wash-out).

The river is navigable for 295 km upstream, as far as the Usina hidrelétrica de Balbina. Between February and August the depth of water is more than 2.10 m and the Cachoeira Morena in it, 260 km upriver, is passable at this time. Interestingly there has been little development so far along this stretch, the settlements and *fazendas* can be counted without running out of fingers.

1. A *S. haraldi* from the right bank of the Rio Uatumã, opposite São Sebastião do Uatumã *(see page 321)*.
2. This specimen, also from the Uatumã, has 11 bars (instead of nine). This discovery was as big a sensation as the "snake-skin" variant *(see page 311)*. It provides supporting evidence that these cutivated varieties derive from a wild discus. I found several like this in the Uatumã. 3-4. Specimens from the Lago Limão. The relationship with *S. discus* in the Jatapu-Uatumã region is similar to that in the Rio Nhamundá-Lago Nhamundá. Two discus species occur – almost overlapping – in adjacent habitats. In this region *S. discus* lives only in the Jatapu drainage (with exceptions – see text), quite the opposite to what has been stated on the Internet and elsewhere. *S. haraldi* occurs only in the Uatumã.

1. São Sebastião do Uatumã is a small town situated where the Rio Maripá joins the lower Rio Uatumã on its left-hand side. As well as the Festival do Tucunaré *(Cichla* fishing festival), St Sebastian is celebrated here once a year. Not only the town, but also the monument showing St. Sebastian hanging bound to a tree and pierced with arrows, is decorated with life-sized aras and herons.
2. The discus collector Boto (= "whale", but his real name is Sr. Augostinho) has his *viveiros* (for holding discus) and containers for transportation by the Rio Maripá, right next to the town.
3. One of Boto's daughters.
4. Boto's daughters at the *viveiro* where they keep their discus in a net fastened to a pontoon in the river (Rio Maripá) (2). Three days before shipping to Manaus they take the discus out and place them in plastic containers (3-5 discus in each) with a high salt concentration. And Boto says that anything that survives that for three days will carry on living.
5. The mouth region of the Rio Jatapu. Truly no small river... There are also several channels in this region, as well as numerous *lagos*. But certainly none of the *lagos* (such as Lago A, K, or similar) listed on the Internet and elsewhere are locally known...

The fish fauna, apparently still rich despite the dam (there are even *en masse* angling trips to the Uatumã offered on the Internet), is almost unbelievable. Be that as it may, there are discus habitats in the lower part, abundant off the mouth of the Jatapu, just like further downstream.

I should mention that in 1996 it was suggested (pangs of conscience?) that an ecological reserve should be created here, and six years later (2002) this actually came into being. The so-called Reserva Biológica de Uatumã includes the *municípios* of Presidente Figueiredo, São Sebastião do Uatumã, and Urucará, with a total of 940,358 ha. Its aim is to conserve the dense tropical forest (below) of the Uatumã-Jatapu region and to maintain its biodiversity, in particular to protect the endemic, rare, and endangered species. In this area the forest also has a dense underbrush. In the open primary forest regions there are unique palms and climbing plants such as *Maximiliana regia,* known locally as the *inajá,* and *Oenocarpus bataua,* which has the Brazilian name *patauá.* The fauna is very typical of the zoogeography of the Guianan Shield. But 14 species are supposedly threatened with extinction (which I can well believe). And there has now been a further piece of forward progress in this region. On the 10th June 2004 the SDS (Secretaria de Estado do Meio Ambiente e Desenvolvimento Sustentável) issued a decree for a new Reserva de Desenvolvimento Sustentável do Uatumã – a conservation and development zone. It will occupy 402,000 ha and be known as RDS-Uatumã (short for Unidade de Conservação do Estado: Reserva de Desenvolvimento Sustentável do Uatumã). It is a victory for the population of São Sebastião do Uatumã (and the discus habitats?), who have been in dispute about it with the IBAMA since 1999. The Secretary of State for the Environment, Virgílio Viana, said in his address that it would offer the potential for the further expansion of tourism, fishing, and the timber activities of the *madeireiros* (whatever that may mean).

However, precautions have been taken upstream of the Balbina hydro-electric dam, in the region of the Rio Pitinga. There are doubts about the Taboca mining company and its activities, and because the risk of a dam breach has been taken into account, a start has been made on keeping fishes in ponds to make sure they are still available for future generations. (Something I myself have been working on for decades, only with smaller fish species).

By now people are also in the process of waking up to the importance of ethno-botany. In 2001 a group of Brazilians performed a census for the region, to obtain exact information for Santa Luzia (Paraná do Amataí), the local Amazon region, the Urucará, São Lázaro (an estuary of the Rio Urucará), Santa Maria (in the *baía* of Uatumã), around Itapiranga and São José

1. The middle and upper Jatapu, all rocks and rapids, without any discus biotopes. 2-3. The Lago Iri (slightly tea-coloured) with Heckel discus biotope (when the water was rising). 4-5. The last left-bank *lago* in the Jatapu before its mouth. Largely clear-felled and cattle pasture. Few habitats, but despite the *fazendas* there are still discus habitats (on the far side, *left in the photo*) in the penultimate (again left-bank) *lago* before the mouth. 6-7. Two typical Heckel discus habitats in the Lago Limão, one with trees (6) and one with *acará-açú* plants (7). 8-9. Further typical habitats in *lagos* in the Jatapu; note that the water is clear, and at most slightly brown. 10. *Nymphaea* sp.

Enseada (on the Rio Urubu), as well as around Silves. They discovered 58 types of fruit tree, of which 34 were native and 24 introduced. The *comunidades* of the *várzea* cultivate 11 native and 12 exotic species. The respective figures for the *terra firme* are 19 and 12, and for city centres 14 and 17.

But let us proceed to the **Jatapu**, the largest (left-hand) branch of the Uatumã. All the other tributaries are discus habitats but not worth mentioning distribution-wise (there are hardly any there), apart from the Urubu, which we will come to later. The Jatapu is more than 300 km long and rises in the Serra Acaraí, in the border region with Guyana. (The Rio Nhamundá and the Mapuera also originate there.) However, it is full of rapids, stones and rocks, and navigable for barely 100 km upstream. The discus habitats end at the small village of Santa Maria on the right bank just before the mouth of the Rio Capucapu.

South of Santa Maria I found *S. discus* (but none of the so-called royal discus that are supposedly found here in large numbers – see below) in the following biotopes: along the right-hand bank, downstream in the mouth regions of the Igarapés Cucuia and Jaturana as well as the Onze de Agosto (= 11th August). Then in the large, almost circular Lago Castanha with its black water; the elongate Lago Badabaxí; the Lago da Velha (= lake of the old); the Lago Macunã (named after an indian tribe, and also written as Makunã); and the Lago Iri *(see page 325),* which contains really black water. (In addition I must comment that my researches came up with very different water parameters and colours for the Uatumã, just as for the Jatapu. Some people say the former contains black water and others that the Jatapu is a river of *água preta*.)

Along the left bank to the south of Santa Maria we come to the Igarapé Boa Vista, which contains discus habitats. Discus also live in the lower Igarapé dos Negros, in the Lagoa Uauacu, then in the *lagos* at the village

Under way in the *voadeira* (with lots of reserve fuel) during a discus exploration trip to the Jatapu-Uatumã region lasting several days. Natasha, already tired of all the sitting *(top)*. Our starting point was the town of Itapiranga, population circa 6,000 *(centre and above)*.

of Panacuré and those at the Usina Remanso (a factory). Further south in the Lagoa Escatici, in the large *lago* at Boca Maracanã (named after the largest football stadium in the world, in Rio), in the Lago Aaçarí, and south of the settlement of Brás in the Lagos do Araçá, Caiuca, and Limão. Discus may also occur in the Lagos Leandro and Leandrinho, which I didn't investigate, but which are very muddy.

It remains to mention, apropos of the Jatapu, that the names Lago A, Lago B, Lago K, etc. are repeatedly seen in the discus trade (as well as on the Internet and in various publications). No *caboclo* or *disqueiro* on the spot knows these names. José, who, with his family, has lived for many years immediately downstream of the first *lagoa* and also eats *acará-disco* bones and all, when he catches them (he says the numerous bones go down easily with *farinha)*, says that maybe his *lagoa* is the Lagoa A, but nobody knows anything about it (well, but a few "discus experts" in Europe and America do...).

Resuming for those who know different: there are no fully-striped individuals in the Jatapu, except for Heckel discus *(S. discus)* with their well known colour pattern. The often mentioned (in the Internet and elsewhere) Royal Blues or fully striped blues *(S. haraldi)* from the Jatapu, are definately not from the Jatapu drainage. They can only come from the Rio Uatumã, or some place else. The Rio Jatapu region is, like the Rio Nhamundá, a Heckel discus habitat.

The water parameters measured in the region also indicate clearly where each particular discus species occurs.

The water parameters in the discus habitats in the Uatumã (see page 321) on 22.01.2004 at 14.00 were as follows: pH 6.48, in 2.5 m of depth 6.31; conductivity 8 μS/cm; daytime temperature (air): 38 °C; near the water's surface:

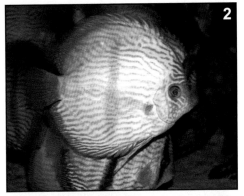

1. A large group of Heckel discus (*S. discus*) that I displayed on a biotope stand at the Aqua-Fisch 2003. The fishes originated from the Uatumã region, specifically from the Lago Iri (lower Rio Jatapu drainage, see text). My intention was to show how these fishes actually live in the wild. There were 50 individuals and an alpha individual rapidly developed from among them *(centre of photo)*. They immediately felt at home in this large group, as that is precisely how they live in the natural habitat.

2. During the four days of the exhibition the alpha individual was constantly busy watching over his group.

3. By the end of the exhibition a new alpha individual appeared to be emerging (as in the wild).

4-5. In the lower Jatapu drainage, as in the mouth region of the Rio Nhamundá, there are *S. discus* variants that look rather like natural hybrids. Here too it can clearly be seen that these two are not the typical Heckel discus (*S. discus*) with which we are familiar.

1. View from the *terra firme* from above a discus habitat on the Lago Saracá (in the background is the Lago Canaraçi or Camaraçi). 2. Returning from fishing in the Lago Canaraçi with Sardes and the *caboclo* Pedro (habitat in the background on the northern lake-shore). 3. Guanavenas Pousada, Hotel de Selva at Silves. One of the first lodges (opened in 1980), and nowadays there is also the Pousada Ecológica Aldeia dos Lagos. Both lie in the jungle, surrounded by this lake region, but nevertheless offer all comforts. 4. During the first hunt for discus here (1983). 5-6. I have often fished in the black Rio Urubu, which has its mouth here. The discus habitat is clearly visible (5). The arrows show the same spot at daytime (5) and at night (6) were I caught a discus on a tree-trunk (it was resting alone on the trunk like a leaf).

7. The map shows (bottom right-hand corner) the mouth region of the Urubu, and to its left the Rio Preto da Eva, which also contains black water although so far I have been unable to find any discus habitats. Then we see a section of the Amazon and Manaus in the lower left-hand corner. 8. Aerial photo of the Urubu's mouth region in the Lago Canaçarí. 9. Rio Urubu sign. 10. The middle part of the Rio Urubu, where Heckel discus habitats can still be found now and then. 11. *Satanoperca* sp. from the mouth region (caught with the blue discus, as were 12 and 14). 12. *Astronotus ocellatus*, wild form: the species is this brilliant red in the mouth region of the Urubu. 13. *Symphysodon haraldi* – a blue discus – from the Urubu-mouth. 14. *Cheatobranchus flavescens* – a particularly lovely colour variant which occurs in the lower Urubu together with the discus.

32.1 °C; in 2.5 m of depth 32.2 °C; oxygen 6.90 mg/l.

Biotope: overhanging trees, roots and tree-trunks; no floating plants. Jungle intact, *fazendas* further away (to the south) on the *terra firme*. The water was almost clear and flowing very slightly. Discus variants: *page 322 (photo 1 & 2)*.

The water parameters in the discus habitat in the Uatumã, in the Lago de Jaquarequara, on 22.01.2004 at 17.00 hours were as follows: pH 6.86 in 2.5 m of depth; conductivity 7 μS/cm; daytime temperature (air): 33 °C; near the water's surface: 33.6 °C; in 2.5 m of depth 30.1 °C; oxygen 5.61 mg/l O_2.

Biotope: overhanging trees, roots and tree-trunks. Floating plants: *Utricularia foliosa* – in large quantities. Jungle intact, *fazendas* on the *terra firme*. The water was almost clear, and motionless. Discus variants: *see page 207*.

In October 1989 (no date) I recorded measurements during the dry period in an this Rio Uatumã disus habitat of pH 6.65 and conductivity 70 μS/cm!

For comparison, in the centre of the Rio Uatumã (definitely not a discus habitat) the water parameters on 22.01.2004 at 18.00 hours were as follows: pH 7.95 in 35 cm depth; daytime temperature (air): 32 °C; at 35 cm of depth 32,7 °C.

Biotope: no floating plants, only (almost) clear, flowing water. Surely effected from the dam.

The water parameters in the discus habitat in the Jatapu, Lago Iri *(see page 325)* on 23.01.2004 at 12.50 hours were as follows: pH 5.36; in 2.5 m depth 5.03; conductivity 9 μS/cm; daytime temperature (air): 35 °C; near the water's surface: 33.6 °C; in 2.5 m of depth 29.8 °C; oxygen 4.56 mg/l.

Biotope: overhanging trees, tree-trunks and branches in the water; no floating plants, but sometimes lots of grass in the water. Jungle intact – primary forest. The water was clear to tea-coloured, motionless. Discus variants: *see page 327*.

Additional measurements in the Lago Limão *(see page 325)* on 23.01.2004 at 10.30 hours were as follows: pH 5.57 in 2.5 m depth; conductivity 6 μS/cm; daytime temperature (air): 35 °C; near the water's surface: 32.3 °C; in 2.5 m of depth 29.5 °C; oxygen 5.97 mg/l.

Biotope: overhanging trees and tree-trunks. No floating plants. Water clear to slightly brown, motionless. Jungle intact. Discus variants: *see page 322, photos 3-4*.

I have additional parameters measured earlier in the Jatapu (October 1989), but no date: pH 5.4 in 35 cm of depth; conductivity 10 μS/cm. Again in the Jatapu, at the same depth and with a slight current, pH 5.2 on one occasion, and another time pH 5.5 (and it was really not as hot as 2004).

Discus habitats extend as far as Itapiranga in the south-west, in the Lagos Jarauçu and Jabutí *(see page 321)*. Although there are numerous *fazendas* there, almost all around, the *caboclos* say that in the dry season there are *acará-disco* everywhere. Even among grasses. This is *S. haraldi* habitat. In the past I have found half-striped individuals and others almost without pattern colour. The water is clear, but cloudy during the rainy season. Unlike the Paraná de Itapiranga, which contains white water all year and offers a direct link with the Amazon and with Manaus (some 15 hours journey by *recreio*). Itapiranga is a peaceful little place with a hotel and a square *(see page 326)*. A *cooperativa de pescadores* (fishermen's cooperative) and a road link to Manaus, though unfortunately the last part of it is so bad that in 2003-2004 no buses and taxis would use it any more.

Rio Urubu: One can also get to Silves from Itapiranga (by car or by boat); it lies on the Lago Saracá, which is connected to (or rather, forms a unit with) the enormous Lago Canaçari *(see page 328)*. Silves is an island and is washed by the Rio Sanabani and the outflow of the genuinely blackwater Rio Urubu.

The (white) population of the region began in 1660 with the founding of the Missão do Saracá by Frei Raimundo of the order of the Mercês, still resident today in the neighbouring town of Itapiranga. Unfortunately its foundation was a bloodbath. The Portuguese killed off the indians, thousands of whom died in 1663 at the mouth of the Urubu. In 1759, when at long last peace reigned again, the little *aldeia* of Saracá was built at the Vila de Silves.

I have found discus in the northern Saracá region, also along the north-western Lago Canaçari area *(see page 328)*. As in the mouth region of the Urubu there are *S. haraldi,* often all-over striped or individuals with some degree of patterning. The southern part is influenced by the Amazon, and contains mainly murky water and no discus habitats.

The water parameters in the discus habitat *(see page 328)* on 12.10.1986 at 14.00 hours, measured by my companion at that time, H. J. Mayland, were as follows: pH 6.4, in some 30 cm of depth; conductivity 16 μS/cm; daytime temperature (air): 32.5 °C; at the water's surface: 31.6 °C; oxygen 6.2 mg/l.

Biotope: overhanging trees and tree-trunks; no floating plants. Jungle intact, *fazendas* further away on the *terra firme*. The water was clear and moving slightly.

Discus variants: *see page 329*.

I have often explored in the lower and middle region of the Urubu *(see pages 328-329)* and have already reported on it extensively elsewhere. There are *S. discus* (Heckel discus) habitats upstream to the Igarapé Jutuarana (also called Jatuarana):

The water parameters on 21.02.2004 at 15.00 hours were as follows: pH 4.67 (ditto at 2 m depth); conductivity 12 μS/cm; daytime temperature (air): 37.5 °C; near the water's surface: 31.3 °C; in 2 m of depth 29 °C; oxygen 4.83 mg/l.

Biotope: overhanging trees, lots of roots and branches; no floating plants. Jungle intact at the habitat, but already numerous *fazendas* in the surrounding area. The water was black, dark tea-colour, almost darker than in the Rio Negro itself and hardly moving *(see page 329)*.

Now, before we go on to the next habitat region, I must not forget to talk about three bitter pills for this region (as regards nature and biodiversity):

1. As regards the Jatapu, there are plans in the offing to build a hydro-electric dam in Roraima (after Balbina is "in place" – see below), and in addition large deposits of iron have been discovered in its upper course.

2. Recently it is no longer the Urucu and Coari *(qv)* that are on everybody's lips, but the Rio Uatumã drainage. The metropolis of Manaus, whose energy supply has been slowly but surely (and literally) drying up, is now banking on the gas reserves discovered by Petrobas in this region. When I was last there, in February 2004, this was the sole reason for the trucks full of building materials thundering along the Manaus-Itapiranga road. The "wells" were discovered three years ago and will supposedly solve all the problems of the capital of Amazonia.

3. Balbina.

Around 2,400 km² of primary forest (plus flora and fauna), including innumerable archaeological sites and a enormous indian region (large parts of the Waimiri-Atroari), were flooded for this power station, commissioned in 1987. It is notorious as the largest ecological disaster in Amazonia. The area is of similar size to that of the Tucuruí hydro-electric dam, but Balbina has produced only 3% as much energy (only some 250 megawatts). And Manaus can neither live nor die on that.

The once black Rio Uatumã has become grey (instead of clear) and there is no improvement in sight. The water level is dropping again, and if something drastic doesn't happen in the coming decades things will no longer look rosy here, either. Interestingly in 2004 the Justiça Federal in Brasilia awarded 27 plaintiffs damages of 320 million R$ (about 120 million U$) against the Centrais Elétricas do Norte do Brasil (Eletronorte), as recompense for inundated land in the region, no longer available for them to use.

Fortunately, at least for the record, an ichthyological census was performed prior to the inundation. 182 species from 28 families and nine orders were recorded from the Uatumã. Including species that to date are known only from the Guianan Shield, Rio Negro, and Solimões-Amazon (INPA, 1993). The work also contains a whole series of stomach contents studies which show insects, fishes, plant residues, detritus, fruits, shrimps, algae, insect larvae, micro-crustaceans, aquatic macrophytes, scales, and fragments of sponge (in that order) as the main components.

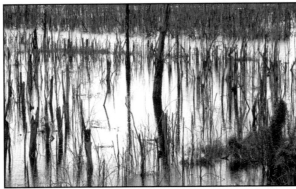
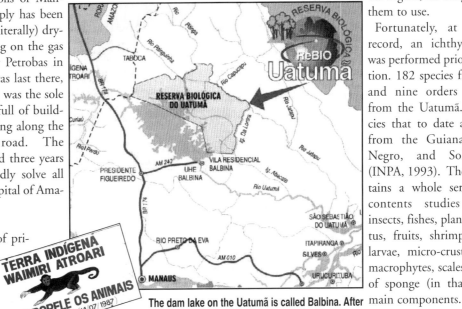

The dam lake on the Uatumã is called Balbina. After the largest ecological disaster in Amazonia had been perpetrated, a biological conservation zone (Reserva Biológica do Uatumã) was created in reparation (?). But how can you conserve what is dead or extinct? *(photos top)*. The Waimiri-Atroari indians are trying to do it using road signs in their reservation *(left)*.

4. Maués region and Rio Abacaxís; Rio Madeira region; Manacapuru region.

Maués region and Rio Abacaxís: We leave the region north of the Amazon where we have advanced as far as the Rio Preto da Eva and cross the Amazon at Urucurituba, which I have already mentioned briefly. But it remains to mention that there are discus habitats not only in the smaller *lagos* to the south and east but also in the large ones to the north-east and north, as well as in the Lago do Arrozal during the dry season *(map, centre)*. Further west lie the Lago Preto, Igarapé do Bacabal, Lago Grande, and Lago Curuçá, which are almost adjacent to the whitewater Paraná Urariá and likewise contain discus habitats. In this region there are individuals that have little pattern colour and are partially striped, as well as others that are plain light-brown (all *S. haraldi*). I didn't record water parameters during my visit, but they are probably not much different from those for the *S. haraldi* habitats discussed below, because of the vicinity of habitats.

As regards Urucurituba itself, it remains to say that recently it has been noteworthy not only for its Festa do Cacau e Feira Cultural (cocoa and culture festival) but also for recent archaeological finds of ceramics and tools from prehistoric times, which will add a further chapter to the fascinating history of the settlement of the Amazon.

The cocoa festival takes place at the beginning of May and was first brought into being in 2002 – despite ancient cocoa tradition – to honour the so-called *cacauicultores* of the region, as their work has made a considerable contribution to the economic upturn in the area. The fruit *(right)*, which was supposedly brought here by pre-Columbian tribes more than 2000 years ago, by now constitutes 80% of the Amazon cocoa production.

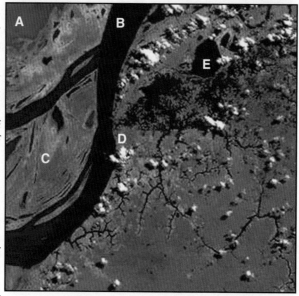

Top: Urucurituba decorated for the second cocoa festival.
Above: Satellite photo of the region (1:50,000): A) Lago Canaçari (light blue = clear water, clouded during the rainy season); B) Amazon (dark blue = white water); C) Ilha do Risco; D) Urucurituba with the great tea-coloured *lagoas* immediately to the north-east; E) Lago do Arrozal (all discus habitats). The numerous *igarapés* (black), huge *fazendas* (light green), and numerous settlements (pink) are also recognisable.

And it is not just the three days and nights of the cocoa festival, headed by the Cocoa Queen, that nowadays attracts around 20,000 enthusiastic onlookers (about double the number of the local inhabitants), but also the unspoiled white sandy beaches, and the clear (to tea-coloured) *igarapés*, *lagos*, and *rios*.

And, of course, the Museu Arqueológico de Urucurituba, which was under construction and yet to be completed during my last visit to the region (2003). The whole benefits from the proximity of Manaus (only 294 km away by river), to which there is a daily *recreio* service.

In addition it is worth trying on the spot the huge variety of cocoa-based luxuries such as jellies, liquers, cocoa bread, cosmetics, butter, honey, and of course, numerous sweetmeats. And to see how every last bit of the fruit is used, for example how the hard shell is utilised as packaging material and even looks attractive. And finally *cacauari*, a drink made from cocoa juice enriched with alcohol (said to be an Inca recipe), is a must.

Because of this cocoa success Governador Eduardo Braga has (July 2004) announced that expansion of production in the region is to be encouraged, as AFEAM (Agência de Fomento do Estado) is financing small producers (who receive 7 R$, about 3 US$, per kilo of cocoa beans). This is part of the so-called *Programa Zona Franca Verde* (programme for the green free zone – recently established in Amazonia).

But the *cacauicultores* also enjoy celebrating. For example, the festival of St Joseph held every year, with fireworks and competitions, in the *comunidade* of Augusto Montenegro. And a three-day festival in honour of St Benedict, now in its 50th year, is celebrated, again in July (2004), in the neighbouring village of Urucarazinho, with the holiday spent fishing on the water and on the beach. And while we're on the subject of fishing, in the last week of August each year there is the Festival do Peixe Liso (the festival of scaleless fishes), brought into being three years ago in order to encourage tourism in the *comunidade* of Vila da Feliocidade-Itapeaçu. Who ever catches the largest catfish wins a prize.

To the south-east of Urucurituba lies **Maués**, which first hit the discus headlines at the end of the 1970s, but has been known as the centre for *guaraná* for thousands of years.

Maués lies, like Manaus, at an *encontro das águas*, on the right-hand bank of the mouth of the Rio Maués-Açú, at the point where it mingles with the muddy Paraná Urariá. It is a similar natural spectacle except that the Maués-Açú is not so "black" as the Rio Negro. (But many people call it *água preta*.) Its *município*, including 165 rural settlements, numbered about 41,000 inhabitants in 2000 and lies 270 km (45 minutes by air) from Manaus. Two airlines, Tapa and Rico, serve it almost daily, as do *recreios*, although the trip takes 22 hours. (You can also, as I did once, drive the 276 km metalled road from Manaus to Itacoatiara and reach Maués from there within six hours via a *recreio* through the Paranás do Ramos and Urariá.) With its 12 hotels and romantic *pousadas*, broad *avenidas* (avenues) in and around the city, kilometre-long sandy beaches (usable for about eight months per year), and more than 50 sometimes unexplored caves in the vicinity (some more than 200 m deep), it is justifiably known as one of the loveliest spots in Amazonia.

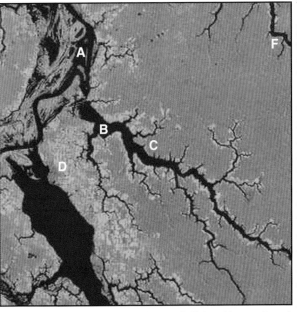

Top: Maués. *Above:* Satellite photo (1:50,000): A) the whitewater Paraná Urariá; B) the clearwater Maués-Mirim; C) the former village of São Francisco, now grown to a city; D) Maués; E) mouth region of the clear Rio Maués-Açú; F) Rio Curuçá. Light green = *fazendas*.

Maués is surrounded by more than a dozen *rios* (see map), an enormous network of rivers such as the Maués-Açú, Maués-Miri (or Maués-Mirim), Apoquitaua, Maraú (or Mamuru), Urupadi, Paraconi (or Pacoval) fed by almost innumerable affluents. However, there are no discus in the latter, which often contain rapids and/or waterfalls (most of them unexplored). In the majority of the others there are habitats where the environment is unspoiled and where I have found some individuals with splendid patterning and others with fewer markings. As well as in a number of *igapós* and in the clear to tea-coloured *lagos*. But only *S. haraldi*, there are no *S. discus* here.

The water parameters were measured by H. J. Mayland during a joint visit to the region on 12.10.1986 around 11.30 hours: pH 6.2; oxygen 6.9 m/l; conductivity 17 μS/cm; daytime temperature (air): 32 °C; at the water's surface 32 °C; and in the habitat 30 °C.

But allow me to say something about the city and its original inhabitants: the first time Maués is mentioned was during a visit to the region by the Jesuit Bettendorf, who named it "Vila dos Maguases". However, in 1759 the Jesuits were outlawed by the Portuguese minister Marquês de Pombal because of conflict with the colonists who wanted to enslave the indians; the missionaries tried to suppress this and stirred up the indigenous people against the white men.

It was Lobo D'Almada, *Governador* of the Capitania do Rio Negro and Grão-Pará, who first managed to get the belligerent Mundurukú, Sateré, and Mawé to live peacefully with the white men and advanced the development of the region. He also played a part in the founding of Canhumã, Juriti, and Luséa. (On the site of the last of these stood an *aldeia* which the Mundurukú called the *uacituba*, which means, roughly, great or fertile earth.) The actual founders of Luséa were, however, the

Portuguese Luís Pereira da Cruz and José Rodrigues Preto, in 1798. They gave the settlement the first and last syllables of their respective names *(lu* and *sé)* and delegated the building work to Capucin monks. The latter achieved this successfully under the Capitania São José do Rio Negro (see Rio Negro) and named the settlement Mawés (= Maués) after the aboriginals. (Others retained the name Vila de Luséa.) The settlement was subsequently the scene of a bloody battle between the *cabanos* and government troops. Not until 1840 did the *cabanos* surrender and swear (what was left of them) allegiance to Luséa's Praça Coronel João Verçosa.

When the Amazon was elevated to the status of province on 5th September 1850, Luséa was officially named Vila de Maués. In 1852 it became part of the Município von Vila Bela da Imperatriz (later Parintins). And a year later the *vila* was raised to the status of *cidade* (city), albeit with yet another new name: São Marcos de Mundurucucami. By this time the majority of the natives from along the Rios Mamurú, Abacaxís, Apoquitauá, and Andirá had already moved away, and retained a land-holding only in the region of Maués. Ever since then they have maintained their culture (to the present day). They are also one of the few tribes to have retained their own language.

In 1865 Maués, as many people still called it, underwent another change of name. Deputado José Bernardo Miquilles, the official in charge, named it Vila da Conceição. Not until 1896 was it all over, and today Maués is one of the few Amazon towns and cities that has (eventually) retained its native name. Translated from the Tupí language, *máu* means intelligent, inquisitive, or talking, and *uêu* parrot. Thus Maués = intelligent or talking parrot. And its inhabitants are certainly intelligent, and also very friendly and ready to help (at least that is my impression to date); and this includes the Sateré-Mawé, who are proud to be *filhos da terra do guaraná* (children of the land of the *guaraná*).

When, in 1664, Padre Felipe Bettendorf visited the region and saw *guaraná* for the first time, he wrote that the Andirás indians had little fruits in their forest which they dried and then crushed to a powder, from which they made little balls which they ate and valued the way white men do gold. This was called *guaraná*. A spoonful of the powder mixed with water produced a drink of such power that they could go hunting for an entire day and night in the forest without a trace of hunger. In addition it healed fevers, cramps, and headaches.

The Sateré-Mawé (photos 1-3 below) are descended from the Andirá and Maraguá and were living in the region between the Tapajós and the Madeira long before white men arrived. There were originally two tribes, and the Sateré lived where the border between the two states of Pará and Amazonas lies today. In 1669 Portuguese Jesuits made first contact with the warlike Mawé in the Maués region. When, in 1825, the Mawé and Sateré allied themselves with the Mundurukú and Mura in the the movement of the *cabanos,* their numbers were reduced considerably by the long years of war. And the few survivors then formed one tribe, the Sateré-Mawé.

In their language *sateré* means fire saurian and *mawé* talking parrot. In 2000 there were 7,134 Sateré-Mawé, and they live along the Rios Mamurú (in 38 villages), Andirá, Urupadi, Miriti, Manjuru, and Uaicurapá, and the *igarapés* in their

reservation. Over a period of more than 300 years the tribes have lost a lot of their original land and their reservation encompasses only 50,000 ha. But they still tend their original *guaraná* regions, the *terras altas*. They have discovered the phytotherapeutic uses of the *guaraná*, and still treat their cultivation of the fruit as a ritual: harvesting; transportation; transplantation of seedlings; special attention to the roasting in a clay oven; heating the oven with special aromatic wood; etc.

As well as this ritual, I was able to witness a second.

When a boy wishes to become a man he must undergo the *tucandeira* (or *tocandira*) test, that of the fire ants *(Paraponera clavata)*, which the Sateré-Mawé call *wty'ama*. (At least, this ritual is practised in Kuruatuba and along the Rios Maraú and Andirá between October and November.)

He is woken very early man to have *jenipapo (Genipa americana)* juice rubbed into his arms; the liquid is yellowish but colours the skin dark blue (this is a widespread method of body-painting among South and Central American indians). Then his hands are scratched with *paca (Agouti paca)* teeth until the blood flows. In the meantime the *sá ari pé*, a kind of glove woven from bast, round or in the form of a fish or bird, will have been prepared. The fire ants, about which I could tell you a tale or two and which not for nothing are also known as bullet ants (because their bite feels like a gunshot wound), captured alive and kept in a *tum-tum* (bamboo tube), are placed in a bowl of the juice of the *cajú (Anacardium occidentale)* leaf to stupefy them. Then, half-asleep, in the *sá ari pé*, *pé* with smoke to make them angry. Meanwhile the *tucandeira* dance will have started, following the rhythm of two steps forward and two back, first to the left and then to the right. Then suddenly, in the middle of the dance, a shrill sound rings out from the *taquara* (a kind of horn) of the *uepy hat* (master of ceremonies), who pulls the *sá ari pés* onto the boy's hands and raises his arms, at the same time relating the origin, the history, and the power of the ritual in the Tupí language. At this point the ants wake up and start to tear the flesh from the boy's hands. (The photos below show the species and its fearful pincers.) The boy beats his hands around as if insane until two men restrain him so that he cannot injure himself further. The dance continues while the boy's tears roll down but he steadfastly clamps his teeth. Eventually he raises his head and thereby demonstrates that he has become a man. The *sá ari pés* are removed. A girl takes him into her hut to look after him. Subsequently he emerges and joins the dance. The boy is now a proud man and can start a family (at the age of 14).

But back to the *guaraná:* the story (one of many) goes that the *guaraná* fruit originated from the eyes of the son of Onhiámuáçabê, who knew all the plants and their uses, and cultivated a great garden full of healing plants, the *noçoquem*, which it was forbidden to enter. One day, without the consent of her brothers, she was made pregnant by a snake that merely touched her on the leg, and bore a strong handsome boy with unusual black eyes. As he grew he preferred to live on fruit. But one day, when he ate chestnuts from a forbidden tree, his envious uncles caught him and had him killed. Because Onhiámuáçabê knew that his body would shortly afterwards be cremated by the shaman in accordance with tribal ritual, she rescued his beautiful eyes which were treasured all over the Maués-Açú-Marau region. She chewed the leaves of a magic plant, washed the eyes in the juice and her saliva, and buried them in the best soil with the words, "My son, you will become the greatest power in nature, to bring good to humans and energy to the weak, and free many from disease, making them whole again."

Legends apart, the fact is that the *guaraná* fruit, once open, really does contain seeds that look like human eyes *(above),* hence the indian name *guaraná* (people's eyes), while in Venezuela they are called *ojo de indio* (= indian eyes). And they do have healing powers and other marvellous qualities (see below).

No less a person than Carl von Martius collected the fast-growing *guaraná* liana (which grows up to 18 m high, supporting itself on other trees) here in 1819 and classified it later as *Paullinia sorbilis*. (However, in in the meanwhile, in 1821, Humboldt and Bonplant assigned the species to the coffee-plant family Sapindaceae and described it from the Venezuelan region

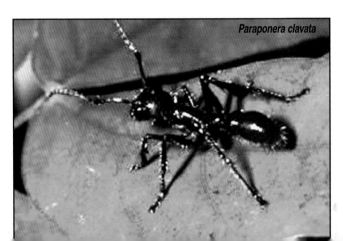

of Cupana as *Paullinia cupana*. Thus Martius' discovery became *P. cupana* var. *sorbilis*.) Martius noted that there was already a regular trade in *guaraná*: ships laden with it travelled up the Madeira to Mato Grosso and Bolivia, along the Iténez and Beni to Santa Cruz de La Sierra and Cochabamba, and returned to Serpa (= Itacoatiara) and Vila Bela da Imperatriz (= Parintins) carrying goods obtained in exchange.

But it is important to be selective because, as with wine, there are considerable variations in quality. The Sateré-Mawé stick firmly to their ancient traditions and cultivate *guaraná* plants only in the best soil on the *terra alta*, the Maraú region. But their harvest amounts to at most two tonnes per year. By contrast an appreciably lower quality is (mass) produced around Maués, on about 6,000 ha, and known as *guaraná de Luséa*.

By the beginning of the 20th century *guaraná* already represented more than 30% of all the alcohol-free drinks produced in Brazil. In 1921 Antarctica, the largest drinks manufacturer in the country, began production locally. Nowadays over 180 tonnes of *guaraná* fruit are harvested annually by more than 3,000 small farmers. They harvest around 1 kg per plant, and this is then roasted in clay ovens, dried, and ground *(below: the roasted fruit in the sack (left), and ground (right))*. The powder has become the basis of a gigantic industry with a local turnover alone of more than two billion US$ annually. And in Brazil as a whole the market value to the drinks manufacturers stands at 3.4 billion US$. *Guaraná* is the number one soft drink in Brazil, and Portugal, Spain, Puerto Rico, Japan, and others are not far behind.

But *guaraná* is not only a delicious drink (I drink it all the time in Brazil) but has numerous other benefits. It is used as an elixir and mixed with other fruit drinks. It contains tannin and phosphorus, and more calcium, magnesium, and potassium than coffee, tea, or *mate*. It is rich in saponin (which is also present in ginseng and, when taken over long periods, is said to have an unsurpassed energy-boosting effect). It is unusually stimulating and not without reason was already being sold at high prices in Europe by the middle of the 19th century as an aphrodisiac and rejuvenation agent. It is prized as an aid to stopping smoking, but the latest craze is available from the merchant Abrahão Leda in Maués, who has invented a mixture that supposedly puts all others, actual and reputed, in the shade. His *kit tesão* or *turbinado* is eagerly consumed by the locals as the Viagra of the forest. It is a mixture of *guaraná* powder with a root extract known locally as *mirantã*. Abrahão (= Abraham) says that he is already selling more than 1000 *kits* (= packs) a month all over Brazil, and three times as much during the annual Festa do Guaraná in Maués (which took place last year for the 20th time). All that is required is to stir a tablespoon of *mirantã*, one of *guaraná* (drink), and a teaspoon of *guaraná* powder in a glass of water and the day is *garantido* (= guaranteed). Abraham also says that *mirantã* has been taken regularly for centuries by the Sateré-Mawé and that he has never known any of them suffer from cardiovascular problems, let alone that one died of angina or a heart-attack. When I first tried it my heart did in fact start to beat faster after a few minutes and I felt a real energy boost (unfortunately I can't offer any comparison with Viagra). The additional energy supposedly causes the blood to circulate better and fat to be broken down in the artteries, and increases strength overall. The *kit tesão* can be purchased in the Bar Turbinado and the Barraca Sateré in Maués. But it is not the only thing you should try in this lovely city: there are also splendid fish dishes such as *moqueca de surubím* or *calderada de tambaquí*. Not to mention the tasty north-eastern specialities such as *buchada de bode* and *galinha á cabidela* which have long since reached here. And you must also taste the home-made liquers, jellies, fruit juices, and *batidas* based on regional fruits, and exotic drinks such as *xexuá* (a kind of wine distilled from wild roots) and *aluá* (a pineapple fermentation).

On the Internet you will find around 472,000 pages on *guaraná* (at www.altavista.com), but if you are thirsty in the Brazilian Maranhão and want it, then please say *"Me vê um Jesus, aí!"*, as there the drink is called Jesus (and perhaps it really is divine). But back to our fishes.

Apart from the discus mentioned the area remains little explored ichthyologically to the present day. In 1997 the Swede Kullander described a cichlid, *Aequidens mauesanus*, from the Maués-region Igarapé do Rio Maraú (it is also found elsewhere). Apart from discus I have found interesting variants of angelfishes *(Pterophyllum)* and a few others.

There are no discus habitats in the Paraná Uariá itself *(see page 345)*, which contains white water all year and is used by shipping. It provides a direct connection (shortcut) from the

Amazon to the Madeira. Down the Paraná do Ramos, into the Paraná Urariá to the Paraná Canhumã, and over to the Rio Madeira *(see page 343)*.

At Santissima Trinidade, further down the Urariá, a *paraná* flows into the latter via a network of whitewaters, and at high water links the gigantic Lago Mirituba with the Amazon via the Lagos do Moura and Porção (and, depending on the time of year, can flow in either direction). So far no discus have been found here, and my researches indicate there aren't any habitats either. They are not found again until the **Lago do Campo**, the **Lago do Araçá**, and the **Rio** and **Lago Curupirá**. Without exception I have found just *S. haraldi* almost completely brown, as well as part-striped and reddish individuals *(photo right)*.

The water parameters I measured in the above mentioned areas varied according to the circumstances on different occasions and habitat (1986-2002): pH 6.1-6.5; conductivity 26-35 µS/cm; daytime temperature (air): 28-35 °C; the water in 1.5 m of depth: 26 °C (early morning) and in 1.5 m of depth towards 16.00 hours: 29 °C. Biotope: clearwater habitats with no aquatic vegetation.

Rio Madeira region: But the "jam-packed" area for discus is the Canumã (Canhumã)-Marimari-Abacaxís region, or at least it has been for the past two decades where collecting is concerned. And the latter is based (almost) exclusively on **Nova Olinda do Norte**. This is a small town of around 14,000 inhabitants on the right-hand bank of the Madeira, and for a long time the departure point for so-called "Madeira discus".

However, no discus has ever been seen, let alone caught, in the Madeira itself (but nevertheless there is popular literature that refers to more than 38 variants from the Madeira as well as almost innumerable Internet sites with "Madeira discus"). They all come from the slightly tea-coloured Rio Canumã and its affluents such as the Mapiá and Acari (left-hand tributaries), the Rios Jaburu and Pacová and the mouth regions of a number of *igarapés* upstream to Foz do Canumã (right-hand tributaries). A huge number of different variants of *S. haraldi (see page 343, and Chapter 4)* come from these habitats, which are almost completely intact and little disturbed. Only now and then are individuals imported from the clear Rio Aripuanã, a right-hand branch of the Madeira *(see below)*.

Nova Olinda do Norte can be reached from Manaus in about 16 hours via the *recreio* that runs every other day. But you can also fly from the Aeroclub in less than an hour via Taxi-Aereo Cleiton for around 300 R$ (about 200 US$). The owner, Cleiton, is a first-rate fellow and helpful in every respect. On my last expedition he immediately called Nova Olinda from Manaus and arranged a meeting with a Mundurukú named Edivaldo dos Santos who was experienced at collecting *acará-disco*.

Cleiton arranged a hotel (the Jardim Paiva) and *voadeira* and recommended the Restaurant Aruanã where Natasha and I ate a delicious *pirarucu ão molho* with rice, black beans, *farinha*, and beetroot late that evening. And here I experienced something that still makes my flesh crawl: little green men! Do they really exist? Why were they visiting this remote Madeira town?

I had hardly finished eating when Donna Fatima, the proprietress and cook, excitedly called me to her balcony next to the kitchen, from where there is a good view of the Madeira. She pointed to a UFO-like object (a flying saucer?) circling high up in the sky over the Ilha Canhumã *(see*

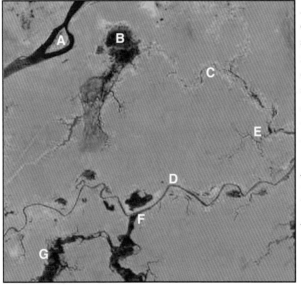

Top: In 2002 I portrayed a habitat from the Lago do Campo with its discus variants at the International Discus Championships in Duisburg.
Above: Satellite photo (1:250,000): A) Ilha de Urucurituba in the Madeira (not the one in the Amazon); B) the clear Lago do Campo (no connection to the Madeira); C) the clear Rio Curupirá links the Campo with the Paraná Urariá (D); E) Lago Curupirá; F) mouth region of the blackwater Abacaxís; G) the Marimari, which flows into the Abacaxís.

page 343) off Nova Olinda. For eight years, she said, it had repeatedly turned up now and then at night, but so far no-one had been able to identify it. People had even gone up in planes without finding anything, although it was in sight from the ground ! We too saw it quite clearly, but when I raised my camera with the 800mm lens to photograph it, there was nothing to be seen through the lens – but it could be seen with the naked eye! I wouldn't believe it if I hadn't seen it for myself. Be that as it may, next time I plan to check out the island of Canhumã more closely and pitch my tent there. Perhaps I will be able to arrange an interview !

As well the UFO there are two hotels, a church, the Bar Paraíba, and even a small supermarket in the little Madeira town (all firmly in the grip of the seven Paiva brothers). And Nova Olinda aims high, with or without UFO, with a *Garantido-Caprichoso*-like festival, held every year exactly two months after that in Parintins. The interval is so that the specialists from Parintins can come and work on the fantastic costumes and on the story for the festival. They call it the Festa da Mandióca. As in Parintins two groups join in "battle". The rivals in this case are *Rosa Vermelha* and *Ciranda de Nova Olinda*. Manioc is the main theme, and its potential uses are reflected in the festive costumes. In 2003 the festival, which lasts three days and nights, attracted an audience of more than 12,000 in the Raimundo Maciel Neto amphitheatre. It was a resounding success.

As well as this event at the end of August there are others such as the Festejos de São José in mid-March, the Festival de Frutas in April, the Festival da Cultura with theatricals and music, the Femucrinon – the festival of Christian music, Baile do Garoto e Garota Bronze (dance of the bronze girls and boys), the last two both in October, and many more (several every month). But back to discus.

Top: Natasha hanging my net to dry on the balcony of the lovely Jardim Paiva Hotel (with Madeira view) in Nova Olinda.
Above: Cleiton's air taxi transports the valuable (discus) freight.

In contradiction to what is generally stated and published (including in scientific work), the three notable discus rivers, the clear Canumã and the black Marimari and Abacaxís, are not branches or tributaries of the Madeira. There is no outflow to the Madeira, only a fluvial connection in the rainy season via the whitewater Paraná Canhumã to the white Urariá channel. From Nova Olinda one must first travel overland (barely 10 km) to reach the Paraná Urariá and then embark in a boat to seek out these habitats *(see map on page 343)*.

The Rio Marimari flows into the Abacaxís before they together feed the Paraná Urariá; these both are true blackwater rivers, like their tributaries (see below for exceptions). And their water parameters are comparable with those of the Rio Negro, Rio Nhamundá, and Rio Jatapu habitats and are true Heckel discus regions. Here there are only *S. discus*, although there are overlaps (such as in the mouth region of the rios Nhamundá and Jatapu), a few of which I will include here *(see pages 341-342)*.

The Marimari, which I have travelled upstream to where it becomes very narrow, is sparsely inhabited and unspoiled. Apart from 10 small (and one larger) Mundurukú indian settlements and a *caboclo* family near the mouth, no-one lives there. The densest jungle, totally undisturbed, borders its banks and extensive Heckel discus habitats can be seen, extending into the Igarapé da Paca and the Miracãoeira, although there is an overlap between two species here *(S. discus x S. haraldi)*. Right in the Igarapé Bem Assim, which forms a link from the middle Marimari to the Rio Canumã (via the Paraná Uraiá) but only in the rainy season

1. *From left to right:* Cleiton, the owner of the only airline to fly once a day to Nova Olinda do Norte from Manaus, with Director Mauro and myself (2004). 2. The airport building in Nova Olinda and the only ground transport (VW bus). 3. Nova Olinda, like Parintins, has its own festival. The photo shows the miniature dolls of the rival groups *Rosa Vermelha* (left) and *Ciranda de Nova Olinda* (right). In the background Linda, Cleiton's secretary. 4. The church of Nossa Senhora de Nazaré e São José dominates the town and there is no lack of aerials for satellite TV. 5. As in many Amazon towns and cities there are hardly any private cars, but lots of motorbikes and nowadays even motorbike-taxis (almost everywhere). 6. The Madeira was formerly noted for its huge gold reserves, and run-down houses like that of M. Joaquim (who buys up everything – sign, inset) are silent witnesses to its heyday.

This page to 342 are intended to provide a glimpse of habitats, variants, and overlap zones for the two species *(S. discus and S. haraldi)* from this region. *Above:* Downstream of the settlement of Mucajá the Marimari is black like the Rio Negro (for water parameters, see text). Here, as in the confluent Abacaxís, there are only true *S. discus*. In the background a *viveiro* for Heckel discus, which are particularly beautifully marked here. *Below:* Typical Heckel discus habitat in the Rio Abacaxís. Aquarists are very keen on individuals from this section of river as they are often very blue with red fin edgings. The aquarium shows the black water.

Above & below: Typical discus habitats in the upper Marimari. The water is here isn't black *(see aquarium),* and there are no true Heckel discus *(S. discus)* here. The Mundurukú indians say there are *acará-azul* (Blues), but I found natural hybrids (2) *(S. discus x S. haraldi)* in those habitats. Perhaps at some time *S. haraldi* (or hybrids) made their way from the Igarapé Bem Assim *(p. 342)* into the Marimari and then upstream to this area. The water parameters are typical for *S. haraldi.* True Heckel discus (1) and habitats can only be found in the lower Marimari. The two natural hybrids below (3 & 4) are from the mouth of the Igarapé Bem Assim. *Leaf: acará-açú.*

The Igarapé Bem Assim *(above & map (= B))* is an overlap habitat for two species. In the rainy season it carries water from the Rio Canumã region to the Rio Marimari (during the dry season from the Marimari to the Canumã area, or is dry), and *S. haraldi* come (came?) here to spawn (November-March). Close to its "mouth region" lies the Marimari blackwater Heckel discus habitat. Many of these various Bem Assim colour variants, like some of those in the mouth region of Jatapu and Nhamundá, are natural hybrids (note the high dorsal fin in the runted specimen above). The water is murky following rainfall (February).

Above: Satellite photo at high water showing the influence of the Madeira (K). Only the white water Paraná Canhumã (F) flows here out, past the mouth of the clear/brown Rio Canumã (C) and an island (E), into the Paraná Urariá. The latter eventually joins the Paraná do Ramos and the Amazon (creating the enormous white-water-washed Ilha Tupinambarana). A) Igarapé da Paca (left-hand Marimari tributary), is *S. discus*-habitat. B) Igarapé Bem Assim (under clouds); G) Igarapé Maypá is *S. haraldi*-habitat. H) Nova Olinda and Ilha Urucurituba (I), the UFO island; J) Lago Sampaio (otherwise clear). *Below:* Rio Canumã with *S. haraldi*-habitat and water.

(November-March). Apparently at this time (in the breeding season) *S. haraldi* are able to penetrate from the clear Canumã (and/or its lagos) to the immediate vicinity of the blackwater Marimari. On the other hand it may equally be the case that a Heckel discus ventures into the water that flows in at this time of year, finds a mate from the other species and "falls in love". In the overlap habitat I found a huge variety of colour variants *(see page 342)*. Sometimes neither *S. discus* nor *S. haraldi,* mostly natural hybrids (see also photo 3, below confirmed by DNA-test) and individuals with unnormal forms and sizes. Typical signs of inbreeding. A second interesting point is that only so-called *acará-azul* (Blues) occur in the upper Marimari, ie no Heckel discus. This was confirmed by the Mundurukú fish-collectors from the Aldeia Mucajá, headed by Nilson Cerrão, a very friendly Tuchaua chief who rules over 300 indians who go to church every Sunday (even though the parson visits it at most once a year). He also told me that the collectors caught discus only at night.

Because I hadn't recorded any water parameters previously I measured them once again during the spawning period, as well as specifically in the overlap zones:

1. In the upper Marimari, at the habitat in the vicinity of the Aldeia Sorval (named after a tree that has almost died out through exploitation for chewing-gum manufacture), the water parameters on 05.02.2004 at 07.45 hours were as follows:

pH 5.69; in 2.5 m of depth: 5.7; conductivity 6 μS/cm; daytime temperature (air): 27 °C; at the water's surface: 29.2 °C; in 2.5 m of depth: 28.0 °C; oxygen 3.11 mg/l. Biotope: roots, branches. No floating plants; jungle intact only around the areas cultivated by the indian settlement of Sorval. The water was relatively still and clear (slightly tea-coloured), and rising. Bottom: muddy. Discus variants: so-called blues, possibly naturally-occurring hybrids (*S. haraldi x S. discus*) or an uncommon *S. haraldi* variant *page 341*.

2. At the Heckel discus habitat downstream of the Aldeia Mucajá *(photos 1-2 below)* (named after the fruit of a local palm that tastes very good but has little flesh) the water parameters on 05.02. 2004 at 10.40 hours were as follows:

pH 4.97; in 2.5 m of depth: 4.85; conductivity 7 μS/cm; daytime temperature (air): 32 °C; at the water's surface: 29.3 °C; in 2.5 m of depth: 29.0 °C; oxygen 1.90 mg/l. Biotope: lots of roots, branches, trees, and overhanging *acará-açú*. No floating plants; jungle intact only around the areas cultivated by the indian settlement. The water was relatively still and tea-coloured (black water), and rising. Discus variants: only *S. discus* – *page 340*.

Note: this habitat lies only some 5-600 m from the next!

3. At the mouth of the Igarapé Bem Assim in the Marimari the water parameters on 05.02.2004 at 15.15 hours were as follows:

pH 6.3; in 2 m depth: 6.3; conductivity 40 μS/cm; daytime temperature (air): 32 °C; at the water's surface: 29.3 °C; in 2.5 m of depth: 29.0 °C; oxygen 2.98 mg/l. Biotope: lots of roots, branches, overhanging trees and bushes, plus lots of washed-in grass and *acará-açú* scrub; floating plants; jungle intact. The water was relatively still and murky, flowing very slightly towards the Marimari or motionless Discus variants: only what I believe to be natural hybrids (*S. discus x S. haraldi*) – *page 342*.

Notes: In the rainy season water flows in from the Canumã, then for two to three months it flows in the opposite direction but only for a short period, and for more than six months it is

Above: The confluence of the Paraná Canhumã with the Rio Canumã is a real spectacle. The permanent white water of the *paraná*, which here comes from the left round the curve *(see page 343, E)*, mingles with the black-looking water of the Rio Canumã flowing from the south-west at the point where the settlement of Foz do Canumã stands on the right-hand side *(inset, left)*. From here on it is called the Paraná Uariá, whose shores have been largely turned into *fazendas (inset, right)*. *Below:* Crowds of fishermen waiting at the confluence of the Rio Abacaxís and the Paraná Urariá, at the village of Abacaxís, at the beginning of February 2004, to block the river with several 100 m long nets in order to catch the *jaraquís* (*Semaprochilodus* spp.) migrating to spawn.

almost dry. Further into the interior lies the so-called Laginho (small lake) where typical Canumã discus *(S. haraldi)* occur. The water parameters there were identical to the latter.

These measurements demonstrate yet again that true Heckel discus practically always live in water with a pH of less than 5.0.

The Rio Abacaxís, into which the Marimari flows, is appreciably more populated, and there are numerous *fazendas, sitios,* and *caboclo* huts along its shores, mainly in its middle and lower course. Only far upstream is the environment still in good order.

In the *comunidade* of São João, during my last visit (2004), I met a *caboclo*, Leite de Souza, who was busy roasting his *farinha*. I wanted to talk to him as I had been told that he was now the chief collector here. Leite (= milk) is 36, has eight children, and has been collecting for six years. About 2,000 individuals every *safra* (= season, normally July/August-October). At present he is paid 0.70 R$ (0.30 US$) per discus (the indians, by contrast, get 5.00 R$ = 2 US$) and he sometimes catches 150 individuals in a single night. Not including *rekel-disco* as Heckel discus are known here. He also confirmed that I was right about only Heckels being found in the Abacaxís region.

Here too I took measurements in the habitat: **In the vicinity of the *comunidade* of Camarão the water parameters on 05.02.2004 at 12.55 hours were as follows: pH 4.99; in 2.5 m depth: 4.9; conductivity 8 μS/cm; daytime temperature (air): 35 °C; at the water's surface: 29.6 °C; in 2.5 m of depth: 28.9 °C; oxygen 3.4 mg/l. Biotope: roots, branches, *maracarana* scrub everywhere. No floating plants; jungle only partially still intact. The water was genuinely black (Rio Negro), relatively still, and rising. Bottom: sand. Discus variants: *page 340*.**

Two further sets of water parameters for *S. haraldi* habitats in the Canumã and its side-arms (as they are found well upstream, as far as the Igarapé Capinarana):

**In the Igarapé São Domingo (a right-hand tributary of the Canumã) the water parameters on 05.02.2004 at 16.45 hours were as follows: pH 5.35, in 1 m depth: 5.4; conductivity

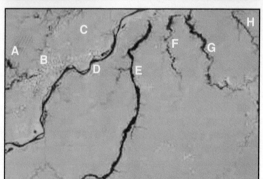

Top: Lago Juma (A) with the nearby whitewater Paraná do Madeirinha (B) (light green = *fazendas*). *Above:* the satellite photo (1:250,000) shows that no "discus water"(A,E,F,G,H) flows into the Madeira: A) Juma; B) Paraná do Madeirinha; C) Ilha da Madeira ("island" because it is washed around by the white water); D) Madeira (at Borba); E) Canumã; F) Marimari; G) Abacaxís; H) Paraconi.

40 μS/cm; daytime temperature (air): 30 °C; at the water's surface: 31 °C; in 1 m of depth: 29.8 °C; oxygen 3.07 mg/l. Biotope: lots of roots, branches, overhanging trees; no floating plants; jungle intact. The water was flowing slightly or still, and tea-coloured. Bottom: white sand. Discus variants: typical Canumã discus *(S. haraldi) – page 343.***

At a discus habitat in the lower Rio Canumã, another Mundurukú indian reservation (like Marimari) the water parameters on 05.02.2004 at 04.45 hours were as follows: pH 5.55; in 2 m of depth: 5.34; conductivity 7 μS/cm; daytime temperature (air): 31 °C; at the water's surface: 31.5 °C; in 2.5 m of depth: 29.3 °C; oxygen 3.39 mg/l.

Biotope: lots of roots, branches, overhanging trees. No floating plants; jungle intact; water flowing slightly, tea-coloured. Bottom: muddy. Discus variants: *S. haraldi – page 343.*

A glorious area. Full of cormorants, herons, and unspoiled vegetation, as far as the eye can see.

Now before we cross the Madeira, a little more on this gigantic and most water-rich affluent of the Amazon: at 1,380,000 km² and occupying more than 20% of the entire Amazon basin, the Madeira is by far the largest of its tributary basins. And at 3,352 km it is also the longest and most water-rich tributary (contributing around 15% of the entire volume of water and sediment in the Amazon). Not for nothing did Orellana, who discovered it on the 10th May 1542, call it the Rio Grande (= big river; only the Amazon has been given a more impressive title: Mar Dolce, = freshwater sea). But there are definitely no discus in it, they would be eaten or die within minutes by large preadors in the murky water *(page 344)* where you can't see your hand in front of your face. The southernmost discus distribution I heard was the clear Rio Atininga, a right-hand tributary to the north of Manicoré. The mayor told me that there were *acará-disco escuro* (dark discus) in it.

Manicoré has more than 38,000 inhabitants (in 2004) and also holds a

1. The Madeira is a muddy whitewater river that rises near Cochabamba in Bolivia, at an altitude of around 3,000 m. 2-5. The gold rush has almost completely destroyed many aquatic life-forms in its reaches. Even children pan for gold, here from the river sand upstream of a section of rapids (2). But the greatest damage has been done by the boats which for decades have pumped up millions of tonnes of substrate in search of gold (3). In the mid 1980s I once saw 187,000 such boats upstream of Porto Velho. The pumping and the mercury residues have made an incalculable hole in the aquatic flora and fauna. Hundreds of species extinct. In the section of river from Guajará-Mirim to Porto Velho I had difficulty finding any fishes at all (5). The majority of the boats are now rotting away (4) as the gold is exhausted. There are rapids and other barriers to fishes off Manicoré, but Teotônio (6), upstream of Porto Velho, is the end of the line for (almost) all species. I have often investigated the Madeira (7) and caught also 80 cm *Rhaphiodon vulpinus (insert)*. 8-9. Things look bad around Porto Velho. 10. The first railway (1912) ran from here to Guajará-Mirim.

RIO MADEIRA REGION: MADEIRA   BLEHER'S DISCUS   **347**

big Festival Folclórico in July, with lots of dancing, costumes and history of the aboriginals, the Mura. As in 2001 an area on the upper Rio Manicoré was declared a reservation, a *terra indígena*, for them. (However, to date no discus have been found in the clear Manicóre, or in any of the 213 *comunidades riberinhos*.)

It is worth mentioning that the mayor has done his best to add to the damage caused by gold prospectors, *fazendas*, and a ship containing an unknown quantity of sulphuric acid which sank here in 2002 and caused an enormous fish-kill: because there is no road to Manicoré, in 2002 he started to hack a 270 km route to the Transamazonica through virgin primary forest and indian territory, without obtaining permission. Luckily, however IBAMA managed to stop him.

On the right-hand side of the Madeira there are still discus habitats in the clear Aripuanã. A lighter-coloured (yellowish) variant with little pattern, upstream to Rio Juma and the Ilha do Mamão. The Juma itself is full of rapids, which also begin further upstream in the Aripuanã. In 1975 and 1976 I visited the Alto Aripuanã, making my way there from the Transamazonica, and discovered lots of new fish species (but no discus), including. the blue royal tetra, subsequently described as *Inpaichthys kerri*. And it was here that I discovered that the cichlid *Chuco axelrodi* is a synonym of *Hypselecara coryphaenoides* (shortly before death it becomes pitch black), the swordplant *Echinodorus inpai (centre, below)* and lots more. In those days the region was cloaked in the densest jungle, but since then vast *fazendas* have been established, plus three *pousadas* (lodges) for anglers: the Pousada da Liga de Eco; then another upstream, near the Cachoeiras da Sumaúma (although you have to walk for 30 minutes through the forest to get there); and still further upstream the Pousada do Rio Maracanã, on a branch of the Aripuanã. This far upstream there is also the huge *jaú* catfish *(Zungaro jahu)*, up to 1.70 m long (TL) and weighing about 100 kg, as well as huge *pirarara* catfishes more than a metre in length. And back then I also found numerous loricariids, some of which are now included in the L-numbers list (but not all of them, by far).

Today the majority of this region belongs to the state of Mato Grosso (Municipio Aripuanã), but it was once the territory of various indigenous tribes such as the Mundurukú, Apiaká, Kayabí, Iamé, Nambikwára, Tupímondé, and Arára, the survivors of whom today call themselves the Tupímondé or Cinta Larga and Arára.

***Top:** In 1975 I found this gorgeous tetra in the Alto Aripuanã, and in 1977 Géry and Junk described it as **Inpaichthys kerri** in honour of the INPA institute and its then head, Kerr.
Above: discus habitat in the middle Aripuanã.*

Virtually untouched by white men until the beginning of of the 20th century, the whole area changed when Brazilian and Peruvian entrepreneurs penetrated high up the Madeira and its tributaries. The adventurous installation of the telephone line by the General Rondon (1908) meant little to them, likewise the *seringueiros*, brazilnut collectors, and fur hunters.

As regards the entrepreneurs: in 1928 a band led by Júlio Torres under the command of the Peruvian Alejandro Lopes (who for decades controlled the rubber harvesting along the Alto Rio Aripuanã and built the settlement of Aripuanã) reached the area and massacred almost one entire *aldeia* of the Iamé. (Its few survivors subsequently called themselves the Cinta Larga (= broad belt)). And so it continued for a long time, and hardly anyone paid any attention to what was going on. Not until the 1950s, when the so-called *"Operações limpar a área"* (cleansing of the region) began and destroyed practically all the Iamé villages between the Rios Juruena and Aripuanã. This gigantic massacre did attract attention and went down in history as "the massacre on the 11th parallel".

When FUNAI came into being in 1966 there were still almost 5,000 indians living in so-called *cidades de palha* (palm cities), but today only 160 Arára and barely 1,000 Cinta Larga (= Tupí-mondé) survive in the Aripuanã region *(right)*. Their land-holding, including the Parque Indígena Aripuanã created for them (originally 3.6 million ha), has continued to shrink, and today only a fraction remains.

I am always enchanted by the river, its almost snow-white

sand, its clear water, and the almost undisturbed environment. But not by its mouth region and the continually expanding Nova Aripuanã, which by 2000 had more than 15,000 inhabitants. The lofty trees of the primary forest, with species such as *Triplaris surinamensis, Piranhea trifoliata, Copaifera martii* – discus spawning sites during the high-water period – and *Alchornea castaneaefolia*, which towers over everything on the *terra firme*, have me under their spell. Trees with a height of 45 m are no rarity, and it was in such giants that only recently (1998) Dutch biologists (Marc and Tomas van Roosmalen, and Russel Mittenmeier) discovered two new species of monkey, *Callicebus bernhardi* and *C. stephennashi* (drawings above). And consider: in the 1990s alone more than 15 new monkey species were discovered in the Brazilian jungle, bringing the total number of species for Brazil to 97. (It thus has the most monkey species of any country on earth.) And this is said to be just the tip of the iceberg.

Another (almost) endemic species is *Polygonanthus amazonicus,* a species of tree that is found here and around Maués. The Aripuanã also marks the westernmost distribution of the huge *Dinizia excelsa*, and the easternmost limit of the natural range of the cocoa tree *Theobroma cacao*. And last but not least, I never cease to marvel at the eye-catching violet-flowered *Physocalymma scaberrima* trees in the region. Unfortunately they are also the source of a much-prized, splendid red hardwood that is (too) popular for furniture and parquet floors.

The only water parameters I have measured, in a habitat near the left-hand mouth of the Rio Arauá on 01.07.1995,

chinodorus inpai

Drawings *(top): Callicebus bernhardi* (left) and *C. stephennashi (right)*, the new monkey species from the Aripuanã region. The photos (taken in 1972) show the Cinta Larga (= Tupí-mondé). They are called Cinta Larga (= broad belt) because they wear broad belts of bast *(centre)*.

are as follows: pH 6.6 (in approx. 30 cm of depth); conductivity 15 µS/cm; night time temperature at 22.00 hours (air): 28.5 °C; water 28.3 °C (in approx. 30 cm of depth); oxygen 3.08 mg/l. Biotope: tree-trunks, overhanging trees, sometimes bushes. No aquatic vegetation. The water was clear, flowing slightly, and falling *(see page 348).* **Bottom: white sand. Discus varieties:** *S. harldi* **(partly striped).**

On the right-hand side, further up the Madeira, there are no more habitats and no records of discus to date. Nor around Borba, a town of more than 25,000 people, again on the right bank and already a well-known stopping-off place back in Natterer's day, where he stayed for a long time (but without finding any *Symphysodon*).

I'm not going to say much about Borba's festival (you can look it up on the Internet), just that in June last year St Antony of Borba attracted more than 40,000 people; but I will mention the Malaysians, who bought up 657,000 ha of jungle to fell for timber (apparently Malaysia has sold out). However, when the clear-felling started there were expropriate. (This does sometimes happen in the Amazon region nowadays.)

We will now "jump over" the Madeira to the gigantic network of waters between it and the Purus (as until Nova Olinda there is no further information, discus-wise). The entire region has been little researched as regards *Symphysodon*. I have found them in the Lago and Rio Juma (there's one of these here too, not just in the Aripuanã drainage) *(see page 346).* A *S. haraldi* variant, which has a pink to reddish base colour and a degree of patterning. The Juma flows into the Paraná do Marmori *(see page 346),* which, depending on the time of year, sometimes contains white water. While the Paraná do Madeirinha (also known as Autaz-Açu) is white almost the entire year and has created the enormous so-called Ilha Madeira. I was unable to find any discus there, but there used to be Blues *(S. haraldi)* in the Lagos Castanhal and Taciuã (whose eastern part is called Lago Grande and the western Lago Manianrã), in the lake region of the Rios Igapó-Açú and Tupanas, as well as in the Lagos Castanho (not Castanhal) and Manaquri. A specimen from the large Ilha do Rei to the north was recorded by science in 1987, but I have no personal evidence. The island is inundated by the muddy waters of the Amazon for most of the year.

Manacapuru region: For decades this name has been bandied about in the discus-world, and discus bearing this name have repeatedly been offered for sale that don't actually come from this region at all. But in the first place – for those who are unfamiliar with the indian word – there are actually four Manacapurus *(map, top):* (1) The Rio Manacapuru with a length somewhat in excess of 50 km; which flows into (2) the gigantic Lago Grande de Manacapuru, more than 50 km long and up to 10 km wide; which narrows at its south-easternmost end and is again called the Rio Manacapuru until it widens again to form (3) a new lake, the Lago Manacapuru, narrower but much longer, creating a vast *lago-igapó-várzea* region more than 70 km long and 1-2 km wide (in some places more than 10 km). On the south-eastern shore of this lake, where it flows into the Solimões (the name of the Amazon from the border with Peru to the mouth of the Rio Negro), and opposite the often white-water Lago Cabaliana, lies (4) the city of Manacapuru *(centre).*

Manacapuru has become an important trans-shipping centre for Amazon shipping, at least since the advent of the metalled road to Manaus. In addition it is the starting point for expeditions and trips up the rivers from the south – the Rio Purus, the Solimões, the Japurá, and many more. Manacapuru is, moreover, the only place west of Manaus that can be reached by car (except that you have to cross the Rio Negro via the hourly ferry).

From this region, as well as from the Purus, somewhat to its west and east, came the majority of the so-called blue discus formerly assigned to the subspecies *Symphysodon aequifasciatus haraldi.* Although there is a different race/population/colour form in every geographically separated region (river or lake), including in the Manacapuru region.

The upper Rio Manacapuru has not yet been explored, and I have no details (yes, there are still a few "white areas", fish-wise, on my map of the world). Only for the two lakes and the middle Rio Manacapuru, where there are two variants, easily separable visually *(see page 352).* Even so, I would like to mention once again a point that is (almost) always overlooked:

Top animals with pronounced markings – especially stripes running all over the body – which were labelled "Royal Blue" *(see page 68)* as long ago as 1965 – are extremely rare in the wild and extremely difficult to catch. They are hardly ever netted, even by the most experienced discus collectors. I have been

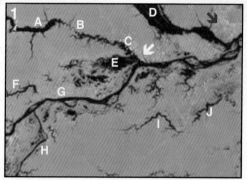

1. *Satellite photo (1:250,000):* A) Lago Grande de Manacapuru; B) Rio Manacapuru; C) Lago Manacapuru; D) Rio Negro; E) Lago Cabaliana; F) Lago Anamã; G) Rio Solimões; H) Rio Purus; I) Lago Castanho; J) Lago Marmori. *Arrowed:* Manacapuru (white), Manaus (red).
2. The new cultural centre of Manacapuru.
3-4. Old Portuguese buildings.
5. Manacapuru's promenade, with prostitutes.
6. You can get a complete meal here for 1 R$ (= 0.30 US$).

fortunate enough to land a few of these unique beauties. Lead-fishes or dominant individuals, as I term them. Just one specimen among hundreds or even thousands, which leads the group and protects them from enemies by destracting the predator from the group. And within a shoal of discus, as in many other groups of animals, it is the strongest, fastest, and often largest of the company. And usually the most beautiful.

The Lago and Rio Manacapuru have been fished since my discovery of the Royal Blues *(see page 68)*. During the collecting season (August-November) *disqueiros* bring specimens to Manaus which are often exported as Purus discus. As far as the layman is concerned they are virtually indistinguishable as the markings of the dominant individuals are similar. In particular, the anal fin markings are almost identical (for further details see Bleher & Göbel, 1992: *DISCUS)*.

However, to the present day discus from the Lago Grande de Manacapuru have rarely been collected. My old friend Anildo Macedo from Manaus, who for years went fishing with me in the Amazon region and sent fishes to Frankfurt once or twice weekly over a period of 20 years, said: "...*você esta doido, para ir no Lago Grande de Manacapuru precisa muito peito*." (...you're crazy, it takes guts to get to the Lago Grande de Manacapuru.) But that never stopped me. From the 1960s to the beginning of the 1980s I was repeatedly able to find a number, albeit small, of fabulous specimens there. And only individuals from there had these most attractive of colour forms *(see page 69)*. "Had", because in recent years no-one has gone that far upstream (the costs are enormous as there are no roads), or no more have been caught. It was here too that I found a variant *(see page 69, photo 8)* which otherwise occurs only in the Trombetas. Only these two have lines running totally straight, almost horizontally, across the body to the outermost tip of the anal fin.

Now, before we proceed to the Rio Negro region, a few words on habitats, spawning behaviour, feeding, syntopic fishes, and water parameters in this region, on the following two pages.

The road from Manaus to Manacapuru is metalled and nowadays surrounded by *fazendas* on both sides. 1. There is now also a dirt road from Manacapuru to Novo Airão (on the Rio Negro), along which everything has likewise been burned or clear-felled for cattle pasture. Erosion everywhere. 2. The *igarapés* have suffered too, and all that remains is a hot, lifeless soup – good enough for cattle, but death to all fishes. 3. Manacapuru's floating tourist hotel terminal *(see sign)* and the Solimões in the background. 4. The harbour at Manacapuru with the water rising.

On this page I have tried to give a small glimpse of the habitats explored on various trips to Manacapuru: 1. Very typical discus habitat in the Lago Manacapuru. 2. Typical variant from the lake, fresh from the net and hence slightly scratched and dark-coloured (fright coloration). 3. *Mesonauta* sp., caught here along with discus. 4. The natural food of discus (seeds and fruits) in the *igapós* at high water (spawning time – photographed in the natural habitat). 5. Probably *Amazonsprattus scintilla,* a member of the Engraulidae, which Roberts first described in 1984 from the Jufaris, a blackwater biotope to the north of here. I have only ever caught them at the time when the water is rising. 6. A typical so-called brown discus *(S. haraldi)* from the Rio Manacapuru at Capella. 7. *Acaronia nassa* from the discus habitat in the Lago. 8. A splendid *Apistogramma* sp. *(agassizii* group) that I found in the Igarapé Macumerí. 9. Discus variants from the Lago Grande de Manacapuru in the net (here croped). Note the upper, juvenile specimen, am individual approaching Royal Blue but the stripes will undoubtedly never extend all-over (see also text, *p. 68).* 10. *Pomacea papyracea,* a hand-sized snail from the *igapó.* 11-13. In the Manacapuru region parent discus lay their eggs in the cracks in these trees in the *igapós* (photographed here during the dry season) and then guard them. In this region this occurs from December to March (photo 13 shows one of the cracks). 14. The water has already risen in the *igapós* and the fishes are spawning, but this is impossible to photograph because the water is cloudy (15) and they flee as soon as you approach, faster than the eye can see.

I measured two sets of water parameters here at spawning time, at Macumerí and Doema. They were very similar to other collecting sites such as Capella, Macuaçú, and Rosario. The parameters were at the mouth of the Macumerí on 02.02.2004 at 8.00 hours: pH 6.49 in 1 m of depth; conductivity 80 µS/cm; daytime temperature (air): 30 °C; at the water's surface: 29.8 °C; in 1 m of depth: 29.2 °C; oxygen 4.33 mg/l. Biotope: spiny palms, roots, fallen trees and branches. No aquatic vegetation but lots of green algae (see left); water slightly cloudy (usually clear to tea-coloured); little movement. Bottom: leaves and sand. At Doema on 02.02.2004 at 9.00 hours: pH 6.77 in 1 m of depth; conductivity 77 µS/cm; daytime temperature (air): 31 °C; at the water's surface: 30.2 °C. in 1 m of depth: 30.2 °C; oxygen 4.54 mg/l. Biotope: roots, fallen trees and tree-trunks. No aquatic vegetation; large amounts of green algae; water rising and moving slightly.

5. Rio Negro region; Rio Branco region.

Rio Negro Region: No other river has been discussed so much, no other river is mentioned so often, and no other river on our planet contains such an enormous mass of black water. The natural spectacle of the confluence of the muddy Amaru Mayu (snake mother of the Earth, as the Amazon was originally called by the Manaús tribe) with the Rio Negro (as Orellana christened it), is one of the most documented natural wonders on our planet *(right)*.

It is truly a mighty river, not only on account of its length of around 2,550 km, but even more so because of the gigantic mass of water it conveys, which constitutes 13-14% of the total volume of the Amazon (it is the most water-rich river in the Amazon basin after the Madeira). And on account of its width, which can measure up to 80 km in some places during the high-water period – wider than any other river anywhere. Its basin is larger than France and half of Germany combined (around 700,000 km^2).

The Rio Negro rises in Colombia, where it is known as the Guainía, before joining with the Casiquiare on the border with Venezuela. But there are no discus so far upstream. The distribution of the Heckel discus *(S. discus)*, the only species that lives in its black water, begins well downstream of São Gabriel da Cachoeira. But more of that anon.

On the left hand side of the Rio Negro mouth region lies Manaus, the capital of the state of Amazonas and today the jungle metropolis in the middle of the largest forest on Earth. However there was also a sort of metropolis or centre here much earlier, long before the arrival of the white man, as when Orellana landed here on the 5th June 1542 he found a large settlement with a central square and a gigantic wooden sculpture. The latter was a city in itself, with gates, towers, and windows, and stood on two large jaguar feet carved in relief. Between the huge jaguar figures (which Orellana called lions) there was an opening into which the *chicha* drink was placed in honour of the sun god. A native supposedly explained that the sculpture was a tribute to the female rulers (Amazons), whom they also honoured with colourful *ara* and parrot feathers. In another square stood the ceremonial house of the shamans, decorated with feathers of the same type and festive sacrificial and dance costumes. Today there are no longer any shamans in Manaus (only *macumbeiros* – kind of voodoo), but the belief in the Amazons remains deep-rooted. It is impossible to get away from the name of the myth of the women with no right-hand breast.

The Manaús (the aboriginals – also called Manaós by some people), were astonished when in 1669 the Europeans erected a *forte* (fortress) next to their settlement. It was called the Forte São José da Barra do Rio Negro and was intended to keep out the constantly intruding Dutch from Surinam. During its construction the legendary Manaú leader Ajuricaba resisted enslavement and chose to die rather than lose his freedom.

Until the middle of the 18th century the major part of Amazonia was actually in Spanish hands. The Spaniards and Portuguese were both seeking for treasure and spices. Not until the Tratado de Madri of 1750 (were they devided South America) did the region officially fall under the dominion of the Portuguese. But until

Top: The world-famous natural spectacle, the *encontro das águas*, where the black Rio Negro meets the Amazon. Manaus can also be seen *(right)* and part of the huge island, the Ilha do Careiro, with *fazendas (left)*. *Centre:* Nowadays large *lanchas* are converted to transport discus. *Above:* In 1974 they were a lot smaller.

1. The Ilha do Careiro with its great Lago do Rei is washed by the Amazon just downstream of the mouth of the Rio Negro. During the dry season the *lago* is filled with clear water, but during the rainy season it is inundated by the sediment-rich waters of the Amazon.
2. The Rio Negro water can be seen flowing for many kilometres along the left-hand side of the Amazon. The photo also shows how far Manaus has spread out along the Amazon.
3. Since the 1970s the colonial buildings have been disappearing into the sea of skyscrapers.
4. At the end of the 1970s and in the 1980s gigantic discus transportation expanded greatly. For example, the exporter Sardis had 16 m long *lanchas* constructed with space for 15,000 discus in the hold. 5. The *lancha* in Werner Herzogs film *Fitzcaralldo*. 6. 1983: Sorting my discus at the harbour in Manaus.

the 19th century the fort remained surrounded by indian settlements and the area roundabout was known as Lugar da Barra, or just Barra. From 1832 onwards it was called Vila da Barra, until 16 years later, on the 24th October 1848, it became Cidade da Barra de São José do Rio Negro. Not until the 4th September 1856 was the name changed to Cidade de Manaós.

The intention was to honour the aboriginals, whom the Franciscans called Manaós, but their correct name was Manaú and signifies "God's mother". It is often spelled variously Manou, Manau, Manao, Manaó, Manaha, Manave, Macnal, Manouh, Manouâ, and also Manáos. The name originates from the Aruaque group of languages, and interestingly the spellings *manouh, manou, manu, and mani* are also abbreviations of the Hebrew word *manouchyaka* and its variants. Furthermore, an scholar from Manaus translates the Tupí word *houcha* as "god of the indians".

Manaus with a "u" appears for the first time in a journal of 1862, in an item by Antonio David Vasconcellos Canavarro on the problem of the *cólera-morbus* (deadly cholera), but it was the *governador*, the poet and professor Álvaro Botelho Maia, who first made it official on the 19th March 1937. The former Barra had finally become Manaus.

A certain amount has already been written about the Amazon metropolis in Chapter 1, but there are a few additional interesting stories about the place that all discus discoverers (and researchers) have visited – or passed through – that I would like to relate here: Towards the end of the 19th century, at the time of the rubber boom, the people here had very clear views and exerted the greatest care regarding expansion and the planting and felling of trees. Green areas were required and the forest had to be preserved (something that does hardly apply today, except on paper and in the media). The legal code of 1896 established that the felling of fruit trees, valuable hardwoods, and construction timber was forbidden and punishable by law. The *governador*, Constantino Nery, declared: *"...a possibilidade de irem se tornando cada vez mais pobres os mananciais que fornecem água para o abastecimento de Manaus, em virtude do corte de madeira pouco moderado para uso de particulares e fabricação de carvão, lembra a necessidade de uma realidade de uma medida de proteção às mattas que cobrem os mananciais"*. (...that there was a possibility that uncontrolled wood-cutting could lead to the impoverishment of the waterways and the town...) For this reason thousands of European trees were imported and planted around the town for its beautification *(below)*. People were environmentally aware and looked to the future. They built the first university in the country, and the town was the second

At the time of the rubber boom only the best was good enough. The most elegant shops and clothes (1); an opera-house intended to outdo La Scala in Milan (3); painted ceiling and rigid curtain showing the *encontro das águas* (2).

to have an electricity supply and a tramway. The latter was authorised in 1882 to provide public transport and was described as *"carros americanos sobre trilhos"* (American cars on rails) or *"...railways sobre trilhos de sistema Bourgois para carga de passageiros"* (...public trains on rails for the carriage of passengers). (All these innovations vanished with the rupper trade collapse, the trams last of all, in 1957.) People had so much money that the wealthy thought it amusing to light their cigars with notes. The finest shops and cafés were built, and, of course, a unparalleled opera house was an essential. And thus the Teatro Amazonas was inaugurated in 1896 with a performance of *Don Quixote*. The neoclassical structure took 17 years to build and the paintings, seating (for 700) and other furniture, and the parquet floor comprising around 12,000 pieces of hardwood, are still *in situ* in their original condition. Nowadays, following renovations in recent years, there are frequent performances, with an opera festival every May and a film festival in December.

During my researches into the Teatro Amazonas I came across a few interesting details, for example there was Crispim do Amaral, a Brazilian by birth, who had studied at the Academia de San Lucas in Italy. The academy (nowadays a museum of art in Rome) was already world-famous as the university of painters by the Middle Ages, and was where people like Titian, Moretto, Bronzino, Sebastian del Piombo, Rubens, Van Dyck, Guercino, Ribera, Raphael, and many others studied. But Crispim was not only a talented painter, but also a musician, actor (including at the Comédie Française), caricaturist, and writer. He spent three years in a Parisian prison (for publishing a political tract) before coming to Manaus, where in 1883 he was responsible for almost all the artwork and other decor, the interior decoration, and the construction of the opera-house facade. One of his glorious paintings can be seen on the ceiling of the auditorium, and another one on the specially constructed (for Crispim's painting) rigid stage curtain, the *encontro das águas* – *where* the Rio Negro meets the Amazon *(see page 356)*. It shows Yara, the mother of the waters, and two groups (3). Only the finest was imported, including bath-tubs (5), and Art-Nouveau houses were erected, like that of the German rubber baron Scholz, nowadays a cultural centre (6). *1965 photo:* The Hotel Amazon, the first skyscraper in Manaus (4).

Photos from around 1900: The second tramway in Brazil was opened in Manaus in 1896 (1). The planting of European trees (2) – in the middle of the largest forest on Earth – was typical of the country. The Teatro Amazon (7) brought in international

deities, one with red hair (= the Rio Negro) and the other blonde (= the Amazon) in the waters. Crispim was a genius, who was absorbed in his work and apparently regarded the opera house as his life's work, although he also brought into being the magazine *O Malho* (= The Wicked), famous all over Brazil. After the completion of the opera house he remained in Manaus, but disappeared (died?) very suddenly at the age of barely 53 years.

The ironwork in the Teatro Amazonas – stairs, balustrades, balconies, seats, columns, statues (interior and exterior), chairs and tables, and iron girders, as well as the framework of the unique cupola – was made in Paris by Koch Frères, at Rue Martel 6. Only the steel girders were made in Glasgow. The bronze sculptures came from Rome, and the numerous mirrors and glasswork were created in Venice (by Murano) and Paris (by Lalique). But only the frames of the mirrors remain, and none of the numerous vases imported from France, China, and Japan survive. Likewise the made-to-measure damask drapes, with threads of real gold, for the governor's box, as well as the Persian carpets, velvet hangings, silks and embroideries from Persia, are history. The original electrical installation has long since been replaced. Back then (March 1896) the *engenheiro* Vicente José de Miranda received R$ 262,563,840 (= US$ 45,584) for the work and purchased all the materials in New York.

The official opening celebrations were scheduled for the anniversary of the independence of the Província do Grão-Pará (5th September), but it wasn't until the 31st December 1896 that the Companhia Lírica, who had travelled from Italy, made their appearance under the *maestro* Joaquim de Carvalho Franco. At the turn of the year there was a performance of La Gioconda by Amilcar Ponchielli. From 1896 to 1937 the opera-house management entered into agreements with the best-known lyricists and performers at home and overseas, for example – to name but a few – the Companhia Lírica Italiana, the Companhia Tomba Italiana de Óperas, Operetas e Óperas Cômicas (1897-1906), the Companhia Dramática Dias Braga, from Portugal (1897 and 1900), the Spanish Companhia de Zarzuelas (1898 and 1905), the Companhia de Operetas Italiana Coniglio & Valla (1898/1899), and the Grande Companhia Dramática Italiana (1899 and 1902). Then there was the Companhia Dramática dos Pigmeus from Rio Grande do Norte, with real dwarves – none of the performers stood more than 90 cm tall. The Companhia Lírica Francesa (1906 and 1907) was one of the last from overseas, as from 1910 onwards, with the beginning of the rubber crisis, there were just a few performances by local groups. During the First World War everything came to a halt, and the final performance was by the Companhia Álvaro Pires, in 1937. And at the end of this "performance" I mustn't forget to mention that world-famous opera singers such as Enrico Caruso and Sarah Bemard never appeared at the Teatro Amazon. These are inventions by the media, for example Caruso in Werner Herzog's film *Fitzcaralldo*.

However, not only the opera house, but also the Rio Negro Palace should be mentioned. A German "rubber baron", Waldemar Scholz, had this splendid residence built for himself at the end of the 19th century, and in 1918, after the collapse, it was converted into an administrative building. Nowadays it is the cultural centre of Manaus, with a permanent exhibition on the life of the *caboclos ribeirinhos* and the domestic culture of the indigenous people, including their villages, *guaraná* harvesting, and methods of processing manioc and sugar-cane.

Between 1913 and 1920 Manaus was just a shadow of its former self. The lights went out – literally! Rubber exportation, which had increased from 31 (1827) to 39,000 tonnes (1913), collapsed. The Brazilians were asking 389 £ per tonne while in the same year (1913) a much larger quantity (47,000 tonnes) came from Asian plantations for 101 £ per tonne. (Around 1931 more than 700,000 tonnes were exported from Asia at a price of only 27 £ per tonne.)

Later on the president, Getúlio Vargas, said that the Brazilians should bother with only three things: *"aço, petróleo e Amazônia"* (steel, petroleum, and the Amazon). This strategy also moved a group of officers to found the *Suframa* in the middle of the 20th century. An organ that was to bring the isolated state back to the forefront, and eventually, just when I started importing discus from there every week during the short collecting period – the tax-free-trade zone was brought into being in 1967. Nowadays Manaus as once again become a world-famous metropolis and is wealthier than back then, even without the rubber. There are now around 110 hotels (23 in the 5-star category), 12 jungle lodges (Amazon Village, Amazon Ecopark, Ariaú, and the floating Jungle Palace in the immediate vicinity, which also hold festivals – photos of the last two jungle lodges *page 360 – top),* 30 tour operators, 12 car-hire companies, four tourist information bureaux, 17

Left-hand page: The coat of arms of the state of Amazonas *(centre)* shows the confluence of the Rio Negro and the Solimões (= Amazon); the eclipse represents the sky of Brazil; the stars stand for peace and progress; the white ship in front of the confluence is a symbol of the loyalty of Amazonia to the republic; the green field the jungle, and the feathers and arrows on it modern civilisation (with jungle?); framed with symbols of navigation and with the inscriptions 22nd June 1832 (when Amazonas became independent) and 21st November 1889 (when it joined the republic). Above everything stands the Amazon eagle, a symbol of size and strength. To the right in the coat of arms are the symbols of industry and to the left those of the tradesmen and agriculture. The drawing *(below)* represents the last Manaú, from the extinct tribe.
This page: 1. Every morning hundreds of *voadeiras* carrying fresh-caught fish arrive in the harbour at Manaus.
2. After his marriage to his wife Robine, Willy Schwartz established himself in this house, at Miranda Leão 106 and traded in animals. From the 1960s in fishes as well. In 1963, as Aquario Rio Negro, he was really involved with discus (from 1965 with Aquarium Rio).
3-4. When the Rio Negro rises higher than normal, Manaus is underwater, including toilets (4).
5-6. Erio Peretti (5), an Italian who started with fishes in São Paulo in 1952, opened his Amazon Aquarium (an export station) in 1962 in Manaus, and thereafter continually expanded. He built a fleet of 10 large boats (20 x 4.5 m) for transporting discus and extensive holding facilities (6) in the 1970s-80s.
7. The third major collector and exporter of discus and other aquarium fishes was my friend Anildo, with whom I travelled over almost a 20 year period and from whom I received fishes weekly. He built holding facilities on the Negro.
8. Nowadays Asher Benzaken, who followed in the footsteps of Schwartz, expanded to be the largest and most modern ornamental fish and discus export business in South America. There are 1,600 vats with 2000 l each, thousands of aquariums and fish trays, and 30 workers..

banks, and 28 consulates, and every Sunday between 07.00 and 14.00 hours the FAPA (Feira de Artesanato e Produtos do Amazon) takes place in the Avenida Eduardo Ribeiro, in the city centre. In 62 of the recognised restaurants in the municipality one can enjoy unique specialities. Fish is a must, whether in soup *(calderada)* or the way the *caboclos* often prepare it, as *peixe moqueado*, where the fresh-caught fish is rubbed with *pimenta* (pepper) and *sal* (salt), wrapped in a banana leaf, and the whole buried in the ground. A fire is then lit on top, and left to burn until it goes out. Then the ashes are scraped to one side and the cooked fish dug up. It tastes quite incredible.

But back to the Rio Negro. Orellana was the first white man to discover it, and he also gave it the name. (The local indigenous peoples called it *Iquiari* and *Ipixuna* (both signify blackwater).) And a Brazilian of Portuguese descent, the naturalist Alexandre Rodrigues Ferreira (1756-1815), was the first to explore it ichthyologically upstream to Barcelos during his *Viagem philosophica*, the expedition to the *Capitanias do Pará, Rio Negro, Mato Grosso e Cuyabá* between 1783 and 1792. Perhaps he was also the first white man to see a discus, but that cannot now be proven. The majority of his natural history collections, including the fishes, were carted off to Paris from the Cabinet d´ Ajuda in Lisbon by the French naturalist Étienne Geoffroy Saint-Hilaire, following Napoleon's invasion and occupation of Portugal. There were supposedly 139 fishes, but only 97 arrived in Paris. Eighteen marine specimens (of species already known) were left behind and 24 were never seen again. In Paris his new species were described by Cuvier around 1816 and subsequently by others as well. They included the largest freshwater fish in the world: *Sudis gigas* (= *Arapaima gigas*), which Ferreira had already described in 1787 as *Paranenses pirarucu* but without publishing his description.

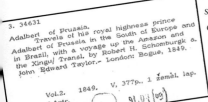

Its common name *pirarucu* comes from the language of the Grão-Pará indians who helped him with his collecting around Santarém (and later at Barcelos), and he wanted to honour them. They called the species *pirá-urucú*. *Pirá* = fish and *urucú* is a tree *(Bixa orellana)* that produces red seeds *(see below)*. (In the adult fish it's rear is red = fish red.) Indigenous peoples have used them for thousands of years for body-painting. The Aztecs (in Náhuatl) knew them as *achiote* and added these fat-rich seeds to their drinking chocolate, as it was reminiscent of blood and had a cult significance. In Europe too the seeds are used as a chocolate additive – but nowadays mainly to colour butter and cheese. Note that the French have corrupted *urucú* to *le roucou* and the Germans call it *Orleanstrauch* (= Orleans shrub), as it was originally obtained from the French town. And the Anglo-Saxons call the tree the lipstick tree. But the term most often used is *annatto*.

The "thief" Geoffroy was honoured in the names of the finest mammals collected by Ferreira: *Inia geoffrensis* and *Saguinus geoffroyi* (the Amazon dolphin and Geoffroy's marmoset), while Ferreira didn't receive any such tribute until 150 years after his death, with *Osteoglossum ferreirai*.

After him came a Prussian. This renowned German performed scientific researches here around 10 years later during his South America expedition (1799-1804), and collected in parts of the Rio Negro, making his way downstream from the Orinoco.

Humboldt

Adalbert

Bates

Wallace

Spruce

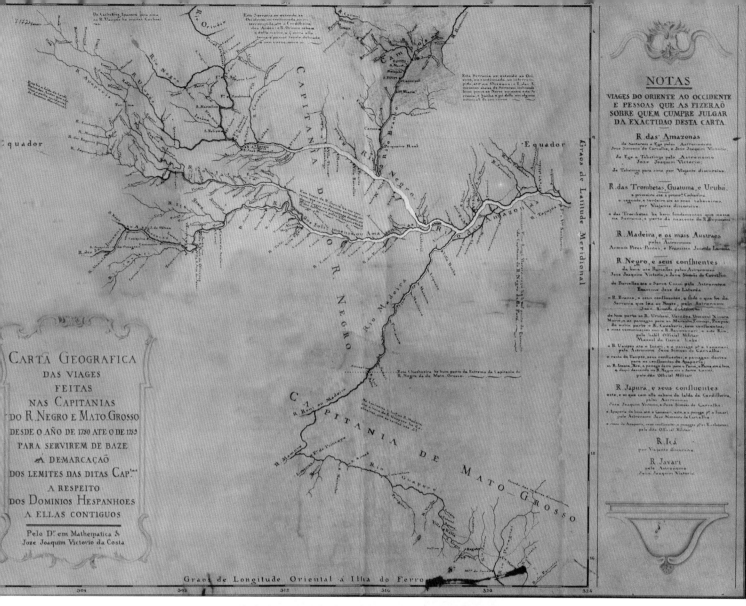

This page: The map from 1780 shows the extent of the Capitania Rio Negro and Mato Grosso at that time. La Condamine was the first (around 1740) to discover that the Rio Negro is connected to the Orinoco – he thought via a channel. Humboldt established in 1800 that it was in fact the Río Casiquiare. A Venezuelan blackwater stream, which is a headwater of the white Orinoco and feeds the Rio Negro via a bifurcation (= where a flowing water branches in several directions where there is a watershed in a flat region). The map was an essential for the second Rio Negro explorer, Alexandre R. Ferreira *(left)*, who first collected fishes there. He studied nature on behalf of Portugal, checking the boundaries between Spanish and Portuguese territory and looking for mining opportunities en route. The fishes he collected were some of the first recorded by science. *Left-hand page below:* A sequence of Rio Negro explorers *(from left to right)*: Humboldt; Adalbert; Bates; Wallace; Spruce (Natterer, Schomburgk, and Edwards not shown); *above:* two lodges in the Manaus area.

Today his name adorns the maps of countries and regions that he never travelled, including a crater on the moon. He appears on the labels of olive oil, wine, and beer. Innumerable (in fact actually counted by science) roads, villages, towns and cities, mountain chains and peaks (I was recently on the "Gumbolt" peak in Kyrgyzstan), funicular railways, parks and nature reserves, ocean currents and bays, institutes, schools, universities, pharmacies, even filling stations and pens bear his name, and Charles Darwin described him as "the greatest scientific traveller who ever lived". I refer to Alexander von Humboldt (1769-1859), or more correctly, Friedrich Wilhelm Heinrich Alexander, Freiherr (Baron) von Humboldt.

His South American expedition with Aimé Bonplant was the most updated expedition of its day, recording latitude, longitude, and contours and making maps, and collecting 60,000 plants (6,300 unknown), fishes (new species), and much more. He introduced plant geography and was the first to recognise correctly the river system of the Orinoco and Rio Negro. As regards the fishes, his observations on the electric eel *Gymnotus electricus* (= *Electrophorus electricus*) and the piranha *Serrasalmus caribe* (which he termed *poisson caribe* and today is *Pygocentrus cariba*) are particularly noteworthy. As well as his book *Ansichten der Natur* (1808), in which he combines descriptions of nature with scientific explanations, and his work on fishes (in 1811 he described a number of species by himself, and others later (1833) with Valenciennes in *Recherches sur les poissons fluviatiles de l'Améric équinoxiale*, including around 30 from the Rio Negro and the Orinoco), it was his magnum opus, *Kosmos – Entwurf einer physischen Weltbeschreibung*, in five volumes (1845-62), that gained the widest distribution.

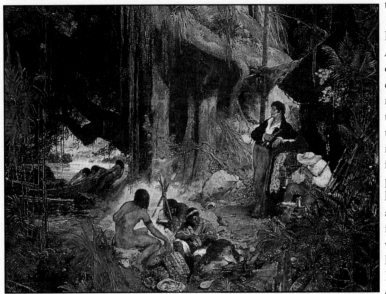

Many branches of natural science regard Humboldt as their founder, *inter alia* physical geography, climatology, and the study of high mountain ranges. His researches in plant geography, the measurement of isotherms, and his contributions to the study of the earth's magnetism are of continuing importance; his works in the field of terrestrial science have gone down in scientific history. In his efforts to disseminate scientific knowledge he remained true to the precept, *Mit Wissen kommt das Denken und mit dem Denken der Ernst und die Kraft in die Menge* (With knowledge comes thought, and with thought quantities of determination and strength). And he had the determination and strength to push through a law in 1857, to the effect that *"jeder Sklave,* der *Preußen betritt, frei ist"* (any slave that enters Prussia becomes a free man). The South American freedom-fighter Simon Bolivar said of him, *"Alexander von Humboldt hat Amerika mehr Wohltaten erwiesen als alle seine Eroberer"*.(Alexander von Humboldt did more good deeds in America than all its conquerors.) That is undoubtedly true, and not only there.

However, the best known explorers of the Rio Negro region were Wallace *(right)* and Bates, both Englishmen, who landed together in Pará (Belém) in 1848. But the Prussian Prince Heinrich Wilhelm Adalbert (1811-1873) is often forgotten; in 1842 he collected in the Rio Negro mouth region on his way upriver. The prince seems to have had a lot in common with myself, apart from nationality: he appears to have had the same lust for exploring, travel, and collecting; he was the first to explore the Xingú (I was the first to explore its largest tributary, the Rio Iriri); in 1866 he was honoured in France *Pour de Mérite* for his work on natural science (my humble self likewise more than 100 years later); Adalbert had "blue blood" in his veins, his grandfather was Friedrich Wilhelm II, King of Prussia (my grandmother also had noble blood). In 1849 the work *Voyage up the Amazon and its Tributary the Xingú* was published in England, translated from the never-published German text of the prince by Sir Robert H. Schomburgk and John E. Taylor. Adalbert had just had a few copies made for his friends. One of the translators, the Quaker Taylor, a former editor of the *Manchester Guardian*, died five years before the work was published, and Schomburgk was by then the British Consul General in the Dominican Republic.

Robert Herrmann Schomburgk (1804-65) was born in Freyburg an der Unstrut and the 200th anniversary of his birth was

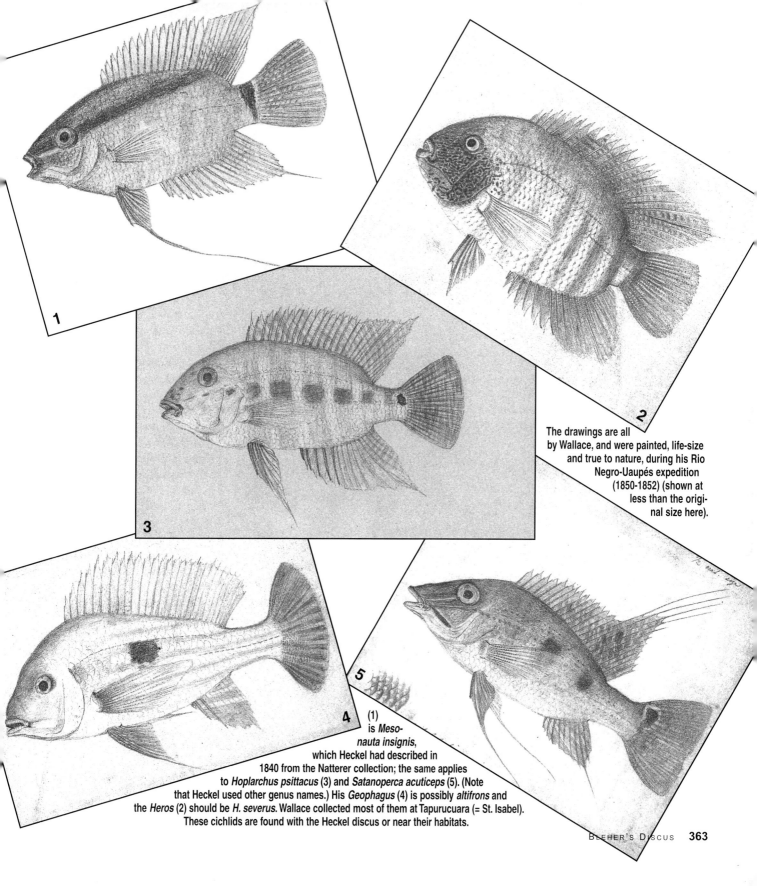

The drawings are all by Wallace, and were painted, life-size and true to nature, during his Rio Negro-Uaupés expedition (1850-1852) (shown at less than the original size here).

(1) is *Mesonauta insignis*, which Heckel had described in 1840 from the Natterer collection; the same applies to *Hoplarchus psittacus* (3) and *Satanoperca acuticeps* (5). (Note that Heckel used other genus names.) His *Geophagus* (4) is possibly *altifrons* and the *Heros* (2) should be *H. severus*. Wallace collected most of them at Tapurucuara (= St. Isabel). These cichlids are found with the Heckel discus or near their habitats.

celebrated on the 5th June 2004 by simultaneous exhibitions in Germany and Guyana. The German geographer, explorer, and naturalist explored what was then British Guiana first for the Royal Geographical Society (1835-39) and later again (1841-43) with his brother Richard – but this time for the British crown – and established boundaries, for which he was knighted in 1844. During the last part of his first expedition Schomburgk covered an almost incredible distance in only around seven months. He negotiated with Amerindian tribes and the Portuguese, measured, marked, collected, sketched, and studied. In September 1837 he paddled from Georgetown far up the Essequibo, then down the Rupununi to Pirara (a former indian settlement, which lies on the current border of Guyana with Brazil). From there overland to the Rio Tacutu, then downriver to the fort of São Joaquim (now in Brazil) which still did stand at that time. Up the Rio Uraricoera, full of rapids and waterfalls, hoping to find the source of the Orinoco in Venezuela but missing it by a few kilometres.

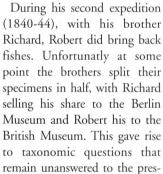

The German ethnographer Theodor Koch Grünberg likewise sought the source of the Orinoco 83 years later, in the company of the first Amazon film-maker, Silvino Santos, and the American explorer Hamilton Rice. They even had an aquaplane. But during this expedition Grünberg died of malaria and they found only the Uraricoera and Parime headwaters. Like Schomburgk they got lost, but Schomburgk made it to the Rio Caroni and then reached Esmeralda on the upper Orinoco, where Humboldt had also been 39 years previously. He made his way up the Orinoco and through the Casiquiare to the Rio Negro, paddled via Barcelos up the Rio Branco, and from its upper course back up the Tacutu, to return the same way that he came to what is today the capital of Guyana. It was also there in British Guiana that on the 1st January 1937, on the Berbice river, he discovered *Victoria regia* (now *V. amazonica*), the huge waterlily which adorns the national coat-of-arms of Guyana although the Brazilians regard it as "their" plant.

Around 80 fish species were subsequently published in 1840, in the two- volume *Fishes of Guiana,* illustrated with 60 fantastic colour engravings, immortalising fishes that had never before been shown in colour *(centre,* see also pages 365 & 381). He had caught more than 40 species in the Rio Negro, the Rio Padauiri and the Rio Branco but didn't bring any specimens back from his travels.

After his return Schomburgk claimed that during his measurements in the then disputed border region between British Guiana, Brazil, and Venezuela he had discovered the *El Dorado* sought in vain by Ralegh (Chapter 1). He wrote about it in an article on Ralegh's expeditions of around 250 years previously. The English version was translated into German by his brother Otto, and *Reisen in Guiana und am Orinoco in den Jahren 1835-39* (Travels in Guiana and on the Orinoco during 1835-39) was published in Leipzig in 1841, with a foreword by Humboldt.

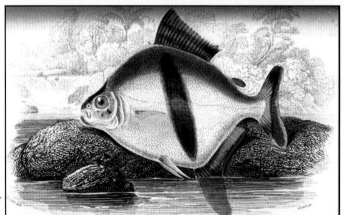

During his second expedition (1840-44), with his brother Richard, Robert did bring back fishes. Unfortunatly at some point the brothers split their specimens in half, with Richard selling his share to the Berlin Museum and Robert his to the British Museum. This gave rise to taxonomic questions that remain unanswered to the present day, as no specimen can be atributed to any exact locality.

Robert, who died young, was honoured not only with *Myleus schomburgkii (centre)* from the Rio Negro (a species known in the aquarium hobby), but also with the orchid genus *Schomburgkia,* as well as the by now extinct Siamese deer *Cervus schomburgki* Blyth, 1863 (on the IUCN *Red List of Threatened Animals* since 1996 – but extinct since 1938). During his final years (1857-64) he was British Consul General in Bangkok.

And it was another Consul General, again a German, Georg Heinrich von Langsdorff (1774-1852), who represented the

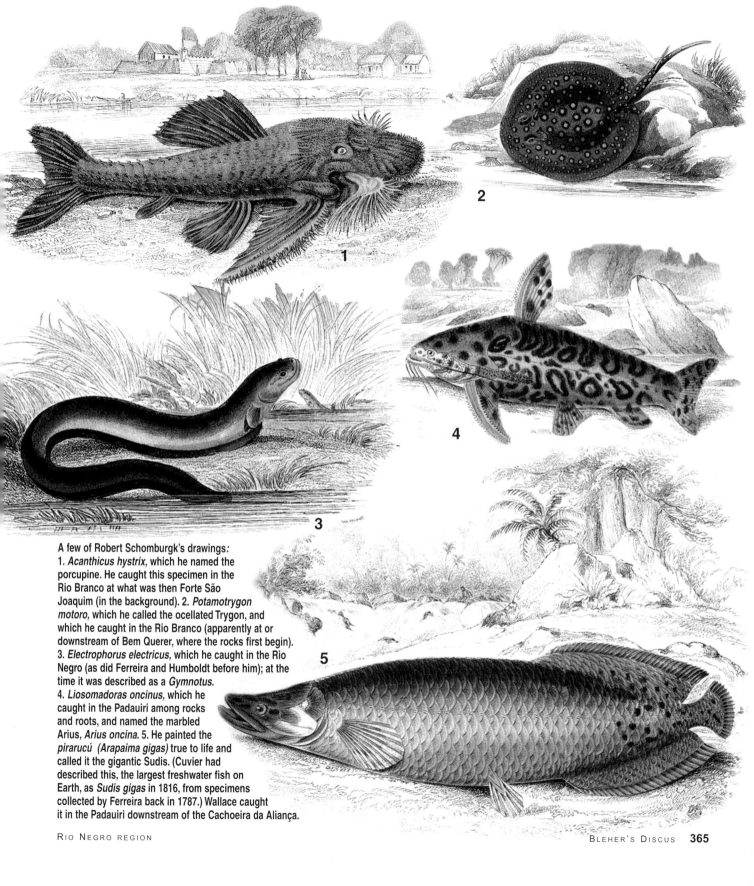

A few of Robert Schomburgk's drawings:
1. *Acanthicus hystrix,* which he named the porcupine. He caught this specimen in the Rio Branco at what was then Forte São Joaquim (in the background). 2. *Potamotrygon motoro,* which he called the ocellated Trygon, and which he caught in the Rio Branco (apparently at or downstream of Bem Querer, where the rocks first begin). 3. *Electrophorus electricus,* which he caught in the Rio Negro (as did Ferreira and Humboldt before him); at the time it was described as a *Gymnotus.* 4. *Liosomadoras oncinus,* which he caught in the Padauiri among rocks and roots, and named the marbled Arius, *Arius oncina*. 5. He painted the *pirarucú (Arapaima gigas)* true to life and called it the gigantic Sudis. (Cuvier had described this, the largest freshwater fish on Earth, as *Sudis gigas* in 1816, from specimens collected by Ferreira back in 1787.) Wallace caught it in the Padauiri downstream of the Cachoeira da Aliança.

Russian Tsar in Rio de Janeiro from 1813 on. He was a learned physician and member of the scientific academy in St. Petersburg, who nourished an ambition for exploration and contributed to the study of natural history in Amazonia via an excellent biological as well as ethnological collection. His explorations from 1821 to 1829 have gone down in history as the Russian-Langsdorff Expedition. It was overshadowed by problems, by sickness and death among the participants and his collapse on reaching the upper course of the Rio Juruena. Three of the participants created fantastic paintings, but one of them, the French painter Adrien Taunay drowned in the Rio Guaporé in mysterious circumstances (like the Graf von Horn 150 years later – see Bleher, A. 2005). The second painter, the Italian Hercule Florence remained in Brazil, and the third, the German Johann Rugendas gave up.

Apart from the painters, he was accompanied by the German botanist Luís Riedel and the zoologist Christian Hasse, as well as the Russian astronomer Nestor Rubtsov. They travelled the Tietê river from São Paulo in very primitive canoes. In the Tietê, moreover, they caught the largest member of the characid family, S*almo dourado* (up to 120 cm TL), a specimen of which Valenciennes described in 1840 as *Salminus maxillosus* (it was thought to be the same species that Alexandre found in the Cuiabá and Cuvier named *Hydrocynus brasiliensis* in 1816). They penetrated into the Pantanal, through the Mato Grosso, down the Guaporé, downriver past the rapids into the Madeira and up the Amazon to Barra (at the mouth of the Negro). There they apparently collected only the *peixe-agulha*, which Ferreira also found and Cuvier described in 1816 as *Hydrocynus lucia* (now *Boulengerella*). From there they followed the Amazon to Pará (Belém).

Baron Langsdorff, better known as Gregori Ivanovitch, covered a similar distance to Natterer and like him achieved an extraordinary amount; but he didn't find any discus, though he did discover 32 species, some new, including the *pacu* in the Rio Cuiabá. He had this fish designed, but only in 1887 was it described by Holmberg as *Myletes mesapotamicus* (later placed in the genus *Piaractus*). In 1988 the three volumes of this expedition containing the splendid paintings of the three artists, including landscapes, people, plants and animals (including the fishes in vol. 3) was re-published in Portuguese.

There is another Rio Negro explorer about whom very little is known. His name was José Solano and around this time he led a natural history expedition up the Orinoco and thence into the Rio Negro drainage. I have been unable to find out anything about the fate of him or his his collections. But back to the Englishmen.

Henry Walter Bates (1825-1982) was, like Edwards an insect enthusiast, and was working as a book-keeper at a brewery when he met the naturalist Alfred Russel Wallace (1823-1913) and with no further ado decided to go with him to South America. Starting in Pará (Belém) they collected together in the Tocantins to Santarém; then Bates travelled on alone from Barra (Manaus) to Ega (Tefé) while Wallace explored up the Rio Negro. When, in 1859, the book-keeper returned to England he took with him more than 14,000 specimens (including 8,000 unknown species) and much more. In 1863 he published his book *A Naturalist on the River Amazon*. Wallace, by contrast, suffered a terrible malaria attack on the Uaupés, but was rescued by his fellow-countryman, the botanist Richard Spruce (1817-1893) (who was plant-hunting along the Rio Negro and Uaupés) and returned in 1852 (see Römer, U. *et al.* 1995: *Uaupés*). A year later he published a book, *A Narrative of Travels on the Amazon and Rio Negro*. His rescuer remained in the Amazon region until 1864. He mapped new rivers and collected seeds of the quinine or China-bark tree along the Río Pastaza and Río Puyo on behalf of the British Foreign Ministry. He became embroiled with native head-hunters but escaped and shipped out 100,000 young plants which were transported to India (where the production of quinine – for treating malaria – was to become the most important source of income). When he returned to England, with 30,000 botanical samples and a knowledge of 21 different native languages, he was physically a wreck. He published nothing further apart from an important work on South American mosses. Long after Spruce was dead this was rectified by his friend Wallace, who in 1908 published *Notes of a Botanist on the Amazon and Andes*. Which also mentions Spruce's Negro-Uaupés-Casiquiare researches (on fishes).

1. The Rio Negro, still in the dry season (January), at the Archipelago de Mariuá. The water colour is more like amber than black over the white sand. Ferreira mentions this, only he has various versions of where the colour came from. However, one was close to the truth: a native told him that it came from the decomposition of the riparian vegetation (they still say this today). While digging in the fine sand in blackwaters, which generally flow over fine white sandy bottoms, I have come across immense layers of leaves and twigs like this one in middle of the Rio Negro *(left-hand page).* 2-3. Two shots of the immense Archipelago das Anavilhanas – said to be the largest archipelago in any river. 4. But the Mariuá archipelago may well be larger.

Schomburgk collected and drew only a few fish species in the Rio Negro, Wallace, by contrast, collected intensively and painted what was an almost incredible number for his day. A total of 212 species which were first published recently (2002) in the book *Peixes do Rio Negro/Fishes of the Rio Negro (see page 363)*. Unfortunately, during the return voyage his ship, the *Helen*, caught fire on the 6th August 1852 and the entire collection was burnt. He saved only a single metal box containing notes, sketches of palms, and the drawings of fishes, which he had reproduced very accurately. He also opined that there were undoubtedly 500 fish species in the Rio Negro. (Today more than 500 species are known.)

Now, neither Wallace nor any of the other collectors mentioned, no more than Cousteau during his monumental Amazon film series (1982), nor any of the authors of the to date most detailed work on the fish fauna of the Rio Negro, *Rio Negro, Rich Life in Poor Water* (Goulding, Carvalho & Ferreira, 1988), managed to find discus. Only Natterer, which shows how difficult it has always been to catch discus.

As well as his pictures, Wallace left behind numerous publications and provided the stimulus for the theory of evolution (he wrote to Darwin about it but was ignored see Bleher, H. 1993c).

To limnology, Amazonia, and ichthyology he bequeathed, albeit in brief, the distinction between the different types of water: "whitewater, clearwater (or bluewater) and blackwater rivers". Whitewater he attributed to the layers of sediment that are transported downstream from the Andes; clearwater he related to a combination of the lixiviated layers of stone in the rocky massifs of the headwaters and (rocky) sediment-poor watercourses; and blackwater he attributed to the decomposition of leaves and other vegetable material *(see page 366)* but he changed this in his autobiography.

On the other hand Humboldt who was surprised that blackwater and whitewater habitats often occur in the immediate vicinity of one another *(see page 375)*, was told by missionaries that the tea-coloured water could be traced back to the roots of certain plants, for example the sarsaparillas. The sarsaparillas belong to the lily family (formerly Liliaceae, now Asphodelaceae), and their genus, *Smilax*, contains about 200 species, found especially in moist areas in the tropics and sub-tropics. Two species grow in Europe, the black-fruited *S. nigra* and the violet-fruited *S. aspersa (see right-hand page)*. The roots of the sarsaparillas contain saponin and saponin-like substances, essential oils, starch, and a bitter resin (no wonder they colour water!). Interestingly, how-

In a typical discus habitat in the middle Rio Negro we also find the *acará-açú* plant (in this case *Licania stewardii*) favoured by the "King". The spiny branches prevent most of the predators from reaching the steeply shelving bank zone, where a large group of discus almost always lurks during the day. Let alone a human or mammal. It also serves as a source of food: flowers drop off, and with them often terrestrial insects (mainly ants); later on the seeds (inset). A perfect symbiose.

The sarsaparilla is a component of many proprietary blood-purification treatments. But it is much more. The number of medications on the Internet that contain (10) is almost boundless. Note the pointed spines along the stems (9); at high-water these provide discus with ideal protection as well as food.

Eschweilera tenuifolia (Lecythidaceae) grows in Rio Negro *igapós* (1), and is known locally as the *macacaricuia* (= monkey pods). Its hard seed pods (3) are cracked open by monkeys or *tambaquís (Colossoma macropomum)*, but the disintegrating tannin-rich seeds (4) form part of the Heckel discus diet. The same applies to another member of the nut-tree family *Lecythis barnebyi* (2). It grows on the *terra firme* and along its banks in the Rio Negro region. Its green (11) or ripe seeds (12) must first be cracked or softened. Not so its flowers (8), which are known as *macucu*. The fruits of the sarsaparilla (a collective term for numerous *Smilax* species) also form part of the discus diet. In Amazonia there are around 20 species which are often found at discus habitats, as on the Rio Içá for example (6), and on the Rio Negro. By the Rio Nhamundá they are called *jaranduba,* and elsewhere *japecanga, japucanha,* or *japicanga-miúda,* and *salsaparrilha* in medical jargon. The fruits are not only similar within the genus – including the European species *S. aspera* (5) – but also resemble those of the Amazonian *Myrciaria dubia* (7). Only the colour of the fruits is different and the so-called *camu-camu* bush (family Myrtaceae) grows also along the Rio Negro bank and in swampy regions.

ever, for thousands of years sarsaparilla species have not only provided discus (and others) with food, shelter, and the means to survive, but one species has even saved the lives of many people. This is how it came about: when, in the spring of 1493, Christopher Columbus the newly discovered islands of the West Indies, he had aboard the *Nina* not just a few *"indios"*, parrots, maize seeds and paprika, but also syphilis. His helmsman and several other sailors were carrying the pathogen (the bacterium *Treponema pallidum*) that causes the disease, previously unknown in the Old World, back to Europe. There it became widespread in just a few years (mainly via French soldiers, giving rise to the name *morbus gallicus* (= the French disease)). A cure was sought in vain until years later when one was found among the "indian" medicines.

The indians utilised a combined syphilis therapy which as a rule led to a cure as the disease affected them less severely than the Europeans. They took decoctions of sarsaparilla root (*Smilax regelii* as well as other species) in combination with a steam-bath and fasting. The steam-bath was targeted at the genitals. (According to US researches the pathogen dies in 30 minutes at 41°C.) Moreover, by around 1898 already 34 *Smilax* species were recognised as medicinal plants – and nowadays even more. In the Caribbean they even brew a wedding drink from *Smilax havanensis* and its "chaney root", which grows in Jamaica and Cuba.

However, as already mentioned, it was not only in these spiny liana-like plants that the origin of the water colour was to be found. Much research has been performed on the composition of blackwaters, but to the present day there is no definite answer. In the case of the Rio Negro it is known that the high content of humic and fluvo-acids are washed out by the rain from the already greatly lixiviated sandy soils of the *terra firme* in its drainage region. And that the colour comes from the roots of plants, from leaves and branches, and many other sources, but there are no satisfactory answers and even fewer to the question of why this phenomenon occurs in various geological regions worldwide.

Likewise very little is known about the natural diet of discus. But the following is of interest in this respect:

numerous plant families such as the Euphorbiaceae, Chrysobalanaceae, Lecythidaceae, Bombacaceae, Myrtaceae, Rubiaceae, Palmae, and Leguminosae are widespread in Amazonia. They are also found in the Rio Negro and the indundation zones (*igapós*). Their species contribute to the tea-coloured hue of the water, and also to the survival of many fish species and a whole chain of aquatic life-forms – their flowers and seeds (or the flesh of their fruits), their leaves and even terrestrial insect "passengers". This has been demonstrated by stomach-contents analyses in around 100 fish species – and undoubtedly there are many more *(see also Discus nutrition in the wild)*.

Munguba trees (*Pseodobombax munguba*, family Bombacaceae) grow at Novo Airão (1) in the Heckel discus habitat (2), and the flesh of their soft, fibrous seeds (3) is also a discus delicacy, as are the softened seeds and flowers (4) of *Hevea spruceana*. In addition the bark of the *seringa barriguda* (= *H. spruceana*) is sold as *catahua* as a treatment for respiratory problems. When brewed to make a tea it is also beneficial for stomach, kidney, and liver pains. This Heckel discus (5) is from the biotope shown in (2).

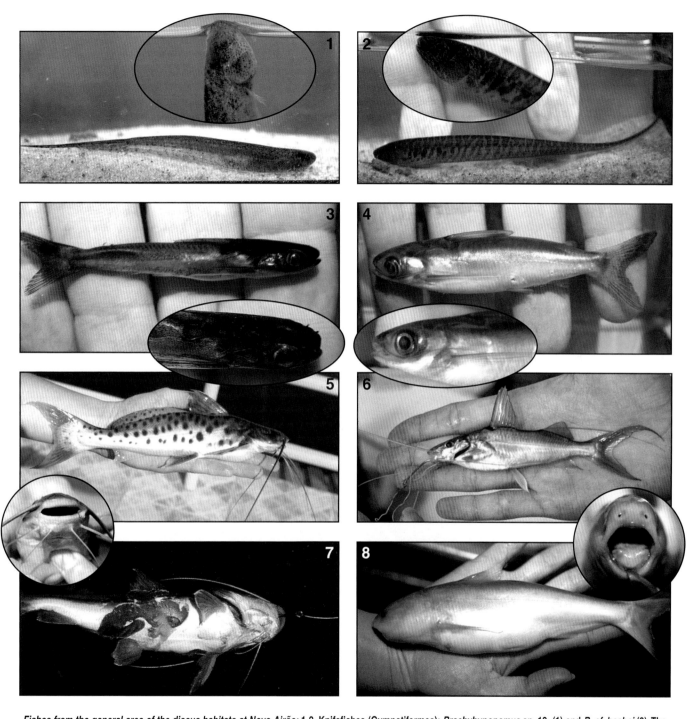

Fishes from the general area of the discus habitats at Novo Airão: 1-2. Knifefishes (Gymnotiformes): *Brachyhypopomus* sp. 12 (1) and *B. cf. beebei* (2). The two inset photos show how they take in atmospheric air. 3-4. Peculiar catfishes: *Centromochlus cf. existimatus* (3) and *C. aff. heckelii* (4), with two unusual, extremely long, filamentous barbels below the eyes *(inset photos)*. 5-7. Pimelodids: *Calophysus macropterus* (5) with its down-curved mouth *(inset)*; *Pimelodus* sp., metallic green; and a larger *Pimelodus* sp. being bitten by whale catfishes. 8. A wahle catfish, *Cetopsis coecutiens,* is an aggressive predator which uses its mouth *(inset)* to rip pieces of flesh from larger fishes (7). They often attack in groups.

The oft-cited *acará-açú* plants are widespread along the Rio Negro and its drainage region, in bank zones and places where discus habitats are found. They belong to one of the above-mentioned families (the Chrysobalanaceae) and at least three species are found here: *Licania hispida* and *L. undulata* in the upper course, and predominantly *L. stewardii* in the middle and lower Rio Negro region. I have found these and 10 similar looking species such as *L. angustata, L. ansisophylla, L. annae, L. celativenia, L. ferreirae, L. krukovii, L. oblongifolia, L. octandra* ssp. *grandiflora, L. stenocarpa*, and *L. teixeirae*, at various discus habitats in Amazonia, including in Peru and Colombia. If the water parameters and the biotope are right, then you can expect to find discus beneath them.

Before we leave Manaus and make our way upriver, a brief note on the existence of habitats near the largest jungle metropolis. In 2000 the city had 2,840,889 inhabitants, but since then the number has risen to well over three million. In 1874, when a fire broke out – started by a firework during the solstice celebrations – and completely destroyed the old fort constructed of wood, around 50,000 souls lived there; after the collapse there were barely 20,000; when I arrived in the 1960s there were still well below 100,000. One can well imagine how, given this population explosion, natural discus habitats in the region have largely disappeared. Skyscrapers are springing up from the ground like mushrooms and the waters are suffering, just like the habitats of the bank regions. Until the 1970s I could still find *S. discus* not too far from Manaus and have data from that time, as well as later. However, from 1990 on there have no longer been any discus to be found, apart from much further upstream.

The Rio Jauaperí is fairly clear in its upper course, and clouded only after heavy rain. But in the blackwater *lagos* of its mouth region *(above)*, where Rio Negro parameters and discus habitats are to be found, there is fine-striped *S. discus* variant *(left)*.

The water parameters measured by me near Manaus on 12.10.1972 at 13.00 hours were as follows: pH 4.70 (in 1.5 m depth); conductivity 8.2 µS/cm; daytime temp. (air): 32 °C; near the water's surface: 27.7 °C. Biotope: intact. No floating plants; water tea-coloured with very little movement. Discus variants: the typical Rio Negro Heckel discus.

It should be noted, in addition, that in the period 1967/68 INPA measured a pH of on average 5.04 and a conductivity of 8.7 µS/cm in a bay 20 km up the Rio Negro, and in the years 1975/76 a pH of 4.46 and a conductivity of 12.1 µS/cm in the Rio Tarumã (a former discus habitat).

In mid-September 1986 at 12.00 hours I measured a pH of 4.42, and my most recent Manaus habitat measurements, on 10.10.1996 at 15.00 hours, were as follows: pH 5.0 (in 1.5 m of depth); conductivity 8.0 µS/cm; daytime temp. (air): 36 °C; at the water's surface 29.6 °C. Biotope: destroyed. Skyscrapers under construction along the bank; no floating plants; the water tea-coloured with little movement.

And further upstream, on the right-hand bank at Paricatuba, the water parameters on 28.09.1998 at 16.30 hours were as follows: pH 4.6 (in 1.5 m of depth); conductivity 10.0 µS/cm; daytime temp. (air): 37 °C. at the water's surface 30.3 °C. **Biotope: still intact; no floating plants; barely any movement.**

Today the southernmost distribution of *S. discus* in the Rio Negro system is to the south of Novo Airão *(see page 370)*. This is the first town upstream and stands on the right-hand bank. Only a few decades ago it was just a handful of huts, but today about 15,000 people live there. Here too is the mouth of the Igarapé da Freguesia where in recent years I was able to find the typical Rio Negro variant of *S. discus* in peaceful *lagos*.

The water parameters on 20.06.-2000 at 11.00 hours were as follows: pH 5.30 (in 50 cm of depth); conductivity 18 µS/cm; daytime temp. (air): 37 °C; at the water's surface: 28.7 °C. Biotope: intact, with *acará-açú* scrub; water tea-coloured with very little movement. Discus variants: page 370.

Novo Airão lies roughly in the centre of the Archipelago das Anavilhanas – said to be the largest archipelago in any river on Earth – and can even be reached via a dirt road from Manacapuru. (It is also the only Rio Negro town that can be reached overland from Manaus.) The archipelago consists of a labyrinth of more than 400 islands, which were declared an ecological station back in 1981. It is the place where manatees *(Trichechus inunguis)* used to be slaughtered in their thousands. The Portuguese called these aquatic, grass-eating mammals *peixe-boi* (= fish-cow). These easily-harpooned animals were regarded as a delicacy (and are still eaten by *caboclos* in the interior, and processed by them for oil and soap). They are now protected(?) in Amazonia and every year Novo Airão celebrates the Peixe-Boi

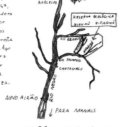

In 2000 a new school was sponsored by the Associação Amazônia in the *communidade* of Sumaúma in the Xixuaú-Xiparinã indian reservation *(see text)*. Another was created right by the Xixuaú in 2002, where not only the children can learn, but also adult indians in evening classes. I was fantastically impressed with the book of the jungle *(livro da selva)* which the indian children (Alda, Artemisia, Deni, Dila, Leni, Lielma, and Marcia) had illustrated in 1997. In it you can read all about the *caboclos* as well as *igapós* (and *terra firme*), the reservation and its ecology, hunting with bows and arrows and traps, fishing with hook and line and various nets, and aquatic creatures that live underwater *(em baixo de água),* including well-known fish species with their names given (an un-named one may be a discus); plus terrestrial animals and birds; what indians grow; what they collect in the forest; and much more (not illustrated).

Festival in honour of the endangered *mamirauá* (the indian name for the Amazon or river manatee).

Across the river is the mouth of the Rio Apuaú, on whose bank live just 16 *caboclo* families, in a *comunidade* controlled by IBAMA. Their leader, João Batista, is the oldest man in the village though he doesn't know exactly how old he is. They are allowed to remove from the wild only what they require for their own needs. But they live not only by fishing and hunting, but carve huge quantities of *churrasco* sticks for a supermarket in Manaus. Most members of the families are now involved and can carve up to 2,000 sticks per day. A dealer transports them to Manaus and pays 6 R$ per thousand. In Manaus they sell for that amount per hundred. I was informed that "*rekel*" (Heckel discus) occur here, upstream to the Igarapé Ambrosio. After that the Apuaú becomes very narrow and the Terra do Jaguar begins – a totally undisturbed and uninhabited region which forms part of the 3,500 km² Estação Ecológica de Anavilhanas reserve. In the left-bank affluents further south and closer to Manaus I had no luck in the Igarapé do Ariuaú (also written Ariãu or Ariuãu), but in the lower Rio Cuieiras, where a German lady researcher has been living for years to study sloths. Here I found specifically individuals similar to the variant from the Rio Quiuini *(see page 381)*.

The Rio Canamaú, a left-hand branch of the Rio Negro further to the north, forms the approximately 20 km long and 1-3 km wide Lagoa Curiuaú in its lower course. Here I have been able to find Rio Negro Heckel discus upstream to the mouth of the Rio Curiuaú. The next large left-hand affluent is the Rio Jauperí (also Jauperí or Jaupery) which rises far upstream on the Roraima plateau and contains lots of rapids. Every time I have investigated its upper course the water was either clear or muddied by rain.

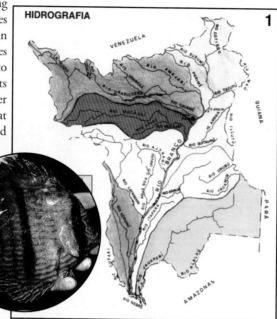

The coloured map (1) shows Roraima with the river basins of the Rio Branco region: White is the Branco itself (almost exclusively whitewater); light green the clear/black, Jauaperí; light brown the black Xeruiní, where indians (5) and *caboclas* (3) still live, there is a lodge (4), and this Heckel discus variant (2). Red and orange are clear, rocky waters.

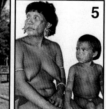

The water parameters on 26.09.1998 in the late afternoon were as follows: pH 5.90; conductivity 14 μS/cm; water temperature 30.5 °C. But there are no discus there. Only in the lower part of its mouth region, where it flows through several *lagos* fed by blackwater *igarapés* and offers ideal Heckel discus habitats, did I find (in 1996) a very finely striped *S. discus* variant (see page 372). There are habitats in blackwater *lagoas* almost as far upstream as the Igarapé Xixuaú – the Lagoas Dipari, Camixixi (near Sumaúma – named after Brazil's largest tree, *Ceiba pentandra),* Coxo and Coxoinha and the *igarapés* of the same names (all at some distance from the right-hand bank); as well as in the two large *lagoas*, Grande and do Tanauaú.

The Xixuaú, a very transparent river, marks the start of the 172,000 ha Xixuaú-Xiparinã indian reservation. This largely unexplored area of primary forest was declared a reservation in 1992 and since then has been the scene of a number of research projects by organisations such as INPA (Instituto Nacional da Pesquisa da Amazonia), the Universidade da Amazonia, FNS (Fundação Nacional da Saúde), the Universitá la Sapienza di Roma (Italy), the University of Birmingham (UK), the Institute for the Quality of Life (Denmark), the Colorado Springs High School (USA), Embrapa Roraima (Brazil), and the Universitá di Salerno and Universitá di Siena (both Italy). The latter, along with the University of Amazonas and the Associação Amazônia, and financed by celebrities such as Mikhail Gorbachev, Lady Madeleine and Selina Kleinwort, Luigi Fabbro and Paolo Roberto Imperiali, university professors, actors, intellectuals, and explorers, has recently established a "health post" in order to alleviate the severe ravages of malaria and the likewise continually increasing problem of hepatitis. Schools have been built and already several TV films, both national and international, have been shot here. The entire region is inhabited by just 59 families (about 570 persons), *caboclos* and indians, 70% of whom are less than 15 years old.

The Rio Branco is the only whitewater river in the Rio Negro drainage region. Where the two rivers meet *(centre)*. To me this spectacle is much greater than the world famous *encontro das águas*, where the Rio Negro joins the Solimôes. In the lower Rio Branco region, close to its mouth *(above)*, there are numerous *lagos* and *lagoas* containing black water, and *S. discus* are to be found in almost all of them – but they never enter the whitewater zone. Often the extremely acid blackwaters with their Heckel discus habitats lie only a few metres from the white, alkaline, sediment-laden waters of the Rio Branco *(left)*.

Further up the Rio Negro, again on the lefthand side the only whitewater river in the entire region enters its black waters (a unique spectacle), the Rio Branco. And here lies an isolated (as is so often the case with discus habitats) clearwater *lagoa* known as Cureru. It is the only biotope in the entire Rio Negro drainage region where I have found a second species of discus. It is almost completely isolated and I was able to reach it only by hydroplane. The discus in question is an interesting colour variant of *S. haraldi* (see page 390), which I reported at the beginning of the 1980s.

Rio Branco region: The Rio Branco belongs to the most northerly state of Brazil, Roraima (see page 374), and is formed by the Rios Tacutu and Uraricoera about 30 km to the north of the capital, Boa Vista. The muddy Uraricoera rises far to the west on the border with Venezuela, at an altitude of more than 1,000 m. It flows over six waterfalls and countless rapids in a rocky stretch more than 500 km long through the reservation of the Yanomami indians before joining the clear Tacutu. The American ichthyologist Paulo Petry is the only person to date to investigate this fast-flowing stream and discover a number of new fish species there. But neither here nor in the Tacutu nor in the 548 km of the Rio Branco are discus to be found.

The course of the Branco is divided into three sections. The first 172 km downstream to the Cachoeira Bem Querer are known as the Alto Rio Branco; from there the 24 km stretch to Vista Alegre is the Médio Rio Branco; and the remaining 388 km to the mouth is called the Baixo Rio Branco. Its main affluents are the Cauamé, Mucajaí, Ajaraní, Anauá, Água Boa do Univiní, Catrimari (the last two are clearwater rivers and my friend Manuel Torres, a discus collector of many years, says that he has caught Blues there), and the blackwater Xeruiní.

The Rio Branco has been renowned for centuries, albeit not for discus but as a centre of attraction for the first Portuguese colonists. Not only in order to capture indian slaves, but also to barter weapons and tools for the much-prized Brazil nuts, ginger and sarsaparilla roots. The Dutch were the first to penetrate here from Guyana, followed by the English, and the Spaniards travelled via Venezuela down the rocky Uraricoera, even though the Portuguese defended the "white river" (Rio Branco) and maintained sovereignty. In 1775 they built a fortress, the Forte São Joaquim, at the confluence of the Tacutu and the Uraricoera. Unfortunately nothing of it remains today. The first settlements were also founded at that time: Senhora da Conceição and Santo Antônio on the Rio Uraricoera, São Felipe on the Rio Tacutu, and Nossa Senhora do Carmo and Santa Bárbara on the Rio Branco. And it was at this time that the Marquês de Pombal initiated the enslavement of the indians in the region, and ordered them to learn the Portuguese language.

1. Aquarium set-up in Russia, included here to show the true *Hemigrammus rhodostomus* (which occurs only in the lower Amazon region).
2. *H. bleheri*, discovered by myself in 1964/65 in the Jufarí. 3. *H. bleheri* in the aquarium. 4. Piranha species *(S. gouldingi?)* from the Jufarí.

The indigenous people didn't mix with the colonists, but ignored their orders and rarely allowed themselves to be seen. In 1789 members of the Macuxi tribe visited a Portuguese settlement for the first time. A few years previously the commandant, Manuel da Gamma D'Almada, had brought in cattle and horses in order to ensure the survival of his fellow-countrymen. (And thus began the zebu invasion of Amazonia.) Until 1869 there were only two *fazendas* here – by 1885 there were over 80.

One of those, Fazenda São Marcos, which was erected close to the fort and still stands today, has long been in the hands of the Macuxi, along with the descendants of D'Almada's original cattle. However, the most tragic chapter in the region's history came with the collapse of the rubber industry. As mentioned earlier, since the end of the 15th century this region had been rumoured to be the site of Eldorado, and this saw a revival with the invasion of the *garimpeiros* in 1912. Gold had been discovered in what is now the Municipio Uiramutã and soon afterwards diamonds in the Yanomami region, in the Serra Tepequem. And in 1980 at Santa Rosa, in Pacaraima, again extending into the Yanomami area. The regions of Uraricoera, Apiaú, Mucajaí, Serra Couto Magalhães, Palimiú, Erikó, Jundiá, Catrimani, Paapiú and Auaris were affected. At the peak of the gold-rush period, in the mid-1990s, more than 40,000 *garimpeiros* were digging away in the region.

Until 1943 the region was part of the Província do Amazon, but was later declared the Território Federal do Rio Branco, and Roraima from the 13th December 1962. And, as is now known worldwide, the Yanomami have lived in the north-east of Roraima (and Venezuela) for more than 10,000 years. When I was here in 2002, hunting for a new swordplant (*Echinodorus heikobleheri* Rataj, 2004), I encountered a raggletaggle group of this tribe hanging around. They asked for money. I would rather have given them pocket knives, food and drink, but they told me that the Brazilian government had granted multi-national interests the prospecting and mining rights for what was supposed to be their reservation; that these rights were being exploited on a large scale; that *fazendeiros* were continually intruding and FUNAI is the one who has the saying

1. Satellite photo: A) The Aracá (or Demini) shortly before it joins the Rio Negro (some people say that the Aracá flows into in the Demini, others vice versa). B) Large lake region with discus habitat. C) The Aracá-Demini confluence. D) The *S. discus* distribution ends in the Aracá to the south of the *comunidade* of Curupira. E) The Demini to the south of Tabocal. F) Rock massif (grey-white) – boundary of the discus distribution. 2. Heckel discus from the Demini. 3. The middle Demini during the dry period. 4. Water colour of the *S. discus* habitat.

– reminding me of the words of Rosa Borōro in 1913 (from his own language translated): "Never trust the white men out of your sight. They are people who domant the thunder, who have no home but instead wander around in order to satisfy their lust for gold. Then they are friendly towards us, as they need us, because the land that they are trampling, the meadows and the rivers they are fighting over, belong to us. Once they have got what they want they are treacherous and untrustworthy!" Tragic, and till true today. And the popular saying *"Quem bebe a água do Rio Branco sempre voltará"* (anyone who drinks the water of the Rio Branco will always come back) is history.

The mouth region of the Branco is the main collecting region for Heckel discus in the Rio Negro. Collectors from Barcelos come to the numerous *lagos* and blackwater tributaries that contain discus habitats. The majority of these are in the waters along the right-hand (lower) bank of the Rio Branco upstream to the Rio Xeriuiní (spelt in various ways).

In the Xeriuiní I first discovered the so-called blue-headed Heckel discus in the 1980s and again in the 1990s *(see page 168)*. Unfortunately I didn't record any data during these trips, but the water **parameters measured by Rolf Geisler at São Angelo on 20.10.1971 were as follows: pH 4.40 (in 50 cm depth); conductivity 11.8 μS/cm; temperature at the water's surface: 29.7 °C.**

But I have data for the adjacent water types in the mouth region of the Rio Branco. **The parameters measured here, in** the Heckel discus habitat (blackwater), on 25.09.1998 at 13.00 hours were as follows: pH 4.40 (in 1.5 m of depth); conductivity 11 µS/cm; daytime temperature (air): 35 °C; near the water's surface: 30.1 °C; in 1.5 m of depth: 29.3 °C; oxygen 2.5 mg/l.

Biotope: intact; no floating plants; water with very little movement. Discus variants: Heckel discus, the typical Rio Negro form.

And the parameters in the Rio Branco (whitewater), barely 50 m away, also on 25.09.1998 at 13.50 hours were as follows: pH 6.40 (in 1.5 m of depth); conductivity 21 µS/cm; daytime temperature (air): 35.5 °C; at the water's surface: 30.6 °C; in 1.5 m of depth: 30.3 °C; oxygen 21.5mg/l.

Rio Negro region: Further upstream, after the Rio Branco, we come to the mouth of the Rio Jufarí (or Jufarís) on the left-hand side. Only a few kilometres earlier there is a *paraná* that runs parallel to the right-hand bank of the Rio Branco, forming a link between the mouth region of the Xeriuiní and the Rio Negro. The Jufarí flows into into a *lagoa* about 30 km long and up to 5 km wide, before the latter in turn joins the gigantic Archipélago da Mariuá which extends up the Rio Negro from here. (Some people maintain that Mariuá is larger and contains more islands than Anavilhanas – estimated at 1,400.)

The Jufarí has a very special significance for me. Not only because I have several times found Heckel discus, the well-known typical Rio Negro form, in the lower course and the *lagos*, but also because in 1964-5 I discovered a tetra here which took the (aquarium) world by storm. Nowadays 2-5 million of them give pleasure to people worldwide every year. More than 20 years later (1987) this fish was described as *Hemigrammus bleheri* by Jacques Géry and Volker Mahnert. Neither Willi Schwartz, to whom I showed it first, nor the scientists would believe me that it was a new species. They all thought it was *H. rhodostomus*, described in 1924 by Ahl, and available under the name "rummy-nose tetra" in the trade. This species was collected at the beginning of the 1920s near Belém, and occurs only there. However, no wild-caught specimens had been exported from Belém since the mid 1950s. It wasn't until a lot later, when, along with Renato Takase (now deceased), I managed to catch some again and thus provide new proof of the considerable differences between the two species *(see page 376)*, that the mills of science began to grind. I found the new fish, known as the *Rotkopfsalmler* (red-headed tetra) in Germany (in English Géry suggested: brilliant rummy-nose) far upstream, in shallow blackwaters where there have been no discus for a long time, but where there are still cardinals *(Paracheirodon axelrodi)*.

The water parameters measured on the spot by Rolf Geisler on 20.11.67 were as follows: pH 5.10; conductivity 5.40 µS/cm; temperature near the water's surface 26.0 °C.

A project has been brought into being in this region, the so-called Projeto Jufarí. It encompasses an area of 50,000 ha, including the rivers Jufarí, Demini, and Padauiri. A joint venture by the *municípios* of Caracaraí (Roraima) and Barcelos (Amazonas). Only around 500 people live in the region, in the *comunidade* of Caicubi and along the Igarapé Caicubi, a more than 100 km long arm of the Jufarí in which I have found Heckel discus. They live almost exclusively from collecting ornamental fishes, mainly cardinals and blue neons, the brilliant rummy-nose tetra, pencil fishes, hatchetfishes, mailed catfishes, Scolof's bleeding-heart tetras, and checkerboard and other dwarf cichlids. Naturally they also catch fish for food and sell Brazil nuts, and about 100 ha are used for agriculture. There is even a generator to provide lighting, a little hospital station, and a primary school. The majority of the region (about 80%) is, however, cloaked in dense jungle, and on the *terra firme* we find trees such as *castanha* (Bertholletia excelsa), angelim-pedra (Dinizia excelsa), guariúba (Clarisia racemosa), sucupira (Dilotropis sp.), pau-d'arco (Tabebuia spp.), matamatá (Eschweilera sp.), the valuable itaúba (Mezilaurua itauba), acariquara (Minquartia guianensis), seringueira (Hevea brasiliensis), and the gigantic *sumaúma* tree. In the almost innumerable *igapós*, where the water level rises and falls by 5-8 metres annually, grow buriti (Mauritia spp.), açaí (Euterpe oleracea), bacabas (Oenocarpus spp.), and patauá (Jessenia bataua). And the families Leguminosae, Euphorbiaceae, and Sapotaceae are strongly represented. The whole area is washed by black water, with the exception of a few stretches of the Demini that are inundated by white water at certain times of the year. The region also contains *campinaranas* or *caatingas,* the name given to a special type of swampy area found only in the Rio Negro basin.

Like almost all Rio Negro regions, this too is virtually inaccessible except by water. We are by now in the Guiana Shield with an average temperature of 26.5 °C, which can drop to at least 18 °C

Left-hand page: I found this Heckel discus variant in the Igarapé Irauaú (the smaller fish is a juvenile from October).
This page: These two Heckel discus habitats (with the relevant variants inset) also lie along the left-hand bank of the Rio Negro upstream at Barcelos. Close to the habitat in the Igarapé Tukano *(top)* there are *igapó* woodlands with *Eschweilera tenuifolia* (Lecythidaceae), *Hevea spruceana* (Euphorbiaceae), *Macrolobium multijugum* (Fabaceae), *inter alia,* which flower between November and April, ie in the dry period adn when the water is rising (photograph taken in January). The fruits fall into the water just at the time when the Heckel discus fry stop feeding from their parents. In addition, as the water rises the remaining seeds are dispersed far and wide. In the Igarapé Bui-Bui (above) we find endless *acará-açú* bushes in which lurk not only groups of discus but also crocodiles.

once a year. The region is noted for its very high biodoversity, with more than 300 fish species, 120 mammals, and around 500 different birds recorded, as well as more than 120 reptiles and amphibians. The Jufarí in part forms the boundary between the states of Amazonas and Roraima, and discus habitats are to be found only in its middle and lower course and in its tributaries.

Shortly before the next large river mouth along the left-hand shore of the Rio Negro there are additional *igarapés* known to contain Heckel discus habitats, for example the Igarapés Irauaú and Bui-bui (often written Boi-boi), *inter alia*. Then we come to the Demini.

The **Rio Demini** is noted for its many names and its numerous meanders; that apart, it rises in the Parima Mountains on the border with Venezuela and is 664 km long to where it joins the Rio Negro. I was able to find Heckel discus only upstream to the area around the *comunidade* of Tabocal; further upstream the rapids and rocky massifs begin. Something few people know, and which was first established by scientists, is that there is a link via various waterways between the Demini and the Xeruiní. The scientists in question were studying bees and wasps in the region and were so terribly stung by a *Caba* species (attracted by their outboard motor) that they had to flee, and thus discovered the link.

It is also noted for numerous aquarium fishes, including dwarf cichlids. (Apropos of which, I have been able to find the as yet undescribed *Dicrossus* species, the so-called "Rio Negro", only here in the middle and upper Demini and never in the Rio Negro.) Its main tributary on the right-hand side is the Aracá *(see page 378 & left)*, which is 551 km long (to where it joins the Demini), and here too there are discus habitats, but only upstream to the Paraná do Marium and Paraná do Calado. Here, often beneath *acará-açú* plants, there lives a variant that usually has wavy lines. Further upstream the river is rocky, with rapids and waterfalls. The Aracá has sandy beaches and is fed by numerous *igarapés*, and has a rich, to date little-studied flora and fauna along its practically uninhabited tributaries such as the Marari and Curuduri.

I also found habitats with the typical Rio Negro Heckel discus in the Igarapés Peixe-Boi, Ariaú, Tukano, and Zamula, and of course, in the gigantic mouth region of the Rio (or Igarapé) Andairá, which is more than 50 km long and up to 20 km wide. The Ereré is a further left-hand arm of the Rio Negro, and its mouth region again consists of several long and apparently endless *lagos*. I found Heckel discus in these as well – again the typical form.

The next river upstream, another left-hand affluent, is the Padauiri (Padauarí, Padaueri), where Schomburgk fished back around 1837, and later Wallace. Discus occur only in the mouth region, as the *rio* is full of rapids and waterfalls, the

1. Rio Aracá with Heckel discus habitat. 2. Rio Aracá – the colour of the water in the discus habitat. 3. The true *Pterophyllum altum*, which is found only in the upper Rio Orinoco affluents. And for comparison: 4. *Pt. scalare*, which is very often sold as the "Altum" or "Rio Negro Altum". The latter is found with the Heckel discus.

first substantial one upstream being the Cachoeira da Aliança. Here in the mouth region I found a lovely Heckel discus variant (see Bleher, H., Supplement 1993a). Although rich in rubber trees, it is only sparsely inhabited. It lies in a little explored paradise with unspoiled affluents such as the Preto and Icié-Mirim.

Discus habitats become increasingly fewer from here on up the Rio Negro, and my most northerly find was off Vista Alegre. There are still Heckel discus here in the Lago Bacururu, behind a large *praia* sandy beach), but I have never managed to find any further upriver.

It is worth noting that one of the three jungle lodges run by the tour operator *Amazon Queen* lies upstream of the mouth of the Branco. (The other two are on the Xeruiní, the Macaroca Lodge *(see page 374)*, and a new one on the Unini.) These lodges are advertised on the Internet mainly to attract anglers from the farthest corners of the Earth and guarantee them large *tucunarés* (peacock bass = *Cichla* spp.) to catch in the Unini, Caurés, Jufarís, Aracá, and Cuiuni. The fishing season extends from October to March, as it is easier to fish during the Rio Negro dry period. (The same applies to discus collecting – although that often ends as early as January.) For around 3,000 dollars you can spend eight days there (six are devoted to angling) trying to hook one or more *tucunarés*, then let the fish go again or cook it (depending on the type of fishing license). You are photographed with your big (or small!) *tucunaré* and put on the Internet (by now there are thousands of anglers holding a *tucunaré* there). Travel to and from Manaus is not included in the price.

It is unknown whether there are Heckel discus around the Ilha Grande, but further down the Rio Negro, along the right-hand, there are. In the Rio Uneuxi – which also represents the westernmost known distribution of the species, I found specimens in the typical habitat. Likewise in the subsequent affluents such as the lower Urubaxí, the lower Rio Arariá (which runs parallel to the Rio Negro for almost 30 km, and in the Cuiuni (also written as Quiuini). Here there are very large numbers of *tulia* palms *(left)*. These offer Heckel discus an ideal home, as, unlike many other palm species, they can live for 7-10 months with trunk and roots underwater. They provide the discus with food and shelter. The discus variants here are very beautiful *(top),* almost metallic light blue (this may depend on diet). Unfortunately I have no data on water parameters.

The Rio Negro is still little settled and in large part totally abandoned to nature. The settlements along the river can be counted on the fingers of

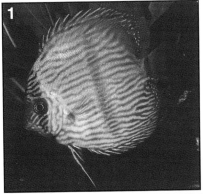

1. *S. discus* from the Quiuini. 2. The *tulia* palm *(Leopoldiana pulchra)* in the Quiuini habitat.

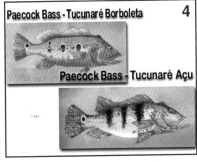

3-4. There is a lot of advertising to attract anglers from all over the world for the four tucunaré species in the region. Five are known to science *(Cichla intermedia, C. monoculus, C. ocellaris, C. orinocensis,* and *C. temensis)* and there are about ten synonyms. There are undoubtedly a lot more valid species in South America. 5-10. Schomburgk painted six *Cichla* species back in 1839.

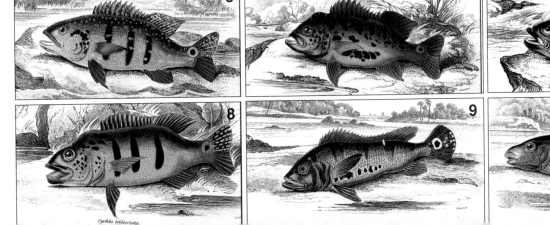

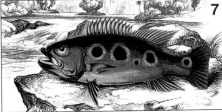

both hands, and many that once existed, such as those of the missionaries (ie mission stations) have disappeared. One of these is Moreira (formerly Moreré at the Rio Moreyra, on 1858 maps), to the north of Barcelos on the right-hand bank. Although this settlement is still sometimes found on current maps (IBGE 1998), it is no longer there. But it has gone down in discus history. The story goes that Natterer caught his discus at Moreré (= Moreira). But Heckel, who described it, actually gives Moreré as the name of the fish *(Chapter 2)* and the locality as "Barra do Rio-negro" (= Manaus). The wording of the description has, however, given rise to misunderstandings and the locality has been variously described and incorrectly interpreted in scientific works. In the old Brazilian literature from around 1900, moreover, it is stated that Natterer collected the fish at Moreré (when he was living in Barcelos) and from then on – until around the middle of the 20th century – local (fish) writings never refer to *acará-baru* (as it was also called) or *acará-disco*, but only to *o peixe de Moreré*, or just *moreré* for short. The only small *comunidade* in the region today is Baturité, and nobody knows anything about Moreré or Moreira any more.

But let us move on to the discus habitats and Barcelos, in whose vicinity, in the Baruri, I found a very light-coloured Heckel discus variant near *acara-açú* plants.

Barcelos was the capital of Amazonia long before it rose to be the Brazilian ornamental fish metropolis – nowadays the majority of its approximately 24,000 inhabitants make their living directly or indirectly from the collection and sale of the *peixe ornamental*. When, in 1750, negotiations with the Spaniards over the colonised territories were in full swing, steps were taken to ensure that the Amazon region also fell to the Portuguese. The *governador* of Maranhão, Francisco Xavier de Mendonça Furtado, was put in charge of establishing boundaries. In the north from the Rio Negro to the Japurá and in the south from the Madeira to the Javari (nowadays the border with Peru). In addition, in 1754 he used the Carmelite mission at Mariuá on the Rio Negro as his main seat of operations. The Carmelite Frei Matias São Boaventura had founded the Missão de Nossa Senhora da Conceição de Mariuá in 1728, and in 1739 Frei José de Madalena had the chapel of São Caetano built, and in 1744 the Nossa Senhora de Santana. However, just two years later Mendonça changed the name of the settlement to Barcelos, after his home town in Portugal. And founded there the first *comunidade de pesqueiros* (nowadays *piabeiros*), to provide settlers on the Rios Negro, Branco, and Solimões with (fish) protein. He imported seed and got agriculture under way. It is also thanks to him that Portugal authorised a new *capitania,* which became separated from the Capitania do Grão Pará, and named São José do Rio Negro.

On the 27th May 1758 it was made official. After Barcelos had been elevated from the status of *aldeia* to that of *vila* on the 6th May 1758, the Capitania de São José do Rio Negro received its permanent seat in the Vila de Barcelos and its first *governador*, Joaquim de Melo e Póvoas. At this time the new *capitania* already had 45 *aldeias* under its control, including Silves, Serpa (now Itacoatiara), and São Paulo de Olivença, which all likewise received the status of *vila*. The Portuguese offered an incentive to colonise the area as far as the Spanish border region, promising to levy no taxes for 16 years.

The new *governadores* of Rio Negro, initially João Pereira Caldas and Manuel da Gama Lobo d´Almada, worked hard in the last quarter of the 18th century to improve both social and economic conditions. Between 1780 and 1820 small additional industries were created, for example the manufacture of cotton material, rope, pottery, candles, and turtlebutter, Coffee, tobacco, cotton, rice, corn, cocoa, manioc, and *cana-de-açúcar* (sugar-cane) were cultivated, and pig and cattle breeding introduced. In the final years of independence exports were worth on average 50,000 Pound Sterling per year (a third of the total exports from Pará). But in 1791 d´Almada transferred the government from Barcelos to São José do Rio Negro at the *encontro das águas* (todays Manaus), when it was realised that the link to the port of Belém would be shorter and quicker. However, in 1799 the government moved back to Barcelos and remained there until 1806, when Barra became the final seat of the *governador do capitão de mar - e - guerra* (governor of sea and war), José Joaquim da Costa. In 1816 the latter

1. The promenade and quay in Barcelos. 2. The Projeto Piaba buildings in Barcelos, rented from the church. The project is a questionable organisation. 5,000 dollars from a patron were spent on a laboratory (backed by the US-government) that doesn't exist. And nobody seemed to care about the existing aquaria. 3. Some of these contained the wrong (and emaciated) discus – not from the Rio Negro (no *S. discus*) but from faraway Uatumã; in others there were corpses *(above)* or live fishes in far too small containers (eg stingrays that couldn't move around); yet others were empty. In addition only one aquarium was correctly labelled. The director in charge came round once, at most twice, a year. I was told. A volunteer worker has to pay R$ 10 (US$ 3.30) from his own pocket to obtain a few *cardinais* (cardinals).

4. The local authorities put aquaria on display, but only for the Festa dos Peixes Ornamentais, in the renovated old town hall (2004). However, some of these fishes were likewise not from the Rio Negro. 5. *Caboclos* come from far and wide for the festival. 6. As well as the *lanchas* of the food-fishermen *(far right)*, the *lanchas* large *(centre)* and small *(near left)* of the *piabeiros* tie up, as well as water-taxis (the two *far left)*. 7. Some people even charter planes from Manaus to Barcelos to attend the Festival do Peixe Ornamental. 8. Once a year there is a meeting of the *piabeiros* (= *pescadores*) where the IBAMA chief, Dr. Henrique dos Santos Pereira *(centre)*, the mayor of Barcelos, José Ribamar Fontes Beleza, and others hold discussions with the *piabeiros*. Problems are discussed and suggestions made.

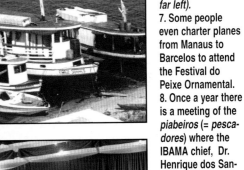
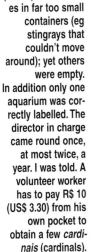

had all the buildings in Barcelos demolished, sparing only the palace, a church (page 383), and the land registry office. When Natterer arrived there Barcelos was still only a tiny place with three stone buildings and a few huts. Little remained of Almada's grandiose edifices, and there was no longer a workforce due to the move to Barra and the outbreak of an epidemic. In addition, part of the indigenous population had been taken away by force to the distant borders for work on the boundaries. Only a few members of the Baniwa tribe managed to escape and return to their lands on the Içana. The complete disappearance of the agriculture and small industries followed in short order. On the 30th March 1876, when it was raised to the status of *comarca*, there was something of a revival. But the eventual breakthrough and economic upturn didn't come until the *piabeiros* around 100 years later, through their hard work and the export of their ornamental fishes. When I visited the place in the 1960s it was still very poverty-stricken and nothing to write home about. Nowadays it is a sizeable, attractive town with vast tourism potential.

Barcelos lies 40 m above sea level, 405 kilometres by air (50 minutes flight) and 496 by boat (about 28 hours journey) from Manaus. It is said that the temperature never rises above 34.3 °C, and never drops below 19.3 °C. That the *município* encompasses 123,120.9 km² and had 20,128 inhabitants in 1999. Rice, beans, maize, and bananas are grown in the poor soil, but the main agricultural crop is manioc. In addition pigs and cattle are bred, as well as poultry. Even so, the majority of provisions have to be imported. *Pacú, surubim, curimatã, jaraquí,*

1. *Recreios* and *lanchas* come from all over Brazil for the fish festival and the spectacle of the discus and cardinal groups. 2. Asher *(right)* and the author on arrival at the airport in Barcelos (2004). 3. *Acará-disco* festival costume.

tucunaré, and *matrinchã* are caught as foodfishes, but the real money comes from collecting ornamental fishes, which yields more than 60% of total income and provides thousands of people in the region with work, food, and a measure of prosperity.

At the peak of the ornamental fish industry (1980s and 1990s) up to 20 million cardinals were caught here annually and exported. Between 40,000 and 60,000 *S. discus.* But thereafter there was a dramatic downturn. In 2002 the sales figures for *Paracheirodon axelrodi* had dropped to 1,258,900 at home and 9,112,492 overseas. By 2003 there was a gradual improvement, with 1,412,100 cardinals despatched within Brazil and 12,288,464 exported. In 2002 14,762 specimens of Heckel discus from Barcelos were sold in Brazil and 26,042 were exported overseas. In 2003 the corresponding figures were 16,835 and 26,841. The largest customers in 2003 were Germany, then Japan, followed by the USA, Holland, and Taiwan. In terms of numbers, *H. bleheri* was in second place with a total of 1,571,150 fishes sold (but traded as *H. rhodostomus* or *Petitella,* as IBAMA doesn't produce accurate permit lists).

Another thing for which Barcelos is noted is its natural surroundings. Splendid sandy beaches along the archipelago, natural resources such as the *seringueiras, castanha do brasil, copaíba, maçaranduba, piaçava* in huge numbers, *palmito* from the *jauari* palm (also called coquillo palm – *see page 388*) and the *pupunha* palm; as well as medicinal plants (some already mentioned), only a fraction of which are exploited. The fauna still includes *onças, antas, queixadas, caititus, veados, jacaré, tracajá* and an incredible wealth of fishes, which are not in any way endangered, including the manatee. Only with additional discovery of gold deposits in the Padauiri, or discoveries such as iron and titanium, *inter alia,* in the Serra do Tapirapecó.

It is said that the nearby Arquipélago Mariuá, which comprises 545,395 ha and more than 1,200 islands (up to 1,400), and

Above: The entry of the *acará-disco* (discus group) on the first night. Their theme (2004), *Um Grito na Floresta, Preservar é Preciso*. A pregnant woman (= Mother Nature) stands on a *Victoria amazonica* leaf under the protection of the water goddess Yara, ringed by discus. (Inset: *acará-disco* banner). *Below:* The cardinal group with the theme: *O Rio comanda a vida*. An indian woman on a butterfly (inset: *cardinal* banner).

Discus group:
1. Discus collector with torch and net.
2. A pregnant woman (= Mother Nature), with wildlife painted on her naked belly.
3. The shaman bewitches the animal hunter.
4. Baniwa indian woman.
5. The indian god Tupã in a unique feather costume.
Inset: Acará-disco T-shirts.

Cardinal group:
6. Cardinal collector with typical net.
7. Indian "head-woman" with thematic banner.
8. They dance indefatigably in their huge feather costumes.
9. Blue and red are the typical colours of the group.
10. Cardinal wood carving.
11. Dancing kids with cardinal mask.

where back in 1996 the *Área de Proteção Ambiental de Mariuá* (Mariuá conservation zone) was created, is the largest on Earth. And although we are in the low-lying Rio Negro region Barcelos is blessed with three nearby waterfalls: the Cachoeira da Alemanha – about two hours away by boat (and 1.50 m high); the Cachoeira do Aracá, the first in the Aracá – 12 hours by boat (1.50 m high); and the previously mentioned Cachoeira da Aliança – 15 hours away by boat (4.60 m high).

Apart from the archipelago reserve, this region (the *município*), which is larger than Portugal, also contains the Parque Estadual Serra do Aracá and large parts of Brazil's largest conservation area, the Parque Nacional do Jaú *(see below)*. But as well as the possibilities for ecotourism and the active sporting fishery (the Rio Negro Lodge and the Aracá River Camp have an office on the spot), in June there is a festival of dance with *bois-bumbás* and *quadrilhas*, and something that puts everything in the shade and must be the second largest festival in Amazonia (after Parintins) – the Festival do Peixe Ornamental.

It began as a celebration of the ornamental fish collectors, the so-called *piabeiros*, and has evolved into an annual spectacle with entertainments (mock battles) that have to be seen to be believed. Thousands upon thousands of people come from all parts of Brazil and even travel from all over the world to participate in this unique spectacle for three days and two nights. And I can't avoid to describe it, specially because it involves the groups representing the *acará-disco* and the cardinal.

Above: The *coquillo* palm (*Astrocaryum jauari*) is widespread in the middle Rio Negro region and grows on pure sand – here in the lower Rio Unini – and generally in the water. It normally starts to flower in August/September (food for adult discus) and the bears fruits by next May-June and provides food for the semi-adult. **Left:** *S. discus* with an extremely broad stripe pattern, to date known only from the Unini.

Local people and members of the two groups work for a whole year on the new theme, writing scripts and composing songs. In addition they create the most fantastic costumes, each one relevant to the theme, with the discus group wearing black and yellow as their main colours and the cardinal group red and blue. Everyone wants to be the best and win.

Here is an extract from the theme of the *acará-disco group* at the 2004 festival: *Um grito na floresta, preservar é preciso!* (A cry in the forest, conservation is essential!)

After the introduction of the *cardume preto e amarelo* (the black and yellow fish shoal – note that they are aware that discus are shoaling fishes), comes the message of the presentation (in case there is anyone who doesn't know it already): that forests are being felled and burned illegally; plants and animals are threatened with extinction; lakes, rivers, and seas are visibly polluted; waste gases are increasingly contaminating the air we breathe; Amazonia has become the goal of bio-piracy. That every day the media show us the thoughtless destruction of the environment (I don't agree with that); Amazonia is noted worldwide as the home of the majority of the surviving indigenous peoples; it contains the largest continuous area of jungle with the largest biodoversity on our planet and a wealth of metal ores; it harbours the most important reservoir of water on Earth with an almost incredible ichthyological fauna. Concerned for this natural heritage, the *acará-disco* group brought the theme *"Um grito na floresta, preservar é preciso"* to the XI Festival Peixe Ornamental.

Reference is made to Brazil's Law 225, which states that every-

On the right in the large photo is a *lago* which is discus habitat. By contrast they don't occur where the Rio Jaú flows, in the upper part of the picture. 1. Nor in the rapids or their vicinity. 2. The water is tea-coloured (= black water). 3. In the headwater region petroglyphs have been discovered that are thousands of years old and indicate early settlement. 4. The Cachoeira do Rio Jaú forms a natural boundary not only for humans but also for discus.

1

2

3

4

one has a right to enjoy an ecologically stable environment, but equally a duty to protect this inheritance for current and future generations.

Next deceased members are remembered with a minute's silence, and there is a tribute to Anildo Macedo, who was the first to work in Barcelos and initiated the collecting and exporting (and supplied fishes from Barcelos practically exclusively to my Aquarium Rio for around 17 years). The musicians march to the tune of *Supremo Acará* (= super cichlid) and dancing girls, dressed as indians in fabulous costumes, join in singing, *"O Índio calma, a floresta reclama, a fauna chora, e a flora soluça, um grito na floresta, é tempo de preservar, somos filhos da Amazônia, que nos faz sonhar tão verde, o verde Amazônia; um grito na floresta, pedindo socorro, um enorme gemido pedindo um choro, é hora de preservar, pois é nossa maior missão, a acará-disco veste o verde, o verde Amazônia, vamos preservar, o acará-disco vem mostrar, vamos preservar, tudo essa cultura popular; Minha alma lamenta o verde perdido, o animal ferido, o rio sufocado, não adormecerei a essa insensatez, não me calarei em meio a tanta negociata, do verde Amazônia."* (Which translates as, "The indian is silent, the forest is moaning, the fauna is crying, the flora is coughing, a cry in the forest, it is time for conservation, we are the children of Amazonia, which makes us dream of green, the green of Amazonia; a cry in the forest, begging for help, a mighty weeping, crying and praying, it is time for conservation, that is our greatest mission, the *acará-disco* clothes itself in green, the green of Amazonia, let us protect, that is what the *acará-disco* signifies, let us protect this entire culture; my soul cries out for the lost green, the injured creatures, the polluted rivers, I cannot sleep because of this inhumanity, I cannot rest while the green of Amazonia is being treated thus.")

Ten people dressed as discus come dancing into the stadium, with groups in incredibly beautiful green costumes representing the natural splendour of Amazonia and bearing slogans such as "Gigantic rivers with the richest ichthyological fauna on Earth"; "Fishes such as the *tucunaré* to eat or the *acará-disco* for ornament"; "Attracts tourists and brings work for the *caboclos* of the region, but these activities must be kept within bounds to avoid environmental damage"; "The discus is important and responsible for the survival of many families"; "A peaceful freshwater fish which is very shy and can grow up to 20 cm long and 3 cm thick" (not true of Heckel discus!); "Its scientific name is *Symphysodon discus*"; "It belongs to the family Cichlidae and was called *discus* because of its shape"; "It cannot tolerate water pollution and requires natural surroundings in the aquarium"; "Its colour ranges from yellow to dark brown, and if it turns black it is feeling bad".

Amazonia is also a region rich in social values, culture, and religion, giving rise to fantastic myths and legends (for example that of Yara, the mother of the waters, who is half fish and half woman), and encouraging popular stories and religious faith. And then in to the stadium comes the *mãe natureza*, a (eight months) pregnant woman clad in green, covered in plants and flowers with her naked belly protruding, and on it a painting of intact Nature *(see page 386);* she is carried by Yara on an enormous *Victoria amazonica* leaf *(see*

The Lago Cureru *(above)*, which lies to the left of the Rio Branco mouth region, is a clearwater lake, and here, at the beginning of the 1980s, I found this interestingly patterned discus *(left)*. Definitely a *S. haraldi*, but the only one to date in the Rio Negro region.

page 385) and out of the giant Yara (the structures arms) comes large fireworks...

The show carries on almost all night, with continually more new costumes, in each case illustrating a further aspect of environmental destruction; the risk of the extinction of many species unless a halt is called; the main problem area for humankind, that of water; that the continual pollution of the lakes and rivers may jeopardise the survival of fishes, plants, animals, and people; that water doesn't need us, but we do need water to survive; that we (Amazonia) who live amid this wealth of water must protect this inheritance, especially when we consider that other people in Brazil suffer water shortages and that by 2025 only

São Gabriel da Cachoeira with the first rapids in the Rio Negro; these prevent further navigation and form a natural boundary for discus. The town (1), founded in 1758 (?) by the Portuguese *comandante* Lobo d´Álmada, was also known as Uaupés (4) for a while. I have even explored here in a helicopter (2) with no doors (6) and penetrated to remote regions in order to map the distribution of discus and other fishes.

I had myself dropped off (3) and collected alone for days in the remotest wilderness, in pouring rain (5) here on the nameless *igarapé* that feeds the Igarapé Bonvina (a left-hand tributary of the Rio Cauaburi). But I couldn't find any discus in the entire region. When the helicopter didn't come back, I tried to march through the jungle – without success, it was impenetrable. Spines (8) and millions of mosquitos (9). But a few days later he came...

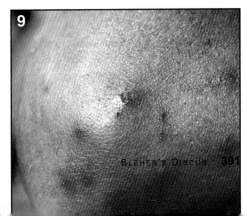

one in three people on this planet will have access to potable water; that water is the greatest asset of human kind.

A series of gigantic floats illustrate that Nature is the source of all life and, of course, there are the *piabeiros* with their *força e coragem, e arte da sobreviência* (strength and courage and survival skills). They dance in their hundreds with torches and hand nets *(see page 386)* – which made me very proud. We are told how difficult it is at night; that the *piabeiro* receives R$ 1.00 (= US$ 0.40) for an *acará-disco*, in Barcelos they are sold for R$ 1.50 to the dealer, who in turn sells them for R$ 5.00-10.00 to the exporters in Manaus, and the latter get US$ 3.00-5.00 for them (or maybe not); that this needs rethinking. Then there are further songs about the *acará-disco* Baniwa indian men and women appear, dancing in splendid costumes; they seize savage animals and take them to the *pajé* (shaman), who dedicates them to the omniscient and all-seeing god Tupã in a ritual on the mountain, while the latter commands their veins to transform into roots, so they can do mo more harm to nature...

It was morning before the two groups – the *cardinal* with the theme *O Rio Comanda a Vida* (The river governs life) – left the stadium, and in the evening the second round began. But it was already predictable that the "King of Amazonia", the *acará-disco*, was again favourite.

Before we leave Barcelos, a few more parameters from Heckel discus habitats in the region.

The water parameters in the Igarapé Tukano (or Tucano) on 31.01.2004 at 11.00 hours were as follows: pH 4.82; in 2.5 m depth 4.73; conductivity 9 µS/cm; daytime temp. (air): 37.5 °C; at the water's surface: 31.9 °C; in 2.5 m of depth: 31.7 °C; oxygen 2.14 mg/l.

Biotope: an intact Rio Negro *igapó* biotope with typical vegetation *(see page 379);* black water with no aquatic or floating plants; water with very little movement. Discus variants: *p. 169*.

The water parameters in the Igarapé Bui-bui on the same day at 14.30 hours were as follows: pH 4.91; in 2.5 m depth 4.89; conductivity 9 µS/cm; daytime temp. (air): 37° C; near the water's surface: 31.8 °C; in 2.50 m depth: 31.7 °C; oxygen 3.85 mg/l. Biotope: intact, full of *acará-açú* scrub; no floating plants; water tea-coloured with very little movement. Discus variants: *page 379*.

An interesting point is that, of course, the parameters for Rio Negro discus habitats differ only slightly – apart from the oxygen content among *acará-açú* plants! And that discus live there not only for protection and food, but also essentially because the oxygen concentration is higher there at daylight.

After Barcelos there are further discus habitats in the Baía do Caurés, in the lower Unini – albeit only far downstream of the first rapids – where I found a very broad-striped Heckel discus variant (the broadest stripe pattern to date). Here it also lives in a habitat full of prickly *coquillo* palms *(see page 388)*, which can readily survive for up to 300 days per year standing in water. Their *tucumã* fruits are a delicacy for blue-and-yellow macaw *(Ara ararauna)*. And very often pieces of the yellow flesh drop into the discus habitats and are greedily eaten (I once saw this quite clearly). In the Jaú (only in its mouth region where it joins the Carabinani, and in its *lagoas*) there is again the typical Rio Negro variant, likewise in the Rio Puduari (not to be confused with the above-mentioned Padauiri).

The habitats and Heckel discus further to the south have already been covered at the beginning, but I would like to say something further about the Jaú National Park (JNP), and, to conclude our discussion of the Rio Negro region, a word about INPA.

Brazil's largest nature park begins about 200 km north-west of Manaus and lies in the *municípios* of Barcelos and Novo Airão. It begins at the confluence of the Jaú with the Rio Negro and extends along the right-hand bank of the Jaú to the mouth of the Rio Carabinani. Thence up the Carabinani, again along the right-hand bank, as far as its headwaters. Then along the border line of the Igarapé Açú, between the Jaú and the Cunauaru, Igarapé Timbó Titica, and Igarapé Sebastião. Then along the Igarapé Maruim and the left-hand banks of the Paunini and Unini, following the latter all the way to its mouth into the Rio Negro, forming here the northern boundary. It lies precisely between 1°00' and 3°00'S and 61°30' and 64°00'W. People live there too, mainly *caboclos*, no indigenous peoples. The majority on the Unini (112 families) and a smaller number on the Jaú (56 families). Seven families on the Rio Carabinani (in 1998) – although here, in the upper course, which contains up to 14 waterfalls or rapids depending on the time of year, the *garimpeiros* are now busily at work. In 2000 this 6,096,086 ha park in the central

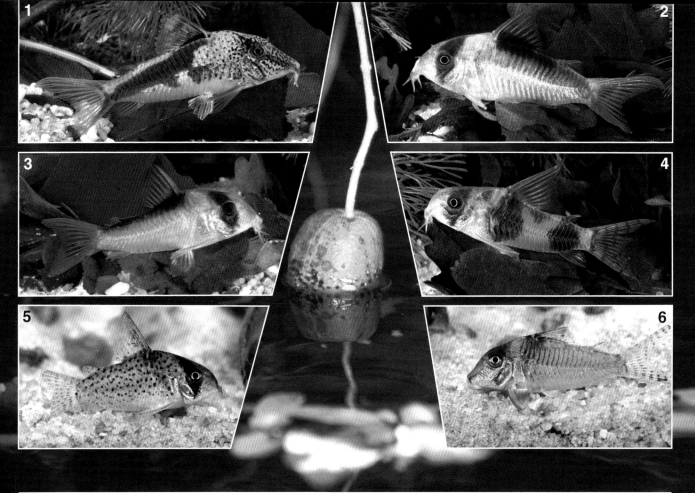

Fishes from the Rio Negro System. *This page:* 1-4. I found these four mailed catfish species living sympatric in the Rio Tiquié (a right-hand Uaupés tributary) in 1996. One of them (1) looks like *Corydoras cortiae*, which is, however, found in the Rio Aguaytia (Peru). The second is an undescribed species (2); *C. adolfoi* (3) also occurs at São Gabriel; *C. tukano* (4) was described in 2004 (there is also a long-snouted form). 5-6. These two species also live sympatric, but in the Demini. One is *C. kanai* (5) and the other (6) undescribed. 7-10. It is less well known that the Rio Demini (a left-hand Rio Negro tributary) is also home to *Apteronotus albifrons* (7), this *Rineloricaria cf. heteroptera* (8), *Paratrygon aiereba* (9), and *Anostomus ternetzi* (10). The center photo shows a *Hevea spruceana* seed and how it slowly disintegrate when the water rises and provide food for discus as well as other fishes. *Left-hand page:* Another food providing seed *(center)* is the mungubeira *(Pseudobomax munguba)*. Also *Pseudacanthicus leopardus (below left)* and *Ancistrus hoplogenys (below right)* feed in the Rio Negro from it.

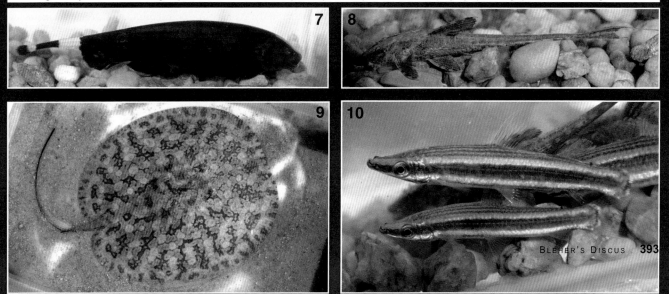

Amazon basin was added to the list of World Heritage sites. It is the largest conservation region in Amazonia and supposedly one of the richest in terms of its biodoversity, with its numerous *igapó* forests, *lagos*, and *paranãs* (all containing black water), which, moreover, are also said to harbour a very large number of electric fishes (Gymnotiformes). Equally unique are the to date little studied petroglyphs *(see page 389)*. However, it is known from local tradition that long ago there was a regular trade with the Incas here (and possibly with pre-Columbian cultures). It is known that at the time of the *Wayna Kapaq* Inca (1495-1528) a regular cultural exchange took place. The Incas brought gold jewelry, coca leaves, and vicuna hides, which they exchanged for c*urare* arrows, wooden seats, baskets, and the famous pottery with negative designs. When, in the 1970s, Padre Casemiro Beckstá showed Tucanos near the Pari-Cachoeira a painting of the Inca ruler *(tawantinsuyu)* Tupa Yupanki (1470-1493), they immediately recognised his finery and ornaments, using the original Quechua names. (Scholars say that the Quechua language adopted by the Incas originated in Amazonia.) And this after he had been dead for almost 500 years. Moreover he was the only Inca who tried to annex Amazonia. But he fell foul of the Anti tribe, epidemics, and malaria in the jungle, and wasn't heard of for months until he finally returned to Cuzco with just a handful of his warriors.

But this is not the only fact that is known. In addition, 17 archaeological sites, with numerous petroglyphs have recently been discovered by the mouth of the Rio Negro alone. They have so far not been dated, but point to two different cultures. This also applies to the ruins at Airão, which have been abandoned since the 1950s, but are now being further studied by the Instituto do Patrimonio Histórico Brazileiro (IPHAN) and may perhaps be restored. This may have been the first settlement in the region in the 17th century.

The previously mentioned dynamic President of Brazil, Getúlio Vargas, cast an eye over Amazonia and in 1952 passed a law founding INPA (Instituto Nacional da Pesquisas da Amazônia), based in Manaus. The official opening took place on the 27th July 1954, but it wasn't until the 1970s that a building, the Campus da Ciência, was started, along with the irregular publication *Acta Amazonica*, still current today, in which the results of research are published. A start was made on the formation of groups to undertake evaluations and study the influence of major projects on nature and the environment, in particular the construction of hydro-electric plants (Balbina, for example). Since 1994 INPA has established new guidelines and intends to promote scientific and technological knowledge, as well as populating the region in harmony with its natural resources and the environment. Apart from its administrative role, INPA has the following 12 c*oordenações de pesquisas* (research departments): Aquacultura (CPAQ)(= aquaculture), Biologia Aquática (CPBA) (= aquatic biology), Botânica (CPBO) (= botany), Ciências Agronômicas (CPCA) (= agronomic sciences), Ciências da Saúde (CPCS) (= health), Ecologia (CPEC) (= ecology), Clima e Recursos Hídricos (CPCRH)(= waterway), Entomologia (CPEN) (= entomology), *Produtos Florestais* (CPPF) (= forest products), *Produtos Naturais* (CPPN) (= natural products), *Silvicultura Tropical* (CPST) (=tropical arboriculture), and *Tecnologia de Alimentos* (CPTA)(= food science). Its main offices occupy 379,868.41 m^2, and it has three *reservas florestais* and two *biológica*s, four research stations, two floating bases, one floating laboratory, and one research vessel. The floating bases are *Catalão*, on the *encontro das águas*, and *Tarumã* on the Rio Negro. The floating laboratory *Herald Sioli* – at the Ilha da Manchanteria – is named after the only recently (2004) deceased (at the age of 94) German professor of limnology and the discover of the cardinal tetra, and the research vessel is the *Amanai II*.

There are 213 *pesquisadores* (23 of them foreigners) currently active, and a total of 775 staff, 169 retired, and 332 freelance collaborators, 117 of them volunteers. Collaboration and cooperation with overseas institutions is currently (2004) continuing only with Jica, Orstom, Dfid, Cirad, Max-Planck, OMS, SUL-SUL, the University of Washington, and the Smithsonian Institution. The permitted research projects must be undertaken by Brazilians, and present an annual summary of their progress and/or their findings and discoveries.

Sioldi worked here for decades and published trail-blazing findings. He headed the Max Planck Institute for Limnology in Plön from 1957 to 1978, as well as the department of tropical ecology, nowadays under Professor Wolfgang J. Junk. Junk too worked for decades in Manaus and published numerous scientific works, chiefly on the ecology of tropical inundation zones, focussing on Amazonia and the Pantanal in Brazil. On the structure and function of inundation zones in relation to the flooding cycle, sediment loading, and geochemical conditions; the adaptations of organisms to the changes between the aquatic and terrestrial phases; primary production and photosynthesis in relation to the flooding cycle; the breakdown of organic material, food networks, and bio-element cycles; biodoversity; and much more.

In addition INPA also has a manatee project in progress, and this endangered species has also repeatedly been bred.

But Manaus, which lies precisely 1,713 km by river from Belém, is also the Amazonian seat of IBAMA; but I will cover that in detail in Chapter 10.

In the upper photo the Rio Solimões can be seen (upstream) at the mouth region of the Rio Purus *(see also page 399)*. *Above:* a bend in the meandering Purus (surpassed only by the Rio Juruá) and its sediment-rich muddy waters. Nevertheless, in the extreme dry season it may contain clear water in its lower course *(inset aerial photo, above left)*. One of my recent expeditions (2004) took place on John Chambers beautiful triple-decked *lancha*, the *Alyson*, seen here at the ultra-modern floating Petrobras filling station on the Tarumã (Manaus), taking on 2,000 litres of fuel for the Purus trip.

6. Purus region; and Tapauá region.

Before we now travel up the Solimões from Manaus and turn into the Rio Purus *(see page 395)*, a little more about the Rio Negro mouth region and the *lagos* in this area.

I have already written about the Ilha do Careiro and the huge Lago do Rei, and that B. de Moreira collected a discus *(S. haraldi)* of 128 mm SL there, although the water was high and turbid (April 1987). And that there is no longer any habitat there today (pasture land). That in the mouth region itself, on the Manaus side, D. Merlin and others collected 14 specimens in November 1923, when the habitats still existed (until the 1970s) and that they were Heckel discus, measuring between 78.7 and 123.9 mm SL.

And I would like to mention yet another ichthyological sensation from this region, first discovered in 1999 by a Swiss INPA employee, Ilse Walker. An eel-like, barely 15 cm long, carnivorous creature with a head and respiratory organ similar to those of the predatory characins of the family Erythrinidae, such that (like the erythrinids) it can survive for months (years?) out of water – as long as its skin doesn't dry out. The creation of a new family and new monotypic genus for this primitive creature, found at Manaus, is currently in progress.

On the tongue of land between the Rio Negro and the Solimões lie the Lagos Victoria Regia (a tourist destination which contains black water, but no discus habitats), Caldeirão and Baixinho (clear water), the blackwater Lago do Limão, the whitewater Iranduba, and the blackwater Lago do Miriti (just before Manacapuru), all with no evidence of discus. I was not a little surprised when on one occasion I saw a fisherman travelling by on one of the floating grass islands in the middle of the Solimões *(above)*.

On the right-hand side of the Solimões lie the large Lagos Grande and Janauacá, which contain white water for the majority of the year, and the Lagos Janauari and Marmori. The latter (or parts of them) contain clear water for longer and only occasionally display signs of the influence of the Solimões. Nowadays there are tourist destinations here, with splendid lodges, including the *hotel ecológico* Amazon Lodge and Village as well as Boa Vida (= good life) and floating restaurants. They have even constructed treetop walkways so that people can marvel at the jungle from above. Discus habitats are to be found, but I have found discus themselves only a few times there, years ago. The large region of lakes south of the Janauacá, which includes the Lago Careiro (and the village of the same name; note that the *lago* is also called Castanho) and that to the west – extending to the mouth of the Purus – are *várzea* areas, which are "controlled" by the Solimões (and the Purus). No habitats – this is mainly pasture land. Likewise along the left-hand bank of the Solimões *(see page 397)*.

After leaving Manaus, **Anamã** is the first city upstream on the Solimões. It lies on the Paraná de Anamã, around 1 km from the left-hand bank of the Solimões. From here it is 5 km to to the Boca de Anamã where the Paraná do Arariá begins and the Lago Anamã (not to be confused with the Lago Amanã, located further west) ends.

The city lies 28 m above sea level and can be reached only by water (a day's journey – 188 km – from Manaus). Its history is linked to that of Manacapuru and the villages of Anori and Codajás, further upstream. After making peace with the native Mura from this region (5,540 of whom still survived in 2000) in 1775, the Portuguese established a headquarters at the chief Mura village (today Manacapuru). The natives were allowed to continue to live there, but were later relocated to the Rio

Left-hand page: In the upper photo the two dominant floating perennial grasses can be seen. The fisherman is standing on the denser *Paspalum repens* and to the left of it is the lighter (and taller) *Echinochloa polystachya* grass. There are, of course, countless other floating plants, mainly *Pistia stratiotes*, plus *Eichhornia* and *Salvinia* species. Roots, branches, and tree-trunks float down the rivers. The lower photo shows part of the Lago Janauari with a floating restaurant and numerous other *lagos*.

This page: 1. Travelling on the Solimões by *recreio* can often be quite an adventure. The rain, powerful gusts of wind, and tall waves are reminiscent of being on the ocean. Vessels seek shelter along the bank, or creep along behind one another in the wind-shadow. 2. A souvenir of the rubber period on the left-hand bank of the Solimões, shortly before Manacapuru. Here and elsewhere the relentless erosion along the riverbanks can be seen, especially where there is clear-felling. 3. There is no longer anything to retain the soil in place, and the small church and the house next to it soon disappeared into the Solimões. 4. Here, again on the stretch upstream near the mouth of the Purus, thousands of square metres of land have vanished into the river and the process continues unabated (probably to the edge of the trees, and if further clear-felling and/or burning takes place, the erosion will continue...). 5. And if not stopped, the Solimões be probably only a few metres deep in a century or two's time, like here... 6.-7. At one spot where we sought shelter from the waves, three of the 37 *Cyperus* species (Cyperaceae) of the *várzea* could be seen together on a single square metre of eroded soil. 8. Collecting at night in 2004, I was no longer able to find any discus in the lower Lago Anamã. The *ajú* (sudden receding water) had resulted in lack of oxygen, which had killed thousands of fishes. A tragic sight which is repeated every year (in November/December). From January/February the water is again black.

Manacapuru (in order to supply fishes from there to the then capital, Barcelos). And in 1785 the commandant then in charge ordered their removal to the spot where Anamã lies today. In 1938 it was elevated to the status of *vila*, and since 1982 has been a city with a *município* encompassing 2,465 km². In 2004 2,383 of the 7,567 inhabitants voted for Luiz Guedes Brandão as *prefeito*. The people live by agriculture, predominantly manioc, maize, water-melons, jute, and black beans. The surrounding *várzea* has been transformed into manioc fields. The remaining primary forest provides a harvest of *madeira de lei* (precious hardwood), *cipó-titica* (a liana species), *borracha* (rubber), *óleo de copaíba* (a palm oil), *castanha* (Brazil nuts), *camaru* (wild fruit), and *malve* (mallow) for the markets of Manacapuru and Manaus. There is no fish-breeding, or rather, no regulated aquaculture with an appropriate structure. But in the dry season many fishes are caught here to supply the markets mentioned above, as well as for shipping to other *municípios* and to Colombia. I have never managed to find a hotel, but I have become familiar with the Festival de Música Anamãense (28th-30th August) and the Festejos de São Francisco de Anamã (which lasts for several days and nights...).

The Lago Anamã is around 40 km long, but relatively narrow (up to 2.5 km wide). There are no longer any habitats in the lower part, just cattle country, clear-felled or burned – the *várzea* regions converted to fields. The last time I was there, at the beginning of December 2004, there had just been an *ajú* and thousands of fishes were floating dead on the surface of the water *(see page 397)*. A ghastly sight, which is repeated every year (and not only here). I caught a few attractive tetras *(centre)*, but no discus, not even dead ones.

The last evidence I found of them was years ago in the tributary region of the Igarapés do Anamã Grande, Mato Grosso, and Alexandre, which contain black water. Likewise in the Paraná Arariá, which, however, like large parts of the Lago Anamã, is muddied by the Solimões during the rains and periods when the water is rising (November/December to June/July), so that no discus are to be found at those times. The colour forms are very similar to those from the Manacapuru region (= *S. haraldi*), hence there was very little commercial discus-collecting here.

The water parameters measured by me here on 2.11.2004 at 23.45 hours were as follows: pH 6.98; conductivity 135.0 µS/cm; daytime temp. (air): 27.0 °C; water 29.1 °C; oxygen 3.31 mg/l. Biotope: water turbid (indundated by the Solimões) and at this particular time *ajú*; covered in floating grass, *Pistia stratiotes* in quantity, as well as lots of *Salvinia auriculata*. Discus variants: *below*.

Purus region: Before we now follow the river upstream, a few data: it rises at an altitude of barely 500 m in the Peruvian Contamana mountains and after 3,325 km flows, full of murky sediments (except in the final section in the extreme dry season), into the even muddier Solimões. As regards its overall length, interestingly it is only around 1,450 km as the crow flies from its source to its mouth. Its serpentine windings (meanders) are surpassed (worldwide) only by those of the very similar Rio Juruá. The Purus is navigable for much of its length: from the mouth to Cachoeira (1,740 km); Cachoeira to Boca do Acre (810 km); Boca do Acre to Rio Iaco (290 km); and even as far as Bolivia via the Rio Acre, one of its small tributaries. During the dry season it may be only 0.80 m deep in places, but during the rainy season it can rise more than any other Amazonian river – up to 24.5 m above its low-water level. On average the water rises by 12-15 m (November to June).

There are only a very few villages along this long river (and nothing worthy of the name of town): Beruri, Tapauá, Arumã, Novo Tapauá (Foz or Boca de), Nova

Fishes from the Lago de Anamã: 1. A *Chalceus* aff. *erythrurus* (the genus has recently been revised and expanded from two to six species). 2. A possibly undescribed *Aphyocharax* species, with splendid colours. 3. This *Prionobrama* species may likewise be new. *Right:* Anamã discus.

1. At the mouth of the Purus, where it joins the Solimões, there are three or four floating houses along the right-hand bank. There is no village far and wide.
2. Upstream we come to the first large river island, the Ilha do Anamã, where the river is more than a kilometre wide (elsewhere it is mainly less).
3. The first (and only) large *praia* (= sandbank) comes into sight. 4. For the voyage on the *Alyson* we needed a *pratico* (guide), Jozuí from Anamã, who knew the Purus and its quirks. 5. Along the lower course there are *fazendas* that are often regularly indundated, unlike the floating houses with their floating gardens.

Olinda, Canutamã, Lábrea, Porto Luzitânia, Pauiní, and Boca do Acre. None has a harbour, at most a floating jetty. Lábrea and Boca do Acre are the only ones that can be reached overland. Provisions are brought in largely by *lanchas*, which transport Brazil nuts and rubber, but mainly all sorts of fishes for the insatiable market in Manaus, back downstream. *Lanchas* belonging to *piabeiros,* carrying hundreds of thousands of mailed and small suckermouth catfishes, also come from the Purus region, likewise the majority of discus exported, although discus have been found in this region only upstream to Canutamã.

At 375,000 km² – more than 10% of which is inundation-zone forest – the Purus basin is the sixth largest in the Amazon region (after the Madeira, Tocantins, Rio Negro-Branco, Xingú, and Tapajós). Its main right-hand tributaries are the Jari, Ipixuna and Itaparaná, Jacaré, Mucuím, and Ituxi, and on its left-hand side the Tapauá and the Cunuiá (Cuniuá, Cunhuá, Cunuhá) – not to mention the *igarapés* and *rios* in its upper course.

The Rio Purus first appears in the literature in 1852, in the *Dicionário Geográfico do Brasil*. The search was on for *drogas do sertão* (drugs, spices, etc) in the mid 19th century. In May 1852 an expedition set out from Manaus under the leadership of Tenreiro Aranha with the aim of finding a link between the Purus and Madeira. But without success, just like Manuel Urbano da Encarnação (the founder of Canutamã) later on (1861). In 1854 Frei Pedro Coriana founded the first mission for the natives on the Purus, including members of the Mura, Cauinici, Mamuru, Jamadi, and Puru-Puru, and called it São Luís Gonzaga. In 1869 the first immigrants *(caboclos)* arrived from Ceará, and at the end of 1871 a larger contingent from Maranhão under the *commando* of Coronel Antônio Rodrigues Pereira Labre. He settled them on the *terra firme* at Amaciari (subsequently Lábrea). The rubber boom (1890) brought more than 50,000 people to the Purus region – more than live along the entire Amazonian part of the long river today; around 1905 it was the Mecca of the *seringueiros*.

Ichthyologically practically unexplored, part from the English geographer and naturalist William Chandless (1829-1896), who, under the commission of the Geographic Society of London, penetrated to the Alto Purus in 1861, 1864-65, 1867, and 1869, studied several native peoples, and discovered fossils (from the Cretaceous) of the fish genus *Phractocephalus* (which still exists today), *inter alia*. Then an expedition in 1963 and my collections, on which I report below. But since the 1970s it has been famous as the collecting area for *acará-disco, Corydoras, Otocinclus,* and *Dianema,* and at the same time for numerous plagues – illnesses, mosquitos, and the dreadful little gnat known as the *pium* (which can pierce anything).

A quite different plague was discovered among one tribe of the many aboriginal peoples that have lived here for thousands of years. The Puru-Puru (who are now extinct) suffered from a skin disease with white spots. Depending on the tribe, the natives called the river the Cuchiguará, Cuchivara, Cuxiuara, Cochinuára, Wainy, Pacajá, or Pacyá. The Spaniards named it the Beni, and in 1833 the Portuguese Cerqueira da Silva named it Rio dos Purus, referring to the Puru-Puru, who lived there (but in their language it means "painted"). Among the white men, it is also worth mentioning the ethnologist Paul Ehrenreich (1855-

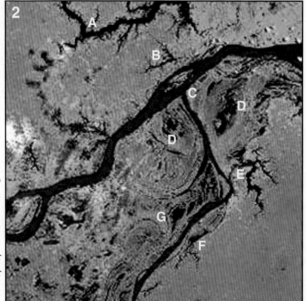

1. Beruri, the first settlement up the Purus. 2. On the satellite photo (1:250,000) we can see: A) Lago Grande de Manacapuru. B) Lago Anamã. C) The mouth region of the Purus, with offshore islands in the Solimões. D) *Várzea* regions. E) Lago Surará (sometimes black water). F) Lago Grande de Paricatuba. G) Lago Aiapuá *(for details see p. 402)*.

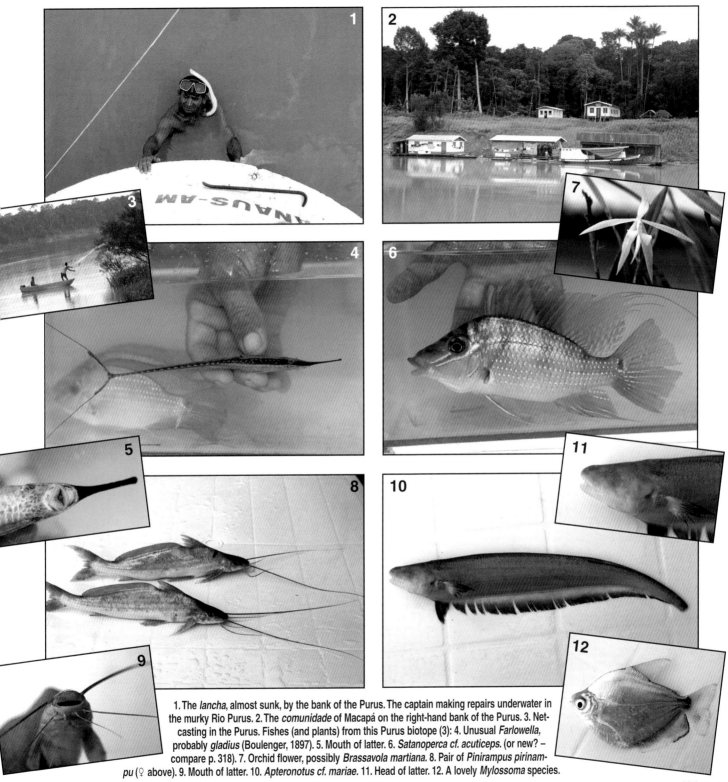

1. The *lancha*, almost sunk, by the bank of the Purus. The captain making repairs underwater in the murky Rio Purus. 2. The *comunidade* of Macapá on the right-hand bank of the Purus. 3. Net-casting in the Purus. Fishes (and plants) from this Purus biotope (3): 4. Unusual *Farlowella*, probably *gladius* (Boulenger, 1897). 5. Mouth of latter. 6. *Satanoperca cf. acuticeps*. (or new? – compare p. 318). 7. Orchid flower, possibly *Brassavola martiana*. 8. Pair of *Pinirampus pirinampu* (♀ above). 9. Mouth of latter. 10. *Apteronotus cf. mariae*. 11. Head of latter. 12. A lovely *Mylossoma* species.

1914), a doctor by education who turned to the study of peoples as an independent researcher. During several research expeditions he busied himself mainly with the ethnology of Brazil, and in particular that of Amazonia and the Rio Purus. In later years he concentrated predominantly on the study of comparative mythology. He was the first to point to the astonishing similarities between the indian myths of North and South America. Around six decades before Claude Lévi-Strauss and his concept of transformations he made the observation that these myths represented a large "organic unit".

Another German was Gustav Wallis, who has gone down in history as a famous botanist; he was born in Lüneburg in 1830 and died at the age of 48 in Cuenca (Ecuador). He collected plants in the Philippines, but predominantly in South America for the Belgian company Linden. And eventually on his own account. He discovered more than 1,000 new species. Among others from the Rio Negro area, one of the most elegant bromeliads, *Guzmania musaica*, the lovely orchid *Cattleya eldorado*, and *C. maxima*, which he rediscovered. He has been immortalised in a number of names, for example *Spatiphyllum wallisi*, a marsh plant which aquarists like to plant underwater and which he probably found in the Purus *(right)*, where it occurs in hundreds of thousands in the *várzea*. (In addition, around 35 *Spatiphyllum* species are known from Latin America and one or two from Malaysia). And *Dracula wallisi*, a truly unique orchid with filigree blossoms. But let us travel on upstream.

Along the left-hand (and right-hand) bank of the Purus the gigantic *várzea* region with its endless *lagos* extends as far as the gigantic Lago Aiapuá. This area has no discus habitats, it is a

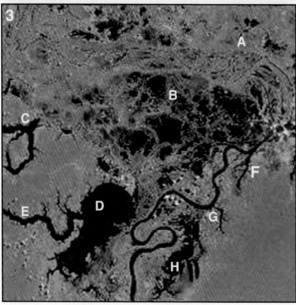

várzea inundation zone, part of the 240 km wide Purus inland delta, and contains white water most of the time *(see below)*.

On the right-hand bank, after around 30 km we come across **Beruri**, whose history is linked with the Muras. Not until 1939 was it raised to the status of *vila*, and since 1981 it has been the *novo município* of Beruri with an area of 17,326 km² (and the tiniest of hamlets such as Sururá, Itapuru, Cuiuanã, Paricatuba, and Supiá, each of which consists of just a few houses). It lies 65 m above sea-level and 192 km (by river) from Manaus. Very little is grown there, at most for personal use; there is a saw-mill, furniture production, a brickworks, and ice-making. I discovered a bank, but no hotel. The *município* is said to have 12,744 inhabitants (in 2004), the town barely half that.

By comparison the Lago Surará is large. It lies somewhat further south, again on the right-hand bank, and most of the time contains white water from the Purus. Only during the dry season is it black, and only then have I been able to find discus. But in the adjacent inundation zone, the Estirão do Surará, there are definitely none. I have already mentioned the following *lagos* – do Ipiranga, Ipiranginha, do Matias, Caviana, and Água-Fria *(see page 157)*; some of these are affected by the turbid water of the Purus, others are isolated and contain habitats. Further along the left-hand bank, where again hardly a soul lives, discus habitats begin south of the Paraná Cuianã (Cuiu-anã), which changes its name and a constitutes a (white-water) link with the Solimões. Here, after a large lake region, we come upon the already-mentioned Lago Aiapuá (or Ayapuá). It too is virtually uninhabited

aiapuá = Maniok und Name des Sees

1. One of the tiny *comunidades* on the Lago Aiapuá. These settlements often consist only of palm-thatched huts, and are called Carapaná, Furo do Pirarara, Furo do Rodela, Lago Rodela, Lago do Veiga, Lago Marajá, and Parana Fortaleza. 2. The *caboclos* have always given me a friendly welcome, including during a visit by aquaplane (1989). 3-4. The majority live in floating wooden houses, as the water level rises and falls enormously. 5. I came across discus habitats beneath spiny palms which offer shelter and food, especially at high water as here. 6. The *bacuripari-liso*, *bacuri*, or *bacu* (*Rheedia brasiliensis*, family Clusiaceae), is a delicious fruit – and not only for discus. The tree bears fruit during the high-water period (January-May), ideal for the fishes. Each tree yields up to 1,200 of these 3-4 cm diameter fruits.

Spatiphyllum wallisi

and designated an *Área Indígena*, and an around 6,100 km² *Reserva Ecológica National* borders the *várzea* mentioned above. An almost totally undisturbed region with optimal discus habitats south of the *várzea* (the whitewater influence).

Note that *aiapuá* is a native word for the manioc plant *(manioca or cassava)*, the Euphorbiaceae of the rainforest. Long before the arrival of the European colonists manioc had spread from central Brazil across South America, Mexico, and the Antilles. The Portuguese took it to Africa and Asia. The up to three metres high plants have a bushy habit of growth and bear greenish-yellow flowers. The crop consists of the starch-rich, up to 8 cm thick and 90 cm long, tuberous roots *(see page 403)*. All parts of the plant contain poisonous linamarine, a glycosid of prussic acid, in their milky sap. The indians have developed a method of extracting this sap by squeezing out the pulverised tubers with the aid of woven hoses, and finally roasting the residue. The manioc flour thus obtained is used mainly in the production of flat-breads, porridge, sauces, soups, and alcoholic drinks *(kaschiri)*. The best-known merchandised product is tapioca, pure starch from the manioc tubers, but by far the most widespread basic form of manioc flour is *farinha*, which is obtained from the ground or grated manioc tubers. It constitutes an essential staple of the daily menu, at every mealtime. Two sorts of manioc are recognised

(on the basis of their linamarine content): the bitter *manioca* with a high linamarine component and the sweet *aipím* with a small component. The *aipím* stores the poisonous glycosid in the outer layers of the skin of the tuber; hence it is sufficient to peel the latter. Because the poison is eliminated by leaching or the effect of heat, safe-to-eat meals can also be produced from the bitter manioc. In Brazil *Manihot esculenta* is also known as *macaxeira*. So much for manioc.

The discus variants in the Aiapuá region are similar to those from the Lago Jari *(see page 410)*, mainly so-called Blues, with stripe patterning on the upper and lower parts of the body. I was also able to find them in the lake complex opposite (on the right-hand side of the Purus), albeit only in the southern part on account of the considerable influence of the Purus: in the Lago Grande de Paricatuba, which is more than 40 km long and up to 10 km wide, and in the blackwater discus habitats in the Lagos São Francisco and Calafate.

I didn't record any parameters during my three Aiapuá research trips, but the water parameters in the **Paricatuba habitat (and they are undoubtedly all similar) on 3.12.2004 at 18.05 hours were as follows: pH 5.72, in 2.5 m of depth 4.69; conductivity 40.0, in 2,5 m of depth 52.0 μS/cm; temperature (water) 29.4 °C, in 2,5 m of depth 27.1 °C, (air) 32.0 °C; oxygen 0.93, in 2,5 m of depth 1.50 mg/l. Biotope: water black, but slightly turbid; spiny palms, grass, *Pistia stratiotes* and *Salvinia auriculata*; roots and trees.**

Interestingly, among the discus (*S. haraldi*) I found four different piranha species (and a few of the latter were rather shredded).

This large region is uninhabited apart from a floating hut at the mouth of the Purus, where I was told of a small *comunidade* that supposedly exists far inland. I didn't find it.

During one of my trips, on the *Alyson*, the luxurious *lancha* of the Englishman John Chambers, there were problems with water coming in *(see page 395)*. We would have foundered were it not for the presence of mind of the *capitano* Almir Viana da Silva (who bore more than 50 knife scars all over his upper body, the result of adventures involving women, or so he said) who headed the *Alyson* full tilt for the bank, so that she sank in only a metre of water.

During the day's halt we caught fishes immediately along the bank zone (see page 401). Later, but still early in the morning, I beat my way up the 12 m high escarpment, through endless lianas, into the *várzea-igapó* region, where I intended to look around for fishes. But I hadn't reckoned on Wayne, who suddenly appeared while I was photographing fungi, flowers, and plants among the undergrowth *(right)*. Wayne had been invited on the trip by Chambers. For more than 25 years he had run an aquarium shop in Yorkshire, then two years ago he came to Manaus, met a pretty *amazonense*, got a divorce, said adieu to the shop, and established himself in the Amazons capital with a new child (nine months later). Now he wanted to run a horticultural business and asked me the names of every plant possible. And because of his questions I didn't attend to where we were going. We found ourselves in an unspoiled inundation zone,

Left-hand page: Hypopomus sp. 2 (*above*) and *Rivulus* cf. *ornatus* (*below*), from a tiny waterhole in an lower Purus *igapó*. *This page:* A glimpse of the *várzea* and *igapó* region where the *Alyson* sank, north of Macapá (see also text): Where ever one steps there are giant seeds (1); popular aquarium plants *Spatiphyllum* sp., rooted in lianas (2); tropical fungi in all imaginable colours and forms (3-6), and even tiny ones barely 5 mm in diameter (see the 4 below); the strangest of flowers (7, 8 and 10); *Heliconia* sp. (9) and almost unbelievable flowers (*Herrania nitida*) sprouting from the tree-trunk *(left and right)*.

untrod by human feet, and every few metres there was something new to discover. I lost all track of time and space, and couldn't find the way back again. And then it began to rain in buckets (an cold tropical downpour). I had left behind my GPS, compass, etc, and started to mark the path we took the way the indians do, but it didn't help. For hours we wandered around lost in the jungle, and every liana, bush, tree, and other plant looked exactly like the next. I couldn't find my route-marks any more. Our tracks on the jungle floor were invisible among the millions of leaves and hundreds of thousands of *Spatiphyllum* plants, or long since washed away by the rain.

By 1600 hours we had certainly covered more than 20 km and had even caught fishes in a water-hole *(see page 404)*. The situation appeared desperate. Something of the sort had never happened to me, not even as an eight-year-old in the Mato Grosso, when I was out hunting all day with the natives and got lost. As we had been here already an entire day without food, I began to keep an eye open for fruits and small wood grubs (especially the fat white ones) which I had often previously eaten to survive. And then the sun came out briefly before setting. But that was enough. I immediately knew the right direction to go as we were east of the Purus bank and, as everyone knows, the sun goes down in the west. Practically running, Wayne loaded with plants and I with fishes, we found our way to the boat. When we were only a few metres from it John and his wife Anna sounded the boat's horn for the first time...

The entire region north of the *comunidade* of Macapá, where a church (the only fixed wooden building) stands on the bank *(see page 401)*, is *várzea* and *igapó*. There we were told that there were seven blackwater lakes in the area (and we fished one of them), but hardly anyone had ever visited them. There were typical discus spawning places in the *igapó (photos left and below)*. They live in the *lagos* further east and come here when the water rises.

Before we come to the best-known Purus discus region, the Paraná do Jari, at Arumã, a brief note on my knowledge of left-hand bank region.

Far in the interior, north-west of the Aiapúa, lies the more than 60 km long Lago Uauaçu (see page 402). An ichthyologically unexplored and (almost) isolated lake where, as in the Piraiauara and the isolated Miraiucá, there are supposedly Blues. Likewise in the Miuá and Baixo, although they periodically receive turbid Purus water via the Paraná do Joari. And in the isolated blackwater Caapiranga, the half-moon-shaped Comprido, Reis, and Bacuri II, which are accessible only by canoe via the same *paraná* when the water rises. They are all uninhabited apart from Bacuri I and the *igarapé* of the same name, where there is an Apurinã village. Totally isolated and likewise uninhabited are the Lagos Pitú Grande, Azul, Preto, Tapagem, and do Moreira. Discus (*S. haraldi*) habitats are to be found everywhere, but they have been neither exploited commercially nor studied scientifically. This also applies to the more than 20 km long Lago Itaboca, which is, however, accessible year-round.

At the beginning of the high-water period the Itaboca is influenced by the

Top: This tree is one of the largest of the Amazon forest and can attain a height of 55 m. Termites (a delicacy much enjoyed by discus– and other fishes) climb high up the trees (as here on a *samaúma* tree) when the high water period begins. In this case in the lower Purus region.
Above: In the dry season these termites build their nests on the ground. *Below:* Typical discus spawning site in the *igapó* (and thousands of *Spatiphyllum* in the underbrush).

1. Discus habitat among the *acará-açú* plants along the west bank of the Lo. Itaboca. 2. At twilight search for discus in the northern Lo. Itaboca, along with Sondomar from Barcelos and Jozuí from Anamã. In the discus habitat we found interesting characins of the family Hemiodontidae: 3. The rare *Pterohemiodus cf. atrianalis* (some regard this genus as a synonym of *Hemiodus* - this is the first known photo of a live specimen). 4. A *Hemiodus* of the *immaculatus* group. 5. One of the *H. semitaeniatus* group. 6. Looks like Fowler's *Anisitsia amazona (Anisitsia = Hemiodus)*, but the anal fin is long, so it could instead be a *Bryconops*.

Purus, but only in its downstream part. Here there are beautiful blue discus with an almost light blue stripe pattern.

The parameters measured in the habitat on 4.12.2004 at 18.30 hours were as follows: pH 5.50 (in 2.5 m of depth 5.45); conductivity 20.0 (19.0) µS/cm; temperature (water) 29.3 °C (29.1 °C), (air) 25.0 °C; oxygen 0.93 (1.20) mg/l. Biotope: water black, slightly muddied (from the Purus); *acará-açú* plants almost everywhere; roots, fallen trees and branches; no floating plants.

It was formerly uninhabited, but in 2004 I spotted seven houses on a tongue of *terra firme* and others under construction. And I noted a number of clear-felled areas. As well as discus, I found splendid *Apistogramma* and four sympatric characins, three of them from the family Hemiodontidae *(see page 407)*. Including a species of the never photographed genus *Pterohemiodus*, whose type locality is the Río Ucayali, which lies in a valley which it shares with the headwaters of the Purus and Juruá. (Maybe a link exists there?)

But back to Arumã, a settlement of around a dozen houses where the Paraná do Jari joins the Purus. If you travel up the latter by *voadeira* for around two hours then you come to the Rio Jari (Jarizinho) and the Lago Jari. Most of the time the Paraná do Jari contains white water, as it is fed by the Purus at Tapira (which exists only on maps), makes a loop to the south, and flows back into the Purus again around 80 km later, at Arumã *(see page 411)*. On one occasion when I travelled this stretch I came across floating Apurinã huts *(centre)* en route. They also call themselves Apurinhã, Ipurinãn, Kangitê, or Popengaré (the last two are also names for their language). They are an indigenous people belonging to the Arawak (Aruaque, Aruák) language group, and their population has increased in recent years, from 2,000 in 1994 to around 2,800 by 2004. They live widely scattered across the Purus region from the Rio Branco (in the state of Acre) to Manaus.

Years ago I visited the Apurinã-do-45 Association near kilometre marker 45 on federal highway 317 (Rio Branco to Boca do Acre). In 1999 a handicraft project was brought into being there in conjunction with a non-government organisation (Pescare-Grupo de Pesquisa e Extensão em Sistemas Agroflorestais do Acre) from the Rio Branco. The project is intended to ensure a basic living for the Apurinã with the aid of their their traditional artistic handicrafts, so that the people can retain their traditional way of life. An integral component of the project is the sustainable use of the tree species in the tribal region, which provide the raw materials for their handicrafts. This excludes any possibility of over-exploitation of this resource. The Apurinã ornaments *(below)* consist exclusively of processed plant seeds and contain no artificial components. They use *tucumã, inajá, jarina, murmurú,* and *açaí* seeds for this purpose. The strings and catches are manufactured from the bark of the *carapicho,* a small tree. If the bark fibre is beaten for a long time, soaked, and then dried it acquires a wire-like structure of

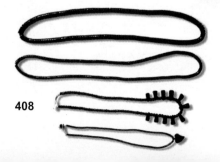
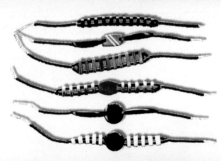

Left-hand page: Apurinã indians on the Paraná do Jari (2004), at the *comunidade* of São Raimundo (which consists of four floating palm-thatched huts). Some 80 Apurinã live here and on the Lago Jenipapo I. They had just traded dried fishes, *farinha,* and jewelry *(below)* for drinks, pots, plastic containers, and confectionery from a *regatão* (travelling merchant) and gave me a friendly welcome in Portuguese. They still hold festivals to honour their dead.

This page: 1. A floating Apurinã hut on the Paraná do Jari.
2. I was able to find discus habitats there as well, but only in the dry season (the photo shows the *paraná* at high water - turbid)
3. In 2004 the *caboclo* family of Eduardo Batista Gomes (aged 39) had been living in the mouth region of the Boca do Jenipapo for nine years.

He catches discus now and then, but lives mainly by agriculture.
4. There are only three small *comunidades* (and a few *caboclo* huts) on the gigantic Lago Jari. This is Santo Augustinho (not shown on any map), which consists of just eight houses with around 40 people.
5. Along the eastern Jari bank lie countless pastel-coloured laterite formations.
6. Discus habitat in the Igarapé Castanha (a Jari tributary) at rising water.
7. Discus habitat in the Igarapé Flecha (a Jari tributary). Full of *acará-açú* and *camu-camu* plants.
8. It is impossible to see the other end of of this gigantic lake.
9. Daytime collecting in the discus habitat in the Igarapé Mari. But they aren't usually collected this way (see *collecting*).
10. Blue discus *(S. haraldi)* immediately after capture in the Lago Jari.

Large photo: A discus habitat can be seen by the eastern bank of the Lago Jari. *Left:* A selection of the variants from the habitat. Note: the pattern doesn't vary much, although the base colour does. All-over-striped specimens (= alpha individuals) are very rare. *Right:* satellite photo (1:2500.000) showing: A) The Lo. Jari, more than 60 km long. A1) The northernmost *comunidade* of Santo Antonio do Jari and the Ig. Flecha (discus habitat). A2) The *comunidade*, Santo Augustinho do Jari, and the Ig. Castanha (discus habitat). A3) The *comunidade* of Mari and the Ig. Mari (discus habitat). B1) Rio Jari. B2) Rio Jari (the *lago* outflow, which flows into the whitewater Paraná do Jari a few km later). C1) The Paraná do Jari, which branches from the Purus and after more than 80 km rejoins it at Arumã (C4). D) Lo. Juçará (accessible only at high water). E) Lo. Jenipapo I (left-hand lake) and the Lo. Jenipapo II (right-hand) which links the Paraná do Jari with the Boca do Jenipapo. F1) The Paraná da Elba is navigable only at high water and cuts out many of the meanders of the Purus (F3) via the Paraná Tata-putauá (F2). There is another short-cut via the Paraná Macaco (F3-F4). Above F2 there are two *lagos*, on the right Pereira, which dries up, and to the left the permanent Baixo (and Piranha in the centre above). G) Lagos Supiá, Paletão (which dries up), Cabeceira Grande, and do Espeto. H) Lo. Itaboca. H1) The entrance to the Lo. Campina. I) Lo. and Ig. Bacuri I (with Apurinã settlement). J) Lo. Bacuri II. K) Lagos Caapiranga, Comprido, Reis, Miuá, and Baixo, Paranás Joari and Taboqinha.

incredible strength. They call it *envira*. Their design is based on ancient traditions and is a symbol of their culture (to ward off the evil eye). Meanwhile, a number of young Christians along the Purus are waiting impatiently to hold their very own New Testament in their hands. A translation into the Kangitê language has been made, with publication scheduled for 2005.

Their floating huts were made from palm fronds and their two *lanchas* were "parked" nearby. The chief told me about his village on Lago Jenipapo I, which had been renamed Nossa Senhora do Carmo, and that his community comprised a total of 80 people. And about the Lago da Cobra, where numerous *acará-disco* live, but also a large snake. That the *lago* was very deep and the snake lived right at the bottom, but anyone who wanted to see it could. Now, I have never seen any giant anaconda, either there or anywhere else. The largest, which my mother caught when I was only a child, was barely 8 m long. All subsequent ones were smaller. One in the Paru region measured around 6 m *(see page 244)* and one I caught myself in 1982 just over 3 m *(see page 99)*. There is no incontrovertible evidence of specimens of more than 9.5 m, just a *Python reticulatus* of almost 10 m which Fred Cochu obtained from Sulawesi more than 50 years ago. So much for giant snakes.

To get to the Lago Jenipapo II you have to turn from the Paraná do Jari into the Boca do Jenipapo, which is full of fishes (the last time I was there an *Anodus* immediately jumped into the boat). After around 800 m you come to the home of the only *caboclo* family in the region (see page 409). Eduardo Batista Gomes has lived there practically autonomously for nine years with his wife and two children, in a wooden hut sited high on the *terra firme*. He told me that he more than got by on his *roça* (manioc), *banana, abacate, laranja, lima, tangerina*, his pigs and poultry, plus piranhas and *tambaquí*. Also that he was born on the Jari and had spent all his 39 years in the region. In the two to three months when there were lots of *acará-disco* he sometimes went fishing with Zézinho, an Apurinã native, who sold them to dealers from Manaus. Only 14 days ago the dealer Clovis had gone back to Manaus with 17,000 discus, towing the fishes in *viveiros* behind his *lancha*. (Just like the first discus collector here, my friend Manuel Torres, *see page 421)*.

I have found discus in the Lago Jenipapo Segundo (II). It has beautiful habitats and is uninhabited. The indians live on the Primeiro (I), where discus supposedly also occur. Here **in the Segundo they are found mainly among** *acará-açú* **plants.** Sometimes all-over-striped (Royal Blue) like the C 4.2 variants (Bleher & Göbel, 1992) and also specimens with less blue striping.

The parameters measured in the habitat on 9.12.2004 at 08.30 hours were as follows: pH 6.15; in 2.5 m depth: 5.98; conductivity 36.0; in 2.5 m of depth: 28.0 μS/cm; daytime temperature (air) 25.0 °C; at the water's surface: 25.9 °C; in 2.5 m of depth: 26.1 °C. Biotope: unspoiled; water black (or rather, clear), but slightly turbid (from the Paraná do Jari/Boca do Jari); *acará-açú* **plants widespread; fallen trees; no floating plants. Discus variants: see above.**

The Lago Jari has only three *caboclo comunidades (right,* see also pages 409 and 411), and they all lie on the *terra firme* on the eastern bank of this gigantic lake. The Lago Jari is so vast that you often feel like you are on the open sea when you paddle around on it. Or you may be happy to get away from it, as I was on one occasion after two hours of storm and rain with waves breaking over the *voadeira*, no flesh left on my sit-upon from the continuous jolting, soaked to the skin, and shivering from head to foot. At around 60 km long and up to 5 km wide, it seems endless, and the other bank is often not visible.

In the lake there are discus habitats only in certain zones around the edge, mainly in *igarapés* such as the Sangue, Flecha, Castanha, Mari (also called Marimari), at their mouths and far upstream. The habitats here are very variable, from fallen trees *(see page 410)* via a mixture of trees, branches, and grass to pure *acará-açú* plant habitat. During the extreme dry season discus are more often to be found in the lake, but as soon as the water is rising to some degree almost only in the *igarapés*. Further up the Rio Jari there are additional discus habitats. That area is practically uninhabited (apart from a few indian settlements). The *caboclo* Adino Nacimento Rodrigo, who accompanied me on one occasion and who was born and grew up by the Lago Jari, told me during the journey: *"Tem é muito acará-disco aqui, o Rio ate acíma, só o boto vermelho vai mais para cima que o acará-disco. E todos os igarapés são de água preta."* (There are discus far upstream in the Jari and only the red dolphin goes further upriver. And all its *igarapés* are black.) But I established that a few of its headwaters contain white water, which, however, becomes lost in the black. And a turbid *igarapé*, the da Pedra, also flows in at the mouth of the Lago Jari. It was a real adventure with Adino. We arrived in a *mutuka* (a small mosquito species) region, where we were almost eaten alive. And during this attack, as we tried to cover ourselves up with anything we had, he told me with a smile about his father. The latter had remarried, at 82, taking a 19-year-old wife, and after she had borne a child she had murdered him in order to inherit his lands and cattle.

The parameters determined at three different Jari habitats were as follows: 1. **The parameters on the east bank north of Santo Augustinho, on 9.12.2004 at 12.30 hours were as follows (all water parameters in 2 m of depth): pH 5.32; conductivity 24.0 μS/cm; daytime temperature (air): 26.7 °C; water: 29.4 °C.**

1. There are only three *comunidades* on the Lago Jari, with 40-50 people in each (and each with four to eight huts). They are all on the eastern bank of the northen half of the lake *(see page 411)*. There are none on the western bank, and just as few in the upper part of the lake, not to mention up the Rio Jari. (It takes 8-10 days by *lancha* to reach the headwaters, and you encounter only indians). Mari, which lies on the Igarapé Mari, is the southernmost of the three *comunidades*. I always received a very friendly welcome there, though the children were often shy (1) or suspicious (2), but that is understandable as most of them have hardly ever seen a white person (and some visibly have indian blood). In Mari I received a hearty welcome at the home of the *caboclo* Pedro Alves Bizer da Silva. One of his nine children had just prepared a *peixe boi* (manatee) (3-4). It had to be boiled for 30 minutes and was then roasted. Tastes like fish. In addition Pedro himself had roasted 10 *tracajá* (aquatic turtles) on the open fire and immediately chopped a few up for us (5). Pedro, who thinks he is 50, had by then lived there for 19 years and told me that discus were first collected there commercially 20 years previously. Collecting had ceased for a few years, but had started again in 2003. He didn't catch any himself, but his son Marcio, by then 20, was well versed and had repeatedly helped me. Marcio had also recently accompanied a Dutchman (who had already discovered various new monkey species in recent years in Amazonia) far up the Rio Jari, where he was able to find additional new species. The Jari is rich in fishes; tonnes of them go to Manaus every year, either on ice or dried (6).

Biotope: natural and unspoiled; water black (or clear), only slightly turbid (from the Paraná do Jari, which was starting to rise due to the influence of the Purus); fallen trees and branches; no floating plants and no grass; substrate laterite, loamy reddish. Discus variants: *page 411*.

2. The parameters at the habitat in the Igarapé Castanha (east bank) on 9.12.2004 at 14.30 hours were as follows (all water parameters in 2 m of depth): pH 5.33; conductivity 23.0 µS/cm; daytime temperature: (air) 27.9 °C; water: 28.1°C. Biotope: unaltered (apart from a *fazenda* under construction opposite, near the mouth; water black; large fallen trees, old branches and roots; no floating plants, but some grass; bottom covered in leaves.

3. The parameters at the habitat at the mouth of the Igarapé Lontra (west bank) on 9.12.2004 at 16.45 hours were as follows (all water parameters in 1.5 m of depth): pH 5.92; conductivity 23.0 µS/cm; daytime temperature: (air) 27.7 °C; water: 29.2 °C. Biotope: water black (or clear), slight turbidity starting (from the Paraná do Jari, which was starting to back up against the Rio Jari); full of *acará-açú*; fallen trees and branches on the beach; no floating plants; bottom just white sand.

The Jari area remains little explored to the present day. You need to spend months in the region to catalogue just a fraction of its fishes. Every time I landed new species, the majority so far undescribed – seven dwarf cichlid species alone, unknown characins and catfishes, and much more. I was able to find discus in yet other *lagos*, including that at Arumã – again a blue form.

I have a few parameters from the upper Lago do Arumã and the *igarapé* of the same name on 7.10.1996 towards noon (all water parameters in around 50 cm of depth): pH 6.32; conductivity 41.0 µS/cm; daytime temperature: (air) 26.7 °C; water: 28.4 °C. Biotope: natural and unspoiled; water clear (the *caboclos* term it *preto)*; fallen trees and branches; *acará-açú* plants; no floating plants; substrate sandy.

Further up the Rio Purus we find just the two tiny *comunidades* of Bé-a-Bá alto and Bé-a-Bá baixo, which are separated by the outflow of the narrow Igarapé Bé-a-Bá *(below)*. Not 12 houses on both sides of the *terra firme*, and barely a dozen floating offshore. A church and school to serve both. Around 80 people who live by fishing and an element of agriculture. Shortly thereafter, likewise on the left-hand bank of the Purus, the Reserva Biológica do Abufari begins at the Lago Campina; inaugurated in 1982, the reserve comprises 288,000 ha and extends almost to **Tapauá** *(right)*, the capital of the *município* of the same name and supposedly the third largest province on Earth. And Tapauá is also the name of a river that flows into the Purus, but more than 200 km distant from the town, further upstream.

The town is enthroned high up on the *terra firme* on the right-hand bank of the Purus, just where the black Rio Ipixuna flows into its white flood. The *município* includes a number of additional very small settlements such as Membeca, the above-mentioned Be-a-Bá, Ponta da Nova Alegria, Piranha, Porto Artur, Jacaré, Baturité, Nova Olinda, Boca do Tapauá (or Foz), Nova Aliança, Boca do Jari, Cassiam, and Seringal do Tambaqui (a rubber plantation). Now, when discussing Tapauá (the town) it is impossible to overlook the name of Daniel Albuquerque. Born in Porto Velho in 1926, he travelled the Purus at the age of just 20, and by 25 had his own *regatão* (trading vessel). He bought himself estates on the Itaparaná, traded in rubber and Brazil nuts, and organised fishing for *pirarucu* in the area. When, at the end of 1955, the state of Amazonas divided his land up into lots of *municípios* (there are by now 62) and removed Tapauá from Canutamá, a suitable place was sought for the seat of government. This was intended to be Boca de Tapauá at the mouth of

1. Bé-á-Bá alto is a small *comunidade* of 30 families on the left-hand bank of the Purus, and is part of the *municipio* of Tapauá. It is separated by a small *igarapé* of the name Bé-á-Bá from Bé-á-Bá baixo, which is even smaller and had no houses on the *terra firme*.
2. The people live by fishing (typical styrofoam boxes by the floating house).

3. A fisherman had caught two *Phractocephalus hemiolipterus* with a *tarafa* (cast net). There are lots of these catfishes in the murky Purus water, and a number of fossils of this genus from the Cretaceous have been found in the Alto Purus.
4. Vegetables are planted in wooden boxes on the raft next to the house.

1. Tapauá is the capital of the *município*. Here a view of the confluence of the Rio Ipixuna with the Rio Purus in 2002 (note the black water flowing into the white and the number of floating houses). 2. Almost the same view in 2004 (the number of floating houses has increased considerably). 3-4. Posters and the annual Festival do Pescador are used to promote tourism. 5. A cross towers high above the main square, overlooking the *encontro das águas*. 6. The Igreja de Santa Rita in the square. 7. The "in" bar, Veranda Tropical, owed by the Hungarian Lorenzo. Top-notch service.

the river after which the new *município* had been named and which flowed through it with for more than 500 km before debouching into the Purus. But because there was no *terra firme* by the *boca*, giving rise to concerns that a "Venezia Brasileira" (Brazilian Venice) would develop there, after a long search the current site was found. It was uninhabited primary forest, which began to be cleared in mid 1956. And because the site overlooked an *encontro das águas* (a view that not even Manaus enjoys), they planned to call the place Estrela do Purus or Pérola do Purus (= the star or pearl of the Purus).

At that time the only people living there were Daniel Albuquerque, João Eliseu Torres (who fished for him), and the floating merchant Milton Rosas. By the 11th December 1957 the first house, the *comune*, was *in situ*. There followed a school and administrative offices. The *tijolos* (bricks) was fetched from Coari in Milton Rosas' *lancha*. At the end of 1959 Daniel was elected the first *prefeito* (mayor) and he held that position for three more terms of four or six years. Today (2004) his son (from his first marriage to a native woman of the Mura tribe) is already serving his second term in the same office.

When, in the last week of October 1963, the aquaplane of the SIL (International Linguistics Society, formerly the Summer Institute of Linguistics) Mission landed here bringing three white men, H. Schultz, H. R. Axelrod, and F. Terofal, there were just eight houses and 234 inhabitants in Tapauá, and Daniel placed one of the former at their disposal. He helped where he could, was a gentleman then as now, but he had never seen the neon tetra often cited by Axelrod as supposedly occurring there. Neither in 1963 nor in the more than 50 years he had lived there and saw daily people collecting fishes. This was also confirmed by his fishermen and acquaintances such as Ciso and Francisco *(below)*. Be that as it may, this first fish expedition was a success. They discovered, *inter alia*, *Corydoras schwartzi*, collected 114 species in four weeks, and were the first to find discus (brown and blue) there – but they thought they were on the Rio Tapauá...

Mailed catfishes are found not far away, just 30 minutes downstream by *voadeira*, in a small left-hand *igarapé (right, above)*. On one occasion I came across the *pescador de peixe de aquario* (= collector of aquarium fishes), as Francisco is known, who had been collecting aquarium fishes for 19 years (this was in 2002). He told me that in the past he had also caught discus, but it was no longer worth the effort. The *gringos* were breeding so many and didn't want to pay anything. Moreover, they were interested only in *pintados* (= painted, ie all-over striped), which were found maybe one in a thousand. He knew all

1. The former mayor of Tapauá, Daniel Albuquerque, aged 77 years, and myself.
2. Albuquerque (34), at the start of his first term of office in 1960.
3. The fisherman Ciso Saraiba da Silva (aged 63), who had lived for 40 years here on the Jiboia, and caught fishes – day-in, day-out – all over the Tapauá region. He looked at the photos of the neon tetra in *Aquatic News* and confirmed that this fish wasn't found there.

4. Likewise the *pescador de peixe de aquario*, Francisco Evilon Fernandes Picanço (42), an aquarium-fish collector here for 20 years, had never seen the neon. He is seen here sorting mailed catfishes (there were 10,000 in the canoe). 5-6. Below decks his *lancha* has been converted to accommodate around 1,800 litres of water in plastic-lined containers, ea. holding 100,000 *Otocinclus* (5) or 100,000 *Corydoras*.

Above: The *igarapé* with hundreds of thousands of *Corydoras* near Tapauá (1). In addition to *C. schwartzi* there are another eight different variants (species?) living sympatric here. Four long-snouts (2-5), possibly belonging to the *C. blochi* group. Individuals with round snouts (6-9), one of them traded as *C. julii* (6) though it has to be *C. trilineatus*. Another as *C. reticulatus* (7), and two as *C. punctatus/agassizii* (8-9), though none of them belongs to those species. *Below:* A few days before my researches food-fishes had been caught in the Igarapé Piranha (15) using *timbó;* these had already all been gathered up, with just the catfishes, mainly doradids, being left to their fate. With gills agape they gasped for air (10), but the majority were dead (11-12). A rotting mess (13), with dead plants (16). Only aquatic bugs (17), an *Ancistrus,* probably *A. dubius* (14), and a beautiful *Pterygoplichthys* sp. (18) were still alive.

Discus and habitats (Itaparaná region): **1-2.** Braço Morto do Itaparaná after the *repiquette* (after the first rise the water recedes again, but soon afterwards rises once and for all). **3.** Blue discus from the habitat (1-2), slightly damaged through daytime collecting. **4.** Discus habitat in the Itaparaná. **5.** Ditto in a *braço morto* (dead branch) at Variae. **6.** Blue discus from this habitat (5) with reticulated markings from the seine and healed piranha bites. **7-9.** On the bank next to the habitat there was a large caiman nest (7). The 20 youngsters that had already hatched remained nearby (9). I stroked one with my index finger, whereupon it immediately rolled its eyes (8). **10-12.** *Virola* sp., an indundated terrestrial plant from the habitat (1) – shelter and food for discus at high water.

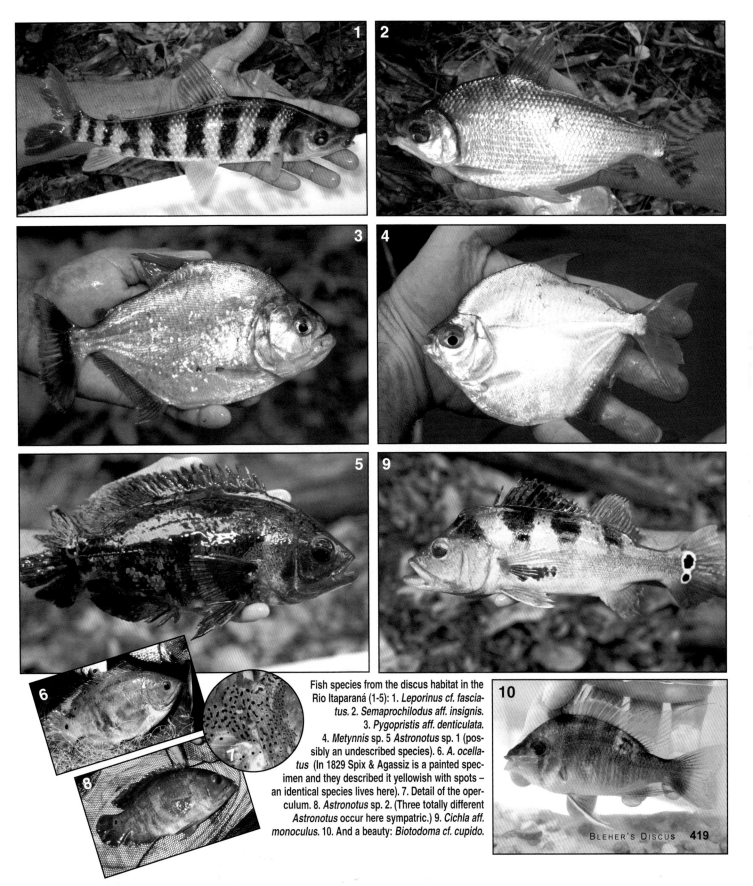

Fish species from the discus habitat in the Rio Itaparaná (1-5): 1. *Leporinus cf. fasciatus*. 2. *Semaprochilodus aff. insignis*. 3. *Pygopristis aff. denticulata*. 4. *Metynnis* sp. 5 *Astronotus* sp. 1 (possibly an undescribed species). 6. *A. ocellatus* (In 1829 Spix & Agassiz is a painted specimen and they described it yellowish with spots – an identical species lives here). 7. Detail of the operculum. 8. *Astronotus* sp. 2. (Three totally different *Astronotus* occur here sympatric.) 9. *Cichla aff. monoculus*. 10. And a beauty: *Biotodoma cf. cupido*.

the categories from the area, for example the *disco-comun* (brown), *cigana* (gypsy), *azul* (blue), *branco* (without pattern colour), *listado* (striped), and the *disco-verde* (green) from Tefé. But he had long since specialised in *corredora* and *otocinclo,* as he called them. Converted a *lancha* for the purpose and fitted out the hold, with plastic-lined wooden boxes *(see page 416).* Each box measured approximately 2.50 x 1.20 x 0.50 m and could contain around 100,000 *Otocinclus* (100 fishes per 1.5 l, without aeration, for a journey of around a week).

Tapauá lies 30 m above sea level and its *município* is, at 96,500 km², larger than Portugal. It is 450 kilometres by air, but 1,176 by river, from Manaus. Naturally the people grow *mandioca, juta,* and *feijão,* and cultivate *abacate, banana* and *laranja.* But more than 40% of the 18,326 inhabitants (2004) live by fishing (for food fish), as there is no cattle and pig breeding worth mentioning in the region. And because fishing is their livelihood it was for this reason that in 2002 the first Festa do Pescador (29th-31st August) was brought into being, in order to encourage tourism and honour the *pescadores* of the region. The largest fish caught wins. In this case a *tambaquí* more than a metre long *(below).* The town is growing, and a single ice-making factory has now become three. The food-fish business is booming. I was there when the fishermen were catching *jaraquí de scama grossa e fina* (two *Semarochilodus* species, large- and small-scaled) that were spawning in the white water and migrating downstream. There must have been at least 5,000, all full of roe. If only they had at least allowed them to spawn and then caught them a week later on the way back...

Brazil nuts are still collected, rubber and palm oil harvested – but for a long time not as much as in the past. Timber is in demand and there is a large sawmill. When I arrived in 2002 the airport had just been surfaced with tarmac – but there were still no buildings (for this reason air traffic was stopped in 2004) and the first hotel was still being erected. Nowadays the owner, Raimundo Araujo de Souza, has competitors. Only the Veranda Tropical bar, owned by the Hungarian Lorenzo, who came here and remained, has none. His bar is still *the* place to meet.

Discus habitats are widespread in the Tapauá region. The first is up the Rio Ipixuna (also written Pixuna, Parapixuna, Paranapixuna) from Tapauá. Here in the blackwater river, mainly along the left-hand bank, there vast quantities of *acará-açú* plants. I found Blues, individuals ranging from those with very little stripe pattern to some with stripes almost all over the body, and some with an irregular pattern (the above-mentioned "gypsy").

And the parameters measured at two different times in the habitat here were:

1. On 5.10.1996, at midday (all water parameters in approximately 50 cm of depth): pH 6.30; conductivity 18.0 μS/cm; daytime temperature: (air) 30.7 °C; water: 31.3 °C; oxygen: 29.1 mg/l. Biotope: natural and unspoiled; water black; *acará-açú* plants; no floating plants; substrate sandy.

2. Near the Igarapé Santa Maria on 4.8.2002 at 14.20 hours (all water parameters in 20-30 cm of depth): pH 6.07; conductivity 10.0 μS/cm; daytime temperature: (air) 35.0 °C; water: 28.3 °C. Biotope natural and unspoiled; water black; *acará-açú* plants; roots, trees, branches, leaves; no floating plants; no grass; substrate = sand.

Somewhat further upstream the Rio do Jacinto debouches on the right-hand side; I haven't explored it, but the same variants supposedly occur in its black water. On the left-hand bank the Igarapé Giboia (Jiboia) (the native name for *Boa constrictor)* debouches shortly before the enormous lagoon – at least 10 km in diameter – where the two *rios,* the Ipixuna and the Itaparaná, meet. It too is said to contain black water, but in my view its water is more like clear. Here lives Ciso *(see page 416),* who has been collecting (food) fishes for more than 40 years, and who specialises in *pirarucú.* There are Blues here, under and around *camu-camu* plants (a *camu-camu* species known locally as *araçá)* which bear splendid fruits when the water starts to rise, and at this time these can easily be "picked" by the discus and other fishes. When the water rises it is possible to reach the Lago Cantagallo, further to the north-west, by canoe, via the Paraná Giboia *(see page 423)* and the Furo de Cantagallo.

The parameters in the Lago Cantagallo habitat on 6.12.2004 at 12.30 hours were as follows: pH 5.68; in 2 m of depth: 5.60; conductivity 21.0; in 2 m of depth: 20.0 μS/cm; daytime

Left-hand page: A gigantic *tambaquí* (*Colossoma macropomum*), more than a metre long, from the Purus at Tapauá. Individuals up to 1.5 m were caught 100 years ago, but nowadays those seen in the markets – this is the chief food-fish of Amazonia (or rather, Latin America) – measure at most 50 cm. The longest museum specimen is 99.5 cm TL.

This page:
1. An interesting feature of the discus variants of the Purus region is that their base colour becomes lighter the further west you go. This is a specimen from the region near the blackwater Lago Tambaquí (where I didn't see any *tambaquís*). At extreme high water there are links with the Rio Mamiá (see map *(centre)* and text).

2. Satellite photos of the Tapauá region (1:250,000): A) Purus. B) The town of Tapauá, at the mouth of the Rio Ipixuna (Itaparanás – see below) where it joins the Rio Purus. C) Rio do Jacinto. D) The vast lagoon where the two blackwater *igarapés* meet and then flow combined past Tapauá (B) to end in the Purus. D1) Rio Ipixuna (Pixuna, or even Parapixuna). E) Rio Itaparaná. E1) Cachoeira in the Itaparaná; the habitats cease downstream of here.

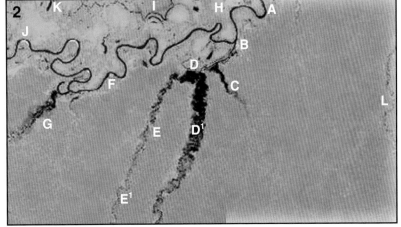

F) The small village of Baturité on the right-hand bank of the Purus. G) The Rio Jacaré, with the large Lago Arimã, before it enters the Purus. H) Reserva Biologica do Abufari, with the Lago Scopema. I) The Igarapé Pauapixina, which is linked with the Rio Coari and Rio Mamiá. J) The *comunidade* of Jaburú on the left-hand bank of the Purus. K) Lago do Tambaquí. L) The eternally long Rio Jari (see page 411). And endless jungle in between.

3-6. Photos (from 1994) of discus being transported from the Rio Cunuhá (Cuniuá) in the Purus region to Manacapuru (and then by truck to Manaus). This was undoubtedly the largest and most unusual method of supplying discus in history. For months the well-known Manuel Torres caught blue and brown discus. Built *viveiros* for his discus (5) and then attached the 19 *viveiros* (3) with a total of 94,000 discus – around 5,000 in each *viveiro* (6) – to his *lancha*. In this way Manuel towed (4) the *viveiros* all the way downriver in 18 days and nights, and he says that only a few of the fishes died.

temperature: (air) 32.7 °C; water: 30.5°C; in 2 m of depth: 29.3 °C. Biotope: water rather clear; completely isolated, uninhabited, natural and unspoiled; *araçá* plants; no plants in the water; substrate = leaves, branches, and sand.

Discus habitats are encountered extensively in this region, including in most of the *lagos* that have no permanent connection with the Purus as well as those that are considerably influenced by its turbid water. I was able to find them in the Lagos Comprido, Preto, Castanho, Onçina, Ancorí, Jatuarana, and, further up the Purus, in the Arimã (right-hand bank) and Jaburú (left-hand bank). In the Lago Solitario, which can be reached only overland, likewise the Lago do Marajá *(see page 424)* with a lovely gypsy variant, and in the Lago do Serrão. Note that the last of these lies further up the Purus in the *comunidade* of Curupati, on land belonging to Zémanão. I had asked him for permission to enter. Natasha and Leonardo accompanied me on the overland trek. It was quite a adventure. Past a giant *bacuri* tree with 6 m long aerial roots extending more than a metres from the ground. Craning our necks looking up at the 50 metre high *assaçú* trees towering straight as a die into the heavens. Sometimes along muddy paths (despite it being the dry season) and through pools, until after around 6 km we reached the isolated lake, more than a kilometre long and 100-200 m wide.

The parameters in the discus habitat there on 3.8.2002 at 15.30 hours were as follows (all water parameters in approximately 50 cm of depth): pH 6.40; conductivity 30.0 µS/cm; daytime temperature: (air) 35.7 °C and 32.0 °C in the shade; water: 32.0 °C in the sun and 25.5-26.0 °C in the shade. Biotope: natural and unspoiled; water black; fallen trees; no floating plants; substrate = sand.

As well as blue discus we also discovered splendid *Chaetobranchus semifasciatus (see page 425)* while a few caimans watched us from the bank. Only when I went closer did they beat a retreat into the water. Interestingly, in this region I always found the swarms of mosquitos and small stinging flies that are encountered everywhere, despite the black water. It is as black as in the Rio Negro, but there are none there. The answer lies in the degree of acidity, and goes to show that not all black water is the same.

Further up the Rio Ipixuna I was able to find lots of blue discus habitats, although the river is settled as far as the BR 230 (Transamazonica) and beyond. It is more than 300 km long and rises north of São Carlos, which lies on the Rio Madeira (but there is no link) south of the BR 319 (Manaus to Porto Velho). The Rio Itaparaná, by contrast, is little populated. There are just three small *comunidades* (the largest had 50 inhabitants in 2004) along its more than 200 km length. It flows almost parallel to the Ipixuna, rises south of the BR 230, and descends over several *cachoeiras* (rapids and waterfalls). I have been able to find habitats and discus upstream almost as far as the first *cachoeira (see page 421)*. In the Lagoa 26, where I found splendid cichlids in the habitat *(see page 425)*, in the Braço Morto do Itaparaná *(see page 418)*, at the *comunidade* of Nova Olinda, and in the Lago do Caissiana. Along with with Natasha and Leonardo I once spent an entire night searching in the last of these and lying in wait with a large bait-fish. There was said to be a more than 50 metre long snake living there, which had already swallowed several people – or, at least, in every case they had vanished here. But although it only came out at night, nothing. Although I did discover in the process that pike-characids *(Boulengerella cuvieri)* build circular sand-nests, a metre in diameter, at night and spawn in them. These individuals were more than 40 cm long and chased away anything that came near them.

Now a few parameters from habitats in this region:

The parameters at Arrombada do Toari/Itaparaná on

1. Cisno the fisherman had lived here on the Jiboia for 40 years. The twin houses stood on stilts, but had no walls. 2. He produced his own hand-made throwing spears, with which he killed the largest freshwater fish on Earth, the *pirarucu*. 3. Capitano Almir, covered in knife scars. 4. Leonardo (who on two occasions helped me explore in the Tapauá region) with a red piranha at a discus habitat (behind). 5. Moth. 6. The tip of Cisno's throwing-spear.

1. A discus habitat *(right in the photo)* in the Rio Itaparaná, in a still bay (discus are very rarely found in flowing water).
2. Discus habitat in the Lago Cantagallo, north-east of the town of Tapauá.
3. In the Rio Ipixuna there are not only *acará-açú* plants but also (even more frequently) large numbers of *camu-camu*, a favourite discus habitat, especially when the water is rising.
4. The Paraná Giboia is, like the *igarapé* of the same name, named after the South American giant snake, the *giboia* (Boa constrictor). During the high-water period it provides access by canoe to the Lago Cantagallo and to the Abufari region.
5-6. *Camu-camu* plants are widespread in the central Purus region (in the black- and clearwater areas). Their fruit (inset) is a food source for many creatures and has a high vitamin C component. And the day-flies are a protein food *(below)*.

Giboia

Hunting for discus at night, and habitats during the rainy and dry seasons (Purus region): 1. Collecting discus at night with Leonardo in the Lago do Marajá in the Rio Ipixuna region. When the water rises the discus also spawn here, in crevices in the large trees. 2. Lago Marajá, the same spot at low water. 3. The Lago do Marajá can be reached only by land. The canoe has to be carried several kilometres from the bank of the Ipixuna so it can be used for fishing in the isolated lake. Only at extreme high-water is it linked to the *lagos* Comprido and Preto and sometimes also to the Rio Ipixuna. 4. A typical habitat in the Lago do Serrão, with the water rising, again at night. 5. And the Lago do Serrão at low water, the same habitat. 6. Natasha often had to walk through swamps and mud at night, as seen here, in order to reach the Lago do Marajá 3 km away. 7. This nocturnal discus hunt was unsuccessful, but there were lots of the catfish *Agamyxis pectinifrons* in the tree-trunk (8). 9-10. When the water recedes from the *igapós* (dry season), the (uneaten) seeds start to sprout.

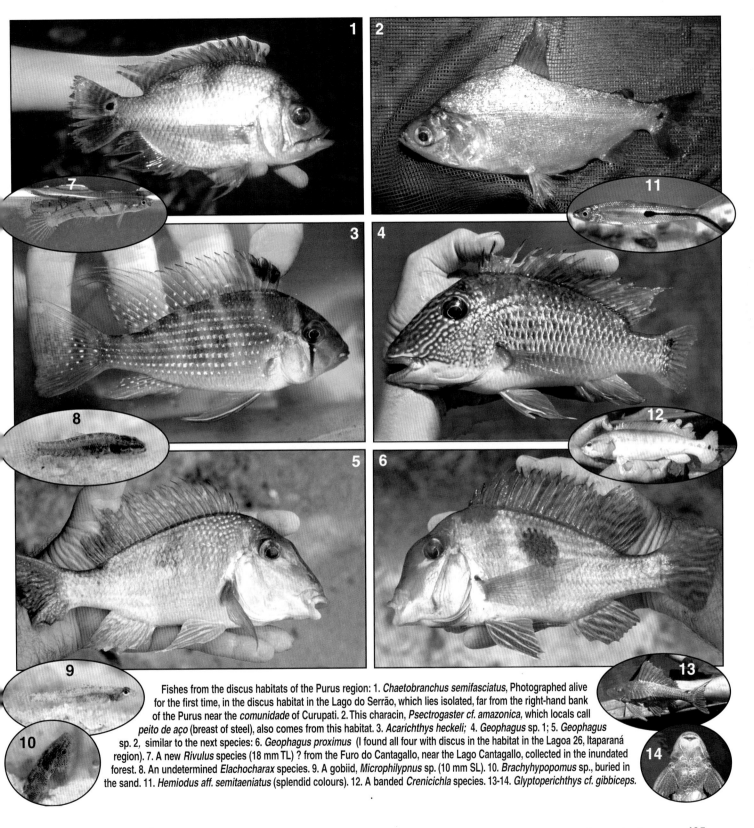

Fishes from the discus habitats of the Purus region: 1. *Chaetobranchus semifasciatus*, Photographed alive for the first time, in the discus habitat in the Lago do Serrão, which lies isolated, far from the right-hand bank of the Purus near the *comunidade* of Curupati. 2. This characin, *Psectrogaster cf. amazonica,* which locals call *peito de aço* (breast of steel), also comes from this habitat. 3. *Acarichthys heckeli;* 4. *Geophagus* sp. 1; 5. *Geophagus* sp. 2, similar to the next species: 6. *Geophagus proximus* (I found all four with discus in the habitat in the Lagoa 26, Itaparaná region). 7. A new *Rivulus* species (18 mm TL) ? from the Furo do Cantagallo, near the Lago Cantagallo, collected in the inundated forest. 8. An undetermined *Elachocharax* species. 9. A gobiid, *Microphilypnus* sp. (10 mm SL). 10. *Brachyhypopomus* sp., buried in the sand. 11. *Hemiodus aff. semitaeniatus* (splendid colours). 12. A banded *Crenicichla* species. 13-14. *Glyptoperichthys cf. gibbiceps*.

6.8.2002 at 08.30 hours were as follows (all water parameters in 50 cm of depth): pH 5.80; conductivity 10.0 μS/cm; daytime temperature: (air) 28.0 °C; water: 26.5 °C. Biotope: completely isolated and uninhabited – totally undisturbed; water black; *araçá-açú* plants; a little grass; no aquatic plants; bottom sandy, with leaves and branches.

The parameters in the Lagoa 26 on 5.8.2002 at 10.30 hours were as follows (all water parameters in around 50 cm of depth): pH 5.80; conductivity 10.0 μS/cm; daytime temperature: (air) 29.0 °C; water: 28.0 °C. Biotope: completely isolated and uninhabited; natural and unspoiled; water black; fallen trees; no plants in the water; bottom covered in leaves with sand everywhere.

The parameters in the Braço Morto do Itaparaná on 7.12.2004 at 12.00 hours were as follows: pH 5.80; in 1,5 m of depth: 5.92; conductivity 16.0; in 1,5 m of depth: 16.0 μS/cm; daytime temperature: (air) 36.5 °C; at the water's surface 30.5; in 1,5 m of depth 29.5 °C. Biotope: completely isolated, uninhabited; natural and unspoiled; water black; fallen trees with forked branches; no aquatic plants; only inundated terrestrial plants in flower *(see page 418)*; bottom covered in leaves, branches, sand. Discus variants: *page 418*.

There had been a *repiquette* and the water level had dropped more than a metre. The discus habitat was now in only in 1.5 m of depth. This happens once a year. Always after the first rise – often a month later – the water drains away for around eight days before rising massively until the end of the rainy season/high water period.

In contrast to the Tapauá region, where *camu-camu* plants are found in the discus habitats (likewise to some extent in the lower Itaparaná), a *Virola* species is present here; one of the more than 60 South American species of the family Myristicaceae, a number of which not only yield exceptional general-purpose timber but also contain a hallucinogenic resin which for thousands of years has been used by the aboriginal peoples of Amazonia, mainly in their ceremonies. (So not only fishes benefit from it...) I learnt about it among the Paumarí indians, nomads who nowadays also live in the Rio Tapauá region, one of the additional discus habitats further upstream and formerly the second most important collecting region.

The Paumarí (also called Pamoarí, Palmarí, Pammarí, Pamarí, Wayai, Yija´ari, or Kurukurú), who belong to the Arawá language group and around 870 of whom still survive (in 2000), call it *kavabo*. They use the inner layer of the bark which is rich in the red resin *(centre)*, ad which they boil up in water to create a thick syrup. This syrup is then dried, ground to a fine powder, and used as "snuff" without any further processing. They use this hallucinogenic "snuff" in the diagnosis and treatment of illnesses, in soothsaying, and for other ritual purposes. Paumarís also believe that it is possible to communicate with the spirit world under the influence of the drug. They formerly used hollow bones to blow it up one another's noses, but I saw how they now instead use empty ball-point pen tubes. But the women don't do this. Instead they boil up the inner layer of bark and drink the resulting tea – likewise intoxicating – or use it to treat inflammations.

The Paumarí use *Virola elongata*, but I noted that the Yanomami use *Virola theiodora*, which occurs in northern Amazonia, and call their drug *epená*. And the Maku in the Uaupés region boil the bark to make a tea which is used to treat malaria as well as stomach and intestinal problems.

One Paumarí village is called Santa Rita and lies on the Lago Marahã, where there are also discus habitats *(right)*. In the past, until 1974, Paumarís lived on the Purus in the *comunidade* São Clemente and usually, like the Apurinã, in floating palm-thatches on rafts. Their traditional nomadic way of life contributed to some of them surviving the European colonisation. They also had their rafts on the Rio Jacaré, another blackwater river, which, like the Itaparaná, rises in the south. There I found very similar discus variants to those from the Itaparaná, often lurking beneath *camu-camu* plants at high water. Nowadays the Jacaré is practically uninhabited. The Paumarí moved into the Rio Tapauá region a long time ago, settling by the above-mentioned Lago Marahã and the Lagos Manissuã and Paricá, as well as along the Rio Ituxi. Only in the Reserva do Caititu do they live by themselves, in other places they have to share with several other indigenous groups. Some people believe they are the descendants of a sub-group of the extinct Puru-Puru, who until the mid 19th century lived by the Lago Jari, the Ipixuna, and in the mouth region of the Ituxi. The other sub-group was supposedly the Juberi (Jubirí, Yuberí) from the lower Tapauá (likewise extinct), who also had villages on the Lagos Abonini and the Mamoriá-Açú.

The Paumarí have been in contact with white men for around 200 years. In 1862, during his botanical expedition to the Purus, Gustav Wallis saw the first *maloca* on the Rio Jacaré. And was present on the Arimã when around 600 Paumarí and Juberi were hired by Manoel Urbano da Encarnação to clear-fell a large area for the construction of the church of Pedro da Ceriana and its mission. But it was Chandless who first published in detail on these cheerful indians, noted for their frequent singing. He labelled them an aquatic people who practiced little agriculture, growing just *manioca*, *aipím*, and *banana*. But stated that they were skilful fishermen, who used bow and arrow to catch the fishes and turtles that were their main food. On one occasion he saw 60 canoes going turtle-hunting (as late as 1955 the same happened to me in the Mato Grosso). Interestingly, according to his account in each canoe a woman paddled and a man kept a lookout for turtles. During the dry season they lived in palm-thatches on the sandy shore and in the rainy season in the centre of the *lagos* on their rafts, in order to escape the mosquitos *(below)*.

The largest settlement on the Purus, Lábrea, which today (2004) has 28,931 inhabitants, was founded on Paumarí land in 1873. Colonel Labre had the streets and houses there lit with lamps burning turtle butter and oil, which the Paumarís had to provide for the purpose. But there are no longer any discus in the Lábrea region, which lies 1,672 kilometres from Manaus by river. And if there were any they would have no future. In this region clear-fulling is in full swing. *Fazendas* are spreading everywhere, and now the BR 230 (Humaitá to Lábrea) is under construction (costing R$ 70 million over the period 2004-2007). Once it is finished it will be usable year-round, and then nobody will be able to stop "progress" any more.

Before I now move on to the Rio Tapauá and the Cuniuá, I would like to make a few more, hopefully interesting, remarks on Jaburu, as it is, apart from Baurité, the only *comunidade* between Tapauá and Foz that is worth mentioning. It too has barely more than 30 inhabitants and mainly floating houses, but, unlike Baurité, lies on the left-hand bank. In the hinterland lie the Lago da Lontra and the Lago do Chapado, where I found discus habitats

1. Today the Paumarí live in the central Purus region and no longer follow their former nomadic lifestyle, but nevertheless still live for the most part in their floating houses on rafts. Here on the Lago Marahã, with discus habitat. At high water this lake too starts to be influenced by the turbid Purus.
2. A typical Paumarí house on stilts, here at Santa Rita on the Lago Marahã.
3. A Paumarí family at Santa Rita.
4. Paumarís eat fish every day. Here a woman in the Araçá village on the Rio Ituxi.

Even so they grow a lot on the *várzea* and *terra firme*. The anthropologist Peter Schröder has established that they cultivate around 28 manioc varieties as well as 30 other plants such as sweet *cassava* and sweet potatoes, *taro*, maize, beans, pumpkin, lots of fruits and palm-trees, and many more.
5. Small, isolated, palm-frond "tents" are woven for the young Paumarí girls, where they spend the night alone until they get married.

in the black water (but no unusual colour variants), as well as in the huge Lago do Cachimbo *(see p. 429)*. I haven't investigated the Lagos Mororó and Socó, but I have fished in the Igarapés do Cachimbo and do Tapají, where *"cardinais"* occur! However, they are not our familiar cardinals, but *Dianema urostriatum,* which actually live in the Rio Negro (!), and because they were caught here in quantity for the exporters in Manaus, and because it was known that cardinals are popular fishes for the masses, these too were simply christened cardinals. Nowadays they are not caught much any more, not only because the demand has diminished, but because timber and food fishes are more profitable. Before entering the mouth region of the Rio Tapauá on the right-hand bank of the Purus we find ourselves in the Foz or Boca de Tapauá, the "Venezia Brasiliana" (worthy of a sizeable book in itself). The ruler of the floating settlement is Cloves, a *commerciante* (merchant) and manager. A splendid blackwater *igarapé* flows through his seringal, full of interesting fishes – but no discus.

Upstream in the Rio Tapauá we soon come upon the floating Paumarí village of Maniçuã (Manissuã), consisting of four wooden huts with palm-thatched roofs, but open on two or three sides (2002). Here the discus habitats begin, even though the river is gigantic, in places more than a kilometre wide, with innumerable *igapós* and flooded woodlands along a large part of its more than 500 km length. Now there is much debate over where it starts or stops, and almost every map shows something different. After passing the right-hand Igarapé Citari, which is more than 100 km long and has discus habitats in its lower part, we come to a larger left-hand tributary further upstream, the Igarapé Capitão, where I have no personal expe-

rience of discus, though I have found them in its first right-hand tributary, the Igarapé Minuá (some people say that the Capitão is instead a tributary of the Minuá). The next left-hand tributary is known locally as the Rio Cuniuá (also written Cunuiá, Cunuhá, or Cunhuá), but on IBGE maps it is called the Rio Tapauá, while the right-hand affluent is the Rio Pinhuã. Be that as it may, both extend for hundreds of kilometres, are virtually uninhabited, and flow through a gigantic area occupied by native peoples, where barely 20 *caboclo* families live.

Upstream in the Cuniuá, which can be reached in a morning from Foz by boat after several typical discus habitats and unspoiled wilderness we come to the tiny Paumarí *comunidade* of Açaí, with the Lago Açaí and discus habitats. And the further up the Cuniuá we travel, the deeper we find ourselves in another world (only in the Nhamundá have I experienced anything similar). Wildlife is no rarity along the eternally green shore, whether it be caimans, turtles, monkeys in the trees, herons, parrots, aras, toucans, or other birds. And given the numerous *lagos* and *igapós* it is also no wonder that people come here from Manaus (less often in recent years) in order to catch discus in large quantities. Around five days upstream by *lancha* we come to Cidadezinha, the largest *comunidade* on the Cuniuá, which is at the same time the first Deni settlement, with around 15 houses and some 150 natives. All the buildings are on stilts. Interestingly I discovered that a few Denis have become bee-keepers and are producing a lot of very good honey from stingless bees. Further upstream is the village of Marrecão. It takes another five hours to get there, and it lies not on the riverbank, but around three minutes inland on foot. There is also an airstrip, as a plane belonging to the Missão Novas Tribos do

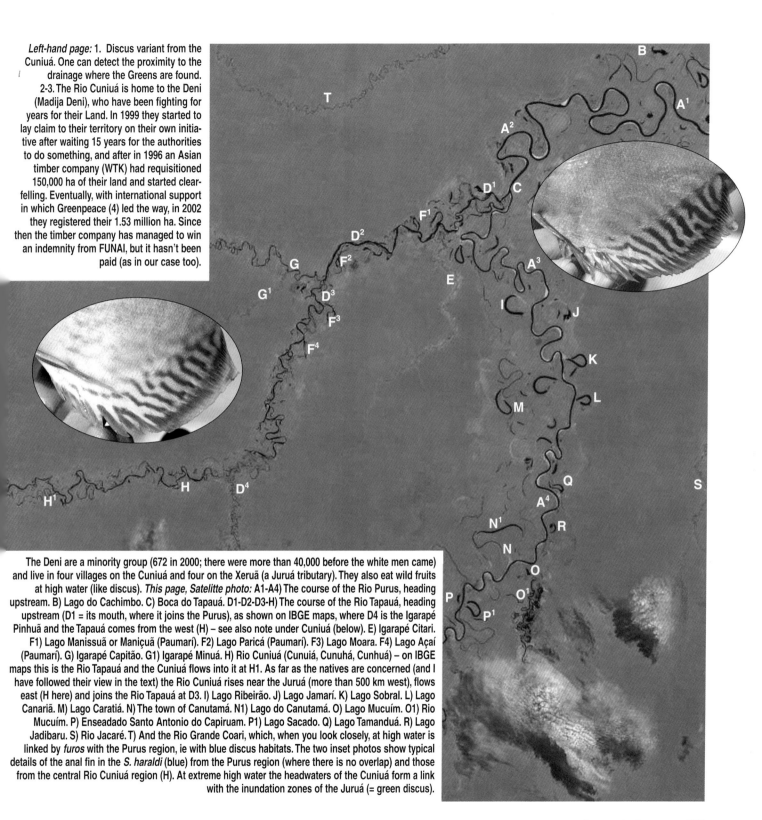

Left-hand page: 1. Discus variant from the Cuniuá. One can detect the proximity to the drainage where the Greens are found. 2-3. The Rio Cuniuá is home to the Deni (Madija Deni), who have been fighting for years for their Land. In 1999 they started to lay claim to their territory on their own initiative after waiting 15 years for the authorities to do something, and after in 1996 an Asian timber company (WTK) had requisitioned 150,000 ha of their land and started clear-felling. Eventually, with international support in which Greenpeace (4) led the way, in 2002 they registered their 1.53 million ha. Since then the timber company has managed to win an indemnity from FUNAI, but it hasn't been paid (as in our case too).

The Deni are a minority group (672 in 2000; there were more than 40,000 before the white men came) and live in four villages on the Cuniuá and four on the Xeruã (a Juruá tributary). They also eat wild fruits at high water (like discus). *This page, Satelitte photo:* A1-A4) The course of the Rio Purus, heading upstream. B) Lago do Cachimbo. C) Boca do Tapauá. D1-D2-D3-H) The course of the Rio Tapauá, heading upstream (D1 = its mouth, where it joins the Purus), as shown on IBGE maps, where D4 is the Igarapé Pinhuã and the Tapauá comes from the west (H) – see also note under Cuniuá (below). E) Igarapé Citari. F1) Lago Manissuã or Maniçuã (Paumarí). F2) Lago Paricá (Paumarí). F3) Lago Moara. F4) Lago Açaí (Paumarí). G) Igarapé Capitão. G1) Igarapé Minuá. H) Rio Cuniuá (Cunuiá, Cunuhá, Cunhuá) – on IBGE maps this is the Rio Tapauá and the Cuniuá flows into it at H1. As far as the natives are concerned (and I have followed their view in the text) the Rio Cuniuá rises near the Juruá (more than 500 km west), flows east (H here) and joins the Rio Tapauá at D3. I) Lago Ribeirão. J) Lago Jamarí. K) Lago Sobral. L) Lago Canariã. M) Lago Caratiá. N) The town of Canutamá. N1) Lago do Canutamá. O) Lago Mucuím. O1) Rio Mucuím. P) Enseadado Santo Antonio do Capiruam. P1) Lago Sacado. Q) Lago Tamanduá. R) Lago Jadibaru. S) Rio Jacaré. T) And the Rio Grande Coari, which, when you look closely, at high water is linked by *furos* with the Purus region, ie with blue discus habitats. The two inset photos show typical details of the anal fin in the *S. haraldi* (blue) from the Purus region (where there is no overlap) and those from the central Rio Cuniuá region (H). At extreme high water the headwaters of the Cuniuá form a link with the inundation zones of the Juruá (= green discus).

Brasil, who have a house and a small shop there, comes once or twice a month.

During my visit a Deni group had just come back from collecting *copaíba* oil, which they exchanged for goods from the *regatão* that comes by at irregular intervals (the Deni are unfamiliar with money). They were very excited as they had seen a *banihada* herd again. None of them could explain exactly what *banihada* were, but they were supposedly larger than *onças* and lived in the treetops. They had seen these creatures years ago and shot one. But it hadn't been killed and the entire herd came after them. It appeared that the Denis were scared of these *banihadas* (the only other thing they are frightened of is the jaguar). The missionaries had also heard of *banihadas*, but didn't know what they were. They thought they might be anthropoid apes, which reminded me how as a child in the Guaporé, south of here, I once saw a strange, huge, gorilla-like animal in a tree. I will never forget it. And the fact is that more than 15 new monkey species have been discovered in the Amazonas alone since 1990, there is undoubtedly much still to discover in the jungle. The only school in the village is open from 8-11 o'clock, but every third day the Deni teachers have to go fishing, and then it remains closed, and the 10 schoolchildren, aged from five to 15, have the day off. Upstream there are two further Deni villages, Viagem (Visagem) and Samaúma, but the discus habitats are long gone.

I have only one set of data from this region, from a habitat in the Rio Tapauá, where the parameters on 5.10.1996 at 13.00 hours were as follows: pH 6.30; in 1.2 m of depth 6.15; conductivity 18.0; in 1.2 m of depth: 18.0 μS/cm; daytime temperature: (air) 34.5 °C; near the water's surface: 30.4; in 1.2 m of depth 29.9 °C. Biotope: water clear (but also termed black); an isolated uninhabited region; primary forest with fallen trees; no plants in the water, apart from inundated *Virola elongata*, but with no flowers (at this time of year); substrate = leaves and sand.

Further up the Rio Purus, I was able to record habitats only along the right-hand bank, and for that I had to rely on statements from the *ribeirinhos*. These habitats are supposedly in the isolated (and generally accessible only overland) Lagos Jamarí, Sobral, Canariã, and Tamanduá, and in the likewise blackwater Jadibaru (see page 429).

The Rio Mucuím, with its large, around 5 km long Lago Mucuím, flows in on the right and was a favourite discus collecting site in the 1990s. Less so more recently. Here I was still able to find beautiful striped specimens (but not with an all-over stripe pattern). At the end of 2002 FUNAI designated a piece of land (73,000 ha) here as the Terra indígena Apurinã do Igarapé Mucuím, extending for 131 km along the *igarapés*, and in 2004 this was legally confirmed. In addition to their main work fishing and cultivating *mandioca*, the Apurinã here also collect lots of wild fruits in the untouched forest, and produce their much-prized *caiçuma* wine from *açaí* and *buriti*. They also still regularly hold their traditional *xingané* festival in honour of their dead *(see page 408)*. When a member of the tribe dies, one of the children is responsible for organising three *xingané* festivals in succession. The Apurinã believe that only with the third festival will the spirit of the deceased be free. They hunt and fish all day long. The fishes are paralysed in the biotope using an ichthyotoxin which they obtain from the so-called *tingui* liana.

One thing I found very interesting. A youth had to be treated for a swollen leg, the result of an insect "bite" (moth?). Through a tiny opening an approximately 2 cm long "worm" (larva) could be seen moving inside, but couldn't get out. The mother blew a stream of smoke into the wound and then covered the latter with a slimy mass of tobacco. About 15 minutes later she pulled out the anaesthetised larva with a pointed splinter of wood. It hurt. I know this because I've been through in myself, on several parts of my body. In Bolivia moths had laid their eggs under my skin, and after the larvae hatched they had to be extracted in precisely the same way. I even had to have a 4 cm long one "surgically removed" from the back of my head.

This entire region is inhabited by a large variety of native tribes. Many have died out, but there are still villages of Jarawará with barely 160 souls, that of the Hi-marimã with even fewer, but they have no contact with white men. A Banawá (Banawá-Yari) village with 120 and that of the mixed-blood Suruaha with 139 people (in 2000). Of these peoples, the Hi-marimã and the Suruaha are not even listed among the 210 tribes registered in Brazil. The Suruaha live by the Igarapés Riozinho and Coxodoá, tributaries of the Rio Cuniuá. They belong to the Arawa language-group and are a remnant of the long-extinct Masaindawa, Kuribidawa, Eidahindawa, Sarukwadawa, Adamidawa, Aijanema, and

Jukihidawa tribes, who lived in parts of the Purus region that were easily accessible for the whites. They abandoned their lands in the first half of the 20th century, after the *seringueiros* and fur-hunters arrived here and made short work of anything and anyone that got in their way (only the Paumarís made friends with them). The majority were massacred or decimated by the diseases of civilisation. A minority fled to the remote *igarapés*, where they still live today, and weren't rediscovered until 1980. But they have no future. Since the massacre of their tribes, suicide has become a ritual among them, reflecting their racial trauma and serving as a protest against the intolerable events of the past.

I found their view of the cosmos very interesting. They say that the universe is composed of three entities:

Adaha: the Earth, the world of the living.

Adahabuhwa: the underworld (which starts beneath under the earth) where the *kurimie bwadahazy* spirits live. Every spirit manifests as a plant, in which it resides.

Zamzama: the sky, the region beyond the sun, moon, and stars. Populated by the *kurimie namhaze* and *agabuji karuji* spirits. The former are characterised by the birds, and the latter are responsible for the wild fruits.

One can learn a lot from the aboriginal peoples, but unfortunately by far the majority of them have disappeared.

But I would also like to mention the Paraná Bela Vista, which is linked with numerous *lagos* and, according to the *seringueiros* I asked (mainly in the Seringal Axioma), harbours *acará-disco*. But that was the furthest south that I was able to find out information from the right of the river. And there are supposedly discus in the *lagos* even further south on the left-hand side of the Purus, some of which are extremely isolated and accessible only by trekking for days overland, for example the horseshoe-shaped Lago do Ronca, *inter alia*.

Thus the southernmost boundary of discus distribution effectively lies around the city of Canutamá, which in 2004 had a total of 10,067 inhabitants. The *município* includes the *comunidade*s of Ribbeira, Belo Monte, Jamunduá, Glória do Rouca, Novo Ariá, Nova Colônia, Forte Veneza, Moará, Seringal do Jaburu, Caburité, Porto Alegre, Paxiuba, and São Francisco. Its name derives from an indian who collected aquatic turtles here under Manuel Urbano da Encarnação, and who on cutting his foot shouted *"canutama, canutama"*, which in his language means roughly "cut foot". It too lies only 30 m above sea level and its *município* comprises 24,027 km². Here, and the further south you go, there are *fazendas*. But, economically speaking, timber still holds first place, followed by Brazil nuts and rubber. There is a hotel and a supermarket, but no hospital, and the *comunidades* are without schools. During my last visit were the wages of the state employees were eight months in arrears, and the *prefeito* Raimundo Amorim had been accused of corruption. The leader of the commission conducting an investigation, *vereadora* Marlete Brandão of the PSDB party, had received death threats. The previous *prefeito*, Raimundo Sampaio da Costa, had already been prosecuted for embezzlement and been fined around 12,000 dollars. But one thing is certain, namely that aquatic turtles lay their eggs here every year and the population finances an IBAMA conservation project.

Canutamá lies 1,320 kilometres by river from Manaus and, like Lábrea further to the south, can be reached every two weeks by a six days and nights journey by *recreio* (the *Barco Santos Reis II*). Or you can fly to Lábrea (twice a week) or Canutamá (once a week) with Rico airline, though the small planes are almost always full. From both places you can then travel downstream by rented *voadeira* to the discus habitats. But there are no longer any commercial collectors in the region.

And now let us go further up the Rio Solimões.

The fate of the Juma, like that of the extinct natives in some cases mentioned in the text, is tragic. Between 1940 and 1965 systematic attempts were made to eliminate the Marmori, Katukina, and Ximarimã along the Cuniuá, the Jamamadi on the Rio Pauini, and the Juma on the Mucuím. In 1948 even by means of a special Peruvian force, which created a *tabula rasa* along the Jacaré. When two whites were attacked by Jumas on the Trufari in 1959, because of the trespass on their land, the population of Canutamá mobilised an army to wipe them out, even hacking down children with machetes. The worst massacre took place in 1964 on the Igarapé da Onça. By 2000 just seven Juma women and children still survived (left). Those responsible for all the massacres were never brought to justice.

7. Coari region; Tefé region; Rio Japurá; and Rio Juruá region.

If we now follow the Rio Solimões further upstream, on the left-hand bank we come to Anorí, which, like Anamã, lies at a distance from the Solimões, around 10 km inland. A relatively small *município* of 6,247 km² and 11,316 inhabitants (in 2000). It lies on the northern shore of the Lago do Anorí, which is fed by blackwater *igarapés* such as the da Luzia, Grande, and Anoríaçú and is famous for its scenic beauty. Cruise ships such as the German *MS Vistamar* and *MS Bremen* with their 7,500 or more GRT (gross register ton) bring tourists here, before they travel on to São Paulo de Olivença, Leticia, and Iquitos, or in the opposite direction to Manaus, Parintins, Santarém, and Belém. Undoubtedly because the lake is in large part still surrounded by unspoiled, natural regions (even though the *fazendas* are continually expanding). Here one can travel along the Paraná do São Tomé, where it really is still possible to enjoy (unchanged) nature to the full and to be sure of seeing Capucin, howler, and other monkey species, birds and *Morpho* butterflies, and much more.

The Lago do Anorí is rather clear, and rarely influenced by the Solimões. Discus have not been found here, but the *ribeirinhos* say they are there.

On the right-hand bank of the Solimões we are still in the Purus inland delta, and almost opposite Anorí begins the *paraná* which here is called the Furo do Mastro, but later becomes the Paraná Cuianã and ends in the Purus. To the south-west of this lies the boundary of the gigantic inland delta, at Paraná do Salsa. In addition, at extreme high water this *paraná* is connected with the around 20 km long and otherwise isolated Lago do Salsa, a clearwater lake, fed by several blackwater *igarapés* and one containing white water, the Água-branca. The *lago* is uninhabited and unexplored, but there have to be blue discus there.

Further up the Solimões we come to the huge Ilha Caxuará, which, like the islands on the recently covered stretch since Manaus, consists almost entirely of pasture land or jute plantations. And the three small adjacent islands likewise. To the north of here the Lagos do Mueru and do Pacu, which contain blackwater and discus habitats, lie practically isolated in the interior.

Next comes **Codajás** on the left-hand bank of the Solimões. It lies 47 m above sea level, was elevated to *vila* status in 1875, and in 1892 to city. The colonisation of the village of the Cudaiás *indios* (now extinct) didn't begin until 1862, when its name was changed to Freguesia de Nossa Senhora das Graças de Codajás. The *município* had 18,753 *habitantes* (= inhabitants) at the end of 2004, has an area of 18,988.4 km², and is famous for its Festa do Açaí, the Amazonian festival of the little lilac-coloured coconut. It is said that the fruit from here tastes better than that from Pará *(see page 218)*. This incredible festival was held for the 17th time in the year of my last visit (2004), from the 29th April to 1st May. The revelry lasted for three days and nights and the theme was *Açaí de Codajás na era da industrialização* (açaí from Codajás in the age of industrialisation). The people are proud of the *açaí*, as 1,200 to 1,500 tonnes of it are harvested per year. This can yield 540 tonnes of *polpa* (pulp) when it has all been processed. And the plan is to increase production to 1,200 tonnes of *polpa*. *Inter alia*, a group from INPA has studied 15 *açaí* regions in Amazonia – from where Martius described the palm species *Euterpe aleracea* more than 150 years ago – and brought some interesting facts to light. The *açaí* variety from Codajás yields the smallest pulp component at only 0.8 g, while the average is 1.4 g (the Tabatinga variety yields 2.0 g), and its seed component is, at 77% of the fruit, the largest, exceeding the Parintins variety at 76%. The Codajás *açaí* thus has a very high energy component at 66 kcal, exceeded only by the Barcelos variety at 83 kcal. Its mineral and fibre components are also

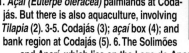

1. *Açaí (Euterpe oleracea)* palmlands at Codajás. But there is also aquaculture, involving *Tilapia* (2). 3-5. Codajás (3); *açaí* box (4); and bank region at Codajás (5). 6. The Solimões and Anorí, which lies on the Lago do Anorí – pasture land all around and along the road to Mato Grosso. 7-8. The Solimões at rising water, full of *Pistia stratiotes*. 9. Codajás on the Solimões *(below left)*.

CODAJÁS-ANORI REGION

high, and vegetable fats (fatty acids) as good as non-existent. In addition, the stems of the young plants are good to eat.

But not only *açaí* cultivation, also a new venture, aquaculture, is on the increase. Amazonia doesn't (now?) have enough fishes to fill all the hungry mouths. For this reason an IBAMA plan to expand *piscicultura* in Amazonia is in full swing, and includes Tilapiine fishes *(see page 432)*. During the last meeting of interested parties from the region it was suggested that huge *viveiros* should be established in order vastly to expand fish-breeding in Codajás.

When you look down from the plane you can see the clear-felled region around and to the north of Anorí along the only road in the region, which leads to Mato Grosso at the Lago Anamã *(see page 433)*. And the pasture-land around Codajás, along the Solimões, and north along the line of the proposed road (already marked on the maps) to the Igarapé do Abreu. But there is still also unspoiled nature to marvel at, and, if you are lucky, a few Codajás goldenwinged parakeets *(Brotogeris chrysopterus)* as well. Discus and suitable habitats are found only in the huge **Lago Miuá** to the west of Codajás, again on the left-hand bank of the Solimões and connected to that turbid river all year. For this reason discus habitats are found predominantly in the north of the lake, where it is fed by blackwater *igarapés* such as the Miuá and Arapari. It is said that blues and browns

occur here, and that they are similar to the variants from the Manacapuru region.

Across from Codajás lies the mouth of a branch of the Solimões *(see page 433)* which washes around the huge Ilha dos Corós, and the latter harbours clearwater *lagos* which as far as I know remain unexplored. This also applies to the Lago Gabriel *(see page 436)* between Anamã and Anorí and a number of others in this region. Only the islands of Paratari and Arraia (also called Mundurucú), the Ilha Ajaratuba, and others in the Purus inland delta are pure *várzea*-whitewater islands and, as already mentioned, are used for jute cultivation *(left)*.

Still further to the west of Codajás there begins another gigantic lake region extending for hundreds of kilometres and vast almost beyond imagination. The northern lakes alone, for example the Lago do Acará with a diameter of around 40 km, or the Lagos Badajós, Piorini and Amanã, each more than 90 km long so that it is often difficult to tell where they begin and end. Not to mention the southern, such as the Lago Mamiá, that in the gigantic Coari mouth region, the Lago Caiambé, and the Lago Tefé. But back to our progression upstream.

Further up the Solimões, after the narrow mouth of the Lago Miuá we come to that of the huge **Lago Badajós**, whose long outflow is called the Paraná Badajós (some call it Rio, but it opens into the lake). Heading up this *paraná* we pass close to the two small isolated Lagos

Left-hand page: 1-5. On the *várzea* island of Paratari in the Solimões jute is harvested at low water and the seeds scattered at the same time (1). These germinate and are automatically fertilised by the rising water. The jute stems are softened by immersing them in the rising water (2) and weighing them down with mud (3). Eight days later the women go into the warm muddy soup to fetch the softened jute (4). Then it is dried before being processed (5).
This page: 6. The clear Lago Badajós with discus habitat (southern bank). 7. Catching a *cascudo* (*Liposarcus* sp.) in the habitat in the Lago Badajós. 8. Transporting manioc soaked in the Paraná Badajós back to the *comunidade*. 9. Freshly killed manatee *(Trichechus inunguis)* in the Lago Badajós. The *comunidade* could now enjoy a feast.

Jutaí and Arpãouba, which lie further to the north. Both are unexplored. Then, near the *comunidade* of Badajós, we come to the gigantic lake, which is little known – the city of Badajós on the Rio Capím in the state of Pará, and the dog breed Badajos-Labrador, are both better known. And, of course, lots of people are familiar with the Spanish province of Badajoz in the Extremadura, where on the 6th April 1812 one of the great battles of European history came to a bloody end. After Lord Wellington had captured Ciudad Rodrigo in January 1812, he went on to attack the impregnable fortress of Badajoz, which was defended not only by Iberian troups, but also by the French (who were friendly and loyal to the the Spaniards), as well as by Germans of the Darmstadt Regiment from Hessen. Wellington fought in vain from the 17th March to the aforesaid 6th April. On the previous day alone he had lost 2,000 men storming the fortress walls, but on the 6th one of his divisions managed to scale the walls, and the end came the next day (to be precise, at two o'clock in the morning on the 7th April 1812). Wellington suffered losses totalling 5,000, including men of all ranks. It was also in Badajoz that the Tratado de Badajós was ratified in 1801, whereby Spain gave up the ownership of Missões, and Portugal ceded its rights in the colony of Sacramento to the Spaniards.

But let us return from the world stage to the lake, whose name also dates back to Iberian times. It contains clear water (but is locally termed *preto*) and harbours *acará-disco* – blues – which have never yet been collected commercially here *(right)*.

During one trip I met an *agricultor* (farmer), Antônio Santana, on the *flutuante* (floating house) of a *ribeirinho* at the *comunidade* Ubim. He and his wife, Raimunda Costa, as well as their six children, had no birth certificates. Antônio told me that in the absence of their own registration he couldn't register the children – apparently an everyday situation in Amazonia.

The **Lago do Aracá**, accessible only via the Paraná do Aracá, and otherwise totally isolated and uninhabited, is a glorious, unspoiled, clear lake. Here one can marvel at the hundreds of cormorants (*Phalacrocorax brasilianus*), known here as *biguás*, spreading their wings to dry at sun-up. As well as the grey herons (*Ardea cocoi*) known as *garça-moura*, and the bird-of-paradise-like heron (*Pilherodius pileatus*) called the *garça-real*, with its blue mask and long crest feathers. I once saw a brown and white buzzard (*Busarellus nigricollis*), which the locals call the *gavião-belo*, carrying a *jaraquí* from the lake. On the same day I also spotted a large *caracará* (*Polyborus plancus*), the crested caracara, several times. Firstly as it rose from the canopy with a snake (but dropped it), and later, as it snapped up an iguana from the overhanging branches just as it was about to jump into the lake. But this is not only a paradise for birds and reptiles, but monkeys – in particular squirrel, howler, Capucin, and woolly monkeys – are often to be seen here too.

The **Lago Piorini**, which lies still further west and again north of the Solimões, and is connected directly with the Lago Badajós via the Paraná Piorini year-round, was new territory as

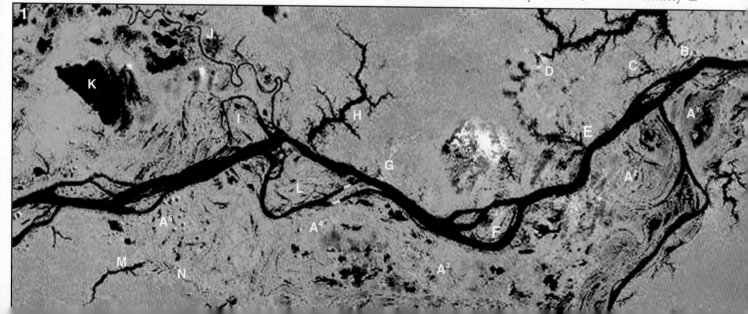

Left-hand page: 1. Satellite photos (1:100,000) of the Solimões upstream from (right to left) the Purus mouth region to the Badajós region: A1-A5) The Purus inland delta. B) The settlement of Anamã with its lake. C) Lago Gabriel. D) Mato Grosso on the Lago Anamã. E) The settlement of Anorí on the Lago do Anorí. F) Ilha Caxaurá. G) Codajás. H) Lago Miuá. I) Ilha do Matrinxão. J) Paraná (or Rio) Badajós. K) Lago do Acará. L) Ilha dos Corós. M) The isolated Lago and Paraná do Salsa, to the west of the Purus inland delta (N). *Fishing for food-fish in the region:* 2. The Boca do Mamiá during the capture of around 2,000 *brasileirinhos* – as *jaraquí* (*Semaprochilodus taeniurus*) are called here – during their spawning migration (from black to white water). Here there are abundant blue *(S. haraldi)* habitats, just as in the Rio Coari Grande drainage *(see page 440). This page:* 3. An almost 2 m long *surubim (Pseudoplatystoma tigrinum)* in the Lago Coari (with a specimen inset). 4. Thousands of *sardinhas (Triportheus* spp.), likewise caugth during their spawning migration (from white to black water). *Inset: S. haraldi* from the Lago Badajós.

regards discus. In November 1998 this gigantic lake had almost dried up and no discus were to be seen. Normally it rises at this time and then contains so much water from December to May that the location of its banks can be discerned only with difficulty. The surrounding area is virtually uninhabited, and, apart from the *comunidade* of Liberdade with its tiny airstrip, there is not much to be seen along the hundreds of kilometres of shoreline. Let alone along the approximately 200 km long Rio Piorini, which forms the western boundary of the Jaú National Park. A totally unexplored region and unspoilt paradise for *peixes-bois* (manatees), which find all they need in the relatively shallow lake with its plenteous grass, although in the dry season they are easy prey. In 1995 IBAMA confiscated 648 manatees killed in the region, and were of the opinion that this represented a mere 40% of the actual *animais abatidos* (animals killed). The Piorini is sought out by professional hunters who fly from Manaus to Codajás and from there travel by *lancha* along the *paraná* for 10 hours to the Lago Badajós, where the manatees are long gone. From there it is another six hours to the Piorini. They are well prepared for a surprise visit from IBAMA, whose swift *voadeiras* can be heard from far off, so they can simply hide their entire booty in the undergrowth. And nobody says a word when the *matança* (slaughter) takes place. The irony is that the *governador* Amazonino had financed motor boats for the *ribeirinhos* of the region.

Sixtytwo-year-old Plácido Pereira told me about it when I came across him with a spear in his hand. He was sitting very still in his boat and staring at the water's surface, in order to detect any manatee coming up for air. If he wounded one he would drag his booty out onto the bank and finish it off with a club or push pieces of wood up its nostrils so that it suffocated. When I mentioned this in Manaus, the 46-year-old IBAMA worker Leland Baroso said that manatees were also being exterminated by commercial fishermen, in that they got caught in their dragnets and suffocated. Although its sale is officially banned, the meat is offered at R$ 3.50 (U$ 1.50) per kilo in the markets. Numerous *recreios* smuggle this "hot" commodity through Amazonia, but most of it is consumed in the *comunidades ribeirinhas*. The shortage of fish in Amazonia forces them to kill other creatures as well, and a *caboclo* family can live for 20 days on the flesh of a single manatee. After boiling *(see page 413)* it will keep for months in its own fat, even in the heat. They call this process *mixira*. I was told this by the Piorini fisherman Hildebrando Fernandes, who at the age of 74 was still hunting manatees and had taught his four sons the skill too.

But it is not only the manatee, but also the real "cow of Amazonia", the *pirarucú*, that is being hunted more than ever *(see below and right)*. The largest freshwater fish on Earth is threatened with extinction (except in Mamirauá – *qv*). And a breeding

Catching *pirarucú* (*Arapaima gigas*):
Left-hand page: 1. In the Lago do Inferno (Lake of Hell) there are (blue) discus habitats – these can be seen in the background. The three *caboclos* standing with their harpoons on the floating grass (at rising water) are waiting to kill the *pirarucú* when it comes up for air.
This page: 2. In the Lago Paraoá, where I have been unable to find any discus, is a favourite *pirarucú* habitat. Here a *caboclo* (in 1996) throwing his harpoon at the giant fish as it surfaces. 3-4. The *pirarucú* is very powerful, can weigh up to 300 kilos (rarely nowadays), and is difficult to haul into the canoe. 5. The Lago do Inferno lies isolated and can be reached only by land almost throughout the year, so the fishes have to be carried for many kilometres.

station in the Rio Jarí (Pará) is unlike to remedy the situation. A wealthy businessman from Belém has recently purchased the long disused, 12 km long canalisation system, including a 184 ha settling pond, which the multi-billionaire Ludwig had constructed to prevent toxic effluent from his factory from getting into the environment and the Jarí *(see page 234)*. The canals and pond are now home to *pirarucús* which are supposedly breeding, and IBAMA has granted the businessman a CITES permit to sell them.

The *caboclo* Nei Gomes da Silva (aged 25) told me that he was after *pirarucú* with his harpoon, but wouldn't let a *peixe-boi* get away. He had learnt how to hunt them from his father, who said that between 1935 and 1964, when manatee hide was used for the manufacture of sewing-machine drive-belts, more than 200,000 individuals were killed every year. Following the ban in 1964 the number remained at around 50,000 per year up until 1973.

Between the Piorini and the Solimões and, of course, westward upstream to the Rio Japurá, there are almost countless *lagos*, the majority of which contain clear water with discus habitats, but are unexplored. Likewise the Lago do Davi, the isolated Lagos Atravessado and do Soró, and the Lago Copeá, which empties into the Paraná Copeá, which provides a more than 120 km long connection between the Lagos Moura and do Paçu and the Solimões at Presidente Médici. The latter is a small *comunidade* diagonally opposite Porto Solimões – the harbour where natural gas and mineral oil from the Urucu are embarked for Manaus. The embarkation port lies west of the mouth of the Coari on the right-hand bank of the Solimões, and is part of the *município* of Coari, nowadays the richest in Amazonia.

At the end of the 1970s, along with Anildo Macedo, I brought back the first Greens (*S. aequifasciatus*) from the **Coari region**. Then, at the end of 1983, several alpha individuals, whose photos travelled all around the world, and which were first bred by Schmidt-Focke. From the mid 1980s at the latest these had risen to the most popular of all discus variants, and today their descendants win almost every championship as "Red-Spotted". But as often happens in this life, whenever someone discovers (invents, shows to the world for the first time, or develops) something, this attracts a host of envious people who do everything they can to steal the glory – or to ruin the whole thing. As regards this discovery, I have never stated in any publication that this variant is found in the Rio Coari or in the Rio Coari Grande. Simply (as with Alenquer or Purus variants) named the region, or, as in the book *DISCUS* (Bleher & Göbel, 1992) given the locality as Lago Coari. As who has ever heard of the Igarapés Açú and do Cerrado, which lie in the mouth region of the Rio Urucu, where I found this "Coari" variant. And they all count as residents of the Lago de Coari *(see left)*.

None of the envious has noted or mentioned that the Urucu and the western part of the Lago Coari form the easternmost boundary of the distribution of the green species. They have searched in the Rio Coari Grande drainage and found the blues that occur there, in waters where no *S. aequifasciatus* live.

The city of Coari lies an the relatively narrow mouth of the gigantic Lago Coari (a few centuries ago it was a lot narrower), on the right-hand bank of the Rio Solimões, around 40 m above sea level. Belgium could fit almost twice into its *município* of 57,230 km². The coordinates are 40°6'22" E and 63°3'21" S, and the green discus begin at 63°6'21" S. The city lies 363 kilometres by air and 463 by river from Manaus and is the second largest in the state of Amazonas. A few years ago its population was estimated at 75,850 inhabitants, but by the beginning of 2005 there were already more than 100,000 people in the *município*, which

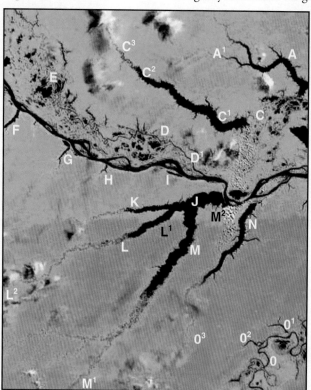

Satellite photo (1:250,000) of the Coari region (with the Solimões in the centre; light blue = white water): A) Lago Badajós. A1) Rio Badajós. B) Lago do Acará. C) Paraná Piorini. C1) Lago Piorini. C2) The *comunidade* of Liberdade. C3) Rio Piorini. D) Lago Copeá. D1) Paraná Copeá. E) The Lagos Moura (left) and Paçu (right). F) Lago Caiambé. G) Lago Catuá. H) Lago Ipixuna. I) Lago Apaurá. J) Lago de Coari. K) Rio Aruã. L) Rio Urucand. L1) Locality for the Coari discus. L2) Urucu station. M) Rio Coari Grande. M1) The connection with the Purus system at high water. M2) The town of Coari. N) Lago Mamiá. O) Rio Purus. O1) Lago Itaboca. O2) Lago Campina. O3) The connection between the Rio Mamiá and the Purus system at high water.

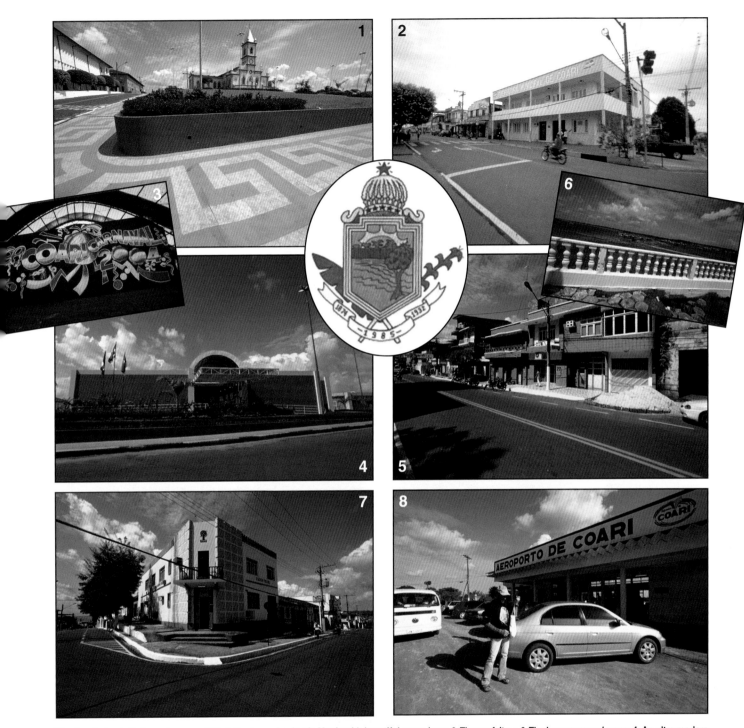

The Coari *pico bello* (*see also page 443):* 1. The imposing Igreja Matriz with beautiful green lawn. 2. The *prefeitura*. 3. The huge *carneval* arena. 4. An ultra-modern hospital. 5. The lovely Hotel Alex Flats (upper floors under construction in 2004). 6. Promenade with view of the *encontro das águas*. 7. Coari's radio and TV station. 8. The airport, the most modern on the Amazon after Belém and Manaus. And *(centre)* Coari's very own *brasão* (coat of arms): The crown symbolises the "Queen of the Solimões" (Coari); the star represents the seat of government of the *município;* the seven crosses the origin of the town under the sign of Christ, and the number seven from the Bible; the five small stars on the lower edge of the crown are the five districts of Coari; the two esses (S) on their sides are just for decoration. In the centre are the most representative features of the region: sun, (blue) sky, primary forest, Brazil nuts, and water (oil and natural gas are missing). In the lower part: right, a banana leaf; left, a jute leaf; and centre, a ribbon showing the most important years (for Coari): 1874 - 1985 - 1932.

borders on those of Codajás, Tapaúa, Anorí, and Tefé. Here the rainy season normally begins in November and continues until February, and July to September are the only months when the water level drops so low that discus can be found.

Although the soil in the Coari region is largely sandy, it is rich in minerals and ores: as well as gold, iron, diamonds, and copper, significant reserves of oil and gas were discovered in 1986 on the upper Rio Urucu. In the very area where back in 1983 I was able to find green discus and also discovered a connection by water between the upper Rio Tefé and the upper Rio Urucu. Where at the end of 1988 on average 60,000 barrels of oil began to be extracted every month, for which Petrobras paid Coari R$ 1.5 million (US$ 750,000) per month royalties. (In 2003 this amounted to more than R$ 19 million, so much that on the 22nd April 2003 the mayor, Adail Pinheiro, emptied the account at the Banco do Brasil and disappeared, after more than R$ 18 million had already gone missing in 2002.)

Initially an oil pipeline was laid to the nearby Rio Tefé and from there the oil was transported downriver. But since March 1997 a new pipeline has run from the upper Urucu to the harbour at Coari, from where the oil is then shipped, and gas pipelines since January 1999. The initial development of the gigantic Urucu facility was completed at the end of 1998 *(below)* but it is continually being extended. By 2001 110 km of roads (71 km surfaced) had been completed, 60 boreholes up to 3,000 m deep were in operation, and 103 km of pipelines had been laid. However, because the gas reserves greatly exceed those of oil (72.42 million barrels) the emphasis was placed on gas, and it was established that the latter amounted to almost 100 billions m^3, which it was thought would support a daily production of 4-5.5 million m^3 (900 to 1,000 megawatts of electricity) for 50 years. (Subsequently the prediction was revised to 25 years.)

When I was last there, in December 2004, permission had been granted by IBAMA and President Lula for the long-disputed construction of a pipeline to Manaus, despite the vast expense involved (estimated at US$ 350 million, with US$ 1 billion of foreign investment already promised). It is supposed to be ready to go into operation in December 2006. And the battle conducted for years in the media against the construction of the 522 km long gas-pipeline from Coari to Porto Velho, through several indian reserves and unspoiled wilderness, likewise appears to be lost. Petrobras has received a permit from IBAMA for that as well.

Soon 2.3 million m^3 liquid gas will flow every day through possibly the most fantastic region of Amazonia, past more than 4,000 aboriginals, including the practically extinct Juma tribe. And a (hitherto) unspoiled wilderness. The watchword is *avança Brasil* (Forward Brazil), and will bring an endless wave of *madeireiros* and *fazendeiros* – integration – with it. A second Rondonia is coming into being, with areas laid waste, dirt and effluent, which no one knows anzhow where to place it and most of it will end up in the river, as it already has for several years in Coari (see page 443). But amid all the *avança* (= forward progress), nobody seems to have thought of that.

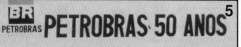

The Petroleum Cross (1) stands on Coari's Pça. G. Vargas, opposite the statue of Christ. The sign (2) says (translated), "Christmas trees and domestic goods depend on a sure and steady production of oil and gas from the boreholes". The station on the upper Rio Urucu (2) is being expanded. Pipeline (3) is being laid to Coari. From there the liquid gas goes by ship (6) to Manaus. But soon the pipeline will go all the way. Petrobras is celebrating (5).

Above: The pipelines from the Urucu station pass through the primary forest for around 50 km from the station to the Rio Tefé, and 285 km to Porto Solimões (and soon a further 417 km to Manaus and 522 km to Porto Velho). *Left and below:* Coari is a clean town *(see page 441),* but rubbish ends up in the harbour and river. Although there are notices everywhere and IBAMA is asking for it to stop. And even sponsored posters, showing a fish with a mask and the slogan (translated) "The eyes don't see, but the rivers feel. Don't dump rubbish into the river". But in vain. The only beneficiaries are the *urubus* (vultures – *above*).

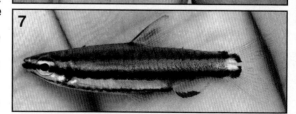
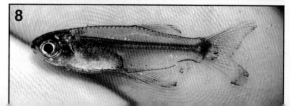

This page: Discus (blue) habitats in the drainage of the Rio Coari Grande *(see also map on page 440)*: discus habitat in the Igarapé Gibian, a right-hand tributary of the Rio Coari Grande (1). The Igarapé do Pau-furado constitutes a year-round link between the Rio Mamiá and the Igarapé São José (Purus region). Here at high water (2). Discus habitat deep in the Igarapé Gibian at rising water (3) and detail (4). Note how the *Virola calophylla* flowers are submerged and the provide food for discus *(see also page 447)*. The author searching for blue discus habitats (5). Small characins from Coari igapós: *Gnathocharax cf. steindachneri* (6); *Nannostomus aff. marginatus* (7); and a brilliant blue *Tyttocharax* sp. (8). *Centre:* A blue discus (*S. haraldi*) from the Igarapé Gibian habitat.

This page: A number of (green) discus habitats from the Rio Urucu region:

The Rio Urucu is nowhere linked to the blue habitats (Rio Coari Grande and eastward). And the parameters are lower here. A bay in the central area with lots of *camu-camu* plants in the habitat, at rising water with washed-in grass (1). The Igarapé Açú harbours beautiful unspoiled green discus habitats (2-3). It is broad where it joins the Urucu (8). In the Igarapé Cajú, north of the *caboclo* hut of old Senhor Cajú, the black water in the *acará-açu* discus habitat can clearly be seen (4). Fishes from the *igapó* region along the Urucu: *Copella* sp. (5); *Apistogramma* sp. from the *agassizii* group (6) and a *Crenicichla* sp. from the *regani* group.
Centre: Green discus *(S. aequifasciatus),* which I found here in 1983-84 and which others say is from the Rio Coari...

While the hopeless discussion of these pipelines continues, Petrobras has already contracted with Confab for the purchase of 400 km of pipe for R$ 242 million (about US$ 100 million). Confab, along with its parent company Tenaris, based in Luxemburg with subsidiaries in Argentinia, Canada, Italy, Japan, Mexico, Venezuela, and even Brazil, is the largest steel-pipe manufacturer on Earth. A network of 14,000 employees extends all over the world and in 2003 sales of pipe amounted to at US$ 3.18 billions.

But they are busy at the Urucu station for the environment, insofar as is possible in the face of *avança*. For example, they have built an orchid nursery there, and the staff have only recently discovered two new species during a census in the area which is to be cleared to make way for housing. A *Peristeria* species and an *Octomeria* which is similar in appearance to *O. serpens*. These are now being studied in the Museu Emílio Goeldi in Belém. Already 76 orchid species have been found in the oil- and gas-drilling region alone, and only seven of them have been identified. Naturally plant seeds are being collected as well, and shown in a huge nursery where there is room for 125,000 plants, including the orchids and also bromeliads. When they have been grown on they will be introduced into re-afforested areas. An area of 20 ha has been allocated to the nursery.

But back to Coari and the habitats. The huge clear- and blackwater mouth region actually consists of four almost parallel *rios* flowing from the south, each of them forming a lake many kilometres wide in its lower course. The most easterly, the Mamiá, is connected with the Coari Grande via the Igarapé do Jacaré at high water, but it empties its masses of water directly into the Solimões to the north-east of Coari.

The Mamiá is linked year-round with the Purus system via the Igarapé São José *(see pages 156 and 440)*. This is undoubtedly the reason why the discus variants there are in part similar to those from the Purus region *(see page 444)*. The Mamiá drainage region is virtually uninhabited. There is a *comunidade* called São José da Boca do Paranã Mamiá at its mouth ("Paranã" because the Lago Mamiá outflow is named thus). And the tiny *comunidade* called Japini, 6 km before the end of the only road leading from the city of Coari to the Lago Mamiá. This dirt road, which isn't shown on any map, is around 30 km long and negotiable only during the dry season.

I was able to find blues (*S. haraldi*) and habitats up the Mamiá to the Igarapé do Pau-furado, which can be navigated via the already mentioned Igarapé São José to the Lago Campina and the Purus. That is around 120 km upstream, while the virtually unexplored Rio Mamiá is more than 200 km long. The habitats and discus variants in the Rio Coari Grande and its *igarapés* are very similar. I last explored the region at the beginning of February 2004 and travelled there with Sabá (Sebastião's nickname), after our

It takes hours or days (depending on the boat) of travel from the town of Coari via the Lago de Coari (1), often passing freshwater dolphins on the way (2), to reach the Rio Urucu mouth region (3). Here at rising water, full of floating grass islands. In addition to my discovery of new green discus variants in the upper Urucu, Petrobras has found new orchids in the area (4). So far the unspoiled blue discus habitats in the Rio Mamiá remain little explored (5).

Rio Urucu (green discus) region: green discus habitat in the Igarapé Cerrado (1). Here grow masses of spiny creepers *(Smilax* species), whose yellow flowers (2) provide the discus with food and their thorns with protection. And, of course, their roots contain saponin and saponin-like substances, essential oil, and starch, and their bitter resin contributes to the water here being so black (3). The mouth region of the Urucu, where it enters the Lago de Coari, is gigantic, the waves are often metre-high, and the other bank cannot be seen (4). In the immensely wide central and lower Urucu (5-6) huge numbers of grass islands are encountered when the water is rising in January-February. A typical Urucu blackwater green discus habitat (8) with *aracá-açú (Licania angustata)* in flower (7) and *Virola calophylla* (9), which also occurs in the Coari Grande *(see page 444),* branches in between, tree-trunks and virtually still water. There are also habitats with floating *Utricularia foliosa* (10) and *Azolla cf. microphylla* (11). If you walk overland in the *igapós* in the dry season, then you will pass discus spawning trees *(left)* and perhaps, when the first rain falls, see numerous fruits and seeds (13) on the ground; if these don't germinate then they provide a rich food supply for the fishes. And then too the survival specialists among the fishes, such as *Hoplias* sp. (14), come out of their mudholes where they have spent months without water.

visit to the Coari Tilapiine *piscicultura* at the *terreno* (estate) of João Martin (even though there are plenty of edible fishes in the area – *see page 437*).

The Rios Coari Grande, Urucu, and Aruã together form the gigantic Lago de Coari, before the latter opens into the Solimões close to the city. The Coari Grande is the largest at around 500 km in length, with countless meanders. About 200 km upstream it is connected by water during the rainy season with the Igarapé Pauapixuna, which ends in the black Abufari (Purus drainage) *(see page 156)*. From here upstream I have as yet to explore the Coari, but I don't believe that discus go that far upstream. In addition, as good as nothing is known about the flora and fauna above and below water (apart from the discus in the lower course). As in the past there are three small *comunidades* in its mouth region. After 4 hours journey in Sabá's 85-horse-powered *voadeira* across the Lago de Coari, where waves broke over our boat, we reached the first of them, Vista Alegre. Here 87 people live by a small bay on the right-hand bank. Some 10 minutes upstream lies Boarí, with around 80 *caboclos*; and the last, in this case on the left-hand bank, is Samaúma, with not many more inhabitants. They live by *pesca* and *roça* (fishing and cultivation). Here (and elsewhere) it is essential to obtain permission from the village elders to explore any of the *igarapés*, as otherwise you may end up with a bullet through the head.

Aldemir from Vista Alegre accompanied Sabá, Natasha, and myself on this expedition. I wanted go to the Igarapé Gibian, where recently not only I had searched for discus but also other *gringos* whom Sabá had guided there. Right at its mouth on the right-hand Coari bank, around halfway along the vast lower section of the Coari

Grande, stands the wooden hut of Donna Luiza. She told me that the previous year (2003) people had come there from Manaus to collect *acará-disco-azul* in her Gibian. They had declared 3,000 individuals but there were actually 7,000 and when it finally came to it they had given her nothing at all. She was very friendly, gave me a *cafezinho*, and her eldest son said that at the height of the dry season thousands of blue discus were to be seen in the *acará-açú* scrub right there in their little bay. And he confirmed that at high water it was possible to get through the Gibian to the Igarapé Jacaré and into the Lago Mamiá.

The parameters in the habitat on 07.02.2004 at 08.00 hours in 2.5 m of depth were as follows: pH 5.40; conductivity 11 μS/cm; daytime temperature (air): 28 °C; water: 27.6 °C; oxygen 1.71 mg/l; unspoiled wilderness, full of *acará-açú* plants; water tea-coloured with very little movement; no floating plants; bottom sand.

Further up the Igarapé Gibian same day at 08.45 hours in 0.30 m (3.0 m) of depth: pH 5.89 (5.68); conductivity 18 (17) μS/cm; daytime temperature (air): 26 °C (very shaded), water: 25.2 (25.1) °C; oxygen 1.33 mg/l at 3.0 m of depth; primeval wilderness, with fallen trees and branches, *acará-açú-* and *maracanã* plants and *Virola calophylla* in full (violet) flower; water brownish with no movement, but the water level had risen; no floating plants but full of flowers and floating leaves; sandy bottom.

In a typical habitat in the Rio Coari Grande on 07.02.2004 at 11.30 hours in 0.30 (2.5) m of depth: pH 5.70 (5.32); conductivity 10 (11) μS/cm; daytime temperature (air): 30 °C, water: 29.5 (27.6) °C; oxygen 1.62 (1.73) mg/l; unspoiled wilderness, full of *maracanã*

plants; water tea-coloured with some movement; lots of grass islands; bottom sand.

For discus variants *see page 444*.

To get from here to the green discus you have to travel back down the huge Lago and then cross the "freshwater sea", as I call the Lago de Coari *(see page 446)*. Normally a day's journey to the first habitats, but with a 85-horse-powered outboard motor on a *voadeira* it can be done in 3-4 hours barring rain and storms. And then you arrive in the Rio Urucu. Apart from the *comunidade* of São João da Moura (with 80 plus *caboclos* and a school) there are only a few further *caboclo* huts along the more than 300 km long river and then nothing more further upstream until the previously mentioned Urucu station with its oil- and gas-drilling towers. The latest news is that a town is being built in the deepest jungle – the only other place I've seen anything similar was in Timika (Indonesia). The Rio Urucu has suffered badly around the harbour at the station, where the black gold was formerly embarked. There are no longer any green discus habitats here. Nowadays the only traffic is supply ships. In the dry season these can't get through, and at the beginning of the high-water period – February – it is equally impossible as the lower Urucu metamorphoses into a gigantic meadow of grass *(below)*. Discus habitats have in part disappeared along the way, in places where the river has been dredged and large regions clear-felled. No doubt *fazenda*s will spring up there, or may already have done so. Likewise the area between the station and the Rio Tefé no longer looks like it did in the past. I can count myself lucky to have seen it as it was.

On these pages a few of the Greens that I found here in the Urucu, where my researches date back to 1977 *(photos 1-2)*. However, those with the entire body red-spotted are in fact rare, and anyone who thinks he can go there and encounter such fabulous specimens everywhere is in for a disappointment.

I brought back data from the middle and lower Urucu habitats that differ from any I had ever recorded before. Here are a selection:

The parameters in the Igarapé Cerrado habitat on 07.02.2004 at 15.35 hours in 0.30 (2.50) m of depth were as follows: pH 5.20 (5.13); conductivity 11 (10) μS/cm; daytime temperature (air): 31 °C, water: 28.4 (26.2) °C; oxygen 1.48 (1.75) mg/l. Unspoiled wilderness full of *maracanã* plants; water black with some movement; *Utricularia foliosa* and *Azolla cf. mycrophylla*, and lots of grass islands; bottom sand.

For discus variants *see page 447* (alpha individual).

In the Igarapé Açú habitat on 07.02.2004 at 16.45 hours in 0.30 (2.50) m of depth: pH 5.22 (5.35); conductivity 15 (17) μS/cm; daytime temperature (air): 30.5 °C, water: 27 (24.6) °C; oxygen 1.28 (1.35) mg/l. Biotope unaltered, full of fallen trees, branches, and roots, with *maracanã* plants, spiny creepers (*Smilax* sp.), and violet-blossomed *Virola calophylla;* water moving slightly; bottom sandy.

In the Rio Urucu habitat on 07.02.2004 at 17.45 hours in 0.30 (2.50) m of depth: pH 5.07 (5.07); conductivity 9 (10) μS/cm; daytime temperature (air): 31 °C, water: 29.4 (29.3) °C; oxygen 2.67 (1.00) mg/l. Wilderness with *acará-açú* and spiny creepers; water black with some movement; bottom sand.

Interestingly the above-mentioned *Smilax* species, a liana-like spiny plant, was wide-spread in the second of these habitats. As I have already mentioned, its roots contain saponin and saponin-like substances, essential oil, starch, and a bitter resin which contributes to pronounced blackwater coloration; its spines provide protection from predators at high water and its huge flowers and seeds are a source of food (including for fishes), as is also the case with *Virola calophylla*.

I had also previously collected in the Igarapé Cajú, named after the *caboclo* who has lived there for along time. Cajú lives at a distance from the three huts on the right-hand bank of the Urucu. Unfortunately at the time of my last visit he was lying in his hammock with a severe attack of malaria and couldn't go with me.
I had

Left-hand page: green discus *(S. aequifasciatus):*
1. A so-called Coari, a green discus from the Rio Urucu.
2. Green, again a Coari (alpha) individual from the Urucu drainage.
3. Green from the Lago Tefé.
Double spread: Rio Urucu mouth region.

a long chat with him and gave him my Halfan (an anti-malarial) tablets. And I hope he recovered in due course.

I have not personally investigated the third and last Lago de Coari affluent, the Aruã (also known as the Arauá), but it supposedly contains green colour variants such as I found further west. *Ribeirinhos* told me this. A few, fishermen who I meet on a previous field trip, resting by the Aruã river mouth, told me that once in a while they catch *acará-disco-pintado* (=greens) and make a *calderada* (fish soup) from it.

Before we now travel further up the Solimões, a little more on the discovery of this region: Carvajal (along with Orellana) was the first white man here. He named it Província de Machiparo and remarked that more than 50,000 primitive people lived there and that the word Coari signified Rio de Ouro or Rio dos Deuses (river of gold or river of the gods). He was so taken with the natives and especially the Omáguas and their skillfully-made ceramic wares (which he compared with those of the Chinese) that I will quote his words on the subject (in translation): "On the 12th day of May we reached the heavily populated province of Machiparo, which borders on another large region, that of the Omáguas, who are of a friendly disposition among themselves but battle together against invaders from the interior who attack their houses on a daily basis. When we were around two *léguas* from the region we saw that it was full of villages. We tried to remain unnoticed as further upstream we spotted a large number of canoes in battle position." (Carjaval wrote that they were between Catú and the Purus = Lago Coari; it was in 1542.)

"There followed a perilous battle, as there were a lot of them, on the water and on land, everywhere possible, so it was a hard fight. Our archers went ashore and fought so hard that the *indios* fled. The *capitão* then ordered 25 men ashore to reconnoitre the area. They found large amounts of supplies such as turtles, and meat, fish, and biscuit in large earthenware containers. Enough to feed more than 1,000 men for a year. The *capitão* then decided to travel further downriver. We continued to fight indefatigably from the raft against the *indios*, it was unimaginable how many were forever arriving along the shore and in canoes. The battle lasted more than two days and nights and we had no time for a moment's rest. It seemed there was no end to this Machiparo region, which we remarked extended for more than 80 *léguas* and was inhabited throughout, one village following another practically without interruption. And some of these villages extended for more than five *léguas* and there were no gaps between the houses along the entire stretch, a wonderful sight to see." (1 légua is 6 km, so the housing area was 480 km long...)

"On the Sunday following that of the resurrection of our Lord, we again saw endless villages in the beautiful land of the Omáguas, but there were too many for the *capitão* to stop... But in the afternoon we came to a village on an elevation, and because it looked small, the *capitão* gave orders to capture it... There was a *casa de reuniões*, where we found various domestic wares: huge pots of more than 25 *arrobas* smaller pots, dishes, bowls, plates, and cups made of porcelain such as the world had never seen; not even that from Málaga is comparable, as it is glazed all over and enamelled with all the colours of the rainbow, so lifelike that it was almost frightening, and in addition the decoration included painted figures just like those dating from Roman times." (*Casa de reuniões* = meeting hall and 25 *arrobas* = 275 litres).

Judging from Carvajal's writings, Amazonia was more densely populated at that time than today. And its people more advanced than Europeans in some ways. Nevertheless they were viewed as primitive and savage, and were slaughtered *(right)* or enslaved.

Only one thing is missing from Carvajal's detailed observations: that the Omáguas altered the shape of their heads by binding them up in childhood *(centre)*.

The first mission here was called Santana de Coari and was founded at the end of the 17th century by no less a person than Samuel Fritz. He united the Catauxi, Iriju, Juma, Jurimagua, Auapes and Puru-Puru tribes. Samuel Fritz (1654-1725) was a native of northern Bohemia, from the village of Trutnov, and in 1672 began to study philosophy in Prague. Within a year later he went over to Jesus and from 1680 to 1683 studied theology at the University of Olomouc. At the age of barely 21 he travelled to the New World as a missionary and for almost 20 years worked from the upper Amazon outwards among the Omáguas. At that time too they were still very numerous and little affected by the European influx. Today they are long gone from Brazil (although in 2000 156 supposedly survived as Kabemba). In Peru the Omágua population in 1976 was estimated at between 10 and 100 individuals.

Fritz founded nearly 40 more missions and in 1689 travelled down the Amazon, falling seriously ill during the journey. He sought a cure in Grão-Pará (Belém) but the *governador* held him for a spy and imprisoned him for 18 months. After his release through the intervention of the Portuguese crown (1691) he returned to his mission, became an advocate of indian rights, and published the first precise map of the Amazon *(above)*. For this purpose he clambered more than 4,000 m up into in the central Andes, to the Lago Lauricocha. He regarded this frozen lake as the source of the Río Maranõn and thus of the Amazon, and this was accepted worldwide until the middle of the 20th century.

He noted (translated from the Spanish), "With much toil and sweat I have prepared this geographical map,

in order to obtain greater knowledge of the rivers Marañon and Amazon, and have personally travelled the greater part of it. Insofar as it was negotiable at all. And although there are already various maps, I must comment, without criticising anyone, that none of them was made with any degree of accuracy, as none of their authors had personally seen or travelled the region, or because their work was based on the confused statements of others."

And I can well sympathise with him, as his map of the Amazon was like my decades of on-the-spot research into the distribution of the discus species. Fritz has gone down in history as the *Apostolo do Amazon*.

The original founders of Coari were the valiant Jurimauás. Nomads, who lived at Paratarí, then on the lha de Guajaratuva (Guajaratuba) between the Lago Taracajá and the Rio Uamori, and were resettled in their current home by the Carmelites Antônio de Miranda and Maurício Moreira. In 1758 Francisco Melo de Povoas changed the name Aldeia de Coari to that of the Portuguese village of Alvelos, and in 1833 it became the Freguesia

1. Original print of the first precise geographical map of the Amazon (1691) by S. Fritz. 2. Things went badly for Orellana and his companions against the natives of Amazonia during the journey downstream (in 1541-42). 3. Amazons, as popularly envisaged. 4. Water-colour by Requeña y Herrera, showing the Spaniards involved in peaceful dealings with Omáguas by the Rio Mesay, Japurá region, while the former were establishing borders for Spain and Portugal (1778-1785).

de Alvelos, with the Nossa Senhora Santana as its patron saint. In 1874 it was granted the status of *vila* and the name was changed back to Coari. Then in 1932 it became a town. At the time of my first visits (late 1960s) it was still a small town of 6,000 inhabitants.

Meanwhile, the first half of December 2004 saw the III Festa do Gás Natural and XII Festa de Banana taking place, with a huge selection of banana costumes on display and more than 20,000 people celebrating all day and night.

Since 2001 the Banana has been poisoned with the Gás, and since 6th February 2005 the bulldozers have been working flat as never before out in the Coari region to make way for the gas pipelines to Manaus. Hundreds of square kilometres of primary forest – and heaven knows how many life forms – will soon be no more than history. Hence a few wildlife data for posterity, from the time before *avança Brasil* in this region, an area where the fauna is particularly endangered.

Of the 171 mammal species alone counted in this eco-region in 1999, 120 occur in a small region on the upper Rio Urucu (= the oil and gas area). These include species that migrate there at high water, for example, monkey species *(Saimiri sciureus, Cebus albifrons, Lagothrix lagothricha)* and the peccary *(Tayassu tajacu)*. But the majority cannot "jump" rivers, which are a natural barrier for them. This by itself is the reason for the biodiversity here, and not only among the primates, but also the insects and many others. They are just as endangered by the rampant biotope destruction as the ant-eater species *(Cyclopes didactylus, Tamandua tetradactyla, Myrmecophaga tridactyla)*, the sloth *(Bradypus variegatus)*, the jaguar *(Panthera onca)*, and the puma *(Puma concolor)*. Likewise the deer *(Mazama americana* and *M. gouazoubira)*, as well as the giant tapir *(Tapirus terrestris)*. And the more than 554 bird species that are found here, including seasonal migrants such as toucans *(Ramphastos cuvieri)*, parrots *(Amazona* spp.) and aras *(Ara* spp.), and, of course, the majority of the sedentary species such as tanagers *(Tangara* spp., *Tachyphonus* spp.), woodcreepers *(Xyphorhynchus* spp.), quetzals *(Pharomachrus pavoninus)*, wattled curassow *(Crax globulosa, Nothocrax urumutum, Mitu tuberosa)* and tinamous *(Crypturellus* spp., *Tinamus* spp.).

This immense region between the Juruá and the Purus lies in the low-lying western Amazon basin (the highest point is less than 60 m above sea level), whose soft sedimentary soils were laid down relatively late on, in the lower Tertiary (2-5 million years ago). Traversed by an almost uniform pattern of waterways, comprising only a few large but thousands of small rivers, all subject to highly seasonal fluctuations in water level. The driest month here is July, with barely 100 mm of precipitation. Its towering forest giants (2.5 m diameter) include *Cariniana decandra, Osteophloem platyspermum, Piptadenia suaveolens, Brosimum* sp., *Eschweilera blanchetiana,* and *Sclerobium paraense*. And the most important plant families are the Sapotaceae, Lecythidaceae, Moraceae, Chrysobalanaceae, Lauraceae, Myristicaceae and the Leguminosae. Their commonest members here are the tree species *Eschweilera alba, E. odora, Pouteria guianensis, Vantanea guianensis, Jessenia bataua, Ragala sanguinolenta, Licania apetala*, and *Iryanthera ulei*, as well as the four palms *Astrocaryum tucuma, Jessenia bataua, Maximilliana regia*, and *Socratea exorrhiza*. Interestingly the Rio Juruá, which we have yet to reach, is the westernmost limit of the natural distribution of the Brazil nut tree.

We are thus in the land of the Green, which shares no river and no habitat with either of the two other species.

Upstream from the Coari region, on the right-hand bank of the Solimões, we come to the isolated Lago Apaurá, which

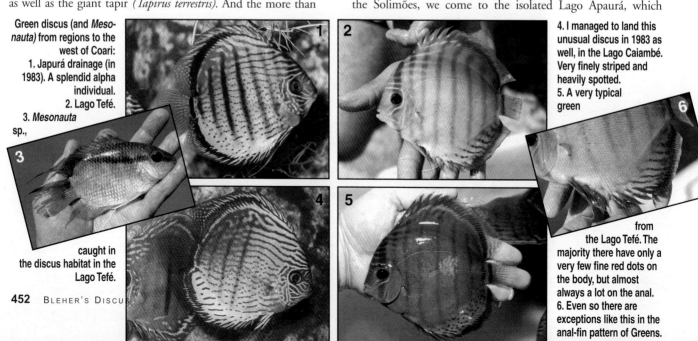

Green discus (and *Mesonauta*) from regions to the west of Coari:
1. Japurá drainage (in 1983). A splendid alpha individual.
2. Lago Tefé.
3. *Mesonauta* sp., caught in the discus habitat in the Lago Tefé.
4. I managed to land this unusual discus in 1983 as well, in the Lago Caiambé. Very finely striped and heavily spotted.
5. A very typical green from the Lago Tefé. The majority there have only a very few fine red dots on the body, but almost always a lot on the anal.
6. Even so there are exceptions like this in the anal-fin pattern of Greens.

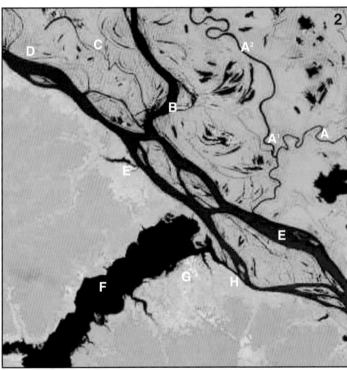

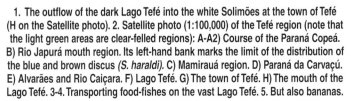

1. The outflow of the dark Lago Tefé into the white Solimões at the town of Tefé (H on the Satellite photo). 2. Satellite photo (1:100,000) of the Tefé region (note that the light green areas are clear-felled regions): A-A2) Course of the Paraná Copeá. B) Rio Japurá mouth region. Its left-hand bank marks the limit of the distribution of the blue and brown discus *(S. haraldi)*. C) Mamirauá region. D) Paraná da Carvaçú. E) Alvarães and Rio Caiçara. F) Lago Tefé. G) The town of Tefé. H) The mouth of the Lago Tefé. 3-4. Transporting food-fishes on the vast Lago Tefé. 5. But also bananas.

remains unexplored. Although I have several times found greens in the Lagos Ipixuna, Catuá, and Caiambé, further on *(see pages 440 and 452)*.

I have not investigated very far up the Igarapé Ipixuna, navigable only at high water, and the same applies to the Catuá with its single *comunidade*, São João do Catuá.

The black **Lago Catuá** (= cat in Basque – perhaps because someone found jaguar tracks in the sand?) is around 10 km long and sandbanks appear around its margins during the dry season. More than 150 years ago Henry Bates, who spent 14 days here, noted the robbing of turtle nests taking place here even then, as it is an ideal nesting place with a "concealed entrance" from the Solimões. Even so, at the height of the dry season, when the lake can be reached only by land, there are no discus habitats to be seen – you have to go to the Igarapé Catuá for them. And there is no (longer) any need to worry about crocodiles; the numbers that Bates encountered can now only be seen in films. Note that Jules Verne stopped off at Catuá in *800 leguas on the Amazon*.

I know nothing about the Igarapé Jutica (Juteca). Then, further upstream, we come to the already-mentioned **Lago Caiambé**. This too contains blackwater, and is around 20 km long and 1-2 km wide. The river feeding it is around 100 km long, and its outflow is again narrow and barely 300 m long. When I came here for the first time in the 1980s, the region was an absolute dream. I arrived through the straits from the Solimões and suddenly a glorious spectacle opened in front of me (I think Bates must have felt something similar). Nowadays there is a *comunidade* of the same name there (in the 1980s a few huts, in Bates's day just one), with three schools and around 1090 voters, which has been elevated to the status of *vila*. Before long it will be yet another Amazon town, like so many others. By the time of my visit to the region in 2004 the Governador da Amazonia, Eduardo Braga, had brought in legislation for the piping of water and the surfacing of the roads. Back then there was dense forest, but now much of it has been cleared and large areas of the *terra firme* are today *fazenda*s.

At the the fifth World Parks Congress in Durban (in 2003) Brazil (in order to impress?) presented a vast nature conservation project for Amazonia to all the representatives of the world press: a total of 3.8 million ha, in regions where the greatest biodiversity on Earth is encountered, was to be placed under protection. Specifically the Rio Urubú State Reserve; Cuieiras State Park; Cujubim Sustainable Development Reserve; and Catuá/Ipixuna Extractive Reserve. High time for the last of these in particular – if and when it actually happens. Unfortunately Caiambé was not included, as recently I have been able to find green discus habitats surviving only at the upper end of the lake.

The splendid (alpha) variant from here *(see page 452)* has appeared in all the discus publications on Earth. Today (20.08.2005) if you look under "Caiambé" on the Internet, you will find (eg at google.com) 156 references, more than half of them mentioning the Green. None of the site owners has ever been there (and none is a wild discus expert), and hardly anyone mentions the discoverer of the Green from here, let alone the title of his work...

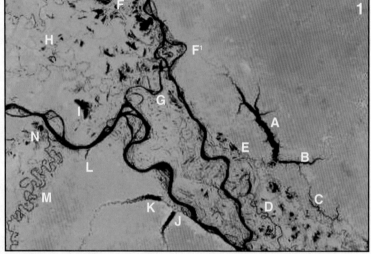

Caimabé lies 2,100 km from the mouth of the Amazon, shortly before Tefé, at 3°30' S and 64°26' W.

We now come to the **Tefé region**, which I have already mentioned in Chapter 1. Officially founded in 1759, but there is evidence dating from 1686 that missionaries named the place Nogueira, which was subsequently changed to Ega (Eda) and then Teffé, before finally becoming Tefé. It is even possible that the *Pueplo de la Loza* marked on Orellana's maps (1542) was Tefé. Be that as it may, in 1718 a population increase took place around the various missions along the Solimões, and this also affected the lake region – most of the settlers were adventurers who had come looking for indian slaves.

Tefé lies 47 m above sea level, set back from the right-hand bank of the vast lake just before the latter empties into the Solimões. The *município* comprises 23,808.9 km² and can be reached only by water or by air (several times daily). The few kilometres of road that exist lead to *comunidades* or *fazenda*s. Two seasons of the year are recognised, the *inverno* (February to June) and the dry *verão* (July to October). As in the past the main activity is fishing for food, followed by agriculture and trade. But light industry is taking root.

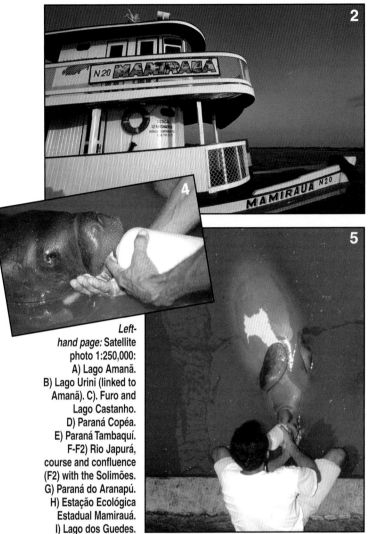

Left-hand page: Satellite photo 1:250,000:
A) Lago Amanã.
B) Lago Urini (linked to Amanã). C). Furo and Lago Castanho.
D) Paraná Copéa.
E) Paraná Tambaquí.
F-F2) Rio Japurá, course and confluence (F2) with the Solimões.
G) Paraná do Aranapú.
H) Estação Ecológica Estadual Mamirauá.
I) Lago dos Guedes.
J) The settlement of Uarini and the mouth region of the Uarini.
K) That of the Rio Cobacá. L) Lago Uará.
M) Rio Juruá. N) Mouth and lake region of the Rio Minerúa. *This page:* 2-3. *Lancha* and terminal at the Mamirauá station. 4-5. There is a manatee conservation project in progress and feed them like babies (4). 6. The posters are intended as reminders that fishes may not be caught during the migration *(piracema).* 7-8. Fish otter young that are caught in the fishermen's nets are reared and then released.

Siluriform

On these two pages we have a world premiere: the migration of siluriform fishes (catfishes) in the Mamirauá region, Rio Japurá drainage, captured in photographs and published for the first time.

In general 34 families from all over the world are assigned to the order Siluriformes, with 14 of them living in Latin America and 10 occurring in the Amazon basin. I was able to observe species from all 10 families during a communal migration from 6th-8th August 1997. By my calculation there were around 2,816,000,000 catfishes on this spawning migration (320 fishes per m^2 *(photo 2),* swimming in 20 layers, one above the next, in a body 4 m deep (in some places even deeper), 20 m wide, and 22 km long). The 10 families were as follows, but I can list only a small selection of the genera and species (with photo number in brackets) here:
Cetopsidae: *Cetopsis* sp.;
Aspredinidae: *Bunocephalus* sp.;
Trichomycteridae: *Pseudostegophilus* sp.;
Callichthyidae: *Dianema longibarbis*;
Loricariidae: *Ancistrus* sp.;
Farlowella sp., *Hypostomus* sp., *Liposarcus* sp., *Loricarichthys* sp., *Parotocinclus amazonica, Rhineloricaria* sp., *Sturiosoma* aff. *brevirostre*;
Pseudopimelodidae: *Microglanis* sp., *Pseudopimelodus* aff. *pulcher*;
Hepapteridae: *Charmocromes* sp., *Goldiella eques, Imparfinis* sp., *Pimelodella* spp., *Rhamdia* sp.;

migration

Pimelodidae: *Cheirocerus goeldii* (15), *Duopalatinus peruanus, Hemisorubim platyrhynchos, Hypophthalmus* sp., *Leiarius marmoratus* (2), *Phractocephalus hemioliopterus* (3), *Pimelodus albofasciatus, P.* aff. *altissimus* (3), *P. ornatus* (4), *P. blochii* (4), *Platystomatichthys sturio, Sorubim lima;*
Doradidae: *Acanthodoras* sp., *Agamyxis pectinifrons, Hassar orestes* (9), *H.* spp. (9), *Hemidoras* sp., *Megalodoras irwini* (2,7), *Nemadoras humeralis* (2), *Opsodoras* aff. *stuebelii* (6), *Oxydoras niger, Platydoras costatus* (9), *Pterodoras granulosus, Scorpiodoras heckelii* (5), *Trachydoras nattereri; T. trachyparia* (9);
Auchenipteridae: *Ageneiosus* sp. (14), *Achenipterichthys* sp., *Auchenipterus* sp., *Centromochlus heckelii, Epapterus dispilurus* (13), *Trachelyichthys exilis*.
There were so many catfish species that it is impossible to catalogue them all. The mass movement was of such dimensions that no species of any other order had room to swim, let alone breathe – they were all gasping for oxygen and pushed right out of the way.
Stingrays, which live exclusively on the sandy bottom, were to be seen at the surface (10); cichlids (11) and characins (12) in the bank zones. Thousands of birds were sampling this spiny migration (1, 8). In a number of places the catfishes grouped into a gigantic snake-like formation – to confuse the birds? And why this huge communal migration?

In Tefé – as everywhere – there are festivals, with the Santa Teresa Festival in October and the *Festa de Castanha* in June being the most important events. But for years Tefé has also been known as the gateway to the huge Mamirauá Reserve and Institute.

I have often visited Tefé, and every time I go back I am disappointed yet again. The destruction of the wilderness is increasing to such an extent that the place is no longer beautiful. On one occasion I was there twice in the space of six months, and in the interval two *igarapés*, in which I had discovered a new characin species, had disappeared. Nature had been obliged to make way for new cattle pastures – only a stinking brew remained where the *igarapés* had once been, and the characins were gone *(see map on page 453)*.

The discus habitats too are disappearing faster than I can describe them. It isn't quite as bad up the Rio Tefé, but Petrobras hasn't done anything to improve the situation. And whether the Floresta Nacional de Tefé in the north, established in 1989 and comprising a total of 1,020,000 ha, has helped to some degree, remains to be seen. In addition, the once most important town on the Solimões, like many others in the state, is rife with corruption. On vacating his office the last *preifeito,* Francisco Hélio Bezerra Bessa, also emptied the public coffers – after failing to pay state employees for the previous four months.

But a few words more on the region in and around the Estação Ecológica Mamirauá. The latter was established by government decree in 1986 and became a state-run entity in 1990. Since 1996 it has officially held the status of RDS (*reserva de desenvolvimento – sustentável* – a region of conservation and sustainable progress). It comprises an area of 1.35 million ha, consistingly solely of inundation land and lying between two great rivers, the Japurá and the Solimões. Former presidents and Bill Gates have visited the station (the latter subsequently gave 40 million dollars to the fight against the greatest scourge of mankind, to malaria research).

There are 5,300 people divided between 61 small *comunidades* in the reserve, and 130 are permanently employed making sure that all runs smoothly, including the infrastructure comprising the 13 floating houses, the four on the *terra firme*, the four pickups, six vessels, 35 *voadeiras*, and a motorbike. In the 10 years to 2000 more than US$ 10 million has been invested, with sponsorship from the Ministério da Ciência e Tecnologia/CNPq, IPAAM (Instituto de Proteção Ambiental do Amazon), the DFID (Department for International Development), and WWF (World Wide Fund For Nature) as well as the European Union. In 2000, when ecotourism began, 210 tourists came here, mainly from Europe. The guides, cooks, paddlers, among other helpers, are all from the *comunidades* of Mamirauá.

The guests are housed in five of the *flutuantes*, each consisting of two suites, and the three days/four nights package (including all meals) costs US$ 80 per person.

The Mamirauá residents

The Rio Japurá forms the boundary between blue and green. On these two pages habitats of both species in the region: 1. Lago Amanã. 2. José's canoe on the *lago* (it sank - *right-hand page*). 3. In order to reach the remote *lagos* and blue habitats during the dry season you have to carry the canoe overland (6-7). 5. Once arrived, with luck a nocturnal search will produce discus like these splendid blues (5) – or a fabulous lizard (4).

1-2. Two green discus habitats west of the Rio Japurá, full of grass (1) at Jarauá and at Jauarauá full of *Utricularia foliosa* and *acará-açú* plants (2). 3-4. I was able to spend the night with José in his hut (3) on the Amanã, which lies north-east of the Japurá. It had a splendid view of the huge lake, which I enjoyed over my müesli (4). 5. The children of Donna Lucieda and José. 6. Their pigs – they are self-sufficient. Their hut lies a week by canoe from Tefé. 7-8. He lent me his canoe for night fishing (7). And the canoe had a leak. When my companion William couldn't bale fast enough we sank (8).

are permitted to fish and practice agriculture, as well as selective felling. They are involved in all the decision-making as well as in the supervisory work in the region. The entire area is dotted with lakes – there are said to be 600 – which are often connected by *paranás* or natural channels. When we consider that the water level in this region varies by up to 15 m between September/October and May/June, then we can well imagine the quantities of water to be found here. To date more than 290 fish and 310 bird species have been recorded, with most of the latter aquatic. The number of amphibians and reptiles has not been recorded.

During the dry period the numerous lakes serve young fishes (eg *tambaquís*) as refuges from predators, and with the coming of the high-water period the extremely shallow *lagos* provide the *pirarucú* with nest sites where they spawn. This is also when the spawning migrations, already mentioned in part, take place. On one occasion I even witnessed a vast siluriform migration here, involving more than 2 billion catfishes from 10 families *(see page 456)*. And of course, as already mentioned and illustrated, this is also when the majority of Amazonian fish species really begin to feed. The table is bountifully spread as long as the *várzea* and *igapó* tree populations are present. And not only the discus, but food-fishes such as *matrinchá (Brycon* spp.*), pacus (Mylossoma* spp.*, Myleus* sp.*,* and *Metynnis* spp.*), pirapitinga (Piaractus brachypomus), sardinhas (Triportheus* spp.*),* and *aracús (Leporinus* spp.*, Schizodon* spp. *and Laemolyta* spp.*),* and millions of fry cram their bellies full.

Fish is the protein of the *ribeirinho*. At least 500 g per person per day is consumed, and this alone requires around 400 tonnes of fish annually from Mamirauá. More than in Tefé itself. The out-lying *ribeirinhos* dry and salt the fish in order to be able to sell it to the *regatões*, while the *caboclos* living near to urban centres offer their fish in the market immediately. Only time will tell how all this will develop further. According to the statistics the species biodiversity, including that of the trees, in Mamirauá's *várzea* and *igapó* regions is greater than elsewhere. More than 250 tree species more than 10 cm in diameter have been identified. But many of them, for example s*amaúmeira (Ceiba pentandra),* have already been considerably reduced by the selective felling. The other giants, *assacu (Hura crepitans), muiratinga (Maquira coriacea),* and *ucuuba (Virola surinamensis)* – all white woods required in quantity for the plywood industry in Manaus and which continue to be felled – as well as hardwoods such as *louro-inamui (Callophylum brasiliense)* and *mulateiro (Calycophyllum spruceanum)*, are seriously endangered. The latter are used for boat building and furniture manufacture. And only recently the demand for *envira-vassourinha (Xylopia frutescens)* has increased, with freemasons paying high prices for their furnishings in Tefé.

When I was here prior to 1986 there were less than 10 *comunidades* and barely 1,000 *ribeirinhos,* but in the meantime the number has increased more than five-fold and everyone needs to live. All these activities are difficult to control and there is no question but that the natural flora and fauna, and in particular the aquatic, will continue to suffer increasingly, especially in this region where almost all of it currently survives.

Without fruits and seeds there will be no discus and hardly any other fishes – and the rare endemic white uakari *(Cacajao calvus),* which feeds almost exclusively on *envira-vassourinha,* and other native species will be increasingly more endangered. Except around Tefé, there have never been many aquarium fishes collected commercially in this region. The small number of discus from the Mamirauá and Japurá regions are not worth mentioning. And if, as briefly mentioned in a Mamirauá publication, the discus in the *reserva* are now fewer in number, then that must be entirely the result of the loss of trees (and thus food).

As already mentioned, greens have to date been found only to the right of the Japurá. Along its left-hand side, where the water is instead clear, there are blues, and I found this to be particularly the case in the huge **Lago Amanã** and its long thin offshoot, the Lago Urini (see pages 458-459).

The Amanã was studied scientifically for the first time at the end of the 1970s by the Canadian biologist Robin Christopher Best (1949-86), who was seeking a peaceful, isolated place for his work on the Amazon manatee, and who until his premature death battled for this region too to be given protection. And his efforts weren't in vain. In August 1998 the state of Amazonas issued a decree for the establishing of a reserve – the RDSA (Reserva de Desenvolvimento Sustentável Amanã). It comprises 2,350,000 ha and a population of around 4,000 in 45 *comunidades.* Here too it remains to be seen how things turn out. Of course this means that a "protected" block of forest of around 5.7 million ha, such as can be found nowhere else on Earth, has now been created by this reserve, along with that of Mamirauá and the previously mentioned Parque Nacional do Jaú.

Note that Amanã is the name of the indigenous Guayanas' water goddess or Siren (= manatee). The perfect name for the lake – as Best may well have thought as well. On the other hand, among the Yanomami *amana* means caiman.

My last visit to the Amanã and the surrounding discus habitats was in 1997 (see page 458-9), before all this had taken place.

The parameters I measured at the Lago Amanã habitat on 07.08.1997 at 14.00 hours in 0,5 m of depth were as follows: pH 5.24; conductivity 15 μS/cm; daytime temperature (air): 24.5 °C, water: 23.5 °C (a *friagem* (sudden drop in temperature) had taken place). Intact wilderness full of *acará-açú*

1. The Rio Juruá downstream of Eirunepé – as in the Purus, there are again no discus habitats here in the river itself. 2-3. I trekked here, covering more than 100 km through unspoiled jungle (in the background) in the central Rio Juruá and Jutaí region, in order better to understand the distribution of discus. 4. Pure wilderness everywhere, butterflies *(inset)* and termites (good fish food). 5-7. I gave clothes and presents to the Kanamari, also called the Katukina, who were on migration through the deepest jungle. With spider monkeys (7) – their ancestors? (see text). They mark the trees.

BLEHER'S DISCUS   **461**

plants·; water tea-coloured with very little movement; fallen trees and roots, no floating plants; bottom sand.

Green habitats are widespread on this side of the Japurá. I found them in the **Lago dos Guedes** and south to the Jarauá, including in the **Lago Jauaruá** *(see page 459)*. Often beneath floating *Utricularia foliosa* and among masses of the various *acará-açú* plants (3-4 *Licania* species) that occur there.

In the *comunidade* of Jarauá, believe it or not, there is even selective rubbish collecting. Once a week the children children go out with push-carts to collect and sort the rubbish in order to sell it. This is coordinated by the teacher, Antônio Daniel de Carvalho, known as Seu Toniho. "We want to make Jarauá an example for others", said the 59-year-old environmentally conscious schoolmaster, who had seen his parents despatch 18 *peixes-bois* in a day. He said it was the fault of their ancestors that nowadays hardly a child along the Amazon knew what a manatee was. In another *comunidade*, Barroso, I found that the teacher, Manuel João de Souza Filho, was unable to give his 36 pupils any lessons for two months because the school was flooded to a depth of almost a metre.

Before we now leave Tefé and make our way along the right-hand bank of the Solimões past Alvarães and Uarini to the Rio Juruá, a few more parameters from green discus habitats in this region and the Japurá. Firstly, from the **Lago Tefé** habitat at three different times:

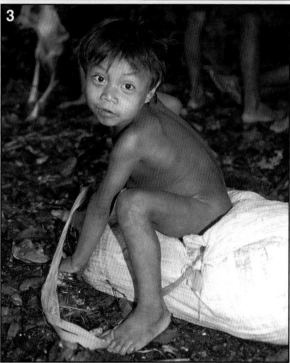

1. Fonte Boa on the right-hand bank of the Solimões. 2-3. Kanemari children, always smiling and cheerful. One boy tied up my net bag with a liana (3). New tiny characins *(right)* from Fonte Boa, and *(left)* the kernels of the seeds on the ground in the *igapó* are a very good fish food.

On 07.11.1985 at 10.00 hours in 0.5 m of depth: pH 5.85; conductivity 15 µS/cm; daytime temperature (air): 32.5 °C, water: 29.9 °C.
On 06.11.1995 at 12.00 hours in 0.5 m of depth: pH 6.7; conductivity 20 µS/cm; daytime temperature (air): 30.5 °C, water: 29.3 °C.
On 06.05.1997 um 14.00 hours in 0.3 m of depth: pH 5.7; conductivity 6 µS/cm; daytime temperature (air): 32.5 °C, water: 30.1 °C. Wilderness still intact at this time, full of tree-trunks, branches and roots; water clear, scarcely any movement; no floating plants; bottom sandy.

In the Rio Tefé habitat on 09.11.1985 at 20.30 hours (10.11.1985 at 07:00 hours), in 0.5 m of depth: pH 5.45 (5.20); conductivity 11 (10) µS/cm; daytime temperature (air): 26.5 (24.0) °C, water: 24.9 (25.0) °C. Water black with minimal movement; fallen trees, lianas, roots, no floating plants; bottom sandy.

Interestingly, the water temperature at night in green habitats is often lower than in those of the other two species.

It is also worth mentioning that I got to know the first discus collector in the region, Raimundo Nonanto de Oliveira Silva. He began to supply the exporters in Manaus at the end of 1962. Fished during the day, collecting up to 30,000 in a season. After 1965, when I introduced night-fishing, he went over to that like the others. When the Nova Colônia de Pesca (new fishermen's colony) was founded in 1976, the discus collector Yoshinary Esashika, nicknamed Zé Japonês (he really was Japanese), was elected its first president. I often collected lots of green discus with him – and later on with his son. Nowadays the Colônia de Pesca alone lands around 10 tonnes of food-fishes in Tefé every day, but never discus.

Alvarães is the capital of the *município* of the same name amd has 13,636 inhabitants (in 2004). It was named after the Portuguese town on the Costa Verde. The black Rio Caiçara, which joins the Solimões here, has not yet been explored, but I have seen typical habitats with huge *acará-açú* thickets of the species *Licania ferreirae*. There may be Greens here. And undoubtedly in the region of **Uarini**. Like Tefé, this place owes its origin to Samuel Fritz. Under the administration of Tefé, it first achieved independence as a *município* in 1981, and borders those of Alvarães, Juru, Fonte Boa, and Maraã. It comprises more than 9,850 km², and lies 50 m above sea level and 727 kilometres by river from Manaus. Here too agriculture is practised, and the *farinha de Uarini* is prized all over Brazil, and even mentioned in cookery books such as *Receitas de pratos que combinam com cerveja* by Maria José Rios and Lizete Teles de Menezes as an addition to the Pato no Tucupí. And also the Hotel Pousada Uacari.

The two rivers that meet here, the Rio Copacá (shown on IBGE maps as the Rio Uarini) whose mouth lies immediately opposite Uarini, and the Rio Uarini, adjacent to the village, have a certain amount in common. In each case the lower course ends in a large, black *lago* of almost identical length (15-20 km long and 1-2 km wide). As in the past there are green discus habitats (Tefé variants), but water melon cultivation is taking over and reducing them. More than 35 ha have been clear-felled for the purpose, 3,000 fruits eventually harvested, and the intention is to expand. But the good news is that on the 19th April 2004 President Lula approved not only the Terra Indígena Porto Prais (4,769 ha) for a Ticuna population here in the Municipío Uarini (this had been outstanding since 1973), but at the same time the Terra Indígena Tupã-Supé in the *municípios* of Alvarães and Uarini, as well as the Terra Indígena Igarapé Grande in the *município* of Alvarães. And even a Terra Indígena Juma for the seven surviving Jumas *(see page 431),* in the *município* of Canutamã.

It should be noted that as long ago as 1929 the Terra Indígena Méria (585.49 ha in what is now he *município* of Alvarães) was founded in this region for the Miranha, and in 1982 the Terra Indígena Miratu (13,198.78 ha, *município* of Uarini) by FUNAI, and that in 1998 they were promised the Terra Indígena Cuiú-Cuiú (38,310 ha, *município* of Maraã) on the Japurá. In 1999 there were still 613 Miranha, who speak only Portuguese, living in Brazil. In Colombia, by contrast, there are said to be almost double that number, who retain their own language, which is similar to that of the Bora but belongs to the Uitoto language group. The name Miranha dates from the colonial era, when the

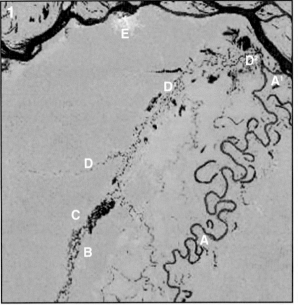

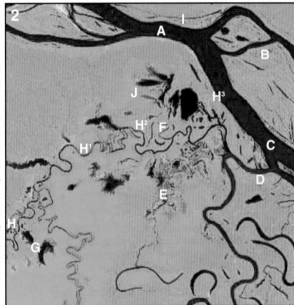

Two satellite photos of the Juruá region *(the lower in detail).*
1. (1:250,000, light blue = white water) The western Juruá region: A-A1) The lower Rio Juruá and its confluence with the Solimões (at A1). (For details see 2). B) Rio do Breu. C) Rio Mineruázinho. D-D2) Course of the Rio Mineruá. E) The town of Fonte Boa.
2. (1:100,000, blue = white water): A) Rio Solimões. B) Ilha Taiaçutuba. C) Mouth of the Rio Juruá. D) The Lago Tamaniquá region. E) The *paraná* that links the southern blackwater lakes of the Mineruá (F) with those of the Purus. G) Unexplored. H-H3) Course of the black Rio Mineruá (H3 is Arara, near the mouth). I) The *comunidade* of Vitória. J) The gigantic blackwater lake regions north of the Mineruá.
Left: *Rivulus* species, Igarapé Preto – Juruá region.
Right: *Hydrolycus* cf. *scomberoides* from the Rio Jutaízinho – Juruá region.

name was devised for the resident hostile indigenous people.

With regard to the Rio Japurá, it remains to say that, with a length of around 2,800 km, it is the fifth longest tributary of the Amazon. It rises in Colombia, in the Cordillera Oriental, around 250 km south of Bogotá, and is known there as the Río Caquetá. Its water is mixed, often turbid and full of sediment, but because of its numerous clear- and blackwater tributaries it is also clear at some times and in some places. Discus have never yet been found in its course, either in Colombia, where for the first two thirds of its length it passes almost countless waterfalls and rapids, rocky massifs and gorges, or in Brazil, where for its final 733 km (as the Japurá), along with the Solimões, it forms part of the largest Amazonian inundation region. As already mentioned there are discus habitats in the numerous *lagos* and *igapós* along its left-hand bank – possibly as far upstream as the Apaporis (*qv*) where I have also found blues. On the right-hand bank Greens are to be found only for the last 100 to 150 km. The limit of their distribution may extend along the Tonantins (*qv*). The northernmost distribution for blues I was able to find along the Japurá was at Jacitara.

I measured the following parameters on 04.11.95 at noon in 0.5 m of depth: pH 7.35; conductivity 29 µS/cm; daytime temperature (air): 33.5 °C, water: 29.6 °C. Water clear, but clouded by steady rising (although the main rise isn't until May-July), with minimal movement; fallen trees, roots, no floating plants. Discus: blue, with slight striping.

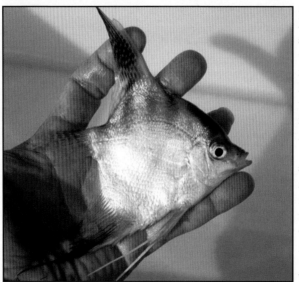

Above: An angelfish *(Pterophyllum* sp.), possibly a new species, from the Rio Mineruá discus habitat. *Right-hand page:* Three greens *(S. aequifasciatus)* (1,3,5) from a Mineruá habitat, and three (2,4,6) from the Lago Tamaniquá, Rio Juruá mouth region.

In addition it is navigable (but only with specialist equipment) for 721 km to the *vila* of Bittencourt, on the Brazilian side. It has only a slight drop (7.5 cm/km), and is shallow, barely 3 m deep – only 1.2 m from July to February. This long river too has a long history behind it and was once the home of a large population, as evidenced by countless finds of ceramics with fascinating zoomorphic decoration (ducks, bats, etc.), which have gone down in history as the Yapurá phase. But to date we know very little more about these pre-Columbian civilisations.

Rio Juruá Region: The next large white-water and sediment-rich affluent on the right-hand side is the most meandering of all rivers on this Earth. With a length of 3,350 km the Juruá is definitely longer than the Purus, although they are very similar and rise quite close to one another in Peru. This river too is in flood from February to April and it is only from July to September that sandbanks and shallow zones are in evidence. And for this reason it is also a lot less navigable – only in the middle and lower reaches, ie from its mouth to Eirunepé (1,850 km) and from Eirunepé to Cruzeiro do Sul (1,270 km), though even so during the dry season its depth decreases to 0.40 m. The *capitão* Almir told me that on one occasion he was aground for 72 days in the middle of the river bed. During the high-water period the river rises more than 10 m.

Discus have never been found in this river, and in this too it resembles its cousin to the east. I have visited the region several times, and it is important to be aware of the distances involved. Just to get from Manaus to, for example, Eirunepé you have to cover a distance somewhat greater than that from Moscow to Lisbon. Around 3,000 kilometres by river.

A unique river, and not only on account of its almost countless meanders, but also because of its numerous names. Juruá is thought to derives from the language of the Guarani, who call it Hiuruá, meaning "the river with a wide mouth" (which it isn't). Orellana noted Yuruá, but said that it was the Río de Cusco (as he thought that was where it originated). Nowadays the natives call it Yorouhá (just as they call the Japurá Yupurá, the Jutaí Yutahy, and the Javari Yavary). In the literature since the 16th century it can be found under names such as Juruná, Jurura, Iuruá, Yuruhá; La Condamine called it Yuruca; then there are Yurva, Yuruba, Yuruá, Yuoroá, Hyurna, Hyuacú, Hyurbia, Hyuruá, Hyuruhá, and many more.

The gigantic Juruá basin lies in a part of the state of Amazonas that to the present day remains for the most part unexplored indian country. This was once the home of 49 different tribes – more than in any (river) valley anywhere else in South America. They belong to the Arawak (Aruak) language group. Unfortunately during my last visit (in 1997) I established that only a fraction of them still survived, specifically around 200 Ararawas,

Mineruá-Japurá

3,964 Kaxinawás, 813 Kampas, 403 Poyanawas, 289 Katukinas (Kanamari), and 2,318 Kulinas. I have had contact with the last two of these (Bleher, H. 1999a). They live isolated in a difficult to traverse, dense primary forest region and continue their nomadic life to some degree. I once encountered Katukinas, far away on the Jutaízinho where I found no discus, but discovered the first mouthbrooding *Apistogramma*. Because I always have some clothes and other presents in my rucksack I was able to clothe an entire family, which they hugely enjoyed *(see page 461)*.

I have never managed to find discus so far upstream, despite optimal black water. Not once south of Juruá, let alone at Caurauari or Eirunepé. Note that Juruá remains the largest *comunidade* with 7,062 inhabitants (in 2004). All the others are small, and essentially this long river is virtually uninhabited except in the state of Acre and its capital Cruzeiro do Sul on the Juruá. Lots of white areas on the maps. From here to the Peruvian border is undoubtedly the least explored region in South America or anywhere else in the world. Of the around 54 indigenous tribes of Amazonia who to date have had little or no contact with the white man, the majority supposedly live in this inaccessible territory.

Green discus habitats are numerous in the lower Juruá region. In its *lagos* large and small, which often lie isolated and can be reached only by canoe at high water – or overland *(see page 463)*. I found them in *lagos* along the right-hand bank, in the Lago Tamaniquá *(see page 465)* beneath countless *Licania krukovii* (*acará-açú* plants), likewise in the Uará and south to the mouth region of the Igarapé Caapiranga. Along its left-hand bank in the huge lake region extending to the lower Rio do Breu.

Further up the Solimões we come to the mouth region of the Rio Mineruá, on the right-hand bank immediately before the huge island of Taiaçutuba. Again a blackwater river, more than 200 km in length with large lakes to the right and left of its lower course *(see page 463)*. Minerva (Minerua-ae) is, as everyone knows, Athena, the tutelary deity and patroness of handicrafts, trade, and medicine. Symbol of wisdom. Whether this has any relevance to the river and the Greens I found there *(see page 465)* is beyond my power to find out.

Even further up the Solimões on the right-hand bank we come to **Fonte Boa** and its *município*, where 36,150 people are said to live (or maybe 25,627, or 31,472 – all three figures date from 2004). Since 1891 it has held the status of *vila*, and it lies 62 m above sea level It is purportedly12,165 km^2 in area and is 1,011 kilometres by river from Manaus. The name originates from a town in Portugal, where the international Animal Production and Environmental Management Congress took place from 2nd to 7th May this year (2005). And both Fonte Boas even have an airport. Except that the Amazon one, has only once a week a flight and that is usually fully booked - month ahead...

Discus have been caught here only once or twice (from the Mineruá), but *pirarucú* is everyday sport, under the auspices of the *prefeitura*. And because the results (of the fishing) were so successful in 2004 everybody enjoyed better Christmas and New Year festivities. Every fisherman received R$ 3,000 (US$ 1,200) extra, as a bonus, to buy clothes, outboard motors, or a freezer, or to put the money in a safe place, towards building himself a house. It is said that the fishing success and the improvement in the quality of the fishes is to be attributed to the Mamirauá project. The *pirarucú*, which in 2003 was still threatened with extinction here and its capture prohibited, is said to be swimming around in large numbers again, and even more so since 2004, because of the mass slaughter (?). And so the 2005 hunting season has just begun...

This all reminds me of my experiences here regarding the men with tails. No, this is not an April-Fool's joke. There is supposedly a tribe who are all said to be descendants of one (human) mother with a male *coatá* (black spider monkey, *Ateles paniscus*). As long ago as the 15th October 1768, Friar José da S. Thereza Ribeiro, of the order of Nossa Senhora do Monte Carmo, set down a written affidavit because he couldn't believe his own eyes and wanted to avoid any recriminations. In this he stated (on page 42) that the indian who had come to him had a tail, *grossura de hum dedo polegar* (as thick as a thumb) and *meio palmo* (half a hand) long, but without hair. And that the indian had told him that every month he had to cut a bit off the tail as it grew too quickly. Ribeiro continued that the precise place where the tribe lived could not be ascertained, only that it was in the Juruá region. I was told this too. Interestingly, I heard a very similar story in Irian Jaya, only with the difference that there the tails were longer (Bleher, H. 2001c). Also, the Frenchman Docteur Henri Bouquet wrote about the men with tails from this region in 1929, in *Les Hommes à queue*, and likewise the naturalist Francisco de Castelnau (1812-1880), in his book about his journey from Rio to Lima *(De Rio à Lima.1843-47.* vol. V. p. 105) the following:

1. there is no physical reason to prevent the human species from having a tail.

2. Many of the *indios* he questioned along the Juruá confirmed the story, even down to the tail being *palmo e meio* (a hand and a half) long.

3. The (abovementioned) affidavit from Ribeiro.

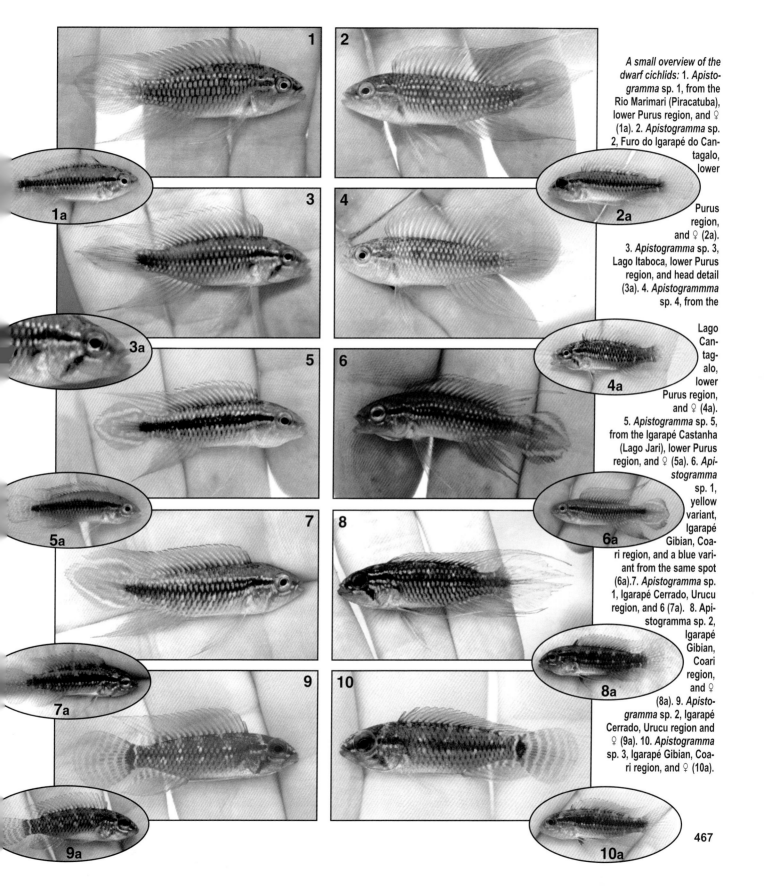

A small overview of the dwarf cichlids: 1. *Apistogramma* sp. 1, from the Rio Marimari (Piracatuba), lower Purus region, and ♀ (1a). 2. *Apistogramma* sp. 2, Furo do Igarapé do Cantagalo, lower Purus region, and ♀ (2a). 3. *Apistogramma* sp. 3, Lago Itaboca, lower Purus region, and head detail (3a). 4. *Apistogrammma* sp. 4, from the Lago Cantagalo, lower Purus region, and ♀ (4a). 5. *Apistogramma* sp. 5, from the Igarapé Castanha (Lago Jari), lower Purus region, and ♀ (5a). 6. *Apistogramma* sp. 1, yellow variant, Igarapé Gibian, Coari region, and a blue variant from the same spot (6a). 7. *Apistogramma* sp. 1, Igarapé Cerrado, Urucu region, and 6 (7a). 8. Apistogramma sp. 2, Igarapé Gibian, Coari region, and ♀ (8a). 9. *Apistogramma* sp. 2, Igarapé Cerrado, Urucu region and ♀ (9a). 10. *Apistogramma* sp. 3, Igarapé Gibian, Coari region, and ♀ (10a).

8. Rio Jutaí; Tonantins – Rio Içá; São Paulo de Olivença; Tabatinga-Benjamin-Constant Region; Leticia-Putumayo (Colombia and Peru); Nanay Region (Peru).

Our next stop up the Solimões is **Jutaí**. On both sides the land of the green discus; although the smaller *lagos* in the interior are to date little explored they nevertheless harbour habitats and no doubt discus. Next, at the mouth of the huge, more than 300 km long Rio Jutaí, we come to the town of Jutaí (some people also call it Foz do Jutaí). It lies 48 m above sea level and is the capital of the *municipio* of 69,857 km² with 22,251 inhabitants (at the end of 2004). In 1996 there were still only 19,325, so it is growing, like all Amazon settlements.

The origin of Jutaí is again to be attributed to the Jesuít Samuel Fritz, at the time when he established missions in Tefé and elsewhere. Until the second half of the 19th century the Spaniards and Portuguese disputed this region. Not until the advent of peace, in 1875, did the *Comendador* Pimenta Bueno take any notice of the place and start to settle the indian region. (Today remnants of the Catuquinas (Katukinas), Marauás, Ariaceus, and others still live in the region.) In 1928 it was raised to the status of *comarca,* and in 1955 it was split from Fonte Boa and turned into a separate *município* with the *subdistritos* of Mutum and Curuena and the seat of government in Boa Vista. Today it borders on the *municípios* of Fonte Boa, Juruá, Carauari, Itamarati, Eirunepé, Benjamim Constant, São Paulo de Olivença, Amaturá, Santo Antônio do Içá, and Tonantins.

Jutaí can be reached only by boat – there are neither roads nor airport – and is 1,072 kilometres from Manaus by river. Apart from limited agriculture and cattle-breeding, the people live from food-fishes such as *matrinxã, tambaquí, jaraquí, sardinha,*

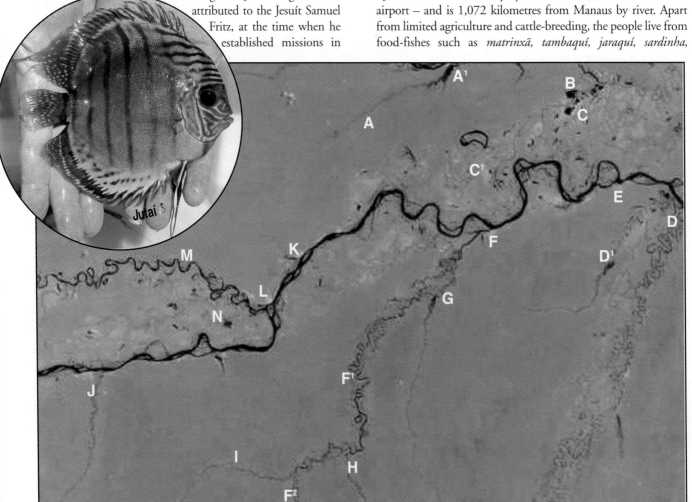

Satellite photo 1:250.000. Rio Purus to Rio Jandiatuba (blue = Solimões). *Green discus region:* A-A1) Rio Mapari (A) and Lo. Mapari (A1). B) Lago Inambe. C-C1) Lo. Sapateiro (C) and Atiparaná (C1). D-D1) Mouth of the Purus (D) and Mineruá (D1). E) Fonte Boa. F-F2) Foz de Jutaí (F) and course of the Rio Jutaí (F2-F). G) Riozinho. H) Rio Mutum. I) Rio Curuena. J) Rio Jandiatuba. K) Mouth of Rio Tonantins and the town. L) Santo Antonio do Içá. *Brown-Blue region:* M) Rio Içá. N) Lo. Jacupará.

1. The black Rio Jutaí is up to a kilometre wide at the *foz* (mouth) and well upstream. 2-3. Copatana is the only other (small) settlement on the Jutaí. upstream. It has almost no electricity, but it does have solar telephone, a *flutuante* (2. = floating shop), and a church (3). It lies on the right-hand bank almost opposite the mouth of the Igarapé Copatana. 4. *Viveiros* for discus. 5. The alpha individual *(arrowed)* stands out clearly in in the middle of this group of green discus. 6. On average, the pattern colour of greens from the Jutaí is almost uniform *(see also page 470)*.

1.-4. Green discus *(S. aequifasciatus)* from the habitat in the Rio Jutaí (in the background). This male specimen (4) is an alpha individual and demonstrates that not only red-spotted individuals are alphas. 5. The detail shows how if an individual (3) receives an injury to the flank. the scales grow back in a circle. closing the wound. 6-7. *Heros cf. efasciatus* from the Jutaí discus habitat: male (6) and female (7). Other fishes (the larger ones) from the same habitat: 8. *Tetragonopterus aff. argenteus.* 9. *Satanoperca jurupari.* 10. *Caenotropus cf. labyrinthicus.* 11. *Leporinus* sp. 3. 12. *Pterophyllum* sp.. 13. This detail is unknown from any *P. scalare* population. 14. *Peckoltia brevis.* 15. *Hypostomus cf. carinatus.*

pirarucú, pescada, tucunaré, and *pacu*. And the majority goes to Manaus or Tabatinga, which are also the transshipment ports for valuable hardwoods such as *cedro* and *jacareúba*, Brazil nuts, rubber, and *sorva* from Jutaí. *Sorva* (*Coumo utilis*, family Apocynaceae), also known as *kumã-uaçú* or *sorva grande*, provides not only rubber sap but splendid fruits with sweet flesh, which I have often eaten in the Amazonian jungle. Jutaí has a restaurant and hair-dresser, sawmill, bakery, and shops, for which the Flutuante Tuchaua acts as supplier.

It is said that their *Festa de São José*, which takes place from 10th to 19th March, is a sight to see – but if tourists wander here they marvel at the river. Its splendid black water is similar to that of the Rio Negro, and the area truly remains virtually unspoiled. Large parts of the Jutaí region are *TI* (indian reserves) and the natives value their environment.

I have found discus habitats extensively here – exclusively with brilliantly-coloured Greens *(S. aequifasciatus)*. I recently visited the sole discus collector, Isaac Perreira Correia, who has by now been collecting in the region for 20 years and lives upstream in Copatana. Isaac told me that there are numerous greens in the largest *TI*, through which flow the long right-hand Jutaí tributaries, the Rios Mutum and Biá. And that the Jutaí tributaries are black, except that the left-hand Curuena is white and no discus live there. Accompanied by the *homen do acará* (cichlid man), as Isaac is known locally, I travelled up the Jutaí to various habitats.

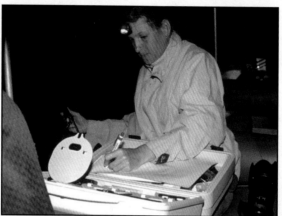

When you travel up the Jutaí from Foz to Copatana you still occasionally see one or more huts – some even with corrugated-iron roofs – and pasture land, but by contrast from Copatana upstream there is hardly a house to be seen. The Jutaí and its *lagos* and *paranás* are bordered by overhanging trees, often full of bromeliads, ferns, succulents, and orchids. You see lots of birds, and here and there a caiman, in this totally unspoiled paradise. Isaac said that 14 days are required to travel up the Mutum, which I had never explored, and that there was no likelihood of meeting another human.

The broad lower course of the Jutaí is largely surrounded by inundation forest as its waters rise to 12 m or more. On the other hand, in the dry season lots of splendid, almost snow-white sand-banks are exposed. But during this period the water is often no longer such a lovely black, but instead rather murky. In one habitat I dived down to 3 m and swam in the almost pitch-black water. I could hardly see my hand in front of my face, but immediately spotted a huge group of Greens, led by a fantastic alpha individual, disappearing into the underbrush.

The parameters on 26.01.2004 at 10.50 hours in 0.30 (3.0) m depth were as follows: pH 5.15 (5.02); conductivity 9 (9) μS/cm; daytime temperature (air): 37 °C, water: 28.3 (28.1) °C; oxygen in 3 m of depth: 0.88 mg/l. Unspoiled wilderness with *acará-açú* bushes; water black with very little movement; trees, branches, and twigs, no floating plants; bottom sand.

Discus variants: *page 473*.

And in the habitat *(right)* in one of the almost innumerable *paranás* in the area, the parameters on 26.01.2004 at 13.00 hours in 0.30 (2.5) m depth were as follows: pH 5.06 (5.05); conductivity 8 (8) μS/cm; daytime temperature (air): 29 °C (in the shade), water: 27.8 (27.7) °C; oxygen in 3 m of depth: 0.92 mg/l. Wilderness intact, indundation forest; water black with slight movement; full of trees and branches, no floating plants to be seen; bottom sand. Discus variants: *page 470*.

When I expressed an interest in the fishes that lived in company with the discus *(see page 471)*, Isaac said that he never bothered with such "by-catches" and always threw them back right away. Nobody wanted them.

Additional Jutaí parameters were little different to those given. I was able to find further habitats in the Lago Jararuá, and in the Rio Pati (a left-hand affluent of the Jutaí further upstream) upstream to the mouth of the Rio Curuena. And even in the Riozinho, the lower right-hand tributary of the Jutaí.

I had good times with the cichlid man. On one occasion he invited me into his wooden house for coffee and cake; told me about his eight children, all grown up, and that he had to live all year on the proceeds of the short collecting period (from the end of August to the beginning of November); how his eight brothers all lived in Manaus and two of them took his discus, but no specimen with large or white eyes, and only those with spots all over; and that he never fed the discus, they found enough in the *viveiros*. He also confirmed my spawning observations: when the water rises they spawn in pairs on branches and tree-trunks and keep the young with them for a long time. In September he had often netted 4-6 month-old discus with a diameter of 6-8 cm.

Further up the Solimões from the mouth of the Jutaí, near the tiny *comunidade* of São Miguel on the left-hand bank, there is a

Two green alpha individuals from a habitat in the middle Rio Jutaí *(background)*. Photo taken during rising water. The upper, a male, has become the new alpha individual of the group, the old alpha individual *(right)* – a female – has had a piece bitten out of her dorsal fin (probably by a large piranha) just where the prolongation begins. Undoubtedly during a diversionary manoeuvre during which she fled while the group made for safety. *Left-hand page:* Natasha testing water parameters at night.

paraná which forms a direct link with the Atiparaná *(see page 468)*, which passes through various *lagos* (Inambé, Sapateiro, etc) and *furos* to the Rio Japurá. Albeit not in the extreme dry season. Along this stretch, as well as in the *lagos*, which all contain black water, there are Greens similar to those in the Jutaí. Isaac told me this, and that there is a *furo* from Copatana to a *paraná* which joins the Solimões more than 10 km west of the mouth of the Jutaí. We followed this route and came to a Ticuna village. (There is a *TI* reserve here as well.) But they were so busy with their game of football that none of them noticed us. I explained to my companions, Natasha and João, that there was a great debate as to whether the Mayas or the Amazon indians invented soccer. One thing is sure, it wasn't the English and the game has been played for thousands of years.

We also passed tiny *comunidades* on the left-hand side of the Amazon, for example São Raimundo, Alegria, Beija-flor, Florianopolis, União, Belezas, and Liberdade, before reaching Belo Horizonte with its splendid black *igarapé*, which is even visited by tourist boats. Along the right-hand bank, except for cleared areas and banana plantations, there is only jungle. The next place we reached on the left-hand bank of the Solimões was Santa Luzia – which even has a landing strip – followed by Tonantins.

Tonantins is the capital of the *município* of the same name, which comprises 6,433 km² and has 15,506 inhabitants. Here we are 40 m above sea level and during my last expedition I found the place had expanded again, but not as far as the tourism infrastructure was concerned. There may have now been a hotel, but there were cockroaches everywhere so I told the boatman to continue. During my visit in 1985 I was told that 5,001 people lived here and the *município* was just four years old, but the place was mentioned as long ago as 1813, when the *igreja do Divino Espírito Santo* (church of the Divine Holy Ghost) was built. For a while under the jurisdiction of Tefé, Tonantins subsequently belonged to São Paulo de Olivença and eventually to Santo Antônio do Içá. No roads lead to Tonantins and it lies 1,109 kilometres from Manaus by river. The main industry is timber felling and the sawmill, followed by rubber and Brazil nuts. Cattle are bred and fish caught for personal use, but overall there is nothing to write home about. Or at best the green discus – Jutaí variants – that occur in the lower part (around 100 km long) of the black Rio Tonantins and its *lagos*. When travelling down the Solimões (as I did during my first research), it can also be reached via the Paraná das Panelas before you get to its mouth. On that occasion I saw three snakes along the short-cut in the space of half an hour, which is unusual in Amazonia (contrary to general opinion). One dropped into the water close to the canoe, a second swam away quickly from in front of it. And the third was in an overhanging tree, hunting

1. Children on the Jutaí at Copatana, on the floating discus *viveiros*. 2. The water of the Jutaí. 3. Discus habitat at rising water. 4. The *ribeirinhos* live mainly from fishing and manioc. Here peeling the manioc near the riverside. 5. Isaac, the discus collector on the Jutaí, with the author.

for eggs, as it looked. There are totally undisturbed green discus habitats in the Tonantins – apart from natives, hardly anyone lives in the region.

In a lower Tonantins habitat the parameters on 26.01.2004 at 18.35 hours in 0.30 (2.5) m of depth were as follows: pH 4.88 (4.83); conductivity 7 (7) µS/cm; daytime temperature (air): 30 °C, water: 27.5 (27.4) °C; oxygen in 2.5 m of depth: only 0.56 mg/l. Wilderness intact, sometimes *acará-açú* bushes; water black with some movement; overhanging trees, branches, and twigs, and a few *Pistia stratiotes;* bottom sand.

When I was back in Tonantins on my last trip, looking for fuel, a *caboclo* at the harbour told me that the previous year, on 18th April 2003, the ship *Itapuranga* had been attacked nearby, en route from Manaus to Iquitos. Shortly before sunrise two *voadeiras* with 10 masked men, armed to the teeth, had captured the vessel. Immediately destroyed the radio equipment, shot one of the ship's boys and hit two others over the head with the butts of their weapons. The crew and 220 passengers had been robbed – including one taking wages to a camp for bio-pirates and bio-pharmacologists. From the captain they took all his passenger receipts.

I think it is also worth mentioning that in 2001 there was no plan of the town or its houses registered in the *Cadastro imobiliário;* no body knew anything about the *favellas* (slums) there; and nothing about local economic requirements; and when I asked for the court house they looked at me as if I was totally stupid – but there are 1,380 state-paid officials in the town (almost 40% of the inhabitants).

The next stop is interesting: **Rio Içá**. Its (discus) inhabitants have puzzled numerous "specialists". Nobody can dispute that browns (and blues) occur here. Only by analysing the distribution, the water parameters, and the biology of the species can we better understand this phenomenon. Which is, of course, also the purpose of this book. But more of that below.

On the journey there up the Solimões, the ship passes several islands and then reaches the town of **Santo Antônio do Içá** on the left-hand bank. It lies almost 10 km before the mouth of the Rio Içá itself. By 1996 its *município* numbered 23,037 inhabitants. The original name of the place was Boa Vista, and it is no longer possible to establish when it was founded, except that this is said to have been before 1831. It is also said that between 1763 and 1772 Fernando da Costa de Ataíde Teve Souza Coutinho transported 340 families from Mazagão in Morocco and settled them in areas at Macapá, Belém, on the Rio Negro and near the mouth of the Rio Icá, and had the Forte de São Francisco (no longer standing) erected here.

But not until the end of 1935, when São Paulo de Olivença was officially declared a *município*, were Tonantins and Boa Vista (= Içá) again mentioned, as *distritos* of São Paulo de Olivença. Thereafter Santo Antônio do Içá received its current name

1. It is difficult to imagine how cold it can be on the Amazon, or, as here, on a *paraná* off the Jutaí. The author, freezing, holding his favourite jungle food: banana sandwich. 2. The floating jetty at São Antonio de Içá during sunset. 3. The huge mouth of the Rio Içá in the grey of morning. *Left:* the *Pirarucú* fountain in São Antonio de Içá, at the main square.

and Boa Vista vanished. A year later the *distrito* became a *município*, and autonomous in 1955, with 12,038 km². Içá is composed of the *comunidades* of Paraná do Tarará, Tarará de Baixo, São Francisco, Bela Vista, Tarará do Meio, São Sebastião, São João, Porto Alegre, São José do Amparo, Santa Cruz da Nova Aliança, Guarani, São João da Liberdade, Cuiava (not Cuiaba, as often stated), Boa Vista and Paraná do Pinheiro, Floresta, and Pinheiro do Meio. It borders on Jutaí, Amaturá, São Paulo de Olivença, and Tabatinga, on Colombia, and on the *municípios* of Japurá and Tonantins. The town stands 70 m above sea level, has no road links, and is 1,199 kilometres from Manaus by river. There is arable farming – with manioc the main crop – cattle and pig breeding. The people eat little fish and lots of meat – nowadays the norm in Amazonia. Fishing is nevertheless a factor in the economy, and mainly involves supplying Tabatinga with *peixe liso* (scaleless fishes). It is the only export apart from timber, the main source of income for the region. There are brickworks and sawmills, as well as bakeries and two hotels.

The Hospedaria lies near the harbour, but was still under construction at the beginning of 2004 and a real shambles – despite which they tried to rent me a room with a wooden bed with no bedclothes and rats underneath. When I asked about the *banho* (bathroom) I was shown a filthy communal toilet at the end of the corridor. The second, the Hotel Fogaça, was better, only the rooms had no windows! It stands on the main street opposite the best restaurant in town. It is managed by Donna Fogaça, who served us freshly roasted chicken with rice, beans, *farinha*, salad, and a *guaraná baré* as well. The town also has an ice-cream parlour, which is located at the main square facing the snow-white fountain decorated with three *pirarucús (see page 475)*. Loud rap music droned out into the pitch-dark night and the local church, still festooned with Christmas decorations, rustled as we bought ourselves *maracujá* ice-cream. A new temple is under construction nearby. There is a bank, a garage, and a mission and hospital station, where the American sister, Sherry Skirrow, for 15 years lovingly cared for the sick in the Amazon Baptist Hospital. Unfortunately she died of cancer in the USA on the 11th February 2004, shortly after my last visit, at the age of only 48. The entire town was waiting on the bank to greet her when she came back for the last time in 2002 – a bigger crowd than for the *governador* two weeks previously.

The entire hinterland of Içá is a huge *TI* reserve, the Ticuna Betânia I, often visited by missionaries from Santo Antônio do Içá who convert the Ticunas and preach in their churches. As already mentioned in Chapter 1, the Ticuna group, with more than 30,000 people, is the most powerful in Amazonia. They are flexible and intelligent, and a number have already become involved in politics. In the *município* elections of 2000 alone, one Ticuna was elected *vereador* (delegate) for Santo Antônio do Içá, one for Benjamin Constant, and three for Tabatinga. One can even learn about these election-results on the Internet. But, if one looks in the web for the tax revenue (*tributos arrecadados*) of Içá only the years 2000-2002 are mentioned (in 2005), and by clicking on those, one will find nothing.

1. Habitat in a *paraná* in the Içá region. Browns are to be found here at night, in the still, tea-coloured water beneath the *acará-açú* and two species of creeper, but only at greater depths during the day. 2. The same *paraná*, untouched but the creepers are killing the forest. 3. Boca do Jacupará.

Above: Typical brown discus habitat in the Boca do Jacupará (Jacurapá). A completely uninhabited region – 100% wilderness. *Below:* It was in a biotope like this were in 1976 I caught the "Rio Içá" discus *(inset, below)*. Browns live here, but individuals with a pronounced central (fifth) bar – or very red coulored ones – are extremely rare, both here and in the entire distribution region of the browns and blues. Further to the north the Swede Hongslo likewise caught brown discus in 1970 *(see page 478)*.

In 1971 the Swede Hongslo caught two brown juveniles (66.5 and 73.6 mm SL) near the *comunidade* of Cuiavá, in the Lago de Cuiavá and in Cocha Comprido respectively. And in December somewhat larger specimens (72.2-87.5 mm SL) (one pictured below), again in Cocha Comprido and in the Igarapé Comprido. The Thayer Expedition (James & Talisman) made a collection (043) in the Içá but it didn't include any discus. In 1976 I investigated the Içá for the first time and discovered the subsequently world-renowned Içá discus *(see page 477)*. However, the striking red individual with a central, Heckel-discus-like bar was a one-off. And I must also mention something that is unknown to many "discus experts": discus with a pronounced central bar are extremely rare, but widespread in the brown/blue discus (*S. haraldi*) distribution region. My friend Manuel Torres, for example, found a single specimen among 20,000 in the Cuniuã (Purus). And I too have only ever netted a single specimen after that first one in the Içá, specifically in the Maués and Alenquer region.

In 2004 I again visited the locality *(see pages 476-477)*, but there had recently been a *ressaca* (a sudden drop in water level after the first rise) in the Paraná Bucaçú. As a result the oxygen content had sunk so far (to 0.05 mg/l) that the majority of the large fishes had died. For several kilometres the surface of the *paraná* was strewn with dead fishes *(below)*. Only in the shallow bays was I still able to find juveniles of a few genera *(right-hand page)*, but no discus (living or dead).

The around 1,600 km long Putumayo-Içá is said to rise in the Andes in Ecuador, in the province of Napo, at a height of about 4,000 m; flows through parts of Colombia (as the Putumayo) and Peru (where it forms the border of those countries for 500 km); then (as the Içá) for its last 250 km crosses the Brazilian state of Amazonas, to finally empty into the Solimões. Its mouth consists of a huge delta with two main branches, with three islands (Mueru is the largest) in between. These branches of the delta are also known as *paranás*. This river is the tenth longest of the Amazon tributaries and its little explored, narrow basin is the eleventh largest at around 148,000 km^2.

Browns and blues are found well upstream in the Içá. I found them in the Lago Jacupará (also called Jacuapá and Jacurapá), in numerous *lagos* and tributaries up to the Rio Pureté (Puratá or Purití), as well as in the Cuvirá. *Ribeirinhos* say that they occur as far as the border, at Visconte de Rio Branco, again blues with an interesting colour pattern. But only on the Brazilian side. I have never yet found any evidence of blue discus in the Colombian part of the Içá, known there are the Río Putumayo. Just greens (see Putumayo).

In the habitat in the Boca do Jacupará the parameters on 27.01.2004 at 08.00 hours in 2.5 m of depth were as follows: pH 5.89; conductivity 27 µS/cm; daytime temperature (air): 31 °C, water: 28.1 °C; oxygen 0.24 mg/l. Wilderness intact, with some *acará-açú* bushes; water clear with some movement; overhanging trees, branches, and creepers everywhere, a few *Pistia stratiotes;* bottom sand.

In the habitat at Betânia the parameters on 27.01.2004 at 06.15 hours in 2.5 m of depth were as follows: pH 5.84; conductivity 14 µS/cm; daytime temperature (air): 30 °C,

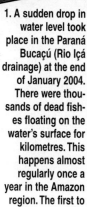

1. A sudden drop in water level took place in the Paraná Bucaçú (Rio Içá drainage) at the end of January 2004. There were thousands of dead fishes floating on the water's surface for kilometres. This happens almost regularly once a year in the Amazon region. The first to die are the adult fishes with a high oxygen requirement, mainly bottom-dwelling Gymnotiformes such as *Apteronotus albifrons* (2), *Rhamphichthys* sp. (4), and Siluriformes such as *Pimelodus* sp. (3). Also species of Characiformes, such as *Schizodon fasciatus* (6) and *Serrasalmus* sp. (7) or the surface-living *Rhaphiodon vulpinus* (5). 8. A brown from Cuiavá (Içá), collected by Hongslo in 1970.

water: 28.6 °C; oxygen 3.81 mg/l. Wilderness intact, with some *acará-açú* bushes; water clear with some movement; overhanging trees, branches, and creepers everywhere; bottom sand.

These are typical parameters for browns and blues, and have nothing to do with the Heckel discus or greens.

Further upstream on the right-hand side of the Solimões lies the town of **Amaturá** with 6,338 inhabitants (1996). The origin of Amaturá is again to be found with the Jesuíts in the 17th century, when it was called São Cristóvão. Later it was for a time known as Enviratiba Castro d´Avelos and then Amataurá. Because it lies close to the border region, in 1968 it was declared a security zone, and in 1981 it became the autonomous *município* of Amaturá, bordering on the *municípios* of Jutaí, São Paulo de Olivença, and Santo Antônio do Içá.

It lies 97 m above sea level, encompasses 5,808 km^2, and is 1,307 kilometres from Manaus by river. Here too the people live primarily from manioc, but in addition lots of lemons and avocados are cultivated, numerous pigs bred, and children reared. The food-fishes are mainly shipped to Colombia. But the main export consists of valuable hardwoods, the largest source of employment.

There is only one *comunidade* at Amaturá, namely São Sebastião on the Igarapé Amaturá. And in its black water I found Greens (we are on the right-hand bank of the Solimões, still in the land of the Greens), very similar to those from the Jutaí. Nowadays anyone who travels up the Igarapé Amaturá will see practically nothing but clear-felled cattle lands along its right-hand bank. Along the left-hand shore all is still right with the world. The habitats resemble those of the Rio Tefé, almost down to the last detail.

In the Green habitat at São Christóvão (also known as São Francisco) the parameters on 25.01.2004 at 17.30 hours in 0.30 (2.5) m of depth were as follows: pH 5.30 (5.84); conductivity 8 (8) μS/cm; daytime temperature (air): 31 °C, water: 28.9 (28.3) °C; oxygen 2.40 (2.48) mg/l. Wilderness intact, with some *acará-açú* bushes; water black with virtually no movement; overhanging trees, branches, and creepers; bottom sand, covered in leaves.

Further along the same bank of the Solimões we encounter the long and meandering **Rio Jandiatuba**. A blackwater river more than 400 km in length and a little explored green discus region. I don't know whether anyone (apart from myself) has found discus here. The colour variants are again similar to those from the Jutaí. I found them in company with large *Tetragonopterus* and *Leporinus*, similar to the Jutaí species. Plus a lovely *Mylossoma* with a large, black-edged, red anal fin.

In the habitat, some 5 km upstream, the parameters on 25.01.2004 at 15.35 hours in 0.30 (2.5) m of depth were as follows: pH 5.28 (5.20); conductivity 9 (8) μS/cm; daytime temperature (air): 34.5 °C, water: 28.3 (27.3) °C; oxygen 2.60 mg/l. Wilderness intact with tall primary forest and *Virola* bushes; water black, with hardly any movement; overhanging trees, branches, and creepers, no floating plants; bottom sand.

During my last visit I met a fisherman, Paulo, who had just been catching piranhas, pacus, and sardinhas (*Triportheus* spp.) for food. He too said that discus are to be found only between September and November, with September the best month by far. That he had seen them in most of the *lagos* along the Jandiatuba, but never in shallow water. Many of the *lagoas* are accessible only by land or across an *igapó*. He rarely catches discus, and when they do turn up in his nets he let them go.

The upper Jandiatuba is part of the Vale do Javari *(see page 493)*, where as yet uncontacted indian tribes live. One of these is thought to comprise around 300 people who live along its head-

All the adult fishes may have died (or disappeared) in the drainage region affected by the *ressaca* (the sudden drop in water level – *left-hand page*). but I was still able to find fry and small (juv.) species extensively. Here just a few of them:
1. *Biotodoma wavrini*, juv.

2. *Semaprochilodus taeniurus*.
3. *Moenkhausia dichroura* (?).
4. *Triportheus* sp. – note the appendage on the lip. This develops only under extreme hypoxia and the fish uses it to locate any residual oxygen.
5. *Leporinus* sp.
6. *Anchoviella* cf. *jamesi* (Engraulididae).

waters, the Igarapés São José and Uchoa, as well as those of the Rios Itacoaí and Jutaí. They supposedly live as they have from time immemorial – just hunting, fishing, manioc, bananas, *pupunha*, and collecting turtle eggs. It is said that they have light skins, wear *tangas,* and the men have long beards. They paint their mouths black and are known as *"Barbados"* (bearded people).

Only recently (2004) President Lula granted a 180-head Ticuna population in the region the *TI* Nova Esperança (20,003 ha) on the right-hand bank of the Jandiatuba, at the outflow of the *furo* from the Lago do Adriano.

Heading upstream along the right-hand bank of the Solimões we come to **São Paulo de Olivença.** In 2004 it had 27,607 inhabitants. Here too Samuel Fritz founded a mission back in 1689, and called it São Paulo Apóstolo. But in 1691 the Portuguese for the first time drove out the missionaries (sent by the Spanish government) by force or took them prisoner. Even so they defended themselves and fought back, until in 1708 the Portugese *capitão,* Inácio Correia de Oliveira, had the villages evacuated. Then Padre João Batista Lana fled to Quito (Ecuador) and recruited an army, which he brought back down the Marañon to the Solimões, recapturing the *aldeias* and freeing the prisoners. But the *governador* of Grão-Pará didn't wait long before sending *Sargento* José Antunes da Fonseca, who avenged the Portuguese and freed those who had been taken prisoner. He placed the *aldeias* of São Paulo Apóstolo and São Cristóvão under the supervision of Portuguese missionaries and changed the names of these settlements to São Paulo dos Cambebas and Enviratiba Castro d´Avelos. The mission at São Paulo dos Cambebas was founded at a site where Cambebas and Ticunas had already lived for a long time. In 1759 the name Vila de Olivença (the name of a Portuguese town) appeared for the first time, but disappeared again until eventually, in 1882, Vila São Paulo de Olivença was established. Since 1968 it has been a town with a *municípo*, and is part of the *área de segurança nacional* (national security zone). Its 19,922 km² border on the *municipios* of Amaturá, Jutaí, Benjamin Constant, Tabatinga, Santo Antônio do Içá, and Alto Solimões. São Paulo de Olivença lies 96.42 m above sea level, and can be reached once a week by plane from Manaus, or once or twice per week by *recreio,* a journey of 1,235 kilometres by river.

The form of agriculture here is similar to elsewhere: *mandioca, arroz, abacaxí, cana-de-açúcar, feijão, milho, melancia, banana, cacau, abacate, laranja, manga,* etc, and again, as almost everywhere, cattle-breeding is on the increase and expanding endlessly, just like the timber felling. But not the fishery.

São Paulo de Olivença lies high up – around 500 steps have to be climbed to reach the elevation on which it stands. There are buildings to both left and right all the way – including two hotels. One of them, the Marques, has a beautiful view over the Solimões *(see page 483)*. When you reach the top you come to the culture square, where the annual festival of the Red Aras versus the Blues is held. With the church, which overlooks it all and past which people roar continuously on motorbikes – the latest mode of transport in the Amazon towns. During our last visit Natasha and I had a meal in the Restaurante Tucunaré – a splendid *pirarucú ão molho* with rice, spaghetti, black beans, lettuce, and a tasty *goiaba* juice as well. Two pretty girls run the business, which has only three tables but a view of the Solimões. Unfortunately a tractor was at work outside, trying to empty the totally clogged up "canal" – an

1. The town of Amaturá lies at the confluence of the Igarapé Amaturá with the Solimões. 2-3. Green discus habitats in the Igarapé Amaturá, a blackwater stream. Here photographed at night/twilight. The plants in the biotope are *Virola* sp.

open channel at the edge of the road, carrying sewage and rubbish to be spewed into the Solimões.

The *município* has only three *comunidades*: Vendaval, Santa Rita do Weill, and Nova Congregação. And a nearby *igarapé*, the Oté *(see page 483)*. This is the easternmost boundary of the distribution of the *piaba-brilliante*, the neon tetra *(Paracheirodon innesi* – see Bleher, H. 2004a). But they are to be seen only from September to October. This also applies to the green discus in the Paraná Cuiavua, whose mouth lies almost opposite the Oté and Isaac said that Jutaí-type individuals live there.

Because it is almost impossible to get a seat in the little plane that flies weekly to São Paulo de Olivença, those with the necessary money fly by chartered plane from Manaus. And this is how Cleiton, who has often flown me, came to arrive there on 29th July 2003, as he tells the story. On board he had four *France-Press* reporters, and he thought it odd when they wanted to fly on by night. So he made himself scarce. He thought that they wanted to involve him in freeing the Colombian former delegate Ingrid Betancourt, who had been taken hostage 17 months previously. Much too dangerous for his liking.

Heading up the Solimões you come to a floating control post in the middle of the river and have to stop. Here the Policia Federal (federal police), IBAMA, and a government team are at work making detailed checks on all vessels travelling up- or downstream.

Further upstream we come to the last right-hand blackwater Solimões tributary before the border, the Igarapé Camatiá *(see page 482)*. Isaac said this was supposedly the westernmost Brazilian boundary for the green discus. It has habitats, I spend a day investigating it on one occasion. Unfortunately the water level was too

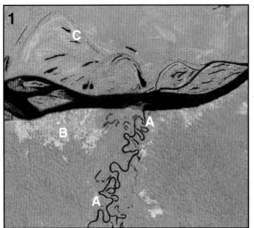

1. Satellite photo 1:50.000: A) The Rio Jandiatuba. B) São Paulo de Olivença. C) Solimões drainage. 2. The Rio Jandiatuba is almost uninhabited and full of habitats. 3. Typical green discus habitat in the Jandiatuba at high water (turbid).

high and I didn't have time for a nocturnal search. So I have no conclusive evidence.

From here there are repeated Ticuna villages, along the Solimões and in the *igarapés*, most of which flow in on the left-hand bank. I have explored there and was unable to find any discus, for example in the Igarapé Cajari, at whose mouth a Ticuna community (Cajarí I) stands, with a second (Cajarí II) further upstream. There are no habitats to be found.

Along the right-hand bank of the Solimões – and further inland – there are no more discus habitats as far as the border with Peru (see below) – almost exclusively permanent whitewater affluents. On the left-hand bank, upstream in the Igarapé de Belém, there are blue discus habitats. Near to its mouth lies the Ticuna village of Belém do Solimões (not Belém in Pará). I came across 14 *aldeias* along the Igarapé de Belém, where a total of 3,359 natives were living in 2004. All under the administration in Tabatinga.

The Ticuna here call the discus *acará-indo*. They are blue, often dark individuals, very similar to the colour form from Terra Santa (B7.10 and C8.6 in the book *DISCUS* Bleher & Göbel, 1992) from the Nhamundá region, but no striped individuals.

Unusually, on my last visit (in 2004) I met a Ticuna in Belém who told me about his *calderada* (fish soup), which he had made just the day before from *acará-indo*. I had never previously heard anything like that from a native.

There are no habitats far upstream in the Belém, but I have found Blues in the adjacent *lagos*. August and September are the best months to see them among branches and overhanging trees. Upriver as far as the Igarapé Tacana, with which it is connected at high water, has also discus habitats in its *lagos* and upper course. The Tacana drainage is also the area where

Rabaut found the first neon tetra (Chapter 1: *Fred Cochu*). But they live only in the small *igarapés*, tiny tributaries, and here too they are known as *piaba-brillante*. By contrast the Igarapé de Belém, which the local natives call the Igarapé Preto, is the type locality for the little tetra *Hyphessobrycon copelandi*, which is often misidentified in the aquarium literature *(see page 485)*.

The parameters in the discus habitat in the vicinity of the Ticuna settlement of Nova Esperança on 25.01.2004 at 12.30 hours in 0.30 (2.0) m of depth were as follows: pH 5.93; conductivity 20 µS/cm; daytime temperature (air): 32.5 °C, water: 27.1 (26.7) °C. Biotope undisturbed; water black with virtually no movement; fallen trees, branches, and leaves in the water, and floating plants – *Pistia stratiotes* and *Salvinia auriculata*; bottom to deep to see.

I could find no habitats, in the Paraná Ribeiro, also home to Ticuna; likewise in the lower Tacana, where I found nine *caboclo comunidades* but no habitats. The parameters were different to those in which discus normally live (eg pH 6.5, conductivity 62 µS/cm; on 25.01.2004 at 12.00 hours noon time).

Caboclos call the Tacana the Igarapé Preto, but the Ticuna living further upstream say Tacana. Manuel, aged 34, had been the *tuchaua* (chief) of one of the Ticuna villages for four years, and was a very intelligent sort. He told me in Portuguese that today more than 61% of the total population of this region of Amazonia are Ticuna. As well as his own language he also spoke the *língua general* or *nheengatu*, a universal language which the missionaries introduced in Amazonia on 30th November 1689, so that everyone could understand one another, as at the time there were more than 300 languages.

The Igarapé Tacana (or Takaná – as a hotel in Tabatinga also writes its name), at whose mouth stands the Ticuna village of Tacanaburgo II (Tacanaburgo I lies on the right-hand bank of the Solimões), is actually a *rio* rather than an *igarapé*, with a width of more than 100 metres in its mouth region. On my last visit I found innumerable *Pistia stratiotes* – whole sections of the river were covered in them.

The Tacana is a bit of a headaches for ichthyologists, as "Igarapé Preto" is the type locality cited for the lovely tetra described in 1965 by Géry as *Poecilocharax weitzmani* (honouring one of his colleagues, also a well-known expert on characins). In the description it says "some 60 km from Leticia". But that could equally be the Igarapé de Belém (also known as Igarapé Preto), whose mouth lies only a few kilometres to the north. When Harald Schultz collected the specimens in December 1960 he noted that Belém was close by, but in those days it was the only settlement in the area. Schultz collected a total of 41 species in the "Igarapé Preto" and these included four new taxa. I was able to find some of these species (in large numbers) in both Igarapé Pretos, but not *P. weitzmani*, which Schultz caught in the biotope of *P. innesi* (the neon). Perhaps this time I was there at the wrong time of year. But interestingly, if one looks at the original drawing of *P. weitzmani* and the type, and compares those with the photos published everywhere in the popular literature as *P. weitzmani*, it is clear that they must be two different species. Moreover the species, as described, lives among neons in almost neutral water, and the

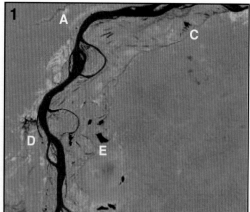

1. Satellite photo (1:100.000): the upper Solimões: A) Rio Puretá. B) São Paulo de Olivença. C) Igarapé Camatiá and *lagos* (unexplored). D) Igarapé Cajarí (no discus). E) Lago Jurupari (unexplored). 2. The Igarapé Camatiá is inhabited along its right-hand bank upstream to the fork, there is even a mission station: Babtist II, with a hydroplane. 3. Discus habitat on the left-hand bank of the Camatiá. 4. In the middle of the Solimões near São Paulo de Olivença at rising water (Jan./Feb.).

1. The steps climb up between houses to the town. 2. São Paulo de Olivença even has a court. 3-4. The view from the heights is beautiful, so they have even created a small park there (4). 5. But around the town and along the Solimões there are cattle farms without end. 6. Only in the Igarapé Oté, where there are neon tetras far upstream, is all right with the world. 8. And in the town, when the big festival of the *vermelho* (red aras) against the *azul* (blue aras) takes place in the Praca da Cultura. 9. Even the pygmy marmoset *(Cebuella pygmaea)* watch.

aquarium species a long way away among cardinal tetras in extremely acid water. *(See right-hand page.)*

The name Tacana presents a further puzzle, as it refers to a different indigenous group to that which lives there. The Tacana tribe enjoyed its golden age long before the advent of the white man, in the Bolivian lowlands *(see page 491)* between the Ríos Beni and Madre de Dios, and today some 3,500 still live in the *departamentos* of La Paz, Beni, and Pando in Bolivia. In 2003 I visited Ixiamas there and once again met a number of members of the tribe (the first time was in 1955). They belong to their own language group (Tacana), unlike the Ticunas living here now. When the Incas spread out into the lowlands they supposedly made first contact with the Tacanas. So how is it that a river in Brazil is named after them? Was it the Franciscans who first penetrated that far back in 1680? (And founded their first missions there in 1731.) Or those searching for quinine, rubber, and gold in the 19th century, who oppressed and decimated them? Or did they once extend this far? Nobody knows.

The Tacana are tall (1.70 m) slim people with delicate facial features, noticeably different from other local ethnic groups. Their beliefs and traditions remain a part of their daily life to the present day. Their shamans celebrate the traditional feasts only on the days most important for agriculture. The *tata janana* and *baba tcuai* as they are called, or *baba cuana* in Pando, are not only medicine men, but guardians of the well-being of the Tacana community and their universe.

Unfortunately I wasn't allowed to attend one of these ceremonies, which take place in huts

in places hidden deep in the jungle. In the Alto Madidi National Park, where I found interesting new characins (but no discus), I saw what wonderful masks they make. Unfortunately their ancient traditions, culture, dancing, and music are not being practised and cherished by the younger generation.

It is worth mentioning that there is a 4,092 m high Tacaná volcano, dormant since1988, on the border between Mexico and Guatemala. And in Guatemala, where in 1983 I investigated the fishes in the Rìo Coatàn, I came across the town of Tacaná at an altitude of 2,410 m, which now (2005) has 53,568 inhabitants. Could there be a millennia-old inter-continental link? Remarkably there is also an ancient Tacaneco language in Guatemala, classified as a Quechuan-Mamean language and again suggesting early contact between the Andean Quechuas and the Tacanas – but who knows?

We do know that *tacaña* signifies "mean" in Spanish and that following the terrible *tsunami* catastrophe in Indonesia, Thailand, Sri Lanka, and India. Latinos labeled President Bush *tacaño* (= skinflint), as he promised only US$ 350 million in aid, less than 0.25% of his expenditure on the Iraq war up to the time of the *tsunami*.

Next comes **Benjamin Constant** on the right-hand bank (or branch) of the Solimões. Although there are no discus here (and undoubtedly never were), as all the waters are sediment-laden and white, this place was cited as the type locality for the subspecies *Symphysodon aequifasciatus haraldi* Schultz 1960.

Not only the nephew of Gentil Rocha, the first and until his death the oldest fish collec-

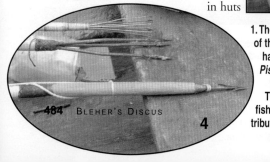

1. The mouth of the Igarapé Tacana. which is also the site of the village of Tacanaburgo I. 2. A typical blue discus habitat in the Igarapé Tacana at rising water, full of *Pistia stratiotes*. 3. There I found juvenile *Schizodon fasciatum* and *Megalechis cf. thoracata*. 4. The Ticuna in the region have an arrowhead for every fish. 5. The Igarapé Capacete is a right-hand Solimões tributary and, like practically all the southern rivers here. contains muddy water without discus habitats.

1. The Ticuna village of Betânia on the Igarapé Betânia (2) with brown discus habitat. 3-4. Igarapé de Belém (or Igarapé Preto) with Blue habitat (4). 5. Branch of the Igarapé de Belém. 6. Discus biotope in the Igarapé do Belém with lots of branches – collecting possible only at night. August-September. Turbid after rain. 7. *Poecilocharax weitzmani*. described from the Igarapé Preto. 8. The *P. weitzmani* of aquarists – a totally different species. 9. Juveniles of three sympatric species: *Serrasalmus* sp. (a), *Phenacogaster* sp. (b), and *Tetragonopterus cf. chalceus* (c) – who profits from whom? 10. The true *Hyphessobrycon copelandi*. photographed at the type locality. 11. Young *Brycon melanopterus* with prolonged lower lip for oxygen uptake. 12. *S. haraldi* from this region, undoubtedly from the Igarapé de Belém habitat.

tor in the area (Rocha collected for Cochu for almost 40 years and later for Tsalikis), but also Gentil himself, with whom I often travelled, confirmed the results of my researches: nobody has ever seen discus in this region, neither in the Rio Javari nor in other south-bank tributaries of the Solimões.

The nephew *(see page 487),* Renato Felix da Rocha – at the time of writing over 60 – worked for his uncle from the age of ten onward. Unfortunately Gentil died five years ago (I saw him for the last time in 1994) and Renato now works at Elektronorte.

In 1996 Benjamin Constant had precisely 23,633 inhabitants, but by now there are twice as many. The town is named after the mentor of Cândido Rondon, who at the age of 23 assisted him in ensuring that the transition from monarchy to republic on the 15th November 1889 took place peacefully. Benjamin Constant Botelho de Magalhães was also responsible for slavery finally ending on the 13th May 1888. As the first Minister of War he introduced the educational system plus the postal and telegraph services (Rondon subsequently laid the cables through the deepest jungle). He added the motto *"Ordem e Progresso"* to the Brazilian flag designed by the Empress Leopoldine, but because, he was unable to get his suggested actions and improvements enacted despite pertinacious campaigning, he departed the political stage. This idealist and benefactor of Brazil died in abject poverty on the 22nd January 1891 in Jurujuba, Niterói.

The development of the current town began in the 18th century. Around 1750 there was an *aldeia* Javari with Ticunas, founded by the Jesuits, nearby at the mouth of the Rio Javari. Here the seat of government of the *capitania* was established by the Portuguese government in accordance with the *Carta Régia* of 18th July 1755, under the supervision of the *governador* of Grão-Pará, Mendonça Furtado. Although, for a number of reasons, the administration was transferred to Mariuá (= Barcelos) on the Rio Negro. São José do Javari became a military station and border post. But not for long, as conditions on the left-hand bank of the Solimões, where Tabatinga stands today, were considerably more favourable. The land higher and a better view for controlling border traffic. This was decided by the *sargento,* Mor Francisco Franco, who named the new settlement São Francisco Xavier de Tabatinga. Shortly afterwards he had a fortress constructed (no longer standing) and brought the military to Tabatinga. It was and still is the border crossing to Peru. And since around 70 years ago to the "tongue" of Colombia at Leticia as well.

By 1854 only traces of the *aldeia* of São José do Javari still remained. The Rio Javari became the border between Brazil and Peru. It flows through a region (the Vale do Javari) that remains largely unexplored and where then, as now, blood often flows. When the border was established by mutual agreement at Tabatinga on the 28th July 1866, a commission travelled up the Javari, and during this expedition the geographer Capitão-Tenente José Soares Pinto was killed during an attack by *indios*. During a new expedition in 1874, led by the future Barão de Tefé,

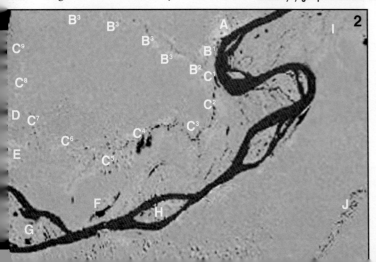

1. Ticuna children from the Ticuna village on the Igarapé de Belém (3). 2. Satellite photo (1:50.000) of the border region: A) Belém do Solimões. B1-B3) Course of the Igarapé de Belém. C1-C9) Course of the Igarapé Tacana (navigable in the high water period at C7 from Leticia (D) onwards and at C9 to Igarapé de Belém as well). D) Leticia and E) Tabatinga (grown together). F) Also the habitat of blues. G) The island of Armacá and H) the Island Arariá (both without habitats). I) Igarapé Camatiá and J) Jandiatuba (green habitats).

1. The harbour at Benjamin Constant.
2. The nephew *(left in photo)* of Gentil Rocha, the first ornamental fish collector in the Benjamin Constant region.
3-9. The method of *açaí* processing in Benjamin Constant: first the fruits (4) are washed, then water is boiled (3) to soften them. Next they are placed in the

pulper (now electric) (5), where the *polpa* (pulp) is separated from the seeds (7) and remains behind. The *polpa* is then sieved (8) and the concentrate packed in plastic bags (9) for further processing (for juice, ices, drinks, etc).
Inset: The Festa da Padroeira – just one of the festivals here.

Capitão de Fragata Antonio Luiz Hoonholtz, the latter's brother Carlos von Hoonholtz met his end. To the present day it is purportedly the Korubo, Flecheiros, and other as yet unknown tribes *(see page 493)* that are responsible for repeatedly killing intruding *gramipeiros* and *madeireiros*. But in general the opposite is the case. (See also Chapter 10.)

Quite a distance west of this region – or so I was told by local Ticunas – are two clearwater lagoons or lakes (the only ones?), in which Blues *(S. haraldi)* supposedly live: the Peruvian Lago Jatimana, which is around 20 km long and whose affluent rises at the edge of the nearby mountains (or in them) and empties into the muddy Javari. And if that is correct (and Ticunas are very precise in what they say), that is the westernmost natural distribution of the genus. The second Ticuna testimony to blues relates to the Lago Contrabando, a smaller lake that lies south of the *comunidade* of Paumari, also in the Javari drainage region.

But to return briefly to the man Benjamin Constant, who was honoured with the town name seven years after his death. In 1968 BC, as some call it, too has formed part of the *área de segurança nacional* because of its proximity to the border, and since the end of 1981 its own *município*, separate from Tabatinga and sharing a border with São Paulo de Olivença, Jutaí, Eirunepé, Ipixuna, Atalaia do Norte (to which the only road leads), and Peru. It lies 65 m above sea level and 1,628 kilometres from Manaus by river. Here too manioc is the most important agricultural crop, but *açaí* cultivation is slowly increasing *(see page 487)*. Cattle breeding has not yet increased to any serious degree, but there is an expanding commercial fishery, and not just in the local market *(below)*. They claim to be one of the most important transhipment ports for fish in the state. In the past there were aquarium fishes, but never discus. Timber processing and export constitutes the lion's share. There are several hotels and restaurants (one floating) – not at all bad – supermarkets, barbers, beauty salons, and clockmakers. Small industry and trade are in full swing. Hence it also has banks, and, of course, festivals such as those of Santo Antônio and São João.

Almost opposite lies **Tabatinga**, with just two islands in between, one of which has diverted the course of the Solimões in the last 50 years. The 60-year-old Renato from Tabatinga, who has often transported me in his boat, told me that the geography, as he terms it, is constantly changing here. During his lifetime has already seen the Solimões dramatically alter its course several times. And his father also told him about it as a child.

Once you are in Tabatinga, it is impossible to tell where the

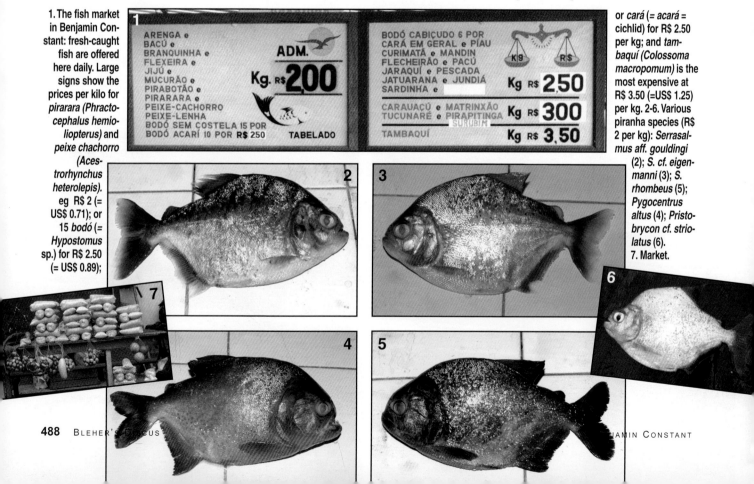

1. The fish market in Benjamin Constant: fresh-caught fish are offered here daily. Large signs show the prices per kilo for *pirarara (Phractocephalus hemioliopterus)* and *peixe chachorro (Acestrorhynchus heterolepis)*. eg R$ 2 (= US$ 0.71); or 15 *bodó* (= *Hypostomus* sp.) for R$ 2.50 (= US$ 0.89); or *cará* (= *acará* = cichlid) for R$ 2.50 per kg; and *tambaqui (Colossoma macropomum)* is the most expensive at R$ 3.50 (=US$ 1.25) per kg. 2-6. Various piranha species (R$ 2 per kg): *Serrasalmus aff. gouldingi* (2); *S. cf. eigenmanni* (3); *S. rhombeus* (5); *Pygocentrus altus* (4); *Pristobrycon cf. striolatus* (6). 7. Market.

town begins and ends. It is completely grown together with Leticia (Colombia). There is no border crossing point. Only a small, inconspicuous sign on the wall of a house to indicate that you are entering Colombian or leaving Brazilian soil. On my first visit (in 1966) the towns were around 20 km apart, on the last (in 1996) there were officially 32,009 inhabitants and in 2004 precisely 37,719. I suspect there are more than double that, or perhaps three times as many.

Back in the mid 17th century the Jesuits founded an *aldeia* here, and in 1766 this became the military and border station mentioned earlier. Tabatinga comes from the Tupí language and signifies "white mud" (on account of the muddy Solimões bottom) and in Tupí-Guraní "small house". Like Benjamin Constant, Tabatinga too first received its *município* status in 1981, with 3,225 km² bordering on São Paulo de Olivença, Benjamin Constant, Colombia, and Santo Antônio do Içá. It lies 1,607 kilometres from Manaus by river, 65 m above sea level, and since 1968 has also been an *area de segurança nacional*. And a lot goes on there *(see page 493)*, the war on drugs is in full swing. Since February 2004 all three of the bordering countries have been cooperating in the fight against the vast amount of cocaine smuggling and bio-piracy. Natasha and I, like all travellers, were (virtually) strip-searched, every bagg and suitcase, no matter how small, was opened at the airport and the contents examined in the finest detail – a bio-pirate would undoubtedly have been locked up. But how can hundreds of small boats, and innumerable jungle paths (with no border crossings) covering thousands of kilometres, possibly be monitored? They are tilting at windmills.

In Tabatinga *açaí* is the second most important crop after *mandioca*, followed by bananas. Cattle play a subordinate role – there are too many large *TI* (Ticuna reserves), which is to the benefit of the primary forest. But the fishery cooperative, like that of Benjamin Constant, is regarded as one of the largest in the (fish) trade in Amazonia.

The town is well provided with hotels, in particular the Takaná and the Tarumã. I finally came to rest in the latter, in existence for only 6 years, and discovered once again what a small world it is. The owner, Claudio Batista de Aguiar, whose wife is Peruvian, is not only the brother-in-law of Isaac the discus collector, but had also worked for Mike Tsalikis in Leticia as a young man. And Mike, an American of Greek ancestry, is a whole story in himself. When, in the 1950s, Tsalikis was collecting animals in Belém (Pará) and exporting them illegally, the police were hot on his heels on precisely the day that Fred Cochu stopped off in Belém.

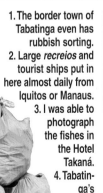

1. The border town of Tabatinga even has rubbish sorting.
2. Large *recreios* and tourist ships put in here almost daily from Iquitos or Manaus.
3. I was able to photograph the fishes in the Hotel Takaná.
4. Tabatinga's steps down to the Solimões, at rising water.
5. There are often drinking-water shortages here on the most water-rich river on Earth, and it has to be purchased in plastic bags.
6. Charcoal is a popular fuel.
7. The *taxista fluvial* (river-taxi stand).
8. Only a few decades ago the Solimões flowed to the left of this tongue of land, but now to the right.

Mike looked for Fred in Madame Peréz's brothel, where he had rented a room from a prostitute for 20 dollars as there were rats running around in the Grand Hotel. The *puta* (the local word for whore) was delighted, as she received at most 50 cents from a client. Fred was there to collect a consignment of fishes from old Cesar and wanted to leave at five o'clock in the morning. He took Mike with him to Florida and thus saved him from a Brazilian jail. During the flight Fred suggested Mike to export animals from Colombia where it was easy to get a permit. At that time Fred was also collecting a DC3-load of fishes from Leticia every week, from Gentil Rocha and Raffael Wandurraga.

Thereafter Mike established himself in Leticia and became world-famous for his animal collecting and (staged) display fights with giant alligators and anacondas (Claudio once saved his life when a snake coiled itself so tight round Mike's neck that he turned blue and almost suffocated). He appeared in TV programs and documentary films. Opened the Tarpon Zoo in Florida (hence the discus name "Tarzoo"), exporting to it every week; supplied jungle creatures to research laboratories, animal businesses, and wealthy private individuals in the USA. The demand was high and for a few dollars Tsalikis purchased the island of Santa Sofia further upstream, renaming it Isla de los Micos. The Capuchin monkeys on the island bred like rabbits and had only to be collected. He built a luxury lodge (the first in Leticia), provided money for the hospital, flew Capuchin monks to the missions on the rivers in his Cessna – and was universally popular. And Fred, whom he had to thank for all this, had laid down only one condition – "You never work with fishes" – with which he had promised to comply. But soon Fred discovered that Mike was nevertheless selling fishes to his competitors in Miami. Furious, he immediately flew to Leticia, but was then duped by Mike, who provided Fred and the TV team he had brought with him with free accommodation and trips into the jungle in his boats, and said that he had no money and had sent fishes only once – and Fred left.

When I met Mike for the first time in the mid 1960s at his pool in Leticia, he tried to sell me fishes too. He would send them to Florida with his regular consignments of animals in his own DC6 plane. In Miami they would be transferred to a flight to Frankfurt, as I was in the process of opening my fish quarantine station there. It sounded good, but Ross Socolof, the former owner of Gulf Fish Farms in Florida, where I worked and learned for two years, warned me, "Don't trust the man, he does not know his fishes and will cheat you". Instead I came to an agreement with old Raffael Wandurraga in Leticia, an honourable and 100% honest indian, who had worked with Fred for decades. Raffael was also the first to bring green discus from the Putumayo.

By the time of my second visit to Leticia (in 1967) Mike was the first honorary American Consul in Colombia. He set the tone in the 1970s in Leticia, and married an attractive Ticuna indian. The Ticunas to whose family he now belonged supplied him with caiman skins and macaw feathers.

Then tourism arrived. Tsalikis founded the Parador Ticuna and had a community of Yagua indians resettled from Peru to Colombia. These former shrunken-head-hunters, who had formerly pursued animals with their blowpipes, now vegetated as a tourist attraction in Ticuna territory, wearing skirts made of bast and serenading the visitors with melodies played on miniature panpipes. By then the animal trade was also illegal in Colombia as the result of pressure from international conservation organisations. But instead of taking things easy – the fish business was now run by his brother George – he had new ideas and, as is often the case with the rich, he could never have enough. Mike started his finally lofty enterprise, entering the cocaine business and transporting the white powder to the USA by the tonne and ship-load, concealed in large tree trunks. But in the course of transporting a consignment worth a quarter of a billion dollars via São Paulo de Olivença, he was betrayed and arrested in Miami and is now doing more than 20 years in clink. And the aquaplane *La Mirañita*, carrying George and a discus consignment from Tarapacá on the Rio Putumayo to Leticia, crashed shortly before landing in Leticia in 1980. George didn't survive, nor did any of the thousands of green discus.

Mike has left behind memories in Leticia – even postcards of his fights. And on one I saw recently is a blown-up detail from a one of his photos from the 1970s, where an anaconda is lying on the load platform of a pickup. Its head twisted and secured to the cab with heavy chains. Its mouth open. In the belly of the snake there is a large bulge. It looks like the outline of a swallowed peccary, but was supposedly a 12-year-old indian boy whom the snake had devoured in the vicinity of Amacayacu.

But that isn't all. The spot where three countries meet is already attracting 10,000 tourists per year (but hardly fish collectors any more, for exceptions see below). Around 15% Brazilian, 40 % from Colombia, and the rest from Peru, the USA, and Europe. They visit the Isla de los Micos and the Parque Nacional Natural Amacayacu (where Wandurraga's ornamental fish facility, built by Cochu, once stood, on the Lago Yacu). And the tourism is being expanded. In April 2003 the director of the department of drugs, Coronel Luis Alfonso Plazas, let it be known that Mike's monkey island with the 48-room lodge and the Parador Ticuna, with a value of 110,000 million pesos, should be made available for the *turismo ecológico*.

Leticia was founded in 1867 by Benigno Bustamante, but it

Repeated finds recently indicate that in the past Amazonia was much more populated and had far more buildings than previously thought. Only 50 km east of Trinidad (Bolivia), where Tacanas also once lived, the remains have been found of pyramidal structures which the Sirionó indians call *ibibaté*. They are supposedly considerably older than those of the Mayas in Central America, and 500-1000 people are thought to have lived in each building *(top)*. In addition they constructed an enormous system of canals for transportation, more than a metre deep and up to 5 m wide *(above)* and laid out like roads. Only recently was made this sensational discovery (Baleé & Erickson. 2005).

was not until the *tratado* (agreement) of Rio de Janeiro in 1937 that the *trapecio Amazónico* (Amazonian trapezium – because of its shape – *see page 493)* finally became Colombian territory and Leticia the capital of the Departamento Amazon with an area of 109,665 km^2. Leticia lies 82 m above sea level and in 1993 it had a total of 35,513 inhabitants – today there must undoubtedly be twice that. Although timber is the main export, here too people live by fishing, and even some from aquarium fishes, but mainly *aruanás* (or *arowanas, Osteoglossum bicirrhosum* and *O. ferreirai),* which are exported in large numbers (most of those originating from Brazil, where fishing for them is banned or allowed only to a limited extent).

The capital city Bogota lies far from the rest of Colombia, cut off by 600 km of rainforest, and Leticia is in the main a town of white Colombians, called *blancos* by the indians. A number originated from the metropolises of Bogotá, Medellín, and Cali. They were small traders, dealing in spirits, electrical goods, and outboard motors in this tax and duty haven. Their number was increased by adventurers and gamblers, drawn by the magic of the jungle life like moths to a flame. Philosophers and fantasists such as have existed since the first days of the *conquista* and in whose veins a few drops of the blood of a Lopéz Aguirre or an Ulrich von Hutten flowed. Have-nots driven across the country by economic necessity combined with a sense of adventure. People who are at home anywhere that money is to be earned from gold, emeralds, timber, coca, or the latest commodity, human organs. Since autumn 2002 corpses have been discovered here in the rainforest, their heads missing and their insides plundered of heart, liver, and kidneys – valuable raw materials for the illegal market in organs for transplantation. And these people seem to die only on days when there is a plane leaving for Bogotá or overseas.

Of course discus habitats are to be found only further to the north of here. As already mentioned, it is possible to travel from Leticia by canoe to the Igarapés Tacana, de Belém, and Jerônimo (known as the Calderón in Colombia). Interestingly some Colombians may call the Igarapé de Belém Calderón, but the border town of Calderón lies on the river known in Brazil as the Jerônimo (where Jobert caught a discus?). There are none in the Colombian part of the Río Pureté (Purutê, Purui). Undoubtedly there are no more until the green discus already mentioned in the Putumayo drainage – and there only in blackwater habitats. In between lie mountains as well as the huge 390,000 ha Parque Amacayacu. The Río Putumayo is, so to speak, an "island distribution" for greens *(S. aequifasciatus),* as in actuality they occur only in its Colombian and Peruvian drainage, and disappear abruptly at the border with Brazil – or rather, shortly before in the last right-hand tributary before the border, the Río Cotuhé. The other distribution region (already discussed) for greens is practically continuous, as can clearly be seen from the maps in Chapter 3.

Here on the Brazilian side, where it is called the Rio Içá, we find the border settlement of Ipiranga with around 600 souls, but the majority of these are the military, who also provide an air service twice a month to Tapatinga. I have never found habitats here. The Colombian border settlement is Tarapacá, with around 1,200 inhabitants, who live mainly from fishing, commerce, and timber sales.

It is said that the Río Putumayo/Içá has its origin in the Rio Guamués in the vicinity of Pasto in Colombia, which in turn has its source in the Lago La Concha. (Some say it rises in Ecuador *(see page 478),* depending on which river is regarded as the main headwater.) The Guamués flows south-east past heavily-wooded lowlands to Puerto Asís. From this point on it is known as the Putumayo, and the border between Perú and Colombia and

1. The *flutuantes* of the fish exporters in Leticia. They have aquaria and wooden boxes lined with plastic (3) in their floating houses, while they themselves live upstairs (1). I recently visited Afonso Cuellar and his son (1 and 3). who formerly collected discus in the Putumayo, but now export almost exclusively arowanas. While IBAMA grants only limited permission for the export of arowanas (*Osteoglossum*

bicirrhosum and *O. ferreirai)* from Amazonia, there are no limitations on export from Colombia. I even found a completely white variant in his tanks. He called it a white arowana and had one specimen (4) together with *O. ferreirai*. They come from a different collecting region (maybe a third species?). They are not albinos.

2. A small tributary of the Solimões in Colombia, which during the high water period forms a link with Calderón (on the Igarapé Jerônimo) via the Igarapé Tacana-/Igarapé de Belém.

Left: Since February 2004 the military of the three adjacent countries – Brazil. Colombia. and Peru – have for the first time been fighting together against the drugs Mafia and bio-piracy. Or at any rate. everything has been put in motion and strict controls imposed. The Vale do Javari is unregulated and has often seen bloodshed. A vast (3.338.000 ha) indian reserve, extending into the *municípios* of Atalaia do Norte. Benjamin Constant, Estirão do Equador, Jutai, and São Paulo de Olivença, has repeatedly been plagued by loggers and gold prospectors. And farming is also on the increase. There is constant friction and bloodshed. The region is so large and impenetrable that, like former FUNAI president Sydney Possuelo. I could write a book about it. My expedition is marked in blue (dashes = on foot) and took place in 1996. Possuelo, for years a FUNAI expert on isolated tribes, has travelled there several times, the last in 2001 (in red), to seek out undisturbed tribes such as the Mayá, Korubos, and Flecheiros, *inter alia* and ensure that they remain. He now believes – unlike in the past – that only thus can they survive. In 2001 Possuelo made contact for the first time with the Tsohon-djapas, nomads who live between the Jutai and Jandiatuba. (*Inset photo*: Possuelo with Korubos.)

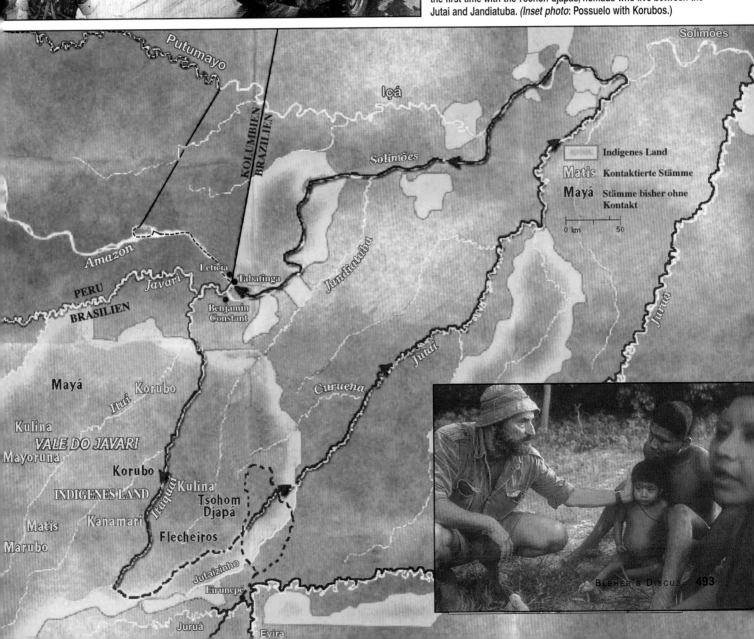

between Ecuador and Colombia runs along its centre. The Colombian part lies in the Departamento Putumayo, established in 1991 with 25,684 km², which by 1993 had precisely 166,679 inhabitants. Since then the number has increased considerably just through the expansion of agriculture, cattle and food-fish breeding, significant oil deposits, and, of course, the gigantic cocaine plantations, against which Bush has campaigned for years without success. With the anti-drug pact and 3.3 billions dollars the American president has tried to show the world that he is ready for the Colombian drug barons. People, and rare animals and plants, have been mercilessly destroyed in order to get rid off the cocaine growers. For around four years cocktails of untested herbicides and chemicals have been sprayed to and fro across the woodlands and fields of Putumayo from small planes under the strict supervision of military helicopters. The irony of it all is that the growers state that they are losing at most 15% of their coca-leaf harvest after a spraying, and they simply replant the area. A number have moved into the national park or still deeper into the jungle, where the chemicals may not be used *(see page 496)*. In addition, Ecuadorian, Peruvian and Bolivian growers have expanded their cultivation as the demand on the streets of US cities and elsewhere has increased even further following Bush's "cocktail party". And the price. The pact is undoubtedly the greatest American disaster in recent times after the Iraq war, and demonstrates yet again that evil cannot be combated at the root.

Unfortunately I haven't been back to the Putumayo since then and cannot say what effect all this has had on the discus world, as there has been no commercial collecting for a long time now. I know from food-fish breeders that they have lost almost their entire stocks after the spraying. *Tambaquís, pirarucús,* and other valuable edible fishes have died in thousands *(see page 496)*.

Before the spraying I was still able to find discus and habitats in the Rio Cotuhé and its *lagos* (all blackwater) – mainly red-spotted individuals with a more bluish instead of green pattern colour. (The "blue discus" described in 1960 as *Symphysodon discus tarzoo* also undoubedly came from here – see Chapter 2: *comments on taxonomy*.) In the Puerto Pipa region (origin of the well-known honeycomb toad *Pipa pipa,* which I have often encountered) on the Peruvian side and also in the Lago Guapapa there. In addition it was in this lake that the old (oldest?) discus collector in Colombia, Noberto dos Santos Perés – who died in 1995 at the age of 86 – once caught 12,000 discus for George Tsalikis. He supplied George until the plane crash (and it was his fishes that died as well), mainly in the 1960s and 1970s. I was told only recently by his three grandsons, Oscar, Argemiro, and Ferid, and his granddaughter Clara, who run the Aquario los Delphines, that no more are collected today. The customers want only specimens with red eyes, and the Lago Guapapa, for example, has none, plus they are generally rare in the Putumayo region (even so I was able to find some – *see page 496)*. The small number they require are brought to them by Isaac from the Jutaí or from Manacapuru to Leticia.

There are said to be greens up the Putumayo as far as the blackwater Río Algodón and further, or so the aquarium-fish collector Alfonso maintains. But to date I haven't been able to secure any evidence myself. The Putumayo (and its Brazilian part, the Içá) contains clear water (except when it rains hard and sediment is washed in), but many of its tributaries are black and contain green habitats. Hopefully still. My last on-the-spot researches were 20 (1986) years ago. (For 2005 collection *see page 638*.)

Further north, after the nearest mountains, we come to the whitewater Río Puré and the black Aguanegra, which join at the border with Brazil to form the Rio Puruí, a right-hand tributary of the Japurá, and in whose lower course I have found blues. There are no discus in its Colombian tributaries. Only to the north in the lower Rio Apaporis.

The **Rio Apaporis** is an 805 km long river, not only little known – as regards fishes – but also virtually unexplored in other respects. It rises in the southern part of Central Colombia, and its clear water (some call it *preto)* flows south-east over numerous rapids and at least one fantastic waterfall, the Raudal del Jirijirimo, before joining the Caquetá/Japurá at Vila Bittencourt. (The right-hand bank is still called the Río Caquetá at this point, and the left-hand, where the Brazilian border post lies, the Rio Japurá, until further south in Brazil where it retains the latter name and finally ends in the Solimões.) For its final approximately 50 km the Apaporis forms the border between Colombia and Brazil. And in this last section, where it flows through a vast inundation region full of *lagos,* there are said to be blue discus. And in the lower Igarapé Preguiça (= sloth), its last left-hand tributary. I haven't yet been able to confirm this myself, but again I obtained the information from Alfonso, who claims to have collected discus in the region for decades.

That means that this must be not only the northernmost point in the distribution of the genus, but also the highest. The discus habitats lie 100 m above sea level and in the region with the most rainfall in Amazonia – up to 4,000 mm precipitation per year (as

The northernmost distribution of the genus *Symphysodon* is shown (blue) on the IBGE map *(right);* this is the Blue population in the lower Rio Apaporis (the map on page 161 doesn't extend that far upstream). The isolated population of greens *(S. aequifasciatus)* in the Putumayo is shown again to demonstrate how this single population is practically hemmed in by the mountains along the Peru-Colombia border. With only one road (shown in red) in the entire region – but it is as good as non-existent.

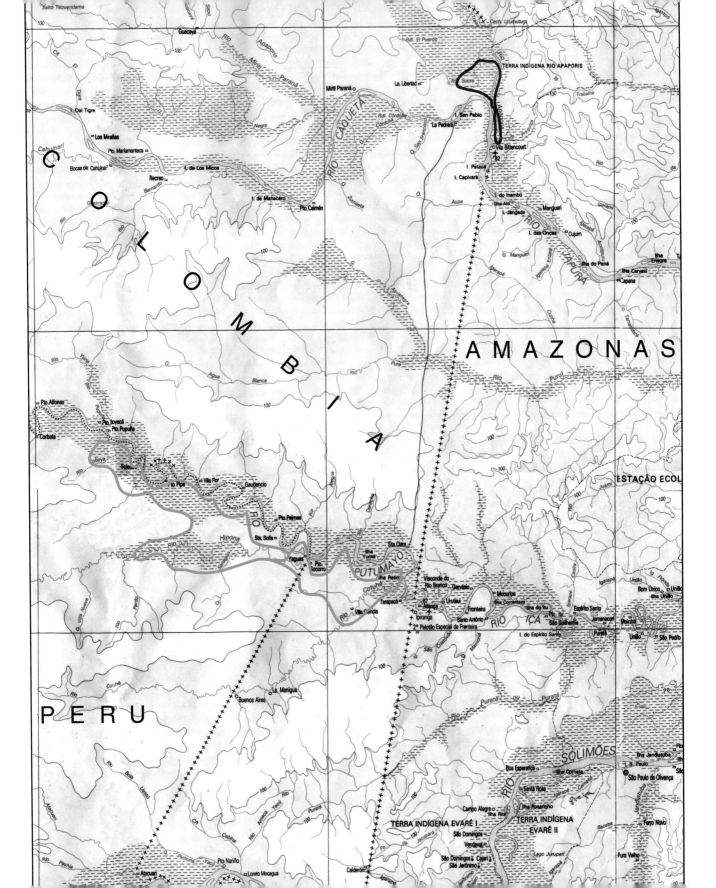

opposed to 2,000-3,000 mm elsewhere along the Amazon). In addition, the high-water months here are June to August, while in the Putumayo the water level is highest from February to May.

The Brazilian border post of Vila Bittencourt is situated at the mouth of the "sloth river". It is a sleepy military post with 360 people. Once, or at most twice, a month, a (supply) plane comes from Tabatinga or a *regatão* up the Japurá, but the latter are rare as the Japurá is virtually uninhabited along its approximately 750 km in Brazil. When Martius travelled up the river he called the Yupurá, in January 1820, there were still a number of native settlements, but these vanished without trace a long time ago, along with their inhabitants.

Things are quite different in La Pedrera. Barely 50 km upstream, past an arduous stretch full of rapids and waterfalls at the Cerros de Yupatí, lies this, the first Colombian town on the right-hand bank of the Caquetá. Until 1987 it was, like Vila Bittencourt to the present day, a tiny military border post. The inhabitants lived by fishing and agriculture, and looked forward to the small propeller planes that now and then brought edibles and medications from civilisation. But in 1988 all this changed at a stroke. The yellow metal was discovered in the nearby Sierranias del Taraíra and within a short time the place was bursting at the seams. Today the gold-miners' town is totally built-up, overpopulated, and in a desolate state. All the rubbish and sewage lying around in the open or spewed into the river. There is no electricity, no hospital or first aid – hygiene is a foreign word. The entire surrounding area (more than 17,000 ha) has been clear felled, and cattle breeding is expanding out of hand. There is an airstrip, just dirt which is supposed to be going to be surfaced one of these days, where a small plane lands once a week. The road marked on maps of this gigantic region – supposedly leading to Santa Clara on the Putumayo – doesn't exist *(page 495)*. Nor are there any discus, as I have been able to confirm on several occasions. The only thing really worth seeing is further upstream – a stretch of rapids some 6 km long, hemmed in by 80 m high, blue-grey granite cliffs. Colossal masses of water descending some 20 m over the steps of rocky falls. A fabulous biotope, called the Canón de Araracuara or Arara-Coara (= ara hole). If only there weren't so many *piúms*.

I can't leave the Apaporis, this remote river with some 16 rapids and at least one fantastic waterfall, without saying something about the people, including two Germans and an American, who have travelled or explored it in the most adventurous way. In addition it should be mentioned that it that flows through the

1-2. Green discus *(S. aequifasciatus)* from the Putumayo drainage. One (1) is from the Río Cotuhá region. where they often have a (light) blue pattern colour. 3. The colours of the flag of Colombia within the outline of the country (the tongue of land at the bottom = the trapezium). 4. An aqua-- culture installation. where all the fishes were killed by Bush's spraying to destroy the coca fields (6). 5. Collecting discus in a Putumayo tributary (in the rain). 7. The oil-wells in the Putumayo belong to the Canadian company Petrobank. A pipeline leads to the Pacific oil terminal.

Departamento Vaupes, which encompasses 54,000 km² and, with 22,000 inhabitants (85 % of them native people, who live mainly in the 36,000 km² protected zone), is one of the most sparsely populated regions the world.

One of the above-mentioned people was Theodor Koch-Grünberg (1872-1924), from the little town of Grünberg in Hessen, Germany. He was a quite extraordinary naturalist. Following his education under the world-famous ethnologist Karl vom Steinen, Koch penetrated into the unexplored regions of northern Brazil and Guyana in the course of several expeditions. In the process he learned to value the free life of the indians and described it in gripping books such as *Roraima* and *Zwei Jahre unter den Indianern* (Two years among the indians). Koch was the first explorer to see masked dances and celebrations of the dead, and made the very first sound and film recordings of South American indians. And said (translated) afterwards:

"This journey took me to regions where before me no white man, let alone researcher and explorer, had ever set foot, and to indian tribes who, far from all the demoralising influences of so-called 'civilisation', have remained true to their ancient traditions and customs and are in fact 'better people' than we are. During my months of dealings with these often reviled (usually unjustly) children of nature, as the result of which I became totally adjusted to them and not only lived as the only white man among them, but also shared their lives, so-to-speak as an 'indian among indians', I was able to obtain a deep insight into their entire perception of the world, and, as well as numerous excellent photographs, amass a splendid ethnographic collection of very unusual pieces, which has just recently made the long journey back to Berlin."

In 1903-1905 Grünberg travelled up the Rios Negro and Uaupés and into the rivers Içana, Aiarí, Curicuriarí, Tiquié, and Cuduiarí in that region, and in the Caquetá basin along the Pirá-Paraná and the Apaporis, where he was without doubt the first European. In addition he said of black and white water that they were a curiosity, as he found the two types of water barely 100 m from one another, but they were flowing through the same forest and over the same substrate, and yet were nevertheless quite different. And of the Tiquié, which he knew in detail, having travelled its entire length several times, that it changed colour three times in its course according to whether the tributaries contained white or black water. Also that the indigenous peoples regarded the black waters as having a healing effect, while the white brought fever with them.

1. The almost uninhabited region, with the *tepuí*-like rock massifs rising from the jungle. through which flows the Rio Caquetá *(centre)* and the Rio Apaporis arrives from the north-west (background). 2. The fascinating ara hole – the *Acara-Coara* rapids in a Caquetá tributary. 3. A Tacana Indian. To the present day the old people live from and with their medicinal plants in the Madidi region (Bolivia). 4. An aboriginal from the Apaporis (Carayuru). 5. The Apaporis still has a rich flora and fauna. including *Boa constrictor constrictor* (and an endemic, *Crocodilus crocodilus apaporensis*, not shown). 6. The extinct 25 m long *C. purusaurus* from south of here.

The German Huebner (1) was a close friend of Theodor Koch-Grünberg and an exceptional photographer. who made his home in Manaus and opened a studio (6). He was the first to photograph the ethnic peoples of Peru, back in around 1880 (2-3). He erected the most spectacular structures in the jungle in order to shoot his photos from the right angle (4). And later even used authentic surroundings for indians in his photographic studio in Manaus. Almost unique for the time.

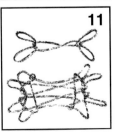

The Danis. an ethnic group of more than 50.000 have lived for thousands of years at an altitude of around 3.000 m in Irian Jaya. New Guinea. And the men practice a penis-sheath cult. using a plant seed for the purpose (1). Precisely the same tradition is practiced by ethnic groups such as the Macuna. Yababana. and Yabuna on the Apaporis. with whom Koch had himself photographed in 1905 (2). In 2001 there were supposedly only 168 Macuna still living in the region. but it is not known whether they maintain their ancient tradition. The method of attaching the sheath and the type of seed used were exactly the same as in New Guinea. It can clearly be seen in the photo (8) shot by Koch on the Rio Aiary as a Kauá applied his 2.5 m long *cerbatana* (blow-pipe). Koch published works of lasting interest. for example that about the Orinoco (4). Made lots of drawings. like this one of the Maloca on the Apaporis (5). and also took photos such as this one of a Maloca on the Alto Rio Negro with it's beautiful interior (6). One can see how big they are. The American Schultes. who has gone down in history as the father of ethnobotany. surveyed the Apaporis for rubber trees for the US government prior to the Second World War and spent years among the natives in the region. Here with Macuna children on the Apaporis in 1940 (3). Later. between 1941 and 1953. he returned repeatedly to the Amazonia region and collected more than 24.000 plants. Here with a waterlily (7). Schultes warned in 1994: "The Indian people and their knowledge are disappearing even faster than the plants themselves." Which is undoubtedly true. In the past – as I have seen for myself - indian children of five or six were masters of bow and arrow and daily went hunting for fishes (9) or wild animals (10). Nowadays this is rare. Koch also learned other things from them. eg the rope game (11).

BLEHER'S DISCUS   499

His journey to the Apaporis began on the 6th February 1905, when he set out from San Felippe on the Rio Tiquié. Accompanied by a few *indios* he followed the river upstream until it was impossible to go any further except on foot. After two days carrying the canoe on their shoulders they reached the Igarapé Yauakáka, which the *indios* said led to the Rio Japurá. Though they didn't mention that it was hundreds of kilometres with obstacles – gradients, fallen trees, rapids or waterfalls – blocking the way every 50 metres. After four days of the severest toil the *igarapé* became easier to negotiate and the first indian settlements appeared. Tribes who had never before seen a white man. On the 15th May he reached the somewhat larger Rio Pirá-Paraná, which was also full of rapids; in one of these he lost his entire supply of salt. The Tiquié indians soon abandoned him as they didn't want to run into a warlike tribe. Koch stayed behind with one helper, and it was supposedly another four to 14 days to the Japurá. They reached the Apaporis on the 21st March, and as luck would have it he encountered another tribe there, who helped him to negotiate the final large waterfall, the Jirijirimo, with his boat. On the 16th April he finally came to the mouth region of the Apaporis.

When, in 1924, he set off on another expedition to Brazil, he didn't realise at the time that it was to be his last. He died a painful death from malaria on the 8th October 1924, in the Amazonian jungle village of Vista Alegre (Rio Branco region).

Koch had often visited the German photographer Georg Huebner (1862-1935) in Manaus and they were friends for 20 years. Huebner had come to photography through his fellow-countryman Albert Frisch. Frisch shot the very first photos of Brazilian indians, from 1865 onwards on the Rios Negro and Solimões, where he also encountered Agassiz and his colleagues when they were collecting Amazonian fishes during the Thayer Expedition. In 1898 Huebner settled permanently in Manaus, which he had already visited 15 years previously during his first photographic expedition to the Peruvian indians, and again in 1894, when he travelled upstream to the Rio Branco. He was the pioneer of the photography of ethnic Peruvians such as the Campa, Mayonisha, Caxibo, Cunivo, Pito, Xipido, and others. Likewise a number of tribes on the Putumayo and Apaporis (nobody knows how he got there). This is fortunate, as his photos are the only ones of long extinct groups that have survived for posterity.

In the heyday of Manaus he opened the studio Photographia Allemã *(see page 498)* and was in demand everywhere. There followed an offshoot in Belém, where photographed not only

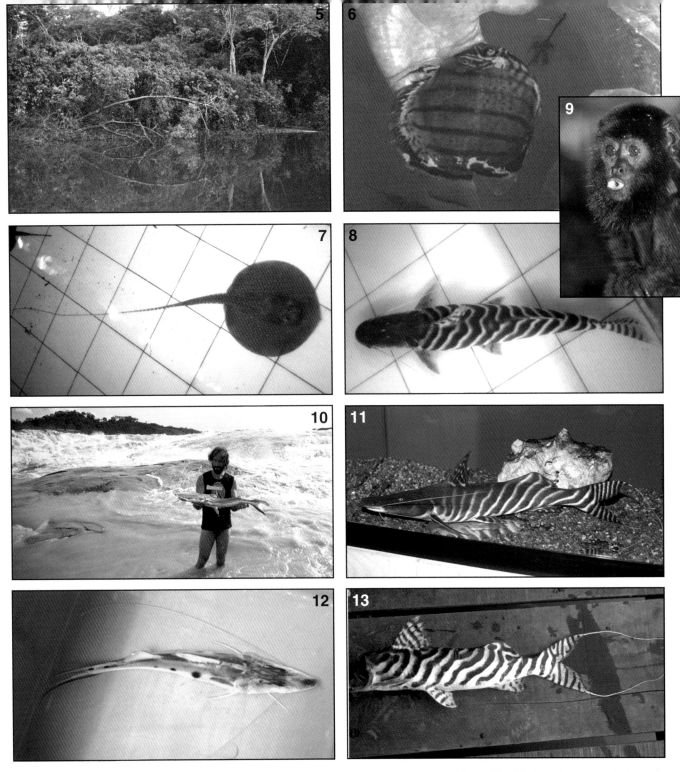

1. Iquitos officially has 400.000 inhabitants, but if the thousands of floating houses are included it must surely be a million. It is the main export centre for aquarium fishes in South America. 2. Fishing on the Rio Nanay. 3. *Pterophyllum scalare* from the Nanay. 4. I have presented authentic biotopes – how fishes live in the wild – at international exhibitions. Here an authentic Nanay biotope at Aqua-Expo 2001 in Belgium. 5. Green discus habitat in the Nanay. 6.Tefé-type discus, caught in the Nanay. 7. Likewise this *Plesiotrygon iwamae*. 8&11. This catfish, collected in the Nanay, is labelled *Merodontotus tigrinus* in the aquarium hobby, but that is a different species. 9. Red howler monkeys in the Nanay trees. 10. A *Sorubimichthys planiceps* in the hand. Only in the Rio Madeira (Brazil) have I been able to find the true *M. tigrinus* (its type locality) among large pimelodids like this, below the waterfalls of Teotonio. (Note: *Merodontodus* has been placed in *Brachyplatystoma* recently.) 13. This is what it looks like (unfortunately the head is missing). 12. A rare catch in the the Nanay: *Platystomatichys cf. sturio*.

eminent figures in the community but also indians in his studio. Unfortunately, like almost everything else, he too had to close down in 1920 with the collapse of the rubber boom. Thereafter he devoted the rest of his life to the flora and private photography. He discovered, described, and photographed numerous Amazon plants.

The third to travel the Apaporis and numerous other Amazon regions was the father of modern ethnobotany, Richard Evans Schultes (1915-2001). His first contact with the greatest rainforest region on Earth came when the Japanese over-ran Asia and the Americans could no longer obtain sufficient rubber. He was commissioned by Washington to count the *Hevea* trees along the entire length of the Apaporis, walking inland from the bank for up to 1,000 metres. His final count was 1.5 million ready-to-harvest rubber trees. And just imagine: at least 1,600 km up the Apaporis and back down again, repeatedly negotiating the countless rapids and waterfalls, beset by malaria, *piúms*, hunger, and loss of equipment, wandering through totally undisturbed dense jungle, waiting months for a chartered plane, and much more. And finally paddling for 40 days back to Manaus, all by himself and delirious with fever, in order to regain his health. I could write a song about it.

And in the end it was all to no purpose. With around one rubber tree per acre (4,047 m²) it was uneconomic to tap the trees and would never have provided enough. In the meantime the Americans had turned to recycling their old tyres and improved synthetic rubber to such a degree that its manufacture was increased.

However, as a result of his years in the jungle Schultes became interested in psycho-active and toxic plants and how they influenced he lives of the indigenous peoples, for example the Kofáns. He was undoubtedly the first white man to recognise the sheer quantity of different medicinal and hallucinogenic plants.

When the Second World War broke out, Schultes went back to Harvard University and accepted a chair in order to study the arrow poison used by the native peoples in the north-west of Amazonia. On the first day after he landed in Bogotá (in 1941) he discovered a new species of orchid, barely 2.5 cm tall, at the edge of a city street. It was later described as *Pachyphyllum schultesii*. And was just one of many that were to follow.

During the years 1941 to 1953 he indefatigably combed the jungle of Colombia. For him the region was the land "where the gods reign", which was also the title of a book. During this period he collected more than 24,000 plants, including many new discoveries. And, like Raleigh, La Condamine, and my mother, he paid special attention to the *curare* toxin and and hunted for its source, as in the meantime it had been discovered in the western world that it had important uses in medicine, for muscle relaxation and surgery.

The fish world of the Apaporis is practically unknown. Apart from the people mentioned above, nobody has explored in this enormous region. A caiman subspecies, *Caiman crocodilus apaporiensis*, was discovered and described by Medem in 1955. Because of its limited distribution region, restricted to the upper course of the Rio Apaporis between the waterfalls of Jirijirimo and Puerto Yaviya, it is threatened with extinction (conservation status: WA I appendix A). The rare black *uakari* (*Cacajao melanocephalus*) (along with numerous other primates) has been discovered here. Its population of around 600 individuals is restricted to the lower Apaporis, to an area of 14,450 ha round the Lago Taraira. Other endemics include the emerald *chiribiquete* (*Chlorostilbon olivaresi)*, the grey-legged tinamou *(Crypturellus duidae)*, as well as the *tamarin (Saguinus inustus)*. The endangered giant otter *(Pteronura brasiliensis,* listed in CITES Appendix I), also occurs in the lower Apaporis, in the area where the blue discus live. And because of the latter I will investigate further.

There are no more *Symphysodon* occurring naturally to the west of here. There is, however, an exception in the Rio Nanay in Peru. But this is not a natural discus distribution, although this is often stated in both popular and scientific literature. It is now known that more than 25 years ago the fish exporters (or one at least – my friend Bustamante) became tired of continually making the difficult (and expensive) journey to the Putumayo or to Tefé to bring back Greens. So why not introduce these fishes "on their doorstep", as the water parameters and all other relevant conditions in the Nanay are identical to those in the Lago Tefé? And it actually worked. After a few years the population from Tefé was breeding and establishing itself in the Nanay, and to a not inconsiderable degree. Every breeder knows how many young a pair can bring into the world, and in the wild they are always "ready for sale" after a year.

I won't go into detail regarding nearby Iquitos, which was a tiny place when I arrived there for the first time but is now a city of more than one million (if the thousands of floating houses are included there must certainly be that many, although officially only 400,000). It has no further relevance to discus habitats. Only those mentioned, in the Nanay, whose mouth lies to the north of the city. This river city is flanked by two additional *ríos:* the Marañon (= Amazon) on the eastern side and the Itaya to the south.

On the following pages I will illustrate how discus normally live in the wild during the day and at night, as well as their spawning behaviour. And finally, a summary of the parameters that can also be used to differentiate the species.

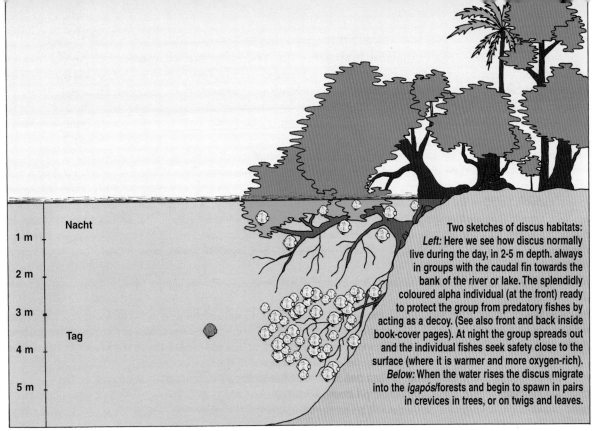

Two sketches of discus habitats:
Left: Here we see how discus normally live during the day, in 2-5 m depth. always in groups with the caudal fin towards the bank of the river or lake. The splendidly coloured alpha individual (at the front) ready to protect the group from predatory fishes by acting as a decoy. (See also front and back inside book-cover pages). At night the group spreads out and the individual fishes seek safety close to the surface (where it is warmer and more oxygen-rich).
Below: When the water rises the discus migrate into the *igapós*/forests and begin to spawn in pairs in crevices in trees, or on twigs and leaves.

Water, Types, Chemical Parameters, and Temperatures – a summary

The data summarised here – for each *Symphysodon* species – clearly demonstrate that each of the three species lives and spawns only in its own special water parameters. A characteristic that nobody has utilised previously. Many of the parameters were measured only recently (2004) and the earlier ones – in part measured by colleagues – are in each case given below.

The equipment used for the measurements (in 2004) was from the German (and international) company WTW, specifically the professional multimeter system MultiLine P4. Mayland and Bleher also generally used WTW equipment for earlier measurements. At the end of the data for each species its average parameters are summarised and repeated below for the purposes of comparison of the average parameters for all three species. All the data are from the habitats where the discus live or spawn. Where a value is followed by a second in brackets then the parameters were also measured at the water's surface at the same spot, for comparison. Where available the following data are given: **Date:** the date of the measurements; **Time of day:** (in brackets); **Location no:** from authors' records; **pH:** degree of acidity; **Temp. °C:** temperature in degrees **Celcius** of the discus habitat at the relevant depth; **Oxy. mg/l:** oxygen content in mg/l; **Conductivity μS/cm:** conductivity in microsiemens/cm; **Biotope:** discus habitat – name of the *rio (R.), igarapé (Ig.), lago (Lo.), lagoa (La.), paraná (Pa.),* or *furo (F.);* **Region:** major river or the nearest settlement; **Water type:** the three generally-recognised types of water: B = black, C = clear, W = white; B/C, B/W, or C/W = in each case indicates a mix (= M) of the two water types indicated.

On the basis of measurements taken in the natural habitat at various times of year the three species can be summarised as follows (figures in brackets are for 2004):

1. **The Heckel discus** lives on average at **pH** 4.78 (4.87); a **temperature** of 28.6 (29.6) °C; and a **conductivity** of 9.2 μS/cm (8.3); in black water.
2. **The green discus** lives on average at **pH** 5.42 (5.18); a **temperature** of 28.2 (27.4) °C; and a **conductivity** of 10.5 μS/cm (9.4); in black water.
3. **The blue/brown discus** lives on average at **pH** 6.14 (6.07); a **temperature** of 28.8 (28.9) °C; and a **conductivity** of 31.4 μS/cm (31.0); in clear water.

In other words, the three species can be differentiated by the water parameters:

1. *Symphysodon discus* – Heckel discus: black water; pH less than 5.0; temperature around 29 °C; conductivity less than 10 μS/cm.
2. *Symphysodon aequifasciatus* – green discus: black water; pH around 5.5; temperature around 28 °C; conductivity greater than 10 μS/cm.
3. *Symphysodon haraldi* – blue/brown discus: clear water(+M); pH greater than 6.0; temperature more than 28 °C; conductivity greater han 30 μS/cm.

A view of the 3 (main) water types in Amazonia: After the *encontro das águas* the black water of the Rio Negro flows for quite some time along the whitewater Amazon, past the Ilha do Careiro which contains clearwater habitats (light blue = clearwater; beige = whitewater; brown = blackwater).

Symphysodon discus

Heckel discus

Parameters measured by Natasha Khardina and Heiko Bleher during the breeding season, January/February 2004, in the biotope (at the spot where the discus were found/collected).
All values at 1.5 to 2.5 m depth (in brackets: at the water's surface) in the discus habitat (day/night).

Date (Time of day)	Location no.	pH	Temp. °C	Oxy. mg/l	Conductivity µS/cm	Biotope	Region	Water type
05.02.04 (12:55)	28	4.99	28.9(29.6)	3.40[1]	8.0[1]	Rio Abacaxís	Rio Abacaxís[1]	B
05.02.04 (10:40)	26	4.97	29.0(29.3)	1.90	7.0	Rio Marimari	Rio Abacaxís[1]	B
05.02.04 (09:00)	24	4.85	29.3	3.14	6.0	Rio Marimari	Rio Abacaxís[1]	B
05.02.04 (07:45)	25	5.70(5.69)	28.0(29.2)	3.11	6.0	Rio Marimari	Rio Abacaxís[2]	B/W
01.02.04 (16:00)	6	4.03[3]	29.8	4.56	9.0	Rio Ariau	Rio Negro	B/C
31.01.04 (11:00)	20	4.73	31.7[4]	2.14	9.0	Paranã do Rocky	Rio Negro	B
23.01.04 (13:00)	6	5.03	29.8	4.56	9.0	Lago Iri	Rio Jatapu[1]	B/C
23.01.04 (12:50)	–	5.03(5.36)	29.8(33.6)	4.56	9.0	Lago Iri	Rio Jatapu[1]	B
23.01.04 (10:30)	–	5.57[3]	29.5(32.3)	5.97	6.0	Lago Limão	Rio Jatapu[1]	B
21.01.04 (15:00)	2	4.67	29.0(31.3)	4.83	12.0	Rio Urubu	Rio Urubu[1]	B

Parameters measured by H. J. Mayland, R. Geisler, W. J. Junk et al., & H. Bleher, 1967-2000

Date (Time of day)	Location no.	pH	Temp. °C	Oxy. mg/l	Conductivity µS/cm	Biotope	Region	Water type
20.06.00 (11:00)	–	5.30	28.7	–	18.0[5]	Ig. da Freguesia	Rio Negro	B
06.12.98 (23:00)	8	4.23	27.5	–	11.0	Rio Nhamundá	Nhamundà[1]	B
28.09.98 (16:30)	–	4.60	30.3	–	10.0	Paricatuba	Rio Negro	B
26.09.98	–	5.90	30.5	–	14.0	Rio Jauaperí	Rio Negro[2]	C
25.09.98.(13:50)	–	6.40	30.3(30.6)	21.50	21.0	Rio Branco[7]	Rio Branco[7]	W
25.09.98 (13:00)	–	4.40	29.3(30.1)	2.50	11.0	Rio Negro	Rio Branco area	B
10.10.96 (15:00)	–	5.00	29.6	–	8.0	Manaus	Rio Negro	B
10.89	–	5.40	–	–	10.0	Rio Jatapu	Rio Jatapu	B
24.09.86 (18:00)	1	4.30	–	–	11.0	Anavilhanas	Rio Negro	B
17.10.86 (07:00)	7	4.42[6]	27.1	7.24	15.0	São Gabriel	Rio Negro[6]	B
15.09.86 (12:00)	–	4.42	–	–	–	Manaus	Rio Negro	B
11.11.83	–	5.10	25.0[4]	–	3.4[5]	Rio Araça	Rio Negro	B
15.04.82	–	4.90	26.5	4.60	8.4	Rio Curicuriaí	Rio Negro	B
08.04.82	–	4.30	26.4	3.70	12.5	Rio Cuiuní	Rio Negro	B
1975/76	VIII	4.46	–	–	12.1	Rio Tarumã	Rio Negro	B
12.10.72 (13:00)	–	4.70	27.7	–	8.2	Manaus	Rio Negro	B
20.10.71	–	4.40	29.7	–	11.8	São Angelo	Rio Negro	B
1967/68	Vb	5.04	–	–	8.7	Manaus (20km N)	Rio Negro	B
20.11.67	–	5.10	26.0	–	5.4	Rio Jufaris	Rio Negro	B
27.11.67	–	4.90	27.0	–	6.8	Rio Padaueri	Rio Negro	B

Average values, Heckel discus: pH 4.78 (4.87); temp. 28.6 (29.6) °C; conductivity 9.2 µS/cm (8.3); blackwater

1. The values are from regions outside the typical (former known) Heckel discus distribution.
2. The measurements from the Marimari (Sorval), where during the rainy season there is initially a link with white water, are for natural hybrid rather than *S. discus* habitats. The Rio Jauaperi measurements likewise do not relate to discus habitats and have not been taken into consideration.
3. <u>The lowest and highest pH values measured to date in *S. discus* habitats.</u> 4. <u>The highest and lowest temperatures to date.</u>
5. <u>And the highest and lowest conductivity values.</u>
6. To date no discus have been found south of the cachoeira (oxygen much higher).
7. The Rio Branco contains neither discus nor habitats. Included only for purposes of comparison.

Symphysodon aequifasciatus

(the former subspecies *S. aequifasciatus aequifasciatus*) green discus

Parameters measured by Natasha Khardina and Heiko Bleher during the breeding season, January 2004, in the biotope (at the spot where the discus were found/collected).
All values at 1.5 to 2.5 m depth in the discus (daytime) habitat (in brackets: at the water's surface).

Date (Time of day)	Location no.	pH	Temp. °C	Oxy. mg/l	Conductivity µS/cm	Biotope	Region	Water type
07.02.04 (17:00)	35	5.07	29.3 (29.4)	1.00 (2.67)	10.0 (09.0)	Rio Urucu	Rio Urucu (Coari)	B
07.02.04 (16:45)	34	5.35(5.22)	24.6 (27.0)[2]	1.35 (1.28)	15.0 (17.0)	Ig. Açú	Rio Urucu (Coari)	B
07.02.04 (15:35)	33	5.13(5.20)	26.2 (28.4)	1.75 (1.48)	10.0 (11.0)	Ig. Cerrado	Rio Urucu (Coari)	B
26.01.04 (18:35)	14	4.83(4.88)[1]	27.4 (27.5)	0.56	7.0	Rio Tonantins	Tonantins	B
26.01.04 (13:00)	13	5.05(5.06)	27.7 (27.8)	0.92	8.0	Paraná-Jutaí	Rio Jutaí	B
26.01.04 (10:50)	12	5.02(5.15)	28.1 (28.3)	0.88	9.0	Rio Jutaí	Rio Jutaí	B
25.01.04 (17:30)	11	5.84(5.30)	28.3 (28.9)	2.48 (2.40)	8.0	Rio Amaturá	Amaturá	B
25.01.04 (15:35)	10	5.20(5.28)	27.3 (28.3)	2.60	8.0(9.0)	Rio Jandiatuba	Rio Jandiatuba	B

Parameters measured by H. J. Mayland & H. Bleher, 1985-97

Date (Time of day)	Location no.	pH	Temp. °C	Oxy. mg/l	Conductivity µS/cm	Biotope	Region	Water type
10.08.97 (20:00)	5	5.70	31.0[2]	–	6.0[3]	Lago Tefé (N)	Tefé	B/C
06.08.97 (18:00)	1	4.89	29.1	–	38.0[4]	Rio Tefé	Tefé	B
30.07.97 (12:00)	8	5.30	28.5	–	40.0[4]	Lago	Juruá area	B
06.05.97 (14:00)	–	5.70	30.1	–	6.0[3]	Lago Tefé	Tefé	B
06.11.95 (12:00)	–	6.70[1]	29.3	–	20.0[3]	Lago Tefé	Tefé	B
10.11.85 (07:00)	–	5.45	24.9	–	11.0	Rio Tefé	Tefé	B
09.11.85 (20:30)	–	5.20	25.0	–	10.0	Rio Tefé	Tefé	B
07.11.85 (21:00)	2	5.45	31.0	–	15.0	Rio Tefé	Tefé	B
07.11.85 (12:00)	1	5.90	29.0	7.47	15.0	Lago Tefé (S)	Tefé	B
07.11.85 (10:00)	–	5.85	29.9	–	15.0	Lago Tefé	Tefé	B

Parameters measured by H. J. Mayland, 1987

Date (Time of day)	Location no.	pH	Temp. °C	Oxy. mg/l	Conductivity µS/cm	Biotope	Region	Water type
01.11.87(14:00)		5.26	–	2.80	6.0	Rio Nanay[5]	Iquitos/Peru	B

Average values, green discus: pH 5.42 (5.18); temp. 28.2 (27.4) °C; conductivity 10.5 µS/cm (9.4); black water

1. The lowest and highest pH values measured to date in *S. aequifasciatus* habitats.
2. The highest and lowest temperatures to date.
3. The highest and lowest conductivity values to date – that for the Igarapé Açú (Urucu) was measured at extreme high water (breeding season) in 2004, and that for the Lago Tefé at the extreme of the dry period in 1997.
4. I have been unable to confirm these two conductivity values from Mayland et al., so they have not been taken into consideration.
5. The green discus in the Nanay (Peru) were introduced there more than 20 years ago, from the Tefé region.

Symphysodon haraldi

(the former subspecies *S. aequifasciatus haraldi* and *S. aequifasciatus axelrodi*) blue/brown discus

Parameters measured by Natasha Khardina and Heiko Bleher during the breeding season, January/February, and December 2004, in the biotope (at the spot where the discus were found/collected).
All values at 1.5 to 2.5 m depth in the discus (day and night time) habitat (in brackets: at the water's surface).

Date (Time of day)	Location no.	pH	Temp. °C	Oxy. mg/l	Conductivity µS/cm	Biotope	Region	Water type
09.12.04 (16:45)	–	5.92	29.2	–	23.0	Ig. Lontra (Lago Jari)	Purus	C
09.12.04 (14:30)	–	5.33	28.1	–	23.0	Ig. Castanha (Lago Jari)	Purus	C
09.12.04 (12:30)	–	5.32	29.4	–	24.0	Lago Jari (centre)[1]	Purus	C/W
07.12.04 (12:00)	–	5.92(5.80)	29.5(30.0)	–	16.0(16.0)	Rio Itaparaná	Purus	C
06.12.04 (12:30)	–	5.60(5.68)	29.3(30.5)		20.0(21.0)	Lago Cantagallo	Purus	C
04.12.04 (18:30)	–	5.45(5.50)	29.1(29.3)	1.20(0.93)	19.0(20.0)	Lago Itaboca	Purus	C
03.12.04 (18:05)	–	5.72	27.1(29.4)	1.50(0.93)	40.0(52.0)	Paricatuba	Aiapuá (Purus)	B/C
02.12.04 (23:45)	–	6.98	29.1	3.31	135.0[4]	Ig. do Anamã	Lago Anamã	C/W
17.02.04 (10:30)	–	6.25	28.2(28.9)	–	22.0	Lago de Maicá	Santarém	C
16.02.04 (11:00)	50	4.78	26.0	–	14.0	Ig. Tapajós*	Santarém[6]	C
16.02.04 (10:00)	49	6.38	28.9(29.5)	–	17..0	Rio Tapajós	Santarém	B/C
16.02.04 (05:30)	48C	6.57	26.4(27.2)	–	22.0	Lago Grande	Alenquer	C/W
15.02.04 (18:00)	–	6.43	30.2(32.0)	–	74.0	Lago Cuipeuá	Alenquer	C
15.02.04 (15:30)	48B	6.37	30.3(30.7)	–	26.0	Rio Curua (Pacoval)	Alenquer	C
15.02.04 (11:00)	48A	6.81	30.2(30.7)	–	26.0	R. Curuá (Barra Mansa)	Alenquer	C/W
15.02.04 (06:30)	–	6.53	26.3(26.9)	–	19..0	Lago Grande	Alenquer	C/W
14.02.04 (14:00)	45	6.39	32.4(31.9)[3]	–	25.0	Rio Curuá	Alenquer	C
13.02.04 (08:30)	–	6.10	25.4(27.9)	–	27.0	Lago Mamaurú	Óbidos	–
12.02.04 (18:00)	46B	6.30(6.36)	30.0	2.58	26.0	Rio Trombetas	Rio Trombetas	B/C
12.02.04 (17:00)	46	5.90(6.05)	30.0(30.5)	0.60	23.0	Lago Curupira	Rio Trombetas	B/C
12.02.04 (14:15)	42	5.52(5.70)	30.0(30.4)	0.87	12.0	Lago Acarí	Rio Trombetas	B/C
12.02.04 (13:00)	41B	5.95(5.83)	30.4(30.5)	1.17	14.0	Lago Batata	Rio Trombetas	C
12.02.04 (09:00)	43	5.81(5.50)	29.0(29.9)	0.00	20.0	Água Fria	Rio Trombetas	B/C
11.02.04 (13:20)	–	5.60(5.75)	28.6(28.8)	2.72	17.0	Lago Nhamundá	Nhamundá	
11.02.04 (13:00)	38	5.75	28.6	2.72	17.0	Tigre	Lago Nhamundá	C
11.02.04 (09:00)	37	6.36	30.0	2.58	26.0	Caburi	Nhamundá area	C
09.02.04 (17:00)	36B	6.11	29.5(29.6)	1.51	59.0	Ig. Maximo	Parintins	C
09.02.04 (16:15)	36A	6.59(6.27)	29.4(29.5)	2.09	31.0	Lago Maximo	Parintins	C
07.02.04 (11:30)	31C	5.32(5.70)	27.6(29.5)	1.73(1.62)	11.0(10.0)	Ig. Gibian	Rio Coari Grande	B/C
07.02.04 (08:45)	31B	5.68(5.89)	25.1(25.2)	1.33	17.0(18.0)	Ig. Gibian	Rio Coari Grande	B/C
07.02.04 (08:00)	31A	5.40	27.6	1.71	11.0	Ig. Gibian	Rio Coari Grande	B/C
07.02.04 (05:00)	32	5.18[2]	28.4	3.10	9.0	R.Coari Grande(centre)[1]	Rio Coari Grande	C
05.02.04 (17:00)	29	5.40(5.35)	29.8(31.0)	3.07	40.0	Ig. S. Domingo	Rio Canumá	B/C
05.02.04 (15:15)	–	6.30	29.0(29.3)	2.98	40.0	Rio Canumá	Rio Canumá	C
05.02.04 (11:00)	27	4.85	29.0	1.90	7.0	upper Rio Marimari[5]	Rio Abacaxís	B/C
05.02.04 (04:45)	30	5.34(5.55)	29.3(31.5)	3.39	7.0[4]	Rio Canumá	Rio Canumá	B/C
02.02.04 (09:00)	23	6.75	30.2	4.54	77.0	Doema	Manacapuru	C/W
02.02.04 (08:00)	22	6.49	29.8	4.33	80.0	Macumeri	Manacapuru	C/W
27.01.04 (14:00)	17	6.48	27.9	0.00	73.0	Rio Camatiá	Camatiá (Solimões)	B
27.01.04 (08:00)	16	5.89	28.1	0.24	27.0	Lago Jacupara	Rio Içá	B/C

Date (Time of day)	Location no.	pH	Temp. °C	Oxy. mg/l	Conductivity µS/cm	Biotope	Region	Water type
27.01.04 (06:15)	15	5.84	28.6	3.81	14.0	Ig. Betânia	Rio Içá	C
25.01.04 (12:30)	9	5.93	26.7(27.2)	0.30	20.0	Ig. Belém	Belém do Solimões	B/C
25.01.04 (12:00)	8	6.30	27.2	0.24	62.0	Ig. Tacaná	Belém do Solimões	B/C
22.01.04 (18:00)	–	7.95	32.7(32.0)	–	–	Rio Uatumã (centre)[1]	Uatumã	B
22.01.04 (17:00)	7	6.86	30.1(33.6)	5.61	7.0[4]	Lagoa Jaquarequara	Uatumã	C
22.01.04 (14:00)	7	6.31(6.48)	32.2(32.1)	6.90	8.0	Rio Uatumã	Uatumã	C

Parameters measured by H. Bleher. 1995-2002

Date (Time of day)	Location no.	pH	Temp. °C	Oxy. mg/l	Conductivity µS/cm	Biotope	Region	Water type
08.08.02 (17:00)	16	6.20	28.5	–	48.0	Lago Solitario	Purus	B/C
06.08.02 (08:30)	–	5.80	26.5	–	10.0	Rio Itaparaná	Purus	B
05.08.02 (10:30)	–	5.80	28.0	–	10.0	Lagoa 26 (Itaparaná)	Purus	B
04.08.02 (14.20)	3	6.07	28.3	–	10.0	Ig. Santa Maria	Purus	B/C
03.08.02 (15.30)	–	6.40	26.0(25.5)	–	30.0	Lagoa 26 (Itaparaná)	Purus	B
24.07.02 (14:00)	2	6.55	28.5(27.9)	–	21.0	Victoria do Xingú	Xingú	C
07.02 (17:00)	1	5.98	29.3(27.9)	–	65.0	Rio Taquanaquara	Portel	C
07.10.99 (15:00)	3	5.98	28.5(27.9)	–	24.0	Rio Parurú	Tocantins	B/C
07.10.99 (08.00)	2	6.55	27.5(26.9)	–	43.0	Rio Tocantins	Tocantins	C
09.99 (12:00)	1	6.30	27.6	–	–	Ig. Grande	Breves	B/C
03.10.98	–	6.20	30.0(34.0)	4.9	18.0	Lago Sapucurú	Trombetas	C/W
07.08.97 (14:00)	–	5.24	23.5[3]	–	15.0	Lago Amanã	Amanã	B
07.10.96	–	6.32	28.4	–	41.0	Lago do Arumã	Purus	C
05.10.96 (13:00)	–	6.15(6.30)	29.9(30.4)	29.1	18.0(18.0)	Rio Ipixuna	Tapauá (Purus)	B
01.07.95	–	6.60	28.3(28.5)	3.08	15.0	Rio Arauá	Madeira	C

Parameters measured by H.-J. Mayland and H. Bleher. 1976-1989

Date (Time of day)	Location no.	pH	Temp. °C	Oxy. mg/l	Conductivity µS/cm	Biotope	Region	Water type
10.89	–	6.65	–	–	70.0	Rio Uatumã	Uatumã	B/C
19.06.89 (09.00)	–	5.65	26.9(31.0)	–	16.0	Lago Jacaré	Rio Trombetas	B/C
15.06.89 (11:00)	–	5.90	27.5(33.0)	–	18.0	Lago do Farias	Rio Trombetas	B/C
13.03.87 (14:00)	–	6.72	–	3.0	116.0	Lago Tarapoto	Leticia	C
25.06.86 (15:00)	–	5.78	–	–	20.0	Ig./Cametá	Rio Tocantins	B/C
25.06.86 (10:00)	–	6.60	–	–	40.0	Rio Tocantins	Rio Tocantins	C
12.10.86 (14:00)	5	6.40	31.6(32.5)	6.20	16.0	Lago Canaçari (Urubu)	Silves	C
12.10.86 (11:00)	6	6.20	32.0	6.9	17.0	Rio Maués	Maués	C
06.10.86 (07:00)	4	6.30	30.3	–	23.0	Lago	Rio Branco	C
03.10.86 (15:00)	3	7.67[2]	31.3	–	63.0	Lago Beruri	Purus	B/C
03.10.86 (07:00)	2	6.76	29.0	–	40.0	Lago Manacapuru	Manacapuru	C
08.76 (10:30)	–	5.50	28.4(29.3)	5.8	25.0	Rio Curuá-Una	Curuá-Una	B/C

Average values, blue/brown discus: pH 6.14 (6.07); temp. 28.8 (28.9) °C; conductivity 31.4 (31.0) µS/cm; clear water, or mixed water

1. "Centre" denotes the area where measurements were taken: in the middle of the river – with no discus habitats – where the current is strongest. 2. The lowest and highest pH values measured to date in *S. haraldi* habitats. 3. The highest and lowest temperatures to date.
4. The highest and lowest conductivity values to date.
5. Heckel discus are found in the black water of the lower Marimari, and Blues in the black/clear water of its upper reaches (hence the extremes of pH) and infertile natural hybrids occur in the overlap zone.
6. The values are from the igarapé (to which the brown discus retreat) where it joins the Tapajós.

To conclude the habitats section, a few additional comparisons of Amazon waters, plus night fishing and reminiscences of helpful friends: 1-2. 1975: black water of the Rio Curua-Una (1). The author eat his meal in the habitat to cool off. 1965: in the Rio Negro (2), with discus baskets in the canoe guarded by an indian girl. 3. 2004: clear Guaporé water (3) with 8-20 m visibility – in the Tapajós, Xingú, Tocantins 3-4 m some places up to 10. 4. 2004: Madeiras white water visibility is less than 5.5 cm. 5. 1985: Sardis *(left)* and Mayland *(centre)* helped me with measurements. 6. 1986: while hunting for discus at night you have to watch out for poisonous centripede...

DISCUS NUTRITION IN THE WILD

In this part of chapter 5 I would like to discuss my experiences with the food intake of discus in the wild. What each species feeds on in the natural habitat, what they depend on, and how they manage to survive on only a small amount of food (during extremely dry periods) in the wild state.

My researches and experiences relate to decades of observation in the wild (I have already mentioned a little of this, eg on pages 291, 309, 340-41, 368-69, 370, 393, 403, 418, 447, and 476) and on various gut and stomach contents studies performed on specimens caught by me. These were anaesthetised immediately after capture using MS 222 and I then injected them with 4% formaldehyde solution via the anal opening *(p. 513)*, so that I could analyse the contents of the unevacuated gut and stomach later in the laboratory. In addition a number of specimens preserved in this way were the subject of gut and stomach contents by Dr. Efrem Ferreira of the INPA Institute *(p. 513)*.

As regards food intake in the wild, it is important to note – as well as the type of food involved – that in discus, as in the majority of larger fishes in Amazonia, the time at which they eat to satiety and undergo strongest growth coincides with the high-water period, when the water level rises and the fishes disappear into the flooded forest. (For this reason practically no discus juveniles, let alone fry, are brought to land by discus collectors, as they grow on during the periods when collecting is impossible.)

At this time not only do many fish species find shelter where they can spawn practically undisturbed (as is the case in discus), but Mother Nature has also arranged things so that many trees bear flowers, fruits, or seeds at precisely this time. And far and away the majority of Amazonian fruits become (edibly) ripe during the high-water period and fall onto the surface of the water in the *igapó*, the *várzea*, or along the bank regions. In addition, mammals or birds feeding on them often cause them to fall. Moreover, when the water level rises discus (and other fishes) have the opportunity to nibble directly on natural, growing, vegetarian foods, including leaves and flowers *(see, for example, p. 370)*.

After years of painstakingly performing gut contents analyses on 8,563 fishes (around 450 species, although they caught no discus) from the Rio Negro region, Efrem Ferreira and Michael Goulding have shown that the majority of the fish species feed predominantly on detritus, fruits, seeds, or flowers and leaves, as well as various types of algae (Goulding, Carvalho & Ferreira, 1988). Of course, the majority of the larger species and the predatory species are piscivorous (consuming fish flesh, fins (predominantly the caudal fins of other fishes), scales) or feed on larger invertebrates. But they don't include the discus. To date I have never – not in the wild, not from gut or stomach contents analyses, let alone in the aquarium - come across a discus that fed on other fishes.

In the wild I have often watched them underwater as they busy themselves searching for food among the roots and the accumulated leaf litter on the bottom. And especially in the sandy substrate (above which they usually live/occur). They literally blow into the fine sand to stir up any micro-organisms and other lower life forms, and then suck these in. Discus also very often pick up sand and "chew" it thoroughly in the way familiar to aquarists from a number of geophagine cichlids. (Note that I have recorded this type of feeding in discus in large aquaria where I was keeping them in groups of 50 or more individuals – *see p. 327* and *p. 513*.)

In addition discus are not disturbed from this and similar methods of feeding by single *Heros* and *Mesonauta*, larger groups of *Crenicichla* or *Geophagus* (often *altifrons* and *proximus*), nor by *Leporinus* species or *Pterophyllum scalare* (*P. leopoldi* is only seldom found with discus, and *P. altum* never), that traverse the terrain with them. They often live together with *Uaru amphiacanthoides,* apparently their closest relative (according to DNA study) and consume the same types of food. Adult *Myleus* and *Tetragonopterus* species are also visitors to their territory and feed with them. And, like the majority of other diurnal fishes, they are continually on the search for food during the day, if they are able to and feel secure.

As already mentioned the discus breeding season is from December to late February or March, depending on the region and the water level. At this time, when the bushes and trees on the shore are standing in water, they migrate into the *lagos*, now accessible to them, and spawn on leaves or tree-trunks (and branches) *(p. 503)*. The water has practically no movement and a rich food supply is present in the places where they shelter (see below).

But let me tell you about the results of my decades of study of the natural foods of each individual discus species, and the findings of other authorities. This is the first (Discus)publication of its kind.

The list below indicates the different types of food in order of precedence:

1. *Detritus;*
2. *Vegetable material (flowers, fruits, seeds, leaves);*
3. *Algae and micro-algae;*
4. *Aquatic invertebrates;*
5. *Terrestrial and arboreal arthropods.*

Above (this photo) a typical blue discus habitat in the Lago Jari region (Purus region) during the dry season. It shows the feeding behaviour of discus and other fishes that feed in their company (eg *Uaru*, *Crenicichla*, and *Mesonauta* species), searching mainly for detritus, micro-organisms, and invertebrates in an environment poor in food. Often in the sand. *Below* (again in the Jari drainage, but during the high-water period). Some changes can now be seen in the feeding behaviour of the fishes. The "table" is richly spread, and as well as the foods mentioned above there are plenty of other things to eat, such as numerous fruits *(camu-camu, acará-açú, inga, pupunha)*, flowers, and leaves.

1. Detritus

The above-mentioned Goulding, Carvalho, and Ferreira found that in 46 individuals of *Uaru amphiacanthoides* (which often lives and feeds together with Heckel discus in the Rio Negro system) measuring on average 149 mm SL from a habitat in the Anavilhanas the stomach contents included 68.2 % detritus. In the same habitat they also found 33 on average 193 mm SL *Geophagus altifrons*, whose stomach contents contained 66.4 % plant material, as well as 17 *Mesonauta insignis* measuring 112 mm SL on average, whose stomachs contained 78.5% detritus, and 107 *Heros severus* of on average 148 mm SL in which 78.8% of the stomach was filled with detritus.

In 97 *Symphysodon discus* studied, from various parts of the Rio Negro region, with an average length of 120 mm SL, I found as much as 79.5 % of the gut and stomach contents to be detritus.

Thus detritus was (and is) far and away the main dietary component in Heckel discus during the dry periods.

Also of interest are the findings of Goulding *et al.* from the lower Rio Urubaxí, where I have again found and studied Heckel discus. At this location they studied juveniles of *Uaru amphiacanthoides* in February 1980 (rising water) measuring on average 34 mm SL (ie some 4-6 weeks old), a total of 11 specimens, whose stomach contents contained in total 70.5% vegetable material (aquatic and terrestrial) and detritus. The 18 adults that they found, measuring on average 180 mm SL, had 61.1 % of the same mixture of foods in their stomachs. And in 11 *Mesonauta insignis* measuring on average 53 mm SL the stomach contained as much as 93.2% vegetable material and detritus.

I have never in fact managed to catch young discus in order to investigate their stomach contents, but these figures should basically be applicable to young Heckel discus as well. In particular in comparison with the *Uaru* fry, as they are very similar. They spawn there during the same period as Heckel discus; like discus the young feed on parental secretions during the first days (and weeks); don't take the first small foods (detritus, micro-algae and vegetable material) until weeks later; and even look very similar to discus fry in the first stages. Now, some of you may be wondering what detritus actually is, and what detritivores are.

In a dietary context, detritus is a collective term for any accumulation of material that is ingested in a state of decomposition. The word means literally "an aggregate of loose fragments". In biology it is often used to denote solid organic waste, including material from decomposing dead plants or animals. Detritus is fine-structured organic and inorganic residues; plant remains in an advanced stage of decomposition; and/or other coarser fragments in the initial stages of decomposition. When detritus is removed from/studied in a fish's stomach it is normally found to contain quantities of fungi, bacteria, protozoa, and algae. As far as I know there has so far been no study undertaken on detritivorous fish groups to establish whether they differentiate between the components of the detritus or obtain nourishment from only a particular group of micro-organisms.

Of the three components listed above it has been a mixture of fine-structured residues and plant remains in an advanced stage of decomposition that I found most frequently in Heckel discus stomach contents. Coarse fragments have also been ingested, but less frequently. The findings of Goulding *et al.* should be mentioned for comparison: of 269 fish species from the Rio Negro system whose stomach contents were studied, 132 species contained only fine detritus, 100 decomposed vegetable material, and only 37 species had coarse fragments in their stomachs.

And where does detritus come from?

The energy component of this diet (that of Heckel discus and other discus species) undoubtedly comes from the jungle vegetation: the vegetable residues and the coarse fragments. And the bulk of the fine-structured organic material derives from these two components. This has already been demonstrated a long time ago (Irmler, 1975, 1976; Irmler & Furch, 1980).

At least one of the three components mentioned above is practically always present in the substrate of an *igapó*, a *lago*, or in the flooded forest of the Rio Negro region (and also in the other Heckel discus blackwater regions such as, for example, the Rio Nhamundá system, and the Abacaxís, Jatapú and Urubu regions). The majority of the saprophytes (plants that grow on rotting material, heterotrophic organisms which derive the organic nutrients they require from dead substrates) that are largely responsible for the decomposition of wastes in the flooded forests of clear- and blackwater zones in central Amazonia are known. They are chiefly saprophytic fungi, which convert organic material (predominantly the "rubbish" of the rainforest) into fine detritus. Algae and bacteria play only a secondary part. Some authors (Walker, 1975, 1978; Walker & Franken, 1983) have gone as far as to suggest that detritivores feed mainly on fungi. This can

A few photos of feeding behaviour, the gut and stomach analysis process in discus, and some of detritus:

1. Here, in a 2500 l aquarium housing 60 wild-caught Heckel discus, we can see the fishes searching the bottom for micro-organisms and/or detritus. This is very typical behaviour, chiefly during the dry season (see also p. 511).

2. Dr. Efrem Ferreira from the INPA Institute, Manaus, during the gut and stomach analysis of various fishes caught by me and preserved immediately. In this case Efrem found more than 90% detritus, vegetable material, and algae, and barely 10% aquatic invertebrates and arthropods, in 10 adult (12-13 cm SL) *S. aequifasciatus* specimens from the Rio Tefé.

3. The preserved discus is slit carefully from the anus forwards, the gut and stomach are removed with fine forceps, and then examination under the microscope begins.

4. The photomicrograph shows the detritus – coarse material and residues – and other materials: the dark spots are the detritus itself; the green is algae; the coloured material is the remains of fruits. (On the left-hand page is a further photomicrograph of detritus, from another region. Here too the dark particles are detritus and the green algae.)

5. After the freshly-caught discus have been anaesthetised with MS 222 in the plastic (in front of the author), they are taken out to check that the gut hasn't already evacuated (through fright or stress).

6. Next a 4 % formaldehyde solution is injected via the anal opening. Only in this way can one be sure of obtaining an unevacuated gut and an accurate analysis of gut and stomach contents.

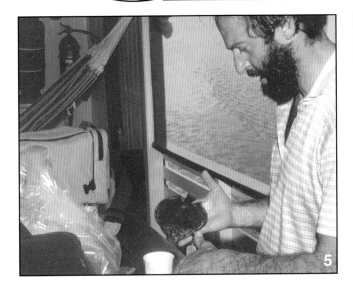

hardly apply to all detritivorous fishes, however; fungi are not the main component of the detritus eaten by Heckel discus.

It has been demonstrated that decomposition of organic remains in the Rio Negro region differs from the norm in other central Amazonian clearwater and blackwater habitats. It takes place in the forest inundation zones mainly during the dry periods, via the agency of arthropods, fungi, and bacteria. However, much still remains to be learned about the processes of decomposition. We can nevertheless extrapolate that the agents mentioned above are all three extensively involved, with the extent of that involvement varying dependent on the location and the individual habitat.

Apropos of detritus, it is interesting to note the following: I was able to capture *Symphysodon discus* during the dry season in a *lago* near the Urubaxí and investigated their gut and stomach contents (12 specimens of 110-135 mm SL), which included more than 50% fine detritus. Goulding *et al.,* who investigated the stomach contents of 89 other fish species, apparently in the same habitat or nearby, found that 52 of them had fed on fine detritus, while 33 species had vegetable residues in the stomach, and only 4 coarse fragments.

The extensive supply of detritus in lakes during the low-water period originates from the surrounding forest inundation zones. The organic debris accumulated there – organic sludge, leaves, and wood in various stages of decomposition – is washed into the lakes by the receding water.

To this I would add my own observations. While spending time underwater in such lakes (and *igapós*) I have often seen (not easy in generally murky visibility) how discus pick deliberately at fallen tree-trunks, branches, and leaves, or take in food from them by sucking motions. And it wasn't long before I discovered that they were often feeding on detritus, washed by the receding water onto not only the bottom, but also any wood and leaves lying in the way, where it then clung. Of course micro-organisms also cling to these surfaces in the same way. The same takes place on the generally sandy substrate – and this is why discus also regularly blow into the substrate, to stir them up and suck them in. This method of feeding by suction also takes place on the wood and leaves *(see p. 511).* A feeding method of this kind supports the fact that Heckel and other discus hardly ever live in a habitat where the water is flowing, let alone flowing fast or with a strong current, as the detritus would be washed away.

In addition it has been shown (Junk (Ed.), 1997) that of 91 fish species in black water, 41 fed on detritus when the water was rising, 55 during the high-water period, and as many as 67 species consumed detritus when the water was receding. At extreme low water it was again 41.

2. VEGETABLE MATERIAL (FLOWERS, FRUITS, SEEDS, LEAVES)

But the Heckel and other discus don't feed only on detritus. Vegetable material constitutes a possibly almost equally important dietary component. Terrestrial vegetation, as true aquatic vegetation – apart from algae – occurs as good as never in the Rio Negro and other Heckel discus habitats. Aquatic vegetation, apart from *Utricularia foliosa,* a number of floating plant species, and occasionally algae, is likewise not encountered in the habitats of the other discus species either.

Now there are a very large number of plants, but no specific analyses as to which types of flowers, fruits, seeds, or leaves are preferred by Heckel or other discus. Nor for any Amazonian fish

species apart from the *tambaquí,* the main food fish of Amazonia, whose diet has been the subject of a number of publications. Nor can I provide any percentage data regarding how much a discus eats of the flowers, fruits, seeds, or leaves of a particular plant genus or species. But I would like to detail my observations as regards what fruits, seeds, flowers and leaves they take at various times of the year in the discus habitats of the river systems, *lagos,* and *igapós* where I have studied their feeding behaviour, and, in the case of some species, the ascertained nutritional value.

Below is a list, in alphabetical order, of the families, genera, and species, which I have been able to identify in the habitat during my researches and from gut and stomach analyses. However, I haven't always been able to determine the species as the flowers, fruits, and seeds within a genus are often similar and correspondingly difficult to differentiate (they are also comparable in nutritional value). The scientific name of each species is followed by a list (in brackets) of the most popular names known to me.

Arecaceae (also Palmae or Palmaceae)

This family has a circumtropical as well as a subtropical distribution with around 3,400 species in more than 200 genera. I have found four of them to form part of the discus diet.

Astrocaryum jauari (tucumã; coquillo palm):
Heckel discus feed on the flesh of the fruits.

The coquillo palm is widespread in the forest inundation zones of the Rio Negro and has adapted to life on the sandy soil *(see p. 388)*. It can grow up to 20 m tall and usually stands in groups of 4-6 palms of differing sizes – all growing from a single rootstock. It can survive for up to 340 days (Piedeade, Parolin, & Junk, 2003) – or longer – in water. Only when the water level begins to fall do they drop their green-yellowish, oval fruits, 3-4 cm in diameter, which then have to float for only a very short time, if at all. This spiny palm produces some four clusters of on average 106 fruits and around 90 cm long, per year. At present there is no evidence that the fruits are eaten by any vertebrates other than fishes.

The Brazilian name *tucumã* is a very far-reaching term. The principal variants are *tucumã-do-amazonas (A. aculeatum* – fruit below left) and *tucumã-do-pará (A. vulgare)*, but there are other species and/or fruits that are known as *tucumã*. To date 47 species have been assigned to the genus *Astrocaryum*; they are distributed north as far as Mexico and their fruits are almost all similar to those of the coquillo palm.

Until 1998 the much-prized *palmito* (palm heart) was obtained from this palm species alone in Amazonia – at the end of the 1970s my friend and supplier of fishes for decades, Anildo Macedo, even erected a large *palmito* factory on an island in the Rio Negro and thereafter exported palm hearts as well as aquarium fishes. But recently the *palmito* from the palm *Bactris gasipaes*, which is known as *pupunha* and has a greater yield, has gained precedence.

I have seen Heckel discus feed on the flesh of the (sinking) fruits during the high-water period. The shells had split and the flesh was protruding. I made this observation several times in the Rio Quiuini region when the water was receding. Undoubtedly young, growing discus also nibble at this food. To my personal knowledge there are 17 fish species that consume this fruit.

Nutritional value (as percentages of the dry weight of a fruit): 8.0% protein; 78.5% carbohydrate; 8.0% lipids; 5.5% ash; and energy 6.988 g/kJ.

Bactris gasipaes (pupunha; pejibaye; pijuayo; peach palm):
Heckel discus, blues, and browns feed on the flesh of the fruits.

The majority of the palms of the genus *Bactris* grow up to 25 m tall (three to five often grow together) and their trunks are clad in long pointed spines. They are widespread in Amazonia, and the genus of more than 250 species is encountered in many parts of the neotropical zone, for example in Ecuador, Bolivia, Peru, and north to Honduras. Its flowers grow in panicles and ripen into trusses of fruits protected by a wooden shell 40 to 80 cm long. This shell is covered in fine, needle-sharp spines (about which I can tell a tale or two, as I have often trodden on the fallen shells on the jungle floor and in the water). One truss can comprise up to 300 fruits 4-7 cm in diameter. The nature of the flowers is worthy of note: they produce 50 to 1,000 female and 10,000 to 30,000 male, and now and then even hermaphroditic, flowers. The fruits have a hard shell which may be green, yellow, pink with green or orange to completely red *(above)*. The aril, the fleshy covering of the seeds *(below)*, comprises up to 95% of the total weight and is fibrous and golden yellow (also beige or white in some species). They normally ripen in January/February.

I have seen Heckel discus feed on them only in the lower Rio Negro region. Blues and browns in the Alenquer region. But never seen greens feeding from it. The fruit fibre (roughage for discus and others?) is certainly enjoyed, but it has other benefits as well.

Nutritional value (per 100 g fruit flesh in *B. gasipaes*): 45% moisture; 3.5% protein; 27.0 g fat; 23.6% carbohydrate; 0.9% ash; 3.8% fibre; 0.18% zinc; 3.8 µg carotene; and 196 calories.

Remarks: The species *Bactris gasipaes* (also written *B. gassipae* by some) is becoming increasingly more popular for the production of palm hearts, whose consumption is constantly rising, and with Brazil still the world leader with more than 70% of the market. Nowadays this species has also been planted in many tropical regions, and since 1990 (by when annual turnover on the international market had already reached 40 million dollars) even in Hawaii, where it is known as the peach palm. In Brazil the flesh of the fruits is also eaten cooked with salt, and processed to oil and animal feed.

Leopoldinia pulchra (jará; tulia palm):
Heckel discus and blue discus feed on the flesh of the fruits.

This palm, also often known as *jará* and *tulia* (and the genus name is also written as *Leopoldina* by some), generally grows to 10 m tall and is widespread in the Amazon region, but grows almost exclusively at the edges of the banks of rivers and

A. aculeatum

Bactris gasipaes

This painting shows the two Amazon palm species mentioned *(p. 516 and 518)*, *Astrocayum jauari (left in the photo)* and *Leopoldinia pulchra (centre)*. It comes from *Reise in Brasil 1817-1820* by the German naturalists Spix and Martius, and the latter named the second genus in honour of the Archduchess Leopoldine, wife of the first emperor of Brazil (see chapter 1). The fruits of both palms also form part of the discus diet.

lakes, as well as in the *igapós* – flooded forests of black- and clearwater zones. During their fruiting period (at high water) the *tulia* palms are often completely submerged.

Only four species are known: *L. major* and *L. paissaba*, which Wallace described in 1853 and 1855, respectively; *L. insignis* and *L. pulchra*, both described by Martius in 1824.

I was able to observe blue discus in the Trombetas region, in the Lago Jacaré *(p. 291)* (this population is regarded by some as Heckel discus), feeding on the flesh of the fruits. In the lower Rio Jaú region (Rio Negro region) I have seen *ukaris* – a herd of at least 50 individuals of the rare subspecies *Cacajao melanocephalus ouakry* – feast on the fruits, and how many that they bit into fell into the water as a result. And Heckel discus were feeding greedily on the flesh of the opened fruits. I experienced a very similar scenario at a group of *L. pulchra* in the Rio Quiuini drainage, although in this case it was large aras that were busy with the fruits and a lot of those they attacked ended up benefiting the Heckel discus, a beautiful variant *(see p. 381)*. And in the Lago Grande de Monte Alegre I saw brown discus feed on the flesh of the fruits of the *tulia* palm.

Nutritional value: unfortunately I have no precise data, only that a high protein value has been demonstrated (Hahn, 2000).

Mauritia flexuosa *(buriti or buriti-do-brejo; miriti):*

Heckel discus feed on the flesh of the fruits.

This is a palm that often grows in swamps and along the *igapós*, not only in the Rio Negro region but widely distributed in Amazonia and even in Bahiá. It flowers between April and August bears fruits nine months later (precisely during the period when the water is rising and the juveniles of discus are growing on). This 20-25 m high palm produces 5-7 long trusses, each comprising some 400-500 fruits.

Discus feed on the golden-yellow flesh of the fruits *(centre)* which comes into view when the little coconuts with their scale-like structure float on the surface for a short time, or those attacked by birds have fallen into the water. As long ago as the 1960s I discovered that this is a favourite food of the cardinal tetra (*Paracheirodon axelrodi*) and later on observed that Heckel discus also feed on it. But not only fishes benefit from it, but also humans and Amazonian industry. The fruit is eaten with enjoyment raw or as the food called *paçoca*. In the latter the *buriti* flesh is mixed with *farinha* and sugar extract. The flesh of the fruits is often frozen, a delicious ice is made from it, and a spread for bread, and liquor and vitamin-rich juices are mixed with it – to mention just a few of the delicacies. Spix and Martius feasted on the fruits almost 200 years ago.

Heckel discus feed greedily on the fruit in the Rio Negro region – and I have seen browns do so at Breves.

Nutritional value of the fruit (percentages per 100 g of fruit flesh): 65.8% water; 1.8% protein; 11.2 % lipids; 20.4 % carbohydrate; 0.8% ash; 7.9 % fibre; 4.3 % carotene; 0.63 % zinc; and 189.6 kcal. The high carotene content encourages the red colour in discus.

Asphodelaceae (Liliaceae or Smilacaceae in the view of some authors)

A family of lily-like plants (mainly rhizomatous plants, shrubs, climbers, rarely tree-like plants) with some 1,170 species variously assigned to 20 genera. Some authors recognise a separate family, the Smilacaceae, with 370 species and 3 genera (or one), with a holarctic, palaeotropical, neotropical, and Australian distribution. Only *Smilax* is important for discus.

***Smilax* spp.** *(sarsaparilla; zarzaparrilla; catbriar; greenbriar):*

Heckel discus, but predominantly blues and browns, as well as green discus, feed on the flowers and fruits.

The spiny liana generally known as sarsaparilla grows to 20 m high. More than 200 species of *Smilax* are known;

Mauritia flexuosa – buriti

they have a circumtropical and subtropical distribution and a number of them (eg *S. aspera, S. nigra*) are even resident in Europe. *Smilax* are often assigned to the family Smilacaceae, a group of woody climbing plants (rarely upright herbaceous plants or branching shrubs) whose flowers are radial and either hermaphroditic or monosexual.

More than 20 species occur in Amazonia, in discus habitats in the Rio Negro as well as in many other regions. In the *lagos* of the lower Jauaperi (Rio Negro region) I have seen Heckel discus consuming the flowers of the species *S. spruceana*, which looks very similar to *S. papyracea*. But they also feed on the fruits and I have repeatedly found the latter in the net along with Heckel discus *(p. 515, photo 4)*. I have also seen brown discus feeding on sarsaparilla fruits in the Rio Içá drainage *(see also pp. 369-370)*.

There are interesting stories about the sarsaparilla. It is often written that it is resident only in South, Central, and parts of North America, but it was already known in Europe in the

Sarsaparilla in art from the past, *inter alia*:
1. *The Metamorphosis*. Woodcut (A. Rusconi) from 1553 showing how Crocus and Smilax were transformed into plants *(below right and below left respectively in the photo)* in mythology *(see text, p. 521)*.
2. *Smilax ornata* from Jamaica, with flower and fruit; this was one of the first sarsaparillas to reach Europe (in the 16th century).
3. *S. aristolochiaefolia* from Mexico, with fruits and section of fruit.
4. One the first sarsaparilla colour prints (*S. officinalis*, from *Medical Botany*, 1793), with flower and fruit.
5. Members of the animal kingdom have also been described with the specific name *smilax*, *inter alia* butterflies such as *Holocerina smilax* (the subspecies *H. s. menieri* is here commemorated on a stamp from the French colonies), or *Eurema smilax* (not illustrated).
6. *Smilax* fossils: were discus already feeding on them millions of years ago?!

Middle Ages – perhaps even earlier. But one thing is sure, it has been known as a medicinal plant for millennia in the New World and China. And the 5-6 mm thick and around 2.5 m long roots *(centre)* are supposed to work miracles.

In all probability it is the roots of the species *Smilax glabra* and *S. china* (generally called Chinaroot), which have their distribution in Asia and are closely related to the American species, that have been "on the market" as a medication for the longest time. It was introduced there for the treatment of psoriasis and other skin diseases, rheumatic diseases, and kidney complaints. It has been famed since time immemorial for its diuretic and sudorific properties, and also as an aphrodisiac and for its invigorating effect on the entire organism. In the 16th century the American sarsaparilla root advanced from the New World across Europe to Asia as a treatment for the venereal disease syphilis. As a result Tibetan doctors came into contact with it and immediately regarded it as being on a par with the Chinaroot. Today both are prescribed during the treatment of stubborn and/or purulent fevers caused by toxins resulting from, for example, bacterial pathogens or allergens. In the Swiss canton of Appenzell Ausserrhoden they have a formula for a treatment containing sarsaparilla, which is used during the treatment of hay fever and other allergies, acne, and difficult-to-heal skin conditions. There (and on innumerable Internet pages) it is recognised that the root has medicinal properties. And according to the latest research the substances contained in various *Smilax* species are effective against viruses, bacteria, and pathogenic fungi, in particular fungal skin conditions. The Malays chew the roots of various species *(Smilax myosotifolia; Smilax calophylla)* as a aphrodisiac. The first species, known there as *itah besi*, is regarded as more efficacious, while the second species, *itah tembaga*, is taken in combination with Betel to augment the effect (as I saw in Perak). The root of the plant has long been prized as a tonic in Germany where it is known as *Sassafraß* (a name which in the English language is applied to a tree of the laurel family) or *Stechwinde*. Barth. Hinr. Brockes (1680-1747), who published, *inter alia*, a work with the title *Irdische Vergnügen in Gott, bestehend in physikalisch- und moralischen Gedichten*, stated, for example, "*Sassafraß kann nach viel Jahren, diese Kräfte noch bewahren, dass, wenn man ihn gleich nicht rührt, man ihn doch von Ferne spürt.*" ("Sassafras can still be so potent after many years that it can be detected from afar without touching it.")

Sarsaparilla derives from the Spanish *zarzaparilla* and means roughly "bramble-like little vine". Around 1536 the *conquista* brought it back to Spain from Mexico, Honduras and Ecuador, and it soon came to enrich the European pharmacopoeia. During his explorations in America (1799-1804) Alexander von Humboldt wrote extensively about sarsaparilla. He wrote, *inter alia*, "*Die Schlingpflanze wächst in Menge an den feuchten Abhängen der Berge Unturan und Achivaquery. Wir fanden zwölf neue Arten, von denen Smilax syphilitica vom Cassiquiare und Smilax officinalis vom Magdalenenstrome wegen ihrer harntreibenden Eigenschaften die gesuchtesten sind. Da syphilitische Übel hier zu Lande unter Weißen und Farbigen so gemein als gutartig sind, so wird in den spanischen Kolonien eine sehr bedeutende Menge Sarsaparille als Hausmittel gebraucht.*" ("The creeper grows in large numbers on the damp slopes of the Unturan and Achivaquery mountains. We found twelve new species, of which *Smilax syphilitica* from the Cassiquiare and *Smilax officinalis* from the Magdalena river are the most sought-after for their diuretic properties. Because syphilis is so commonplace among both whites and coloureds in the land, in the Spanish colonies a very significant quantity of sarsaparilla is used as a home remedy.")

The main active ingredients in the roots are saponin and saponin-like substances, a certain amount of essential oil, starch, and a bitter resin. In Dinand's *Handbuch der Heilpflanzenkunde* of 1921 it is stated, *inter alia*: "*Die Wurzelabkochung ist schweißtreibend, blutreinigend, gelinde reizend, absonderungsfördernd und wird als gutes Mittel gegen Syphilis, chronischen Ausschlag, Gicht, Rheumatismus, chronische Hautkrankheiten, Hypochondrie, Hysterie und Neuralgie gerühmt.*" ("The decoction of the roots is sudorific, mildly irritant, and laxative, it purifies the blood and is renowned as a good treatment for syphilis, chronic acne, gout, rheumatism, chronic skin diseases, hypochondria, hysteria, and neuralgia.") Today it is an ingredient in many proprietary preparations for purifying the blood.

And saponins are strongly toxic to fishes; in Africa and southeast Asia I have seen the natives use the saponin-rich juice of the roots for fishing.

Nutritional value: I have no nutritional value, only that it consists of saponin; saponin-like substances; essential oil, and starch.

Remarks: The name *Smilax* was already in use before the Greeks (p. 521). But some people derive the scientific name from the English "smile", as if you grasp the liana then the sharp, poisonous spines hurt, the pain is in fact brief but nonetheless effective, and during the moment of pain you should smile instead of crying. "Don't worry – be happy" as we would say today. (See also p. 447.)

Left-hand page: The medicinal root of the sarsaparilla (in this case that of *S. aristolochiaefolia*).
This page: A few *Smilax* species worldwide and comments:
1. The Brazilian *S. laurifolia* at the start of flowering, rising water.
2. *S. megacarpa* from Borneo, with green fruits.
3. The fruits and flower (inset) of the South American *S. walteri*.
4. Typical fruit of an Amazonian sarsaparilla shortly before high water.
Not illustrated: S. ovalifolia and *S. lanceaefolia*, which occur in India, are widely used in medicine. *S. glyciphylla* is the medicinal sarsaparilla of Australia and *S. macabucha* is used in the Philippines against dysentery and other illnesses, while *S. anceps,* the sarsaparilla from Mauritius, has an excellent medicinal reputation. In Iran the asparagus-like young shoots of *Smilax* species are eaten, and baskets are woven from, *inter alia, S. pseudochina*. The Mexicans regard *S. rotundifolia* as sudorific and purificant.
In commerce two sarsaparilla groups are recognised on the basis of their starch content (floury material beneath the bark). The starch-rich species include, *inter alia, S. officinalis, S. syphilitica,* and *S. papyraceae* from Honduras, Venezuela, and Ecuador. The low-starch species come from Mexico, Jamaica, and Peru, and the last two are preferred on account of their acid taste.
And now a little more from the *Smilax* mythology: The story goes that there was a fertility spirit with the name Crocus (or Crocos) – a Spartan, little of whose myth is recorded, but for which there are numerous parallels. Like a number of fertility spirits he had homosexual tendencies and was a lover of the bisexual Hermes. But Crocus was mortal and during a hard game Hermes wounded him. Some say with a discus, others with a quoit, and where his blood fell saffron grew, a flower that retains the colour of his blood. For this reason saffron is purple. Another version of the Crocus myth tells that Crocus was either the liberator of the nymph Smilax (or spurned her), who was associated with a climbing plant containing a resin prized by sorceresses. Surviving fragments also tell how Crocus ignored the advances of Smilax as he was obsessed with hunting. And as a young Spartan he preferred to share a camp with men rather than spend time with a woman. This is supported by the liaison with Hermes. The goddess Aphrodite, finding Smilax forlorn, turned her into a climbing plant with flowers that smelled like carrion and fruits as dark as the night (p. 359). In addition Aphrodite asked Artemis to find the uncaring hunter and changed him into a plant as well – the crocus. The story of Crocus originates from the time before the Greeks and doesn't have Greek roots. There are still other versions, but the only fact is that Crocus and Smilax are immortalised in Renaissance woodcut called *The Metamorphosis* (p. 519); today *Smilax aspera* winds around the ruins of Aphrodite's temple on Cyprus; the sarsaparilla so prized for millennia by discus and humans is still called *Smilax*, and Crocus is no less valued by humans, as *Crocus sativus* (= the saffron crocus), renowned worldwide.

Bombacaceae

The cotton-tree family, with some 180 species, includes trees that grow to 70 m tall and frequently have a trunk that is thickened to store water. They are distributed predominantly in damp regions in Europe and the Americas, as well as in arid regions elsewhere in the tropics and subtropics. Of the 30 genera, two are important for discus.

Pseudobombax munguba (munguba, munguuba, mungumbeira, assacurana, embiriti):

Heckel discus, blues and browns feed on the flowers and the flesh of the fruits.

The tree is one of the giants of Amazonia – with a height of more than 50 m *(p. 406)*. It occurs mainly in the Amazon area and the regions along the borders with Colombia and Peru. Although this species is encountered predominantly in the *várzea* regions (white water), it also grows in *igapós* and along clearwater zones. It possesses relatively large flowers *(centre),* produced upright in pairs (or small groups); these may be up to 8 cm tall, and are 2-3 cm in diameter and fleshy. The elliptical fruits can attain a length of up to a good 23 cm, with a diameter of 12 cm, and beneath the brown-red skin lies a whitish or yellowish mass containing the seeds. It flowers and bears fruit during the high-water period. Munguba trees bear some resemblance to some species of the Malvaceae (and are sometimes assigned to that family).

In the Novo Airão region (Rio Negro) I have seen Heckel discus feed on the flowers and the flesh of the fruits, and blues and browns likewise in the Trombetas region.

Nutritional value: rich in protein and lipids.

Remarks I: It has been found (Horna & Zimmermann, 2000) that during the high-water period (March-July) the lower part of the tree (which is underwater) gives off appreciably more carbon dioxide than during the dry season, with the value rising from 2.0 to 10.0 mg/l, which is, of course, particularly beneficial for the young, growing discus. And there is a yet another factor favouring the discus: the *munguba* tree is home to termites of the genus *Nasutitermes*, which utilise it at this time, climbing up the trunk *(p. 406),* normally to a level where the water doesn't reach – but that does often occur, or they fall down. Either way they constitute an additional food for many creatures. In addition this tree is very often covered in epiphytes.

Remarks II: Pachira aquatica (syn. *Bombax aquaticum*), known as *monguba*, is often confused with *P. munguba*, including because its flowers, leaves, and fruits are similar.

1. *Pseudobombax munguba*, the central *munguba* tree on the bank above a discus habitat.
2. Fruit *(right)* and fruit flesh *(left)*.
3. Flowers of *P. munguba*.

Quaraibea cordata (sapota-do-Solimões; sapota-do-Peru; zapote):

Green discus feed on the flesh of the fruits. This tree species is distributed from the central to the upper Amazon basin, grows to 40 m tall, and almost always has a straight trunk, conspicuous along the river banks. Its flowers are pale pink-red and grow directly from the trunk or branches. Its up to 1,000 fruits ripen between February and May – during the high-water period. They are relatively large, up to 15 cm long, 12 cm thick, weigh 300 to 1,100 g, and inside the brownish skin the orange flesh component represents more than 35% of the fruit (five kernels occupy the remainder) – a lot in comparison with other fruits. It is a tasty fruit, and is also processed into juice.

I have seen green discus feeding greedily on the flesh of fruits floating in the water, in the Tefé and in the Jutaí region, as well as in the Rio Nanay.

Nutritional value (per 100 g fruit flesh): about 75.0% water; 1.0 g protein; 0.4 g lipids; 15.3 g carbohydrate; 0.8 g ash; 5.0 g fibre; 0.13 mg zinc; 2.40 g carotene; and 69.0 kcal.

Caesalpiniaceae (regarded by some as Fabaceae or Leguminosae)

A family of trees, or shrubs, or herbs, or lianas; resinous, or not resinous. About 150 genera and 2200 species are known from tropical and subtropical areas. I have so far found one species to be relevant to discus.

Cassia leiandra (marimari; mari-mari; ingámari; mariri-dá-várzea; seruaia; pachapacta):

Heckel discus as well as blue and brown discus feed on the flesh of its fruits and nibble the softened seeds.

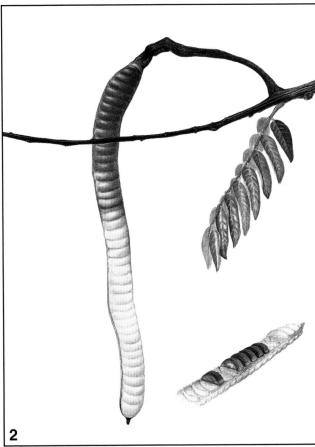

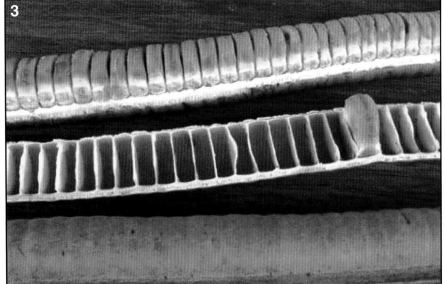

1. *Quaraibea cordata* fruit, with fruit flesh and seeds.
2-4. *Cassia leiandra* pods (2); the rows of seeds embedded in the fruit flesh (3); and the large inflorescence (4).

This tree grows to 8 to 12 m tall and is found predominantly in the central and lower Amazon region, in flooded forests and damp areas along the *igapós* and *lagos*. It bears woody bean-like pods, up to 80 cm long and yellowish in colour, between December and May (rising water and the high-water period), with the majority ripening from January to March. In the pod, which constitutes some 50% of the entire fruit, there are up to 120 seeds in rows, embedded in a greenish, creamy mass of flesh *(p. 523)* which represents around 10% of the fruit as a whole (the seeds themselves provide the remaining 40%). The flesh is greedily eaten by discus (and other fishes) from the open pods (which usually split open, or are opened by animals). It is slightly sweet and the natives like to eat it. In the past I have managed to survive on it in the jungle.

I have seen Heckel discus feed on the flesh of the fruits in the lower Rio Negro, in the Rios Abacaxís and Marimari (the latter gets its Indian name from the tree). Blues and browns consume the fruit flesh in the Alenquer region. Discus and smaller fishes nibble at the seeds only after they have softened, or not at all.

Nutritional value: I have no details available, except that the seeds taste like a cross between chocolate and carob and sucked they dissolve, and that proteins are known to be 15 g per 100 g fruit (seed).

Remarks: I have not yet been able to determine whether discus feed on the fruit flesh or seeds of the species *C. spruceana*, but this may well be the case as it is very similar. That apart, there are 500 *Cassia* species known among the Caesalpiniaceae.

Chrysobalanaceae

A family of trees or shrubs; leptocaul, with a tropical, rarely subtropical, distribution. Some 400 species in 17 genera, but only one genus is (very) important for discus.

Licania spp. *(acará-açú):*

Heckel discus feed on the flowers and the flesh of their fruits; the same applies to greens, as well as to blue and brown discus. A*cará-açú* species must constitute far and away the majority of the vegetable food material consumed by discus.

The shrubs of this genus are widespread throughout the Amazon region and north to Costa Rica, into southern Bolivia, and in Peru and Ecuador. They grow several metres high – I have encountered some 10 m tall *(right)* – and have small, often golden yellow flowers and fruits. This is probably the origin of the German name *Goldpflaumengewächs* ("golden plum tree", from the flower and the plum-like fruit – *see p. 368)*. The up to 30 cm tall cluster of flowers appears and opens at the beginning of the rainy season. During the high-water period some species bear flowers and others fruits. But I have hardly ever seen either during the dry season. The fruit has a usually greyish or yellowish rind that can be 2-3 mm thick, but soon softens or bursts when floating in the water. The flesh of the fruits is whitish to yellowish and slightly pasteurized.

In the central Rio Negro drainage I have seen Heckel discus feed variously on the flowers and the flesh of the fruits of the species *L. stewardii (p. 368)*, as well as on *L. apetala,* and in the Rio Nhamundá *(right)* it must have been *L. stewardii.* I have seen green discus feed on *L. angustata* flowers *(p. 447)* in Rio Urucu tributaries, and on the flowers and fruits of *L. arborea* (it may instead have been *L. longipedicellata)* in the Putumayo drainage. And blue and brown discus feed on the flowers and the flesh of the fruits of *L. parviflora* in the Lago Batata and in the Rio Curuá *(p. 277)*. In *lagos* in the Amazonian region bordering Colombia (Rio Tacaná drainage) I have seen them nibbling the flowers (and fruits) of *L. micrantha (right), L. brittoniana (right),* and *L. blackii.* Although I may be wrong about the last species.

It should be noted that 25 *Licania* taxa have been recorded in Colombia – only in its Amazonian part – and the majority of them look very similar. My identifications are based on the relevant herbarium material. But in the final analysis the composition of the flowers and fruits differs very little within the genus.

Nutritional value: the fruit is very rich in proteins and water-soluble carbohydrates, and lipids. I could also verify that large amounts of sugars are transported as sucrose (no wonder Colombians call it *cana dulce* and eat the sweet pulp fresh – as other American indigenous people do), which might be the reason why discus eat the flowers.

Remarks: Although there are more than 100 taxa in the genus and new ones are constantly being described, as far as I know nobody has to date established that these plants not only provide "homes" and shelter for numerous fishes (and other animals), but their flowers, fruits, and seeds represent an indispensable food source. In all these years I have only rarely discovered a discus habitat without *acará-açú*, and when I have done so, only in the dry season. When the water is rising, and undoubtedly also during the high-water period, these shrubs assure their survival. Undoubtedly this record of my years of research, with the evidence that *Licania* species represent a very important food source for discus and play a role in their survival, is the first publication of its kind. And hopefully not the last.

Rheedia gardneriana
Rheedia brasiliensis

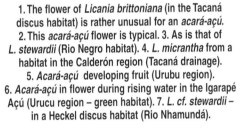

1. The flower of *Licania brittoniana* (in the Tacaná discus habitat) is rather unusual for an *acará-açú*. 2. This *acará-açú* flower is typical. 3. As is that of *L. stewardii* (Rio Negro habitat). 4. *L. micrantha* from a habitat in the Calderón region (Tacaná drainage). 5. *Acará-açú* developing fruit (Urubu region). 6. *Acará-açú* in flower during rising water in the Igarapé Açú (Urucu region – green habitat). 7. *L. cf. stewardii* – in a Heckel discus habitat (Rio Nhamundá).

Clusiaceae (Hypericaceae; Ebenaceae; Guttiferae)

This family of trees, shrubs, climbers, and herbaceous plants, which has a cosmopolitan distribution in the tropics and temperate zones, encompasses 40-50 genera with 1,000-1,200 species (depending on the author), but only one species is of interest here.

Rheedia brasiliensis *(bacuripari-liso; bacuri; bacu; uvacupari):*

Blue and brown discus feed on the white flesh of the fruits.

This tree occurs predominantly in the state of Amazonas, by aquatic habitats in *igapós* and *várzea* regions. It grows to barely 10 m tall and the trunk is 10-15 cm in diameter. Its small, white, scented inflorescence (it can be recognised by sense of smell) contains some 10-15 male and numerous hermaphroditic flowers. The round, golden yellow fruits are 3-4 cm in diameter *(p. 524)* and have a thin rind enclosing a large amount of white, slightly acidic flesh containing 1-3 seeds. And they too are always ripe during the high-water period.

My observations on discus feeding behaviour as regards this fruit go back many years. In the drainage of the Lago Aiapuá blue discus fed greedily on the flesh of the fruits (and so did I – *p. 403*) and in the Monte Alegre region (Lago Grande, at Alenquer) brown discus too. However, I have as yet been unable to ascertain whether Heckel discus feed on the fruits of *R. spruceana* in the central Rio Negro or blues and browns on those of the species *Rheedia macrophylla*, known as *bacuripari* or *bakupari*, in the Alenquer and Xingú regions and elsewhere in the lower Amazon drainage, where they occur.

Nutritional value (per 100 g fruit flesh): 72.3 g moisture; 1.9 g protein; 2.0 g lipids; 7.4 g fibre; 1.0 g ash; 20.0 mg calcium; 36.0 mg phosphorus; 2.2 mg iron; 0.04 mg vitamin B; 105 kcal (and other less relevant values).

Remarks: There are more than 45 species of the genus *Rheedia* and some confuse *bacuripari* with *bakuri*. The latter is, however, *Platonia insignis* (syn. *Aristoclesia esculenta)*, a tree that grows to more than 25 m tall, is often tapped for rubber, and whose fruits are much larger (up to 15 cm long and 900 g in weight).

Cecropiaceae (placed by some in the Urticaceae)

This family, some of whose members provide homes for ants, includes trees, shrubs, and climbers, often with aerial roots, with a neotropical distribution. There are 275 species known in 6 genera, of which one (or two, see below) is of interest:

Pourouma cecropiifolia *(mapati; uva-da-mata; uva-da-amazônia; curucua; umbaúba-de-cheiro; umbaúba(imbaúba)-de-vinho; amaitam; imbaúba-mansa; tararanga-preta; imbaubarana; sucuua).*

Blue and brown discus feed on the fruit.

The *mapati* tree grows to between 3 and 15 m tall and is often confused with the true *imbaúba* tree (*Cecropia latiloba*). Its distribution lies in Amazonia, Peru, and Colombia. It flowers between March and April and normally bears fruits from July onwards (until September/October). Its dark, grape-like fruits *(left & below)* have a diameter of 2-3 cm, and their white gelatinous flesh is slightly sweet. People eat them, but they are also used to make a lot of wine (mainly by natives) and spirits (commercially). This is also the source of the name *uva-do-mato* (grape of the forest). Now apparently a number of discus also have a sweet tooth – or are they in fact actually connoisseurs of wine or spirits?

To date I have seen blue and brown discus feeding on the whitish flesh of the fruits in the Lago Amanã region.

Nutritional value (per 100g fruit flesh): 82% moisture; 0.3 g protein; 0.3 g lipids; 0.9 g fibre; 16.7 g carbohydrate; 0.2 g ash; 34 mg calcium; 10 mg phosphorus; 0.6 mg iron; 0.22 mg riboflavin; 0.30 mg niacin; 0.6 mg reduced ascorbic acid; and 64 kcal.

Remarks: I have found the seeds of *Cecropia latiloba*, called *imbaúba* or *embaúba* in Brazil and assigned by some authors to the family Moraceae, in the net together with blue discus in the Trombetas region, so they too may be food as well *(p. 291)*.

Euphorbiaceae

The spurge family consists of mainly milky-sapped, sometimes succulent trees, shrubs, herbaceous plants, and climbers, sometimes cactus-like with the branches acting as organs of photosynthesis. Some 5,000 species in 300 genera are known from all over the world; two genera are of interest for discus.

Hevea spruceana *(seringa-barriguda):*

Heckel discus feed on the disintegrating flesh of the fruits (during the rainy season).

The rubber tree is widespread in the Amazon region, but predominantly to the north of the great river, often in *igapós* and flooded forests, where it can grow up to 30 m tall. Simultaneously with the rising of the water it

Pourouma cecropiifolia

Left-hand page: Drawings of a *mapati* tree (*Pourouma cecropiifolia*) branch *(below)* and its grape-like fruits.
This page:
1. A *piranheira* tree, *Piranhea trifoliata.* The photo shows clearly the reason for the specific name *trifoliata* (= with three leaves). Meanwhile *Piranhea* comes from Piranha. But why? Now, the well-known vegetarian *tambaquí* and the carnivorous piranhas belong to the same family (Serrasalmidae), and were often confused, especially in the past (Bleher, H., 2003d). And because the "false piranha" (*tambaquí* = *Colossoma macropomum*) loves the fruits and seeds of this tree (apparently even more so than the discus), it was given these popular and scientific names. Green discus often benefit from the fruits when monkeys nibble them and let them drop, or when

they fall during the high-water period, as here in the Mamirauá region (2).
3-4. The fruits of the *seringa-barriguda* (*Hevea spruceana*) are often caught in the rising water, as here in the Rio Negro (3). And then the Heckel discus (and also other discus and fish species elsewhere – *see pp. 291, 370, 393, and 528*) can feed without worry on the ruptured fruits and softening seeds (4). If we consider that in 1953 the water in the Rio Negro rose 29.69 m above sea level and 21 m is almost an annual event, we can well imagine that the fishes not only have plenty of fruits and seeds on which to feed, but also spread them widely – a very important factor in the biodiversity of Amazonia (as long as there are still trees).
5. *Piranhea trifoliata* – a young tree, in which it can be seen how the fruits on the woody twigs ripen at rising water.

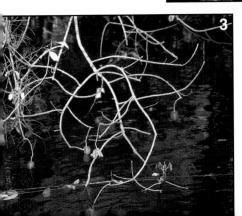

produces its fruits, which are particularly relished by the main food-fish of Amazonia, the *tambaquí (Colossoma macropomum)*. And a lot of the fruit is dropped for the discus to eat, or they feast directly on the flesh disintegrating out of the rind. But so much has been published on this rubber-tree species and the *tambaquí*, in books, in the scientific and popular literature, and on thousands of Internet pages, that I needn't bother.

I have repeatedly seen Heckel discus feeding on the flesh of fruits floating in the water in the lower Rio Negro regions *(p. 527)*. When pulling in the net in the Heckel discus habitat I have often found lots of these fruits in the net as well, along with other fruits and seeds *(p. 529)*. But discus also fed on *seringa-barriguda* flesh in the Japurá and Trombetas region where I caught blues and browns. I cannot say anything about possible links between discus and the fruits of the other 10 *Hevea* species (and variants/subspecies) in the Amazon region.

Nutritional value: very rich in protein and lipids.

Remarks: Indigenous tribes have a tradition of boiling up the tree bark to make a tea and a treatment for liver and kidney ailments. But *Hevea brasiliensis* is preferred for rubber production *(pp. 256-261)*.

Piranhea trifoliata (piranheira; sardina caspi):
Heckel discus appear to feed on this fruit as well, likewise blue and brown discus. Greens definitely.

The *piranheira* is a much-prized, long-lived hardwood, widespread in Amazonia, but very characteristic of the *igapó* regions – black- and clearwater habitats. It too is one of the great trees of Amazonia and bears its fruits right at the time when the water begins to rise *(p. 527)*.

In the central and lower Rio Negro, and in the Tapajós drainage region and in the Caburi (Nhamundá region), I have seen Heckel discus and blues and browns (respectively) picking at the softened flesh of the fruits. But in no instance was I able to ascertain whether they were also eating it. Only in the Tefé-Mamirauá region, where *piranheiras* are often encountered, was I repeatedly able to see green discus actually feed on the fruits.

Nutritional value: unfortunately no details are available.

Remarks: Radio-carbon dating has shown that *piranheiras* are among the longest-living trees of Amazonia, at around 500 years. Only three others live even longer, between 900 and 1,400 years. And this "select group" stand amid very young, short-lived trees in the *igapó*.

Lecythidaceae

This family of nut-trees encompasses around 400 taxa in 20 genera and is widely distributed in the tropics of the New World, less so in Africa and Asia. Predominantly in low-lying *terra-firme* regions, at the edges of lakes and still waters. The genus *Eschweilera*, which is found in the Amazon drainage region and north to Central America, contains more than 120 species, and *Lecythis* more than 50. I have to date found discus food in both these genera.

Eschweilera tenuifolia (macacaricuia; matá-matá, matá-matá-folha-grande):
Heckel discus feed on the flesh of the fruits, seeds, and possibly the young leaves.

This tree, also known as the *matá-matá* ("kill-kill"), often grows *(as shown on p. 369)* in parts of a Heckel discus habitat (but not only there) and towers around 20 m into the sky when it isn't largely submerged. Its fruits, which, as in most species of the family, are nut-like, have a hard outer shell – and hence are also called monkey nuts *(coco de mono)* – but they slowly disintegrate when floating or are cracked by monkeys.

Heckel discus consume the flesh of the fruits and disintegrating seeds in the lower Rio Branco region. I have found pieces of young leaves (possibly those of *E. tenuifolia*) in the stomachs of individuals from this habitat.

Nutritional value: The flesh of the fruits is rich in tannins and protein.

Remarks: Its young leaves are a favourite food of *ukari*. And in addition it (often?) happens that the ripe fruits are full of Microlepidoptera larvae (Sesiidae), which undoubtedly enrich the menu of the discus (and other fishes). The species is sometimes assigned to the genus *Lecythis*, and the natives often use the name *matá-matá* for other *Eschweilera* species as well.

Lecythis barnebyi (macucu):
Heckel discus feed on pieces of the (cracked) seeds and the flesh of the fruits.

This species grows along the banks zones of the *terra firme*, by rivers and lakes. It too attains a height of 20-30 m.

In the Rio Negro region I have seen Heckel discus eating the flesh of the fruits and picking up pieces of softened seed. *(For details see p. 369 and below)*.

Lecythis pisonis (sapucaia; castanha sapucaia; cumbuca de macaco; marmita de macaco; quatete):
Blue discus in the Trombetas region feed on the flesh of the fruits and the seeds.

Left-hand page: "Monkey pot" is the English name for the unusual "packaging" of the nuts in *Lecythis* species (this is *L. pisonis*), as monkeys often open the lid before it falls (1) in order to get at the mass of tasty yellowish flesh (which is also a discus favourite) and the nuts *(see text p. 530)*.

This page: 1. The *L. pisonis* "packaging" always hangs with the lid downwards, so the flesh, along with the seeds, is accessible to the fishes hence for distribution. 2. Green fruits of the *araparí (Macrolobium multijugum)*. 3-4. Flower of *L. pisonis,* closed (3) and open (4). 5. Fruits of *araparí (centre)* and *Hevea spruceana (left)* in the Heckel discus net in the Jaú region.

This tree species can grow up to 30 m tall. Its white flowers come out at the beginning of the rainy season and it bears fruit predominantly during the high-water period. The large shell containing both nuts and fruit flesh (as is usually the case in *Lecythis*) hangs "head-down" and the "lid" falls off automatically when the seeds are ripe. It is also often opened earlier by monkeys, as for many the soft seeds and the surrounding yellowish flesh are a delicacy. Not for nothing are the shells known as *cumbuca de macaco*, which roughly translates as "monkey trap". As when, after opening the shell, they reach inside with their hands to bring out the nuts and the flesh, the hand almost always gets caught in the narrow opening. Only older individuals manage to avoid this *(p. 528)*.

The partly striped blue discus in the Lago Jacaré (Trombetas) (thought by some to be Heckel discus) feed on the fruit flesh. In the lower Rio Xingú I saw yellowish discus feeding on it, and, interestingly, I found *quatete* by the Río Nanay (Peru) and saw how the green discus introduced there had a feast on it. The seeds and flesh of the fruits had been dropped by red howler monkeys.

Nutritional value: high in lipids (more than 51%).

Remarks: 11 Amazonian species of the genus *Lecythis* have now been added to the IUCN list under "threatened with extinction"; they include *L. barnebyi,* but luckily not (yet) *L. pisonis.*

Leguminosae (= Fabaceae)

This is the family with the most flowering species – more than 18,000. I have ascertained that discus feed on:

Macrolobium multijugum (araparí):

Heckel discus, blue and greens feed on the flesh of the fruits.

The distribution of this medium-sized tree species appears to be largely restricted to Amazonian blackwater and clearwater regions, *igapós* and flooded forests. I have found it extensively in the Rio Negro region, in the Mamirauá and Rio Urubu drainages, as well as by the Trombetas.

1. *Malpighia glabra* – the original drawing dates from 1756. 2. *Acerola* (or *cereja-de-Cametá*), as the Brazilians call it – discus food along the Tocantins. 3-4. The delicious flesh of the fruit.

Around 200 species of the genus *Macrolobium* are known. Their fruits and seeds are very similar. They flower towards the end of the dry period and fruits appear during the rainy season. I have also found the species *angustifolium, discolor, gracile, limbatum,* and *punctatum* in the Rio Negro region. In the Jaú region Heckel discus feed on fruits dropped by ukaris (p. 529). I don't know whether they eat the other species. But in the Mamirauá Greens consume the fruits; likewise blue discus in the Urubu and Lago Batata (Trombetas). Unfortunately in the latter region a whole area of the *igapó* vegetation is history because of the 4-6 m high accumulation of bauxite waste.

Nutritional value: unknown.

Remarks: Macrolobium is also assigned to the Caesalpiniaceae or Papilionaceae by some.

Malpighiaceae

The Malpighiaceae, with around 1,100 species and some 65 genera, are small selfsupporting trees or shrubs, the majority of them climbers. The family is distributed in Africa (including Madagascar) and the Americas.

Malpighia glabra (acerola; cereja-de-Cametá; Brazilian cherry; pitanga; cereza de indias; prunha de Madagascar; Barbados cherry; scobillo):

Brown discus feed on the flesh of the fruits. Up to 5 m tall trees that are distributed in the Amazon region and north to the Caribbean, but cultivated in many parts of the Americas and elsewhere. I found them in large numbers along the Rio Tocantins and in the surrounding area (except at the Tucuruí hydro-electric dam, where everything has died out). The fruits, which resemble large cherries, look similar to the *camu-camu*. They too are also initially green and later red. The fruits with their whitish to reddish, somewhat sweetish flesh, are very popular (nowadays almost worldwide). Interestingly on the Tocantins they are known not as *cereja-de-Cametá* (Cametá cherry), but also as *prunha de madagascar* (Madagascar plum), as some people say they originated from

Madagascar. (105 million years ago? But science sees it the other way round.)

However the discoverer of the species, Linnaeus' pupil and one of his 17 "apostles", Pehr Löfling, who found it in 1754 in Venezuela, certainly had no inkling of this. Or that this *acerola*, Barbados cherry, or *escobillo*, would come to be valued as a vitamin C booster, a skin gel, and much more, and would even sold by multi-national companies such as Nestlé in the form of ice-cream, drinks, etc. He cannot have foreseen any of this when he died in 1756 of malaria, at the age of just 26, in San Antonio de Caroní, Nueva Andalucia (now in Venezuela). Or that two years later, in the monumental work that was to become the basis for all scientific descriptions, his mentor would describe the plants, fishes (more than 50 new species from the Orinoco system), amphibians, and reptiles he had discovered. Including *Malpighia glabra* Linnaeus, 1758, with the remark *"cerezas indias"* (p. 530).

Some *Virola* fruits and flowers:
1. *Virola flexuosa* with open fruits.
2. *V. sebifera*, fruits in the process of splitting. 3. *Virola* sp. flower and closed fruit at Calderón, Colombia.

I have seen brown discus feeding on flesh of the fruits in the discus habitat at Cametá and getting a vitamin boost that really made itself felt. But not yet otherwise. I am also unable to say whether Heckel discus feed on further species of the genera *Acmanthera* and *Burdachia*, or the species *Byrsonima spicata*, which I encountered in the Rio Negro region and whose fruits often look similar (that of *B. spicata* is, however, yellowish).

Nutritional value: very rich in vitamin C and vitamin B1. It is estimated that this natural vitamin C has double the effect of artificially produced ascorbic acid. And of all known fruits (worldwide), after the *camu-camu* (see below) it has the highest vitamin C content.

Remarks: The popular name *acerola* is often – on an international level and in homeopathy – ascribed to the species *M. punicifolia*, which is a synonym of *M. glabra* (five *M. glabra* varieties have been described).

Note that the genus *Malpighia* includes more than 250 taxa and that Linnaeus named the genus in honour of the Italian anatomist and professor of medicine Marcellus Malpighi (1628-1694). Malpighi – along with Nehemia Grew (1641-1712) – is regarded as the father of plant anatomy, for which he created the terminology used ever since. In 1661 he discovered the *Rete Malpighi* – the basal layer of the skin (the *stratum basale* of the epidermis), also called Malpighi layer. In addition the Malpighi follicles in the spleen (splenetic follicles) and kidneys (renal follicles), as well as the Malpighian tubules of the glomerula, bear his name. He was also responsible for comparative anatomical studies of the liver and the first accurate description of embryonal development in chickens. His studies of the capillaries in humans also gave him an insight into plant anatomy. And as early as 1669 the diligent professor was elected to membership of the Royal Society in London. Malpighi applied the new techniques of microscopy to comparative anatomical studies, culminating in the publication in 1675 of his work *Anatome plantarum* (The anatomy of plants).

Mimosaceae

The mimosa family includes some 40 genera and around 2,000 species worldwide. They are mainly trees and shrubs, but also include herbaceous plants. One genus is apparently important for discus.

Inga spp. *(ingá):*

Heckel discus, blue and brown discus feed on the flesh of the fruits.

There are 258 *Inga* species, occurring from Mexico to Uruguay. Half of the species predominating in the Amazonian drainage region have adapted to life on the *terra firme* – up to an altitude of 3,000 m – and the other half of the so-called *ingás* to survival in *igapós* and the *várzea*. Their fruits are almost without exception prized as food *in natura* – and nowadays almost worldwide. The *ingá* trees grow to 10 to 15 m tall (only the *ingapéua*, *Inga macrophylla*, remains small). They all have long, woody, bean-like pods (which in some species attain a length of up to a metre – *p. 532)* containing rows of seeds encased in a sweetish white flesh *(p. 530)*. The fruits ripen mainly between March and June (the high-water period) and are often nibbled by monkeys, aras and toucans, and in the process fall into the water. Once they are opened the fishes feast on them.

Ingás: 1-3. *I. edulis* tree standing on the bank of Lago 26 (Tapauá region) when the water was starting to rise (beginning of December), with the almost ripe fruits already hanging down (1) right above the blue discus habitat below (3). This species had flowered here months earlier (2). 4-5. A second *Inga* species (*I. austristellae*), which I found at the green discus habitat in the Rio Nanay region (Peru) (4), also serves the discus there as a much-enjoyed food when the water rises. And often comes complete with the well-known *ingá* ants (5), which provide a tasty additional food. These ants live in and from the *ingá* tree, hence the name. 6-7. Open (5) and closed (6) *I. cinnamomea* from the Abacaxís.

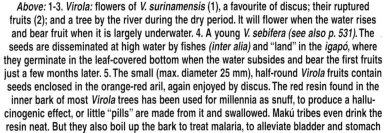

Above: 1-3. *Virola:* flowers of *V. surinamensis* (1), a favourite of discus; their ruptured fruits (2); and a tree by the river during the dry period. It will flower when the water rises and bear fruit when it is largely underwater. 4. A young *V. sebifera (see also p. 531).* The seeds are disseminated at high water by fishes *(inter alia)* and "land" in the *igapó*, where they germinate in the leaf-covered bottom when the water subsides and bear the first fruits just a few months later. 5. The small (max. diameter 25 mm), half-round *Virola* fruits contain seeds enclosed in the orange-red aril, again enjoyed by discus. The red resin found in the inner bark of most *Virola* trees has been used for millennia as snuff, to produce a hallucinogenic effect, or little "pills" are made from it and swallowed. Makú tribes even drink the resin neat. But they also boil up the bark to treat malaria, to alleviate bladder and stomach ailments, to heal wounds and infections. However it has only recently been realised that the agents in the fruits are even more potent *(p. 534).*
Below: Camu-camu: These fruits are another special discus food *(p. 535),* and ripen when the water rises (6) so discus can feed on them directly (7), as here in the Rio Itaparaná (8).

Myrciaria dubia (camu-camu)

Genipa americana (jenipapo)

In the Rio Canumã drainage I have seen blue and brown discus feed on the flesh of the fruits of *I. cinnamomea* (known as *ingá-açú*, *ingá-chinela*, and *ingá-grossa*). Likewise in the Jari area (Purus region). In the blackwater Rio Itaparaná (Tapauá region) blue discus fed on the fruits of the species *I. edulis*, which is also called *ingá-cipó* or *ingá-rabo-de-mico*.

Nutritional value (per 100 g of fruit flesh in *I. edulis* – pp. 530-531): about 82% moisture; 1.0 g protein; 0.1 g lipids; 15.5 g carbohydrate; 1.2 g fibre; 0.4 g ash; and 60 kcl.

Remarks: *Ingas* are cultivated for their fruit in many tropical countries of the world, for example in the Caribbean, USA, East Africa, and Hawaii and other Pacific islands. They are also planted in many places to provide shade in coffee, tea, and cocoa plantations. In addition ice-cream is made from them (hence the English name "ice-cream beans"). And last but not least: *Inga* species are also assigned to the family Fabaceae by some.

Myristicaceae

The nutmeg family consists of trees with a tropical or pantropical distribution. Some 380 species in 16 genera are recognised. To date I have been able to identify only the following as important for discus:

Virola spp. *(bicuiba; bucuva; bucuvcú; sebo; otobo; nu namo).*
All three discus species feed on the flowers, fruits, and seeds.

This genus of large trees is not only prized as a useful wood, but their hallucinogenic red resin has been used for millennia by numerous indigenous peoples. These trees occur almost exclusively by or in the waters of the entire Amazon drainage and north to Central America, and can attain a height of a good 25 m.

Heckel discus feed on the flowers in the Rio Negro region, although I cannot say which of the four species that occur here – *V. theiodora*, *V. calophylloidea*, *V. cuspidata*, or *V. rufula*. As regards *V. calophylla* (*ucuúba*), I know that greens like to feed on their flowers in the Urucu drainage region – I observed this when the water was rising (January-February) *(p. 447)*. Blues feed on the flowers and fruits in the Coari Grande drainage *(p. 444)*, and I also saw this in the Mamiá *(p. 448)*. As regards *V. surinamensis* (a species that occurs north to Central America), in the Alenquer region Blues and Browns feed on the fruits, while Greens consume this species in the Tefé region and Mamirauá. Blue and brown discus were feeding on the fruits of *V. elongata* in the Rio Tapauá when the trees were already submerged (January-February). And blue discus on an unidentified species in the Itaparaná *(p. 418)*. And, again at rising water, green discus were feeding on a *Virola* species in the Rio Amaturá. *(See also pp. 531 and 533.)*

Nutritional value – hallucinogen content: Not only the bark, roots, and leaves but also the seeds and flowers of *V. calophylla* (among other) contain the well-known hallucinogens Dimethyltryptamin (DMT) and 5-Methoxy-DMT. The alkaloid content of the flower tips (a favourite food of many discus) is 193mg per 100 g of dehydrated plant material. 96% of this is DMT. The content of the leaves is 155mg per 100 g of dehydrated leaves, and again 96% of this is DMT. By contrast the alkaloid content of the bark is lower at 9mg per 100 g of dehydrated material, comprising 91% DMT and 9% 5-Methoxy-DMT. The roots contain only 1mg alkaloids per 100 g of dehydrated material, of which 87% is DMT and 13% 5-Methoxy-DMT. The question is, what effect does the high alkaloid content have on discus?

Remarks: As well as the known information on *Virola* (p. 426), I know that indigenous peoples (Puinave and Kuripako) from the Colombian Vaupés (Apaporis region – blue discus) use snuff called *yá-kee* and *yá-to*. They get this from the reddish resin of the inner bark of three *Virola* species: *V. calophylla*, *V. calophylloidea*, and *V. elongata*.

Myrtaceae

The myrtle family consists of trees and shrubs which occur in the tropics, the subtropics, and temperate parts of Australia. Around 3,000 species are known in 140 genera, and one is of interest as regards discus.

Myrciaria dubia (camu-camu; araçá-d'água; caçari; rumberry): Green, blue, and brown discus feed on the fruits.

This is a bushy river side tree, which grows to 8-10 m tall but spends a large part of its existence in (under) water (its roots are invariably underwater for half the year or more). It is widespread

in Amazonia, in the blackwater *igapós* and also in *várzea* regions. It bears its fruits right from the start of each rainy season. The fruits are similar in appearance to those of the European *Smilax aspera (p. 369)*. They are round, dark, reddish or slightly orange in colour, up to the size of a lemon, usually with a diameter of only 2-3 cm. A single tree can bear more than 1,000 fruits per year *(see also p. 367)*. The genus is called *araçá* in Brazil (as is the genus *Psidium*) and it is said that there are as many species of *araçá* in Brazil (including those in Amazonia) as there are beaches (more than 4,000 km).

I found that many discus species/variants – apart from Heckel discus – feed on *camu-camu* fruit, for example Greens in the Rio Urucu (Coari region – *see also p. 445)*, but not only there. Blues and Browns in the Tapauá region (Itaparaná; Tapauá; in various *lagos* such as Lago 26, Tambaquí, *inter alia),* and so on. Almost everywhere that *camu-camu* occurs in (mainly) western Amazonia (where there are no Heckel discus), discus exploit these fruits.

Nutritional value (some of the most important contents per 100 g fruit flesh): 2,605.76 mg vitamin C; 76.53 mg nitrogen; 9.41 mg phosphorus; 106.96 mg potassium; 2.51 mg magnesium; 8.47 mg calcium; 0.40 mg iron; 0.18 mg boron; 0.18 mg zinc; 0.32 mg copper; 0.54 mg manganese; with a pH of 2.86-3.10.

Remarks: The *camu-camu* has the highest natural vitamin C component of any fruit known to date. For comparison: oranges provide 53 mg vitamin C per 100 g fruit, likewise vitamin-C-rich *acerola* fruit has been shown to provide 1,677 mg per 100 g fruit; but *camu-camu* provides up to 2,700 mg vitamin C per 100 g fruit. Compared to oranges *camu-camu* provides 50 times more vitamin C, 10 times more iron, 3 times more niacin, twice as much riboflavin, and 50% more phosphorus. *Camu-camu* is also a significant source of potassium with 711 mg per kg of fruit.

Moraceae

The mulberry family encompasses 40 genera with some 1400 species. They are trees, shrubs, and climbers, rarely herbaceous plants; usually with milky sap, often with aerial roots. They occur in temperate zones and the subtropics, and are widespread in tropical regions.

***Ficus* spp**. *(figo bravo* = wild fig; *apuí; ojé):*

The wild fig trees are numerous and widespread in the Amazon region, found almost exclusively in *várzea* and *igapós*, the forest inundation zones. Some species can attain a good 30 m in height and survive for a long time underwater. There are also large numbers of so-called strangler figs.

I have seen Heckel discus feed on wild figs in the Rio Negro drainage (Jaú, *inter alia*), were they call them *mapimissu;* blues and browns eat the flesh of the fruits in the Nhamundá region (Caburi), in *lagos* along the Trombetas and in the Jari region; green discus in *lagos* along the lower Juruá were feeding on *F. insipida* fruits which had fallen from a tree some 35 m high through the agency of Capuchin monkeys.

Nutritional value: 1.2-1.3 g proteins; 17.1-20.3 g carbohydrates; 0.1-0.3 g lipids; 1.2-2.2 g fiber; 0.48-0.85 g ash; and 80 calories. In *F. insipida* fruits contain the highest concentrations of protein so far known (but not eatable by man).

Remarks: In the upper Amazon the sap from *F. insipida* is collected from trees along the river, in the morning before the sun is out, to cure parasites. And to avoid breaking the equilibrium of forces and invoking the trees bad will, some groups pay the tree back with tobacco, alcohol o dieting.

Passifloraceae

The passionflower family includes around 530 species in 16 genera. They are trees, shrubs, and climbers with a tropical to subtropical distribution. I have found one to be relevant for discus.

Passiflora nitida (*maracujá-do-mato; maracujá-do-rato; maracujá-suspiro, maracujá-do-cheiro):*

Blue and brown discus feed on the flower petals. A climbing plant that is widespread along the rivers and lakes of Amazonia (I have found them mainly by clear waters), but also grows in dry areas, for example along the jungle roads. It often has large flowers (typical passionflower blossoms) up to 10 cm in diameter. The flowers are to be found almost all year round. The fruits, which are often red and almost globular with a diameter of some 7 cm, are seen mainly between January and June *(see on top).*

In the Rio Içá I have seen brown discus feed on the flower petals. But to date I have found no evidence that the fruits are also consumed by discus, although the natives make fruit juice from them or eat them raw (somewhat bitter, but a great thirst-quencher).

Nutritional value: unknown.

Rubiaceae

The madder family, with around 13,000 species of trees, shrubs, climbers, and herbaceous plants in some 650 genera, has a cosmopolitan distribution – predominantly in the tropics. Two species are of interest for discus.

Psychotria spp.:

Heckel discus, green and blue discus feed on the fruit.

A shrub that looks very similar to the coffee bush (including its fruits and their size), and is widespread in Amazonia. More than 700 species are known in the genus circumtropically, but the best-known of them is the species *P. viridis (chacruna)*, which is both naturally distributed and cultivated in the western Amazon region, in Colombia, Peru, and Ecuador. In the Rio Negro region I saw Heckel discus nibbling at the fruits of a *Psychotria* sp. Blues were eating *P. viridis* fruits in *lagos* at Tacaná, and green discus in the Nanay drainage.

Nutritional value: unknown, but principal active biochemicals such as tryptamine alkaloid N, N-dimethyltryptamine (DMT) and *beta*-carbolines are present in the leaves of *P. viridis, P. carthagiensis* and probably other species (see below).

Remarks: P. viridis contains hallucinogenic tryptamine as its main psycho-active ingredient, and has been taken for a long time, mainly in western Amazonia, in the form of *ayahuasca* (in combination with *Banisteriopsis caapi*). The leaves are used for this *(below right)*. Is this why discus like to feed on the fruit? One thing is sure, nobody becomes dependent on it. And the fruit of one species, *P. erecta*, is used by the Karapaná- and Witoto-indians in preparation of one of their forms of curare.

1-3. *Genipa americana*, known as *jenipapo* in Brazil. A tree on the edge of the bank (1), the flower (2), and the *jenipapo* fruit (3), whose flesh can be seen in the open fruit *(left in the photo)*. 4. Fruits of *Psychotria viridis*, widely known as *chacruna*. Discus feed on them, but humans cultivate them for their hallucinogenic effect.

Genipa americana (jenipapo; yenipapa; jeipapeiro; mandipa):

Blue and brown discus sporadically consume the flesh of the fruits.

A tree which grows up to 20 m tall and is found in the north of Amazonia, often along clear waters, and often in the flooded forest. Its distribution continues north as far as Mexico. It flowers between July and September and bears fruit from October to April. Older trees (15-20 years) can produce up to 600 fruits per year. The fruits may be greyish to yellowish and some 10-15 cm long *(right)*. The thin rind quickly softens in the water and the pale yellow flesh with numerous small (0.8 cm long) seeds comes out. The seeds are eaten by larger fishes, the flesh by discus. But the *jenipapo* – called *maluco* (= crazy) in Mexico is also very popular with humans. The fruit flesh is processed into fruit juice, ice-cream, alcohol *(jenipapada* and *huitochado),* compotes, and sweet desserts. According to the Amerindians *huitochado* helps against rheumatism; the stewed fruit is a treatment for asthma and inflammations of the respiratory tract; the fruit flesh numbs toothache; and the juice of the green fruit is effective against stings. The juice is also used as a dye by many indian tribes (see also Remarks).

Nutritional value (per 100 g fruit flesh): water 57.6 g; protein 5.2 g; lipids 0.3 g; fibre 9.4 g; glycerine 25.7 g; ash 1.2 g; calcium 40.0 mg; phosphorus 58.0 mg; iron 3.6 mg; vitamin B 0.04 mg; vitamin B2 0.04 mg; niacin 0.50 mg; ascorbic acid 33.0 mg; amino acids (per g nitrogen [N6.25]); lysine 326 mg; threomnine 219 mg; tryptophan 57 mg; and 113 calories. Tannin levels are extremely high.

Remarks: Coffea arabica is undoubtedly the best-known plant in the family. But a few additional interesting remarks.

In Guyana I have seen ripe *jenipapo* fruits used as bait for fishing; and wild and domesticated animals feeding on fruits lying on the bottom (and the water). In the Darién

1-3. *Ficus* spp.: a mighty *Ficus* tree in the Rio Negro, where it not only provides shelter for discus in the dry season (1), but also fruits when the water rises (2-3). 4-6. *Jenipapo:* A Chocó indian has grated the green fruit with the result that the liquid has already turned greyish (4). He cuts a cane to dip it the liquid and use it as a brush. 5). Nearby lies a green *jenipapo (Genipa americana)* fruit. The next step is the body-painting. Once on the skin, the *jenipapo* dye remains unchanged – pitch black – for around 20 days (4).

peninsula (Panama) I have seen how the colourless juice of the green fruit oxydises in air and then becomes brown, blue-black, and finally pitch black. I have also seen when the indians painted themselves with it the colour still hadn't faded after 20 days; how it is used as an insect repellent, and fabrics, hammocks and baskets are coloured with it; how they eat it as a remedy for jaundice and in large amounts when they have worms, or take it as a diuretic. In Puerto Rico I heard how scientists have discovered antibiotic activity in the fruit. In 1964 Dr. W. H. Tallent isolated and described two new antibiotic cyclopentoid monoterpens from the *jenipapo*.

The juice of the crushed green fruit and the stewed rind are supposed to work wonders when applied to sexual injuries, and also cure throat ailments. And the boiled root is a powerful laxative. If the bark of the trees is incised they exude a whitish, sweet-tasting sticky material (latex). Applied diluted to the eyes this is said to clear cataracts and clarify clouded sight. The juice of the leaves is taken as an antipyretic in Central America. The pulverised kernels of the seeds are emetic and caustic. And I have seen Guatemalans (Guatemaltecos) carry *yenipapa* fruits around as they believe that they bring good luck and protect them from illness.

Verbenaceae (Lamiaceae)

The verbena family includes not only herbaceous plants but also trees and shrubs. The 2,600-3,000 species in around 100 genera have a tropical and subtropical distribution worldwide, but just one is of interest for discus.

Vitex cimosa (tarumã):

Heckel discus feed on the flesh of the fruits.

This species is widespread in Amazonia, predominantly in *várzea* regions and *igapós*. In the flooded forests their nutritious fruits and seeds are a welcome food for numerous fish species. This undoubtedly doesn't apply just to *V. cimosa*, as more than 250 tree and shrub species in the genus are found circumtropically and in the subtropics. A number of species are also prized as commercial timber, especially tree species from Indochina, likewise African species. In North America they have reached pest proportions, especially along the beaches of the Atlantic coast. Not so in Amazonia. Here they flower mainly before the onset of the rainy season and their fruits ripen at just the right time during the high-water period, and until the water recedes again *(right-hand page)*. At that time the fishes find their "table" is copiously laden.

I have ascertained that in the lower Rio Negro drainage the Heckel discus feed on this fruit as well. Unfortunately I haven't yet been able to study any other discus species in this respect. But they undoubtedly feed on it too.

Nutritional value: rich in protein and lipids.

Remarks: Hippocrates, Dioscorides, and Theophrastus mention *Vitex* as having extensive applications in healing, including for stopping post-partum haemorrhage, and that cooked fruits, applied via immersion, were helpful against uterine complaints. In addition it was known very early on that *Vitex* suppresses lust and promotes chastity (hence its English name "chaste tree"). This applies only to the fruit of *V. agnus-castus* which occurs around the Mediterranean. In addition the dried fruit, which has the aroma of pepper and a taste to match (hence also "monks' pepper"), is nowadays a well-known herbal remedy and is even sold as a medication via the Internet, primarily for women's problems such as pre-menstrual syndrome; infertility; fibrous mammary cysts; acne (linked with menstruation); amenorrhoea; dysmenorrhoea; endometriosis; menorrhagia (heavy menstrual bleeding); and as an aid in childbirth and *post partum*.

The only question is, do discus females know this too?! Be that as it may, they do know that there are no contra-indications to eating it – or at most an occasional itchy, urticaria-like skin rash which soon disappears again when it is no longer consumed.

In conclusion: There are many more flowers, fruits, and seeds *(eg pp. 538-539)* that are eaten by discus, I cannot possibly list all those recorded during hundreds of research strips to the Amazon and laboratory studies. Or indeed, how many and which other species and variants also feed on the plant materials mentioned above. In addition I have been unable positively to identify/analyse some of the plant material found during gut and stomach contents studies (nor has the INPA Institute). I have photographed as yet unidentified fruits, flowers, and seeds in various habitats *(eg p. 295, Trombetas region)* in order to be able to study them further. In the process I only recently came across the red seeds of *Bixa orellana* (*urucú; urucum; colorau – p. 360*), family Bixaceae, which Heckel discus eat and which have undoubtedly

Flowers and fruits of Amazonia – discus food *(both pages):* The majority of trees, shrubs, and climbers of Amazonia flower towards the end of the dry period – or at the beginning of the rainy season (2-5). Many flowers then fall onto the water (1), often because of the heavy tropical rain, and provide the first flush of food for the hungry discus (and other fishes). Thereafter the fruits – many still unclassified – ripen as the water rises (10-11). Then fall off when they are ripe, or are "handed out" by animals in the flooded forest, as here at a blue discus habitat in the Lago Cantagallo (Purus), where hundreds of fruits are lying on the surface – the "table is spread" (9). The photo above (5) shows an area of flooded forest in the Rio Negro region with lots of lilac-flowered *tarumã* trees (and a single one, *inset*). Plus detail of the flower of a *Vitex* species, and fruit (6+8).

contributed to the extremely red base colour of some individuals in the Abacaxís region *(p. 175)*.

I have also tried to give an idea of what a wealth of vegetarian food is present during the high-water period, when the majority of medium-sized and small fishes, including all the discus species, can eat their fill and need have practically no fear of piscivores as the latter would have trouble finding them. Leaving aside the fact that a gigantic part of the Amazon basin (plants and trees) is underwater, it is mainly at this time that the plants bear the often-cited flowers and fruits. In addition it is important to realise that in the *igapós* and *lagos* of the black and clearwater zones in the flooded forest regions *(p. 539)* there are up to 50 fruit-bearing trees per hectare that produce massive crops of fruit. Just two examples: *Hevea spruceana* bears on average 55 kg fruits per hectare in a (high-water) season and *Eschweilera tenuifolia* can produce as much as 426 kg of fresh fruit.

According to Junk *et al.* (1992), a full analysis of the fruits and seeds consumed by fishes in Amazonia produces the following average values (in terms of the dry weight of a fruit): 36% carbohydrate; 31% fibre; 19% crude protein; 9.5 % crude fat; 3.5 % ash. The protein component was highest in *Hevea* and *Pseudobomax* with 43% apiece, and lowest in *Psidium* with around 11 % protein. In addition it is worth mentioning the evaluation by Junk *et al.* of gut and stomach contents in fish species occurring in black water compared to those caught in white water: almost without exception the fishes from black water had consumed appreciably more fruits and seeds than those from white water. Of 91 specimens studied, at rising black water 31 had fed on fruits and seeds vs. 18 whitewater fishes, and at high water 40 vs. 24. Only when the water was falling was the figure for whitewater fishes higher. And the growth rate of the fishes was around 60% more rapid at rising water than at falling.

Summary: I have shown that plant material is the second most important nutritional element for discus in the wild – and undoubtedly by far the most important during the high-water period. (See also below under *a summary of the discus diet in the wild,* on page 590, the conclusions about the diet of discus in general.)

3. ALGAE:

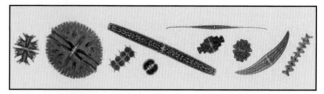

I have found that algae likewise play a part in discus nutrition, albeit again predominantly depending on the season of the year.

But what are algae in actuality?

The true algae are highly variable, photosynthesising plants, without roots, without leafy stems, and without a vascular system. Algae live predominantly in water (fresh and salt) – which covers more than two thirds of the Earth's surface – down to 258 metres of depth. They undoubtedly constitute the major part of primary production on Earth and practically all aquatic life is dependent on their production – including in Amazonia. They are of great importance not only ecologically but also phylogenetically, as it is widely believed that all groups of animals and plants originated in water, as representatives of these ancient evolutionary lineages are still found in fresh and salt water today.

Note that a number of so-called "algae" (the "blue-green algae") are unable to photosynthesise, but are nevertheless classified in the group in everyday parlance, as they are very similar in appearance to the true algae (see Ferrugia, 1993). It is also for this reason that some "algae" are labelled "plants" (true algae) and others "animals" (blue-green "algae").

The algae *(sensu lato,* here and below) are divided, at higher taxonomic levels, into kingdoms, phylum (divisions) and classes of organisms. Of the 34 recognised phylum (Hoek *et al.,* 1993), I have found five that occur in Amazonian waters (but not only there) and are of importance in discus nutrition. These are listed below in alphabetical order, along with further details of each individual group. Unfortunately so far neither I nor anyone else has been able to make any concrete statement as to which species of algae the various discus prefer (or, indeed if they do have any preference). It is also virtually impossible to quantify the amounts of algae that they consume in the course of a year. But I have found that in the dry season discus consume appreciably more algae than during the high-water period. And in the case of some species I have established (on the basis of observations in the wild and stomach and gut contents analysis using specimens preserved in the field) on what phylum (class/order/family/genus/species) they feed. And in some cases the percentage of algae in the stomach contents. The nutritional values (pigments and energy reserves) of each phylum (or class) of algae are given at the end of the relevant section.

Two green algae blooms in the Lago Grande de Manacapuru. *Top:* in the habitat where I found that blue discus *(Symphysodon haraldi)* had eaten a lot of it during rising water (at the end of January) (see text on the following pages). *Bottom:* here the enormous biomass of these green algae can be seen in detail.

This is just one result of years of research into the algae-feeding behaviour of discus and the first publication on the subject.

The "algae" found to play a role in discus nutrition:
– Phylum **Chlorophyta**: classes Chlorophyceae, Zygnematophyceae, generally known as green algae.
– Phylum **Cyanophyta** (=Cyanobacteria): class Cyanophyceae, generally known as blue-green "algae".
– Phylum **Euglenophyta**: class Euglenophyceae, generally known as thread algae.
– Phylum **Heterokontophyta**: classes Bacillariophyceae (= diatoms), Chrysophyceae, Xanthophyceae, generally known as brown algae.
– Phylum **Rhodophyta**: class Florideophyceae, generally known as "red algae".

The **Chlorophyta**, generally known as **green algae**

This phylum (division) consists of photosynthetic organisms, which are largely found in freshwater, some in marine habitats and a few types are terrestrial, occurring on moist soil, on the trunks of trees, on moist rocks, and even in snowbanks. Various species are highly specialized, some living exclusively on turtles, sloths, or within the gill mantles of marine mollusks. The various species can be unicellular, multicellular, coenocytic (having more than one nucleus in a cell), or colonial. Those that are motile have two apical or subapical flagella.

It is generally accepted that early chlorophytes gave rise to the plants. Cells of the Chlorophyta contain organelles called chloroplasts in which photosynthesis occurs; the photosynthetic pigments chlorophyll *a* and chlorophyll *b*, and various carotenoids, are the same as those found in plants and are found in similar proportions. Chlorophytes store their food in the form of starch in plastids and, in many, the cell walls consist of cellulose. Unlike in plants, there is no differentiation into specialized tissues among members of the division, even though the body, or thallus, may consist of several different kinds of cells. The Chlorophyta are divided into in 11 classes and some 18 orders, around 550 genera, and 8,000-10,000 species. The species of three classes (Chlorophyceae, Charophyceae, Zygnematophyceae) are restricted almost exclusively to freshwater habitats. In Amazonia I have to date found two of them in the discus food chain: the Chlorophyceae and the Zygnematophyceae.

Chlorophyceae

Heckel discus, browns, blues, and greens feed on them.

Within this class some four orders, 355 genera and 2,650 species are recognised, by far the majority of them living in freshwater habitats, but with a few in brackish water and terrestrial. They come in a wide variety of shapes and forms, including free-swimming unicellular species, colonies, non-flagellate unicells, filaments, and more. They also reproduce in a variety of ways, though all have a haploid life-cycle, in which only the zygote cell is diploid. The zygote will often serve as a resting spore, able to lie dormant though potentially damaging environmental changes such as dessication. I have repeatedly found species of the order **Chlorococcales** (and lately also species of the order **Volvocales** – see below) in discus stomachs and habitats. The cells of Chlorococcales reproduce by spores and may be solitary or in non-filamentous colonies. Some 215 genera and 1,000 species are known, mainly living in fresh water, but also terrestrial, for example on the sides of trees, rocks, and walls exposed to wind and rain.

In the stomachs of Heckel discus from the Rio Negro I have several times found the crescent-shaped cells of algae of the genus *Kirchneriella* (probably *K. fenestrata,* family Oocystaceae), as well as *Dictyosphaerium pulchellum* (family Dictyosphaeriaceae). Apropos of which it should be noted that 27 green algae species have been found in the lower Rio Negro drainage alone (Rai & Hill, 1980) and Heckel discus undoubtedly also consume other species. I also found another member of this order, the species *Pediastrum duplex* (family Hydrodictyaceae) in the stomachs of three large browns and five blue discus from the lower Arapiuns and Tapajós drainages. *Pediastrum* colonies are circular, flat, and radial in structure *(see p. 543).* Each of the eight discus had consumed a not inconsiderable amount (around 20% of the stomach contents) at the start of the dry period (July). In addition, when the water level was falling I found *Sphaerocystis schroeteri* (family Palmellopsidaceae) and *Treubaria crassipina* (family Oocystaceae – or Treubariaceae according to some authors) in the stomachs of six blue discus in the Rio and Lago Juma area. However in this case these green algae species, which are widespread in Amazonia and undoubtedly also consumed by discus elsewhere, constituted on average only 15% of stomach contents. In the stomachs of nine green discus from the Lago dos Guedes at the height of the dry season I found a whole series of green algae occupying up to 30% of stomach contents: *Ankistrodesmus falcatus* and *A. fusiformis* (family Oocystaceae); *Botryococcus braunii* and *Dimorphococcus lunatus* (family Dictyosphaeriaceae); *Coelastrum cambricum, Scenedesmus acuminatus,* and *S. quadricauda* (family Scenedesmaceae).

Taxonomic note: some assign *Pedistrum* to a separate order Sphaeropleales and *Sphaerocystis schroeteri* to the order Tetrasporales.

Zygnematophyceae

Heckel discus, blues, and greens feed on these green algae.

This class comprises two orders (Desmidiales; Zygnematales) with around 50 genera and about 6,000 species, which live exclusively in fresh water (only the genus *Spirogyra* has also been found in slightly brackish waters). They are unicellular or multicellular.

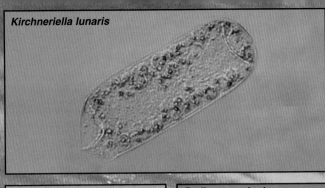
Kirchneriella lunaris

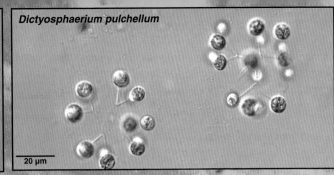
Dictyosphaerium pulchellum

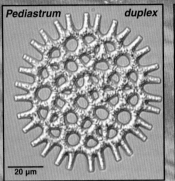
Pediastrum duplex

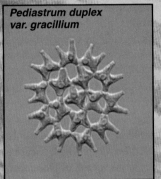
Pediastrum duplex var. gracillium

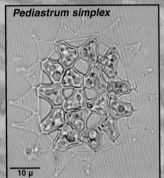
Pediastrum simplex

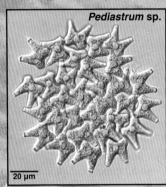
Pediastrum sp.

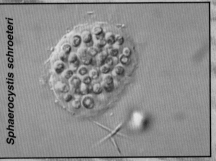
Sphaerocystis schroeteri

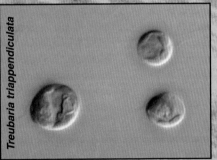
Treubaria triappendiculata

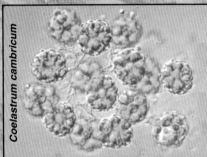
Coelastrum cambricum

Ankistrodesmus falcatus

Ankistrodesmus fusiformis

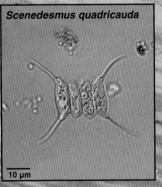
Scenedesmus quadricauda

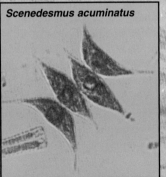
Scenedesmus acuminatus

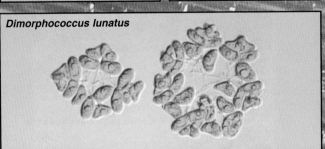
Dimorphococcus lunatus

Botryococcus braunii

CHLOROPHYTA – GREEN ALGAE

The **Desmidiales**, with around 30 genera and some 5,000 species, usually populate still waters with a low pH (between 4 and 7). The normally unicellular and usually solitary Desmidiales (each with a single chloroplast) are probably derived from the multicellular Zygnematales *(see below)*. Only a few genera form loose aggregations of algal strands.

More than 20 years ago 313 species of Desmidiales with very interesting behaviour were already known only from Amazonia (Uherkovich, 1984): they have adapted to low light conditions and are always found beneath shaded banks (discus habitats), well away from strong light, during both the dry and the high-water periods. They are particularly attractive algae, and this is reflected in their apt German names of *Zieralgen* (ornamental algae) and *Schmuckalgen* (decorative algae).

I have found *Closterium kuetzingii* (family Desmidiaceae) in the stomachs of Heckel discus from opposite Barcelos (Rio Negro) in January (still the dry period). The algae component in the stomach of six adult specimens was around 10%. *Closterium* species are, moreover, specially adapted to extremely soft waters and consist of a single curved cell *(right)* with a point at each end, a round cross-section, and no isthmus. I also found a number of desmids (a term for members of the Desmidiaceae) in the stomachs of 10 discus from the Alenquer region: *Hyalotheca dissiliens, Staurastrum hystrix,* and *S. quadrinotatum* representing somewhat more than 10% of the total contents. There was an algal "bloom" at the time (extreme low water). The same applied on one occasion at rising water (January) in the Lago Manacapuru *(see p. 541)* where I netted 15 blue discus with a considerable stomach contents component of *Gonatozygon pilosum* (family Mesotaeniaceae). And in the stomachs of eight greens from the lakes at the mouth of the Juruá I found more than 20% of the contents to be the Desmidiales *Bambusia brebissonii, Cosmarium contractum, C. margaritiferum, Euastrum evolutum, Micrasterias truncata, M. rotata,* and *Staurastrum brachiatum,* and a member of the order Volvocales (family Volvocaceae, class Chlorophyceae): *Eudorina elegans.* This was my record number of species *(see p. 545)*. On one occasion in the Lago Catuá I found a green specimen with a *Xanthidium* species (family Desmidiaceae) in the stomach and gut.

Note: Because the Desmidiaceae are micro-algae and very small (5-500 μm) they are also a basic food for lower animals such as amoebae, water fleas, rotifers *(see p. 545),* and many others.

The **Zygnematales**, with some 18 genera and more than 900 species are found almost exclusively in fresh water (to date only *Spirogyra* has been found in slightly brackish water) with a pH in the range 6 to 8. They take the form of simple, unforked threads or single cells. The order could well be classed as thread algae as they form colonies with almost infinitely long threads.

The genus *Mougeotia* (family Zygnemataceae), 125 freshwater species of which are known, is noteworthy. Each cell has a plate-like axial chloroplast with Pyrenoiden, and the chloroplast can vary its position according to the light: in low light it positions itself with its surface at 90 degrees, and in strong light parallel, to the direction of the incident light. And it is for this reason that the species of the genus are often present in discus habitats. Many years ago, in the Lago Cuipeua near Alenquer, I found 11 *S. haraldi* in which somewhat less than 10% of stomach contents was filled with these algae. But *Mougeotia* species are widespread and undoubtedly provide discus and many other fishes with food in numerous Amazonian habitats during the dry period (from which my data derive), and also when the water is receding, when I have repeatedly found them in places frequented by discus.

I have found the genus *Spirogyra* (family Zygnemataceae) in the stomachs of brown discus from the Tocantins drainage. I got the impression that they sucked these green algae from the submerged leaves to which they were attached. This was at rising water in the month of December. But I have no data on the percentage of stomach contents involved. *Note:* The green algae include the *Caulerpa* species, one of which, *C. taxifolia* from the Indo-Pacific, is notorious in that it invaded the Mediterranean in 1984 (via aquarium water from the Oceanographic Institute in Monaco). No-one had any idea that this tropical algae species could survive in the Mediterranean, but it soon reached plague proportions in many places. It has adapted to Mediterranean temperatures and has no enemies there (see Amsler, 1993).

Taxonomic notes: Some authors regard the Chlorophyta as including only the class Chlorophyceae with 21 orders, others divide the Chlorophyta in 8 classes with the class Chlorophyceae subdivided into 23 orders. Also some do not recognize the class Zygnematophyceae and ascribe the order Zygnematales to Chlorophyceae with the single familiy Zygnemataceae. Again others ascribe the class Zygnematophyceae to the phylum Charophyta and include in the latter the desmids. Others assign the Desmidiaceae and order Desmidiales to the class Conjugatophyceae on the basis of their mode of sexual reproduction (known as conjugation), and the Conjugatophyceae to the Charophyta. Moreover, in the confusion that is algae systematics the class Charophyceae with several orders (including the Zygnematales) is also assigned to the Charophyta. The higher-level taxonomy used in this work is based on the work and systematic divisions of Hoek *et al.*, 1993.

Nutritional value of the Chlorophyta:

The most important pigments are:
- Chlorophylls: chlorophyll a and chlorophyll b;
- Carotenes: β-carotene;
- Xanthophylls: lutein; violaxanthene; neoxanthene.

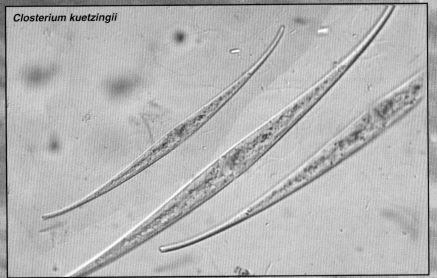
Closterium kuetzingii

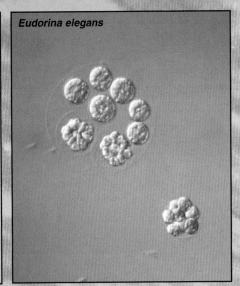
Eudorina elegans

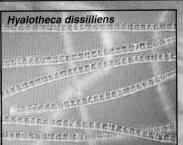
Hyalotheca dissiiliens

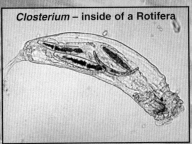
Closterium – inside of a Rotifera

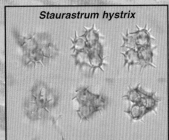
Staurastrum hystrix

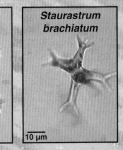
Staurastrum brachiatum

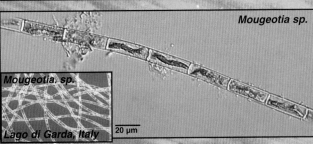
Mougeotia sp.
Mougeotia. sp.
Lago di Garda, Italy

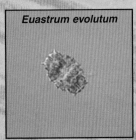
Euastrum evolutum

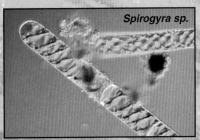
Spirogyra sp.

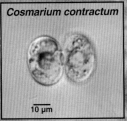
Cosmarium contractum

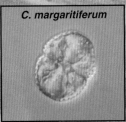
C. margaritiferum

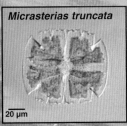
Micrasterias truncata

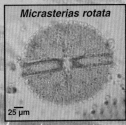
Micrasterias rotata

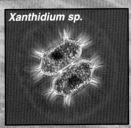
Xanthidium sp.

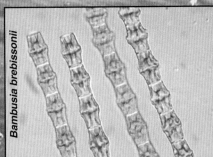
Bambusia brebissonii

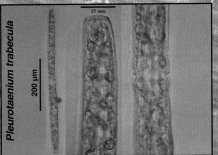
Pleurotaenium trabecula

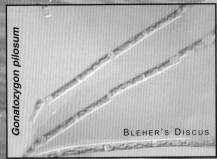
Gonatozygon pilosum

CHLOROPHYTA – GREEN ALGAE

BLEHER'S DISCUS   545

Other pigments:
- Chlorophylls: chlorophyll c_1 ; c_2 ; c_3 ;
- Carotenes: α-carotene; τ-carotene;
- Xanthophylls: zeaxanthene; echinenon; β-cryptoxanthin; antheraxanthin; siphonein; siphonoxanthene.

The most important reserve material is:
- Starch (starch compounds – α-1.4 glucan).

Cyanophyta (= Cyanobacteria), generally known as blue-green "algae"

The life-forms popularly known as blue-green "algae" are actually blue bacteria (= Cyanobacteria). They are among the most primitive of the "algae", as they are organisms without cell nuclei, which are assigned to the Eubacteria (and, together with the Archaebacteria, belong to the Prokaryota). All the other algae (and all other living organisms) with so-called cell organelles are assigned to the Eukaryota. Cyanobacteria were the dominant life on Earth during the long Proterozoic period, which lasted for almost two billion years, and for this reason this time is also known as the "age of the Cyanobacteria". They gradually increased the oxygen content of the Earth's atmosphere and thus paved the way for the obligate aerobic bacteria and the Eukaryota in particular. And some 1.9 billion years ago the first, probably microscopically small, unicellular Eukaryota appeared (following which their evolutionary branching followed a billion years ago).

There are unicellular, colonial, thread-like, and simple parenchymatous blue-green "algae". (The parenchymatous is a type of thallus organization – thalli organized into true tissues composed of several different types of cells. All of the "green plant" groups, e.g. mosses, ferns, gymnosperms, angiosperms, have a parenchymatous construction.) And not one of the blue-green "algae" species possesses flagellated cells at any stage in its life cycle. The Cyanophyceae are the only class within the Cyanophyta.

Cyanophyceae

Heckel discus, blues, and greens feed on these.

The class comprises five orders, more than 150 genera, and over 2,000 species which live in a huge variety of biotopes: in fresh and in salt water, on damp soil, in ice and on glaciers, as well as in deserts and hot (up to 70 °C or more) springs *(centre)*. The majority of species are, however, found in fresh water, in standing or only slightly moving eutrophic waters (= discus habitat), and in these places blue-green "algae" are a frequent component of the plankton. I have found that discus feed on members of three orders, the Chroococcales, Nostacales, and Oscillatoriales.

The **Chroococcales** consist of both unicellular blue-green "algae" and those that form or agglomerations of cells. The latter are held together by polysaccharide mucus. Reproduction takes place via cell division, including the formation of nanocysts, as well as by budding. They occur in salt and fresh water, on soft mud, sand flats, damp soil and gravel, mountain rocks that regularly dry up, on rocks in fast-flowing streams, in pools and freshwater lakes (discus habitats in Amazonia). Some 135 species are known.

Top: blue-green "algae" bloom in the 60 °C Paralana Hot Springs in Australia. *Above: Microcystisaeruginosa* bloom in Europe.

In the Tapajós drainage I found *Microcystis aeruginosa* (family Chroococcaceae) along with *Oscillatoria limosa* (order Oscillatoriales, family Oscillatoriaceae) in the stomachs of two large browns (greyish individuals of almost 22 cm TL). Some 20% of their stomach contents was filled with these blue-green "algae". In the stomachs of greens in the Tefé region (Lago Caiambé) I found two blue-green "algae" genera: *Snowella* (synonym *Gomphosphaeria)*, probably *S. lacustris* (family Chroococcaceae), and *Microcystis* (almost certainly *M. aeruginosa)*. And the quantity in their stomachs was not inconsiderable. In addition, in another individual from the same area I found a considerable amount of an unknown *Aphanothece* species (family Chroococcaceae), cylindrical in form and with large numbers of cells aggregated into gelatinous colonies. And on one occasion, during rising water in February 2004, a gigantic blue-green "algae" bloom, such as I have never otherwise seen in discus waters of Amazonia, was visible in the lower Tapajós *(right)*. This sort of thing normally occurs in the dry period (when I have repeatedly found the majority of discus there to have blue-green "algae" residues in their stomachs). The two species *Microcystis aeruginosa* and *Aphanizomenon flosaquae* (family Nostocaceae) were almost certainly involved, and the bloom was caused by the pollution (poisoning) of the waters by gold-miners in the Itaituba region.

Taxonomic note: The genus *Snowella* is assigned by some authors to the family Merismopediaceae.

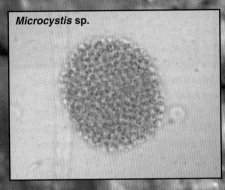
Microcystis sp.

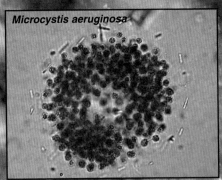
Microcystis aeruginosa

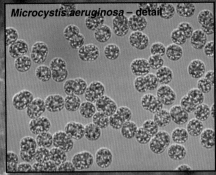
Microcystis aeruginosa – detail

A gigantic poisonous algal bloom in February 2004 in the lower Rio Tapajós and the mouth region of the Arapiuns was almost certainly the blue-green "algae" species *Microcystis aeruginosa* and *Aphanizomenon flosaquae*. Caused by gold prospecting at Itaituba?

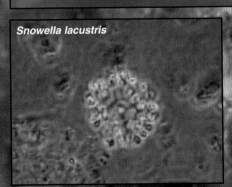
Snowella lacustris

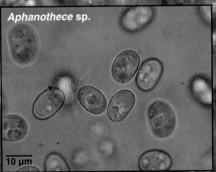
Aphanothece sp.

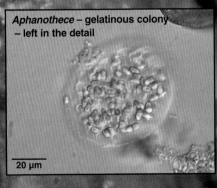
Aphanothece – gelatinous colony – left in the detail

CYANOPHYTA – BLUE-GREEN ALGAE

BLEHER'S DISCUS 547

The order **Nostacales** consists of thread-like blue-green "algae" which reproduce via hormogonium (portion of a filament between heterocysts; detaches as a reproductive body). They occur almost exclusively in fresh water – a very small number of species also on damp soil, on limestone and rock surfaces (where they cause ink-like black streaks in the direction of the water flow), or on the sea-coast. A number of the roughly 235 species are symbionts of other plants.

In the Trombetas region I have seen semi-adult blue discus pecking at *Azolla cf. microphylla* and established that blue-green "algae" were living as symbionts in the leaf cavity of this aquatic fern and were eaten by the discus. They may have been an *Anabaena* species *(A. hassalii* or *A. azollae)* of the family Nostocaceae. And in the Rio Tefé, during the dry season (July-August 1986), I found that up to 20% of the stomach capacity of 10 green discus *(S. aequifasciatus)* was filled with blue-green "algae". In this cas *Aphanizomenon flosaquae* (family Nostocaceae). And two greens examined from the Urucu drainage had again consumed blue-green "algae" of the genus *Anabaena*. Here too I found them in the leaf cavity of an aquatic fern *(Azolla filiculoides) (see p. 549)*. And in the Mamirauá region, between the Rios Solimões and Japurá, I have often caught green discus, which had eaten three blue-green "algae" species of the genus *Anabaena (A. hassalii, A. spiroides,* and one unknown to me) plus *Aphanizomenon flosaquae* (all family Nostocaceae).

Oscillatoriales are thread-like "algae" and reproduce via hormogonium. As in Nostacales, the ramifications in this order are only pseudo-ramifications. Akinetes (thick-walled resting cells) and heterocysts (clear, thick-walled cells occurring at intervals along the filament) are absent. The more than 200 species occur in salt, brackish, and fresh water, in hot springs, waste water, and on muddy sand that regularly dries out. I have found species the genus *Oscillatoria (O. sancta, O. tenuis,* and *O. terebriformis* – family Oscillatoriaceae) in the stomachs of five blues in the Amanã region. The amount of blue-green "algae" species must have been almost 20% of the stomach and gut contents. At any rate 15-20%.

The genus *Spirulina* (family Oscillatoriaceae), a familiar name to aquarists and discus enthusiasts, also belongs to this group. More than 10 species occur mainly in lakes in Peru and Chile, in Australia, in Myanmar, and from the Sahara to East Africa. For more than two decades extensive research has taken place in *Spirulina* "farms" worldwide with a view to making greater use of this cultivated blue-green "algae" genus not only for animals but also for human consumption. Scientists at the Osaka Institute of Public Health in Japan have discovered that *Spirulina* stimulates the activation of natural anti-cancer substances in the body. This was made known at the 30th annual meeting of the Japanese Society for Immunology in November 2000. The first American company to introduce *Spirulina* as a natural foodstuff was Earthrise Nutritionals (=Earth Food)in 1979, thus heralding the "blue-green revolution". In 1982 they produced 1200 tonnes, and nowadays they market *Spirulina* in more than 30 countries. They breed these blue-green "algae" under the most stringently controlled conditions and are the only "farm" in the ISO 9001 Quality Management System. This micro-"alga" contains 60% protein and a very high β-carotene component, its iron content is supposedly equal to that of meat, and it contains vitamin B12 and the rare fatty acid GLA (Gamma-Linolenic Acid). And scientific research confirms that constant intake results in improved health. (This is undoubtedly also why wild-caught discus are much more resistant than tank-breds.) It is even said that the sulfolipids in this blue-green "alga" are active against the AIDS virus.

Nowadays there are around three million pages on the Internet where you can read about *Spirulina* and order products manufactured mainly from the species *Spirulina platensis* (often assigned to the genus *Arthrospira)* and *S. maxima.*

Taxonomic note: some authors assign the family Oscillatoriaceae to the Nostacales and others assign the genus *Oscillatoria* to the family Microcystaceae.

Note: Certain strains of *Anabaena flosaquae, Aphanizomenon* species such as *A. flosaquae,* and *Microcystis aeruginosa* (of which there is only strain) are the cause of toxic "blooms" in water, including in the aquarium, and are toxic. These toxins can cause necrosis and liver haemorrhage.

Nutritional value of the Cyanophyta:

The most important pigments are:
– Chlorophylls: chlorophyll a;
– Carotenes: β -carotene;
– Phycobilins: altophycocyanin; phycobilin; phycocyanin; phycoerythin;

Top: Earth Food in California is the largest Spirulina farm in the world.
Above: *Spirulina* is now also processed into food for ornamental fishes.

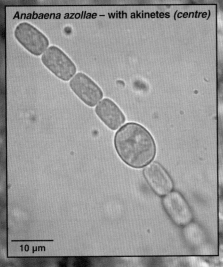
Anabaena azollae – with akinetes *(centre)*

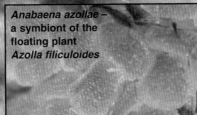
Anabaena azollae – a symbiont of the floating plant *Azolla filiculoides*

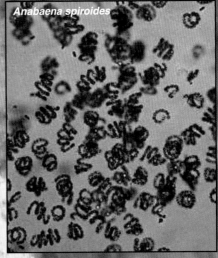
Anabaena spiroides

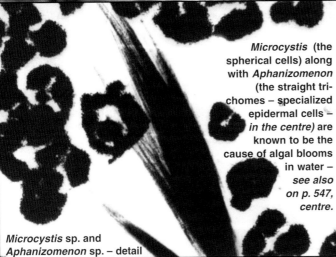
Microcystis (the spherical cells) along with *Aphanizomenon* (the straight trichomes – specialized epidermal cells – *in the centre*) are known to be the cause of algal blooms in water – *see also on p. 547, centre.*

Microcystis sp. and *Aphanizomenon* sp. – detail

Aphanizomenon flosaquae — 30 micron

Aphanizomenon sp.

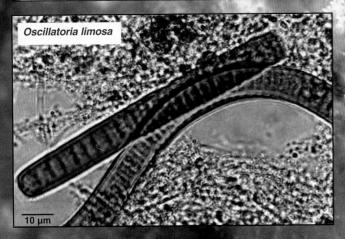
Oscillatoria limosa

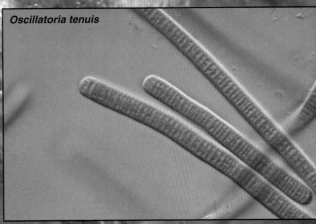
Oscillatoria tenuis

CYANOPHYTA – BLUE-GREEN ALGAE

BLEHER'S DISCUS 549

– Xanthophylls: canthaxanthene; echinenon; myxoxanthophyll; oscillaxanthene; zeaxanthene;
Other xanthophylls:
– β-cryptoxanthene; isocryptoxanthene; mutachrome.
The most important reserve material are:
– Cyanophycine (arginine-asparagine polymer); Cyanophycine starches (starch-like compounds – α-1.4 glucan).

The **Euglenophyta**, also known as "**thread algae**"
The vast majority of the Euglenophyta are unicellular free-swimming flagellate algae. The outer part of the cell consists of a firm but flexible layer called a pellicle, or periplast, which cannot properly be considered a cell wall. Reproduction occurs by longitudinal cell division. The species of this small phylum are placed in a single class:

The Euglenophyceae
Greens, browns, and blues feed on these flagellated algae.

The class is divided into six orders with some 40 genera and more than 800 species. The majority occur in fresh water, a number in brackish water, in the sea, or in the ground. They often predominate in fresh water, especially in eutrophic waters. If present in large numbers they can cause an algal bloom *(centre, in the Rio Tefé)*. Although around a third of the members of the class posses green chloroplasts and are capable of photosynthesis, there is a strong tendency towards heterotrophy within the class. And all the unpigmented Euglenophyceae rely entirely on heterotrophy for their nutrition (can ingest or absorb their food, which is stored as a polysaccharide, paramylon). The majority of its heterotrophic forms are saprotrophic (obtain their nutrients from dead organic matter), but a number are phagotrophic (forms with a special apparatus for capturing algae, bacteria, yeast cells, etc. and a mouth (cytostoma) for ingesting them). I found one order and family in the stomachs of discus.

To date more than a hundred taxa within the order **Euglenales** and the family Euglenaceae have been found in Amazonia alone. They are commonplace in the *lagos* of the clear- and black-water regions of central Amazonia. In addition a number of small crustaceans harbour species of Euglenales. And I have seen them cause algal blooms where dirty waste water empties into the Lago Tefé. I have found a total of 10 species in the stomachs of green discus.

On one occasion I even found several *Euglena* species in green

discus in the Mamirauá region. In the guts and stomachs of seven adult individuals I found the species *E. acus*, *E. oxyuris*, *E. pusilla* and *E. spirogyra*, and these must have represented almost 20% of their stomach contents. *Euglena* species swim with a flowing motion with the aid of their flagellae. In *Euglena* species it is possible clearly to see, at only 100x magnification, the orange-red ocellus which gives the Euglenophyta their name: Greek *euglena* = with lovely eyes. This single, locomotory flagellum is a photosensitive "eyespot" (stigma), which ads the organism in orienting itself towards light.

Phacus longicauda, known in German as the *Herzflagellat* (heart flagellate) has flattened cells which are completely rigid. Posteriorly it bears a terminal spine almost as long as the body of the cell. This alga possesses numerous chloroplast discs. The cells are in itself slightly wound, so that the organism swims with a gently rotating motion. A single alga is on average 85 to 115 μm long and the species' habitat is generally limited to still and slow-flowing waters. To date 10 *Phacus* species are known from central Amazonia alone. I have found *P. pleuronectes*, *P. suecicus*, and *P. longicauda* in the stomachs of various green discus, but they undoubtedly also feed on other species. Likewise *Trachelomonas*, 90 species of which occur in the Lago Camaleão (Confronti, 1993). I have found *T. armata*, *T. hispida*, and *T. volvocina*, as well as an undetermined species, in the stomachs of various green, brown, and blue discus – up to 20% of its contents.

I have found the colourles *Astasia klebsii* in the waste water from the town of Tefé but not in the stomachs of discus. It lives predominantly in heavily polluted waters and is undoubtedly of little interest to discus as food.

Taxonomic note: The Euglenophyta are sometimes known as the Euglenophycota and some authors assign these flagellates to the class Flagellata, phylum Mastigophora. The defining character of the flagellae is that they serve as locomotory organs.

Nutritional value of the Euglenophyta:
The most important pigments are:
– Chlorophylls: chlorophyll a and chlorophyll b;
– Carotenes: β-carotene;
– Xanthophylls: neoxanthene;
Other pigments:

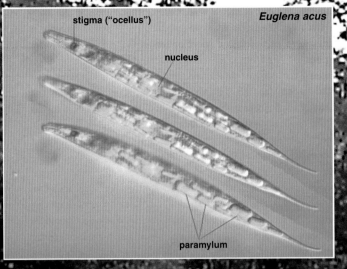

Euglena acus
- stigma ("ocellus")
- nucleus
- paramylum

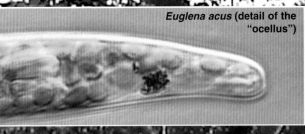

Euglena acus (detail of the "ocellus")

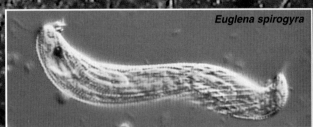

Euglena spirogyra

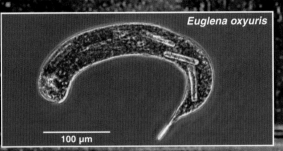

Euglena oxyuris
100 µm

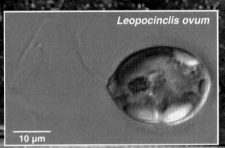

Leopocinclis ovum
10 µm

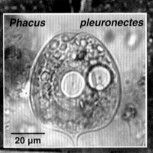

Phacus pleuronectes
20 µm

Phacus longicauda

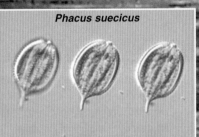

Phacus suecicus

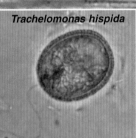

Trachelomonas hispida

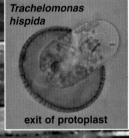

Trachelomonas hispida
exit of protoplast

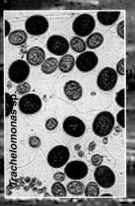

Trachelomonas sp.

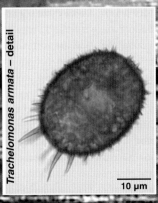

Trachelomonas armata – detail
10 µm

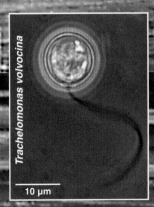

Trachelomonas volvocina
10 µm

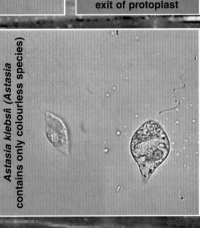

Astasia klebsii (*Astasia* contains only colourless species)

EUGLENOPHYTA

– Carotenes: τ - carotene;
– Xanthophylls: zeaxanthene; echinenon; β-cryptoxanthin; diatoxanthene; diadinoxanthene;

The most important reserve material is:
– Paramylon $(\beta$-1.3 glucane).

The **Heterokontophyta,** generally known as **"brown algae"**.

This phylum includes at least nine classes and it is amazing how precisely its highly complex electro-microscopic structures coincide, which characterizes the Heterokontophyta primarily. The majority of the classes are found predominantly in fresh water, and I have found two in the stomachs of discus, the Bacillariophyceae and the Chrysophyceae.

The **Bacillariophyceae** (= diatoms) Heckel discus, greens, browns, and blues feed on diatoms.

The diatoms are unicellular or colony-forming algae. Each cell is enclosed by a two-part silicified wall which resembles a petri dish with its overlapping lid and is termed the frustulum (or frustule). The class contains two orders – although this is a matter of dispute and opinions differ greatly *(see below)*. As mentioned earlier, I have followed van den Hoek *et al.* (1993) and their classification into some 250 genera and more than 100,000 species, which live in the sea and in fresh water, as well as on damp and dry ground. I have found both of the two orders within the class, the Pennales and the Centrales, in the stomachs of discus.

The **Pennales** are an order of predominantly benthic forms. The oldest known fossil Pennales from freshwater sites originate from the early Tertiary (60 million years ago). The cells are elongate, linear, lanceolate, or oval. The cell wall is set with bilaterally symmetric structures. In some of the Pennales there is a crevice, termed the raphe (longitudinal median line or slit on the diatom valve).

To date I have found more than 12 species in the stomachs and guts of discus. I found an as yet unidentified diatom species of the genus *Pinnularia* (family Pinnulariaceae), which in fresh water usually occurs only in nutrient-poor waters, in the Abacaxís and in the Rio Negro region. In a total of 10 adult Heckel discus from the two regions up to 10% of the stomach and gut contents was *Pinnularia* sp. I found *Tabellaria* (family Tabellariaceae; probably *T. fenestrata),* which is widespread in the stiller, more or less nutrient-rich freshwater regions of Amazonia, in the stomachs of Rio Negro Heckel discus – but only in smaller percentages. And that is typical, as although to date 43 diatom species have been found in the entire central Amazon drainage, in no case was a large biomass of the species present. This also supports my earlier research, where I always found fewer diatoms in the stomachs of discus in comparison to other algae groups.

In the Alenquer region, however, I found a few further species in the stomachs of brown discus: *Nitzschia acicularis* (family Bacillariaceae) and an undetermined *Navicula* species (family Naviculaceae). Both latter genera occur widespread in nutrient-rich (eutrophic) fresh and brackish water, as there are many species of *Nitzschia* and *Navicula* contains more than 350 taxa. The stomach and gut of five adult Alenquer specimens contained less than 10% of these two species. However, in green discus too I have repeatedly come across diatoms of this group. In the Mamirauá, Tefé, and Coari regions, where I found the species *Amphora ovalis* (family Catenulaceae), *Gomphonema augur, G. constrictum* (family Gomphonemataceae), an undetermined *Gyrosigma* species (family Pleurosigmataceae), as well as *Surirella linearis* (family Suriellaceae), *Synedra goulardii var. fluviatilis* and an unidentified *Synedra* species (family Fragilariaceae). I found these only in discus caught in predominantly stagnant waters during the dry season. In all three regions I investigated a total of 15 semi-adult and adult individuals (in each case five per region/collecting site), and the stomach contents of those from *lagos* in the Mamirauá region contained between 5 and 10% diatoms: *A. ovalis, Gomphonema augur,* and *G. constrictum,* and on one occasion *Surirella linearis* as well. In the Urucu (Coari region) green discus I found that around 10% of the stomach contents was diatoms: *Gyrosigma* sp., *S. goulardii var. fluviatilis;* and the stomach contents of specimens from the Lago Tefé included rather more than 10% diatoms: *A. ovalis, Gomphonema constrictum, Gyrosigma* sp., *Surirella linearis,* and *Synedra goulardii var. fluviatilis.*

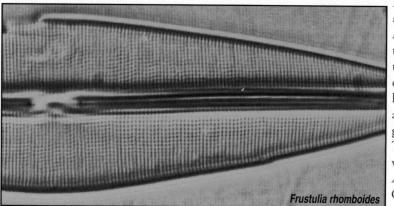

Frustulia rhomboides

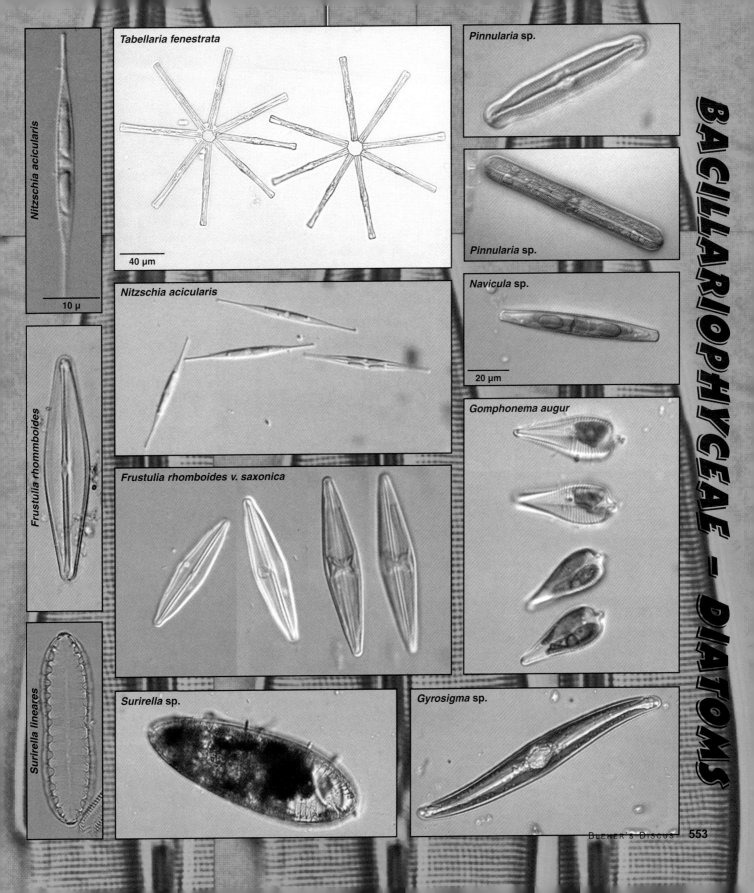

The majority of the above-mentioned taxa, as well as *Frustulia* species (family Naviculaceae) are widespread in central Amazonia. And I have found *Frustulia rhomboides* as well as *F. r.* var. *saxonia* in the blue discus of the Rio Manacapuru. Three adult individuals had both taxa in their stomachs, representing circa 5% of the contents.

The **Centrales** are the oldest fossil diatoms known to date and were present as long ago as the early Cretaceous (120 million years ago). They are often circular to elliptical, but also polygonal with a radial structure emanating from a central point. But the wealth of forms within the Centrales is enormous, and they have just one feature in common: They never have a raphe.

I have repeatedly found the following Centrales had been eaten by green discus: *Eunotia asterionelloides, E. flexuosa, E. triodon,* and a further *Eunotia* and a *Cyclotella* species (family Melosiraceae) which I have not yet been able to identify. In none of the adult specimens examined did these comprise more than 15% of the stomach and gut contents – but often all five algae species mentioned were present, chiefly in the Tefé, Mamirauá, and Fonte Boa regions and in the *lagos* along the lower Juruá. In addition, on one occasion I found the stomachs of four adult blues and browns from the Canumã drainage, as well as two blues from the Manacapuru region, to contain *Aulacoseira* (synonym *Melosira*) *granulata, A. granulata* var. *angustissima, Rhizosolenia eriensis,* and *R. longiseta* (family Rhizosoleniaceae). These are freshwater algae, very commonplace in the plankton of nutrient-rich Amazonian waters, in *lagos* and habitats with little current. In the stomachs examined the bulk of them consisted of *A. granulata* var. *angustissima* – up to 10% of the stomach contents; the others were present in far smaller percentages or in only some specimens.

In the Tapajós and the Alenquer region I found the stomachs of two and three adult brown individuals respectively to contain the species *Actinella brasiliensis, E. diodon,* and *E.* cf. *lunaris,* as well as *Aulacoseira granulata* (family Melosiraceae). Here too *Aulacoseira granulata* was dominant in the stomach contents.

But my researches have been unable to throw much light on the quantities of the individual species present in Amazonia, and it would be almost impossible to make any general statement on this because of the variations in water level.

Note: Considerably more species of the Bacillariophyceae are present in Amazonia. Four taxa have been found in the Lago Castanho, not far from Manaus (Uherkovich & Schmidt, 1974) – in just one lake. Undoubtedly discus also eat many other algae species from this group.

Taxonomic notes: In more recent overviews the diatoms are assigned to the phylum Bacillariophyta and subdivided into three (or more) classes: the Pennales with no raphe are assigned to the class Fragilariophyceae; the Pennales with a raphe to the class Bacillariophyceae; and the Centrales to the class Coscinodiscophyceae. In addition the genera are variously assigned, for example *Tabellaria* to the order Tabellariales, in the class Fragilariophyceae and the phylum Ochrophyta. Some authors also assign the genus *Synedra* to this taxon and the order Fragilariales. *Eunotia* is variously assigned by some authors to the orders Pennales or Eunotiales. *Rhizosolenia* is assigned to the order Rhizosoleniales, in the class Coscinodiscophyceae and the phylum Ochrophyta; likewise *Cyclotella* and *Aulacoseira* are assigned to the order Melosirales, and *Amphora* to the Thalassiophysales, within the last-named class.

Nutritional value of the Bacillariophyceae:
The most important pigments are:
– Chlorophylls: chlorophyll a;
 chlorophyll c_1 ; c_2 ; c_3 ;
– Carotenes: β-carotene;
– Xanthophylls: diatoxanthin;
 diadinoxanthene; vaucheriaxanthene; heteroxanthene;
 Other pigments:
– Carotenes: ε- carotene;
– Xanthophylls: zeaxanthene; β-cryptoxanthin; neoxanthene;
 The most important reserve material is:
– Chrysolaminaran (β-1.3 glucan).

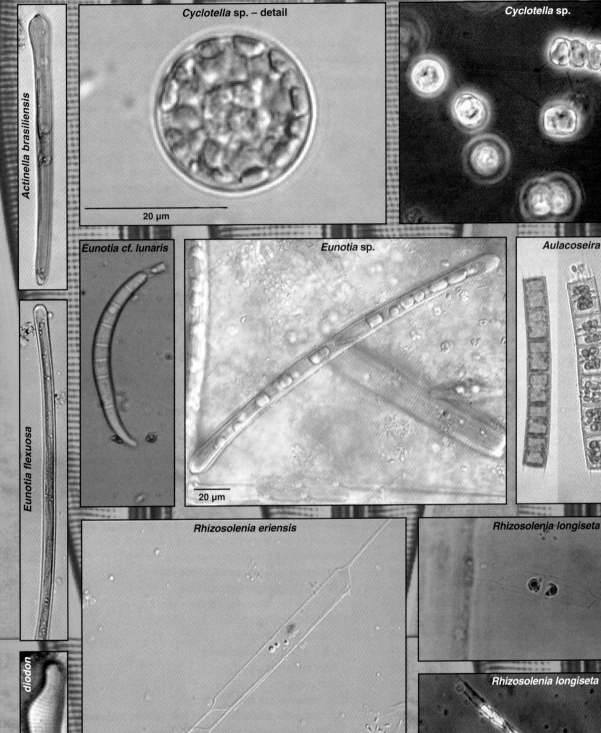

The Chrysophyceae

Green and blue discus feed on the so-called "**golden algae**".

The name of this class derives from the golden yellow to brown colour of the cells of its species (from Greek *chrysos* = gold). The golden or gold-brown algae comprise some 200 genera and the number of species is estimated at more than 1,000. The greater part of this wealth of forms are found in fresh water, but a number of species also occur in brackish and sea water. They predominate in oligotrophic lakes with acid to neutral water (pH 5 to 7.5) and are encountered chiefly in the habitat of green, brown, and blue discus. Three orders are recognised, and I have found two of them relevant to discus.

The **Ochromonadales** have two uneven (monadoide) flagella of uneven length. Some of these unicellular algae have loricae or shells, such as *Dinobryon* (family Dinobryaceae), which is sessile and grows in branched colonies *(see page 557)*. Three species of this unique group of forms I have found in the stomachs of green discus from the Tefé region: *D. bavaricum*, *D. cylindricum*, and *D. sertularia*. The stomach contents of five adult individuals comprised around 10% of these algae. And I found two species, *D. cylindricum* and *D. sertularia*, in the stomachs of two adult blues in the central Purus region (Lago Tambaquí). In these cases the amount was some 8-10% of the stomach and gut contents.

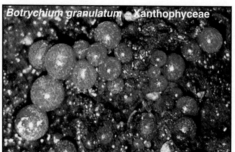
Botrychium granulatum – Xanthophyceae

The **Mallomonadales** are flagellate cells with diatom-scales, and form colonies as well. A typical example is the genus *Synura* (family Synuraceae) whose colonies are round or ellipsoid and consist of pear-shaped cells anchored to one another at the narrow posterior end. *Synura* and related genera such as the unicellular, scaled, flagellate *Mallomonas* (family Synuraceae), likewise found in central Amazonia, differ from other flagellate Chrysophyceae. They contain chlorophyll a and c_1 but no c_2 and they lack the otherwise typical photoreceptor apparatus. These and other differences in the cell structure of the above-mentioned genera are so fundamental that recently a separate class, the Synurophyceae, has been proposed for *Synura* and related genera.

I have found *Mallomonas teilingii*, as well as a second, undetermined, *Mallomonas* species and *Synura uvella*, in the stomachs of discus. In four semi-adult *S. haraldi* from the Manacapuru region I found 10-15% of the stomach contents to consist of these three flagellate species. *Note:* There are supposedly at least six species of a third class, the **Xanthophyceae**, that occur in central Amazonian waters, mainly as freshwater plankton (Putz & Junk *in* Junk (Ed.), 1997). Many species of this class are soil-dwelling algae, which develop rapidly on the moist ground in the inundation forest and undoubtedly provide discus (and other fishes) with a reliable source of food at rising water. These include species like *Botrychium granulatum* and *Botrydiopsis arrhiza*, which I found on the forest floor in the inundation zones and *igapós*. However, I have to date been unable to find them in the stomachs of discus *(centre & p.557)*.

Taxonomic notes: Some recent authors no longer recognise the phylum Heterokontophyta and some elevate its classes to the status of phylum, eg Chrysophyceae = Chlorophyta; or assign the Xanthophyceae to the phylum Ochrophyta. Some also replace the order Ochromonadales with the order Chromulinales. Others separate the class Phaeophyceae (=Fucophyceae) from the phylum Heterokontophyta and place it, along with 11-13 orders, some 250 genera and 1,500 to 2,000 species, in its own division (Phylum), the Phaeophyta.

Remarks to the latter: Phaeophyta is known as brown algae and found almost exclusively in marine environments. Like many red algae they prefer the littoral and sublittoral zones. In clear tropical waters they have even been found at depths of 200 metres. Their main distribution area is the sea-coasts of temperate zones. Brown algae can attain a considerable size *(Macrocystis* and *Nereocystis* up to some 100 m in length); these large forms are known as seaweeds. As a rule they are attached via special holdfasts to solid substrates (rocks, bivalve shells, etc) (benthic existence). A few species (eg *Sargassum*) occur floating (pelagic) in the open surface waters of warm seas (eg the Sargasso Sea). The gametes of the brown algae are generally multiflagellate. By contrast the oocytes and tetrasporangia are – where present – non-flagellate.

Nutritional value of the Chrysophyceae:

The most important pigments are:
- Chlorophylls: chlorophyll a; chlorophyll c_1 ; c_2 ;
  (except family Synuraceae)
- Carotenes: β-carotene;
- Xanthophylls: fucoxanthene;
  Other pigments:
- chlorophyll c_3 ;
- Carotenes: α- carotene; ε- carotene;
- Xanthophylls: zeaxanthene; a-cryptoxanthin; antheraxanthin; violaxanthin; neofucoxanthene; diatoxanthene; diadinoxanthene; neoxanthene;

The most important reserve material is:
- Chrysolaminaran (β-1.3 glucan).

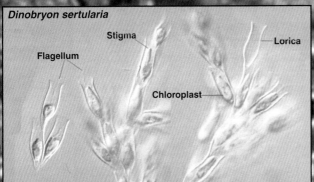
Dinobryon sertularia — Flagellum, Stigma, Chloroplast, Lorica

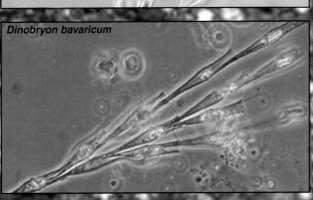
Dinobryon bavaricum

Dinobryon cylindricum
20 µm

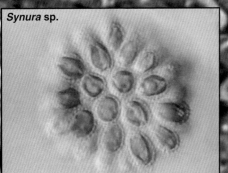
Synura sp.

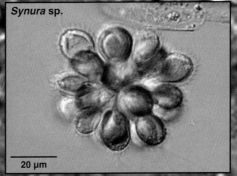
Synura sp.
20 µm

Mallomonas sp.

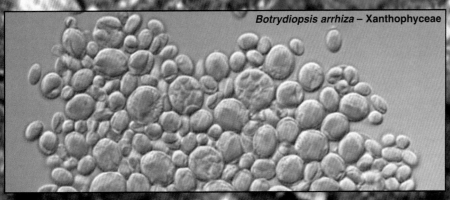
Botrydiopsis arrhiza – Xanthophyceae

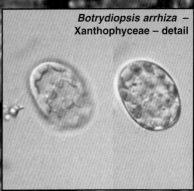
Botrydiopsis arrhiza – Xanthophyceae – detail

CHRYSOPHYCEAE – XANTHOPHYCEAE

BLEHER'S DISCUS

The **Rhodophyta,** generally called "red algae"

The division Rhodophyta includes two classes and 18 orders. The red colour of the thalli derives from phycoerythrin (phycoerythrobilin), and phycocyanin (phycocyanobilin), which is also present, produces a shade of colour verging on bluish *(see page 559)*. Flagellate cells have never been found. The majority of the 5,000 to 5,500 species are marine (and have been found down to depths of 268 m), and only around 150 occur in fresh water. And of the approximately 600 genera only 20 are encountered in fresh water. As a rule red algae are benthic. They usually grow on solid substrates (rocks, bivalve shells, etc) and epiphytic on other algae, sometimes also parasitic on close relatives. I have repeatedly found them on the floor of inundation forests. They attach to substrates via differentiated rhizoid-like cells. Only the class Florideophyceae has so far been found important for discus:

The **Florideophyceae**

Brown and blue discus feed on these red algae

The class contains 13 orders and all the included species consist fundamentally of ramified cell-filaments and these are often united to form more or less complex pseudoparenchymatic thalli. These thalli can be ovate, circular, or flattened (some look like species of the aquarium-plant genus *Myriophyllum – see page 559*). The cells can be mononuclear but are often polynuclear, and the nuclei are often polyploid. I have found species of just two orders in the stomachs of discus.

The orders **Acrochaetiales** and **Batrachospermales** are very distinctive members of the red algae, and two genera of them are particularly to be found in central Amazonia:

Audouinella (Acrochaetiales: family Acrochaetiaceae) and *Batrachospermum* (Batrachospermales: family Rhodophyceae). The former lives as a tiny epiphyte on the latter. Although *Audouinella* species are predominantly marine, a number also live in fresh water, for example *A. violaceae*, and benefit from their association with the genus *Batrachospermum*, which is restricted to fresh water. I have found them both in blue discus in the Uatumã drainage. And the above-mentioned epiphyte was indeed on *B. moniliferum* and four Blues had eaten them. Although there was only a small amount in the stomach contents.

Remarks: Some Florideophyceae species are used directly by humans for food, while cell wall polysaccharides are extracted from others for use as gels, and additives in food and cosmetic products.

Nutritional value of the Rhodophyta:

The most important pigments are:
- Chlorophylls: chlorophyll a;
- Phycobilins: phycocyanin; allophycocyanin; phycoerythin; phycobilins;
- Carotenes: α-carotene; β- carotene;
- Xanthophylls: zeaxanthene;
  Other pigments:
- Xanthophylls: α-cryptoxanthene; β-cryptoxanthene; lutein; antheraxanthin; violaxanthin;

The most important reserve material is:
- Starch (starch-like compounds– α-1.4 glucan).

Summary: Apart from the fact that the biomass represented by algae can be highly variable on account of the huge fluctuations in water level, it is possible to list only a fraction of the algae species that occur in the vast Amazon region and are eaten by discus and other fishes.

To date I have found the species listed above in the gut and stomach contents of a total of 213 discus, and especially during an algal bloom such as I have repeatedly encountered in the Tapajós, the Rio Tefé, and the Manacapuru and Rio Içá regions – and elsewhere. And, of course, during the dry period, when they had eaten larger quantities of algae than in the course of the rising and high water periods.

Among the discus examined (45 *S. discus;* 74 *S. aequifasciatus,* and 94 *S. haraldi)* there were 105 specimens whose gut and stomach contents comprised around 10% algae. In 95 specimens the figure was 15-20%, and 10 discus had consumed as much as approximately 30% algae. Only in three of the specimens was the algae component around 5%.

By comparison, when it came to the algae groups 86 discus had eaten green algae – Chlorophyta, 35 discus blue-green algae – Cyanophyta, 19 discus flagellates – Euglenophyta, 58 discus brown algae – Bacillariophyceae, 11 discus brown algae – Chrysophyceae, and 4 discus red algae – Rhodophyta.

As well as my results, the work of Junk *et al.* (1997) is of interest to anyone curious about the food consumed by other fish species; during the high water period they found algae in the stomachs of 41 out 91 fishes examined, all of them blackwater species. When the water was receding 21 species had consumed algae (but none of them discus, as they did not collect any). And in February 1980, in (or near to) the Urubaxí habitats (Rio Negro), Goulding *et al.* (1988) found algae in *Metynnis* with an average standard length of 142 cm, and these fishes live or travel with Heckel discus. Likewise in many other fishes, but in general smaller species. Algae are thus (almost) the "daily bread" of fishes and without doubt the third most important food source for discus, as well as an important basic element in their nutrition – albeit in variable quantities depending on the time of year.

And not only discus, but also humans have long since learned to appreciate the enormous nutritional value of algae.

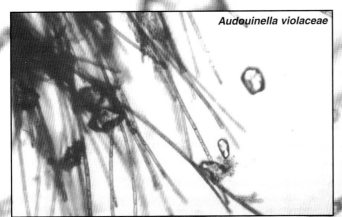
Audouinella violaceae

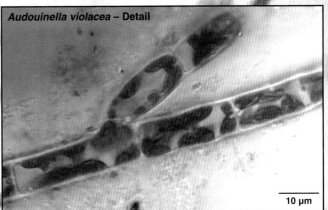
Audouinella violacea – Detail
10 µm

Batrachospermum sp.

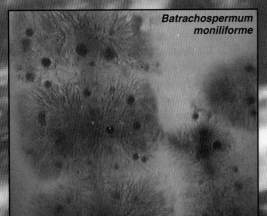
Batrachospermum moniliforme

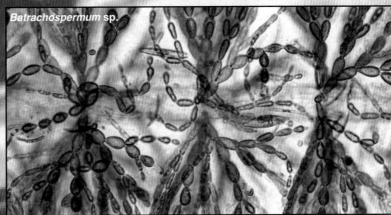
Batrachospermum sp.

RHODOPHYTA - RED ALGAE

4. Aquatic Invertebrates:

This may sound improbable to the discus enthusiast, breeder, dealer, or expert, but discus don't eat many aquatic invertebrates in the wild. This is by virtue of the very nature of things. Discus occur mainly in acid – and nutrient-poor – waters, where invertebrates have less chance of survival than in neutral or alkaline habitats with an appreciably greater food supply. In discus waters aquatic invertebrates must generally feed only on detritus and algae. And because of the huge fluctuations in water level another negative factor comes into play: the likelihood of female and male invertebrates finding one another and a host during the high-water period is reduced to a minimum.

But what are aquatic invertebrates?

An aquatic invertebrates is an animal without a vertebral column and which lives in the water. Generally speaking they are plankton (Greek = that which wanders), communities of organisms with little or no intrinsic locomotory ability, floating in the open water and mostly transported by the current. These planktonic organisms are termed zooplankton – planktonic plants in fresh water are called phytoplankton and the marine plankton is given often different names. The zooplankton of fresh water are termed limnoplankton and consist mainly of unicellular organisms, rotifers, and small crustaceans. And limnoplankton is divided into five groups: eulimno-, heleo-, telmato-, creno-, and potamoplankton. Eulimnoplankton (also called lake plankton) inhabits the open water areas of lakes and is the best adapted to planktonic life. Heleoplankton (or pond plankton) comprises forms that live partly benthic and partly planktonic. The organisms of the telmatoplankton are likewise facultatively planktonic, but live mainly in seasonal pools and tolerate extreme temperature fluctuations. When the water dries up they bury themselves in the mud and aestivate. Specific crenoplankton (spring plankton) and exclusive potamoplankton (river plankton) are unknown.

If planktonic organisms are larger than 5 mm then they are classed as macroplankton, which is found practically only in the sea (eg medusae, larger crustaceans, and jellyfish). Forms measuring 1 to 5 mm in size are classified as mesoplankton (eg the small crustaceans). And organisms ranging from 0.05 to 1 mm in size, barely visible with the naked eye, form the microplankton (eg unicellular organisms, all the rotifers, and some of the small crustaceans). Organisms measuring less than 0.05 mm are termed nannoplankton; they can pass through the finest plankton net and hence can be collected only using centrifuges, filtration, or by letting them set down. By far the majority of the nannoplankton consists of algae, flagellate and other algae together with ciliate organisms, and it isn't generally regarded as belonging to the zooplankton or aquatic invertebrates.

In aquatic invertebrates the main prerequisite for life in water is the reduction of the rate of sinking to a minimum. This is achieved by reducing relative density by storing light-weight substances such as fats, oils, and gases in the organism; by increasing the water content of the tissues; or by enlarging the relative surface area of the organism. The last of these is achieved on the one hand by improving the ratio between the surface area of the body and its volume (the smaller the body, the larger its relative surface area) and on the other by the development of body extensions as in, for example, the small crustaceans.

The transformation of the composition of the plankton in the course of the year is known as cyclomorphosis, and its individual stages as temporal variations. Cyclomorphosis is particularly widespread in water fleas and rotifers and genetically determined.

The horizontal distribution of the plankton is fairly homogenous apart from the formation of "clouds" by the crustaceans and the phenomenon of euplanktonic forms shunning the shoreline. The reason for vertical migrations in the zooplankton is the variation in light intensity during the course of the day. The freshwater crustaceans of oligotrophic lakes (eg *Daphnia* spp.) rise to the surface at night and return to the depths in the morning.

I have often found diverse aquatic invertebrates in the stomachs and guts of discus – as has Efrem Ferreira in other fishes (per. comm.). Below I list the groups, which I have found and which had been eaten. It hasn't been possible to identify every invertebrates species, but usually the genus or family and always the order, class, or phylum to which the species is assigned. As in the previous part of *Discus nutrition in the wild* I will discuss each group only briefly – for detailed information I refer the reader to the often comprehensive scientific literature.

The aquatic invertebrates of Amazonia can be divided into three main groups, the **zooplankton**, the **benthos** (here only benthic animals) and the **perizoon** (Junk & Robertson *in* Junk (Ed.), 1997). The zooplankton consists of the free-swimming aquatic invertebrates. The benthos (Greek, = depth) comprises the faunal communities of the bottom. The perizoon comprises communities found predominantly on rooted and floating plants and on those in inundation-zone woodlands (chiefly in black- and clear-water zones).

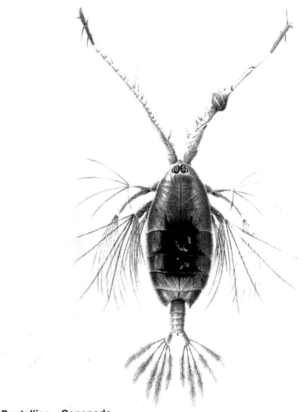

Copilia – Copepoda

Pontellina – Copepoda

Sapphirina – Copepoda

Oncaea – Copepoda

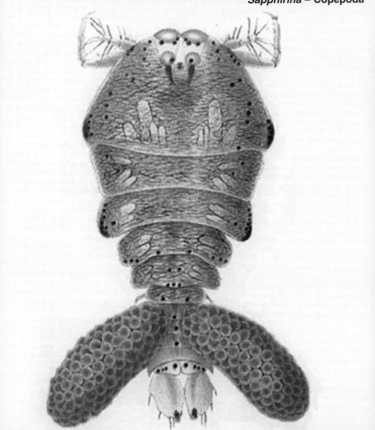
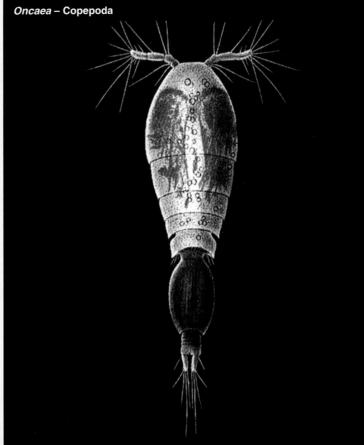

The zooplankton

As long ago as the 1980s it was established (Robertson & Hardy, 1984) that there are at least 250 rotifer species (phylum **Rotifera**), 20 freshwater members of the order **Cladocera** (phylum Arthropoda: subphylum Crustacea; families Macrothricidae and Chydoridae), and around 40 copepods (Crustacea: subclass **Copepoda**; order **Calanoida**) in Amazonia. By now the species count for the Rotifera has risen to more than 300. They also represent the largest percentage of the zooplankton group found in discus stomachs.

The Rotifera (or **Rotatoria**) – rotifers or wheel-animalcules Browns, blues and greens feed on them.

There are supposedly around 2,000 species worldwide, the majority in fresh water. They live sessile, parasitic, or free-swimming, and are around 0.04 to 3.0 mm long. A number of species even live in mosses and also in the sea. And in the dry season they are well able to survive in a dormant stage (Kyroptobiosis). They have a characteristic form: there is a complex masticatory and grasping apparatus (the mastax) on the head in the region of the pharynx, resembling a rotating wheel (hence rotifers, wheel-animalcules, Rotifera, and Rotatoria). It consists of a buccal field (ring of cilia) around the mouth and a circumapical band anteriorly. The "wheel" can take many forms and is used as an identification character. It is used for both food-gathering and locomotion *(centre)*. Under favourable conditions reproduction takes place without fertilisation (parthenogenesis). Under favourable conditions by heterogeny.

On one occasion 152 rotifer species were found among the roots of floating plants in the Tapajós during the dry season (Koste, 1974). There I have also several times found them in discus stomachs or seen them eaten during this period I once found rotifers of the species *Brachionus dolabratus*, *B. alcatus*, and *Keratella americana* in five adult Browns, representing around 10% of their stomach contents. On another occasion around 5% of the gut and stomach contents of four *S. haraldi* again consisted of species of the family Brachionidae (two unidentified *Keratella* species) as well as *Epiphanes macroura* (family Epiphanidae).

And as chance would have it, as I write it is the 6th of January, which in Italy is the day of the *Epifania di Nostro Signore*. This reminded me of the Epiphany festival of the Greek Orthodox Church, celebrating the birth of Jesus again on the 6th January, as the culmination of the Christmas festivities in Greece. Water is blessed so that evil spirits will leave the Earth. The Greek word *epiphaneia* signifies a manifestation or appearance (in this case of Jesus to the Magi) and was also used with regard to Roman emperors: the arrival (appearance) of a ruler, state visit. Although one emperor, Antiochus IV (175-164 BC), the eighth ruler of the kingdom of the Seleucides, adopted the title Theos Epiphanes, = Manifestation of God (Zeus). And his Jewish enemies changed the Greek letter *phi* to *mu* and called him Epimanes (= the crazy one), as he not only captured Jerusalem, destroyed large parts of the holy city and Jewish records, and murdered (in 172 BC) the "Prince of the Realm", the High Priest Onias III (in those days *de facto* the leader of Israel), but also banned the Mosaic rites, the celebration of the Sabbath and circumcision, ordering heathen festivities instead. By 168 BC he had desecrated the Jewish temple in Jerusalem and had erected a focal altar where pigs were sacrificed and a statue of Zeus with the features of Antiochus IV. Theos Epiphanes, the Manifestation of God. In 167 BC Antiochus IV even minted a four drachma piece that may be seen as the epitome of his divine rule and religious politics. On the obverse there is a head of Zeus with the features of Antiochus IV, and on the reverse a Zeus Nikephoros enthroned, with sceptre, Victoriola, and the altar of Zeus, with the inscription "King Antiochus, victorious god in human form" (Theos Epiphanes Nikephoros).

But let us take a further look into discus stomachs, as nobody knows what the German Christian Gottfried Ehrenberg (1795-1876), who in 1828 coined the word *bacterium* (from the Greek *bakterion*= small stick), was thinking of when, in 1832, he chose the name *Epiphanes* in his description (one of around 400) of this member of the zooplankton.

At Alenquer I found an unidentified species of "manifestation" (*Epiphanes* sp.), along with *Brachionus zahniseri* and *Lepadella cristata* (family Lepadellidae), in the stomachs of three large reddish-brown discus, representing around 8-10% of the stomach contents. And in the stomach and gut of two adult blues from the Canumã I found *Lecane bulla* (family Lecanidae) and *Ptygura pendugulata* (family Flosculariidae) albeit not very much. A blue in Maués had eaten a lot of *Polyarthra vulgaris* (family Synchaetidae), almost 5% of the stomach contents. To the west of the Rio Japurá, in *lagos* where I have often found greens, I also discovered a whole series of Rotifera in discus stomachs many years ago – nowadays there are said to be 284 species in the blackwaters of central Amazonia. However, at that time I didn't determine them to species level. On a total of three occasions I have found a number of different species in the stomachs of adult individuals – each time in two specimens that I was conserving. Up to 10% of stomach contents, at the extreme of the dry season.

ROTIFERA

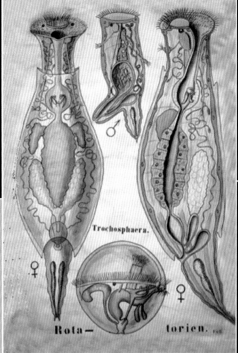

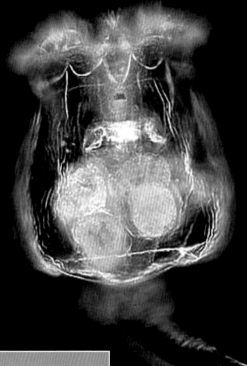

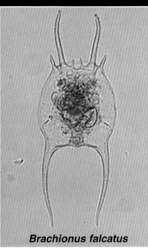
Brachionus falcatus

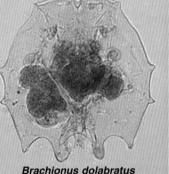
Brachionus dolabratus

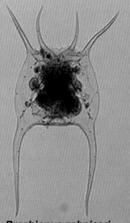
Brachionus zahniseri

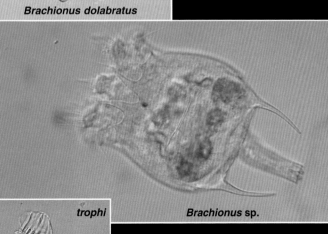
Brachionus sp.

Brachionus sp.

Epiphanes sp. **Epiphanes macroura** *trophi*

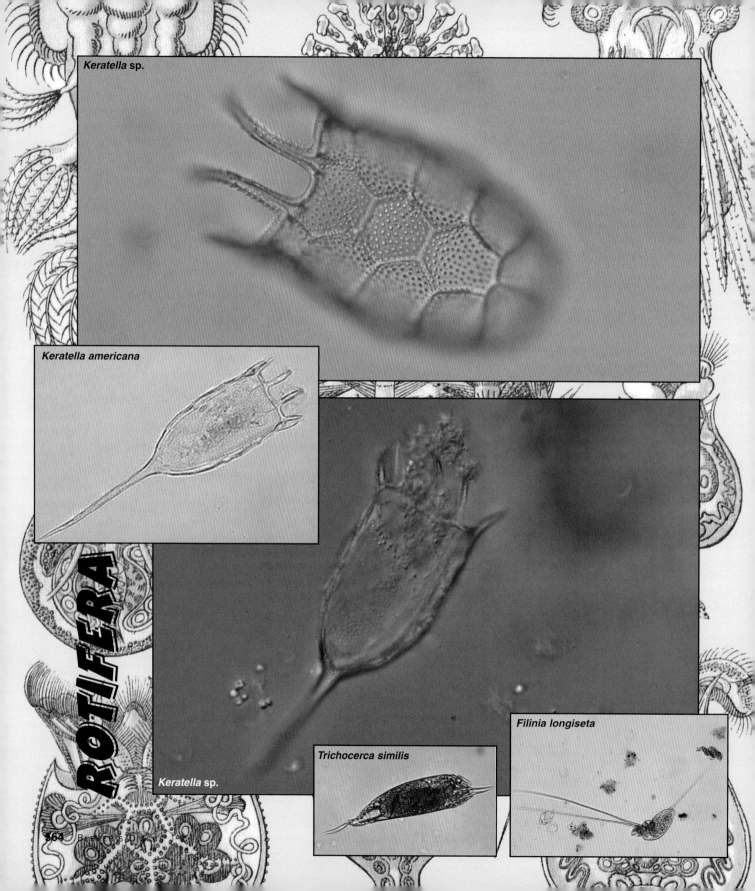

Keratella sp.

Keratella americana

ROTIFERA

Keratella sp.

Trichocerca similis

Filinia longiseta

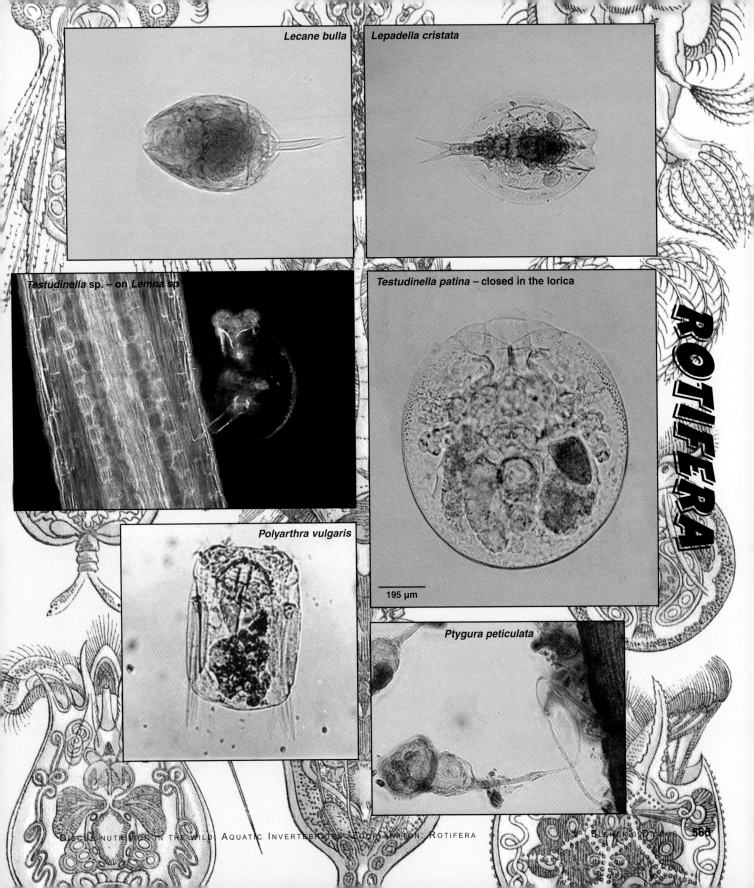

Another genus which I have found in discus has the unusual name of *Testudinella* (diminutive of Latin *testudo* = a tortoise; family Testudinellidae), which it was given because when danger threatens its species can retract head and foot into the round, flat lorica in a split second *(p. 565)*. Browns and blues eat them nevertheless. I found them in the stomachs of three individuals in the Lago dos Campos. They also occur on the roots of floating plants and I have seen them on *Lemna* spp. *(p. 565)*.

I have found the species *Filinia longiseta* (family Filiniidae) and *Trichocerca similis (p. 564;* family Trichocercidae) in discus stomachs at the beginning of the dry season (August) in the Canumã region. The stomach contents of four semi-adult individuals included around 5% of these species.

However, it has been mainly the rotifer genus *Brachionus* that I have found in Amazonia, and chiefly the species *B. dolabratus, B. falcatus, B. patulus, B. zahniseri gessneri,* and *B. z. reductus,* along with a number of *Keratella* species such as *K. americana* and *K. cochlearis*. And discus undoubtedly also feed on them elsewhere. Apropos of which I would like to mention that Goulding *et al.* (1988) mention 50 Rotifera species in the Rio Negro and found these in the stomachs of 27 fish species (but these included no individuals of more than 70 mm SL and no discus). In addition they established that the majority of fishes in *lagos* feed on them. Undoubtedly Heckel discus also feed on them there, only to date I haven't been able to prove it.

Taxonomic remarks: In zooplankton systematics too there is a very high degree of disagreement regarding the status of some taxa, ie whather they are phyla, class, family, or genus. Differences exist in the systematics used by different authors. Some regard the Rotifera as a phylum, others recognise a phylum Aschelminthes or Nemathelminthes with the Rotifera (or Rotatoria, as they were formerly called – and still are by some) as a class – to mention only a few examples.

Cladocera – water fleas

Browns, blues and greens feed on them.

Water fleas, often also collectively known as daphnia (after the genus of that name), are well-known to aquarists, who have been collecting one or another species (there are supposedly more than 90 species overall in the inland waters of central Europe) for their pets for around 100 years.

Water fleas inhabit all types of standing waters, from deep lakes to shallow puddles. They are trophic specialists, with representatives ranging from predatory to strict herbivore with all stages in between. They graze *Aufwuchs* from plants, filter out detritus, sift through the mud of the bottom, or glide along the underside of the water's surface. The majority of species of the families Chydoridae and Macrothricidae as well as some of the Sididae are planktonic and live in various different microhabitats. Predominantly on macrophytes, plant detritus, and organic sediment and swimming only short distances, if at all. They are also classed as meiobenthos and are usually around 1 mm long. Chydoridae species are often encountered in large numbers, with a population density of more than a million individuals per square metre of substrate in 20 or more species. Albeit less in Amazonia. The majority of species of the other eight families of this order are chiefly (or entirely) considered benthic.

Ax (1999) regards the suborder Cladocera as a natural group within the class Branchiopoda. More than 400 of these water-flea species are known, but only eight of them live in salt water. The Cladocera are divided into four infraorders (the Anomopoda, Ctenopoda, Onychopoda, and Haplopoda) and 11 families. I have found species from six families in discus stomachs: *Ceriodaphnia cornuta* and *Daphnia gessneri* (family Daphniidae); a *Bosmina* species and *Bosminopsis deitersi* (family Bosminidae); *Moinia* sp. and *Moinodaphnia macleayi* (family Moinidae); *Macrothrix* sp. (family Macrothricidae); a species of the family Chydoridae (which includes 22 to 32 genera, depending on the author, with well over 100 taxa – around a quarter of all known water-flea species); and *Diaphanosoma* species (family Sididae).

In the Lago Jari region I once found up to 5% of the stomach contents of brown and blue discus to be the predominant *Daphnia gessneri*. I found them in five adult individuals, and on one occasion *Ceriodaphnia cornuta* were also present. The females of *Daphnia gessneri* can grow up to 5 mm long, while the males attain barely 2 mm. In the Lago Manacapuru, in the dry season, I found a *Bosmina* species, as well as *Bosminopsis deitersi,* in blues, but only a few specimens in each of the three adult individuals studied. This also applies to studies of discus stomachs and guts in the Tapajós region, where during the dry season eight adult individuals had eaten only a few specimens of an unidentified chydorid species. In the Tefé region I found again only a few specimens of a *Moinia* species in the stomach of a Green, and on only one occasion. And five blue discus in the Lago Anamã had water fleas of the genera *Bosmina* and *Diaphanosoma* in their guts and stomachs, comprising almost half of the contents. This was the only time that I found that really large numbers of water fleas had been consumed by discus. In addition there is a *Bosmina* species that occurs frequently in Amazonia, but the females attain a length of only 0.3 to 0.7 mm and males are somewhat smaller still.

I would like to add the following: researches have been performed in several *lagos* of central Amazonia *(inter alia* by Carvalho, 1981), and it was found that water fleas were present in the smallest numbers in the zooplankton, around 60% of which

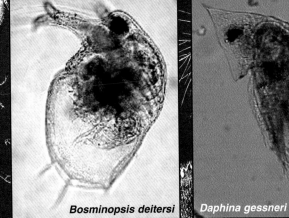
Bosminopsis deitersi

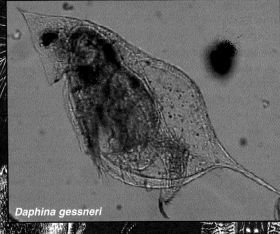
Daphina gessneri

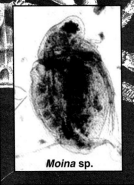
Moina sp.

Macrothrix sp.

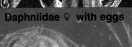
Daphniidae ♀ with eggs

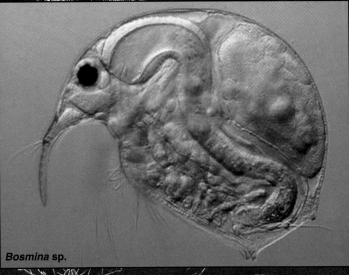
Bosmina sp.

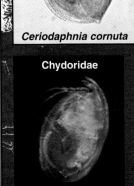
Ceriodaphnia cornuta

Chydoridae

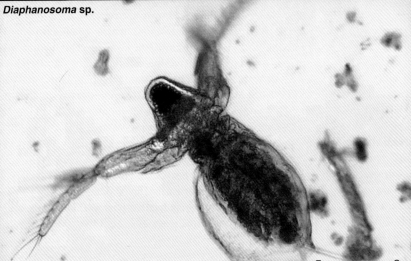
Diaphanosoma sp.

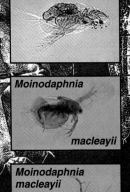
Diaphanosoma sp.

Moinodaphnia macleayii

Moinodaphnia macleayii

CLADOCERA

Discus-Nahrung in der Natur: Aquatische Invertebraten: Zooplankton: Cladocera

consisted of Copepoda *(see below)*, 30% Rotifera, and only around 10% Cladocera. The latter hardly occur at all in the black water of the Rio Negro and other *S. discus* habitats. (Only two species, *Ceriodaphnia cornuta* and *Daphnia gessneri,* were found in the Rio Negro, but never in the stomach of a fish – but in the aquarium Heckel discus immediately pounce on water fleas.) On the other hand, Goulding *et al.* (1988) found a significant quantity (more than 25% of stomach contents) of other Cladocera in the Rio Negro but only in the stomachs of *Apistogramma pertensis, Gymnotus anguillaris,* and a small characid.

Copepoda – copepods

Browns, blues, and greens feed on them.

Copepods occur mainly in the sea, and of around 14,000 species (2,300 genera, 210 families, 10 orders) only 125 live in fresh water. Their size varies between ca. 500 µm and 2.5 mm and the species are difficult to identify. However, depending on the length the antennae and the number of their joints three orders can be readily distinguished:

1. The **Calanoida**: First antennae very long with up to 25 joints *(centre).* Only the first right-hand antenna in males with clasping organ. They are swimming plankton that feed on bacteria and microscopically small algae.
2. The **Cyclopoida**: First antennae with 8 to 17 joints. Both first antennae in males with clasping organs. They are found in small bodies of water and are predators that also feed on carrion and algae.
3. The **Harpacticoida**: First antennae with 8 joints maximum. Both first antennae in males with clasping organs. They are poor swimmers and hence found mainly in the substrate.

The planktonic larvae of these tiny crustaceans are called nauplii (singular nauplius: from the Greek *nauplios* = a type of swimming shellfish) and are very small, often only 20 µm in diameter. They undergo up to six moults. Until the first moults they have only three pairs of limbs *(p. 569)* which serve for locomotion and for capturing and taking in food. After further moults they develop additional limbs; and after the fifth they are sexually mature; but only after the sixth moult is the typical copepod form recognisable. Fascinating creatures, as can be clearly seen from some splendid 19th century drawings *(on page 561; right & centre).*

I have found at least three different unidentified species (p. 569) of the order **Calanoida** in discus stomachs. On several occasions adult Browns and Blues in the Alenquer region had consumed Calanoida species. I once found several specimens in the stomachs and guts of five adult Browns, without being able to record percentage data. On a second occasion, again involving five adult Browns, almost 5% of stomach contents consisted of Calanoida.

I have found two genera of the order Cyclopoida in the stomachs and guts of discus: one *Mesocyclops* species (perhaps *M. leukcarti* or *M. longisetus)* and a *Thermocyclops* species (possibly one of the well-known species *T. decipiens* and *T. minutus,* which predominate in a number of lakes in central Amazonia) (all family Cyclopidae). In each of four Greens in the Mamirauá region I found a large amount of *Mesocyclops* – up to 10% of stomach contents. And in four blues in the Canumã region somewhat more than 5% of the stomach and gut contents consisted of a *Thermocyclops* species.

Remarks: Copepods, especially harpacticoids, can be very numerous in the substrate zone (the benthal), and more than 2000 can live in a single litre of sand (no wonder that discus and numerous other fishes often sift the sand). Also of interest are the planktonic forms which feed on phytoplankton, even picking unicellular organisms, *inter alia,* from the water with their

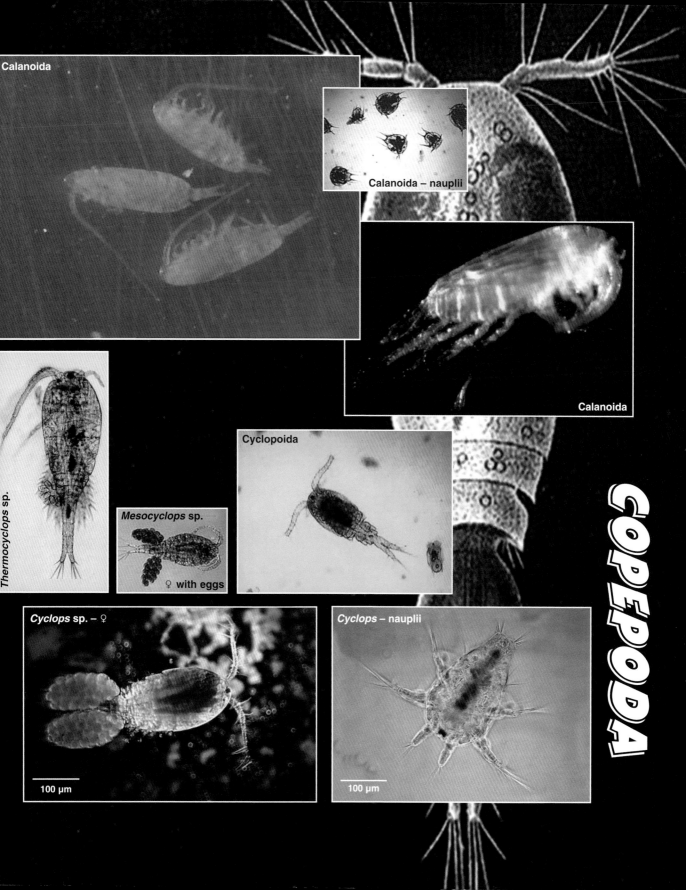

remarkable mouthparts. And by feeding on zooplankton, of course, discus indirectly also take in carbohydrates, which are richly represented in the phytoplankton.

And finally, scientists believe that copepods may constitute the largest biomass on Earth – they vie for this title with the Antarctic krill *(Euphausia superba).*

Taxonomic remarks: The copepods are systematically classified as follows: phylum Arthropoda (arthropods); sub-phylum Crustacea (crustaceans – there are more than 52,000 described species of these alone); class Maxillipoda; subclass Copepoda (copepods).

The Benthos

Heckel discus, browns and blues, as well as greens feed on it.

The benthos is an aggregation of organisms living on or at the bottom of a body of water. (The benthal is the habitat at and in the bottom of a body of water.). The benthic community can be composed of a wide range of plants, animals and bacteria from all levels of the food web. The benthos (or benthon – the life-forms of the benthal) can be divided into three distinct communities: 1. infauna: plantas, animals and bacteria of any size that live in the sediment. 2. epifauna: plants, animals and bacteria that are attached to the hard bottom or substrate (eg. to rocks or debris); are capable of movement; or live on the sediment surface. 3. demersal: bottom-feeding, or bottom-dwelling fish that feed on the benthic infauna and epifauna. The benthos is subdivided into the macrobenthos, meiobenthos, and microbenthos.

I have found the following taxonomic groups in discus stomachs and seen them eaten: rotifers (**Rotifera** or Rotatoria); copepods (**Copepoda**); bear-animalcules (**Tardigrada**); oligochaete worms (**Oligochaeta**); ostracods (**Ostracoda**), the aquatic larvae of dipteran insects (**Diptera**), and sponges (**Porifera**). I have already discussed the relevance of the first two groups to discus.

There are supposedly more than 10,000 species of bear-animalcules *(below),* phylum **Tardigarda**, but only around 750 have been described (35 of them from the sea).

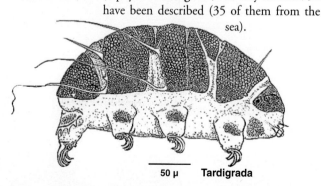

50 μ   **Tardigrada**

They are meiobenthos, 0.05 to 1.2 mm long, eight-legged animals, whose appearance and awkward-looking motion makes them somewhat reminiscent of miniature teddy bears. Depending on their habitat they are described as essentially marine, freshwater, or terrestrial, but the distinction between the last two categories isn't clear-cut, hence the term limnoterrestrial is also frequently used. Although bear-animalcules are sometimes highly resistant to dehydration, they are all designed for active life in a thin film of water. And if their environment dries up, they sometimes congregate and form a little barrel-like aggregation. Some species can survive for many years in this state (which students of the Tardigrada term anhydrobiosis), often at temperatures ranging from –272 °C to +150 °C. They occur worldwide and in all climatic zones, and have been recorded at up to 6,600 m of altitude and down to depths of 4,690 m in the sea (down to 150 m in fresh water). In Amazonia I have found them on algae strands and other aquatic vegetation in discus habitats (but mainly on waterlilies).

At rising water I found a number of bear-animalcules in the stomachs of five blue discus in the Rio Coari drainage, but they represented on average less than 5% of stomach contents.

The subclass **Oligochaeta** (oligochaete worms) belongs to the class Clitellata in the phylum Annelida (annelid worms). There are around 3,500 species measuring from less than 1 mm to around 3 m long. The aquatic forms are regarded as part of the macrobenthos; they occur chiefly in fresh water, in bank zones and all types of land (eg meadows and woodlands) – I have even seen them in bromeliads. A small number of species inhabit the marine littoral. The Oligochaeta also include the genus *Tubifex* (family Tubificidae) and whiteworms (Enchytraeidae), well-known among aquarists worldwide. Although I have only been able to find once representatives of the suborder Tubificina in Amazonia, it is said that members of both families have been found in the Rio Negro. But frequently I came across unidentified species of the genera *Dero,* and *Nais* (family Naididae) in blackwater zones. In the lower Rio Negro I discovered both in the stomachs of five *S. discus* – full house! I have repeatedly seen *Dero* and *Nais* in Heckel discus habitats and especially the distinctive *Dero.* The species of the last-named genus often construct a tube to live in *(p. 571)* from small particles of vegetable origin and epidermal secretions. These tubes can be free-floating or attached to a twig, leaf, or the bottom, and I found one such circa 1 cm long "dwelling" in the stomach of a Heckel discus in the Rio Xeruini. Other discus undoubtedly eat them too.

The **Ostracoda** or ostracods are bivalve shellfish belonging to the phylum Arthropoda, subphylum Crustacea, and form part of the macrobenthos. They have overall the most complete fossil record of all animals; they first appeared back in the Cambrian, diversified during the Ordovician, and are present in abundance in the Silurian. These on average 0.5 to 2 mm long ostracods look like tiny mussels (one marine species, *Gigantocypris agassizii*, grows to 34 mm long and in the Upper Silurian some ostracods there were even 100 mm long). The body is protected by the hard bivalve shell *(p. 573),* whose two halves are hinged by an elastic strip. The seven pairs of extremities include a long and a short pair of antennae, which are also used for locomotion. Using this equipment they can move at an even, in some cases even a very rapid, pace. Depending on the species they feed in different ways, but often on plant remains, diatoms and other algae, diatoms, detritus, and carrion. More than 4,200 species are distributed worldwide and have conquered almost all available habitats. The majority occur in the sea, above all among the benthos of the littoral, but also in the abyssal down to around 6,500 m of depth. The freshwater species all belong to the suborder **Podocopa** and live in lakes, rivers, small bodies of water, ground water, watery places with an extremely high salt content, and hot springs with temperatures of 30-54 °C where they graze algae and bacteria. In Amazonia they are also encountered in holes in trees and in Bromeliacea.

Thirtynine ostracod species are known from the Japurá and blackwater regions, but the majority (97 species) are found in the mixed water of central Amazonia, where brown and blue discus sometimes also spend time at the beginning of the rainy season.

I have several times encountered unquantified amounts of ostracods in the stomachs of discus. On one occasion in two greens in the Mamirauá region; once in three greens in the Urucu; and once in the stomach and guts of three Heckel discus at rising water in the middle Rio Negro region. But I found the numerically largest quantity of ostracods in the gut and stomach of five blue discus in the Canumã. They must have represented almost 20% of the stomach contents. A very nice calcium-boost.

As regards the aquatic insects of the benthos, a whole series of members of the order **Diptera** (two-winged flies) are important in many respects. From a systematic viewpoint they belong to the superclass Hexapoda (insects) and to the subclass Pterygota (winged insects). More than 144,000 dipteran species from 189 families are at present known from all over the world, with around 60,000 species currently assigned to the suborder Nematocera (mosquitoes, midges, gnats) and 84,000 species to the suborder Brachycera (flies). But we are concerned here (in the benthic group) only with their larvae, as aquatic insects are rarely completely aquatic creatures. Mostly they alternate between the aquatic and the terrestrial environment and live part of the time on land or in the air. This variation in habitat is linked to their life cycles and the development of the individual. The stages of their metamorphosis – egg, larva, pupa, and imago – are variably (and not always) adapted to life in water. Recognised developmental cycles include a totally terrestrial way of life, four amphibious variants, and a purely aquatic life cycle.

I have found larvae of the following families in discus stomachs: **Chaoboridae** and **Chironomidae,** and the family **Polymitarcyidae** from the order **Ephemeroptera** (mayflies).

The 12-15 mm long larvae of the family **Chaoboridae** are predatory and don't hang at the water's surface like the larvae of the closely-related mosquitoes in order to take in air via a breathing-tube. They lack this respiratory organ on the posterior body and breath entirely via the surface of the body. For this reason they normally adopt a horizontal body position. They also feed on water fleas, which is a double benefit to the discus. Unlike mosquito larvae, chaoborid larvae are found in blackwater regions with low pH values. I have several times found individual larvae in Heckel discus and in greens at the beginning of the rainy season.

In the Rio Negro region I found in 3-10 larvae in each of 10 *S. discus*, and in the Abacaxís five individuals had eaten 2-5 larvae apiece. And eight greens in the lower Juruá region had even more – up to 12 larvae – in their stomachs.

The much-prized fish food known to aquarists as "bloodworms" consists of the 2 to 14 mm long larvae of midges of the family **Chironomidae**.

During their short lives these midges consume little or no food. After mating, which takes place in a swarm with males, females drop their usually gelatinous clusters of eggs into the water, where they swell considerably and open out into strings and other types of aggregation containing up to 2,000 eggs. (The developmental sequence is egg, larva, pupa, and imago – the last being the fully-formed individual.) The wormlike body of the larva after hatching is similar in form in most species of the family and exhibits a wide spectrum of adaptations to the most varied aquatic habitats. The larvae feed on algae and detritus, but small insects, crustaceans, and worms are also taken. The majority of chironomid larvae construct "houses" which rival those of the caddisflies (order Trichoptera) in their variety. Only the predatory midge larvae (subfamily Tanypodinae) live free-swimming. The pupae, which to date I have never found in discus stomachs, also generally take no food, only oxygen via the body surface.

BENTHOS

OSTRACODA

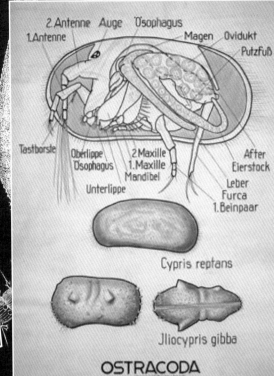

CHIRONOMIDAE

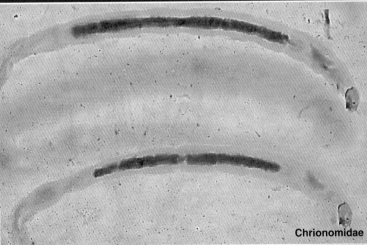

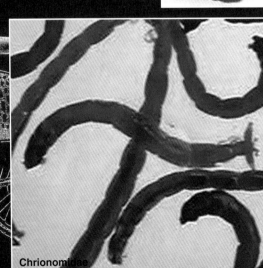

The larvae are regarded as part of the macrobenthos and at least one, *Chironomus paragigas*, occurs only in blackwater lakes (= discus habitats) and other waters. It cannot survive in white and mixed water and in the course of evolution has completely adapted to nutrient-poor acid water with a low conductivity, which is all to the good where Heckel and green discus are concerned. (In much the same way the genus *Diamesa* can survive only in the nutrient-poor, minimal-conductivity water of glacial streams up to 2,400 m in the Alps.) I have also found them in the stomachs of both discus species. Five Heckel discus in the Rio Nhamundá had swallowed between 10 and 20 larvae apiece. In the Rio Negro I found around 10 larvae apiece in the stomachs of 10 adult *S. discus* and around 12 larvae per stomach in three greens in the Mineruá. But this is very interesting, as although few species of the family are known from black water, more than 90 described species have been found in Amazonia to date, and around 400 undescribed.

It is also worth mentioning that as well as my records from the Rio Negro, Goulding *et al.* (1988) found chironomid larvae in the stomachs of 88 fish species, more than in all the other fish species studied. Although there are practically no mosquitoes of the family Culicidae along the Rio Negro, as noted in the past by Spix & Martius (1823), Wallace (1853), and Spruce (1908). This (wonderful) phenomenon along the largest blackwater river on Earth, shared by the Rios Nhamundá, Abacaxís, Marimari, Jatapú, and Urubu (*S. discus* habitats), is to date unexplained. Apparently haematogenous mosquitoes cannot survive in these extremely nutrient-poor, minimal-conductivity waters. But Heckel discus can – and only there.

Interesting is, that there is another chironomid species, *C. gigas*, which can survive only in sediment-rich white water. Also: that chironomids must have existed in a virtually unchanged form for around 170 million years. And that a short time ago chironomids were discovered in fossil of a tropical fauna in Dominican amber on Hispaniola, dating from 15-20 million years ago. (Did discus have more mosquito larvae to eat back then?)

The name **Ephemeroptera** refers to the short life-span of the imagines (plural of imago) (Greek *ephemeros* = living for only one day). There are 11 families in central Europe alone and their larvae are an important source of food for fishes there.

The mayflies (or dayflies) are hunted by hirundines, bats, and dragonflies (see also terrestrial invertebrates). Egg-laying takes place above the water and the eggs sink to the bottom. The larvae develop on and under stones and in the aquatic vegetation, in the mud and sand, or in self-dug burrows. The larval stage can last for one to four years. When they are ready to "hatch" they ascend and look for a place for their final moults so they can make their nuptial flight and mate. The larvae have large facetted eyes and well-developed mouthparts. They live mainly on sediment in the water (saprobes). A striking feature is the paired tracheal gills (usually composed of branchial lamellae) on the first seven abdominal segments as well as the "tails" on the posterior abdomen, which with a few exceptions number three, unlike those of the stoneflies. Mayfly larvae are important indicators of water quality. The final moult of the larvae takes place at the water's surface or on land, with the larva metamorphosing into a winged but not yet sexually mature subimago, which then moults again within a few minutes into the adult insect (imago).

The larvae of the mayfly *Asthenopus curtus (below, left;* family Polymitarcidae) occur only in the vicinity of the water's surface because of their high oxygen requirement. They are widespread in the Neotropics and have powerful mandibles that allow them to bore into tree-trunks lying in the water. (Is this why in Amazonia discus often spend the day beneath tree-trunks and branches?) I have found them repeatedly in the stomachs of brown and blue discus.

Remarks: Interestingly the genera *Dero,* and *Nais* occur in 2-4 m of depth (part of the discus daytime food?), while the species *Asthenopus curtus, Cyclestheria hislopi, Hebetancylus moricandi,* and *Tenagobia melini* live in the upper regions and may be a component of the evening food of discus.

Porifera – sponges

It is difficult to believe, but of the around 5,000 sponge species 3% (about 150) have to date been found in fresh water and a number also in Amazonia. Sponges have a low degree of cell organisation, but nevertheless the cells perform a variety of functions. Sponges live sessile, usually on hard substrates in the sea, but in rivers – and especially in the Rio Negro and other Amazonian regions (particularly in *igapós*) – they grow on trees *(p. 575)* and dry out seasonally but without dying. Instead they produce resistant gemmula which expand again when the water rises. Sponges filter suspended particles from the water and often form a symbiotic relationship with numerous bacteria, algae, and other aquatic organisms. Sponges also produce toxins that perhaps offer them protection against predators. Many of these chemicals are used in pharmacology. They have, for example, respiratory, cardiovascular, gastro-intestinal and antibiotic effects.

Their size ranges from 1 mm to several metres across and they are filter-feeders, ie detritus, plankton, and bacteria are sucked in

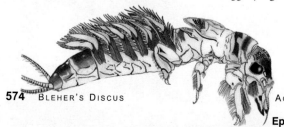

Ephemeroptera – *Asthenopus curtus*

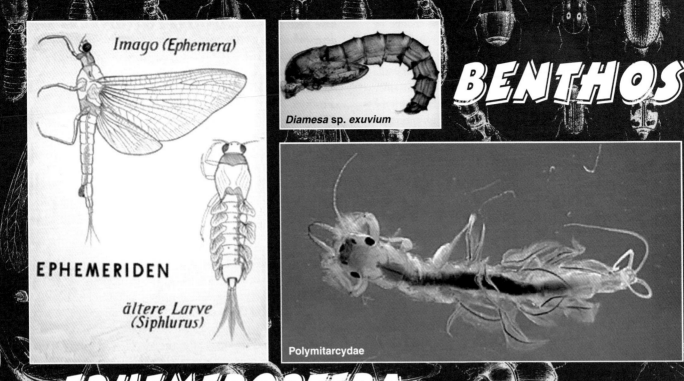

BENTHOS

EPHEMEROPTERA

PORIFERA

by choanocytes and ingested into the body of the sponge by phagocytes. Respiration is also via filtration of the water.

In recent years several new species of the families Spongillidae and Metaniidae have been described from Amazonia, *inter alia* by Volker-Ribeiro, who has been working on the Porifera since 1970 and has also described new genera (eg *Oncosclera*) from the Rio Negro. Goulding *et al.* (1988) found that 67 fish species in the Rio Negro had eaten Porifera. Even the large *Uaru amphiacanthoides, Leporinus* and *Metynnis* species, which all occur there or live there in communities with the Heckel discus, had more than 25% freshwater sponges in their stomach contents. Goulding *et al.* never caught and studied discus, but on one occasion I found Porifera in the stomachs and guts of five Heckel discus, although the quantities were smaller. Apparently the sponge had been ingested with other types of food. Sponges consist mainly of silica and have little nutritional value – but may be effective as roughage.

The Perizoon

As already mentioned, the perizoon comprises communities that occur mainly on rooted and free-floating plants and those in the inundation-zone forest. In Amazonia predominantly in black- and clearwater regions. And it consists almost entirely of the aquatic larvae of insects. To date I have found the following groups in discus: the freshwater mites (**Prostigmata**); shrimps (Crustacea: **Decapoda**); **Diptera** and **Ephemeroptera**. There are, of course, further groups that form part of the perizoon, but to date I haven't found any others to be part of the discus diet, and I have already discussed the last two above.

The systematic classification of the freshwater mites (Hydrachnidia, Hydrachnellae, Hydracarina, or suborder **Prostigmata**) is hotly disputed. Many authors, above all in the older literature, regard them as a probably polyphyletic group, but nowadays the group is often classed as monophyletic, chiefly on the basis of the glands present in the cuticula (Glandularia). Some authors assign them to the phylum Arthropoda (arthropods), the class Arachnida (spiders), and the order Acariformes (mites and ticks) There are exclusively aquatic families. Adults are as a rule predatory, while the larvae usually parasitise insects or bivalves. The freshwater mites have no bristles on the cuticula of the body and in all species two or more legs have been transformed into swimming-legs with long bristles. They have eight legs and often gaudy coloration (eg blue-green, red, black and yellow, or black and white). They are between 0.5 and 4.0 mm long and there are different species – at least 5,000 taxa worldwide – occurring in all possible freshwater habitats. Only a few species have adapted to salt water. Unfortunately the South American Hydrachnidae have been little studied and I have had to rely on the family description. Viets (1954) recorded 104 species – 64 of them new – from various rivers and lakes in Amazonia. Since then hardly any taxonomic work has been done on the group, although the freshwater mites are very numerous in Amazonian waters and constitute a food source for numerous fishes, *inter alia*.

They live mainly on submerse vegetation, among the roots of floating plants large and small, and, of course, many can be found in the gigantic islands of grass that come floating downriver at the beginning of the rainy season. Apropos of these floating islands, Junk (1973) writes that the richest aggregations of aquatic invertebrates in Amazonia are to be found among their root tangles and that almost all invertebrate groups are represented there.

I have also found them several times in browns and blues. In the lower Uatumã, in the Alenquer region, as well as in the Rio Coari. On each occasion I investigated individuals that had swallowed several Prostigmata at rising water. On one occasion in the stomachs of five Blues in the Utatumã, once three browns in the Alenquer region, and again five blues in the Rio Coari.

In the Heckel discus habitat in the Rio Negro region I found five *S. discus* of around 10 cm SL in which more than 25% of the stomach contents consisted of freshwater mites. Goulding *et al.* (1988) found *inter alia Micromischodus sugillatus* (family Hemiodontidae) measuring 10.5 cm SL in the same habitat, and their stomach contents again contained more than 25% Prostigmata.

The shrimps of Amazonia (Crustacea; order **Decapoda**) are represented by 15 species of the family Palaemonidae (palaemonid shrimps) in central Amazonia (Magalhães & Walker, 1986), as well as one species of the family Euryrhynchidae which is relevant to discus. The majority of the small number of Amazonian shrimp species occur in black- and clearwater regions, although the most widespread species, *Macrobrachium amazonicum (p. 577)*, is restricted to whitewater habitats and rarely survives pH values of less than 5.0. It is also one of the largest, at up to 15 cm long. It is of relevance only for predatory fishes and humans, both of whom benefit from its large biomass.

In the stomachs of eight blues in the Lago Manacapuru I found several 5-6 mm long *Euryrhynchus amazoniensis* larvae (they undergo only a single larval stage). And a number of even smaller, but clearly recognisable, larvae of the smallest species, *Palaemonetes ivonicus* (which has two larval stages), in the stomachs of three Heckel discus in the Jatapu. But shrimps are not an important factor in the discus diet, of that I am sure. (See also a summary of the discus diet in the wild.)

PERIZOON

PROSTIGMATA

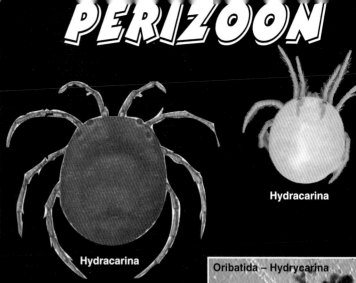
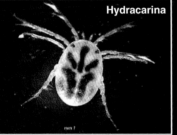

DECAPODA

5. TERRESTRIAL AND ARBOREAL ARTHROPODS:

Terrestrial and arboreal Arthropoda also form part of the discus diet in the wild – albeit almost exclusively insects (superclass Hexapoda: class **Insecta**). Apart from insects, to date I have recorded once mites (order Acariformes), but never, for example, beetles (order Coleoptera), which are frequently eaten by large predatory fishes such as members of the family Osteoglossidae, or by other animals. It all depends on natural circumstances. The regular alterations in the habitat make feeding on terrestrial invertebrates difficult for discus particularly during the low-water period. They cannot take beetles from the trees as arowanas (order Osteoglossiformes) do, and rarely seize food that has landed on the water like many characin species. (The majority of the latter live at or near the surface. Catfishes and knife eels come up to take in air and often snap up insects that have landed on the water. Discus rarely do so. In this respect the study by Goulding *et al.* (1988) is particularly informative: they found terrestrial and arboreal invertebrates almost exclusively in the stomachs of species of the orders Osteoglossiformes, Characiformes, Gymnotiformes, Siluriformes and only three cichlid (order Perciformes) species. During the high-water period too it is not easy for discus to take these foods, as the different invertebrate groups have adapted to the immense fluctuations in water level via a variety of survival strategies. Arboreal invertebrates often undertake vertical migrations to the tree-tops, and many terrestrial forms make horizontal migrations to the *terra firme* – those with wings fly there – when the water begins to rise (or even earlier). But we need to distinguish not only between the terrestrial and the arboreal, but also within those groups, as both contain migratory as well as static forms. A number of terrestrial forms remain in their habitat in a dormant state, underwater or in the ground (in a natural or self-constructed retreat, or as eggs). Then again, other terrestrial forms are active underwater during the high-water period, often on tree-stumps and roots. Hence I will again discuss my personal insect experiences.

It is well known that the insects, with more than 800,000 species, constitute the largest group within the Arthropoda (arthropods); that all insects have a body divided into head, thorax and abdomen; that the first three segments of the abdomen bear a pair of limbs; and that they are characterised by their tracheal system, which is an adaptation to a terrestrial way of life.

Authors usually divide the Insecta into 32 orders, and I have recorded the following orders in the discus diet: **Hymenoptera** (bees, wasps, ants, etc): family **Formicidae**; **Isoptera** (termites); **Diptera** (dipterans or true flies, some of those I discussed earlier): family **Simuliidae**; **Ephemeroptera** (mayflies or dayflies, whose aquatic larvae have been discussed earlier); **Collembola** (springtails), and **Diplura** (two-tailed bristletails).

Formicidae – ants
Blue, brown, and Heckel discus feed on them.

More than 8,800 species in 292 genera have been described within the ant family – but there may well be more than 20,000 (Hölldobler & Wilson, 1990). They have a worldwide distribution and make up 10-15% of the entire terrestrial biomass. But in Amazonia – where the majority undoubtedly occur – the percentage is even higher (Römer *et al.*, 1995). Wilson (1988) found 43 species from 26 genera on a single tree in western Amazonia, and on the *terra firme* more than eight million ants and a million termites have been recorded per hectare of jungle (the two combined, along with their closest relatives, the bees and wasps, make up 75% of the insect biomass in Amazonia).

To date I have found species from six genera in discus stomachs. Repeatedly an undetermined number of the genus *Pseudomyrmex* in the stomachs of five brown discus at rising water in the Alenquer region. I also found them in the stomachs of four blue discus in the Canumá region. There must have been between five and 10 small workers – perhaps of *P. viduus*, a widespread species which occurs from Mexico to Brazil. It is known that the workers of this species are very aggressive and ready to bite (as I well recall from finding a colony while collecting beneath some branches), in contrast to most other species of the genus. *Pseudomyrmex* often live in holes in trees, where they brood their young and "farm" insects of, for example, the families Pseudococcidae *(p. 581)* and Coccidae. But *Pseudomyrmex* are also known for their "ant gardens" *(left)*, which I have seen time and

ant- garden

Daceton sp.

Atta sp.

Isoptera

Isoptera

Insects constitute the largest biomass in Amazonia. Ants *(top)* are a food much enjoyed by fishes when they are available (washed down by heavy rainfall, fallen down when their tree-nests are disturbed by other animals, or overtaken by rapidly rising water levels). In the case of mayflies *(above, left, & right)*, which often fall into the water in their millions after mating the fishes have it rather easier.

again and are to be found on *Coussapoa* trees (Cecropiaceae), *Macrolobium acaciifolium*, and *Pterocarpus amazonum* (Fabaceae), as well as *Ocotea* (Lauraceae) and *Pseudobombax (p. 522)* in Amazonia and in other American regions. Perhaps these ant gardens it is the most complex mutualism between plants and ants, which is an aggregate of epiphytes assembled by ants. The latter bring the seeds of the epiphytes into their carton nests. As the plants grow, nourished by the carton and detritus brought by the ants, their roots become part of the framework of the nests. The ants also feed on the fruit pulp, the elaiosomes (food bodies) of the seeds, and the secretions of the extra floral nectarines. And not only *Pseudomyrmex* constructs these fascinating "ant gardens", but also species predominantly of the genera *Azteca*, *Camponotus*, *Crematogaster*, *Monacis*, and *Solenopsis* – often in the forks of trees *(p. 578)*.

There are other plant-ant symbioses that should be mentioned briefly as again the species involved are consumed by discus. During field studies in Mexico scientists have discovered that some acacias are inhabited long-term by ant colonies. These trees constantly produce nectar from their leaves, as a "payment" to their ants. In return the latter hunt the tree's enemies – eg caterpillars, small beetles, and other insects – and in this way protect it successfully in the long term. It is now known that acacias *(inter alia, see below)* cooperate with ants in a whole variety of ways. While a number of plants attract patrolling ants from the surrounding area, others harbour ant colonies on a permanent basis. The enzyme responsible for this was tracked down by the scientists via some interesting observations. Specifically, the patrolling ant species were found to prefer the nectar of the attractant acacia species, while the colony-forming ants specialised on a particular acacia species like only the nectar of "their" host plant. Chemical analysis shows that the nectars of the attractant acacias contain saccharose (cane sugar), glucose (grape sugar/dextrose), and fructose (fruit sugar/laevulose). By contrast the nectar of the acacia species inhabited by ant colonies contains no saccharose. Enzymological studies of the "resident" ants eventually revealed that they lack the enzyme that breaks down saccharose – but the latter is present in large quantities in the plant of their specific host: the plant can break down the saccharose and make the desired carbohydrate available to the ants in a so-to-speak "predigested" form. This nectar is thus unattractive to "alien" ants, but provides a complete diet for the "specialists".

This system has been found in plant species of the following genera: 1. *Acacia* (Leguminosae); 2. *Cecropia* (Cecropiaceae); 3. *Macaranga* (Euphorbiaceae); 4. *Ochroma* (Bombacaceae); and 5. *Piper* (Piperaceae). And each of the trees has something to offer, respectively: 1. proteins and lipids; 2. glycogen and lipids; 3. lipids, starches, and proteins; 4. lipids, plus some starches and proteins; 5. lipids and proteins. With one exception (4), the plants appear to provide the "specialists" with a complete diet that includes extrafloral nectarin and homes, in return for which the ants act as "police" and protect the plant from invaders. A quite perfect symbiosis, which undoubtedly takes place in many other plants that I have encountered, complete with such ant colonies, above discus habitats. The protectors of the above-mentioned plants are species of the ant genera *Pseudomyrmex*, *Azteca*, *Crematogaster*, *Pheidole*, and *Solenopsis*, and the last three (or four) of these are genera that I have seen discus consume, and/or which I have found in their stomachs.

I have found at least one (possibly two) *Azteca* species –10-15 individuals – in the stomachs of five greens in the lower Juruá region. Species of the genus *Crematogaster (right)* are sometimes known as acrobat ants and measure up to 3 mm long; I have found them in the stomachs of three blue discus in the Manacapuru region. There were 7-10 ants per stomach.

Pheidole, with 524 taxa, is by far the most species-rich ant genus in the New World. Numerous species occur in Amazonia and are difficult to differentiate with the naked eye. I found an undetermined *Pheidole* species, known as the inga ant *(p. 532)*, feeding on the flesh of the *ingá* fruit in the Nanay region and have recorded them in the stomachs of four green discus (but only single specimens). It is also known that *Pheidole* are the main species active in keeping caterpillars and other harmful insects away from *Inga* trees, for example in Costa Rica where *P. biconstricta* performs this service for *Inga densiflora* and *I. punctata* (Koptur, 1984). In Amazonia too this is known in *Inga* species, especially *Inga edulis* (where I am not the only one to have recorded it), and in *Bixa orellana* (Bixaceae), but to date most

Solenopsis sp.

Insecta   Arthropoda   Hymenoptera

FORMICIDAE

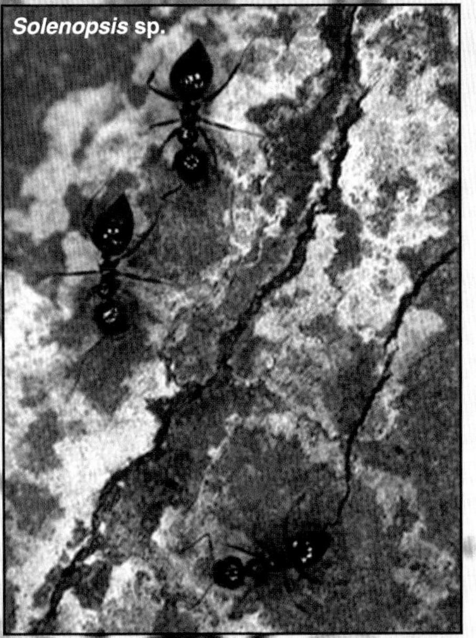

Solenopsis sp.

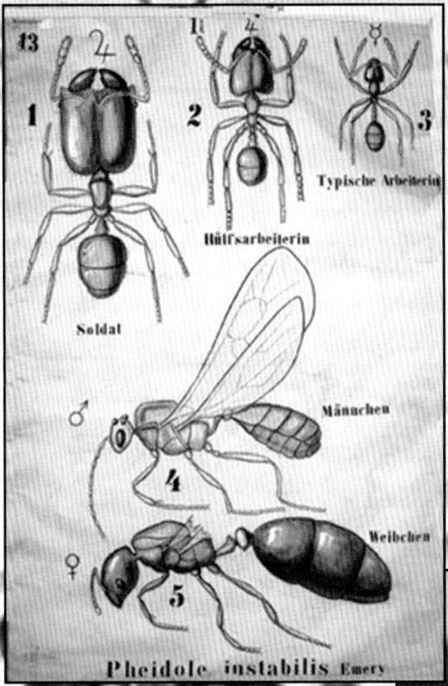

Pheidole sp.

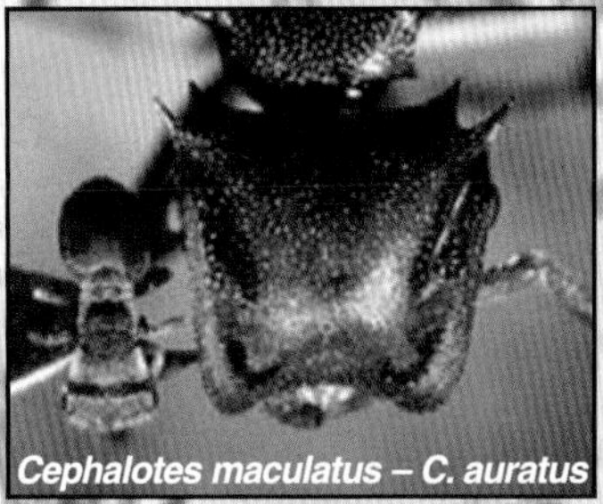

Cephalotes maculatus – C. auratus

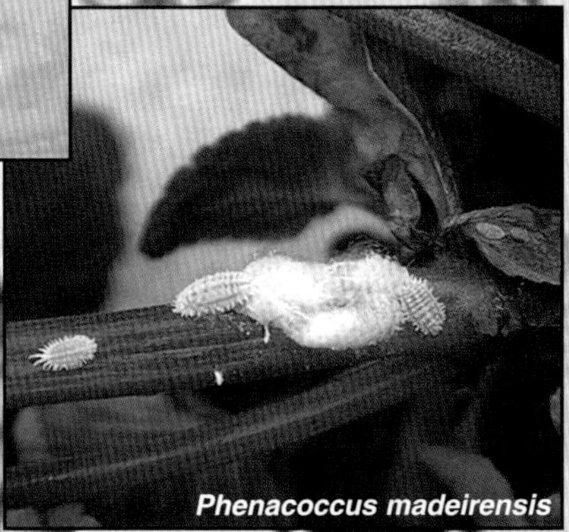

Pheidole instabilis Emery

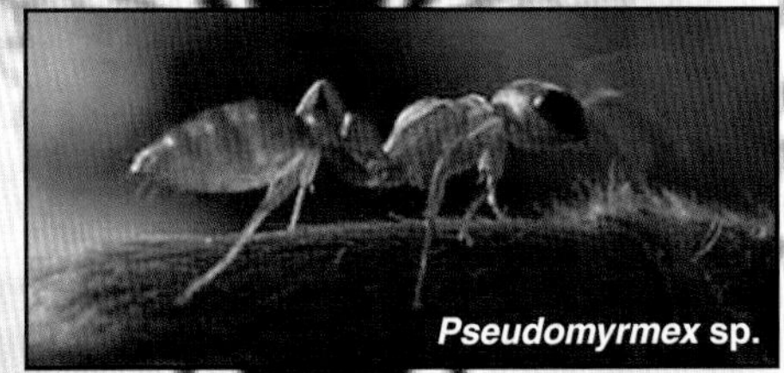

Pseudomyrmex sp.

Pseudomyrmex sp.

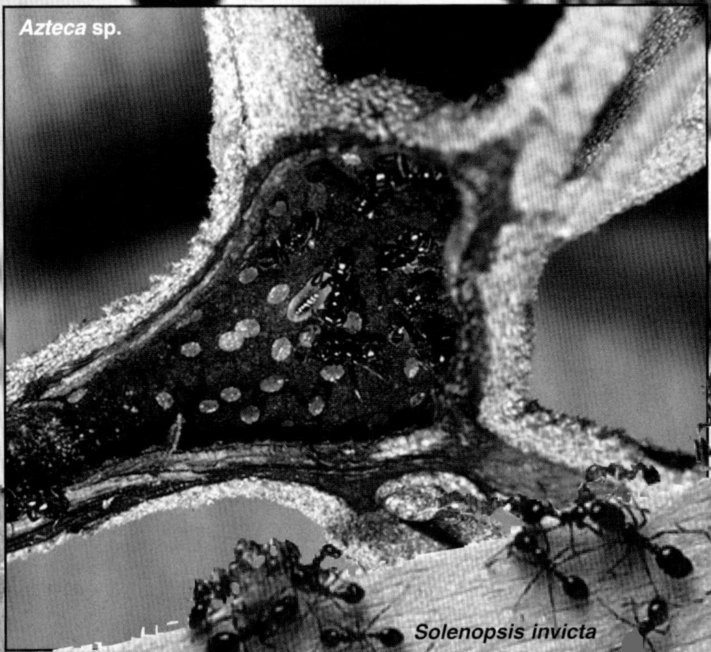

Phenacoccus madeirensis

Azteca sp.

Daceton sp.

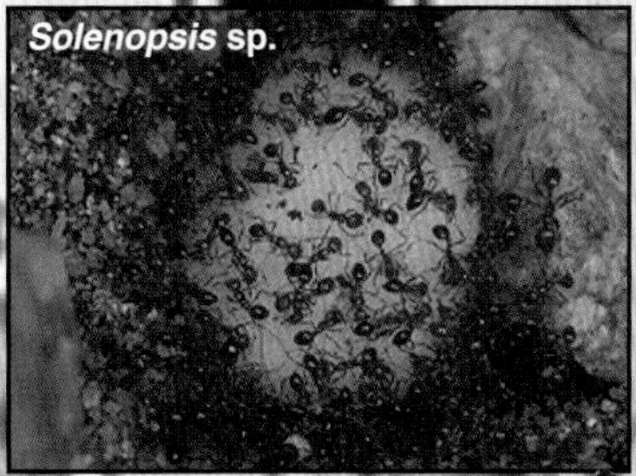

Solenopsis sp.

Solenopsis invicta

of the research into this unique co-evolution of reciprocal assistance has taken place in Central America.

Solenopsis is the genus of the (world) famous fire ants. These little beasts have often made my life hell, tormenting me for hours on end and even more so during my sleep. Including only recently, at the end of 2005, during my 7,500 km expedition through the state of Mato Grosso where over a period of 10 days and nights I fished and researched 28 habitats and catalogued more than 150 fish and 15 potential aquarium-plant species. They included at least three new characins and a new mailed catfish. And it was for the latter that I had to suffer until the early hours of morning. The rainy season had begun and the fire ants were marching in an endless broad column along the bank of the *corregós* (stream) whose course I had followed deep into the jungle, and they were also precisely where shortly after midnight I had discovered a resting *Aspidoras* in the beam of my flashlight. I had to wade through them barefoot and they clambered up my legs as if possessed. In a matter of seconds my body was burning like fire, including where it hurts most! I beat around me wildly, rubbed my legs up and down to get rid of them, tore off all my clothes, but nothing seemed to help. They bite in so hard that often you remove only the body and the heads of these 2-3 mm long ants remain in your flesh. Only when I stood there naked and picked them all off did the pain abate somewhat, but I still didn't have the *Aspidoras*. So I had to go through the whole thing again, this time trying it with shoes – but to no avail. The drama lasted for several hours and they wouldn't let go even in the water. Not until around 5 am, when it was already growing light, and only after spending an hour looking for my car keys (which had fallen in the water), was I able to make my way back and travel onward. But even while driving I was bitten a few more times – in the worst possible places. I have often had similar experiences when catching discus – so often that I prefer not to remember all those agonising hours. But apparently they don't bite discus when the latter eat them. I have found single individuals of this genus in the stomachs of two greens in the Tefé region and four blues in the Coari drainage.

The worst species are *S. invicta (p. 581)*, *S. geminata*, *S. richteri*, and *S. xyloni*. And anyone who had once made their acquaintance can tell a tale or two about it. For a long time now in Australia too, where they have found a second home, and in 14 southern and western states of the USA. They supposedly travelled there as stowaways in ships' cargoes.

Nasutitermes nigriceps soldier

In Amazonia I have recorded two more species of the same subfamily (Myrmicinae) as the fireants, which may also be of relevance for discus. One is *Cephalotes atratus*, which lives on larger trees in the Amazonian inundation-zone forests, but occurs widespread and frequent north to Central America. They move actively up and down the trees via the channels in the bark and I have several times come across aggregations comprising numerous individuals in the rootstocks of orchids, bromeliads, and other epiphytes. But this is a xyloecete species, which builds its nests mainly in holes in trees. The species is spiny and around 1 cm long. Undoubtedly only workers and/or larvae are eaten (if at all) – and then usually *C. maculatus*, which looks similar but is much smaller *(p. 581)*. And the other was *Daceton armigerum* which has been shown to produce a so-called trail and recruitment pheromone. (Recruitment is defined as communication that brings nest mates to some point in space where work is required.)

I am fairly sure that it was these two ant species that I found in the stomachs of three Blues in the Manacapuru region. But only 5-8 workers per Discus.

The distribution of the leafcutter ants of the genera *Acromyrmex* and *Atta* is restricted to the tropical Americas south and north of the Equator. Thirty-nine species are known in the two genera, and it was thought that they fed on the pieces of leaf they cut *(p. 579)* until the naturalist Thomas Belt (1874) first suggested the possibility of fungus culture, a hypothesis that was scientifically confirmed in 1893. Unfortunately I haven't found any of them in discus stomachs, but discus undoubtedly feed on their larvae when the opportunity offers (fallen from trees into the water as ithe result of heavy rain or strong winds, dislodged by other creatures, or washed in by the rapidly rising water at the beginning of the rainy season).

Isoptera – termites

Browns, blues, and greens feed on them.

The termites (sometimes inaccurately termed white ants) belong to the winged insects (subclass Pterygota). They grow to between 2 and 20 mm long *(Macrotermes goliath* can have a wing length of 88 mm, and the queen of *M. natalensis* with swollen abdomen can be up to 140 mm long). According to the latest data there are 283 genera with 2,761 species worldwide, but classification at family level often varies. Engel & Krishna (2004) distinguish the Hodotermitidae (gatherer termites); Kalotermitidae (dry and damp wood-dwellers); Rhinotermiti-

ISOPTERA

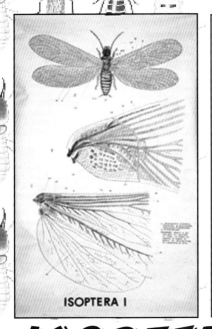

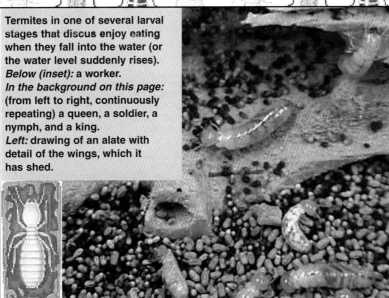

Termites in one of several larval stages that discus enjoy eating when they fall into the water (or the water level suddenly rises).
Below (inset): a worker.
In the background on this page: (from left to right, continuously repeating) a queen, a soldier, a nymph, and a king.
Left: drawing of an alate with detail of the wings, which it has shed.

ISOPTERA I

Termites – workers of the family Rhinotermitidae

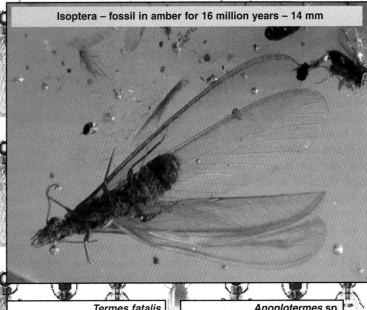

Isoptera – fossil in amber for 16 million years – 14 mm

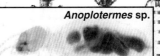

Termes fatalis *Anoplotermes* sp.

dae (hypogean termites); Termitidae (epigean termites); Termopsidae (rotten-wood termites). Termites feed by cultivating fungi or eating dead plant material, in particular wood. Only a minority of the species are completely or partially phytophagous. The gut of termites is home to unicellular flagellates that are able to digest cellulose. Termites make nuptial flights once a year, and these magical spectacles are triggered by rainfall. The comparatively large, winged reproductive forms have the task of carrying the genes of this huge agglomeration of species in this world. When two termites have found one another they shed their wings and look for a hiding-place – a safe place for the founders of a kingdom. Even if only one pair survives there will be a new kingdom with new megafamily, working in perfect cooperation to construct mighty fortress. At the centre of the termite construct lies the reproductive organ of the colony: the chamber of the queen, the only fertile female (and a replacement female is almost always present). Throughout her approximately three years of life the queen operates like a production line, every 3-4 seconds producing an egg around 3 mm in length and which has a two-week incubation period during which it is tended by workers. The majority of termites are blind, they communicate via pheromones, chemical substances with a communication function. The metabolic processes of termites, along with the fungi that they cultivate in the interior of the structure, produce a lot of heat and moisture in the nest, thus creating a climate favourable to the termites, with a temperature of 30 °C and 96-99% atmospheric humidity. If the intrinsic humidity is insufficient they bring in additional water. In desert regions termites utilise water sources that are virtually inaccessible to other animals or humans. Up to 40 m deep shafts have been found beneath termite nests, used to carry water up drop by drop.

In Amazonia I have found four genera of one family (Termitidae) and one species of a second (Rhinotermitidae) that are relevant for discus. *Nasutitermes* spp., whose striking soldiers *(p.582)* I discovered up in the trees where they had their nest *(see also p. 522)*, belongs to the family Termitidae, as do the next three species mentioned here. But the four blues captured in the Manacapuru region had swallowed only several eggs and in each case 2 - 8 larvae. One had even consumed two workers. The termite species in question may have been *N. nigriceps (p. 582, centre)*. Likewise the workers and eggs that I found in the stomachs of two greens in the Rio Tefé may have been the species that Linnaeus (1758) described as *Termes fatale* (from a soldier form from Guyana) and which is widespread in Amazonia *(p. 583)*. I found *Microcerotermes* (larvae and workers), which also build their nests in or on trees *(centre)*, in three Blues from lago Jari at rising water. Apparently the termites were taken by surprise by the rapid rise in water level, to the benefit of the discus, which had consumed between 2-5 larvae and 2-3 workers. Probably *M. arboreus*. I found a species of a further genus, *Anoplotermes (p. 583, soldier)*, in the stomachs of four browns in the Tapajós at the beginning of the rainy season. Workers and larvae. Each Discus had consumed about 5 individuals.

I was able to determine the only other termites I have found in the stomachs of discus only to family level (Rhinotermitidae). The four greens in question, in the Lago Tefé region, had swallowed quite a number of larvae – up to 12 each discus. They were subterranean termites only 3-10 mm long and must have been flooded out by a strong tropical rainstorm. The discus had a feast. In addition, something I have seen other fishes feed on *en masse* (but have no evidence for discus) is termites that died during their vast nuptial flights and fell onto the water – or the wings they shed. And I would also like to mention that wherever I have been in the Tropics, whether in the countries of the Americas, Africa, New Guinea, Australia, or Asia, the local people everywhere have used termites as bait for fishes. The fishes pounce on them when they land on the surface like humans on strawberries and cream. Discus undoubtedly eat a lot more termites when they have the opportunity.

One further comment: termites are native to Europe around the Mediterranean, and also introduced in the regions around Paris and Hamburg, as termites from the eastern USA landed there back in the 19th century. If you live in these areas you need only to "dig around" to find plenty of discus food...

Diptera – dipterans:
Brown and blue discus feed on them.
I have already mentioned *(p. 572)* the dipterans and their partly amphibious and those considered to have a totally aquatic lifestyle. Here it is the terrestrial blackflies (*Simulium* sp., family Simuliidae) that I have recorded in discus. These are small compressed flies, only around 1-1.3 mm in length, with strikingly broad, crystal-clear wings and thickened legs *(right)*. They live only in the open and lay their eggs underwater in fast-flowing streams and rivers, and in rapids (ie not in the discus habitat). In the Tropics they often occur *en masse*, but only the females "sting" and are vectors of blood parasites, poultry diseases, and various worm infestations. The best-known of these diseases is the so-called "river blindness" which affects around 20 million people worldwide, 99% of them in West and Central Africa. During my expeditions to Guinea and

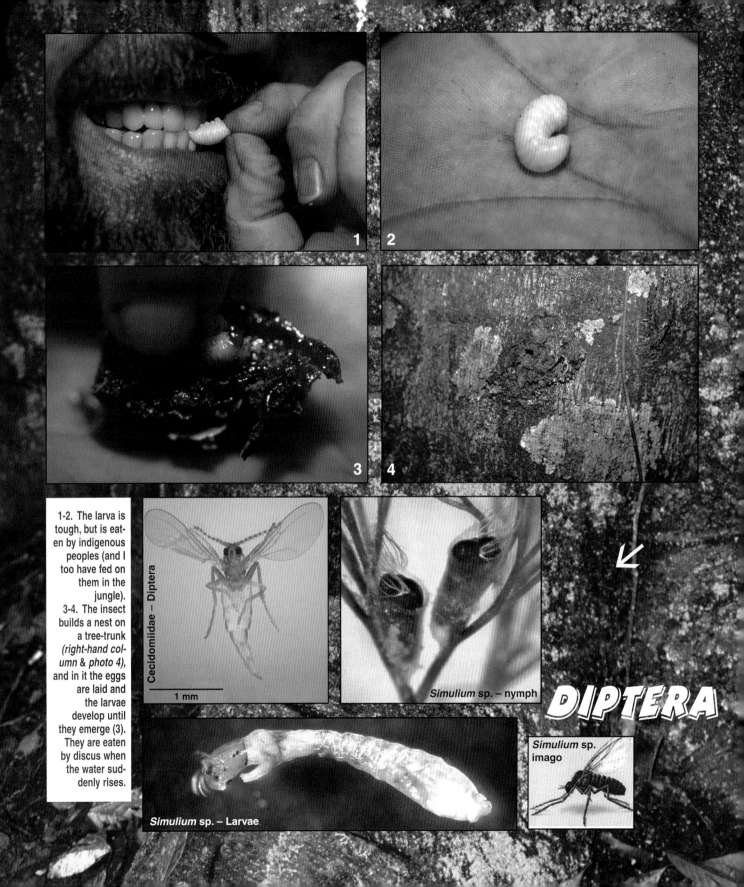

Mali I repeatedly encountered people who were almost or totally blind, making it difficult or impossible for the men to catch fish, their daily bread, in the Niger. In Bamako, the capital of Mali, I more than once came across blind beggars who had formerly been fishermen; and the worst affected were the people of the village of Fougadougou, where the economy and survival itself are dependent entirely on fishing, as practically nothing grows any more in this desert region. But a small white pill should provide relief (or rather, prevention). As in 1987, two years before my last visit the US pharmacological company Merck & Co made the Mectizan pill, which supposedly prevents river blindness, available free of charge to the then approximately 300 inhabitants of Fougadougou. When an improvement was seen (reduction in new cases), the pills were distributed to almost a million people in Mali, under the auspices of a government program and Sight Savers International. Towards the end of 2004 the 50-millionth pill was swallowed in West Africa. Perhaps this will also help improve life expectancy, which is, for example, less than 33 years for a man in Liberia. Although the pill has to be taken annually, after 20 years immunity against river blindness is achieved. During my numerous African expeditions I have been noticeably lucky (apart from bilharzia, which I have come home with three times and undoubtedly still carry), and in Amazonia river blindness is virtually unknown. Including among discus, which occasionally consume these black flies...

I have come across them only twice in discus stomachs: in two browns in the Curuá region (where there are many – or rather very many – of these insects in the tributaries) which had several – fragmented – in their stomachs. And in two browns in the Arapiuns, which had likewise eaten several. Undoubtedly many discus feed on them when they die and drop onto the water. And especially where there is fast-flowing water nearby.

The biology of the blackflies has been little studied. However, 118 diatom species (Bacillariophyceae) have been found in the larvae of three blackfly species. And 50 genera of Bacillariophyceae, Chlorophyta, Cyanophyta, and Euglenophyta have been found in the stomachs of *Simulium incrustatum* larvae. Thus the discus that feed on them indirectly consume a not inconsiderable quantity of algae *(see also p. 554)*.

Ephemeroptera – mayflies or dayflies:
Brown, blue, and green discus feed on them.

The mayflies are thought to be the most primitive of the winged insects. They have been in existence since the Carboniferous (about 300 million years ago). Again they constitute an order within the insects (see also p. 574), with around 2,800 species (around 200 in Europe). The individual species attain a body length of 3-38 mm and a wingspan of 80 mm maximum (genus *Euthyplocia*). The largest European species is *Palingenia longicauda*, which attains 130 mm long overall and occurs only in the Hungarian Tisza area.

The forewings of mayflies are significantly larger than the hindwings. Both wings have a pronounced and clearly visible venation and, unlike in all the more recent winged insects cannot be folded back against the abdomen, but instead are folded upright above the back when at rest. (Butterflies are part of the more recent winged insects and they also cannot fold their wings against the body.) In addition these insects have two or three long abdominal "tails", the outer two being the cerci and the middle one the terminal filum. Also striking are the large facetted eyes *(p. 585),* which are modified into double eyes or turbinate eyes in the males of many species; in such cases the original facetted eye is divided into two parts, one oriented laterally and the other upwards. The anterior legs are also modified in males and have claspers for copulation at the hind end. Adult individuals are characterised by atrophied mouthparts and a gut with no digestive function. When inflated hard with air the central gut serves the insect as a body-stabilising "skeleton", and the connections with the anterior and posterior gut are closed off. Modification of these organs is possible because the adult insects no longer feed. They live for only a few days, which they use for copulation and egg deposition. The final moult of the larvae into adult individuals usually takes place synchronously, and for this reason large swarms of male mayflies often occur. Females fly into these clouds where they are seized by the males, and mating takes place in flight. The females lay their eggs in the water of rivers and streams (less frequently in still waters) after previously flying a number of kilometres upstream (compensatory flight).

I have repeatedly seen this incredible spectacle in Amazonia. Million upon millions of these dayflies whirring around us and dying (after mating) in their millions above the water while the fishes had a field day. The last time was in July 2004 on the Rio Iriri (a Xingú tributary). It was evening and hundreds of thousands came to the surface to fill their bellies to bursting with these mayflies, specifically *Asthenopus curtus* (possibly also *Tortopus harrisi* – p. 5/9; both family Polymitarcyidae). In addition, the species of the genus *Ulmeritoides* (family Leptophlebiidae) are similar in appearance *(below);* they too live in Amazonia but south of the Amazon.

I have recorded these mayflies in the stomachs of four greens in the Juruá region; in four browns in the lower Xingú; and in two browns and two blues in the Lago Jarí. In each case there must have been 3 – 10 individuals (sometimes only fragments) of one (or more) of the species mentioned.

Collembola – springtails:
Browns and greens feed on them.

This is an order of widespread, important, ground-dwelling insects that are involved in the breakdown of dead plant material. They constitute the largest order within the ancient insect group (around 400 million years old), and form part of the class Entognatha. (Some authors ascribe class status to the Collembola.) The Entognatha which are together with the Insecta direct children of the Hexapoda, also include the orders **Protura** and **Diplura**. The Collembola are small, fundamentally wingless creatures, typified by a characteristic springing organ

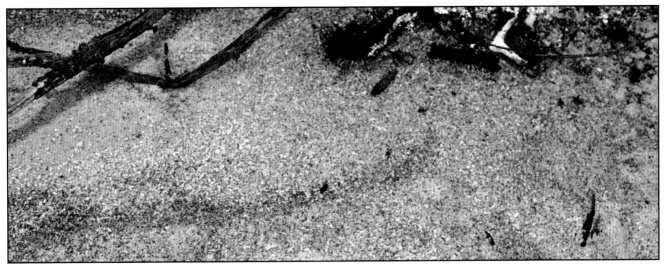

These two photos were taken at night on the Rio Iriri. Minutes after sunset the sky was white with millions and millions of mayflies (perhaps *Tortopus harrisi* – drawing left-hand page, centre). They died soon after mating above the river and covered the water's surface *(top)*, whereupon thousands of fisheshad a feast and snapped them up in a frenzy. The sandy shore was also thickly covered with them *(above)*.

(furcula) on the underside of the body. With the aid of this they can leap large distances relative to their body length. The more than 7,500 species (around 2,000 of them in central Europe) are found worldwide in almost all types of habitat, even above the snowline in mountainous regions, as well as in marginal zones of the Arctic and Antarctic, where they feed on algae and pollen. By far the majority of the species, however, live in the upper layers of the soil, on plant waste, though they are sometimes also found on the water (where fishes can snap them up). They sometimes occur in vast numbers.

Collembola are small, 0.2-10 mm long (usually 1-2 mm) animals. The body is elongate and cylindrical or compressed and spherical. They are normally grey to brown, sometimes also colourless or colourful, and often densely covered in hairs. The mouthparts may be either biting-chewing or piercing-sucking in form. The individual mouthparts are inset into the head capsule (endognathous). The eyes are not compound, and are sometimes atrophied in soil-dwelling species, which may nevertheless be light-reactive, perhaps because of the presence of up to eight occelli. The antennae are variable in length *(p. 587)*.

I have found specimens of this insect order in discus stomachs, but rarely. There is no doubt that access to these unique insects not so easy for discus, although I have encountered them time and again in Amazonia. And Goulding *et al.* (1988) found them in significant numbers in the stomach of only one characin species. I have found two brown discus in the Tapajós region and three greens in the Mamirauá region had consumed a number of individuals. It is very difficult to quantify the amounts consumed in percentage terms. One Collembola species looked like *Pseudosinella dubia* (family Entomobryidae), which was said to be endemic to the USA (Arkansas), but has recently been also recorded from South America. And the second species could have been *Hypogastrura purpurescens* (family Hypogastruridae). The latter has apparently a worldwide distribution, as it is known from Australia to northern Europe and is certainly also found in the Americas.

Diplura– two-tailed bristletails:
Heckel discus and perhaps Greens feed on them.

The order Diplura consists of eyeless insects, as a rule only a few millimetres (rarely 6 cm) long. The abdominal segments bear vestigial extremities (styli) and inflatable bladders. They live in the ground (in Amazonia often in the inundation forest), in plant litter, and beneath stones, and there are around 800 species worldwide (around 50 in central Europe). A number of species are predatory, but the majority feed on fungal hyphae and detritus. I have found species of only one of the eight families in discus stomachs.

The species *Parajapyx adisi* (family Projapygidae) survives underwater, dormant in its silken cocoon, for 5-7 months in the inundation forest – mainly in the *igapó*, the blackwater habitats. I found them in three Heckel discus stomachs at rising water. But only a few individuals (1-3). Two greens in the Tefé-region had also what looked to me like two-tailed bristletails in their guts, but the correct identification is still pending.

Remarks: The terrestrial millipede *Myrmecodesmus adisi* (class Diplopoda: order Polidesmida: family Pyrgodesmidae) can also live for a long time underwater in the inundated areas, on rootstocks in black water, and feeds on algae. Likewise *Hanseniella arborea* (class Diplopoda: order Symphyla: family Scutigerellidae) survives the high-water period. And these are both good candidates for Heckel and green discus to eat.

Acariformes – mites:
Blue discus feed on them.

I won't go into detail regarding this large order of mites with over 20,000 species worldwide, which belongs to the class Arachnida (spiders, mites, scorpions, and other arachnids). Except to mention that I have found one species from the suborder **Oribatida** *(see also Perizoon, p. 576)*, family **Haplozetidae**, in discus stomachs – the only terrestrial arthropods found during my researches that don't belong to the insects.

I have found a number of individuals of this one species, *Rostrozetes ovulum* which is widespread in Amazonia (*R. foveolatus* is a junior synonym). The morphology of this species is very variable between all the areas where it occurs, whether in eastern Peru, Bolivia, Brazil, Venezuela, Panama, El Salvador, Mexico, the eastern USA north to Canada, Angola, Madagascar, the Maldives, or Sumatra, and yet there is only one subspecies, from a Peruvian montane region. Adults feed predominantly on fungi and can survive for a long time underwater. And that is also the time when discus eat them. That is, only when the water rises (and it is impossible to estimate how many they may then consume during the high-water period).

I found several specimens in the stomachs of three Blues as the water was starting to rise. Freshly swallowed.

Summary: As mentioned at the start of this section, terrestrial invertebrates are not widely available to discus in Amazonia. And, apart from a few of the groups mentioned, they are undoubtedly not significant. They constitute a mainly sporadic supplementary food supply which depends on circumstances (and is often a matter of chance).

An overall summary of discus nutrition can be found on the following pages.

COLLEMBOLA

COLLEMBOLA

Hypogastrura purpurescens

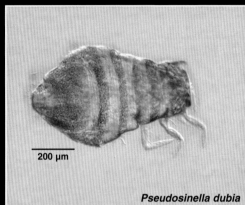
Pseudosinella dubia

ACARI

Rostrozetes ovulum

Hanseniella sp. – Scutigerellidae

Parajapyx adisi

Mymecodesmus adisi

DIPLURA

Parajapyx sp.

A Summary of the Discus Diet in the Wild

Essentially one should be aware that there are, of course, a whole series of significant factors involved in the food intake of discus in the wild, some of which I have mentioned above.

I would now like briefly to summarise the various results of my researches, and show schematically which groups of organisms are consumed by each discus species in the dry season and during the high-water period. However, I suggest that the following should also be taken into account: as well as my results, based on decades of study, those of other researchers who have published material on the diet of various Amazonian fish species, should also be mentioned and taken into consideration (as I have already done in part – and a few further examples are given below). Evaluating all these results together undoubtedly produces an even better understanding of the food intake of Amazonian fishes.

First of all I would like to point out that the low pH and conductivity of the nutrient-poor waters in some regions of Amazonia (and not only there) have obliged the flora and fauna to adapt to these extreme values in the course of evolution in order to survive. But not all life forms, by far, have done so, as should be apparent from the text above. However, one that has is *S. discus*, and it can now survive in the wild only in places where these low values obtain. In all probability this is also a limiting factor in the distribution of *S. aequifasciatu*s. These circumstances undoubtedly contribute to the differences in the diet of the three discus species in the wild.

The numerous studies also show that in white-, mixed-, and some clearwater regions, where the pH is 6-7 or more, and the conductivity more than 25 μS/cm, the biomass and propagation rate of algae and aquatic invertebrates are appreciably greater (in the dry season and at rising water) than in waters with lower values for these parameters. As a result the availability of aquatic invertebrates is appreciably greater for brown and blue discus than for green or Heckel discus. It is undoubtedly also for this reason that browns and blues have a much larger distribution region than the other two species. An increased food supply offers more opportunities for development and reproduction. This is supported by my more than 30 years of feeding hundreds of wild-caught discus almost every day. Freshly collected brown and blue discus have refused all usual dried foods, but have quickly taken live or deep-frozen aquatic invertebrates. (Dried foods are too far removed from what Nature has to offer, while aquatic invertebrates such as *Chironomus, Daphnia, Cyclops*, etc. are still the most similar to some of the natural foods.) Green discus, by contrast, always take several days and Heckel discus the longest (if ever) to accept commercial frozen foods...

My researches show that detritus is consumed by all three discus species and is their main food over much of the year. Only at the beginning of and during the high-water period, as well as in the initial stages of the water level dropping, is this not the case. At the beginning of rising water fruits, seeds, and plant material in general constitute the main diet of discus. Between April/May and September (the time of maximum water level and the water receding) the fruits and seeds of the numerous tree species mentioned are available to discus (and other fish species), and undoubtedly also others such as the widespread species *Cecropia latiloba (embaúba), Vitex cimosa (tarumã),* and *Crateva benthami (catoré),* which bear fruit throughout the entire period (Yamamoto *et al.*, 2004).

My results, as well as those of other workers (on various Amazonian fish species) demonstrate: during the low-water period the fishes – in particular the larger non-piscivorous species – often eat very little (as there is little to eat), or nothing at all. Not only have I repeatedly netted emaciated discus towards the end of the dry season, but numerous other Amazon fishes were half-starved at this time. Mainly other cichlids such as *Heros, Hoplarchus, Hypselacara,* and *Uaru* (and to a lesser extent *Geophagus* and *Satanoperca*, as these "eartheaters" sieve through the sand), or numerous large characin species of the genera *Leporinus, Laemolyta, Schizodon* and *Rhytiodus.* But I have also often seen emaciated stingrays (family Potamotrygonidae), as well as millions of cardinal tetras. (For this reason the latter rarely live more than a year in the wild. But they don't occur in the discus habitat.)

Apropos of which the following study is of interest and worth mentioning. For more than a year the stomach contents of the characin *Triportheus angulatus* in the Lago Camaleão (central Amazonia) were investigated and it was established that during in the high-water period specimens of 8-12 cm (all sizes are SL) had consumed around 60 % fruits and seeds, and 12-20 cm specimens as much as 100%; and, as the water level fell, it was found that all specimens of 4-20 cm or more had consumed practically only fruits and seeds. While during the dry season 67-100 % of all sizes of the 800 specimens studied having empty stomachs at this time. Ie they had consumed no food (Yamamoto *et al.* 2004).

The amounts of food, ie the percentage components of stomach contents, found during the dry season agree with other studies (Lowe-McConnell, 1964; Goulding, 1980; Goulding & Carvalho, 1982) which likewise indicate that the dry season is a lean period for numerous Amazonian fish species. And in addition by far the largest number of empty fish stomachs were found in adult specimens. Juveniles often contained some food during the dry season.

Heckel discus – *Symphysodon discus*
Low-water period

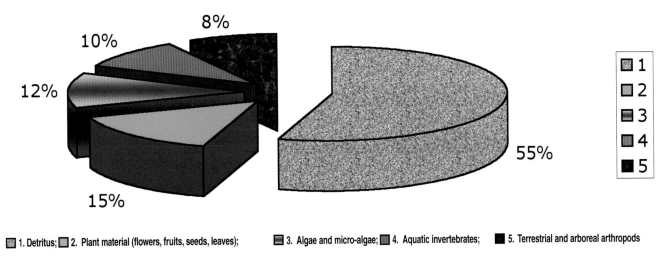

■ 1. Detritus; □ 2. Plant material (flowers, fruits, seeds, leaves); ■ 3. Algae and micro-algae; ■ 4. Aquatic invertebrates; ■ 5. Terrestrial and arboreal arthropods

The schematic shows a cross-section of the components of the Heckel discus diet recorded by the author during the low-water period (up to the beginning of the high-water period). During this period detritus predominates, followed by plant material, although during the dry season only *acará-açú* and a few other plants are available to discus. Of the algae (which are more strongly represented when the water begins to rise), green algae and blue-green "algae" predominate. As regards the invertebrates, aquatic forms were taken (found) fairly regularly, but rarely terrestrial and arboreal arthropods.

Heckel discus – *Symphysodon discus*
High-water period

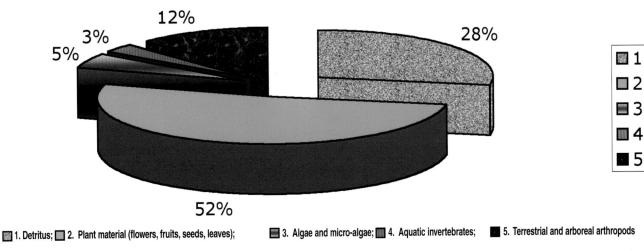

■ 1. Detritus; □ 2. Plant material (flowers, fruits, seeds, leaves); ■ 3. Algae and micro-algae; ■ 4. Aquatic invertebrates; ■ 5. Terrestrial and arboreal arthropods

After the water has risen considerably it is difficult to arrive at a precise evaluation of the food intake of discus. But the values given here for Heckel discus (and analysed by the author during years of research), should be very close to the actual values during flood season. They also reflect what is available during this period in the extreme (Heckel discus) blackwater habitat. Without exception plant material predominates, followed by detritus, terrestrial and arboreal arthropods, and algae, with aquatic invertebrates least represented at this time.

Green discus – *Symphysodon aequifasciatus*
Low-water period

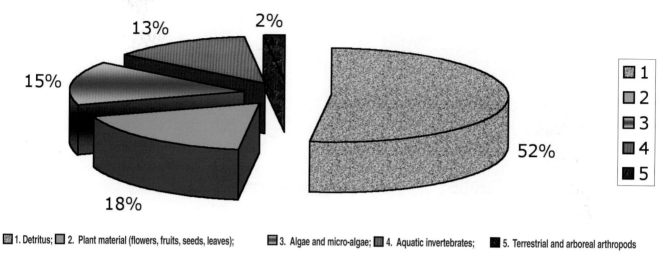

■ 1. Detritus; ■ 2. Plant material (flowers, fruits, seeds, leaves);   ■ 3. Algae and micro-algae; ■ 4. Aquatic invertebrates;   ■ 5. Terrestrial and arboreal arthropods

The schematic shows a cross-section of the components of the green discus diet recorded by the author during the low-water period (up to the beginning of the high-water period). Detritus predominates in the dry season, followed by plant material, although little of the latter is available to discus, apart from *acará-açú*, and in some places *camu-camu, inter alia*. The algae often occur as "blooms", chiefly at rising water. Here too green algae and blue-green "algae" predominate. As regards the invertebrates, aquatic forms were taken (found) fairly regularly, but very rarely terrestrial and arboreal arthropods.

Green discus – *Symphysodon aequifasciatus*
High-water period

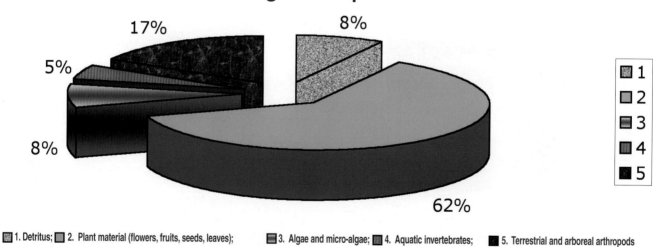

■ 1. Detritus; ■ 2. Plant material (flowers, fruits, seeds, leaves);   ■ 3. Algae and micro-algae; ■ 4. Aquatic invertebrates;   ■ 5. Terrestrial and arboreal arthropods

In green discus too it is difficult to arrive at a precise evaluation of food intake during the high-water period. But the values given here for Greens (and analysed by the author during years of research) should be very close to the actual values during flood season. Again they reflect what is available during this period in the predominantly blackwater habitat of green discus. Here too plant material predominates, but in this Amazon region followed by terrestrial and arboreal arthropods, then algae with detritus, and aquatic invertebrates least represented.

Brown and blue discus – *Symphysodon haraldi*
Low-water period

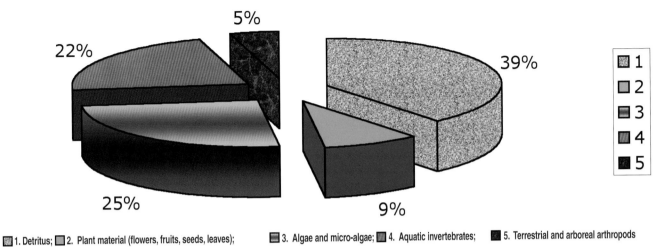

1. Detritus; 2. Plant material (flowers, fruits, seeds, leaves); 3. Algae and micro-algae; 4. Aquatic invertebrates; 5. Terrestrial and arboreal arthropods

The schematic shows a cross-section of the components of the brown and blue discus diet recorded by the author during the low-water period (up to the beginning of the high-water period). Detritus predominates during the dry season, followed by algae (blue-green "algae" and brown algae predominate). But there is an almost equally large amount of aquatic invertebrates, which are richly available in the habitat around this time of year. Very little plant material, as this is largely inaccessible – in the main only *acará-açú* and *camu-camu* are available. Terrestrial and arboreal arthropods are only very rarely available.

Brown and blue discus – *Symphysodon haraldi*
High-water period

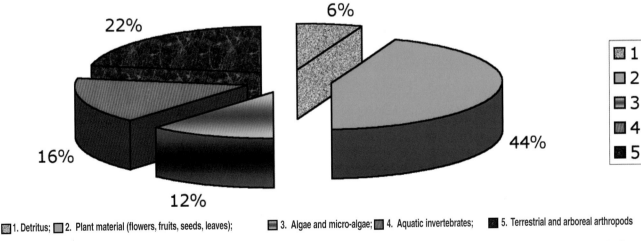

1. Detritus; 2. Plant material (flowers, fruits, seeds, leaves); 3. Algae and micro-algae; 4. Aquatic invertebrates; 5. Terrestrial and arboreal arthropods

In brown and blue discus too it is difficult to arrive at a precise evaluation of food intake during the high-water period. But the values given here for Greens (and analysed by the author during years of research) should be very close to the actual values. Again they reflect what is available during this period in the predominantly clearwater habitat of brown and blue discus. Plant material predominates, but followed by terrestrial and arboreal arthropods and aquatic invertebrates. Than algae and detritus represents the smallest component during this period.

In addition, Silva *et al.* (2000) studied *tambaquí* stomach and gut contents over the course of an entire year, and found that during the high-water period, when there is a rich supply of fruits and seeds, the protein content fell to a relatively low level (between 11 and 15% for the foods consumed) but during the low-water period, when there was little to eat and predominantly zooplankton was consumed, the protein content was much higher, varying between 47 and 57% of the foods consumed).

The latter doesn't necessarily always apply to discus, but, as regards proteins, the much-cited two varieties of *camu-camu* fruit trees (bushes) and their vitamin-C "boost" (60 times more vitamin C than oranges) are available over a very long period for discus (and other fishes – but undoubtedly discus in particular, as they often live there). These bushes are found almost exclusively in the lowest-lying inundation zones and *igapós* that are thus underwater most of the time (often the entire year) and accessible to discus.

It is, of course, also important to take a general look at the vitamins that are significant for discus (and other fishes). Here too an analysis by Silva *et al.* (2003) is very interesting. They discovered that at least 133 tree species in central Amazonia produce fruits and seeds that are consumed by Amazonian fish species during the high-water period and have been recorded during stomach and gut studies, either whole or part-digested. The majority of the fishes concerned, and discus too, consume a variety of fruits and seeds in order to achieve a balanced amount of proteins, carbohydrates, fats, and vitamins in their diet (Araujo-Lima & Goulding, 1998).

In the *tambaquí* alone 40 different types of fruit were recorded (Silva, 1997) and their composition analysed in greater detail. Based on dry fruit material the following nutritional values were obtained: 7.4 ± 3.33% crude protein, 10.94 ± 7.12% lipíds, 50.34±13.08% carbohydrates, 20.57 ± 11.54% crude fibre, 3.3 ± 1.26% ash, and a crude energy of 1851.0 kJ/100 g MS.

The analysis of 14 types of seeds species produced mean values of 11.2±6.5% crude protein, 18.5±16.1% lipids, 49.2±20.2% carbohydrates, 11.4±9.4% crude fibre, 3.0±1.5% ash, and 20.7±4.1 kJ/g MS crude energy.

And last but not least as far as plant material and fruits are concerned, Heckel discus also make considerable use of them, as in the Rio Negro region (Nhamundá-Abacaxís-Marimari, *inter alia*) the majority of plant species that occur along the rivers are the same as those encountered in the *igapós* and *lagos*. Thus Heckel discus have the opportunity to find their preferred fruits, seeds, and perhaps flowers over a wide area almost year-round, albeit to a limited extent during the dry season.

As regards the zooplankton, I would like to add that during the period between the end of the dry season and the beginning of rising water there is a phase when the most zooplankton and terrestrial invertebrates (predominantly insects) are eaten (or available for eating), and not only by discus. In particular terrestrial insects need to be on the move, especially ants and termites with their vast terrestrial and arboreal colonies.

The zooplankton also propagates out of all proportion during this period (Koo, 2000), but is highly dependent on the oxygen concentration in the water. At rising water and during the high-water period the oxygen content in the water is reduced and this limits the reproductive activity of the zooplankton. Fishes appear to be less affected. Discus and numerous other Amazonian fish species hardly at all. Certain species have even evolved special adaptations to counteract the regular reduction in the oxygen concentration. A number change their lip structure (eg species of *Triportheus, Astronotus*, and other genera) in order to be able to take up larger amounts of oxygen. Or to access layers where the concentration is highest. The low oxygen concentration in the water during the high-water period doesn't cause any interruption in food intake by the fishes, either. This has been demonstrated in, for example, characins such as *T. angulatus*, as they had consumed insects, fruits, seeds, and other plant material in the inundation forest as well as in the open water. It was found (as I have discovered in discus, as well as many another fish species) that this characin can adapt to an extremely low oxygen content. At 0.4 mg/l O_2 *T. angulatus* takes up oxygen at the water's surface, aided by a "double chin" – the extension of the lower jaw (Braum & Junk, 1982; Saint-Paul & Soares, 1987; Winemiller, 1989). This type of adaptation permits the fishes not

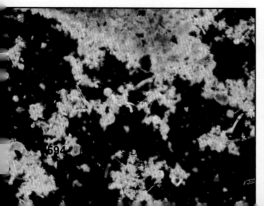

only to survive a period of hypoxia (shortage of oxygen in the water) and anoxia (no oxygen in the water), to, but also to obtain sufficient energy to pursue their normal activities, like searching for food.

The algae rank in third place in the discus diet over the year. They have long been an important nutritional source for humans. Moreover it has been shown that the algae genera *Actinella* and *Eunotia* are indicative of acid water (Patrick & Reimer 1966). Long-term stomach contents studies of the larvae of the blackfly *S. perflavum* (Diptera: Simuliidae), in which it was found that these two algae genera had been eaten, have established that this blackfly species is associated with tropical oligotrophic and prefers acid water (like discus).

Apropos of which, it is also worth mentioning a further finding: various planktonic organisms have been recorded from black, clear (mixed), and white water during long-term studies.

The following comparison relates to planktonic organisms from the clearwater Japurá drainage, mixed waters and the whitewater Solimões region:

Taxon	black water	mixed water	white water
Rotifera	284	23	0
Cladocera	5	29	43
Ostracoda	39	97	29
Calanoida	11	51	66
Cyclopoida	22	49	61
Chironomidae	0	3	3
Acariformes (mites)	0	0	2

And planktonic organisms from black-, white- and mixed-water zones, compared with open water (OW) and *igapó/várzea* (lake regions):

Taxon	black water		mixed water		white water	
	OW	*igapó*	OW	*igapó*	OW	*várzea*
Volvocaceae	42	–	38	–	–	–
Rotifera	87	5	34	–	–	–
Cladocera	6	–	5	–	8	1
Ostracoda	2	11	3	–	7	–
Calanoida	–	23	3	10	–	–
Cyclopoida	5	27	19	1	13	1
Mysidacea	1	–	–	–	–	–
Diptera	–	–	–	–	1	–
Acariformes	–	–	–	1	–	1
Fish larvae	–	–	–	1	–	1

The first comparison clearly shows that fishes (including discus) in black water (and clear water) benefit predominantly from rotifers (Rotifera) (see also p. 562). Water fleas (Cladocera) are virtually absent in black water, and are found mainly in mixed water (and even more so in white water), as are ostracods (Ostracoda) *(p. 572)*. Copepods (Copepoda: Calanoida) are also strongly represented in mixed water, although, along with the Cyclopoida (Copepoda: Cyclopida) they are found mainly in white water and hence are of only very limited availability for all discus species. This is amplified by the comparison of open water and *igapó*, where algae (Volvocaceae) predominate in the open black and mixed water, followed by the rotifers.

These two tables are thus in every respect also strong evidence for discus food intake.

In conclusion, it is interesting to note that Goulding *et al.* were unable to find either whole fishes or fish flesh in any of the numerous *Uaru* stomachs investigated. And I too have not found a single fish, let alone fish flesh, in any of the discus stomachs and guts studied.

This evidence of the diet of discus in the wild should give all enthusiasts, breeders, and even the experts something to think (or rather, rethink) about.

Discus communities – Sympatric species & Predators

In the course of the text overall I have on several occasions discussed the fish species that occur together with discus in the wild or sporadically visit their territory (eg on pages 85, 92-93, 95, 265, 269, 281, 295, 299, 317-318, 329, 352-353, 371, 381, 393, 398, 407, 419, 425, and 464). To avoid repetition, on the following 27 pages I will catalogue only those fish species not previously mentioned that I have encountered in the relevant discus habitats. In each case with a short text. I will mention the discus species that occur in the region concerned, as well as the fishes which I have found to be only "visitors" – welcome as well as unwelcome.

It should be noted that by day discus live together in large aggregations, and as a result there is little room for other species in their biotope. Apart from the groups that occur (live) near (or at) the water's surface, for example, species of the characin genera *Triportheus*, *Chalceus*, and *Bryconops*, and cichlids of the genus *Mesonauta*. And of course, the predominantly bottom-dwelling species that impinge very little on the discus, such as eartheaters of the cichlid genera *Geophagus* and *Satanoperca*, catfish species of the family Loricariidae, and, of course, stingrays, which are often encountered living in the sandy bottom of discus habitats during the dry season. In addition there are the "visitors", such as characins of the genera *Leporinus*, *Schizodon*, *Tetragonopterus* and *Laemolyta*, as well as a number of larger *Crenicichla* species. Small and very small fish species are rarely to be found in the discus biotope. Hardly ever any small characins or catfishes. No *Corydoras*, *Aspidoras*, small *Loricaria* or *Rhineloricaria*, *Ancistrus*, *Peckoltia*, *Hypancistrus* and similar small so-called L-number catfishes. Likewise the very large catfishes, such as *Phractocephalus hemioliopterus*, and species of the genera *Pseudoplatystoma*, *Brachyplatystoma*, *Sorubimichthys*, and *Zungaro* are practically never found there (exceptions are mentioned in the text and on the following pages).

The last-named species are of course, predatory fishes, ie piscivores, but to date I have never heard of them eating discus, or visiting their territory in search of prey. That is primarily the province of the piranhas. (The latter eat almost everything that comes near their mouths if they are hungry, but feed predominantly on the fins of other fishes – discus fins are also a favourite.) Piranhas and predatory characins of other genera, eg large *Acestrorhynchus* and *Hydrolycus* species and *Rhaphiodon vulpinus,* can pose a threat to discus. On several occasions I have watched a piranha slowly approach without being spotted by the "leader" of the discus (some piranhas have a coloration that matches the biotope completely – see photos – and so are almost invisible) and suddenly attack, biting a piece out of an often outstandingly beautiful discus (or the "leader" itself, while guiding the group to shelter), or ripping deeply into the caudal or anal fin. This is also the reason why the most splendid discus specimens often have bits missing, ie bite wounds. It is rare (and I have never yet seen it) for a discus to be torn part, let alone eaten, by piranhas. And piranhas are never shoaling fishes, not a single species of the true piranhas lives in groups. They are solitary like the majority of predatory fishes, except during their spawning migrations. At such times I have repeatedly seen hundreds of thousands of piranhas on migration.

On one occasion – and I will never forget it – this happened in the Rio Guaporé. I was swimming underwater five metres from the bank, in around 3.5-4 metres of water, when I looked up and found myself surrounded by piranhas. They were all *Serrasalmus rhombeus* (the black piranha, purportedly the most dangerous of all the species). There must truly have been umpteen thousand of them, as the water was very clear and I could see a long way – 20 metres or more. They formed a circle around me and swam round and round me for what seemed like hours. Apparently they wanted to discover if I represented a threat to them or their migration. In fact it was only two minutes (I certainly didn't have air for longer) before they continued on their way. But the procession past me really did last for hours, without interruption. Back on the bank unharmed, I followed their migration for a long time and was astonished beyond measure. Firstly that I hadn't been attacked by these, the most predatory of all the piranhas (or rather the most dangerous of all fishes, or so they say – as according to the Hollywood film makers the great white shark is a pussycat by comparison). I haven't tested the hypothesis (and perhaps there isn't enough meat on me!). And secondly, because the shoal seemed endless and I could hardly believe that there could be such a gigantic biomass of black piranhas at all. Especially not in my beloved Rio Guaporé, where as a child I made my first attempts at swimming and at the age of eight got to know the fishes here with my mother.

And now I will let the photos and captions speak for themselves, so that you can get to know and understand the discus communities and enemies. And hopefully in the future maintain discus in a more biotope-correct fashion – in aquaria all over the world. That is what I would like to see. As once you have seen a shoal of 50 or more discus in an aquarium (I don't insist on your doing so in the wild), you certainly won't want to keep just one or two specimens any more.

At the end of the discus community pages you will find a summary of these researches.

Rio Xingú

In lower Rio Xingú discus habitats I have repeatedly come across piranhas, the worst enemies of discus. And there are at least three species there: 1-2. *Serrasalmus rhombeus*, the black piranha (the most dangerous), locally called *piranha de cardume* (it said that these form schools, but I saw only solitary ones – except during migration). 3-4. An as yet undetermined *Serrasalmus* species, which looks similar to *S. rhombeus*. 5-6. According to Géry (per. comm.) this is the true *S. humeralis*. Possibly the largest of all the piranhas. 7. A *caboclo* on the lookout for aquatic turtles. 8. Discus habitat in the lower Xingú.

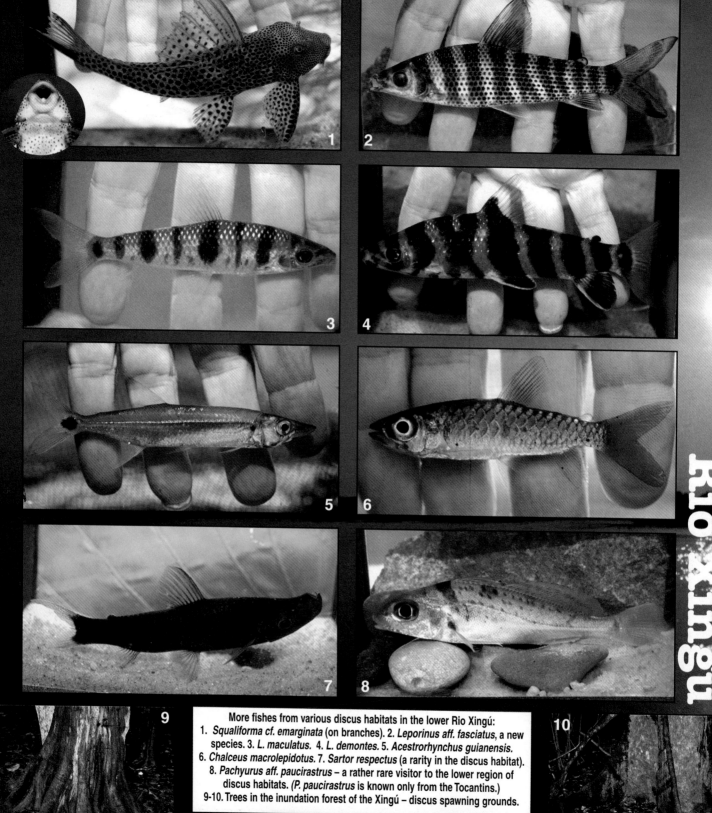

More fishes from various discus habitats in the lower Rio Xingú:
1. *Squaliforma cf. emarginata* (on branches). 2. *Leporinus aff. fasciatus*, a new species. 3. *L. maculatus.* 4. *L. demontes.* 5. *Acestrorhynchus guianensis.* 6. *Chalceus macrolepidotus.* 7. *Sartor respectus* (a rarity in the discus habitat). 8. *Pachyurus aff. paucirastrus* – a rather rare visitor to the lower region of discus habitats. (*P. paucirastrus* is known only from the Tocantins.) 9-10. Trees in the inundation forest of the Xingú – discus spawning grounds.

Rio Xingú

These species too are from discus habitats in the lower Xingú (although I have recorded them all upstream of Altamira as well, where there are no discus):
1. *Myleus cf. schomburgkii* (different to *M. schomburgkii* from the Rio Negro).
2. *M.* sp. (Myleinae C), perhaps a new species; the natives call it *pacu manteiga*. 3. *Satanoperca cf. jurupari*. 4. *Geophagus cf. surinamensis*.
5. *G. aff. proximus*. 6. *G. argyrostictus*. 7. An *Aequidens* sp. here resting close to the water's surface at night, just like discus. 8. Discus habitat in the Xingú.

Rio Xingú

Stingrays are also regularly encountered in the discus habitat of the lower Rio Xingú: 1. An approximately 100 cm diameter ray of the largest species, *Paratrygon aiereba*, in the net. 2-5. Here *Potamotrygon cf. leopoldi* (2) often occurs with interesting markings: you can find almost all the letters of the alphabet, eg J (3), C (4), and O (5). Can they actually write?! 6-7. A lot of fuel and containers (6) are needed for a Xingú (7) collecting trip up river.

ERRATA

Page 12: change Jürgen Weissflog to Günter Weissflog.

Page 13: change *sapsibo* to *spasibo*.

Page 134: missing photo caption: Heiko Bleher and Jacques Géry (from left to right, in Dr Géry's office/home).

Page 136: change Santar8Em to Santarém.

Page 402 missing captions: 1. Lago Aiapuá (black water). 2. The Aiapuá branch. 3. Satellite photo (1:100,000): A) The Parana Cuianã (white water), to the left the Purus and the Solimões. B) Várzea with countless lakes (often white water). C) The unexplored Lago Uauaçu. D) Lago Aiapuá, and E) its branch. F) Lago Caviana. G) The village of Paricatuba and Lago Água-Fria. H) Lago Grande de Paricatuba.

Page 403: change *aiapuá* = Maniok und Name des Sees to *aiapuá* = manioc and name of the lake.

Page 436: change Lago do Aracá to Lago do Acará and change Paraná do Aracá to Paraná do Acará.

Page 440: caption: missing the "**B**" (for Lago do Acará) on the satellite photo in the upper right portion of the photo.

Page 503: change: Nacht to night and change: Tag to day.

Page 505: change:

25.09.98.(13:50)	–	6.40	30.3(30.6)	21.50	21.0	Rio Branco[7]	Rio Branco[7]	W

to

25.09.98.(13:50)	–	6.40	30.3(30.6)	2.15	21.0	Rio Branco[7]	Rio Branco[7]	W

Page 508: change

05.10.96 (13:00)	–	6.15(6.30)	29.9(30.4)	29.1	18.0(18.0)	Rio Ipixuna	Tapauá (Purus)	B

to

05.10.96 (13:00)	–	6.15(6.30)	29.9(30.4)	2.91	18.0(18.0)	Rio Ipixuna	Tapauá (Purus)	B

Page 570: change Tardigrada to Tardigarda.

Page 656: change: General (page 656) (includes everything what is not listed in the following four index-register) to General (page 656) (includes everything what is not listed in the following three index-register).

Page 670: change Robert Doineau to Robert Doisneau.

change throught to through.

MISSING IN REFERENCES

Geisler, R., 1969. Untersuchungen über den Sauerstoffgehalt, den biochemischen Sauerstoffbedarf und den Sauerstoffverbrauch von Fischen in einem tropishen Schwarzwasser (Rio Negro, Amazonien, Brasilien). *Archiv. Hidrobiol.,* **66** (3): 307-325.

Geisler, R., 1970. Des derzeitige Stand unserer Kenntnisse über den Diskusfisch *Symphysodon aequifasciata axelrodi*. *DATZ,* **23**: 9-13, 40-44, 75-78, 131-133, 171-174, 196-198.

Geisler, R., 1972. I. Blaue und Braune Diskus in Amazonien; II. "Echter Discus" *(Symphysodon discus)* im Flussgebiet des Rio Negro; III. Fischtransport vom Urwald zur Sammelstelle Manaus. *TI (Tetra Information),* N. 19-20-21.

Geisler, R., 1981. Les *Symphysodon*. Les biotopes des *Symhysodon*. *Aquarama,* **59**: 37-41; **60**: 34-38.

Geisler, R. & F. Schneider, 1976. The element Matrix of Amazon Waters and ist Relationship with the mineral content of Fishes. *Amazoniana VI,* **1**: 47-65.

Geisler, R. & S. Annibal, 1984. Ökologie des Cardinal-Tetra *Paracheirodon axelrodi* im Stromgebiet des Rio Negro/Brasilien sowie zuchtrelevante Faktoren. *Amazoniana IX,* **1**: 53-86.

Geisler, R., 1996. Ökologie der Diskusfische in *Diskus:* 22-30, Ulmer, Stuttgart.

Géry, J., 2001. DISCUSsion – Le point de vue du systématicien. *Aqua Plaisir* N. 55, mars 2001: 18-20; N. 56, avril 2001: 18-19; N. 57, mai 2001: 18-19.

Junk, W. F., 1985. Temporary fat storage, an adaptation of some fish species to the waterlevel fluctuations and related environmental changes of the Amazonian rivers. *Amazoniana,* **9** (3): 315-351.

About the author

Heiko Bleher was born on October 18, 1944 in a bunker in the ruins of Frankfurt on Main. He was the fourth and last child of Ludwig Bleher and Amanda Flora Hilda Kiel. Amanda's father Adolf Kiel was the well-known "Father of Water Plants", a pioneer of the modern aquarium who established the world's largest plant and ornamental fish farm in Frankfurt. In those early days his adventurous daughter Amanda travelled around the world collecting fishes and plants. She was the first woman to ride a motor bike in Germany, and competing against men, won 148 European Moto-Cross car races, won championships in tennis, table tennis (world vice champion), European skating and ice skating, and was the first woman to fly an aircraft without an engine... Just as Amanda followed in her father's footsteps, so Heiko followed his mother's. At 4, he saw his first discus at an aquarium fish exhibition in the still ruined Frankfurt Zoo. Later he travelled with her to Africa then, aged 6, throughout Europe collecting plants and fishes. When he was 7, his mother took him, his elder brother and two sisters with her on his first discus hunt – a highly adventurous exploration trip deep into the "green hell" of the of South American jungle. They reached areas inhabited by unknown Indian tribes, some of whom had killed and eaten 4 missionaries shortly before. They lived with the natives for over 6 months, sampling 60 new aquatic plant species, countless fishes and many other animals. Still a child, Heiko learned to live like the Indians, eating the same food and collecting fishes and plants in the Mato Grosso, He learned about the life and behaviour of fishes, and became familiar with the amazing variety of fish that exist in unspoiled nature. He also discovered the wimpel-piranha, but no discus at that time.

After two years, Amanda Bleher's return to civilization with her four children made newspaper headlines around the world. In 1959 she decided to settle permanently in Brazil. It was there that Heiko helped build a water plant nursery and fish breeding establishment in the jungle outside Rio. In 1962, he moved to the US and attended the University of South Florida, studying at night and learning more about fishes. He took courses in ichthyology, biology, limnology, oceanography, parasitology and many others. During the day he worked at Elsberry's Fish Farm and later at the Gulf Fish Farm. Two years later he returned to Rio to open Aquarium Rio and start his own collecting in Brazil. He first opened several compounds in the interior, others later in other parts of South America. At the end of 1964 he discovered the first new species to be named after him – *Hemigrammus bleheri*, the brilliant rummy-head tetra, now one of the most widely-sold aquarium fishes. He also discovered the "Royal Blue", his first new strain of discus, now world famous along with many other species. Some years later Heiko explored many new, uncollected areas, and by 1967 he moved his company Aquarium Rio to Germany, returning monthly to Brazil and South America to collect. Over the years, generally alone, Heiko penetrated jungles in all South and Central American countries. He also travelled to the Amazon area as many as 10 times a year in search of discus and others species. In the 1970s he expanded his operations to include Africa, Asia and Oceania (Australia, New Guinea, etc.) and began to give lectures around the world. He made his first Discus-TV film, "Expeditionsziel Aquarienfische" with the German ZDF and made many TV appearances in different countries. His first Discus book was published in 1982 and re-printed 10 times. Since then he has published articles on discus in magazines around the world. His first documentary film "The Wimpel Piranha" was made In 1983, followed by films on freshwater fishes in New Guinea, Australia, Central America and Brazil then four films on discus in the 90s. Until 1997 from Frankfurt he supplied wholesalers world-wide with new species, including new discus variants every year, mostly from his own discoveries. Between 1965 and 1997, besides introducing most of the wild discus variants into the hobby – directly or by means of the breeders – he introduced more than 4,000 aquarium fish species he had discovered (or re-discovered). This includes the variants such as "blue-headed Heckel", "Alenquer", "Red-spotted greens" from the "Coari" and "Japurá" regions and the famous "Rio Içá" discus, and also rainbowfishes such as *Melanotaenia boesemani*, *M. lacustris* and *M. praecox* (most probably now one of the most sold aquarium fishes), angels such as *Pterophyllum altum*, dwarfs such as *Nanochromis nudiceps* and *Steatocranus bleheri* or *Channa bleheri*. Among other fishes attributable to Heiko's explorations are also many loricariids (as many as 800, at the time of printing), new *Corydoras* species, almost countless tetras and dwarf cichlids from West Africa and South America, knife fishes, puffers and flounders. One of his best-known discoveries was the first freshwater sawfish known, in 1982, in a remote northern Australian lake. For his contributions to the hobby Heiko was elected Man of the Year in England in 1993, and later in France. He holds many other titles and has met kings, presidents, ambassadors and senators. He is happiest and at his most relaxed away from it all when deep in the jungle searching for rare or new fishes. In 1992 he created the quarterly magazine "aqua geõgraphia", for Aquaprint (later Aquapress) publishers, a unique publication dedicated to virgin and bizarre habitats, expeditions to new, uncharted places, endangered species, biology, herpetology, botany, myths and aquatic legends, the wonderful world of water, and much more. He is the managing editor of the scientific journal "aqua, Journal of Ichthyology and Aquatic Biology" and still finds time to write numerous articles and has been working for more than 15 years on a forthcoming tome on all fresh- and brackish water fishes. His lectures take him to the five continents each year and he is frequently invited to judge fishes (mostly Discus) in exhibitions worldwide. He also collaborated with and organized the first International Discus Show and Exhibition in 1986 in Tokyo, coordinated the first three Aquarama Exhibitions and Conferences held biannually in Singapore and many others. Heiko continues to travel almost monthly to remote jungle areas to find new fishes, and several times each year to remote, unexplored Amazon habitats in search of discus...

His life is dedicated to fishes.

Photo credits

By far the majority of the photographs in this book have been taken by the author throught the last four decades (and earlier). But the book would not have been what it is, without the additional help of excellent photographers and painters. The following contributors, organisations and museums, have helped to make it such a rich first volume, full of photographic, detailed, information of Amazonia with the aid of extensive maps, satellite photographs, and paintings. The contributors are listed below in alphabetic order (by surname or Institution) with the country and in parenteses the subject to which they have contributed:

Amazon Exotic Imports (discus – Lago Grande).
Aquapress Archiv, Italy (people, country scenes, biotopes, fishes – flora and fauna – macro- and microscopic photos).
Aqua magazine, Japan (discus).
Barham und Coomes (their maps of rubber trees – redrawn).
Bleher, Amanda F. H., Brazil (photos of indians and our field trips in the 1950s).
Brockskothen, Willy, Germany (Discus).
Chou, Sharmon, Japan, Aqualife Magazine (discus).
Cochu, Fred, USA (photos from the early days...).
Fair Wind Co. Ltd., Japan (discus).
Frech, Peter, Germany (biotopes and discus).
Géry, Jacques, France (discus & Harald Schultz's photos).
Hustinx, E., Belgien (discus – Lago Grande & Henry Bak).
Innes, William T., USA (first photos of a discus and colour paintings).
Inge, Jürgen, Germany (large catfishes).
IBGE, Brazil (national institute of maps – photos of maps).
Kahl, Burkardt, Germany (cardinal tetras, brilliant rummy nose tetra and discus).
Kamihata, S., Japan (photos from the author).
Khardina, Natalya Uzbekistan (photos of the autor, fishes and macro-photography).
Kullander, Sven O., Sweden (discus photos - type material).
Leguen, Roger, France (photos of Marajó).
Lukhaup, Chris, Germany (shrimps).
Machado, Jose de Paula, Brazil (Jaú-photos).
Martinelli, Pedro, Brazil (black and white photography).
Matsuzaka, Minoru, Japan, Fair Wind Co. (animal and fish photos).
Mayland, Hans J., Germany (discus).

Mori, Fumitoshi, Pisces, Japan (discus).
Muséum national d'Histoire naturelle, Bibliotheque D' Ichthyologie, France (credit for the photo of Pellegrin by Robert Doineau).
Naturhistorisches Museum, Zentralarchiv, Vienna, Austria (Natterers paintings; map and drawings).
Peret, Joao Américo, Brazil (indians).
Petersmann, Hans, Germany (Heckel discus and orchids).
Pisces Publishing, Japan (discus).
Roggo, Michel, Switzerland (underwater photography).
Rüegg, Peter, Switzerland (discus and habitat).
Schmidt-Foche, Eduard (early Discus and people shots).
Schultz, Harald, Brasilien (different photos, including photos of his discus from the 1950s and early 1960s).
Senckenbergische Forschungsinstitut u. Naturmuseum – department of ichthyology, **Friedhelm Krupp**, Germany (type material).
Shen, Daniel, Wiser Publishing Co. Ltd., Taiwan (discus).
Socolof, Ross, USA (photos from the begin of the aquarium hobby in America, including early photos of Fred Cochu).
Stevenson, Roland W. Vermehren, Brazil (photos of indian paintings).
Taras, Christina, Chile (expedition photos).
T.F.H. Publications, Inc., and **Pres. Glen Axelrod**, USA (old photos of discus from their monthly magazine Tropical Fish Hobbyist – including its cover from 1952 with a discus on (see page 21).
Untergasser, Dieter, Germany (detritus).
Verswijver, Gustaaf, Belgium (indian photos).
Warzel, Rainer, Germany *(Crenicichla* spp., I collected).
Williams, Jeffrey, Smithsonian Institution, Washington, DC, USA (photos of the type material of all the subspecies, of *Symphysodon* from L. P. Schultz's description in 1960).

Special note. The publisher and author are thankful for this contribution, but they want to ad: with almost 5,000 pictures, it could be possible, that one contributor has been omited, but if so, it was definately not done by purpose. In such an event, please inform us, and the credit will be given, accordingly, in volume 2. Thank you.

Tapagem, Igarapé de 292
Tapagem, Lago 406
Tapají, Igarapé do 427
Tapajós 49, 131, 136, 149-155, 194-196, 242, 250-270, 280, 307, 334, 507-509, 528, 547, 549, 554, 558, 562, 566, 584, 588
Tapajós, Igarapé 507
Taparaua 120
Tapauá 120, 157, 211, 216, 396, 398, 400, 414, 415, 416, 420-423, 426-430, 507-508, 532, 534-535
Tapauá, Boca do 414, 429
Tapuiuçu 250
Taquera, Igarapé 230
Taracajá, Lago 452
Taraira, Lago 502
Tarapacá 490, 492
Tarapoto, Lago 507, 508
Tarará de Baixo 476
Tarará do Meio 476
Tarará, Paranã do 476
Tarumã (Tarumá) 151, 300, 395
Tata-putauá, Paraná 411
Tefé 41, 43, 46-47, 64, 67, 75-80, 82, 84-85, 115-116, 120, 124-125, 131-132, 133-136, 139, 142, 145, 159, 161, 178-184, 189, 216, 366, 420, 432, 434, 442-443, 449, 452-454, 458, 459-460, 462-463, 468, 474, 479, 486, 501-502, 506, 513, 522, 528, 534, 550, 552, 554, 556, 558, 566, 582, 584
Tefé, Barão de 76, 486
Tefé, Lago 67, 180-181, 184, 189, 449, 452, 453, 462, 506, 550, 552, 584, 617, 639
Teffé, Lago 120, 125, 135
Teotônio 347
Terra Santa 90, 288, 302, 304, 307, 310, 603, 621,639
Tigre 507
Timbó Titica, Igarapé 392
Tiquié 497, 500
Tocantins 120, 122, 133, 136, 147, 192, 193, 203, 216, 218, 219, 222, 223, 224, 225, 226, 250, 366, 400, 507, 508, 509598, 623, 627, 630, 631
Tonantins 43, 116, 145, 161, 468, 474, 475, 476, 506
Topaiós 250
Trinidad 56, 195, 491
Trinidad den Orinoco 258
Trombetas 145, 151, 161, 164, 172-173, 201-202, 216, 242, 280, 282, 284, 286, 288-304, 310, 507-508, 526, 528, 530, 535, 540, 600-601, 622-632, 634
Trombetas, Forte do 280
Tucumanpijó, Igarapé 230
Tucuruí 531
Tueré, Cachoeira Pilão grande do 230
Tukano, Igarapé 155, 169, 379, 380, 392
Tukanos, Cachoeira 394
Tupaiús (Indig. Stamm) 250
Tupajú (Indig. Stamm) 251
Tupa Yupanki (tawantinsuyu-Inca ruler) 394
Tupí (Indig. Stamm u. Sprache) 298, 306
Tupinambarana, Ilha 36, 312, 343

U

Uabuís (indig. tribe) 310
Uará, Lago 159, 455
Uariá 174
Uariá, Paraná 153, 333, 345, 344
Uarini 455, 462, 463
Uatumã 151,164, 170, 171, 206-207, 288, 321-324, 326, 327, 330, 331,507, 508
Uatumã, Baía von 326
Uauaçu, Lago 406
Uauacu, Lagoa 326
Uaupés (Vaupés) 36, 38
Ubim, Lago 157
Uiramutã 377
Urará, Lago 185
Urariá 164, 177

Urariá, Paraná 332, 333, 337, 338, 343, 345
Uraricoera 364, 376, 377
Urini, Lago 455, 460
Uruará 246
Urubaxí 155, 512, 514, 558
Urubu 505, 507-508, 512, 525, 530, 574
Urubuquá, Lagoa 149
Urucará 322, 324, 326
Uruchi, Lago 149, 272
Urucu (Urucú) 159, 440, 442, 443, 445-450, 452, 467
Urucurituba 332, 337, 343
Urucurituba, Ilha 343
Urucurituba, Paraná 320
Urumu 149
Ururu, Lago 149
Usina Remanso, Lago 326

V

Vale do Javari (Javarí) 479, 486, 493
Vandaluzia (= Andalusia) 276
Vaupés 534
Veiga, Lago do 403
Vajão, Lagoa 151, 300
Velha, Lago da 326
Vendaval 481
Veneza, Forte 431
Verde, Lago 149, 194, 195, 262, 264
Victória do Xingú (Victoria) 147, 240, 507, 508
Victoria Regia, Lago 396
Vila Bela da Imperatriz (Parintins) 334, 336
Vila Bittencourt 161, 464, 494, 496
Vila Curuá 272
Vila Bela da Santíssima Trindade 36
Vila da Barra 356
Vila da Conceição 334
Vila da Feliocidade-Itapeaçu 333
Vila de Araticu 226
Vila de Barcelos 382
Vila de Maués 334
Vila de Olivença 480
Vila de Silves 330
Vila Franca 264, 272
Villa Bella 280, 284
Villa Pauxí 280
Villas Boas 238
Vista Alegre 155, 376, 381, 448, 500
Vista Alegre Médio 376

X

Xingú 43, 59, 102, 116, 136, 147, 216, 218, 234, 236, 238-241, 270, 280, 362, 507-509, 526, 530, 587, 597-600, 623, 625, 630
Xiriri, Lagoa 151, 300
Xixiá, Lago 204, 310
Xixiá-mirim, Lago 603

Y

Yacu, Lago 490

Z

Zé-Açu, Lago 151

Rio Preto da Eva 329, 332
Rio Puduari 392
Río Puré 494
Rio Puretá 482
Rio Pureté (Rio Purutê) 161, 492
Rio Purui (Puruí) 492, 494
Rio Purus 145, 161, 210, 211, 212, 350, 395-396, 400, 401, 402, 414-415, 421, 429, 430, 440, 468
Rio Putumayo 133, 136, 142, 145, 161,216, 178, 189, 468, 478, 490, 492, 494, 496, 500, 502, 524, 639

Río Puyo 366
Rio Quiuini 374, 516, 518
Rio Riozinho 161, 468, 472
Rio Sanabani 330
Rio São Manuel 262
Rio Solimões 76, 78, 131, 135, 153, 155, 246, 350, 359, 382, 395, 431, 432, 463, 595
Rio Surubiú 270, 272, 273
Rio Tacaná 524, 525
Rio Tacutu 364, 376
Rio Tambaí 224
Rio Tapajós 131, 136, 149, 216, 234, 236, 260, 268, 269, 280, 400, 507, 602, 630
Rio Tapauá 157, 416, 426, 427, 428, 429, 430, 534
Rio Taquanaquara 147, 232, 507, 508
Rio Tarumã 372, 505
Rio Tefé 67, 79, 80, 82, 84, 85, 159, 181, 182, 442, 443, 449, 458, 462, 479, 506, 513, 534, 550, 558, 584, 617, 627, 629
Rio Tiquié 393, 497
Rio Tocantins 122, 147, 190, 192, 216, 222, 531, 627, 630
Rio Tonantins 216, 468, 474, 506, 507, 508
Rio Trombetas 39, 68, 72-75, 88, 99, 103, 131, 136, 145, 151, 161, 164, 172, 173, 270, 282, 290, 291, 298, 300-301, 304, 507-508, 518, 522, 528, 530, 600
Rio Tucuranã 242
Rio Tueré 230
Rio Tupai-paraná 250
Rio Tupanas 349
Rio Turuna 290
Rio Uaicupará 151, 319, 334
Rio Uamori 452
Rio Uarini 463
Rio Uatumã (Uatumá) 100, 145, 151, 164, 170, 216, 288, 321-327, 331, 507-508
Rio Uaupés 363, 366, 497
Rio Uiacurapá 312
Rio Umarizal 230
Rio Uneuxi 155, 381
Rio Unini 155, 167, 381, 388, 639
Rio Urariá 337
Rio Uraricoera 364, 376
Rio Urianã 230
Rio Uruará 246
Rio Urubaxí 381, 512, 514
Rio Urubu 120, 121, 122, 127, 129, 133, 136, 139, 151, 161, 216, 288, 326, 328-331, 505
Rio Urubú 454
Rio Urucará 326
Rio Urucu 101, 159, 178, 182, 183, 331, 440, 442, 445, 446, 449, 452, 506, 524, 534, 535, 552
Rio Urucurituba 242
Rio Urumu 246
Rio Urupadi 333, 334
Rio Wainy 400
Rio Xeriuiní (Xeruiní) 101, 168, 377, 381
Rio, Xié 38
Rio Xingú 124, 147, 193, 216, 218, 236, 238, 240, 242, 280, 400, 530, 597-600, 623
Riozinho, Igarapé 430
Risco, Ilha do 332
Rocky, Paraná do 505
Rodela, Furo do 403

Rodela, Lago 403
Ronca, Lago do 157, 431
Rondonia (Braz. state) 62
Roraima (Braz. state) 288, 331, 374, 376, 377, 378

S

Sabina, Igarapé 319
Sacado, Lago do 157, 429
Sacopema, Lago 157
Sacuri, Lago 151, 300
S. Domingo, Igarapé 507
S. Gerônimo, Igarapé 161
Salé, Lago 203, 304
Salgado, Lago 73, 75
Salgado, Lagoa 299
Salsa, Lago do 432, 437
Salsa, Paraná do 432, 437
Samauma, Igarapé 159
Samauma, Lago 149, 272
Samauma I, Lago 151, 300
Samauma II, Lago 151, 300
Sumaúma, Lagoa 374
Sampaio, Lago **343**
San Antonio de Caroní 531
San Felippe 497
Sancta Irena (= Santarém) 252, 253
Sangue, Igarapé 412
Santa Maria, Igarapé 420, 508
Santa Sofia, Ilha 490
Santa Clara 496
Santa Cruz da Nova Aliança 476
Santa Cruz de La Sierra 336
Santa Luzia 474
Santa Maria 326
Santa Maria de Belém do Grão-Pará 216
Santa Maria do Aru 248
Santa Rita do Weill 481
Santa Rosa in Pacaraima 377
Santana, Lago 304
Santarém 43, 47, 116, 124, 120, 125, 131, 133, 135, 136, 139, 142, 149, 194, 195, 196, 202, 216, 218, 242-264, 266-270, 280, 282-283, 286, 288, 304, 316, 360, 366, 432, 507
Santissimo Milagre 253
Santo Antônio 272, 273
Santo Antônio, Cachoeira de 234, 235, 237
Santo Antônio do Içá 468, 474, 475, 476, 479, 480, 489
Santo Antônio do Uruá-Tapera 288
São Angelo 377, 505
São Cristovão 34
São Domingo, Igarapé 346
São Francisco 333, 398, 404, 431, 475, 476, 479, 486
São Francisco, Forte de 475
São Gabriel da Cachoeira 36, 99, 354, 391, 393, 505
São João 346, 476, 488
São João da Liberdade 476
São João do Catuá 454
São Joaquim 38
São Joaquim, Forte 364, 365, 376
São José 334, 338, 354, 356, 382
São José de Marabitanas 36
São José do Amparo 476
São José do Aru 248
São José do Javari 486
São José Enseada 326
São José, Forte de 280
São José, Igarapé 444, 446, 480
São José do Pão Furado 157
São Lázaro 326
São Luís Gonzaga 400
São Marcos de Mundurucucami 334
São Paulo 34-35, 39, 61, 64, 75, 119, 127, 359, 366, 382
São Paulo de Olicença 161, 216, 382, 432, 468, 474-476, 479, 480-483, 488-490, 493
São Paulo dos Cambebas 480

São Raimundo 236
São Sebastião 106, 107, 476, 479
São Sebastião da Boa Vista 230
São Sebastião do Uatumã 15, 321, 322, 323, 324
São Sebastião dos Flores, Igarapé 159
São Tomé, Paraná do 432
Sapateiro, Lago 468, 472
Sapucuá, Lago (Lagoa) 74, 151, 292, 300-301
Sapucurú, Lago 304, 507, 508
Saracá, Lago 301, 328, 330
Scopema, Lago 421
Sebastião, Igarapé 392
Securiti, Furo do 157
Seringal do Jaburu 431
Seringal do Tambaquí 414
Serra Capiranga 151, 306
Serra Couto Magalhães 377
Serra da Escama 280
Serra da Estrella 42, 47
Serra da Lua 246, 247
Serra do Aracá 388
Serra do Mel 242
Serra do Roncador 240
Serra do Tapir pecó 384
Serra do Tumucumaque 245
Serra Formosa 240
Serra Guariba 308
Serra Pataquara 242
Serra Tepequem 377
Serra Tumucumaque 242
Serrão, Lago do 422, 424, 425
Panacuré 326
Sierranias del Taraíra 496
Silves 507, 508
Siqueira 280
Sítio de Taperinha 264
Sobral, Lago 157, 429, 430
Socambu, Igarapé do 159
Socó, Lago 427
Solimões 43, 72, 76, 78, 92, 116, 131, 135, 145, 153, 155, 159, 161, 178, 184, 187, 350, 351, 395-400, 404, 431-450, 453-455, 458, 462-466, 468, 472, 474-475, 479-486, 488-489, 492, 494, 500, 507-508, 622-623
Solitário, Lago 157, 422, 507, 508, 615
Soró, Lago do 440
Sumaúma, Cachoeiras da 348
Supiá, Lago 402, 411
Surará, Lago 157, 400, 402
Surinam 288, 290
Surubiú 270, 272, 273, 274
Sururá 402
Sustentável do Uatumã 324

T

Tabatinga 43, 44, 145, 161, 190, 213, 216, 468, 472, 476, 480-496, 623
Taboqinha, Paraná 411
Tacaná, Igarapé (Tacana) 213, 481-482, 484, 492, 508, 620, 623
Tacaná, Lago 536
Tacanaburgo I 482, 484
Tacanaburgo II 482
Taciuã, Lago 153, 349
Taiaçutuba, Ilha 463, 466
Tamanduá, Lago 157, 429, 430
Tamaniquá, Lago 159, 185, 464, 466
Tamaquerinha, Igarapé 230
Tambaquí, Lago (do) 210, 212, 421, 535, 556
Tambaquí, Paraná 455
Tambaquizinho, Lago 157
Tambor 248
Tanauaú, Lagoa 374
Tanquera, Igarapé 230
Tapacú, Igarapé 230

Rio Branco 28, 99, 145, 155, 161, 167, 168, 169, 209, 216, 266, 286, 364-365, 375-378, 382, 390, 400, 408, 478, 500, 505, 507-508, 528
Rio Caburi 310
Rio Cachorro 290
Rio Cagi 224
Rio Caiçara 453, 463
Rio Camaçá 157
Rio Camaraipi (Camaraipí) 147, 230, 232
Rio Camatiá 507
Rio Canamaú 155, 374
Rio Canumã (Canumá) 153, 164, 177, 208, 337, 338, 343, 344, 345, 346, 507, 534
Rio Capím 436
Rio Capitari 270
Rio Capucapu 326
Río Caquetá 161, 464, 497
Rio Carabinani 392
Rio Cariatuba 147, 232
Río Caroni 364
Río Casiquiare 361
Rio Catrimari 376
Rio Cauaburi 391
Rio Cauamé 376
Rio Caurés 381
Rio Citaré 242, 243
Rio Coari Grande 159, 182, 213, 331, 421, 429, 440, 444, 445, 448, 449, 507, 534
Río Coatán 484
Rio Cobacá 455
Rio Codajás 396
Rio Copacá 159, 463
Rio Cotuhá 189, 496
Río Cotuhé 492, 494
Rio Cucari 149, 246
Rio Cuchiguará (Cuchivara; Cochinuára) 400
Rio Cuduiarí 497
Rio Cuieiras 374
Rio Cuipeuá 278
Rio Cuiuini (Cuiuni; Cuiuní) 135, 166, 381, 505
Rio Cuminã (Cuminá) 284, 290
Rio Cuminapanema 270, 272, 298
Rio Cunauaru 159
Rio Cunhuá (Cuniuá; Cunuhá) 157, 210, 211, 421, 428, 429
Rio Cupai 124
Rio Cupijó 147
Rio Curicuriaí 505
Rio Curicuriarí 497
Rio Curió 230
Rio Curiuaú 374
Rio Curuá (Barra Mansa; Pavoval) 507
Rio Curuá 145, 149, 200, 202, 242, 246, 270, 272, 275, 277, 278, 298, 503, 507, 508, 524
Rio Curuá-Una 246, 248, 507, 508
Rio Curuçá 333
Rio Curuçambá 283
Rio Curuduri 380
Rio Curuena 468, 472
Rio Curumú 272
Rio Curupirá 337
Rio Cuvirá 161
Rio Cuxiuara 400
Rio das Oeiras 147, 226
Rio das Onças 320
Río de Cusco 464
Rio de Janeiro 50, 55, 61, 68, 75, 76, 96, 106, 107, 125, 220, 364,
Río de las Amazonas 216
Rio de Ouro 450
Rio Demini 380, 393, 609
Rio do Breu 463, 466
Rio do Jacinto 420, 421
Rio do Norte 288
Rio dos Deuses 450

Rio dos Purus 400
Rio Erepecuru 284, 290
Rio Esmeralda 364
Rio Grande 346
Rio Grande do Sul 34
Rio Guajará 230, 246, 347
Rio Guamá 216, 219, 220, 221, 230
Rio Guamués 492
Rio Guaporé 36, 366
Rio Guaribas 320
Rio Iaco 398
Rio Içá 101, 132, 145, 161, 198, 213, 216, 308, 359, 369, 468, 475-477, 479, 492, 507-508, 520, 536, 558, 639
Rio Içana 497
Rio Icié-Mirim 380
Rio Igapó 349
Rio Imabú (Inambú) 290
Rio Ipixuna 120, 129, 250, 400, 414, 415, 420, 421, 422, 423, 424, 507, 508
Rio Iriri 362, 587
Rio Itacarará 270
Rio Itacoaí 480
Rio Itanhauã 159
Rio Itaparaná 157, 211, 400, 421, 422, 423, 507, 508, 534
Rio Itapecurú 242
Rio Itatira 230
Rio Itaya 502
Rio Ituquí 149
Rio Ituxi 400, 426, 427
Rio Jaburu 337
Rio Jacaré 157, 230, 400, 421, 426, 429
Rio Jacaré Paruzinho 230
Rio Jacinto 157
Rio Jacundá 147, 228
Rio Janal Grande 230
Rio Jandiatuba 161, 178, 468, 479, 481, 506
Rio Japurá (Japura) 103, 136, 159, 178, 184, 216, 350, 432, 440, 453, 455, 458-459, 464, 474, 494, 500, 528, 562 595
Rio Jarauçu 238, 240
Rio Jari (Jarí)147, 157, 234, 242, 244, 245, 280, 310, 400, 408, 411-414, 421, 440, 534
Rio Jatapu (Jatapú)136, 145, 164, 170, 171, 206, 322, 296, 323, 327, 338, 505
Rio Jaú 389, 392, 518, 535
Rio Jauaperí (Jauapery) 155, 168, 372, 374, 505
Rio Jauari 149, 246
Rio Javari (Javarí) 484, 486
Rio Juami 161
Rio Jufarís (Jufari) 378, 381, 505
Rio Jumá (Juma) 153, 159, 348, 349
Rio Juruá (Jurua) 136, 159, 185, 186, 212, 216, 395, 398, 432, 452, 455, 461, 462, 463, 464, 466, 535, 554, 618, 619
Rio Juruena 262, 348, 366
Rio Juruti (Juriti) 151
Rio Jutaí 103, 136, 161, 188, 216, 246, 468, 469, 471, 472, 473, 480, 506, 621
Rio Litani 242
Rio Macuricanã 312
Rio Madeira 36, 124, 136, 138, 142, 153, 161, 216, 332, 337, 338, 501
Rio Madre de Dios 484
Rio Maicá 262
Rio Maicuru 149, 242, 280
Rio Majari 147
Rio Mamiá 157, 270, 272, 281, 421, 440, 444, 446, 448, 534
Rio Mamirauá 440, 530, 550
Rio Mamuru 151, 153
Rio Mamurú 312, 333, 334
Rio Manacapuru 209, 350, 351, 353, 396, 552
Rio Manacori 208

Rio Manduacari 230
Rio Manicoré 346
Rio Manjuru 334
Rio Mapari 468
Rio Mapiá 337
Rio Mapuera 291, 294
Río Marañon 451, 502
Rio Marapauá 230
Rio Marari 380
Rio Maraú 153, 333, 335, 336
Rio Marimari 153, 164, 176, 177, 296, 338, 342, 467, 505, 507, 524
Rio Marinaú 230
Rio Maripá 323
Rio Maués 136, 333, 335
Rio Maués-Açu 153
Rio Maués-Mirim 153
Rio Mesay 451
Rio Mineruá 187, 455, 463, 464, 466, 468
Rio Mineruázinho 463
Rio Miri 333
Rio Mirim 333, 347
Rio Miriti 334
Rio Mocajatuba 228
Rio Moju 220, 230, 246, 248
Rio Mucajaí 376
Rio Mucuím 157, 400, 429, 430
Rio Mujuá 230
Rio Mujuí 246, 248
Rio Muriapiranga 230
Rio Mutum 161, 468, 472
Rio Nanay 142, 161, 178, 189, 290, 501, 502, 506, 522, 530
Rio Negro 354-394, 396, 607-610
Rio Nhamundá 86-98, 306-311, 326-327, 330, 338, 369, 505, 512, 524-525, 574, 604-605, 607, 629, 631
Rio Novo Airão 166
Río Orinoko 499
Rio Oriximiná 151
Rio Paca 230
Rio Pacajá 147, 230, 232, 233, 400
Rio Pacoval 320, 333
Rio Pacyá 400
Rio Padauarí (Padaueri, Padauiri) 155, 364, 380, 392, 505
Rio Paicuru 242
Rio Paloemeu 242
Rio Panaúba 228
Rio Pará 216, 222
Rio Paracatú 310
Rio Paraconí (Paraconi) 153, 333
Rio Paraguay 70
Rio Paraipixuna 211
Rio Paraná-pixuna 250
Rio Parapixuna 157, 420
Rio Parauaquara 149, 246
Rio Parauarí 153
Rio Parauhaú 228
Rio Paricatuba 372
Rio Parú (Paru) 149, 216, 242, 243, 244, 245, 246, 248, 249, 280, 290, 298, 299, 300
Rio Parú do Oeste 151
Rio Parú Grande 230
Rio Parurú 507, 508
Río Pastaza 366
Rio Pati 161, 472
Rio Pauini 431
Rio Pinhuã 428
Rio Piorini 438, 440
Rio Pirá-Paraná 497, 500
Rio Pirico 230
Rio Pitinga 230, 324
Rio Pixuna 420
Rio Pracupi 230
Rio Pracuruzinho 230
Rio Preto 122, 380

Monacis 580
Monguba 234, 235, 236
Montanhas de Tumucumaque 245
Monte Alegre 197, 246, 247, 270, 518, 526
Monte Dourado 234, 236
Moreira (=Moreré) 110, 135, 381, 382
Moreira, Lago do 406
Mororó, Lago 427
Mosqueiro 230
Moura, Igarapé 292
Moura, Lago (do) 337, 440
Mucajá (Siedlung) 340
Mucajaí 377
Mucuím, Igarapé 430
Mucuím, Lago 157, 429, 430
Mueru, Lago do 432
Muiraquitãs, Lago 264
Museum Emilio Goeldi 264
Myanmar 131, 235

N

Nhamundá 68, 77-101, 145, 151, 153, 164, 168, 170, 202-206, 288, 294, 304-316, 322, 326, 327, 330, 428, 481, 505, 507, 512, 524, 525, 528, 535, 574, 603-605, 607, 622, 629, 631-634
Nhamundá, Lago 89, 136, 138, 173
Nhamundá, Paraná do 304, 307
Nanay, Lago 536
Nanay, Río 161, 216, 468, 501, 502, 506
Napo, Rio 478
Negro, Barra do Rio 43, 76, 116
Negros, Igarapé dos 326
Nossa Senhora de Belém 216
Nova Aliança 414
Nova Aripuanã 153
Nova Canaã 248
Nova Colônia 431
Nova Congregação 481
Nova Esperança 480, 482
Nova Olinda do Norte 337, 338, 339, 343, 349, 398, 414
Novo Airão 351, **370-371**, 372, 392, 522, 607
Novo Ariá 431
Novo Tapauá 398
Nueva Andalucia 531

O

Óbidos (Obydos) 149, 151, 201, 216, 242, 247, 270, 274, 280- 288, 300, 304, 307, 507
Oeiras do Pará 226
Onça, Igarapé da 431
Onçina, Lago 422
Orinoko, Río 361, 362, 364, 366, 380
Oriximiná 72, 74, 75, 88, 151, 201, 216, 242, 284, 286, 288, 289, 290, 292, 300, 301, 302, 304
Otá, Igarapé 230
Oté, Igarapé 483
Ouro, Igarapé do 230
Outeiro 230
Outeiro, Furo do 246

P

Paca, Igarapé da 338, **343**
Pacajaí, Cachoeira Grande do 230
Pacoval 149, 153, 246, 270, 272, 276, 277, 278, 284
Paçu, Lago (do) 432, 440
Padauiri 365
Pajé, Igarapé 230
Paletão, Lago 411
Palhar Grande, Lagoa 292
Panauarí 365
Panauarí, Lago do 304
Panaúba 147, 228
Panelas, Paraná das 474
Pará (= Bras. Staat u. früher Name für Belém) 37, 38, 43, 44, 48, 50, 55, 90, 116, 216, 217, 218, 220, 222, 224, 226, 228, 230, 232, 240, 244, 245, 246, 250, 254, 258, 262, 268, 280, 284, 288, 294, 299, 308, 312, 315, 358, 360, 362, 366, 382,
Pará, Castanha do 90
Pará do Uruará 149
Paracarí, Lago 149, 201, 272
Paraconi 346
Paraiso do Portugal (= Santarém, Portugal) 252
Paranemá, Lago 312
Paraoá, Lago 439
Parque Indígena Tumucumaque 242
Paraquí, Lagoa 151, 300
Paratari, Ilha 435
Parauacú, Lagoa 151, 300
Paricá, Lago 426, 429
Paricatuba 400, 402, 404, 505, 507
Parintins 86, 151, 203, 216, 262, 283, 286, 288, 305, 307, 310-320, 334, 336, 338, 339, 388, 432, 507
Parú 242, 243, 244, 245, 246, 248, 249, 280, 412
Parú do Oeste 290, 298, 299, 300
Parú, Lagoa 300
Patagonia 124, 125
Patos, Lago dos 272
Pauapixina, Igarapé 157, 421
Pau-furado, Igarapé do 444
Pauiní 400
Pauxís, Forte 280
Pedra, Igarapé 292
Peixe-Boi, Igarapé 155, 380
Pereira, Igarapé 230
Pereira, Lago 157
Pimenta, Cachoeira 230
Pinheiro do Meio 476
Pinheiro, Paraná do 476
Pinhuã, Igarapé 429
Piorini 438, 440
Piorini, Lago 159, 434, 438, 440
Piorini, Paraná 438, 440
Piquiá, Igarapé 270
Piracuquara, Cachoeira 230
Piraiauara, Lago 157
Piranha, Cachoeira 230
Piranha, Igarapé 417
Piranhas, Lago das 304
Pirarara, Furo do 403
Pirini, Igarapé 184
Pitú Grande, Lago 406
Planalto Maracanaquara 242
Poção, Igarapé 230
Ponta Alegre 237
Ponta da Nova Alegria 414
Ponta da Sfadeza 322
Ponta, Lago da 157
Porção, Lago 337
Porção Grande, Lago do 304
Porto de (do) Moz 124, 147, 193, 238, 239, 240
Porteira, Cachoeira 164, 292, 293, 301
Portel 230, 232, 507, 508
Portel, Baía de 230
Porto Alegre 248, 431, 476
Porto Artur 414
Porto Bom Jesus 228
Porto Companhia de Docas do Pará 228
Porto Custódio 228
Porto de Altamira 240
Porto Lider 228
Porto Luzitânia 400
Porto Munducurus 228
Porto Novo 248
Porto Prais 463
Porto Trombetas 99, 288, 289, 292, 297, 301, 310
Porto Velho 99, 347, 414, 422, 442, 443
Pousada da Liga de Eco 348
Pousada do Castelo 284
Pousada do Rio Maracaná 348
Pracuí, Baía de 230
Prainha 246, 270
Preguiça, Igarapé 494
Preto, Igarapé 463, 482, 485
Preto, Lago 149, 304, 332, 406, 422, 424
Pucu, Igarapé do 320
Puerto Asís 492
Puerto Pipa 494
Puerto Yaviya 502
Pupunha, Lago 157

Q

Quito 252

R

Ramos, Paraná do 203, 312, 3163, 319, 320, 322, 333, 337, 343
Raudal del Jirijirimo 494
Rei, Lago do 153, 355, 395, 396, 411
Retiro, Igarapé do 303
Ribeirão, Lago 157, 429
Ribeiro, Paraná 482
Rio Abacaxís 131, 136, 145, 153, 161, 164, 168, 174, 175, 176, 216, 296, 332-346, 505, 507, 524
Rio Acangatá 230
Rio Acará 216, 219, 220, 221, 226, 230
Rio Acaraí 240
Rio Acari 337
Rio Acre 398
Rio Açú 333, 335, 349
Rio Água Boa do Univiní 376
Rio Aguaytia 393
Rio Aiarí (Aiary) 497, 499
Rio Ajará 147
Rio Ajaraní 376
Río Algodón 494
Rio Alto Anapú 147, 230
Rio Amaturá 145, 161, 506, 534
Rio Amazonas 270
Rio Anapú 230
Rio Anauá 376
Rio Anauera 147
Rio Andairá 380
Rio Andirá 151, 203, 312, 320, 334, 335
Rio Anori 396
Rio Apacu 300
Rio Apaporis 161, 190, 494, 497, 499, 502
Rio Apoquitauá (Apoquitaua)153, 333, 334
Rio Apuau 155
Rio Apuaú 374
Rio Aquiry 400
Rio Aracá 380, 381
Rio Araça 505
Rio Araguaia 121, 122, 244
Rio Arapiuns 196, 264, 266, 267, 269
Rio Ararirá 155, 381
Rio Ararirá 135
Rio Aratari 230
Rio Araticú 147
Rio Araticus 226
Rio Arauá 349, 507, 508
Rio Ariau 505
Rio Aripuanã 153, 337, 348
Rio Aruá 159
Rio Aruã 266, 440
Rio Aruanã 230
Rio Atauá 230
Rio Atiningá 346
Rio Badajós 159, 437, 440
Rio Banã 230
Rio Barcarena 230
Rio Batovi 238
Rio Bauana 159
Río Beni 36, 484
Rio Biá 472
Rio Bóia 161

Itapuru 402
Itarim 149
Itatingão, Igarapé 230
Itatinguinho, Igarapé 230
Itatuba, Igarapé 319
Itelvina 149
Iténez, Río 61, 102, 336
Ituquí, do 195, 264

J

Jaburú 421, 422
Jabutí, Lago 330
Jacaré 414, 421, 426, 429, 431
Jacaré, Igarapé 159, 448
Jacaré, Lago 151, 173, 296, 302, 507, 508, 518, 530, 601
Jacaré, Lago do 290, 291
Jacaré, Lagoa 90, 292, 296
Jacitara, Igarapé 230
Jacu, Igarapé 319
Jacupara, Lago 507
Jacupará, Boca do 476, 477, 478
Jacupará, Lago 1651, 468, 478
Jadibaru, Lago 157, 429, 430
Jamadis 400
Jamarí, Lago 157, 429, 430
Jamunduá 431
Janauacá, Lago 208, 396
Janauarí, Lago 208, 396, 397
Jandiatuba 468, 479, 480, 481, 486, 493, 506
Japurá 76, 77, 82, 103, 145, 159, 178, 184, 432, 440, 451-465, 474, 476, 494, 496, 500
Jaquarequara, Lago de 330
Jaquarequara, Lagoa 507, 508
Jará, Lago 304
Jaraki 149
Jaraquítuba 149
Jaraquituba, Lago bei 199
Jararuá, Lago 472
Jaranaca, Lagoa 300
Jarauacá, Lagoa 300
Jarauçu, Lago 330
Jari, Boca do 412, 414
Jari, Lago 210, 404, 408, 409, 411, 412, 426, 467, 507, 511, 535, 566, 587
Jarí 147, 216, 237
Jari 216, 228, 229, 232, 233, 234, 235, 236, 237
Jarí, Lago 616, 632
Jari, Paraná do 406, 408, 409, 411, 412, 414
Jatapu (Jatapú) 151, 322, 323-327, 330-331, 512, 574
Jatapu, Lago 171
Jatimana, Lago 488
Jatuarana, Lago 157, 422
Jaú, Cachoeira do Rio 389
Jauaperi, Lago 520
Jauarauá, Lago 184, 462
Jauari, Lago (do) 149, 278
Jauari, Lagoa 246
Jenipapo I, Lago 409, 411
Jenipapo II, Lago 411, 412
Jenipapo, Boca do 409, 411, 412
Jenipapo Segundo, Lago 157
Jereuá, Lago 151, 300
Jerônimo, Igarapé 492
Jibóia, Lago 151, 300
Joari, Paraná (do) 406, 411
Juçará, Lago 411
Juma, Lago 153, 208, 346, 348, 349
Jundiá 377
Jurará, Lago 157
Jurimauás 451
Juriti 333
Jurity, Lago 64, 67, 125
Juru 463
Juruá, Rio 145, 159, 185, 186, 212, 432, 452-468, 506
Juruacá, Lagoa de 300

Juruás 159
Jurujuba 486
Juruparí, Lago (Jurupari) 153, 482
Juruti 203, 288, 302, 303, 304, 306
Juruti, Lago do 306
Juruti Miri, Lago 203, 304
Juruti Velho, Lago 304
Jutaí 145, 147, 149, 161, 188, 468-481, 488, 493, 494, 506, 522
Jutaí, Lago 436
Jutaí, Paraná 506
Jutaí, Foz de 468
Jutica, Igarapé 454

K

K, Lago 326

L

Lábrea 400, 427, 431
La Concha, Lago 492
Lago (Lagoa) A (Jatapu) 323, 326
Lago 26 (Itaparaná) 507, 508, 614, 615, 628
La Paz 484
La Pedrera 496
La Pilata 624
Lauricocha, Lago 451
Laxenburg (Austria) 28
Leandrinho, Lago 326
Leandro, Lago 326
Leticia 55, 56, 120, 126, 131, 133, 135, 161, 216, 432, 468, 482, 486, 488, 489, 490, 492, 494, 500, 507, 508
Limão, Igarapé 230
Limão, Lago (do) 322, 325, 326, 330, 396, 505
Limão, Paraná do 312, 318, 319, 320
Limãozinho, Paraná do 320
Lontra, Cachoeira da 272
Lontra, Igarapé 414, 507
Lontra, Lago da 427
Lucca (Italy) 31
Luséa (=Maués) 334

M

Macaco, Igarapé 230
Macaco, Paraná 411
Macapá 401, 405, 406, 475
Macumerí, Igarapé 353
Macumeri 507
Macunã, Lago 326
Macurani, Lago 312
Macuricanã, Lago (de) 202, 310, 312
Madeira 36, 208, 216, 236, 332-356, 366, 382 398, 400, 422, 501, 507, 508
Madeira, Ilha da 346
Madeirinha, Paraná do 346, 349
Maicá, Furo de 194
Maicá, Lago (do) 149, 194, 264, 507
Maicá, Paraná de 194
Maipá, Igarapé 343
Mamão, Ilha do 348
Mamauru, Igarapé 270
Mamaurú, Lago 149, 283, 286, 287, 507
Mamiá, Boca do 437
Mamiá, Lago 159, 440, 446, 448
Mamiá, Rio 157, 159
Mamirauá 159, 372, 453-460, 466, 527, 528, 530, 534, 550, 552, 554, 568, 572, 588
Mamoré, Río 236
Mamori, Lago 208
Mamurus 400
Manacapuru 68-73, 91, 145, 153, 159, 161, 209, 216, 332, 350-353, 372, 396-400, 421, 494, 507-508, 541, 552-558, 566, 576, 580, 582, 611-622, 630
Manacapuru, Lago 69, 72, 209, 350, 353, 507, 508, 566, 576

Manchanteria, Ilha da 394
Manaos 43, 115, 116
Manaós, Barra de 104
Manaquiri, Lago 153, 349
Manaus (Manaós) **104-105**, 354-361
Mangal 149
Manianrã, Lago 153, 349
Manicoré 153, 346, 347, 348
Maniçuã, Lago 429
Manissuã, Lago 426, 429
Mapari, Lago 468
Mapueira, Rio 151
Maraã 463
Maracanhã, Lago 310
Maracanhã 202
Maraco, Lago 149
Maraguá 334
Marahã, Lago 426, 427
Marajá, Lago do 403, 422, 424
Marajó, Ilha 43, 116, 147, 120, 192, 216, 220, 228-233
Marajó, Baía de 216
Maranhão (Braz. state) 400
Maratuba, Igarapé 230
Maraú 336
Mari, Igarapé 409, 411, 412, 413
Maria Curupira, Lago 153
Marimari 131, 139, 145, 153, 164, 175, 176, 177, 337, 338, 340, 341, **342**, 343, 344, 346
Maripá, Lago 149, 246
Maripauá, Lago 153
Mariuá 486
Marium, Paraná do 155, 380
Marmoré 36
Marmori, Lago 396
Marmori, Paraná do 349
Maruim, Igarapé 392
Mastro, Furo do 432
Mata Limpa 149, 264
matapi 624
Matias, Lago do 157, 402
Mato Grosso 32-40, 62-64, 70, 102, 244, 262, 263, 336, 348, 361, 366, 398, 406, 426, 432, 434, 437
Maués (Maues) 124, 145, 153, 161, 208, 216, 332, 333, 334, 335, 336, 349, 478, 507, 508, 562
Mauritius 521
Mawés (= Maués) 334
Maximo, Igarapé 316
Maximo, Lago 151, 312, 316, 317, 318, 507
Mayonisha 500
Medellín 224
Medicilândia 246
Melgaço 230, 232
Melgaço, Baía do 147, 230
Membeca 414
Meratuba, Igarapé 230
Mesorregião do Baixo 288, 304
Metternich 28
Microrregião de Óbidos 288, 304
Minas Gerais (Bras. Staat) 34
Mineiro, Igarapé 230
Minerúa, Rio 455, 463, 464, 465, 466, 621
Minuã, Igarapé 157
Miniá, Igarapé 428, 429
Mira, Lago do 157
Miranda Leão, Rua 68, 70, 359
Miri, Igarapé 230
Mirim de Alenquer, Paraná 270
Mirim, Paraná 270
Miriti, Lago do 396
Mirituba, Lago 337
Miuá, Lago 159, 411, 434, 437
Moará 431
Moara, Lago 429
Mocambo 282
Moju 216, 220, 221, 230

Cametá, Igarapé 508
Camicha, Lago 300
Camija, Lago 151, 300
Camixixi, Lagoa 374
Campa 500
Campina, Lago 411, 414, 440, 448
Campo, Lago do 153, 337
Campo Grande, Lago 208
Canaçari, Lago 99, 207, 328, 239, 330, 332, 507, 508
Canariã, Lago 157, 429, 430
Candeeiro, Igarapé 296
Candirú, Igarapé 230
Canhumã 333, 337, 338, 343, 345
Canhumã, Paraná 337, 338, 343, 345
Cantagalo 32
Cantagalo, Lago 467
Cantagallo, Furo do 420, 425
Cantagallo, Lago 420, 423, 425, 507, 539
Canumã 164, 177, 208, 337, 338, 342, 343, 344, 345, 346, 534, 554, 562, 566, 568, 572, 578
Canumã, Foz do 337, 345
Canutamá, Lago do 429
Canutamã 157, 398, 400, 414, 429, 431, 463
Capacete, Igarapé 484
Capinarana, Igarapé 346
Capintuba, Lago 149, 272
Capitão, Igarapé 428, 429
Capituba, Lago 272
Capoeirão, Igarapé 230
Caquetá 494, 496, 497
Caracaraí 378
Caranã, Igarapé 266
Carananzaru 149
Caratiá, Lago 157, 429
Carauari 468
Careiro 396
Careiro, Ilha do 355, 396, 504
Carimóseen, Lago 151, 300
Carumbé, Igarapé 230
Carvaçú, Paraná da 453
Casiquiare 354
Castanha, Igarapé 409, 411, 412, 414, 467, 507
Castanha, Lago (Lagoa) 151, 170, 300, 326
Castanhal, Lago 153, 349
Castanheira 248
Castanho, Lago 153, 349, 396, 422, 455, 554
Catuá, Igarapé 159, 454
Catuá, Lago 183, 440, 454
Caurés, Baía do 155, 392
Caviana, Lago 402
Caxiuanã, Baía de 147, 230
Caxuará, Ilha 432
Ceará (Braz. state) 400
Cerrado, Igarapé (do) 159, 440, 447, 449, 467, 506
Chapado, Lago do 427
Chapéu, Lago do 157
Coari 145, 159, 161, 178, 182, 183, 213, 216, 432-467, 534, 535, 552, 570, 576, 582
Coari, Lago de 159, 182, 440, 446, 447, 448, 449
Cobra, Lago da 412
Cocha Comprido 478
Cochabamba 336, 347
Cocoajá, Igarapé 230
Codajás 159, 432, 433, 434, 437, 438, 442
Comprida, Cachoeira 230
Comprido, Igarapé 478
Comprido, Lago 157, 411, 422, 424
Contrabando, Lago 488
Copacabana 106
Copatana 469, 472, 474
Copatana, Igarapé 161, 469
Copeá, Lago 440
Copeá, Paraná 440, 453
Corós, Ilha dos 434, 437
Corréa 149

Corta Corda 248
Costa Verde 463
Coxodoá, Igarapé 430
Cruzeiro do Sul 464, 466
Cuiabá 36, 238, 260, 262, 263, 268, 366, 476
Cuianã, Paraná 402, 432
Cuiava 476, 478
Cuiavua, Paraná 481
Cuipeuá, Lago 149, 198, 200, 272, 278, 507
Cuipier, Lagoa 272
Cuiuanã 402
Cuiuni 155
Cuminá 290, 298, 299, 300
Cuminá, Lagoa 151, 300
Cunhuá 400, 421, 428, 429
Cuniuá 210, 211, 212, 400, 421, 427, 428, 429, 430, 431, 478
Cunivo 500
Cunuiá 400, 428, 429
Cunuris 310
Cureru, Lago 155, 209, 390
Curí, Igarapé 266
Curipera 149
Curipera, Lago 272
Curitiba 35, 36
Curiuaú, Lagoa 374
Curralinho 230, 232
Curuá 200, 202, 242, 246, 248, 270, 272, 275, 276, 277, 278, 281, 282, 284
Curuá, Boca do 270, 272, 278
Curuá do Norte 270, 272
Curuá-Una 149, 507, 508, 509
Curuaí, Lago 264
Curuçá, Igarapé 270
Curuçá, Lago 153, 332
Curuçambá, Igarapé 287
Curuena 468, 472
Cuyabá 360
Curumú, Lago (Curumu) 149, 201, 272, 286
Curumucuri, Lago 304
Curupira, Lago (Lagoa) 151, 153, 295, 300, 304, 507
Curupirá, Lago 337

D
Damiana, Igarapé 230
Darién (Panama) 538
Davi, Lago do 440
Dipari, Lagoa do 374
Doema 507

E
Ega (Eda) 366
Eirunepé 468, 488
Elba, Paraná da 411
Encantada, Lagoa 299
Enseadado Santo Antonio do Capiruam 429
Erepecuru 296, 298, 299
Erepecuru, Lago (Lagoa do) 99, 151
Ereré, Igarapé 155
Esca-Ábidis (= Santarém Portugal) 252, 253
Escatici, Lagoa 326
Esmeralda 252
Esmeralda, Igarapé 230
Espeto, Lago do 411
Espírito Santo, Paraná do 312
Estirão do Equador 493
Estirão do Surará 402

F
Farias, Lago 295
Farias, Lago do 507, 508, 302
Farias, Lagoa de 292
Faro 288, 304, 306, 308, 310
Faro, Lago de (do) 91, 145, 164, 173, 306, 604-605
Flecha, Igarapé 409, 412

Fonte Boa 187, 462, 463, 466, 468, 554
Fortaleza, Paraná 403
Francesa, Lagoa da 312
Freguesia, Igarapé da 372, 505

G
Gabriel, Lago 434, 437
Gállien 276
Garcia de Resende 216
Genuaso, Igarapé 155
Gergoris (Gorgoris) 252
Gibian, Igarapé 159, 448, 467, 507
Giboia (or Jiboia), Igarapé 420
Giboia, Paraná 420, 423
Goiás 34, 36, 70
Grande de Cuminá, Ilha 299
Grande de Manacapuru, Lago 68, 72, 159, 400, 541
Grande de Monte Alegre, Lago 149, 518
Grande de Paricatuba, Lago 157, 400, 404, 613
Grande de Santarém, Lago 149
Grande do Curuaí, Lago 149, 286
Grande do Jauari, Lago 272
Grande do Tapará, Ilha 149
Grande, Igarapé 147, 228, 230, 231, 508
Grande, Ilha 147, 149, 155
Grande, Lago 149, 153, 157, 190, 196, 197, 200, 209, 246-247, 249, 264, 270-278, 286, 332, 349-353, 396, 507, 518, 526, 541
Grande, Lagoa 292, 299, 374
Grão Pará 216, 217, 250, 268, 451, 486
Grão-Pará 358, 360
Grito, Igarapé do 223
Guainía (Rio Negro) 354
Guajará, Baía de 216, 221
Guajaratuva, Ilha de 451
Guapapa, Lago 494
Guaporé, Rio 36, 366, 509
Guayana 56, 72, 87, 288, 290, 294, 324, 326, 331, 538, 584
Guedes, Lago dos 159, 184, 455, 462

I
Içá, Rio 468, 474, 475, 476, 477, 478, 479, 480, 489, 492, 494, †507, 508
Içana, Rio 38
Icoaraci 220, 230
Igreja Matriz e Santo Antonio 272, 273
Inambé, Lago in den 468, 472
Inferno, Lago do 438, 439
Ipaiva, Lagoa 246
Ipiranga 492
Ipiranga, Lago do 402
Ipiranginha, Lago in den 402
Ipiranguinha, Lago in den 157
Ipixuna, Igarapé 159, 454
Ipixuna, Lago 440, 454
Ipixuna, Rio 414, 415, 420-426, 507, 508
Iquitos 142, 161, 432, 475, 489, 501, 502, 506
Iranduba 396
Irauaú, Igarapé (Rio Negro) 169
Iri, Lago (Jatapu) 151, 170, 325-327, 330, 505
Iripixi, Lagoa 151, 300
Itaboca, Lago (Purus) 407, 411, 440, 467, 507
Itacoatiara 333, 336
Itandeua, Lago 271, 272
Itaparaná, Rio 414, 418, 419, 420, 421, 422, 423, 425, 426, 533, 534, 535, 615, 628
Itaparaná, Braço Morto do 418, 422, 426
Itaqui, Forte 280
Itaituba 242, 262, 268, 547
Itamarati 468
Itandeua, Lago de 270
Itapiranga 326, 330, 331
Itapiranga, Paraná de 330

Places

A

Aaçarí, Lago 326
Abacaxís, Rio 145, 151, 153, 161, 332, 333, 337, 338, 340, 345, 346,512, 524, 532, 540, 552, 572, 574
Abaetetuba 99, 224, 226, 227, 230
Abonini, Lago 426
Abufari, Paraná do 157
Abuí, Lagoa 291, 292, 294
Abuí, Paraná do 292
Acagantá, Igarapé 232
Açaí, Lago 428, 429
Açaituba, Igarapé 230
Acapuzinho, Lagoa 151, 300
Acará, Lago do 434, 437, 440
Acarapuxí, Paraná 155
Acarí, Lago 151, 295, 304, 507
Achipicá, Lago 151, 300
Acre (Bras. Staat) 62
Acre, Boca do 398, 400, 408
Açú, Igarapé 159, 392, 440, 449, 506, 525
Adriano, Lago do 480
Aduacá, Igarapé 319
Água Fria, Lago (Lagoa) 151, 157, 284, 300, 402
Aiapuá, Lago 157, 400, 402, 403, 404, 406
Airão 607
Ajaratuba, Ilha 434
Alanquer (=Alenquer) 276
Alãoquer (= Alenquer) 276
Alemanha, Cachoeira da 384
Alenen-Kerk (=Alenquer) 276
Alenquer 101, 142, 145, 149, 161, 195, 197, 198, 199, 200-202, 216, 218, 242, 246-247, 249, 268, 270-279, 440, 478, 503, 507, 516, 524, 526, 534, 552, 554, 562, 568, 576, 578
Alenquer, Igarapé de 270, 273
Aliança, Cachoeira da 365, 380, 388
Algodoa, Lago 204, 307, 310, 603
Almeida, Lago 300
Almeirim 147
Alonquer (= Alenquer) 276
Altamira 240, 599
Alter-do-Chão 264
Alto Solimões 480
Alunquer (= Alenquer) 276
Amaniú, Lago 153
Amanã, Lago 159, 209, 213, 350, 396, 397, 398, 400, 434, 458, 460, 462, 507, 508, 526, 612
Amapá (Braz. state) 147, 236, 244, 245
Amaru Mayu (Rio Negro) 354
Amataí, Paraná do 326
Amaturá 468, 476, 479, 480, 506
Amaturá, Igarapé 479
Ana, Igarapé 230
Anamã 145, 396-400, 407, 432, 434, 437
Anamã, Boca de 396
Anamã, Igarapé do 507
Anamã Grande, Igarapé do 398
Anamã, Paraná de 396
Anapú 230, 232, 233
Ancorí, Lago 157, 422
Andairá, Igarapé 155
Andirá 334, 335, 621
Andirá, Boca do 320
Angelim, Igarapé 230
Angustura, Forte da 280
Aninga, Lago 312
Aningau, Lago 149, 198
Anorí 432, 433, 434, 437, 442
Anorí, Lago do 432, 437
Apé, Lago 151, 300
Apaporis, Río 534
Apaurá, Lago 440, 454
Aquiqui, Paraná do 240
Aracá, Cachoeira do 384
Araçá, Lago 203, 304
Araçá, Lago do 326, 337, 436
Aracá, Paraná do 436
Aranapú, Paraná do 455
Arapiuns, Rio 149, 264, 266-269
Arajá, Igarapé 292
Arajasal, Lago 151, 300
Arapaí, Paraná 149, 272
Arapapá, Lago 320
Arara, Lago 159
Arara, Lago bei 159
Arariá, Igarapé 486
Arariá, Paraná do 396
Araticum, Igarapé 300
Araú, Igarapé 230
Archipélago de Mariuá (da) 367, 378, 384
Archipélago das Anavilhanas 155, 367, 372, 505
Ariaú 606, 607
Ariãu, Igarapé do (Ariaú) 155, 374, 380
Arimã 421, 422, 426
Arimã, Lago 157, 421
Aripuanã 337, 348, 349
Ariuãu, Igarapé do 374
Arpãouba, Lago 436
Arraial de São Vicente 36
Arrozal, Lago do 332
Aruã, Cachoeira de 266, 267
Aruanã 121, 122
Arúba, Lagoa 299
Arumã 398, 406, 408, 411, 414
Arumã, Igarapé 230
Arumá, Lago do 507, 508
Arumã Lago do 157, 414
Atabapo 115
Atalaia do Norte 488, 493
Atiparaná, Lago 468, 472
Atravessado, Lago 440
Ausencia, Lago 157
Aveiro 268
Azul, Lago 406

B

Bacabal, Igarapé do 332
Bacabaú, Lagoa I 151, 300
Bacabaú II, Lagoa 151, 300
Bacuri I, Igarapé, 411
Bacurí, Lago 157
Bacuri II, Lago 411
Bacururu, Lago 381, 608
Badabaxí, Lago 326
Badajós, Lago 159, 434, 435, 438, 440
Badajós, Paraná 435, 436, 437
Bagre 228
Bahiá 518
Baixinho, Lago 396
Baixo, Lago 157, 411
Barcarena 230
Barcelos 38, 99, 135, 155, 165, 166, 169, 360, 364, 377-396, 407, 434, 486, 630
Barbara, Lago 157
Barés 270
Barra 354, 356, 366, 382
Barra Mansa 149, 200, 276, 278, 281
Barracão de Pedra 284
Barreiras, Cachoeira bei 244
Barreirinha 320
Batata, Lago 151, 173, 288-296, 300-302, 507, 524, 530, 634
Baturité 382, 414, 421
Be-a-Bá 414
Bé-a-Bá, Igarapé 414
Bela Vista 476
Bela Vista, Paraná 157
Belém 37-38, 43-44, 48, 50-60, 70, 72, 76, 88, 116, 120-121, 127, 129, 133, 139, 142, 147, 161, 216-246, 260, 264, 270, 274, 283, 288-289, 362, 366, 378, 382, 394, 432, 440-441, 446, 451, 475, 481-482, 485-486, 489, 492, 500, 507-508, 630, 633
Belém, Igarapé de 213, 481, 482, 485, 486, 492, 508
Belo Horizonte 474
Belo Monte 147, 240, 241, 431
Bem Assim, Igarapé 177, **342**, 343, 344
Bem Querer 365, 376
Beni, Rio 336, 484
Benjamin Constant 55, 67, 103, 120-135, 161, 216, 468, 476, 480-493
Beruri (Berurí) 398, 400, 402, 508
Betânia, Igarapé 485, 508
Boa Sorte, Lago 159
Boa Vida 396
Boa Vista 284, 298, 299, 300, 326, 376, 468, 475, 476
Boa Vista dos Ramos 151
Boca Maracanã, Lago 326
Bocas, Baía das 147
Bogotá 57, 492, 502
Bonvina, Igarapé 391
Botos, Lago dos 149, 270, 271, 272
Braço Grande do Arapiuns, Igarapé 266
Branco, Baixo Rio 376
Branco, Boca do Rio 155
Branco, Cachoeira do Rio 304
Branco, Castello (Castelo) 216-217, 228, 235
Breves 99, 192, 228, 229, 230, 231, 232, 507, 508
Buibui, Igarapé (Rio Negro) 155
Bucaçú, Paraná 478

C

Caapiranga, Igarapé 466
Caãpiranga, Lago (Caapiranga) 157, 411
Cabaliana, Lago 72, 350
Cabeçeira Grande, Lago 157, 411
Cabeleira, Igarapé 270
Caburi 507, 528, 535
Caburi, Igarapé 305
Caburi, Lago 304, 307
Caburité 431
Cachimbo, Igarapé do 427
Cachimbo, Lago do 427, 429
Cachoeira da Aliança **365**
Caiambé, Lago (Lagoa de) 79, 101, 159, 183, 440, 452, 454
Caicubi, Igarapé 378
Caipurú, Lagoa 151, 300
Caissiana, Lago do 422
Caiuca, Lago 326
Cajari, Igarapé 481,
Cajari, Igarapé 482
Cajú, Igarapé 159, 445, 449
Cajueiro, Cachoeira der 272
Cajúeiro, Igarapé 230
Cajuti 272
Calado, Paraná do 155
Calado, Paraná do 380
Calderão, Paraná do 307
Calderón 492
Calafate 404
Caldeirão, Lago 396
Calderón 43, 47, 116, 525, 531, 620
Calipso 252
Camaçari, Lago 151
Camaleão, Lago 550
Camaraipi 233
Camarão 346
Camaraçi, Lago 328
Camatiá 481, 482, 486, 507
Camatiá, Igarapé 481, 482
Cametá 192, 222, 223, 530, 531

O

Oliveira, Inácio Correia de 480
Onassis, Aristoteles Onassis 234
Orellana, Francisco de 105, 216, 250, 314, 315, 346, 354, 360, 450, 451, 464
Oscar, Argemiro 494
Otto Schulz 237

P

Paiakan, Paulo 241
Palheta, Francisco de Melo 260
Passini, Tomas Passini 218
Pauxí (Pauxís, Pauxys) (indig. tribe) 280, 282, 283
Pávlova, Anna 218
Paxiuba (indig. tribe) 431
Pedro I. 38
Pellegrin, Jacques 42, 43, 47, 115-139, 164, 178, 179
Pereira, Antônio Rodrigues 400
Pereira Caldas, João 382
Pereira da Cruz, Luís 334
Pereira, Plácido 438
Peretti, Erio 70, 359
Petry, Paulo 376
Pietsch, Hans 56
Priestley, Joseph 254
Perreira Correia, Isaac 472
Petersmann, Georg 52
Petit, M. Georges 116
Picanço, Francisco Evilon Fernandes 416
Plazas, Luis Alfonso 490
Pombal, Marquês de 333
Ponchielli, Amilcar 358
Povoas, Francisco Melo de 452
Praetorius, Wilhelm 49, 50, 52, 262, 264, 262
Prinzen von Holstein-Oldenburg 29
Puinave (indig. tribe) 534

Q

Quadros, Jânio 238
Quilombo 282, 283, 284
Quilombos do Pacoval 272

R

Rabaut, Arnould 52, 54
Rachow, Arthur 56
Raddi, Joseph 31
Ramalheiro, Antônio 248
Ralegh (Raleigh), Walter 44, 45, 46, 76, 502
Rêgo, Maria de 38
Rhome, Romulus J. 264
Ribamar Fontes Beleza, José 383
Ribeira de Niña, Domingos da 282
Ribeiro, Eduardo 360
Rice, Hamilton 364
Riedel, Luís 366
Ridley, Henry Nicholas 258, **259**
Rios, Maria José 463
Rocha, Gentil 484, 487, 490
Rocha, Renato Felix da 486
Rodrigues Preto, José 334
Rondon, Marechal 62
Roosmalen, Marc und Tomas van 349
Roosevelt, Theodore 62
Rouca, Glória do 431
Rubtsov, Nestor 366
Rugendas, Johann 366

S

São Josétegrantes, Frei Cristóvão de 222
Sabrosky, Curt 126
Sacher, Franz 28
Sagratzki, Bruno 48
Santana, Antônio 436
Santos Pantoja, Eloi dos 232
Santos Pereira, Henrique dos 383
Santos Perés, Noberto dos 494
Santos, Silvino 364
Sateré-Mawé (indig. tribe) **334**, 335
Sateré (indig. tribe) 333, 334, 335, 336
Sazima, Ivan 227
Schmidt 61, 64, 66, 69, 77, 96, 101
Schmidt-Focke, Eduard 96, 120, 123, 126, 135-139, 164, 178, 183
Scholze & Pötzschke 48, 49, 53
Schomburgk, Richard 361, 362, 364, 366, 380, 381
Schomburgk, Robert Herrmann 362, **364**
Schreibers, von 30, 31
Schultes, Evans 502
Schultz, Harald 43, 61, 62, 64, 65, 66, 67, 77, 262, 416
Schultz, Leonard P. 66, 67, 119, 120-139, 164, 190
Schwartz, Adolf 120
Schwartz, Willy 359, 378, 626
Segderu, E. 79, 617
Seidel, C. 262
Selina Kleinwort 374
Seuben 276
Siggeklow, C. 48
Silva, Almir Viana da 404
Silva, Cerqueira da 400
Silva, Coreolano Sarrasin da 286
Silva, Olimpio Viera da 306
Silva Rondon, Cândido Mariano da 62, 486
Silva, Silvano 284
Sioli, Prof. Dr. 119
Siqueira Mendes, Manuel José de 284
Soares, José Carlos 288
Soares, José Pinto 486
Sochor 36
Socolof, Ross 57, 490
Solano, José 366
Souza Filho, Manuel João de 462
Souza, José Nicolino de 288
Spix, Johann Bapt. von 31-35, 76, 96, 251, 419, 517, 518, 574
Spruce, Richard 36, 38, 360, 361, 366
Stavros, Niarchos 234
Stegemann, Carlos 61, 64
Steindachner, Franz 41, 124, 125, 318
Steinen, Karl von den 497
Steinen, Wilhelm von den 238

T

Takase, Renato 218, 220, 378
Tallent, W. H. 538
Taras, Christina 77, 78
Taunay, Adrien 366
Taylor, John E. 362
Teixeira, Pedro 268, 280
Ternetz, C. 262
Terofal, Fritz 120, 416
Thayer, Nathaniel 41
Thomson, R. W. 256
Tikuna (indig. tribe) **65**
Torres, João Eliseu 416
Torres, Júlio 348
Torres, Manuel 90, 206, 306, 311, 376, 412, 421
Trajano, Augusto 236
Troschel 109, 318
Tsalikis, George 494

V

Vaillant, M. León 42, 47
Valenciennes 109, 362, 366
Vandalen 276
Vargas, Getúlio 358, 394
Verçosa, João 334
Vespucci, Amerigo 216
Victoria I. 28
Vieira, Antônio 226
Vieira, Celestino 268
Vila Nova, José de 282

W

Wallace, Alfred Russel 284, 360, 361, 362, 363, 366, 368, 380
Wandurraga, Raffael 490
Weddell, M. de 44
Wellesley, Arthur 28
Willliams, J. 120
Wallis, Gustav 402, 426
Wazna Kapaq (Inca tribe) 394
Wattley, Jack 164
Weißflog, Günther 164
Wickham, Henry Alexander 258, **259**
Witoto (indig. tribe) 536
Wolf, C. 35
Wolf, Herb 54

X

Xavier 246, 248, 250, 270
Xipido (indig. tribe) 500
Xokleng (indig. tribe) 64

Y

Yababana (indig. tribe) 499
Yabuna (indig. tribe) 499
Yañez-Pinzón, Vicente 216, 217

Z

Zell 164
Zeus 562
Zo'é (indig. tribe) 298, 299

Carvalho Franco, Joaquim de 358
Castelnau, Francis comte de 466
Castro d'Avelos, Enviratiba 479, 480
Castro, Miguel João de 268
Catrimani (indig. tribe) 377
Cauinicis (indig. tribe) 400
Caxibo (indig. tribe) 500
Caxuíana (indig. tribe) 282
Cesar, Alter 13
Chambers, John 395, 404
Chandless, William 400
Charies, Jacques Alexandre César 256
Chocó (indig. tribe) 537
Cinta Larga (=Tupi-mondé) (indig. tribe) 349
Cochu, Fred 52-62, 68, 119, 218, 220, 262, 412, 481, 484, 489, 490, 630
Cleiton Taxi-Aero 338
Collins, James 258
Coriana, Pedro 400
Costa de Ataíde Teve Souza Coutinho, Fernando da 475
Costa Falcão, Francisco da 268
Costa, José Joaquim da 382
Costa, Lucio 320
Costa, Raimunda 436
Costa, Raimundo Sampaio da 431
Cousteau, Jacques 89
Crocus (Crocos) 521
Cro-Magnon 624
Cuvier 109, 360, 365, 366
Cvancar, J. 48

D
D'Almada, Erst Lobo 333
Danis (indig. tribe) 499
Darwin, Charles 362
Deostede (Tefé) 79
Dias, Bartolomeu 276
D. Manuel 216
D. Merlin 155
Dom João II (Portugal) 216
Dom João VI (Portugal) 30, 32
Dom Pedro 30, 34, 216
Donna Domintila 34

E
Edwards 361, 366
Ehrenberg, Christian Gottfried 562
Ehrenreich, Paul 402
Eimecke, W. 48
Elisa (Bonaparte) 31
Encarnação, Manuel Urbano da 400, 431
Ender, Thomas 34
Epiphanes, Theos 562
Escobar, Pablo 224

F
Fabbro, Luigi 374
Falcone, Giovanni 224
Falcone, Maria 224
Faria, João Barbosa de 282
Ferdinand von Toscana, Großherzog 31
Ferneau, François 254
Ferreira, Alexandre Rodrigues 360, 361, 366
Ferreira, Efrem 77, 248, 296, 510, 512, 513, 560
Ferreira, Raul 226
Ferreira Ribeiro de Lima, Lucio Marçal 320
Florence, Hercule 366
Fonseca, José Antunes da 480
Ford, Henry 260
Franz-Joseph I. (II. des Hl. Röm. Reiches) 28, 30
Friedrich II. (v. Preußen) 28
Fritz, Samuel 450, 463, 468, 480
Furtado, Mendonça 228, 486

G
Gama Lobo d'Almada, Manuel da 376, 382

Gama, Vasco da 276
Garrits, Albert 308
Gates, Bill 458
Geisler, Rolf 505
Géry, Jacques 43, 47, 120, 133, 138, 250
Göbel, Manfred 164, 165
Goeldi, Emil August 55, 230, 240
Gomes, Eduardo Batista 409, 412
Goodyear, Charles 256
Gorbachev, Mikhail 374
Goulding, Michael 147, 240, 510, 512, 514, 558, 566, 568, 574, 576, 578, 588
Greiner, Joseph 237
Greenberg, Albert 54
Griem, Karl 50, 51, 52, 54, 55, 56, 220
Griem, Walter 48, 50, 51, 52, 53, 54, 55, 56
Grünberg, Theodor Koch 364, 497, 498
Guaicaris (indig. tribe) 310
Guajará (indig. tribe) 36
Guarani (indig. tribe) 476
Guimarães, Miguel Antônio Pinto 264
Günther 41, 124

H
Hagmann, Godofredo 264
Hancock, Thomas 256
Hannibal 253
Hans Schmidt 61
Haseman, J. D. 155
Hasse, Christian 366
Heckel, Jakob 28, 36, 39, 40, 43, 47, 109-116, 123, 124, 131, 133, 135, 137, 138, 139, 163
Hermes 521
Holford, William 320
Homann 164
Hooker, Joseph Dalton 258
Hooker, William Jackson 258
Hoonholtz, Anton Ludwig von 76
Hoonholtz, Antonio Luiz 486
Hubbs, Carl Hubbs 119
Huebner, Georg 500
Humboldt, Alexander von 360-368, 520, 636
Huaorani (indig. tribe) 260
Hixkaryána (indig. tribe) 86, 309

I
Imperiali, Paolo Roberto 374
Inge, Jürgen 77
Inkas 250
Innes, William T. 51, 52, 53, 54, 56, 60, 64, 125
Ivanovitch, Gregori 366

J
James I. 45
Jeguí 624
João VI. 30, 32
Jobert, Clément 41- 48, 79, 96, 115, 116, 133, 135, 139, 252, 262
Jozuí 399, 407
Julius Cäsar 253
Junk, Wolfgang J. 394, 505

K
Kamaiurá - Alto Xingú (indig. tribe) **59**
Kanahtxe (Wai-Wai-chief) 95
Karapaná (indig. tribe) 536
Katharina II. 28
Kauá (indig. tribe) 499
Kayapó (indig. tribe) **260**, 261
Khardina, Natasha 240
Kiel, Adolf 48, 53, 60, 62, 636
Kilian, B. 262
Kner, Rudolf 36, 40, 41, 110, 113, 124
Kolumbus, Christoph 370
Kullander, Sven 120, 131-135
Kuripako (indig. tribe) 534

L
Laborde, Alexandre de 29
La Condamine 44, 46, 47, 76, 252, 254, 260, 361, 464, 502
Lacorte, Rosario 119
Ladiges, Werner 48, 49, 57
Lana, Padre João Batista 480
Langsdorff, Heinrich Freiherr von 364, 366
Leda, Abrahão 336
Leopoldine 28, 30, 32, 34, 216, 517
Lepe, Diogo de 216
Lévi-Strauss, Claude 402
Lipman, Hyman 256
Lopes, Alejandro 348
Lopez, Dias Lopez 218
Lorenzo (Tapauá) 415, 420
Lyons, Earl 126
Ludwig 234
Ludwig XIV 43
Ludwig XV 28
Luséa 333, 334, 336

M
Maia, Álvaro Botelho 356
Macedo, Anildo 359, 390
Macintosh, Charles 256
Makú (indig. tribe) 533
Makuna (indig. tribe) 499
Maldonado, Pedro Vincente 46
Maria Theresa fr. Austria 28
Mariavalva, Marques de 30
Marie-Louise (Marie Ludovica) fr. Austria 28, 30, 34
Martius, Philipp Friedrich v. 31, **32**, 34, 35, 76, 251, 252, 335, 336, 496, 517, 518, 574
Mawé (indig. tribe) 333, 334, 335, 336
Mayas (indig. tribe) 250
Mayland, H.-J. 69, 75, 77, 78, 99, 224, 300, 330, 333, 504, 505, 506, 507, 508, 509
Melícola, Gergoris (or Gorgoris) 250, 252
Melo Palheta, Francisco de 260
Melo e Póvoas, Joaquim de 382
Mello, Thiago de 320
Mendes, Chico 254
Mendonça Furtado, Francisco Xavier de 246, 250, 270, 382
Menezes, Lizete Teles de 463
Mertens, Carl 52, 55
Metternich, Wenzel, Clemens **28**, 29, 30, 31, 32, 34, 36
Mewes, Axel 86, **96**
Meyer, Axel 136, 140
Michaux 256
Mikan, Johann Christian 31, 34
Miranda, Antônio de 452
Miranda, Vicente José de 358
Miquilles, José Bernardo 334
Mittenmeier, Russel 349
Montgolfier, Joseph-Michel und Jacques-Étienne 256
Moreira, Maurício 452
Motta, Manoel da 280
Müller & Troschel 109, 318
Mundurukú (indig. tribe) 36, 272,333, 334, 337, 338, 341, 344, 346, 348
Mura 334, 346, 400, 402

N
Natterer, Johann 28, **31**, 32, 34, 35, 36, 37, 38, 39, 40, 47, 61, 76, 96, 106, 109, 110, 135, 164, 165, 217, 218, 284
Nery, Constantino 356
Neto, Raimundo Maciel 338
Nicolaï, Monique 78
Nourissat, Jean-Claude 320

Serrasalmus elongatus (Lo. Piracatuba) **613**
Serrasalmus geryi **623**
Serrasalmus gouldingi **623**
Serrasalmus cf. *gouldingi* (Javari) **488**
Serrasalmus humeralis 597, **623**
Serrasalmus nattereri (Rio Guaporé) **36**
Serrasalmus nigricans (Lo. Jarí) **616**
Serrasalmus pingke **622**
Serrasalmus rhombeus (Rio Xingú) **597**
Serrasalmus rhombeus (Lo. Jarí) **616, 623**
Serrasalmus rhombeus (Javari) **488**
Serrasalmus cf. *rhombeus* (Novo Airão) **607**
Serrasalmus cf. *rhombeus* juv. (Juruá) **618**
Serrasalmus sp. (Rio Xingú) 597, **609**
Serrasalmus sp., juv. ((Ig. Tacaná) **485**
seringa barriguda *(= Hevea spruceana)* 370
Simulium incrustatum 586
Simulium perflavum 554, 595
Simulium sp. 584, 585
Smilaceae 518
Smilax 521
Smilax aspera 369, 518, 521, 535
Smilax aristolochiaefolia **519**, 521
Smilax calophylla 520
Smilax china 520
Smilax glabra 520
Smilax glyciphylla 521
Smilax havanensis 370
Smilax lanceaefolia 521
Smilax laurifolia 521
Smilax megacarpa **519**
Smilax myosotifolia 520
Smilax nigra 368, 518
Smilax officinalis **519**, 520, 521
Smilax ornata **519**
Smiulax ovalifolia 521
Smilax papyracea 520
Smilax papyraceae 521
Smilax rotundifolia 521
Smilax regelii 370
Smilax spruceana 520
Smilax syphilitica 521
Smilax walteri **519**
Snowella lacustris 547
Solenopsis sp. 580, 581, 582
Solenopsis geminata 582
Solenopsis invicta 581, 582
Solenopsis richteri 582
Solenopsis xyloni 582
Sorubimichthys planiceps (Madeira) **623**
Sorubimichthys planiceps (Teotonio) **500**
Sphaerocystis schroeteri 542, 543
Spirulina maxima 548
Spirulina platensis 548
Spatuloricaria sp. (Mapuera) **295**
Spatiphyllum sp. 404
Squaliforma cf. *emarginata* (Xingú) **598**
Staurastrum brachiatum 544
Staurastrum hystrix 544, 545
Staurastrum quadrinotatum 544
Sturisoma cf. *robustum* (Juruá) **619**
Sudis gigas (= *Arapaima*) 365
Surirella lineares 552, 553
Synedra goulardii var. fluviatilis 552
Synedra sp. 553
Synura uvella 556

T

Tabellaria fenestrata 552, 553
Tardigrada 570, 571
tarumã *(= Vitex cimosa)* 536
Tenagobia melini 574
Termes fatale (fatalis) 583, 584
Testudinella patina **565**
Testudinella sp. **565**

Tetragonopterus aff. *argenteus* (Jutaí) **471**
Tetragonopterus cf. *chalceus* juv. (Tacaná) **485** (Jutaí) **471**
Thermocyclops decipiens 568
Thermocyclops minutus 568
Thermocyclops sp. **569**
Thoracocharax securis (Lo. Anamã) *612*
Tortopus harrisi 587
Trachelomonas armata 550, **551**
Trachelomonas hispida 550, **551**
Trachelomonas volvocina 550, **551**
Trachydoras trachyparia (Japurá) **457**
Traira 149, 264
Treubaria triappendiculata 543
Trichechus inunguis 455, **610**
Trichocerca similis 564, 566
Triportheus sp. juv. (Içá) **479**
Triportheus spp. (Lo. Badajós) **437**
Triportheus angulatus (Lo. Anamã) 590, 594, **612**, 622
Tubifex 571
Tubificidae 571
Tubificina 570, **571**
Tucandeira *(Paraponera clavata)* 335
tucumã (=*Astrocaryum jauari*; coquillo palm) 516
Tyttocharax sp. (Coari) **444**

U

Uaru, 511, 590, 595
Uaru amphiacanthoides (Rio Nhmaunda) **604**
Uaru amphiacanthoides (Lo. Solitario) **615**
Uaru amphiacanthoides (Lo. Jarí) 510, 512, 576, 616
Ulmeritoides sp. 587
urucú (urucúm = *Bixa orellana*) 360, 539-540
Utricularia foliosa (Urucu) 447, 514

V

Vandiella chirosa (Trichomycteridae) **227**
Verbenaceae 536
Virola calophylla **444**, 534
Virola calophylloidea 534
Virola cuspidata 534
Virola elongata 534
Virola flexuosa 531
Virola rufula 534
Virola sebifera 531, 533
Virola spp. **533**
Virola surinamensis 533, 534
Virola theiodora 534
Vitex cagnus-castus (= chaste tree) 538
Vitex cimosa 538, 590

Z

zarzaparilla 520
Zungaro zungaro **610**
Zwergotter (Fischotter) 93, 94

PEOPLE

A

Abarés (– Barés, indig. tribc) 270
Ábidis, Prince 252, 253
Aborigines 624
Abreu Castelo Branco, João de 228
Adalbert, Prinz Heinrich Wilhelm 360, 362
Agassiz, Jean Louis Rodolphe 34, 238
Agostinho, Marcos 226
Aguiar, Claudio Batista de 489
Ajuricaba (Manaú) 354
Alcântara, Pedro d' 38
Alenquer, Pedro de 276
Alexander I. Pawlowitsch, Zar 29
Alexander II. 28
Alexandre de Laborde, Graf 29
Álmada, Lobe d´ 391
Alanos 276
Albuquerque, Daniel 414, 416
Alencar 276
Alexander I. 29
Amaral, Crispim do 57
Amorim, Raimundo 431
Andersen, T. B. Andersen 264
Antiochus IV 562
Aphrodite 521
Arrais, Aldo 274
Artemis 521
Assyrer 624
Axelrod, H. R. 54, 64, 67, 119, 120, 121, 122, 123, 125, 126, 127, 129, 133, 135, 139, 173, 416, 428
Aztecs (indig. tribe) 250

B

Bak, Henry 200
Bates, Henry Walter 75, 284, 360-366, 454
Beckstá, Casemiro 394
Benzaken, Asher 25, 87, 359
Bernardi, Enrico 218
Bettendorf, Jõao Felipe 250, 334
Bleher, Amanda Flora Hilda 59-63, 96
Bleher, Familie 60
Bleher, Michael (brother) 626
Blumenkaiser 30
Bolivar, Simon 362
Bonaparte, Napoleon 28, 29, 30, 31, 109
Bonplant, Aimé 362
Bouguer, Pierre 46
Bouquet, Henri 466
Borba, Antonius von 349
Borôro, Rosa 377
Braga, Eduardo 454
Brandão, Luiz Guedes 398
Britski, Heraldo **38**, 39
Brockes, Barth. Hinr. 520
Burkhardt, Jacques 40, 41, 42, 239
Burkhardt, James und Hunnewell 284
Bustamante, Benigno 490
Büttner, Richard 52, 55

C

cacauicultores 332
Cabral, Pedro Álvares (Alves) 216, 217, 276
Caldeira, Francisco 216, 268
Calipso 250, 252
Camões, Luís de 276
Canavarro, Antonio David Vasconcellos 356
Carnevale, Corrado 224
Carreras, José 105
Caruso, Enrico 358
Carvalho, Antonio Albuquerque Coêlho de 280, 510, 512, 566
Carvajal, Gaspar de 250, 312, 314, 315

Myleus cf. *rubripinnis* (Terra Santa) **603**
Myleus schomburgkii (Rio Negro) **364**
Myleus cf. *schomburgkii* (Xingú) **599**
Myleus sp. (Myleinae C) **599**
Mylossoma sp., juv. (Caburi) **305**
Mylossoma sp. (Purus) **401**
Myrciaria dubia **369, 534, 535**
Myrmecodesmus adisi **588**
Myrtaceae 534

N

Naididae 570, 571
Nannostomus aff. *marginatus* (Coari) **444**
Nasutitermes 522
Nasutitermes nigriceps 582, **584**
Navicula sp. 552, **553**
Nemadoras humeralis (Japurá) **456**
Nitzschia acicularis 552, **553**

O

Ochmacanthus sp. (Rio Preto, Juruá) **619**
Odontostilbe sp.? (Cauburi) **305**
Ochroma (Bombacaceae) 580
ojo de indio (=Guaraná) 335
Oligochaeta 570, 571
Opisthocomus hoazin **266**
Opsodoras aff. *stuebelii* (Japurá) **456**
Oscillatoria limosa 546, **549**
Oscillatoria sancta 548
Oscillatoria tenuis 548, **549**
Oscillatoria terebriformis 548
Osteoglossum bicirrhosum (Lo. 26) **615**
Ostarcoda 570, 571
otter 93, 94

P

Pachira aquatica (= monguba) 522
Pachyurus aff. *paucirastrus* **598**
pacu manteiga (Xingú) **599**
Palaemonetes ivonicus 576
Palaemontes ivonicus 577
Palingenia longicauda 586
Palmae (Palmaceae) 514
Pamphorichthys cf. *scalpridens* (Caburi) **305**
Panaqolus spp. (Rio Arapiuns) **267**
Pantera onca 245
Paracheirodon axelrodi 54, 58, 518
Parajapyx adisi 588, **589**
Parajapyx sp. **589**
Paraponera clavata 335
Parancistrus sp. (Rio Arapiuns) **267**
Parotocinclus sp. (Cauburi) **305**
Paratrygon aiereba (Demini) **393**
Paratrygon aiereba (Xingú) **600**
Parischnogaster 582
Passiflora nitida 535
Passifloraceae 535
Paullinia cupana 336
Paullinia cupana var. *sorbilis* 336
Paullinia sorbilis 335
Peckoltia aff. *brevis* (Jutaí) **471**
Peckoltia sp. (Rio Jarí, PA) **237**
Peckoltia sp. (Rio Areapiuns) **267**
Peckoltia sp. **596**
Pediastrum duplex 542, **543**
Pediastrum simplex 543
peixe boi (=*Trichechus inunguis*) 610
Pellona flavipinnis (Rio Nhamundá) **92**
pescada (Lo. Tefé) **617**
Phacus longicauda 550, **551**
Phacus pleuronectes 550
Phacus suecicus 550, **551**
Pheidole biconstricta 580, **581**
Phenacoccus madeirensis 581

Phenacogaster sp., juv. (Ig. Tacaná) **485**
Phractocephalus hemioliopterus (Japurá) **456**
Phractocephalus hemioliopterus **596**
Pimelodus blochii (Japurá) **456**
Pimelodus ornatus (Japurá) **456**
Pimelodus sp. (Novo Airão) **371**
Pimelodus sp. (Juruá) **619, 622**
Pimelodus zungaro (Rio Negro) **610**
Pimelodus aff. *altissimus* (Japurá) **456**
Pinirampus pirinampu (Mapuera) **293**
Pinirampus pirinampu (Purus) **401**
Pinnularia sp. 552, **553**
Piper (Piperaceae) 580
piranha 404, 411, 414, 417, 422
piranha de cardume **597**
Piranhea trifoliata 527, **528**
piranheira 527
pirarucú (=*Arapaima gigas*) **438, 439**
Platonia insignis 526
Platydoras costatus (Japurá) **456**
Platystomatichthys cf. *sturio* (Nanay) **500**
Plesiotrygon iwamae (Nanay) **500**
Pleurotaenium trabecula 545
Pocnemus unifilis 266
Poecilia velifera 54
Poecilocharax weitzmani (Ig. Preto-Typ.) **485**
Poecilocharax weitzmani ? (Jufarís – Aq.) **485**
Poecilocharax weitzmani (Ig. Tacaná) **620**
Polyarthra vulgaris 562, **565**
Pomacea papyracea (Lo. Manacapuru) **352**
Pontederia diversifolia 281
Potamorrhaphis guianensis (Lo. Bacururu) **608**
Potamotrygon aff. *humerosa* (Terra Santa) **603**
Potamotrygon aff. *signata* (Tapajós) **602**
Potamotrygon cf. *scobina* (Tapajós) **602**
Potamotrygon cf. *leopoldi* (Xingú) **600**
Potamotrygon leopoldi (Tapajós) **269**
Potamotrygon motoro (Rio Branco) **365**
Potamotrygon sp. (Tapajós) **602**
Pourouma cecropiifolia (=mapati) 526, **527**
Prionobrama sp. (Lo. Anamã) **398**
Prionobrama sp. 1 neu? (Cauburi) **305**
Pristis perotetti (Trombetas) **300**
Pristis sp. (Óbidos) **300**
Pristobrycon cf. *calmoni* (Terra Santa) **603**
Pristobrycon cf. *striolatus* (Javari) **488**
Pristobrycon cf. *striolatus* (Terra Santa) **603**
Prochilodus sp. (Rio Mamiá) **281**
Psectrogaster cf. *amazonica* (Lo. do Serrão) **425**
Pseudacanthicus cf. *hystrix* (Rio Jarí, PA) **237**
Pseudacanthicus leopardus (Rio Negro) **393**
Pseudacanthicus sp. L 273 (Tapajós) **602**
Pseudanos cf. *trimaculatus* (Lo. Anamã) **612**
Pseudobombax munguba 370, 393, **406**, 522, 580
Pseudomyrmex 578, 580, 581
Pseudoplatystoma **596**
Pseudoplatystoma tigrinum (Coari) **437**
Pseudosinella dubia 588, **589**
Pseudotylosurus 622
Psidium spp. 540
Psychotria carthagiensis 536
Psychotria erecta 536
Psychotria spp. 536
Psychotria viridis 536
Pterohemiodus cf. *atrianalis* **407**
Pterocarpus amazonum 580
Pterophyllum altum (Atabapo) **380**
Pterophyllum eimekei (Zeichnung) **49**
Pterophyllum leopoldi (Xingú) **625**
Pterophyllum leopoldi **510**
Pterophyllum scalare (Mamiá) **281**
Pterophyllum scalare (Tapajós) **602**
Pterophyllum scalare (Rio Nhamundá) **604**
Pterophyllum scalare (Rio Negro) **380, 609**

Pterophyllum scalare (Manacapuru) **611**
Pterophyllum sp. (Jutaí) **471**
Pterygoplichthys sp. (Tapauá) **417**
Ptygura pendugulata 562
Ptygura peticulata 562, **565**
pupunha (= *Bactris gasipaes;* peach palm) 516
Pygocentrus altus **623**
Pygocentrus altus (Javari) **488**
Pygocentrus nattereri **623**
Pygopristis denticulatus **622**
Pygopristis aff. *denticulata* (Itaparaná) **419**
Pyrrhulina aff. *brevis* **281**

Q

Quaraibea cordata 522, **523**

R

Ramphastos toco 245
Raphiodon sp. juv. (Caburi) **305**
Rhaphiodon vulpinus (Madeira) **347**
Rhaphiodon vulpinus (Amazonas) **622**
Reedia bakuri 526
Rheedia bacuripari 526
Rheedia brasiliensis 403, 524, 526
Rheedia gardneriana 524
Rheedia macrophylla 526
Rheedia spruceana 526
Rhizosolenia eriensis 554, **555**
Rhizosolenia longiseta 554, **555**
Rhytiodus 590
Rivulus sp. (Juruá) **463**
Rivulus sp. (alto Trombetas) **296**
Rivulus sp. 1 (Caburi) **305**
Rostrozetes foveolatus 588
Rotatoria 562, 570
Rotifera 562, 570
Rubiaceae 536
Rupicola rupicola 245

S

Salminus maxillosus **366**
Salminus 623
sapota-do-Solimões *(= Quaraibea cordata)* 522
sapucaia (=*Lecythis pisonis*) 528-530
Sarasaparilla 518-520
Sartor respectus (Xingú) **598**
Sassafraß (Sassafras) 520
Satanoperca cf. *acuticeps* (Purus) **401**
Satanoperca jurupari (Jutaí) **471**
Satanoperca jurupari (Lo. 26) **614**
Satanoperca jurupari (Juruá) **619**
Satanoperca aff. *jurupari* (Novo Airão) **607**
Satanoperca cf. *jurupari* (Rio Xingú) **599**
Satanoperca leucostictus (Lo. Maximo) **318**
Satanoperca cf. *leucostictus* (Lo. **Canaçarí**) **329**
Satanoperca cf. *leucostictus* (Novo Airão) **607**
Satanoperca lilith (Rio Nhamundá) **605**
Satanoperca sp. (neu?) (Juruá) **619**
Satanoperca sp. (Lagoa 26) **425**
Scenedesmus acuminatus 542, **543**
Scenedesmus quadricauda 542, **543**
Schizodon fasciatus (Juruá) **618**
Schizodon fasciatus juv. (Ig. Tacaná) **484**
Schizodon vittatus (Juruá) **618**
Scorpiodoras heckelii (Japurá) **456**
Semaprochilodus aff. *insignis* (Itaparaná) **419**
Semaprochilodus insignis (Lo. Tefé) **80**
Semaprochilodus taeniurus (Mamiá) **437**
Semaprochilodus taeniurus, juv. (Içá) **479**
seringa-barriguda (=*Hevea spruceana*) 526-528
Serrasalmus altus (Lago Solitario) **615**
Serrasalmus altus (Ig. Tacaná) **620**
Serrasalmus altus (Javari) **488**
Serrasalmus cf. *eigenmanni* (Piracatuba) **613**
Serrasalmus cf. *eigenmanni* (Javari) **488**

Genipa americana 534, 536, 537
Geophagus argyrostictus (Xingú) 599
Geophagus proximus var. (Rio Arapiuns) 269
Geophagus proximus (Rio Tapajós) 269
Geophagus proximus (Lagoa 26) 425
Geophagus aff. *proximus* (Xingú) 599
Geophagus sp. new? (Lo. Piracatauba) 613
Geophagus sp. (Tapajós) 269
Geophagus sp. (Rio Nhamundá) 605
Geophagus sp. (Rio Mapuera) 294
Geophagus spp. (Trombetas) 601
Geophagus sp. 2 (Lagoa 26) 425
Geophagus surinamensis? (Trombetas) 601
Geophagus cf. *surinamensis* (Xingú) 599
Geophagus thayeri? (Lo. Maximo) 318
Gigantocypris agassizii 572
Glyptoperichthys cf. *gibbiceps* (Lagoa 26) 425
Gnathocharax cf. *steindachneri* (Coari) 444
Gomphonema augur 552, 553
Gomphonema constrictum 552
Gonatozygon pilosum 544, 545
Gymnotus anguillaris 568
Gyrosigma sp. 552, 553, 554

H

Hanseniella arborea 588
Hanseniella sp. 589
Harpacticoida 568
Harpia harpyia 245
Hassar orestes (Japurá) 457
Hassar spp. (Japurá) 457
Hassar sp. 622
Hebetancylus moricandi 574
Hemiodus gracilis (Lo. Bacururu) 608
Hemiodus huraulti (Trombetas) 601
Hemiodus immaculatus group (Itaboca) 40
Hemiodus aff. *semitaeniatus* (Lagoa 26) 425
Hemiodus cf. *semitaeniatus* (Bacururu) 608
Hemiodus semitaeniatus-group (Itaboca) 407
Hemiodus sp., new? (Lo. Maximo) 317
Herrania nitida (Purus) 404
Heros festivus (Guaporé) 36
Heros sp. (Trombetas) 601
Heros sp., juv. (Caburi) 305
Heros cf. *severus* (Lo. 26) 614
Heros cf. *severus* (Lo. Tefé) 85, 617
Heros cf. *efasciatus* (Jutaí) 471
Hevea brasiliensis 528
Hevea spruceana 379, 393, 527, 528, 529, 540
Holocerina smilax menieri 519
Homo sapiens 624
Hoplarchus 590
Hoplarchus psittacus (Terra Santa) 604
Hoplarchus psittacus (Rio Demini) 609, 622
Hoplias malabaricus (Rio Itaparaná) 628
Hoplias sp. (Ig. Cerrado) 447
Hydrolycus cf. *scomberoides* (Juruá) 463
Hyla flavoguttata 223
Hyalotheca dissiliens 544
Hydrolycus wallacei (Lo. Anamã) 612
Hyphessobrycon copelandi (Ig. Preto) 485
Hyphessobrycon eques (Lo. Maximo) 317
Hypogastrura purpurescens 588, 589
Hypostomus cf. *carinatus* (Jutaí) 471
Hypselecara temporalis (Lo. 26) 614
Hypselecara temporalis (Lo. Tefé) 617

I

imbaúba (embaúba = *Ceropia latifolia*) 526
Ina geoffrensis 623
Ingá-Ameisen (*ingá*-ants) 532
Inga astristellae 532
Inga cinnamomea 532, 534
Inga densiflora 580
Inga edulis 530, 532, 534, 580
Inga macrophylla 534
ingá-mari (=*Cassia leiandra*) 522-524
Inga punctata 580
Inga spp. 531
Inpaichthys kerri 348

J

jaranduba (=maracaraná) 95, 369
japecanga 369
japucanha (japicanga-miúda) 369
jenipapo (=*Genipa americana*) 536
Jobertina 43

K

Krebs 293
Keratella americana 562, 564, 566
Keratella cochlearis 566
Keratella spp. 564
Kirchneriella fenestrata (**Oocystaceae**) 542
Kirchneriella lunaris 542, 543

L

Laemolyta 590, 596
Laetacara sp. (Barra Mansa) 281
Lamiaceae 536
Lecane bulla 562, 565
Lecythidaceae 528
Lecythis barnebyi 369
Lecythis pisonis 528, 529, 530
Leiarius marmoratus (Japurá) 456
Leptotila verrauxi 304
Lecane bulla 562, 565
Lecythis barnebyi 528, 530
Lecythis pisonis 530
Leiarius marmoratus (Lo 26) 628
Lemna sp. 565, 566
Leopoldinia insignis 518
Leopoldinia major 518
Leopoldinia paissaba 518
Leopoldinia pulchra 517, 518
Lepadella cristata 562, 565
Leporacanthicus sp. (Tapajós) 602
Leporacanthicus spp. (Rio Arapiuns) 267
Leporinus cf. *affinis* (Trombetas) 601
Leporinus desmontes 598
Leporinus aff. *fasciatus* 598
Leporinus fasciatus (Nhamundá) 605
Leporinus aff. *friderici* (Lo. Anamã) 612
Leporinus friderici friderici juv. (Juruá) 618
Leporinus aff. *granti* (Trombetas) 601
Leporinus maculatus 598
Leporinus sp. neu (Lo. Maximo) 317
Leporinus sp. 1 (Jutaí) 471
Leporinus sp. juv. (Içá) 479
Leporinus sp. (Trombetas) 601
Leporinus sp. (Lo. Piracatuba) 577, 590, 613
Leporinus cf. *tigrinus* (Nhamundá) 605
Leptodoras juruensis (Juruá) 619
Licania angustata 370, 447, 524
Licania annae 370
Licania ansisopylla 370
Licania apetala 524
Licania arborea 524
Licania blackii 524
Licania brittoniana 524, 525
Licania celativenia 370
Licania ferreirae 370
Licania krukovii 370
Licania longipedicellata 524
Licania micrantha 524, 525
Licania oblongifolia 370
Licania octandra spp. *grandiflora* 370
Licania parviflora 277, 524
Licania spp. 524
Licania stenocarpa 370
Licania stewardii 368, 524
Licania teixeirae 370
Liliaceae 518
Liosomadoras oncinus (Trombetas) 296
Liosomadoras oncinus albino (Trombetas) 296
Liosomadoras oncinus var. (Trombetas) 296
Liosomadoras oncinus (Padauiri) 365
Liposarcus sp. (Trombetas) 601
Liposarcus sp. (Lo. Badajós) 435
Lobotes ocellatus (= *Astronotus ocellatus*) 35
Lonchocarpus nicou 260
Lontra 94
Loricaria 596

M

macacaricuia (=monkey pods) 369
Macaranga (Euphorbiaceae) 580
Macrobrachium amazonicum 576, 577
Macrolobium acaciifolium 580
Macrolobium angustifolium 530
Macrolobium discolor 530
Macrolobium gracile 530
Macrolobium limbatum 530
Macrolobium multijugum 379, 529, 530
Macrolobium punctatum 530
Macrotermes natalensis 584
Macrotermes goliath 584
Macrothrix sp. 566, 567
macucu (= *Lecythis barnebyi*) 369, 528, 529
Mallomonas teilingii 556
Malpighia glabra 530
Malpighia puniciifolia 531
Malpighiaceae 530
Malvaceae 522
mapati (= *Pourouma cecropiifolia*) 526
mapimissu (= *Fiscus* sp. - Rio Negro) 535
Maracaraná 94, 95
maracujá-do-mato (= *Passiflora nitida*) 535
marimari (mari-mari = *Cassia leiandra*) 522-524
marmita de macaco (=*Lecythis pisonis*) 528-530
matá-matá (=*Eschweilera tenuifolia*) 528
Mauritia flexuosa 518
Megalechis cf. *thoracata* ((Ig. Tacaná) 484
Megalodoras irwini (Japurá) 456
Merodontotus tigrinus (Teotonio) 500
Mesocyclops leukarti 568
Mesocyclops longisetus 568
Mesocyclops sp. 569
Mesonauta 510, 511, 512, 622
Mesonauta festivus? (Ig. Tacaná) 620
Mesonauta cf. *guyanae* (Rio Nhamundá) 604
Mesonauta cf. *mirificus* (Juruá) 618
Mesonauta insignis (Lo. Piracatuba) 613
Mesonauta aff. *insignis* (Manacapuru) 352
Mesonauta sp. juv. (Caburi) 305
Mesonauta sp. (Lo. Solitario) 615
Mesonauta sp. (Lo. Tefé) 452
Metynnis calichromus (Rio Nhamundá) 604
Metynnis hysauchen (Rio Nhamundá) 604
Metynnis sp. 1 (Rio Itaparaná) 419
Micrasterias rotata 544, 545
Micrasterias truncata 544, 545
Microcerotermes arboreus 584
Microcystis aeruginosa 546, 547, 548
Micromischodus sugillatus 576
Microphilypnus sp. (Furo d. Cantagallo) 425
Moenkhausia dichroura? (Içá) 479
Moenkhausia heikoi (Rio Iriri) 239
Moinia sp. 566, 567
Moinodaphnia macleayi 566, 567
monguba (=*Pachira aquatica*) 522
monkey pot 528-530
monks' pepper (= *Vitex agnus-castus*) 538
Moraceae 535
munguba (= *Pseudobombax munguba*) 522
Myletes mesopotamicus (=*Piaractus*) 366

Arius oncinus (=*Liosomadoras*) **365**
Aspidoras sp. **582, 596**
Asphodelaceae 518
Astasia klebsii 550, 551
Asthenopus curtus 574, 587
Astrocaryum aculeatum **516**
Astrocaryum jauari **516**, 517
Astrocaryum vulgare **516**
Astronotus cf. *ocellatus* (Lo. Solitario) **615**
Astronotus ocellatus (Lo. Canaçarí) **329**
Astronotus ocellatus (Rio Itaparaná) **419**
Astronotus sp. 1 (Rio Itaparaná) **419**
Astronotus sp. 2 (Rio Itaparaná) **419**
Audouinella violaceae 558, **559**
Aulacoseira granulata 554, **555**
Aulacoseira granulata var. *angustissima* 554
Azolla cf. *microphylla* 447, **546**
Azolla filiculoides 548, **549**

B

bacterium 562
Bactris gasipaes **516**
bacuri (bacuripari-liso; bacu =*Rheedia brasiliensis*) **403**, 526
bacuripari (= *Rheedia macrophylla*) 526
bacuripari-liso (=*Rheedia brasiliensis*) **403**
bakterion 562
bakuri (= *Platonia insignis*) 526
Bambusia brebissonii 544, 545
Banisteriopsis caapi 536
Barbados cherry 531
baruca (Indig. name for discus) 290
Batrachospermum moniliferum 558
bentos 284
bicuiba (= *Virola* spp.) 534
Biotodoma cupido (Lo. 26) **614**
Biotodoma cf. *cupido* (Rio Itaparaná) **419**
Biotodoma wavrini (Rio Nhamundá) **92**
Biotodoma aff. *wavrini* (Içá) **479**
Bixa orellana 540, 580
Bixaceae 539-540
Bombacaceae 522
Bosmina sp. 567
Bosminopsis deitersi 566, **567**
Botrychium granulatum **556**
Botrydiopsis arrhiza **557**
Botryococcus braunii 542, **543**
Boulengerella cuvieri (Lo. 26) **614**,
Brachionus dolabratus 562, **563**, 566
Brachionus falcatus **563**, 566
Brachionus patulus 566
Brachionus spp. **563**
Brachionus zahniseri 562, **563**, 566
Brachionus z. reductus 566
Brachychalcinus orbicularis (Lo. Anamã) **612**
Brachyhypopomus cf. *beebei* (Novo Airão) 371
Brachyhypopomus sp. (Caburi) **305**
Brachyplatystoma tigrinus? (Nanay) **500**
Brycon amazonicus juv. (Ig. Tacaná) **485**
Bryconops cf. *giacopinii* (Lo. Bacururu) **608**
Bryconops sp. 1 (Ig. Curuçambá) 287
Bryconops sp. 2 (Ig. Curuçambá) 287
Byrsonima spicata 531
bullet ants (*Paraponera clavata*) 335

C

Cacajao melanocephalus ouakry 518
Caenotropus cf. *labyrinthicus* (Jutaí) **471**
Caesalpiniaceae 522
Callicebus bernhardi 349
Callicebus stephennashi 349
Calanoida spp. 568, **569**
Calanoida nauplii 569
Calophysus macropterus (Novo Airão) **371**
Calophysus sp. **622**

Camponotus sp. 580
camu-camu (*Myrciaria dubia*) 370
Candiru (Rio Preto - Juruá) **619**
Carnegiella strigata (Ig. Tacaná) **620**
Cassia leiandra 522, **523**, 524
Cassia spruceana 524
Catoprion mento (Terra Santa) **603**
Caulerpa taxifolia 544
Cebuella pygmaea 483
Cecropia (Cecropiaceae) 526, 580
Cecropiaceae 526
Cecropia latiloba (= imbaúba) 526, **590**
Centromochlus cf. *existimatus* (Airão) **371**
Centromochlus aff. *heckelii* (Airão) **371**
Cephalotes atratus 582
Cephalotes auratus 581
Cephalotes maculatus 581
Ceriodaphnia cornuta 566, **567**, 568
Cervus schomburgki 364
Cetopsis coecutiens (Airão) **371**
chacruna (= *Psychotria viridis*) 536
Chaetobranchus flavescens (Lo. Canaçarí) **329**
Chaetobranchus semifasciatus (L. Serrão) **425**
Chalceus aff. *erythrurus* (Anamã) **398**, **612**
Chalceus cf. *epakros* (Nhamundá) **605**
Characid sp. 3 (Ig. Curuçambá) **287**
Characid sp. 4 (Ig. Curuçambá) **287**
Characid sp. 5 (Fonte Boa) **462**
Characidium (*Jobertina*) *interruptum* **43**
chaste tree (=*Vitex cagnus-castus*) 538
Cheirocerus goeldii (Japurá) **457**
Chinaknolle (chinaroot) 520
Chironomus sp. 572, **590**
Chironomus paragigas 572
Chlorophyceae 542-
Chlorophyta 542-
Chyrodoridae sp. **567**
Chrysobalanceae 524
Cichla flavo-maculatus? (Rio Demini) **609**
Cichla monoculus (Novo Airão) **607**
Cichla monoculus (Lo. Solitario) **615**
Cichla monoculus (Rio Tefé) **85**
Cichla aff. *monoculus* (Rio Itaparaná) **419**
Cichla temensis (Nhamundá) **607**
Cichlasoma sp., juv. (Caburi) **305**
Clusiaceae 526
Closterium kuetzingii 544, **545**
Coelastrum cambricum 542, **543**
Colomesus asellus (Trombetas) **601**
Colomesus asellus (Lo. Jarí) **616**
Colomesus asellus (Lo. Maximo) **317**
Colossoma macropomum 85, 227, **305**, 527, 528
Copaibapalme 94
Copella sp. (Urucu) **445**
Copepoda 568
Corallus caninus 245
Corydoras adolfoi (Tiquié) **393**
Corydoras blochi group (Tapauá) **417**
Corydoras cf. *cortiae* (Tiquié) **393**
Corydoras cf. *julii* (Tapauá) **417**
Corydoras kanai (Demini) **393**
Corydoras cf. *punctatus*/*agassizii* (Tapauá) **417**
Corydoras cf. *reticulatus* (Tapauá) **417**
Corydoras schwartzi (Tapauá) **417**
Corydoras sp. 1 (Demini) **393**
Corydoras sp. 1 (Trombetas) **296**
Corydoras sp. 2 (Trombetas) **296**
Corydoras sp. 1 (Tiquié) **393**
Corydoras tukano (Tiquié) **393**
Cosmarium contractum 544, 545
Cosmarium margariteferum 544
Coussapoa-tree 580
Crateva benthami 590
Crematogaster 580

Crenicichla cf. *lugubris* (Lo. Solitario) **615**
Crenicichla cf. *marmorata* (Nhamundá) **605**
Crenicichla sp. (Nhamundá) **605**
Crenicichla sp. (Rio Nhamundá) **93**
Crenicichla sp. (Cachoeira Porteira) **293**
Crenicichla sp. (Lagoa 26, Itaparaná) **425**
Crenicichla sp. (La. Maximo) **318**
Crenicichla sp. (Urucu) **445**
Crocus sativus 521
Ctenobrycon hauxwellianus (Caburi) **305**
Curimata vittata 85
Curimata sp. (Rio Mamiá) **281**
Cycla Monoculus (=*Cichla monoculus*) **35**
Cyclestheria hislopi 574
Cyclopoida spp. 568, **569**
Cyclops sp. **569**, **590**
Cyclops nauplii **569**
Cyclotella sp. 554, **555**

D

Daceton armigerum 582
Daceton sp. 579
Daphina gessneri 567
Daphnia sp. 560, 566, 568, **590**
Daphnia gessneri 566, 568
Daphniidae spp. (with eggs) **567**
Diaphanosoma spp. **567**
Dicrossus sp. (Rio Demini) **609**
Dictyosphaerium pulchellum 542, **543**
Dimorphococcus lunatus 542, **543**
Dinobryon bavaricum 556, **557**
Dinobryon cylindricum 556, **557**
Dinobryon sertularia 556, **557**
Dysticus marginalis 281

E

Eigenmannia sp. (Caburi) **305**
Elachocharax sp. (Lo. Cantagallo) **425**
Enchytraeidae 570
Epapterus dispilurus (Japurá) **457**
Epiphanes sp. 562, **563**
Epiphanes macroura 562, **563**
Eschweilera tenuifolia (= matá-matá) 369, 379, 528, 540
Euastrum evolutum 544, 545
Eudocimus ruber 229
Eudorina elegans 544, 545
Euglena acus 550, **551**
Euglena oxyuris 550, **551**
Euglena pusilla 550
Euglena spirogyra 550, **551**
Eunotia asterionelloides 554
Eunotia cf. *lunaris* 554, **555**
Eunotia diodon 554
Eunotia flexuosa 554, **555**
Eunotia sp. 554, **555**, **559**
Eunotia triodon 554
Euphausia superba 570
Euphorbiaceae 526
Eurema smilax 519
Euryrhynchus amazonensis 577
Euryrhynchus amazoniensis 576
Euthyplocia sp. 586

F

Farlowella gladius (Purus) **401**
Ficus insipida 535
Ficus sp. 535
figo bravo (= *Ficus* spp.) 535
Filinia longiseta 564, 566
fire-ant (*Paraponera clavata*) 335
Fluviphylax sp. (Trombetas) **295**
Fluviphylax cf. *simplex* (Caburi) **305**
Frustulia rhomboides 552, **553**
Frustulia r. var. *saxonia* 552, **553**

G

Index

The Index is devided in:
GENERAL (page 656) (includes everything what is not listed in the following four index-register);
FLORA & FAUNA (page 656) (plants and animals; the numbers in bold are for fishes pictured (with its location found in parentheses); discus variants are not listed here – they are all over the book, each variant with its location mentioned;
PEOPLE (page 660) (scientists, researchers, discoveres, exploreres, authors, friends, indigenous tribes, and celebrities mentioned);
PLACES (page 663) (cities, villages, houses, streets, rivers, *igarapés, córregos, lagos* and *lagoas, paranás, furos, canos,* and mountains);
italic names are animals and plants;
bold names are higher taxa;
bold numbers are pages with pictures.

GENERAL

A
Alvarães 453, 462, 463
Amaro Tonico **259**
Amazonica Inc. (USA) 54, 55, 56
Apaes de-Nhamundá-Region 151
Apalaí 242
Apiaú 377
Aquarium Hamburg 52, 54
Auaris 377
Augusta 32
Austria 32
ayahuasca 536
Azteca 580, 581

B
Ballhausplatz 28
banha peixe boi (=Seekuhfett) 610
barbasco (Fischgift) 260
Bauxit 290

C
Cabanagem 38
caguri 624
caiá 624
Cassiam 414
catahua (Indig. Medizin) 370
chicha 354

D
Delectus faunae et florae brasiliensis 34

E
Empire Tropical Fish 52, 53
Erikó 377
ewa 624

F
Festa do Cacau e Feira Cultural 332
Festa do Guaraná (Maués) 334
Festa da Mandióca (Nova Olinda) 338

G
Garantido-Caprichoso (Parintins) 312
Goki's Zooversandhaus 58
Guaraná 333-336
Gulf Fish Farms (Florida) 14

H
haruga 624
Hotel da Bettina (Alenquer) 273
Hotel Jardin Paiva (Nova Olinda) 338
Hotel Tropical (Manaus) 77

hallucinogenic tryptamine 536

I
imiró 624
Inanu 149
INPA 77
jegui 624

K
kasatuti 624
kasaga 624
kit tesão (in Maués) 336

M
matapi 624
menico (fish poison) 260
mirantã 336

N
Naqada I Dynasty 624
Nimurid 624

O
oreja de negro (fish poison) 260

P
Paapiú 377
Palimiú 377
Pando 484
pacará (fish poison) 260
Paramount Aquarium 52, 54
Pito 500
Predynasty Naqada I 624

Q
Quechua (Indig. language) 394

R
ressacar 478
Rosa Vermelha-Ciranda de Nova Olinda 338

S
Sacher-cake 28
Saponin (saponin) 520
Silurifom migration 456-457

T
tamboril (fish poison) 260
Thayer-Expedition 76
Tratado de Madri 354
timbó (fish poison) 260
timbó cedro (fish poison) 260
timbó colorado (fish poison) 260
timin (fish poison) 260
turbinado 336
tucúm 624

U
UFOs 337-338

W
wairó 624
weheku 630

FLORA & FAUNA

A
Acacia (Leguminosae) 580
acará-açú (Curuá) 277
acara-açú (Jatapu) 325
Acanthicus hystrix (Rio Branco) **365**
Acarichthys heckelii (Novo Airão) **607**
Acarichthys heckelii (Lo. Piracatuba) **613**
Acarichthys heckelii (Tapajós) **602**
Acarichthys heckelii (Lago do Serrão) **425**
Acarichthys heckelii juv. (Cauburi) **305**
Acaronia nassa (Barra Mansa) **281**
Acaronia nassa (Lo. Manacapuru) **352**
Acaronia cf. *vultuosa* (Lo. Piracatuba) **613**
acerola (cereja-de-Cametá = *Malphigia glabra*) 530
Acestrorhynchus sp. (Lo. 26) **614**
Acestrorhynchus cf. *falcatus* (Trombetas) **601**
Acestrorhynchus falcatus (Lo. Tefé) **617**
Acestrorhynchus guianensis (Xingú) **598**
Acestrorhynchus microlepis (Lo. Tefé) **617**
Acromyrmex 582
Actinella sp. 554, 555, 595
Actinella brasiliensis 554, 555
Aequidens sp. (Xingú) **599**
Aequidens sp. (Lo. Solitario) **615**
Agamyxis pectinifrons (Lo. Marajá) **424**
Ageneiosus sp. (Japurá) **457**
Amazonsprattus scintilla (Manacapuru) **352**
Amphora ovalis 552
Anabaena azollae 548, 549
Anabaena flosaquae 548, 549
Anabaena hassalii 548
Anabaena spiroides 548, 549
Anacardium occidentale (Cajú) 335
Anchoviella cf. *jamsesi* (Içá) **479**
Ancistrus dubius (Tapauá) **417**
Ancistrus hoplogenys (Rio Negro) **393**
Ancistrus sp. (Rio Preto - Juruá) **619**, **622**
Ankistrodesmus falcatus 542, 543
Ankistrodesmus fusiformis 543
Anoplotermes sp. 583, 584
Anostomus aff. *intermedius* (Juruá) **618**
Anostomus ternetzi (Demini) **393**
Aphanizomenon flosaquae 546, 547, 548, **549**
Aphyocharax sp. neu? (Lo. Maximo) **317**
Aphyocharax sp. neu? (Lo. Anamã) **398**
Apistogramma pertensis **568**
Apistogramma sp. (Barra Mansa) **281**
Apistogramma sp. (Ig. Macumerí) **352**
Apistogramma sp. 1 (alto Trombatas) **296**
Apistogramma sp. (Tigre, Nhamundá) **305**
Apistogramma sp. (Urucu) **445**
Apistogramma sp. 1 (Piracatuba-Purus) **467**
Apistogramma sp. 2 (Cantagallo-Purus) **467**
Apistogramma sp. 3 (Itapoca-Purus) **467**
Apistogramma sp. 4 (Cantagalo-Purus) **467**
Apistogramma sp. 5 (Jari-Purus) **467**
Apistogramma sp. 1 (Gibian-Coari) **467**
Apistogramma sp. 2 (Gibian-Coari) **467**
Apistogramma sp. 3 (Gibian-Coari) **467**
Apistogramma sp. 1 (Cerrado-Urucu) **467**
Apistogramma sp. 2 (Cerrado-Urucu) **467**
Apteronotus albifrons (Demini) **393**
Apteronotus cf. *mariae* **401**
apuí (= *Ficus* spp.) 535
araçá (= camu-camu = *Myrciaria* spp.) 535, 536
araçá-d'água (= camu-camu= *Myrciaria dubia*) 534
Arapaima gigas **365**, **438-439**
araparí (= *Macrolobium multijugum*) 530
Aratinga guaroupa 223
Arecaceae 514
Argonectes cf. *longiceps* (Trombetas) **601**
Aristoclesia esculenta 526

Stölting, K., Salzburger, W., Bleher, H. & A. Meyer, 2006 (in press). Preliminary Revision of the Genus *Symphysodon* Heckel. *aqua, Journal of Ichthyology and Aquatic Biology: Special Publication 2*.

Teixeira, L. M., 1998. Seleção de estirpes de rizóbios para jaranduba (Macrosamanea pubiramea – Leg. Mimosoideae), em dois solos da Amazônia. Monografia da UTAM, Depto. de Engenharia Florestal, Manaus, Brazil, 39 pp.

Toledo, M.a de & P. Ragazzo, 2002. *Peixes do Rio Negro (Fishes of the Rio Negro) Alfred Russel Wallace (1850-1852)*. Editora da Universidade de São Paulo, 517 pp.

Tulard, J., Gengembre, G., Goetz, A., Jourquin, J. & T. Lentz, 1998. *L'ABCdaire de Napoléon et l'Empire*. Flammarion, Paris, France, 119 pp.

Turrill, W. B., 1963. *Joseph Dalton Hooker: Botanist, explorer and administrator*. The Scientific Book Club, London, 228 pp.

Uherkovich, G. & G. W. Schmidt, 1974. Phytoplanktontaxa in dem zentralamazonischen Schwemmlandsee Lago Castanho. *Amazoniana*, 5: 243-283.

Uherkovich, G. & H. Rai, 1979. Algen aus dem Rio Negro und seinen Nebenflüssen. *Amazoniana*, 6: 611-638.

Uherkovich, G., 1984. Phytoplankton. In: Sioli, H. (Ed) *The Amazon – Limnology and landscape ecology of a mighty tropical river and its basin*. Monographie Biologicae, Junk, Dordrecht, Netherlands, 295-310.

Untergasser, D., 1991. *Gesunde Diskus und Großcichliden, Band 1*. Bede-Verlag, Kollnburg, Germany, 136 pp.

Vieira, I., 2000. Freqüência, constância, riqueza e similaridade da ictiofauna da bacia do rio Curuá-Una, Amazônia. *Rev. bras. de Zoociências Juiz de Fora*, 2 (2): 51-76.

Vieira, I. & A. J. Darwich, 2003. Sinecologia da Ictiofauna de Curuá-Una, Amazônia: características hidroquímicas, climáticas, vegetação e peixes. *Acta Limnologica Brasiliensia*, 11 (2): 41-64.

Viets, K. 1954. Wassermilben aus dem Amazonasgebiet (Hydrachnellae, Acari). (Systematische und ökologische Untersuchungen.) Bearbeitung der Sammlung Dr. R. Braun, Arau, und Dr. H. Sioli, Belém. *Schweizer Zeitschrift der Hydrologie*, 16 (1): 78-151; 16 (2): 161-247.

Walker, I., 1978. Rede de alimentação de invertebrados das águas pretas do sistema do Rio Negro. I: Observações sobre a predação de uma ameba do tipo *Amoeba discoides*. *Acta Amazonica*, 8: 434-438.

Walker, I., 1985. On the structure and ecology of the micro-fauna in the Central Amazonian forest stream Igarapé da Cachoeira. *Hydrobiologia*, 122: 137-152.

Walker, I. & W. Franken, 1983. Ecosistemas frágeis: a floresta da terra firme da Amazônia Central. *Ciencia Interamericana*. 23: 9-21

Wallace, A. R., 1853. *Narrative of travels on the Amazon & Rio Negro with an account of the Native Tribes, and Observation on the Climate, Geology, and Natural History of the Amazon Valley.* Reeve and Co., London, 541 pp.

Wallace, A. R., 1905. *My life – a record of events and opinions.* Chapman & Hall Ltd., London. 2 vols., 435 & 459 pp. (fishes of the Rio Negro pp. 285-286, 3 pls, in vol. 2).

Weick, F., 1980. *Die Greifvögel der Welt.* Verlag Paul Parey, Hamburg & Berlin, Germany, 160 pp.

Wheeler, A. C., 1953. Arrival of Pompadour Fish *(Symphysodon discus)*. *Water Life*, 195 pp.

Wichard, W., Arens, W. & G. Eisenbeis, 1995. *Atlas zur Biologie der Wasserinekten.* Gustav Fisher, Jena, Stuttgart, & New York, 338 pp.

Wilson, E. O. & B. Hölldobler, 1988. Dense heterarchies and mass communication as the basis of organization in ant colonies. *Trends in Ecology and Evolution*, 3 (3): 65-68.

Wood, C. A., 1974. *An Introduction to the Literature of Vertebrate Zoology.* Georg Olms Verlag, Hildesheim & New York, 643 pp.

Wurzbach, C. von, 1862. *Biographisches Lexikon des Kaiserthums Oesterreich.* Kaiserlich-königlichen Hof- und Staatsdruckerei, Vienna, 184-235.

Yamamoto, K. C., Soares, M. G. M. & C. E. de C. Freitas, 2004. Feeding of *Triportheus angulatus* (Spix & Agassiz, 1829) in the Camaleão lake, Manaus, Amazonas state, Brazil. *Acta Amazônica*, 34 (4): 653-659.

Yuyama, L. K. O., Aguiar, J. P. L., Macedo, S. H., Gioia, T., Yuyama, K., Fávaro, D. I. T., Afonso, C., Vasconcellos, M. A. & S. M. F. Cozzolino, 1997. Determinação de elementos minerais em alimentos convencionais e não convencionais da região amazônica pela técnica de análise por ativação com nêutrons Instrumental. *Acta Amazônica*, 27 (3): 183-96.

Zerries, O., 1980. *Unter Indianern Brasiliens: Sammlung Spix und Martius 1817-1820.* Pinguin Verlag, Salzburg. 282 pp.

Ziburski, A., 1991. Dissemination, Keimung und Etablierung einiger Baumarten der Überschwemmungswälder Amazoniens. *Trop Subtrop Pflanzenwelt*, 77: 1-96.

Zuanon, J. & I. Sazima, 2004. Vampire catfishes seek the aorta not the jugular: candirus of the genus *Vandellia* (Trichomycteridae) feed on major gill arteries of host fishes. *aqua, Journal of Ichthyology and Aquatic Biology*, 8 (1): 31-36.

Römer, U., 1995. Uaupés. *aqua geōgraphia,* 11: 6-27.

Salazar, A. P., 2004. *Amazônia, Globalização e sustentabilidade.* Editora Valer, Manaus, Brazil, 396 pp.

Sánchez, H., Hernández, J. I., Rodríguez, J. V. & C. Castaño, 1990. *Nuevos parques nacionales, Colombia.* Instituto Nacional de los Recursos Naturales y Renovables y del Ambiente (INDERENA), Bogotá, Colombia, 213 + 25 pp.

Santos, D. d. (Ed.), 1998. *Seres vivos. Vol. 1: Nossas aves, animais da floresta.* Seduc, Maues, Opism, Manaus, 89 pp.

Schomburgk, O.A. (Ed.), 1841. *Robert Hermann Schomburgk's travels in Guiana and on the Orinoco during the years 1835-1839. According to his reports and communications to the Geographical Society, London. Edited by O. A. Schomburgk with a preface by Alexander von Humboldt together with his essay on some important astronomical positions in Guiana.* Leipzig. English translation by Roth, W. E., 1931. Argosy Co., Georgetown. 202 pp + plates & folding map.

Schultz, H., 1959a. A Hunt for the Blue Discus. *Tropical Fish Hobbyist,* 16-30.

Schultz, H., 1959b. Children of the Sun and Moon. *National Geographic,* March, 340-363.

Schultz, H., 1959c. Tukuna Maidens Come of Age. *National Geographic,* November, 629-649.

Schultz, H., 1961. Blue-Eyed Indian. *National Geographic,* July, 65-89.

Schultz, H., 1962. *Hombu.* Colibris Editora Ltda, Amsterdam/Rio de Janeiro, 32 pp.

Schultz, H., 1964. Indian of the Amazon Darkness. *National Geographic,* May, 737-758.

Schultz, H., 1966. The Waurá: Brazilian Indians of the Hidden Xingu. *National Geographic,* January, 130-152.

Schultz, L. P., 1960. A review of the pompadour or discus fishes, genus *Symphysodon* of South America. *Trop. Fish Hobby.,* 8 (10): 5-17.

Schultz, L. P., Herald, E. S., Lachner, E. A., Welander, A. D. & L. P. Wood, 1953. *Fishes of the Marshall and Marianas Islands, Volume I.* Smithsonian Institution, Washington DC. 660 pp.

Schwarz, T. 2002. *Deutsche am Amazonas. Forscher oder Abenteurer? Expeditionen in Brasilien 1800 bis 1914. Begleitbuch zur Austellung im Theologischen Museum, Berlin-Dahlem in Zusammenarbeit mit dem Brasilianischen Kulturinstitut in Deutschland (ICBRA).* Berlin: LIT Verlag, 139 pp.

Sibley, C. G. & B. L. Monroe Jr., 1990. *Distribution and Taxonomy of Birds of the World.* Yale University Press, 1111 pp.

Silva, J. M. C., 1998. *Um método para o estabelecimento de áreas prioritárias para a conservação na Amazônia Legal.* Report prepared for WWF-Brazil, 17 pp.

Silva, S., 1996. *Fruit in Brazil.* Empresa das Artes, São Paulo, Brazil, 230 pp.

Silva, J. A. M. da, Pereira Filho, M. & M. I. de Oliveira-Pereira, 2003. Fruits and seeds consumed by tambaqui (*Colossoma macropomum,* Cuvier, 1818) incorporated in the diets: gastrointestinal tract digestibility and transit velocity. *R. Bras. Zootec.,* 32 (6), suppl. 2: 1815-1824.

Silvano, R., Oyakawa, O., do Amaral, B. A. l. & A. Begossi, 2001. *Peixes do Alto Rio Juruá (Amazonas, Brasil).* Editora da Universidade de São Paulo, Brazil, 296 pp.

Silveira, T. A. & R. Vizeu Lima, 1985. *Roteiro espeleológico das Serras do Ererê e Paituna (Monte Alegre-Pará).* Trabalho apresentado no Congresso Brasileiro de Espeleologia, mecanografadas, Brasília, 64 pp.

Smith, N. J. H., 1979. *A pesca no rio Amazonas.* Amazonas, Manaus, 154 pp.

Socolof, R., 1996. *Confession of a Tropical Fish Addict.* Socolof Industries, Florida, 267 pp.

Sousa, A. C. G. de, 1983. *Síntese da História de Tefé.* Tefé/Am., 31 pp.

Spix, J. B. de, Agassiz, L. & F. C. de Martius. 1829. *Selecta Genera et Species Piscium Brasiliensium.* Typisch, Wolf, Munich, Germany, 138 pp. (reprinted 1995)

Spix, J. B. von & C. F. P. von Martius. 1823-1831. *Reise in Brasilien in den Jahren 1817-1820.* Unveränderter Neudruck des in München in 3 Textbänden und 1 Tafelband erschienen Werkes. Herausgegeben und mit einem Lebensbild des Botanikers C.F.P. von Martius sowie mit einem Register versehen von Karl Mägdefrau. Hanno Beck (Herausgeber), Brockhaus, Stuttgart, Germany. 1. Band pp. 3-412.

Spix, J. B. von & C. F. P. von Martius. 1823-1831. *Reise in Brasilien in den Jahren 1817-1820.* Unveränderter Neudruck des in München in 3 Textbänden und 1 Tafelband erschienen Werkes. Herausgegeben und mit einem Lebensbild des Botanikers C.F.P. von Martius sowie mit einem Register versehen von Karl Mägdefrau. Hanno Beck (Herausgeber), Brockhaus, Stuttgart, Germany. 2. Band, pp. 416-884.

Spix, J. B. von & C. F. P. von Martius. 1823-1831. *Reise in Brasilien in den Jahren 1817-1820.* Unveränderter Neudruck des in München in 3 Textbänden und 1 Tafelband erschienen Werkes. Herausgegeben und mit einem Lebensbild des Botanikers C.F.P. von Martius sowie mit einem Register versehen von Karl Mägdefrau. Hanno Beck (Herausgeber), Brockhaus, Stuttgart, Germany. 3. Band, pp. 887-1388.

Spruce, R., (1817-1893). 1908. *Notes of a botanist on the Amazon & Andes.* Macmillan, London. 2 vols.

Stawikowski, R. (Ed.), 1994. *Amazonas. Datz-Sonderheft.* Verlag Eugen Ulmer, Stuttgart, 74 pp.

Sterba, G. 1959. *Süßwasserfische aus aller Welt, Teil. 2.* Urania Verlag, Leipzig – Jena - Berlin, 688 pp.

Stiassny, M. L. J., 2002. *Ein kompliziertes Netzerk: Cichliden und ihre Verwandten.* www.cichlidae.ch/artikel/ein_kompliziertes_netz.../ein_kompliziertes_netzwerk.htm – 12.08.02.

Federal do Pará, Programa Pobreza e Meio Ambiente, Belém, Brazil.

Monteiro, M. Y., 1975. Um Jovem Cineasta. *Journal do Comércio, Manaus,* 27.

Monteiro, M. Y., 2002. *A Capitania de São José do Rio Negro.* Editora Valer, Manaus, Brazil, 144 pp.

Moraes, R., 1960. *Na Planície Amazônica.* 6a edição. Conquista, Rio de Janeiro, Brazil, 229 pp.

Naito, T. (Ed.), 1994. *Amazônia II.* Fair Wind Co. Ltd., Tokyo, Japan, 235 pp.

Nascimento, E. P. do, Drummond, J. A. (Eds.), 2003. *Amazônia – Dinamismo econômico e conservação ambiental.* Garamond Ltda., Rio de Janeiro, Brazil, 334 pp.

Nowak, R. M., 1991. *Walker's Mammals of the World, Volume I.* Fifth Edition. John Hopkins University Press, Baltimore & London, 642 pp.

Nowak, R. M., 1991. *Walker's Mammals of the World,* Volume II. Fifth Edition. John Hopkins University Press, Baltimore & London, 1007 pp.

Ogawa, Y., 1995. The search for the "Alenquer". *Discus year book '95-'96, Special edition,* 10-22.

Paepke, H. J., 2003. *Die Segelflosser.* Westarp Wissenschaften, Hohenwarsleben, Germany, 144 pp.

Pellegrin, J., 1899. Les Poissons vénéneux. *Thèse de Doctorat en médecine,* Challamel, éditeur, Paris, France, 121 pp.

Pellegrin, J., 1899. Poissons envoyés par M. Jacquot d'Anthonay, vice consul de France à Manaos (Brésil). *Ibid.,* 405-406.

Pellegrin, J., 1902. Cichlidés du Brésil rapportés par M. Jobert. *Ibid.,* 181-184.

Pellegrin, J., 1903. Description de Cichlidés nouveaux de la collection du Muséum. *Bulletin du Muséum d'histoire naturelle,* 120-125.

Pellegrin, J., 1904. Contribution à l'étude anatomique, biologique et taxinomique des Poissons de la familie des Cichlidés. *Mémoires de la Société zoologique de France,* 1903 (wurde erst 1904 pupliziert): 41-399, 42 figures, planches IV à VII *(Thèse de doctorat ès sciences naturelles).*

Pellegrin, J., 1909. Characinidés du Brésil rapportés par M. Jobert. *Bulletin du Muséum d'histoire naturelle,* 4: 147.

Pennington, T. D., 1997. *The genus Inga. Botany.* Royal Botanic Gardens, Kew, London, 844 pp.

Petersmann, H.-G., 2001. Der Nhamundá-Diskus. *Das Aquarium,* 09/01, 20-25.

Petersmann, H.-G., 2003. Blue Heckel Discus from the Nhamundá. *aqua geōgraphia,* 24: 58-63.

Piedade, M. T. F., Parolin, P. & W. J. Junk, 2003. Estrategias de dispersão, produção de frutos e extrativismo da palmeira *Astrocaryum jauari* Mart. no igapós do Rio Negro: implicações para a ictiofauna. *Ecologia Aplicada,* 2 (1): 31-40.

Pires, J. M., 1984. The Amazonian forest. In: Sioli, H. (Ed.), *The Amazon: Limnology and landscape ecology of a mighty tropical river and its basin.* Junk, Dordrecht, Netherlands, 581-602

Porter, D. 1993. On the road to the Origin with Darwin, Hooker, and Gray. *Journal of the History of Biology,* 26 (1): 1-38.

Prance, G. T., 1978. The poisons and narcotics of the Dení, Paumarí, Jamamadí and Jarawara indians of the Purus river region. *Rev. Brasileira de Botânica,* 1: 71-82.

Prance, G. T. & K. S. Brown Jr., 1987. The principle vegetation types of the Brazilian Amazon. In: Whitmore, T. C. & G. T. Prance (Eds.), *Biogeography and Quaternary History in Tropical America.* Clarendon Press, Oxford, 30-31

Prezia, B. & E. Hoomaert, 2000. *Brasil Indígena: 500 anos de resistência.* FTD, São Paulo, Brazil.

Prinz von Preussen, A., 1849. *Travels of his royal highness Prince Adalbert of Prussia, in the south of Europe and in Brazil: with a voyage up the Amazon and the Xingú.* Translated by Sir Robert H. Schomburgk and John Edward Taylor. D. Bogue, London. 2 vols. xvi + 338 pp. & v + 377 pp.

Rai, H. & G. Hill, 1980. Classification of Central Amazon lakes on the basis of their microbiological and physicochemical characteristics. *Hydrobiologia,* 72: 85-99.

Rataj, K., 2004. A New Revision of the Genus *Echinodorus* Richard, 1848 (Alismataceae). *aqua Journal of Ichthyology and Aquatic Biology Special Publication* l: 1-139.

Regan, C. T., 1905. On drawings of fishes of the Rio Negro. *Proceedings of the Zoological Society of London,* 1: 189-190.

Reis, A. C. F., 1934. *Manáos e outras villas.* EDUA, Editora da Universidade do Amazonas, Manaus, Brazil, 144 pp.

Reis, A. C. F., 1967. *As origens de Parintins.* Governo do Estado do Amazonas, Manaus, Brazil.

Reis, A. C. F., 1989. *História do Amazonas.* Belo Horizonte, Itatiaia, Brazil, 145.

Reis, R. E., Kullander, S. O. & C. J. Ferraris, Jr. (Eds.), 2003. *Check list of the freshwater fishes of South and Central America.* (CLOFFSCA) Edipucrs, Porto Alegre, Brazil, 729 pp.

Riedel, D. (Edr.), 1959. *As Selvas e o Pantanal, Goiás e Mato Grosso.* Editôra Cultrix, São Paulo, 314 pp.

Riedl-Dorn, C., 1998. *Das Haus der Wunder.* Verlag Holzhausen, Vienna, 308 pp.

Riedl-Dorn, C., 1999. *Johann Natterer und die Österreichische Brasilien Expedition.* 2000 Petrópolis Editora Index, Brazil, 192 pp.

Rios, M. J. & L. T. de Menezes (Eds.), 1992. *Receitas de pratos que combinam com cerveja.* Editora Marco Zero, São Paulo, Brazil.

Robertson, B. A. & E. R. Hardy, 1984. Zooplankton of Amazonian lakes and rivers. In: Sioli H. (Ed.), *The Amazon – Limnology and landscape ecology of a mighty tropical river and its basin.* Monographiae Biologicae, Junk, Dordrecht, Netherlands, 337-352

Koptur, S., 1984. Experimental Evidence for Defense of *Inga* (Mimosoidae) Saplings by Ants. *Ecology,* **65**: 1787-1793.

Koste, W., 1974. Zur Kenntnis der Rotatorienfauna der "schwimmenden Wiese" einer Uferlagune in der Várzea Amazoniens. Brasilien. *Amazoniana,* **5** (1): 25-59

Kubitzki, K. & A. Ziburski, 1994. Seed dispersal in flood plain forest of Amazonia. *Biotropica,* **26**: 30-43.

Kullander, S. O., 1986. *Cichlid fishes of the Amazon River drainage of Peru.* Swedish Museum of Natural History, Stockholm, Sweden, 431 pp.

Kullander, S. O., 1996. Eine weitere Übersicht der Diskusfische, Gattung *Symphysodon* Heckel. *Diskus – Datz Sonderheft,* October 1996, 10-19.

Kullander, S. O., 1998. A phylogeny and classification of the South American Cichlidae (Teleostei: Perciformes). In: Malabarba, L. R, Reis, R.E., Vari, R.P., Lucena, Z.M.S., & C.A.S. Lucena (Eds.), *Phylogeny and Classification of Neotropical Fishes.* Edipucrs, Porto Alegre, 461-498

Kullander, S. O., 2003. *Guide to the South American Cichlidae.* http://www2.nrm.se/ve/pisces/acara/symphyso.shtml

Kullander, S. O. & R. Britz, 2002. Revision of the family Badidae (Teleostei: Perciformes), with description of a new genus and ten new species. *Ichthyological Exploration of Freshwaters,* **13** (4): 295-372.

Kullander, S. O. & F. Fang, 2004. Seven new species of *Garra* (Cyprinidae: Cyprininae) from the Rakhine Yoma, southern Myanmar. *Ichthyological Exploration of Freshwaters,* **15** (3): 257-278.

Kurella, D. & D. Neitzke, 2002. *Amazonas Indianer, Lebensräume, Lebensrituale, Lebensrechte.* Reimer, Linden-Museum, Stuttgart, 332 pp.

La Condamine, C.-M. 1745. *Relation abrégée d'un voyage fait dans l'intérieur de l'Amérique méridionale. Depuis la Côte de la Mer du Sud, jusqu'aux Côtes du Brésil & de la Guiane, en descendant la rivière des Amazones; Lûe à l'Assemblée publique de l'Académie des Sciences, le 28. Avril 1745. Avec une Carte du Maragnon, ou de la Rivière des Amazones, levée par le même.* Veuve Pissot, Paris.

La Condamine, C.-M. 1754. Mémoire sur l'inoculation de la petite vérole, par M. de la Condamine, *Mercure de France,* June 1754, 64-126.

Ladiges, W., 1973. *Schwimmendes Gold vom Rio Ukayali.* Engelbert Pfriem Verlag, Wuppertal, Germany, 85 pp.

Laudato, L., 1998. *A critica – Yanomami pey këyo: O caminho yanomami.* Universidade Católica de Brasília, Brazil. 326 pp.

Lima, J. 1992. *História dos negros, que através da luta conseguiram libertar-se dos senhores de escravos de Santarém, Pará.* Oriximiná, Brazil, 12 pp.

Lima, C. A. & M. Goulding, 1998. *Os frutos do Tambaquí, Ecologia, Conservaçao e Cultivo na Amazônia.* Sociedate Civil Mamirauá MCT-CNPq, 186 pp.

Linnaeus, C. 1758. *Systema naturae per regna tria naturae, secundum classes, ordines, genera, species, cum characteribus, differentiis, synonymis, locis. Tomus I. Editio decima, reformata.* L. Salvii, Stockholm, 824 pp.

Lindner, H., 1967. Eleven Years Spawning Discus. *Tropical Fish Hobbyist,* **1967** (6): 4-25.

Linke, H., 2000. *Ihr Hobby: Altum-Skalare.* Bede-Verlag, Kollnburg, Germany, 79 pp.

Loefling, P., 1758. *Iter Hispanicum, eller resa til spanska länderna uti Europa och America, forratad Ar 1751 til Ar 1756.* Stockholm, 316 pp.

Lorenz, S. da Silva, 1992. *Sateré-Mawé: os filhos do guaraná.* CTI, São Paulo, 160 pp.

Lüling, K. H., undated. *Südamerikanische Fische und ihr Lebensraum.* Engelbert Pfriem Verlag, Wuppertal-Elberfeld, Germany, 84 pp.

Lyons, E., 1959. *Symphysodon discus Tarzoo.* New blue discus electrify aquarium world. *Tropicals Mag.* **4** (3): cover, 6-8, 10.

Lowe-McConnell, R. H., 1964. The fishes of the Rupununi Savana district of British Guiana, South America. I. Ecological study of fish species and effects of the seasonal cycle on the fish. *Journal of the Linnean Society* (Zoology), **45** (304): 103-144.

Machado de Paula, J., 1991. *The Lower Amazon Tapajós – O Baixo Amazonas.* Editora Agir, Brazil, 160 pp.

Maia, L. M., 1997. *Influência do Pulso de Inundação na Fisiologia, Fenologia e Produção de Frutos de* Hevea spruceana *(Euphorbiaceae) e* Eschweilera tenuifolia *(Lecythidaceae), em áreas inundáveis de igapó da Amazônia Central.* Tese de Doutorado Instituto Nacional de Pesquisas da Amazônia/Universidade Federal do Amazonas, Brazil, 186 pp.

Mayland, H. J., 1987. *Diskusfieber.* Landbuch-Verlag GmbH, Hannover, Germany, 215 pp.

Mayland, H. J., 2002. *Amazoniens Diskusfische,* Dähne Verlag, Ettlingen, Germany, 157 pp.

Martinelli, P., 2000. *Amazônia o povo da águas.* Terra Virgem Editora, São Paulo, Brazil, 263 pp.

Medem, F. 1981. *Los Crocodylia de Sur America. Vol. 1. Los Crocodylia de Colombia.* Colciencias, Bogota, Colombia, 354 pp.

Meggers, B. J., 1971. Amazonia: *Man and Culture in a Counterfeit Paradise.* Harlan Davidson, Chicago, Illinois, USA.

Mendes do Santos, G., Jegú, M. & B. de Merona, 1984. *Catálogo de Peixes comerciais do Baixo Rio Tocantins.* Eletronort/ CNPq/ INPA, 1a edição, Projeto Tucuruí, Manaus, Brazil, 83 pp.

Mikschi, E., 2002. Rudolf Kner. *Datz, Sonderheft Harnischwelse,* **2**: 6-9.

Matsuzaka, M., 1992. *Amazônia.* Minoru Matsuzaka, Japan, 200 pp.

Mitschein, T. A., 1994. *Curauá: Ananas erectifolius.* Universidade

Hölldobler, B. & E. O. Wilson, 1990. *The Ants.* Springer-Verlag, Berlin & Heidelberg, Germany, 732 pp.

Horna, J. & V. R. Zimmermann, 2000. Neotropical Ecosystems. *Proceedings of the German-Brazilian Workshop,* Hamburg, Germany.

Humboldt, F. H. A. & A. Valenciennes, 1833. Recherches sur les Poissons fluviatiles de l'Amerique équinoxiale. In: Humboldt, F. H. A. & A. Valenciennes, 1833, *Recueil d'observ-ations de zoologie et d'anatomie comparée faites dans l'Océan Atlantique dans l'intérieur du Nouveau Continent et dans la Mer du Sud pendant les années 1799, 1800, 1801, 1802 et 1803.* Paris. Vol II. (Zool): 145-216, 4 pls.

Huxley, L. 1920. *Thomas Henry Huxley – a character sketch.* Watts & Co., London, 120 pp.

Innes, W. T., 1935. *Symphysodon discus* (Heckel) – The Aristocrat of the Aquarium. *The Aquarium.* October 1935, 119-122.

Innes, W. T., 1936. The Neon Tetra *Hyphessobrycopn innesi* Myers. *The Aquarium.* November 1936, 135-136.

Innes, W. T., 1948. *Exotic Aquarium Fishes. A Work of General Reference.* Ninth Edition. Innes Publishing Company, Philadelphia, USA, 511 pp.

Irmler, U., 1975. Ecological studies of the aquatic soil invertebrates in the inundation forest of Central Amazonia. *Amazoniana,* 5 (3): 337-409.

Irmler, U., 1976. Zusammensetzung, Besiedlungs-dichte und Biomasse der Makrofauna des Bodens in der emersen und submersen Phase dreier zentralamazonischer Überschwemmungswälder. *Bio-geographica,* 7: 79-99.

Irmler, U. & K. Furch, 1980. Weight, energy and nutrient changes during the decomposition of leaves in the emersion phase of Central-Amazonian inundation forest. *Pedobiologia,* 20: 118-130.

Jardine, W. (Ed.), 1852. *The Fishes of British Guiana.* (Parts I & II). Vols. XXXIX & XL of The Naturalist's Library. W. H. Lizars, Edinburgh.

Jena, D. 1996. *Die russischen Zaren in Lebensbildern.* Verlagsgruppe Weltbild GmbH, Ausgburg, Germany, 556 pp.

Jobert, C., 1870. Recherches pour servir à l'histoire des organes du toucher chez les poissons. *Soc. Philom. Bull.,* 7: 194-206.

Jobert, C., 1875. Recherches sur les organes tactiles de l'homme. *Acad. Sci. Compt. Rend.,* 80: 274-276.

Jobert, C., 1876. Recherches sur l'appareil respirtaoire et le mode respiration de certains crustacés brachyures (crabes terrestres). *Acad. Sci. Compt. Rend.,* 81: 1198-1200.

Jobert, C., 1878. Sur la préparation du curare. *Acad. Sci. Compt. Rend.,* 86: 121-122.

Jobert, C., 1878. Sur une maladie du caféier observée au Brésil. *Acad. Sci. Compt. Rend.,* 87: 941-943.

Jobert, C., 1879. Sur l'action physiologique des strychnées de l'Amérique du Sud., *Acad. Sci. Compt. Rend.,* 89: 646-647.

Jobert, C. & G. Pouchet, 1875. Sur la vision chez les cirrhipèdes. *Soc. Biol. Mém.,* 2: 245-247.

Jobert, C. & G. Pouchet, 1876. Contribution à l'histoire de la visione chez les cirrhipèdes. *Journ. Anat.,* 12: 575-594.

Jordan, D. S., 1922. *The days of a Man.* World Book Company, Yonkers on Hudson, New York, 710 pp.

Junk, W. J., 1973. Investigation on the ecology and production-biology of the "Floating meadows" Paspalu-Echinochloetum on the middle Amazon. II. The aquatic fauna in the root zone of floating vegetation. *Amazoniana,* 4 (1): 9-112.

Junk, W. J., 1993. Wetlands of tropical South America. In: Whigham, D., Hejny, S., & D. Dykyjova (eds.), Wetlands in the Amazon floodplain. *Hydrobiologia,* 263: 155-162.

Junk, W. J. (Ed.), 1997. *The Central Amazon Floodplain. Ecology of a Pulsing System. Ecological Studies.* Springer-Verlag, Berlin & Heidelberg, 525 pp.

Kaestner, A., 1993. *Lehrbuch der Speziellen Zoologie, Band I: Wirbellose Tiere 3. Teil: Mollusca, Sipunculida, Euchiurida, Annelida, Onychophora, Tardigarda, Pentastomida.* Gustav Fisher Verlag Jena, Stuttgart, 608 pp.

Kaestner, A., 1993. *Lehrbuch der Speziellen Zoologie, Band I: Wirbellose Tiere 4. Teil: Arthropoda (ohne Insecta).* Gustav Fisher Verlag Jena, Stuttgart, 1279 pp.

Kalkmann, A. L., Costa Neto, A. N. & D. Kern, 1986. Salvamento Arqueológico na Região de Porto Trombetas (PA). *Relatório da 2ª etapa de campo, novembro/dezembro de 1985.* Museu Paraense Emílio Goeldi, Belém, Brazil.

Keller, G., 1976. *Discus.* T.F.H. Publications, Neptune City, NJ, USA. 96 pp. (Originally published by Franckh'sche Verlagshandlung, W. Keller & Son., Stuttgart/1974 under the title *Der Diskus, König der Aquarienfische.*)

Kasselmann, C., 1995. *Aquarienpflanzen.* Eugen Ulmer GmbH, Stuttgart, 472 pp.

Kesselring, A. K. Jr. & A. de Oliveira, 1992. *Relatório de atividades da Frente de Contato do rio Purus e complementação do projeto de localização e assistência aos grupos isolados.* Funai, Brasília, 53 pp.

Koch-Grünberg, T., 1906. Bericht über seine Reise am oberen río Negro und Yapurá in den Jahren 1903-1905. *Zeitschrift der Gesellschaft für Erdkunde zu Berlin,* 1906: 80-101 + maps.

Koch-Grünberg, T., 1917. *Von Roraima zum Orinoco. Ergebnisse einer Reise in Nordbrasilien und Venezuela in den Jahren 1911-1913.* Reimer, Berlin, 280 pp.

Koch-Grünberg, T. 1910. *Zwei Jahre unter den Indianern. Reisen in Nordwest-Brasilien 1903-1905.* Bd. 1. Wasmuth, Berlin, 186ff., 206.

Koch-Grünberg, T., 1923. *Vom Roraima zum Orinoco. Ergebnisse einer Reise in Nordbrasilien und Venezuela in den Jahren 1911-1913. Bd. 3. Ethnographie.* Strecker & Schröder, Stuttgart. x,446 pp.

Ferreira, L. V., 2000. Effects of flooding duration on species richness, floristic composition and forest structure in river margin habitat in Amazonian blackwater floodplain forests: implications for future design of protected areas. *Biodiversity and Conservation*, 9: 1-14.

Ferrugia, F., 1993. Algae. *aqua geõgraphia*, 7: 106-113.

Figueroa, A. L. G., 1998. *Guerriers de l'écriture et commerçants du monde enchanté : histoire, identité et traitement du mal chez les Sateré-Mawé (Amazonie Centrale, Brésil)*. EHESS, Paris, 585 pp.

Fittipaldi, C., 1986. *A lenda do guaraná: mito dos índios Sateré-Mawé*. Série Morena, Melhoramentos, São Paulo, Brazil, 16 pp.

Filho, J. M., 2004. *O Livro de Ouro da Amazônia*. Ediouro, Rio de Janeiro, 398 pp.

Filho, P., 2000. *Estudos de História do Amazonas*. Valer Editoria, Manaus, Brazil, 239 pp.

Franceschini, D. (Ed.), 1997. *Sateré-Mawé: mowe'eg hap*. Seduc, Maues, Opism, Manaus, Brazil, 105 pp.

Frisch, J. D., 1981. *Aves Brasileiras*. Volume I. Dalgas-Ecoltec Ecologia Técnica e comércio Ltda, São Paulo, Brazil. 353 pp.

Frey, H., 1978. *Das Süsswasser Aquarium*. Neumann Verlag, Radebeul & Berlin, Germany, 265 pp.

Gallois, D. T., 1992. De arredio a isolado: perspectivas de autonomia para os povos indígenas isolados. In: Grupioni, L. D. B. (Ed.). *Índios no Brasil*. Secretaria Municipal de Cultura, São Paulo, Brazil, 121-34.

Géry, J., 1954. Hier et Aujourd'hui – II. L'histoire du "Discus". *L'aquarium et les poissons*, 41: 13-17.

Géry, J., 1977 (1978). *Characoids of the world*. T.F.H. Publications, Inc., Neptune City, NJ, USA, 672 pp.

Géry, J., 1994. Neon. *aqua geõgraphia*, 7: 82-85.

Gomes De Souza, L. A. & M. Freitas Da Silva, 2003. Bioeconomical potential of Leguminosae from the Lower Negro River, Amazon, Brazil. *Lyonia*, 5 (1): 15-24.

Goulart, A., 1968. *O Regatão (mascate fluvial da Amazônia)*. Conquista, Rio de Janeiro, Brazil.

Goulding, M., 1980. *The fishes and the forest: Explorations in Amazonian Natural History*. University of California Press, Berkeley, CA, USA, 280 pp.

Goulding, M., 1981. *Man and Fisheries on an Amazon Frontier*. Dr. W. Junk Publishers, The Hague-Boston-London, 137 pp.

Goulding, M. & M. L. Carvalho, 1982. Life history and management of the Tambaquí, (*Colossoma macropomum*, Characidae): An important Amazonian food fish. *Revista Brasileira de Zoologia*, 1: 107-133.

Goulding, M., Carvalho, M. L. & E. G. Ferreira, 1988. *Rio Negro, Rich life in poor water*. SPB Academic Publishing bv, The Hague, Netherlands, 200 pp.

Goulding, M., Barthem, R. & E. Ferreira, 2003. *The Smithsonian Atlas of the Amazon*. Smithsonian Books, Washington and London, 253 pp.

Greuter, W., McNeill, J., Barrie, F., Burdet, H., Demoulin, V., Filgueiras, T. & D. Nicolson, (Eds.) 2000. *International Code of Botanical Nomenclature (Saint Louis Code). Regnum Vegetabile*. vol. 138. xviii + 474 pp.

Guiry, M. D. & E. N. Dhonncha, 2004. *AlgaeBase version 2.1*. World-wide electronic publication, National University of Ireland, Galway. http://www.algaebase.org

Guter, J., 1999. Fish myths of the World. Part 1. *aqua geõgraphia*, 18: 92-100.

Hahn, W. J., 2000. *Molecular phylogenetics of the Palmae*. Submitted November 2000 to the EMBL/GenBank/DDBJ Databases, at The European Bioinformatics Institute (EBI).

Hartmann, G., 1986. *Xingú – Unter Indianern in Zentral-Brasilien*. Staatliche Museen Preußischer Kulturbesitz, Berlin, Germany, 323 pp.

Hartmann, T., 1982. Artefactos indígenas brasileiros em Portugal. *Boletim da Sociedade de Geografia*, 100 (1–6, 7–12).

Heckel, J., 1840. Johann Natterer´s neue Flussfische Brasiliens nach den Beobachtungen und Mitteilungen des Entdeckers beschrieben. (Erste Abteilung, die Labroiden). *Annalen vom Museum der Naturgeschichte Wien*, 2: 327-470.

Heckel, J. & R. Kner, 1858. *Die Süsswasserfische der Österreichischen Monarchie*. Verlag von Wilhelm Engelmann, Leipzig, 388 pp.

Heil, M., Rattke, J. & W. Boland, 2005. Postsecretory Hydrolysis of Nectar Sucrose and Specialization in Ant/Plant Mutualism. *Science*, 308 (5721): 560-563.

Heil, M., Greiner, S., Meimberg, H., Krüger, R., Noyer, J.-L., Heubl, G., Linsenmair, K. E. & W. Boland, 2004. Evolutionary change from induced to constitutive expression of an indirect plant resistance. *Nature*, 430 (6996): 205-208.

Henman, A., 1982. *O guaraná, sua cultura, propriedades, formas de preparação e uso*. Cadernos de Vida Natural, 10, Global/Ground, São Paulo, 77 pp.

Higuchi, H., 1992. *An updated list of ichthyological collecting stations of the Thayer Expedition to Brazil*. Electronic version (1996) http://www.oeb.harvard.edu/thayer.htm

Hilbert, K., 1988. *Relatório de viagem do Projeto de Salvamento Arqueológico na região de Porto Trombetas (PA). III Relatório*. Museu Paraense Emílio Goeldi, Belém, Brazil.

Hildebrand, M. von & E. Reichel, 1987. Indígenas del Mirití-Paraná. In: Correa, F. & X. Pachón, (Eds.), *Introducción a la Colombia Amerindia*. Instituto Colombiano de Antropología, Bogotá, 135-150.

Hoek, C. v. d., Jahns, H. M. & D. G. Mann, 1993. *Algen*. Revised 3rd edition. Georg Thieme Verlag, Stuttgart & New York, 411 pp.

Cogger, H. G. & R. G. Zweifel, 1992. *Reptilien & Amphibien – Enzyklopädie der Tierwelt.* Jahr-Verlag GmbH & Co. Hamburg, 240 pp.

Collaz, K. G. (Ed.), 2000. *Lexikon der Naturwissenschaftler.* Spektrum Akademischer Verlag, Heidelberg & Berlin, Germany, 505 pp.

Comber, R., 1991. *Vögel – Enzyklopädie der Familie und Arten.* Naturbuch Verlag, Germany, 240 pp.

Comissão Pró-Índio de São Paulo. 1994. *Relatório de atividades: November, 1991 to April, 1994.* São Paulo Pro-Indian Commission, São Paulo, Brazil.

Confronti, V., 1993. Study of the Euglenophyta from Camaleão Lake (Manaus-Brazil). *Rev. Hydrobiol. Trop.,* **26** (1): 3-18.

Cruls, G., 1930. *A Amazônia que eu vi.* Ed. José Olimpio, Rio de Janeiro.

Cunha, E. da, 1902. *Selva e sertão.* Reprinted 2002 in: *Euclides da Cunha,* Casa de Cultura Euclides da Cunha. San José do Rio Pardo, SP & Bibliotheca virtual do estudande Brasileiro: www.literaturabrasileira.ufsc.br.

Cunha, E. da, 1906. *Relatório da comissão mista brasileira-peruana de reconhecimento do Alto Purus.* Reprinted 1995 in: Coutinho, A. & Cunha, E. da., *Obra completa,* 2 vols., Nova Aguilar Editora, Brazil, 1738 pp.

Cunha, E. da, 1908. O Inferno Verde. Reprinted 1995 in: Coutinho, A. & Cunha, E. da., *Obra completa,* 2 vols., Nova Aguilar Editora, Brazil, 1738 pp.

Davis, S. H., 1978. A invasão do parque Indígena Aripuanã: A desintegração das tribos Cintas-Largas e Suruí. In: Davis, S. H., *Vítimas do milagre: o desenvolvimento e os índios do Brasil.* Zahar Editores, Rio de Janeiro, Brazil, 105-17.

Davis, W., 1997. *One River: Science, Adventure and Hallucinogenics in the Amazon Basin.* Simon and Schuster/Touchstone, New York, 537 pp.

Derickx, J. & J. A. Trasferetti, 1992. *No Coração da Amazônia, Jurá o Rio Que Chora.* Vozes, Petrópolis, Brazil, 181 pp.

Diener, P. & M. de Fátima Costa, 1999. *A América de Rugendas.* Kosmos, São Paulo, Brazil, 167 pp.

Derby, O. 1898. O Rio Trombetas. *Boletim do Museu de História Natural e Etnografia do Estado do Pará,* **II** (1-4): 366-82.

Dubelaar, C. N., 1986a. The petroglyphs in the Guyanas and the adjacent Areas of Brazil and Venezuela. An Inventory. *Monumenta Archaeologica,* **12**: 304.

Edwards, W. H., 1847. *A voyage up the river Amazon: including a residence at Pará.* D. Appleton & Company, New York, G. S. Appleton, Philadelphia. 256 (8) pp., (1) leaf of plates, illustrated.

Ehrenreich, P., 1891. Beiträge zur Völkerkunde Brasiliens. *Veröffentlichungen aus dem Königlichen Museum für Völkerkunde, s.l.,* **2**: 1-80.

Ehrenreich, P., 1929. Viagem aos rios Amazonas e Purus. *Revista do Museu Paulista,* Brazil, **16**.

Encarnação, M. U. da, 1900. Carta sobre costumes e crenças dos índios do Purús, dirigida a D. S. Ferreira Penna. *Boletim do Museu Paraense de História Natural e Ethnographia, s.l.,* **3** (1): 94-7.

Engel, M. S. & K. Krishna, 2004. Family-Group Names for Termites. *American Museum Novitates,* **3432**: 9.

Ette, O. (Hg.) 1999. Humboldt, Alexander von: *Reise in die Äquinoktial-Gegenden des Neuen Kontinents. Mit Anmerkungen zum Text, einem Nachwort und zahlreichen zeitgenössischen Abbildungen sowie einem farbigen Bildteil.* 2 vols. Insel Verlag, Frankfurt am Main & Leipzig, 1648 pp.

Esteves, F. A., Bozelli, R. L. & F. Roland, 1990. Lago Batata: um laboratório de Limnologia Tropical. *Ciência Hoje,* **11** (64): 26-33.

Falconer, C. 1938. *The Chemistry and Technology of Rubber Latex.* Chapman & Hall Ltd, London, UK., 2-4

Félix, R. de Cássia. 1987. *Relatório de identificação e delimitação da Área Indígena Paumari do Rio Ituxi.* Funai, Brasília, Brazil, 82 pp.

Ferrarini, S. A., 1980. *Tapauá: sua história, sua gente.* Ed. Calderão, Tapauá, AM, Brazil, 107 pp.

Ferreira, A. R., 1787. *Viagem filosófica ao Rio Negro.* Extrato do diário da Viagem Filosófica, Rodrigues Ferreira resumiu todas às atividades de quatro anos de expedição, desde a saída de Portugal até o fim de setembro desse ano (1787). Museu Paraense Emilio Goeldi, Brazil. (unpublished).

Ferreira, A. R., 1885. *Diário da viagem philosofica pela capitania de São José do Rio-Negro com a informação do estado presente.* Printed from his unpublished manuscript in *Revista do Instituto Histórico e Geográfico Brasileiro,* **XLVIII** (1): 1-234 (1885).

Ferreira, A. R., 1886. *Diário da viagem philosofica pela capitania de São José do Rio-Negro com a informação do estado presente.* Printed from his unpublished manuscript in *Revista do Instituto Histórico e Geográfico Brasileiro,* **XLIX** (1): 123-288 (1886).

Ferreira, A. R., 1887. *Diário da viagem philosofica pela capitania de São José do Rio-Negro com a informação do estado presente.* Printed from his unpublished manuscript in *Revista do Instituto Histórico e Geográfico Brasileiro,* **L** (2): 11-141 (1887).

Ferreira, A. R., 1888. *Diário da viagem philosofica pela capitania de São José do Rio-Negro com a informação do estado presente.* Printed from his unpublished manuscript in *Revista do Instituto Histórico e Geográfico Brasileiro,* **LI** (1): 5-166 (1888).

Ferreira, A. R., 1971(1972). *Viagem filosófica às Capitanias do Grão-Pará, Rio Negro, Mato Grosso e Cuiabá.* Conselho Federal de Cultura, Rio de Janeiro, Brazil, vol. **2**: 246 pp.

Ferreira, A. R., 1974. *Viagem filosófica às Capitanias do Grão-Pará, Rio Negro, Mato Grosso e Cuiabá: memórias (sobre) antropologia.* Conselho Federal de Cultura, Rio de Janeiro, 161 pp.

Bleher, A .F. H. & H. Bleher, 1999. Iténez II. *aqua geōgraphia,* **19**: 36-46.
Bleher, H., 1993a. *DISCUS – Wild-caught and captive-bred forms* (supplements). Aquaprint Verlags GmbH, Germany, 48 pp.
Bleher, H., 1993b. The last Wai-Wais. *aqua geōgraphia,* **4**: 28-43.
Bleher, H., 1993c. Waigeo. *aqua geōgraphia,* **5**: 28-41.
Bleher, H., 1994a. Abunã. *aqua geōgraphia,* **6**: 6-19.
Bleher, H., 1994b. Discus. *aqua geōgraphia,* **8**: 80-83.
Bleher, H., 1994/5. *DISCUS – Wild-caught and captive-bred forms* (supplements). Aquapress, Italy, 48 pp.
Bleher, H., 1995. Africa *Terra incognita, Tunisia. aqua geōgraphia,* **9**: 91-111.
Bleher, H., 1996/7. *DISCUS – Wild-caught and captive-bred forms* (supplements). Aquapress, Italy, 48 pp.
Bleher, H., 1998b. *DISCUS – Wild-caught and captive-bred forms* (supplements). Aquapress, Italy, 48 pp.
Bleher, H., 1998c. Dr. Eduard Schmidt-Focke – A Life Devoted To Discus. *aqua geōgraphia,* **17**: 56-68.
Bleher, H., 1999a. Jungle Walk. *aqua geōgraphia,* **18**: 22-34.
Bleher, H., 1999/2000. *DISCUS – Wild-caught and captive-bred forms* (supplements). Aquapress, Italy, 48 pp.
Bleher, H., 2000. *Keep'em alive! 1 – Der Wimpelpiranha – Brasilien.* Aquapress Video Production, Italy.
Bleher, H., 2001a. Brazil – The world's greatest biodiversity in danger! *aqua geōgraphia,* **21**: 88-96.
Bleher, H., 2001b. *Welt der Diskus IV – Abenteuer Nhamundá.* Aquapress Video Production, Italy.
Bleher, H., 2001c. Yakati. *aqua geōgraphia,* **22**: 6-25.
Bleher, H., 2001d. Aquarium history – Part 1: How it all began. *Nutrafin Aquatic News,* Issue # **1**: 13.
Bleher, H., 2001e. Ethiopia part IV: The Surma and related tribes. *aqua geōgraphia,* **23**: 6-22.
Bleher, H. 2003a. Pre-Columbian. *aqua geōgraphia,* **24**: 190-211.
Bleher, H., 2003b. Nhamundá. *aqua geōgraphia,* **24**: 54-56.
Bleher, H., 2003c. Mozambique. *aqua geōgraphia,* **24**: 212-225.
Bleher, H., 2003d. Genital-Biting Fish Terrorise Villagers of Port Moresby. *aqua geōgraphia,* **24**: 142-143.
Bleher, H., 2004a. Fishes in nature and in the aquarium: Neon, The history of the neon and cardinal tetras. *Nutrafin Aquatic News,* Issue # **1**: 4-5.
Bleher, H., 2004b. Aquarium history Part 4: What happened at the start of 1800? *Nutrafin Aquatic News,* issue # **4**: 13.
Bleher, H. & M. Göbel., 1992. *DISCUS – Wild-caught and captive-bred forms.* Aquaprint Verlags GmbH, Germany, 148 pp.
Bleher, H. & H. Linke., 1991a. *World of Discus – King of the Amazon – History and Care.* Aquaprint Video Production, Germany.
Brandão, F. M., 1997. *Terra Pauxí.* Secult-PA, Bélem, 105 pp.
Brook A. J. & W. Ells, 1987. The Feeding of Amoebae on Desmids. *Microscopy,* **35** (7): 537-540.

Burgess, W. E., 1981. Studies on the family Cichlidae: 10. New information on the species of the genus *Symphysodon* with the description of a new subspecies of *S. discus* Heckel. *Tropical Fish Hobbyist,* **29** (7): 32-42.
Burgess, W. E., 1989. *An Atlas of Freshwater and Marine Catfishes – A Preliminary Survey of the Siluriformes.* T.F.H. Publications, Inc., Neptune City, NJ, USA, 784 pp.
Burnett, D. G., 2000. *Masters of All They Surveyed: Exploration, Geography, and a British El Dorado.* The University of Chicago Press, vol. X. 298 pp.
Cardini, M., 1927. *La vita e l'opera di Marcellus Malpighi.* Pozzi, Rome.
Carneiro da Cunha, M. & M. B. de Almeida (Eds.), 2002. *Enciclopédia da Floresta – O alto Juruá: práticas e conhecimentos das populações.* Editora Schwarcz Ltda. São Paulo SP., 735 pp.
Carvajal, F. G. de, 1992 (reprint). *Relatório do novo descobrimento do famoso rio grande descoberto pelo capitão Francisco de Orellana (1542).* Página Aberta, São Paulo, 55, 57, 63, 65.
Carvalho, M. L., 1981. *Alimentação do tambaquí jovem* (Colossoma macropomum *Cuvier, 1818) e a sua relação com a comunidade zooplanktónica do Lago Grande-Manaquirí, Solimões, AM.* MSc Thesis, INPA/FUA, Manaus, Brazil.
Castro, E. M. R. & R. E. Acevedo, 1991. *Negroes do Trombetas: Etnicidade e história, Belém.* Núcleo de Estudos de Altos Estudos Amazônicos/UFPA, Brazil, 74 pp.
Cavalcante Gomes, D. M., 2002. *Cerâmica Arqueológica da Amazônia.* Editora da Universidade de São Paulo, Brazil, 355 pp.
Chambers, J. Q. N. H, Iguchi, H. & J. P. Schimmel, 1998. Ancient trees in Amazonia. *Nature,* **391**: 135–136.
Chandless, W., 1866a. Ascent of the river Purus. *Journal of the Royal Geographical Society,* **XXXV**: 86-118.
Chandless, W., 1866b, Notes on the River Aquiry, the principal affluent of the River Purus. *Journal of the Royal Geographical Society,* **XXXVI**: 119-128.
Chandless, W., 1868, Notas sobre o Rio Purus. *Separata dos Arquivos da Associação Comercial do Amazonas,* **3** (9): 21.
Chapelle, R., 1982. *Os índios Cintas-Largas.* Edusp, Belo Horizonte, Itatiaia, São Paulo, Brazil, 140 pp.
Chou, S., 1993. *Memorabilia of Discus.* AquaLife Magazine Special publication, Tokyo, Japan, 148 pp.
Cinta-Larga, P., 1988. *Mantere ma kwē tinhin : histórias de maloca antigamente.* Belo Horizonte, SEGRAC/CIMI, Brazil, 132 pp.
Cochu, F., 1993. Interview 2-6-1993, parts 1 and 2. On tape recorded by Ross Socolof. Aquapress, Italy.
Cochu, F., 1993. Interview 2-6-1993, parts 3 and 4. On tape recorded by Ross Socolof. Aquapress, Italy.
Cochu, F., 1993. Interview 7-6-1993, parts 1 and 2. On tape recorded by Ross Socolof. Aquapress, Italy.

References

Abreu, R. L. S. de & M. A. de Jesus, 2004. Natural durability of *Bactris gasipaes* Knuth (peach palm, Arecaceae) stipe II: insects. *Acta Amazonica*, 34 (3): 459-465.

Adis, J., 1992. Überlebensstrategien terrestrischer Invertebraten in Überschwemmungswäldern Zentralamazoniens. *Verh. naturwiss. Ver.*, 33: 21-114.

Adis, J. & W. J. Junk, 2002. Terrestrial invertebrates inhabiting lowland river floodplains of Central Amazonia and Central Europe: a review. *Freshwater Biology*, 47: 711-731.

Adonias, I., 1963. *A cartografia da região Amazônica*. Instituto Nacional de Pesquisas da Amazônia, Rio de Janeiro, 716 pp.

Agassiz, L. & E. C. Agassiz 1868. *A Journey in Brazil*. Translated and reprint 1975 in: *Viagem ão Brasil : 1865-1866*. Edusp, Belo Horizonte, Itatiaia, São Paulo, 323 pp. (Reconquista do Brasil: 12)

Aguiar, J. P. L., 1996. Tabela de composição de alimentos da Amazônia. *Acta Amazônica*, 26 (1/2): 121-26.

Alanis, R., 2000. Sateré-Mawé apostam na força do guaraná. *Amazônia Vinte e Um.*, 2 (7): 15-17.

Alencar, M., 1970. Silvino Santos no Rastro das Amazonas. *Journal do Brasil*, 8th August 1970.

Alho, C. J. R., Carvalho, A. G. & L. F. M. Padua, 1979. Ecología da tartaruga da Amazônia e avaliação de seu manejo na Reserva Biologia de Trombetas. *Brasil Florestal*, 9: 2947.

Alvarez, G. O., 2000. Os Sateré-Mawé. In: Alvarez, G. O., & Reynard, N. (Eds.). *Amazônia cidadã : previdência social entre as populações tradicionais da região Norte do Brasil*. Coleção previdência Social, Série Especial, 1, MPAS, Brasília, Brazil., 78-95.

Amsler, K., 1994. The Green Death – Algae II. *aqua geõgraphia*, 8: 109-113.

Andrade, L. M. M., 1994. O papel da perícia antropológica no reconhecimento das terras de ocupação tradicional: o caso das comunidades Remanescentes de Quilombos do Trombeta (Pará). In: Silva *et al*. (Eds.), *A perícia antropológica em processos judiciais*, Editora da UFSC, ABA/São Paulo Pro-Indian Commission, Florianópolis, 53-59.

Anonymous, 1935. *Das Aquarium. Deutscher Almanach für Aquarien- und Terrarienfreunde*. Verlag Das Aquarium, Berlin, Germany, 224 pp.

Anonymous, 1993. *Memorabilia of Discus*. AquaLife Magazine Special publication, Tokyo, Japan, 148 pp.

Anonymous, 1994. *Amazônia II*. Fair Wind Co., Ltd. Tokzo, Japan, 235 pp.

Anonymous, 1996. *Mamirauá – Management plan*. Ed. Brasília: SCM, CNPq/MCT, IPAAM, Manaus, Brazil, 94 pp.

Anonymous, 1999. *Recuperação do lago Batata*. Environment Report. 2 ed. Mineração Rio do Norte S. A., Brazil.

Anonymous, undated. *Viagem Filosófica: pelas capitanias do Grão, Rio Negro, Mato Grosso e Cuiabá, 1783-1792 Alexandre Rodrigues Ferreira*. Conselho Federal de Cultura, Brazil. Unpublished manuscript, 140 loose sheets: mainly illustrations.

Asp, N., Johansson, C., Hallmer, H. & M. Siljeström, 1983. Rapid enzymatic assay of insoluble dietary fiber. *Journal of Agricultural and Food Chemistry*, 31: 476-482.

Ax, P., 1999. *Das System der Metazoa II. Ein Lehrbuch der phylogenetischen Systematik*. Gustav Fischer Verlag, Stuttgart, Germany, 384 pp.

Baleé, W. & C. L. Erickson, 2005. Intoduction: Time and Complexity in Historical Ecology. In: Baleé, W. & C. L. Erickson (Eds.), *Time and Complexity in Historical Ecology: Studies from Neotropical Lowlands*. Colombia University Press, New York, 432 pp.

Batista, A. O. 1998. *Seres vivos. v. 2: nossos peixes, pequenos animais*. Seduc, Maues, Opism, Manaus, 79 pp.

Bayern, T. von., 1897. *Meine Reise in den Brasilianischen Tropen*. Verlag von Dietrich Reiner, Berlin, Germany, 79 pp.

Behrens, D., 1999. *Curauá-Faser – eine Pflanzenfaser als Konstruktionswerkstoff?* Dissertation, Verlag Dr. Köster, Universität Hohenheim, Germany, 210 pp.

Bernardino, F. R., *Uiara, o boto vermelho. Emoções Amazônicas: Guia fotográfico-sentimental dos ecosistemas amazônicos*, Amazon Multimedia Stock Edições, Brazil, 127 pp.

Bernardino, F. R. & R. de Souza Omena Jr., 1999. *Aves da Amazônia, Guia do observador*, Paper Editora, Manaus, Brazil, 240 pp.

Bertier de Sauvigny, G. de. 1988. *Metternich Staatsmann und Diplomat im Zeitalter der Restauration*. Dederichs, Munich, Germany, 560 pp.

Bethell, L. (Ed.) 1995. *Cambridge History of Latin America*. vol. XI. Cambridge University Press, UK. (18 essays on Brazil).

Bisby, F. A., Ruggiero, M. A., Wilson, K. L., Cachuela-Palacio, M., Kimani, S. W., Roskov, Y. R., Soulier-Perkins, A. & J. van Hertum (Eds), 2005. *Species 2000 & ITIS Catalogue of Life: 2005 Annual Checklist*. CD-ROM, Species 2000, Reading, UK.

Bittencourt, A., 1973. *Dicionário Amazonense de Biografias*. Conquista Editora, Rio de Janeiro, Brazil.

Bleher, A. F. H., 1998. Iténez. *aqua geõgraphia*, 17: 6-19.

Bleher, A. F. H. 2005. *Iténez – Fluss der Hoffnung*. Aquapress, Italy, 287 pp.

water will kill the fishes. Formally only used by indigenous people to collect, today widely used also by white man, although its use is prohibited in Brazil. (See also page 260-261.)

tiba (tiwa - tiua - tuba): abundance; full.

Tijuca (tiyug): foul liquid; mud; swamp.

timburé (ximburé): a species of river fish with black blotches or bands *(Leporinus* sp.).

tinga (ibitinga): white.

Trapecio Amazônico: = the Amazonian Trapezium, the name for the trapezoid area of Colombia where that country borders Brazil and Peru near Leticia.

tratado: treaty, agreement.

tributos arregadados: collected taxes. (Portuguese)

tucum (tu'cum - tecum): a strong, fine thread obtained from the young leaves of the Brazilian palm species *Astrocaryum vulgare,* widely used by indigenous people for making string (eg bowstrings), mats, clothes, etc.

tucunaré: any species of the edible fish genus *Cichla,* widespread in Brazil. The various species are often given supplementary names, eg tucunaré-açú.

tucupi: sap obtained from the manioc (qv) root, highly toxic in its raw state. Once cooked it becomes edible and is added to popular Amazonian foods such as wild duck, *pirarucú, tacacá,* etc.

Tupí: an indigenous tribe that inhabit(ed) northern and central Brazil, along the Amazon and the coast; one of the principle languages in South America, belonging to the Tupí-Guarani *(qv)* language group.

Tupí-Guarani: one of the four major linguistic groups in tropical and equatorial South America; indigenous tribes belonging to this linguistic group.

U

ubá: a canoe (generally made from a tree-trunk); a tree used for making canoes.

una: black.

V

várzea: a general term for whitewater floodplain in the Amazon region. (Portuguese)

vereador: deputy, representative.

voadeira: a boat, normally made of aluminium, very shallow and fast, with an outboard motor. Nowadays widespread in Amazonia.

X

Ximaana: a tribe inhabiting the Rio Javari region on the border between Brazil and Peru.

Ximana (Xumana - Xumane - Jumana): a tribe of the Aruaque group, inhabiting the region of the Rios Japurá and Solimões (western Amazonia).

ximburé: see *timburé.*

Y

Yara (iara): goddess of the waters; mother of water; a mythical woman that lives on the bottom of rivers.

yarapé: a small navigable river. From the Nheêgatu language: *yara* = canoe + *pé* = way.

yawara: jaguar; dog; wolf; cat..

Ytucy: Mother of Waterfalls, an indian name for the Rio Ituxi (a branch of the Purus), because it is full of waterfalls.

pirang arrî: very red, among the Guajararas.

piranha: a large predatory characin (family Serrasalmidae), found only in tropical South America to the east of the Andes, and there in almost every body of water. More than 35 species are known in the genera *Pristobrycon, Pygocentrus, Pygopristis,* and *Serrasalmus.* The name, which originates from the Guajarara *pirãi,* is now used worldwide.

pirapitinga: see *pacu-caranha.*

pitang (pirang): red, in the Tupí language.

pitáng: child, among the Kamayura/Kamayurá.

pium: a tiny midge that can deliver a painful bite through anything, although the pain lasts only a few hours, unlike with mosquito bites.

Polizia Federal: (the Brazilian) State Police – an organisation active nation-wide.

poraquê: the electric eel *(Electrophorus electricus),* probably derived from *muraquê* in the Guajarara language.

porco-do-mato: wild pig *(tazahú* among the Guajarara).

ppuru: a calabash (gourd) for fetching water.

praias: beach; also used in Amazonia to denote a sandbank exposed in the middle or at the edge of a river when the water level recedes. (Portuguese.)

puca (arapuca - puçá): a trap.

puçá: trap for fishes (and other aquatic animals). But today in Brazil (and else where in Portuguese speaking areas) widely used for hand-net, to catch fishes. (Portuguese)

pupunha: a small coconut, *Bactris gasipaes* (see p. 516).

puru: false; lie; liar.

puru-puru: painted.

Q

quilombo: a community of escaped black slaves of African origin (Brazilian Portuguese).

quilombola: a community of the descendants of escaped black slaves of African origin, still found today in certain places in the Amazon basin. See also quilombo.

R

recreio: a large motorboat (lancha, *qv),* the main form of transportation for people and freight, serving almost all places in Amazonia as a river-bus. Also called barco de linha. (Portuguese)

rede: fishing net (also called rede de pesca); hammock. (Portuguese)

restingas: levees (in Amazonia tall banks of silt, usually covered in tall jungle).

ressaca: in the Amazon region, an annual drop in the water level, which takes place once a year, always after the first rise following the start of the rainy season. (See also *ajú.)*

S

seringa: rubber (see also pp. 254-5). The original Portuguese meaning of the word is "injection" (see pages 254-5).

seringueira: a rubber tree (usually *Hevea brasiliensis).*

seringueiro: a rubber collector.

Surui: a tribe in the Parque do Aripuanã, Madeira region, Rondônia State.

T

Tacana: a tribe with a surviving population of around 3,500 in Bolivia. (And Tacaná is an *igarapé (qv)* in the Tabatinga region (see pp. 481-491), some times written also Tacana.)

tambaquí: the usual name in Brazil for *Colossoma macropomum,* a large characoid of the family Serrasalmidae (sometimes further assigned to the subfamily Serrasalminae), better known elsewhere as the *pacu (qv).*

tanga: loincloth (among the indians); a tiny swimsuit (among Brazilians).

tapioca: a food made from starch from *tucupi (qv)* from the manioc *(qv)* root.

tapira: bark, in the language of the Bororo indians.

Tapauá: (1) the largest (left-hand) tributary of the Rio Purus; (2) capital of the province of the same name, situated on the Rio Purus. According to the Encyclopedia of Amazonia the name means *"o que e manso"* = "that which is peaceful", but in the Nheêgatu language it signifies "where the root ends" supposedly referring to the large, branched mouth of the river Tapauá.

Tapuio (Tapoju): the name of the aboriginals who once populated a gigantic region extending from what is now the states of Goias and Mato Grosso to Amazonas and Pará, and along the Rio Tapajós. In 1998 there were still 235 members of a tribe calling themselves Tapuios living in Carretão, Goias. But they are supposedly descended from the Xerente or Xavante.

tatu: armadillo, in the Kayabi/Kayabí/Kajabi language in the Xingú region and on the Rio dos Peixes (= river of fishes).

tép (teto): fish, among the tribes formerly known as the Caiapós meridionais and Caiapós do sul (today the Panarás and Crenhac(ar)ores, respectively). The former also used *topú* and the latter *tépo.*

TI: abbreviation for Território Indigeno = indian reserve.

timbó: fish poison. A root which desolved (beaten) into the

mairá: a species of manioc, typical of the north of Brazil. (Called *mandiocaçu, mandioca açu,* and *mandioca grande* in Portuguese)

manaty: mama or mummy in the Gabilí language group.

Manau: a tribe of the Aruaque group, which once lived in the Rio Negro region.

manauara (manauense): native to, resident in, or relating to Manaus (the capital of Amazonas State).

mandioca: see *manióc.* (Portuguese.)

Maní (Maniva): the goddess of manioc *(qv)* and peanuts.

manióc (manioca- mandioca - aipím - macaxeira): a root which, suitably prepared, has been the principal source of nutrition of the indigenous Amazonian peoples for thousands of years. The word originated among the Guajarara tribe and is derived from Maní *(qv,* a goddess, and *oca* (a house)) and signifies "the goddess Maní, buried in her house, cultivating (creating) the nutritional root".

manioca: see *manióc.*

maracujá: passionfruit.

marimari: make good; value something; also used as a form of address for a respected person. (From the ancient Araucana indian language.)

massau (sa'wi - sagüim - sauim - soim - sonhim - sagüi - tamari - xauim): a small monkey species with a long tail, very common in Amazonia.

matê: a tea-like drink made from herbs (food or nutrition in the Apiacá language).

mato: forest; jungle; weeds. (Portuguese)

milho: maize.

mirim: small.

mocambo: an escaped slave. From Mocambo, a port in Mozambique, from which slaves were shipped to Brazil as long ago as the 16th century. See also quilombo.

moponga (mu'põga - mupunga): an indigenous method of catching fishes by beating on the water (with a stick or hands) to drive them into a trap.

mucúi: two. (Among the Guajarara)

município: province. (Portuguese)

myra: people.

myra puru puru: painted people

O

Oapixana (Vapixiana - Vapixana - Uapixana - Wapixana - Vapidiana - Oapina): a tribe of the Aruaque group from the Alto Rio Branco, on the border with Guyana.

Oapina: see Oapixana.

oca: an indigenous house.

ocara: centre or central square of an indian village.

ocaruçu: a large square (larger than an *ocara* (*qv*)).

onça: puma (*zauára, zauarahú* in the Guajajara language).

onça pintada: jaguar (*zauára-hú piním-piním* among the Guajajara).

P

pacu (pacú): a name given to many large characoids of the family Serrasalmidae (sometimes further assigned to the subfamily Serrasalminae) in South America. But never to the Piranhas. In the aquarium hobby the name is applied mainly to *Colossoma macropomum,* but rarely in Brazil, where this species is normally called the *tambaquí* (*qv*).

pacu-caranha: Piaractus mesopotamicus (family Serrasalmidae); also called *pirapitinga* along the Amazon, and *caranha* in the Araguaia-Tocantins. In many places around the world this species and its congener, *P. brachypomus,* are called simply *pacu* (*qv*).

pacu-piranha: Catoprion mento (family Serrasalmidae) (in Brazil).

palhoça: an indigenous house.

palmito: palm-heart (the heart of certain Brazilian palm trees, noted as a vegetable, today exported practically world-wide from the Amazon).

pará: river.

Pará: a former name for Belém; the Brazilian state in which Belém lies.

para: Halt, Stop. (Portuguese)

Paracanã: an indigenous tribe encountered during the construction of the Tucurui hydro-electric dam on the Rio Tocantins.

paraíba (para-iba - paraiwa): bad river; unnavigable river.

paraíba (parabiwa): irregular (inconsistent) wood.

paraibuna: a small, dark, unnavigable river.

paraná: a channel connecting lakes and/or rivers, whose flow changes direction or stops depending on the time of year. The word is Portuguese but derived from the indigenous Baniua language.

parroto: parrot, hyacinth ara (among the Guajarara the latter is called *azurú*). (Portuguese.)

pé: the/that, among the Kamayura/Kamayur on the Xingú in Mato Grosso, but in Portuguese it means foot.

peixe: fish. (Portuguese)

piná: a type of tall slender palm with a white central "heart", also called *palmito* (*qv*), typical of the Brazilian *mata atlântica,* the former Brazilian coastal primary rainforest.

piracema: a mass spawning migration of fishes.

iapuçá (japuçá - jupuçá - jauá - sauá): a type of monkey.
iba (iwa - iua - iva): bad; ugly; unusable.
iba (ubá): a tree; its wood.
IBAMA: the Instituto Brasilieiro do Meio Ambiente e dos Recursos Naturais Renováveis (Brazilian Institute for Environment and Natural Renewable Resources).
ibi: earth.
ibitinga (tinga): white earth.
ican: head, from the indigenous Apinagé language from the Tocantins region.
ig: see *i*.
igapó: blackwater floodplain, whose lakes and rivers do not normally have any overlap with whitewater inundation zones (*varzea, qv*).
igarapé: a navigable stream. (Derived from the Nheêgatu *yarapé (qv)* meaning "canoe way".)
igreja: church. (Portuguese)
iguaçu: a large body of water; a large lake; a large river.
inã: mother, from the indigenous Apinagé language from the Tocantins region. The derivation of the scientific genus name of the Amazon dolphin, *Ina geoffrensis*.
indaiá: a species of palm.
Ipira (pira): fish, in the language of the Kayabi/Kayabí/Kajabi in the Xingú region and the Rio dos Peixes (= river of fishes).
ira (iracema – irapuã): honey.
iracema (ira tembé - iratembé): honey lips.
irapuã (ira puã): honey stock.
iratembé: see *iracema*.
irupé: the giant Amazonian waterlily, the *vitória régia (Victoria amazonica)*.
ita (itaúna): rock, stone.
itajubá (ita ajubá): yellow rock/stone.
itu: waterfall or rapids, in the language of the Kayabi/Kayabí/Kajabi in the Xingú region and Rio dos Peixes (= river of fishes).

J

jaçanã: a type of tropical bird with very long toes to enable it to walk on floating vegetation; also known as lily-trotters.
jacaré: crocodile, caiman. (From the indigenous Apiacá language from the Tapajós region)
jacaúna: individual with a black chest.
jacu (yaku): a herbivorous gallinaceous bird.
jacuí (jacu): small.
jaguar (yawara): dog; fox.
jaguaracambé: dog with a white head (= a wild dog).
japira (yapira): honey.

jaú (jahu): a local name for the giant catfish *Zungaro jahu*.
Jumana (Ximana - Xumana): a tribe of the Aruaque group, living in the region of the Rios Japurá and Solimões (western Amazonia).
jumbeba (jurumbeba - ju-mbeb): a type of cactus.
jupuçá (iapuçá - japuçá): see *iapuçá*.
juru (ajeru - jeru - ajuru - papagaio): a hardwood tree with edible fruit (fruit-flesh).
jurubatiba: a place full of spiny plants.

K

kaapora (caapora - caipora): a person who lives in the forest (or on poor land).
kabu'ré: see *caboclo*.
kcal: abbreviation of kilogram-calorie or kilocalorie (= 1000 calories), in nutrition, a measure of the heat-producing potential of a food. On average, animal products have a higher calorific value than vegetable matter.
k.k. Naturalien-Cabinet: historically, a natural history museum and research institute, in imperial Austria and other German-speaking areas.

L

lago: lake; in Amazonia a *lago* is not just a normal lake, but a body of water that may be totally isolated some of the time, but normally it forms a unit with other waters (rivers, lakes, *igarapés (qv), paranás (qv), or furos (qv)*) at high water. (Portuguese)
lagoa: normally, but not necessarily, a very large lake. The literal meaning of the word is "bay", referring to (sea) coastal bays, but not in Amazonia, where it is cognate with *lago*. (Portuguese.)
lancha: boat, launch, motorboat. In Amazonia usually applied to a fishing boat or houseboat, for centuries the transport of *caboclos (qv)*. (Portuguese)
lauré (pauetê-nanbiquara): a red *ara* (macaw).
lei: giant snake, anaconda. (Among the Karajá)
Lipids: a generic term for fats, oils, waxes, and related products found in living tissues. They are insoluble in water.

M

macaco: monkey. (Portuguese)
macacão: very large monkey. (Portuguese)
macaíba: see *macaúba*.
macaúba (ma'ká i'ba - macaíba): the macaba tree.
macaxeira: see *manióc*. (Portuguese)
madeireiros: tree-fellers; timber dealers.
mãe: mother. See also *ina*. (Portuguese)

canoa: a canoe hewn out of a tree-trunk (= dug-out). One of the first indigenous words learnt by the Spanish *conquistadores*. Nowadays the normal means of transport for the *caboclos* of Amazonia.

canos: channels linking lakes *(lagos)* (smaller than other channels/*paranás* (*qv*)).

capim (caapii): narrow-leaved grass.

capivara: the capybara (*Hydrochaeris hydrochaeris*), a large rodent (the largest today).

capoeira: dead woodland. (Nowadays also the name of one of the best-known Brazilian dances – see *berimbau*.)

cará: is used today in Brazil for *Geophagus brasiliensis* (family Cichlidae), and also often used for other cichlids (see *acará*). The word originates from the Tupí-Guarani language, where it means root or tuber.

caranha (pacu-caranha): characins of the genus *Myleus*.

carapanã: biting midge; mosquito.

carapeba (acarapeba - acarapeva - acarapéua): *lingua geral* for the marine fish *Diapterus rhombeus* (family Gerreidae). *Carapeba-branca* is a name for the marine fish *Diapterus auratus* as well as for *Eugerres brasilianus* (also family Gerreidae).

cari: white man; race of white men.

cariboca: see *caboclo*.

carioca (kari'oka): the house of a white man; also a term for an inhabitant of Rio de Janeiro.

carrapato: a tick.

cátete: beautiful.

comprido: long. (Portuguese.)

comunidade: settlement, a collection of houses/huts; a small community. (Portuguese.)

curumim (kurumí): boy, young man.

D

dejue: fish, among the Karajá.

Deni: an indigenous tribe of the Arauke language group, barely 300 of whom today live along the *igarapés* in the valley of the Rio Cunhuã (Cuniuá), between the mouths of the Rios Xiruã and Pauini in the state of Amazonas.

departamento: a province or district (in most Latin American countries, but not in Brazil).

detritus (from Latin *deterere* = rub off): rubbish, residual matter. Depending on its origin detritus is classified as organic or inorganic. In aquatic biology, organic detritus is the (usually decomposing) residues of dead plants and animals, as well as faecal matter. Inorganic detritus is eroded rock. Allochthonous detritus enters a body of water from the surrounding area (eg from the inundation forest) while autochthonous detritus originates in a body of water. Detritivores, which may include fishes and invertebrates (eg worms, molluscs, crustaceans) are animals that feed on detritus.

disqueiro: a boat (a *lancha, qv*), used to collect discus; sometimes also a fisherman who collects mainly (or only) discus. See also *barco de pesca*. (It is an Amazonian-*caboclo*-word.)

F

farinha: manioc meal, the "daily bread" of a *caboclo*. (Portuguese)

feijão: black beans, the Brazilian national dish. (Portuguese)

flor: flower. (Portuguese)

flutuante: floating buildings (built on rafts) in Amazonia, which sell everything imaginable. The word is also used, but rarely, for floating dwellings. (Portuguese)

folha: leaf. (Portuguese)

fruto (fruta): fruit. (Portuguese)

fulvic acids: organic acids, formed (as are humic acids) via humification

FUNAI: Fundação Nacional do Indio (the Brazilian national indian authority).

furo: A temporary (water) connection through the jungle between – or to – lakes or rivers. (Portuguese)

G

gaivota: sea-gull. (Portuguese)

garimpeiro: gold prospector. (Portuguese)

goma: a food starch obtained from *tucupi* (*qv*) and used in the manufacture of tapioca, *inter alia*.

governador: governor. (Portuguese)

guaraná: an alcohol-free Brazilian drink (see p. 336).

Guarani: (1) an indigenous tribe of tropical South America. The great nation of the Tupí-Guarani once extended from northern Argentina through central South America to Amazonia. Today only fragments remain, except in Paraguay.
(2) the language group of the Tupi-Guarani, still one of the most widespread indigenous language groups in central South America.

guaraní: warrior; fighter.

H

há: fish, among the indigenous Camacãs; *huá* among the Camacãs in the Bahia region (spoken nasally).

I

i (ig): water; small; fine, thin, emaciated.

iaé (kamaiurá): the moon.

iandé: the constellation of Orion.

anama: large; thick; family; relatives; race; nation. (Also the name of a lake in the state of Amazonas, written Anamã.)
ananã: aromatic fruit; pineapple.
andira: lord of ill-omen.
andirá: bat. (Also a (discus)river name, the Rio Andirá.)
antã (atã): strong.
anta: tapir (which in turn originates from the indigenous Apiacá language, from the Tapajós region called *tapira*).
aondê: owl.
ara: with a few exceptions, generally applied to large parrots, but also to small ones such as budgerigars; to heights; and (rarely) to anything that flies (eg insects).
araçá: a tropical fruit from Brazil.
aracema: the parrot family (budgerigars, parrots, macaws); or another bird group; or a *piracema* (mating flight).
arara: large parrot (macaw); large bird.
araraúna: black ara (hyacinth ara).
ararê: friend of parrots.
area de segurança nacional: national security zone (usually border regions).
ati: small sea-gull.
atiati: large sea-gull.
auá (avá - abá): man; woman; person; indian.
avanheenga (awañene - abanheenga - abanheém): the language of humans as opposed to animals. (*Língua geral* of the Tupís and Guaranis)
avaré (awa'ré - abaré): friend; missionary; catechist.
avati (bati - auati): blonde people; maize.
awañene (abanheém): see *avanheenga*.
ayapuá: see *Aiapuá*.
ayuru (ajuru): a hardwood tree with edible fruits.

B

bacuri: "the river with something warm"; thin wood.
baiacu: warm-blooded animal; pufferfish (Pisces: Tetraodontidae).
baixo: low, below. (Portuguese)
banana: banana (Portuguese); in the indigenous Baniua language the word signifies *paraná (qv)*.
banheiro: bath, toilette. (Portuguese)
banho: in Brazil this can signify a (public or private) bath, but also lavatory or to take a bath. (Portuguese)
barco de linha: see *recreio*.
barco de pesca: a fishing boat (with a large (or several) compartments to store ice), sometimes used for collecting discus, but usually for food-fish.
beijú: a cake made from roasted manioc meal, rolled together.

beîu: bread; cake.
beraba: shine, gleam; radiate.
berimbau: a Brazilian musical instrument, like a bow with one playing edge – the main instrument of the *capoeira (qv)*. (Literally, "hill full of holes".)
biraquera: a place where fishes sleep or rest.
bucui: a river with fine sand.
burití: a species of palm (*Mauritia flexuosa*, see p. 518); its fruit.
buritizal: a group (or plantation) of *buriti (qv)* palms.

C

caá (kaá - caa-hó): woodland; leaf.
cabana: straw hut. (Portuguese)
cabanagem: the revolt, during the 1830s, of the *cabanos (qv)* in Pará.
cabanos: people living in *cabanas* (straw huts), hence the name; more specifically, a group of *cabanos* who in 1832 revolted against the government of Pará and in 1835 took control of the province. However, the revolution was of short duration, being suppressed by 1840, with great loss of life.
caboclo: having white ancestry; a mix of white and indian, the original terms for which, *cariboca (kariboka)* and *carijó*, are old names for indigenous tribes, as are *caburé (qv), tapuio (qv),* and *caboclo. Caboclo* was also formerly the name for the indigenous Guarani from the region between the Lagoa dos Patos (Rio Grande do Sul) and Cananéia (São Paulo). Nowadays, however, it is a general term for people living in Amazonia (mainly along the riverbanks), and/or for immigrant river-dwellers (= *caboclos riberinhos*).
caburé (kaburé - cafuzo - caipira): a stocky person, now also a alternative term to *caboclo. Caburé* (or *caburey*) is also the name for the ferruginous pygmy-owl *Glaucidium brasilianum*. And *kaburé-iwa* signifies owl nest in the tree *Myrocarpus frondosus* (Leguminosae: Papilionoideae). (*Caburé* is also the name of a town in the state of Maranhão.)
cachoeira: rapids, waterfall. (Portuguese)
cacira: a wasp (with a very painful sting).
cafuzo: see *caburé*.
caipira: see *caburé*.
caipora (caapora - kaa'pora): A type of landscape in the *sertão,* the semi-arid north-east of Brazil.
cajú: passionfruit; the nuts of the passionfruit.
calderada: a fish soup, a real speciality of Amazonia (Brazilian).
caminho: path, way. (Portuguese)
camu-camu: a little-known fruit with the highest known vitamin C content (and eaten by discus), nowadays grown commercially in the state of Acre instead of rubber.

Glossary

This glossary contains not only English, but also Portuguese and Brazilian terms (written at the end of eqch text, if Braziulian (Braz.) or Portuguese (Port.) word), plus words used among the *caboclos* and numerous indian words from the languages of various indigenous tribes, chiefly those of Amazonia. Words in brackets immediately following the glossary term relate to variations in pronunciation/spelling, or are other words with the same meaning. The names of indigenous tribes, their languages and/or language groups, are given in only a limited number of cases, as there are around 170 languages in Brazil alone (out of 210 tribes that still exist), but the majority of the words of unspecified provenance are from the Tupí-Guarani language group.

A

abá (avá - auá - ava - aba): man, people, person, be human, indians.

abaçaí: evil genius; persecutor; persecutor of the indians; an evil spirit that haunts indians and drives them mad.

abdomen: in arthropods, the posterior body.

abaetê: a good person; a trustworthy person; a respected person.

abaetetuba: village of good people. (Also the name of a town on the lower Amazon, see p. 224.)

abaité: bad people; impulsive people; trespassers.

abaré (abarê - abaruna): friend.

abati: maize; golden hair; blonde.

abuna (abaruna): priest with black robes.

abunã: an Amazonian dish made from turtle eggs. (Also the name of a Brazilian town and a Brazilian river in the state of Acre.)

açaí (yasaí): fruit that weeps; fruit that emits fluid; small brownish coconuts that grow in clusters on the *açaizeiro* tree (a palm with a slender trunk and thin leaves, and which also produces palm hearts).

acará (acaraú): heron; white bird. Nowadays the Tupí word is widely used for cichlids, eg *acará-bandeira* = angelfish; *acará-disco* = discus; and just *acará* = cichlid; but the indigenous word means nothing of the sort. (There are also towns, lakes and rivers called Acará in Brazil.)

acará-açú: widespread plants in Amazonia – often in discus habitats (applies normally to species of the genus *Licania*).

acarapeba: see *carapeba*.
acarapéua: see *carapeba*.
acarapeva: see *carapeba*.

acre: derived from *áquiri*, the name for the feathered costumes of the Mundukurú. (Nowadays also the name of a Brazilian state and of a Brazilian river.)

açu: large; considerable; long; wide.

água: water. (Portuguese)

aguapé: floating plants; also used for waterlilies, including *Nymphaea* spp. and the *vitória-régia, Victoria amazonica.*

aiapuá: a Nheêngatu (indian) word for manioc – the original form. (Also the name of a lake in the state of Amazonas, see p. 403.)

aimara: a tree, also called *araçá-do-brejo* by Brazilians.

aimará: a tunic made of cotton and feathers, worn chiefly by the Guarani (indians).

aipím: non-toxic manioc; dry root.

ajajá (aiaiá - ayayá): a spoonbill.

ajú: back-flow, drop (of water) – every year, around 1-2 months after the beginning of the high-water period, the water level in the numerous branches of the Amazon suddenly drops by a metre (or more), resulting in a reduction in oxygen levels and the mass die-off of the aquatic fauna, with millions of fishes dying.

ajuru (ayu'ru - ajeru - jeru - juru): a hardwood tree with edible fruits (fruit-flesh).

aldeia: village (usually applied to indian villages). (Portuguese)

alimentos: edibles; the equivalent in the indigenous Apiacá language is *matê* (*qv*). (Portuguese)

ama: mother, among the Kamayura/Kamayurá; *ami* among some other tribes.

amán: rain, among the Kamayura/Kamayurá on the Xingú in Mato Grosso.

amana (amanda): rain, among the Tupí.

amanaci (amanacy): mother of the rains.

amanaiara: wife of the rains, lord of the rains.

amanajé: messenger.

amapá (ama'pá): a tree of the family Apocinaceae *(Parahancornia amapa)*; its bark is bitter and its sap is used for medicinal purposes, eg the treatment of asthma, bronchitis, and inflammation of the lungs, as well as for healing open wounds without leaving scars.

amo: more, among the Kamayura/Kamayurá on the Xingú.

Some variants collected at the end of 2005 and in the begin of 2006: 1-2. *S. aequifasciatus* from the Putumayo region. One can see the typical fine red spots (2), or almost none (1) – except for those always present in the anal fins. Also typical for Putumayo green: no red eyes. 3-5. Three blues from the Canumã region, sold as "Içá-Madeira". Although they are not from the Rio Içá, they still have a very interesting pattern colour (which I found similar in the Alenquer region – see Chapter 4). 6-7. Greens from the Nanay – the typical Tefé variant (introduced from there). 8. A blue variant typical of the Terra Santa region. 9. This one is sold as "Royal-blue" and was collected in the Canumã-area. Also the golden (10) and yellow (11) *S. haraldi*. 12. And a typical Heckel-Discus from the Unini, an alpha animal wth a bright blue head pattern.

At the end a few discus photos of recent collections, on the left, this, and the following two pages: Above a typical Rio Mineruá biotope with greens I collected there. They live in harmony with this beautiful *Pterophyllum scalare* variant. The *S. aequifasciatus* from the Mineruá region have a beautiful orange base colour. I showed them for the first time at the Aqua-Fisch-Exhibition in Friedrichshafen, Germany, March 2nd to 5th, 2006. I displayed 9 authentic biotopes – as fishes live in nature – vistied by 26,000 peole during the four days.

In addition I would like to mention the following: discus won't normally enter a "normal" net of the sort with which angelfishes and many other cichlids can be captured. And discus can never be obtained with plant toxins (they are too cunning and too quick). This has been demonstrated by a number of the numerous research expeditions. For example the INPA ichthyological team, which collected a considerable amount of material during three expeditions in the 1980s and at the beginning of the 1990s and thereby discovered various new species, never had a single *Symphysodon* in the net. They fished almost exclusively with the *timbó* poison mentioned on several occasions. It is even the case that INPA staff performed field studies on the Rio Negro over a 15-year period and in the process catalogued 457 different fish species – but not one discus. I have been given the opportunity to look at this unpublished mammoth work, and *Symphysodon discus*, which is has its stronghold in the Rio Negro and a number of tributaries, and occurs there in its millions. On the basis of INPA collections this species doesn't occur in the Rio Negro. Even rotenone, which ichthyologists worldwide have (unfortunately) been using for decades to perform censuses of underwater faunas, has to date failed to produce any discus. Not one scientist has managed it.

Discus are maddeningly difficult to catch. I can tell a tale or two after my around 40 years of practical experience. I find it particularly amusing when I hear or read about the "exciting expeditions" by so-called discoverers or collectors of discus. It is truly unbelievable what appears in the media worldwide. For example a world-famous breeder of my acquaintance (he really does breed) made a trip to Manaus and came back with the story of how he and his companion sorted through more than 60,000 discus to find only 400 good specimens. Although for as long as discus have been collected (in real commercially numbers only since the mid 1960s), with a single exception *(p. 421)* there have hardly ever been more than 30,000 caught in a season (August to January/February). And then only by hundreds of natives, trained collectors, who travelled for months on their lanchas to catch at most 4,000 discus. So there have never been 30,000 in Manaus at one time. I have been there on several occasions each year and I have worked with fishes and discus all my life, so no-one can pull the wool over my eyes. But there are people who believe it.

Such stories repeatedly confirm what my grandfather, Adolf Kiel, the fish and aquatic-plant breeder, told me: "My boy, don't believe what you read in the papers, at most 10% of it is true..." (And there was no Internet in those days.)

Now you might say, "Then surely that must also apply to this text!" And I leave it up to you to judge, dear reader. But the moral of the story is, don't believe the "discus collectors".

Catching discus is and always will be difficult, as they are highly intelligent creatures, underestimated by many people.

Furthermore, if tourists want to catch fishes in Brazil (likewise in Peru) then they must obtain an amateur licence from IBAMA. Such a *licença de amador* is obligatory for catching fishes and this applies all over the country. It can also be obtained via the Internet at www.ibama.gov.br/pescaamadora/licenca/. Only pensioners and children under 17 can fish without a license, which costs R$ 20 for amateurs fishing from the bank and R$ 60 for those fishing from a boat. Nowadays more than 3,000 foreigners annually visit Amazonia alone for the angling. The income from these permits goes to the PNDPA (Programa National de Desenvolvimento da Pesca Amadora), an institution that holds seminars on environmental education, supports and educates the fishermen, provides them with environmental conservation programmes, and endeavours to use the data they supply to contribute to the conservation of the Amazonian fish fauna.

In conclusion, a note on Goethe's critique of Alexander von Humboldt, which can in fact readily be seen as a compliment. It refers to "instruments" such as almost everyone has, prefers, or uses, and I subscribe to Goethe's viewpoint, which he set down in his *Faust* and which might be described as "anti-instrumentalism":

Ihr Instrumente freilich spottet mein,
Mit Rad und Kämmen, Walz und Bügel.
Ich stand am Thor, ihr solltet Schlüssel sein;
Zwar euer Bart ist kraus, doch hebt ihr nicht die Riegel.
Geheimnisvoll am lichten Tag,
Lässt sich Natur des Schleiers nicht berauben,
Und was sie deinem Geist nicht offenbaren mag,
Das zwingst du ihr nicht ab mit Hebeln und mit Schrauben.

Alexander von Humboldt was undyingly enamoured of his instruments. My instrument is and always will be a fish net ...

On the final pages of volume I, a few more photos of people who have helped me with collecting, and without whom it would have been virtually impossible to collect so many data and fishes. Unfortunately I can show only a few, with it being the natives (indigenous peoples and *caboclos*) who helped me the most: 1. A *caboclo* family on the Lago Batata (Trombetas). 2. Natasha, who has assisted me on numerous research trips, including with collecting. 3-6. Members of the Wai-Wai on the Nhamundá and Trombetas. 7. *Praticos* who helped me reach the remotest regions. 8. Indians who helped during night-fishing and prepared meals. 8. Women who stood by their men, even when mosquitoes were eating them alive. 10-13. A huge variety of indigenous tribes who are at odds with "civilisation" and FUNAI, but stood by my side. 14. And last but not least, it is always a pleasure to meet the always happy children of the *caboclos* in the depths of the jungle, here *acará-disco* boy...

Nowadays discus are sorted on the spot after being collected, and placed in so-called *viveiros* for transportation to Manaus (or Belém): 1. Here Manuel is sorting Blues, Browns, Nhamundá Roses, and fully striped (alpha) specimens by the Lago Nhamundá. 2. The *viveiros* are moored in a bay of the lake. 3-4. The different variants are placed in different *viveiros* (3) and are caught out again for onward transportation (4). 5. Unfortunately it repeatedly happens that discus die in the *viveiros* if the water in them isn't renewed (by through-flow), or through disease. 6. And the water in the transporting containers seethes if the discus are put in carelessly (they dash wildly to and fro). 7. A glimpse into the *recreio* with my discus – three days to Manaus...

The all-night work, cold, wet, with plagues of mosquitoes and other biting flies (although there are none of the latter in Heckel discus regions), certainly make life difficult. Even for a *caboclo*. In addition, they usually have no luck at full moon as the discus (and many other fishes) don't come as close to the water's surface. They can thus spend all night in the wilderness for nothing.

New collecting methods are being devised. And lo and behold, in a number of places they are even succeeding. For example at the Lago Cuipeuá: here, as everywhere, discus are found in groups in their normal habitats, but during the dry period (at extreme low water) among the washed-up grass. (I have otherwise seen this only twice in the Urucu and Trombetas.) For a month (or at most two) the local *caboclos* exploit these "discus assemblies". By day all the men encircle the area of grass *(shown on p. 631)*. The discus have no chance and often hundreds end up in the net. The fishermen then select the desired colour variants. In the Cuipeuá this means reds and browns with a prominent central band (which are extremely rare) – as well as individuals with pattern colour, which are nowadays in demand. The rest are then released again.

In the Nhamundá region the seasoned discus collector Manuel Torres has recently devised a completely new method of collecting at night. The reason is because of the need for selectivity, as the customers now want only striped or extremely colourful discus that can compete with the highly coloured forms cultivated worldwide, which are today also available for very little money. It is, of course, very time-consuming to be selective during normal night-fishing. Especially when the collecting is being done by *caboclos* who have very little idea or awareness of customer requirements – as far as they are concerned a fish is a fish. The result is thousands of discus captured, all of which must be paid for, and this may be only matter of a dollar or two, but still a waste of money when at most 10 or 20 individuals conform to the customer's wishes. (This is also the reason why the majority of discus collectors have given up – it no longer pays.)

At 7 o'clock in the evening (it is dark at 6), when there is no full moon, Manuel goes hunting underwater with mask and watertight lantern. He and his adopted son dive down to around 2-4 metres in the deep water near the Serra Guariba and, by the light of the lantern, literally pick out only the best "sleeping" discus from below in the deep water. He has become shrewd and brings to the surface only the specimens he is sure he can sell (and at a good price). The entire operation takes a lot of effort (and you have to be able to hold your breath a long time), but the method is very effective, as no specimens are injured and there are no longer such quantities as in the past. There are far fewer losses and injuries and hardly any individuals that suffer a miserable end as nobody wants them. I was interested when, during my last visit (at the end of 2004), Manuel told me that ridiculous numbers of discus had been caught here during the heavy collecting following my first finding this discus region (in 1996). And that now the discus have become cunning. They no longer come to the surface at night but remain 2-4 m down. It seems to me that the discus there have followed in the footsteps of the caimans: when I was a child, and even during the 1960s, if I went looking for caimans at night I could always get quite close and literally pick them out of the water. Nowadays that is almost impossible. They have long since swum away. There is a saying, "Nowadays the crocodiles in the Amazon have become so bright they can read and write". Perhaps this is also now the case with the Nhamundá discus...

In the Rio Jutaí region, by contrast, (and not only there) a quite different collecting method is employed. When the water level is extremely low (for just 1-2 months) they pile large heaps of branches all over the underwater sandbanks *(p. 631)* – down to 3 m deep. Then the branches are left lying there for several days. After a while discus – in this case the beautiful Jutaí greens – normally arrive, as they find not only shelter but food there. As discussed under *Nutrition*, detritus and micro-organisms accumulate there, usually washed in by the receding water or by the current. And that is food. After around a week a large seine is used to pull the lot ashore, including all the branches, onto a sandbank, with no wooded bank to get in the way or need to be cleared. Thus this type of collecting is also environmentally friendly. Fishing takes place only by day. And in this way discus (and a number of other fishes – see *Discus Communities*) are captured. As at almost all collecting sites, at most two to four individuals are put in a plastic box and taken to the *viveiros* in the evening.

This collecting method has been used for a number of years. It is (or was) employed in the Lago Nhamundá, the Lago Macuricanã, and the Lago Grande. And nowadays blue discus are sometimes caught this way in the Lago Jari.

Here a few different fishing methods: 1. The Tucano and Tuyuca tribes use the *cacuri* to catch fishes migrating upstream. This basket-trap is set by the bank; it has an opening for the fish to swim in but they can't get out again. But they let discus go again as they are a symbol of fertility in their culture. 2. These kind of traps were set by *caboclos* in the Tocantins region and they too catch discus with them. 3. Indians, including on the Rio Nhamundá, have evolved their fishing methods over the millennia. 4-5. *Caboclos*, by contrast, only recently. In regions such as the Jutaí, Purus (4), and Lago Nhamundá (5) they now swear by the branches that they pile in the deeper water, catching discus there around a week later. It is easier as they can haul the branches, complete with the discus, onto the open sandbank (5). But this can still injure the fishes (7). 6. In the Cupeiuá they encircle huge clumps of grass like this with a large net.

young leaves of the palm *Astrocaryum vulgare*) long before the Europeans. (Identical in form and manufacture with the nets used by primitive tribes in West Africa and those of the natives of New Guinea – as I have established on several occasions.) Those of the Tucano are called *weheku*. But they didn't catch any discus with them, as they could only use them for the night-catch, and the indians don't fish then. In addition to their numerous sorts of plant toxins *(p. 261)*, which they normally use only in the dry period (nowadays other people also use these poisons, even though it is illegal), they also have a very interesting (bait-wise) method of fishing that they use in the *igarapés* and *lagos*. They fence off part of the water (with bamboos or wooden stakes), and impale a termite nest on a long pole with leaves on it (or cut down a tree complete with nest). The pole is then sited in a fenced-off area firmly in deep water so that the termite nest is partially or totally submerged. Then they wait. As soon as the leaves move they know that fishes are digging around the nest to get at the much-enjoyed "white ants". And now the natives need only to shoot their arrows to be sure of hitting a fish. Perhaps this was how the indians caught the first Heckel discus for Natterer.

The collecting of live discus definitely first began in the 1920s (see Chapter 1). The first Europeans were dependent on the natives and later on the seine-nets they brought with them. Collecting was hard work, as standard seine-nets could rarely be used in the habitats (which weren't known at the time). That is also one reason why discus arrived relatively late on in America and Europe. For decades people collected everything they could and freighted it to the Old World and the USA. But never discus. Once the first specimens had (accidentally) turned up in the net discus were deliberately hunted, but always under misconceptions and with the wrong nets.

It wasn't until the end of the 1940s/early 1950s that Fred Cochu managed to send fishermen, under the leadership of Old Cesar, from Belém to the Tocantins and Xingú regions on a fairly regular basis to come back with discus. Fred continually had to bring new net material from the States, as the seines, which were up to 60 m long and 6-8 m deep, never lasted long. A complete section of riverbank could be encircled with such a huge net (I still use one). This was. as mentioned earlier, initially in the Rio Tapajós, then the Rio Tocantins and the lower Xingú. It wasn't until much later that it was discovered that discus often "make their homes" in *lagos*. But because the banks were overgrown and covered in trees, and the water full of branches and other vegetation that had fallen in, the whole stretch of shore had to be encircled and closed off. It then often took a whole day's work to clear the bank region. Only thus was it possible to drag the huge seine ashore *(p. 627)*. This procedure was not only tiresome, but also exhausting and a huge effort – to find perhaps only a few discus – or none at all - in the net.

Discus collecting continued in this way until the 1970s, and in some places persists even today.

When, in the mid 1960s, I went on my own first discus collecting trip with *caboclos (p. 626)* and saw how time-consuming and environmentally destructive *(p. 637)* this collecting method was, and how the specimens captured ended up scratched and otherwise injured, I began to explore at night. And I actually discovered – mainly just after midnight – that discus, like many other diurnally-active fishes, almost always take their rest alone (during the day they live almost exclusively in large groups), close to the surface in the tangle of wood beneath or near a tree-trunk in the water *(p. 629)* or between submerged branches. Hunting for and illuminating them with a powerful flashlight brought success. Blinded by the light, they could be lifted out of the water unharmed – as long as you approached through the tangle and didn't frighten them. Ever since then the majority of discus collectors have used my method. On the Rio Negro, where I started, the *Festival do Peixe Ornamental* has been held for the past 10 years in the town of Barcelos, with groups of discus people showing how it is done *(p. 386)*. And all the *piabeiros*, without exception, have fished in this way there ever since.

In other regions too this method is used today, for example in the Manacapuru and Nhamundá regions, in the Lago Jari region, and among the indians on the Rio Marimari and the *caboclos* on the Rio Abacaxís, as well as by itinerant *disqueiros*. Although rarely using a normal flashlight – they can't afford the expensive batteries. They usually have a lamp with a long cable attached to a vehicle battery. The charge lasts all night or even for the entire canoe trip. Nowadays it is possible to recharge in almost any settlement in Amazonia (via generators) and often even in a *lancha*.

But, as with everything, the discus collectors don't have it easy.

The night-fishing technique that I introduced is still nowadays the most widespread in Amazonia: 1. You go to the habitat on a pitch-dark (moonless) night, as here in the Rio Nhamundá in 1998. 2. The prow of the canoe is pushed right inshore close to the bank, where the Heckel discus are resting near the surface among the branches. 3-4. If you spot a discus, in this case a green in the Rio Tefé (4), then you must carefully lift it out with the hand-net while shining the lamp on the fish (3,7). 5. This is a so-called *rabetta*, an outboard motor for shallow water, ie for getting through to the *lagos*. 6. Clever collectors catching discus in black water fill linen bags with leaves from the habitat and place them in the discus holding containers.

Further collecting methods (and the results): 1. As mentioned in chapter 5, it is often necessary to walk a long way through the jungle to find discus habitats in lake regions during the dry season. 2. If night-fishing proves unsuccessful then they try it by day (here in Lago 26) with gill-nets, into which the fishes are driven. 3-4. The gill-nets are very long: 100-400 m (and 4-16 m deep). The net is rapidly scanned from the canoe for discus, to avoid their being injured. Lots of Pimelodids and Loricariids entered the net, even a *Leiarius marmoratus* (4), but no discus. 5-6. After returning from Lago 26 I tried it in a bay in the Itaparaná, but caught only a large *Hoplias malabaricus* that leapt – complete with the net – against my leg and sank its teeth in.

A series of photos of single daytime discus collecting: 1-2. 1985: A possible habitat was located in the upper Rio Tefé (1). The 60 metre long (and 8 m deep) net was spread around the habitat (2). 3-5. All the large dead tree-trunks and branches had to be hacked away (3) and hauled out onto the bank (4). Often a number remained in the depths and had to be removed (5) before after the day's hard work the net could be hauled to the bank (in the hope that there might be a discus in it). 6. The discus were kept in suspended nets near the collecting site. 7. The same activity in 1974 on the Rio Negro. 8. Searching for discus in the lower Rio Tocantins (1965).

Photos of discus collecting trips: 1-2. 1966, en route in Goias with empty (folded) fish boxes in the VW (1) and my brother Michael *(left in the photo)*. Every 50 km the dust clogged the oil filter (2). 3-4. In the beginning (1965-1966) I often slept on the ground in the habitat after fishing at night (3). Later on in hammocks or in the *lancha* (4) (here at a Rio Negro Heckel Discus habitat in 1976). In Manaus I "boarded" my fishes with Willy Schwartz (here in 1965) (5) before despatching them to Rio (6). In the 1980s we had already developed (with a German plastic manufacturer) these fish transportation containers (here in 1986, with my discus at the harbour in Manaus).
8. 1997, at the Eduardozinho (Manaus airport), with fishes.

Today the indians (all the little "civilised" and the "uncontacted" ones, at least) still fish with bow and arrow, as they did in the beginning *(top left)*. The immigrant *caboclos* in Amazonia have learnt this method of fishing from the natives and practise it just as often *(top right)*. When fishing by day, if you are lucky then after a lot of time and trouble you may find a number of discus in the net, as here on the lower Xingú. Here *(above & left)* the typical Browns of this Xingú region, along with a *Pterophylum leopoldi*.

COLLECTING

I have already in part discussed how discus are collected in this chapter, but I would like to give an overview of how it was in the beginning, how it developed, and how it has also changed to some degree over the course of time (something for which I am in part responsible).

But when discussing catching fish we mustn't forget the Australian Aborigines, as it is possible that they were the first *Homo sapiens* to master fishing and at the same time revere fishes (Bleher, 2002) as the aboriginals of Amazonia still do today. And in particular revere the discus as a symbol of fertility (Bleher & Linke, 1991a). The earliest evidence of fishing in Europe comes from the La Pilata cave in Spain, in the form of drawings by the Cro-Magnon people who lived between 40,000 and 20,000 BC. But what astonished me most during my researches into fishing was something I discovered in the Assyrian culture. But why? They too were fishermen, worshipped their fish god Dagon (half human, half fish), and were among the inventors of writing, with their first "alphabet" having a "fish" as its second sign (letter) (Guter *et al.*, 1999). But it wasn't that, but rather the fact that I came across a disc representing a fish *(centre)* that looks very much like the "King of Amazonia", but dating from the Naqada I dynasty (4000-3600 BC). How can this be explained? As from what we know today, the genus *Symphysodon* is quite recent and did not exist at the time of Gondwanaland, before the continents drifted apart, and so it could hardly have been found in the ancient Middle East. On the other hand the Assyrians were not only accomplished fishermen *(below left)*, but also keen divers and undoubtedly the inventors of the aqualung, something that (as far as I know) nobody else has realised hitherto, although the sculptures, dating from 865-860 BC, left behind in a temple in Nimurid (in the north-west palace) clearly prove this *(below right)*. In other words, perhaps the highly intelligent people of this extremely advanced civilisation had some kind of contact – or cultural exchange – with the Americas long before the time of Columbus, as some African tribes did (Bleher, 2001e) and thus heard about this Amazonian beauty...

But let us return to our Amazonian fish collectors and the distribution of discus today.

Initially, of course, the first Europeans had to improvise and relied on the aboriginals to collect fish for them. Not to mention the fact that the natives were better at fishing (and still are) than any traveller, explorer, or scientist. They grow up with fishing, and the river is their life. They caught (and sometimes still do) fish with bow and arrow – and often have different arrowheads for different fishes *(p. 484)*. They make basket- and other fish-traps such as the *cacuri (right)* and *caiá* (called *wairó* and *ewa* by the Tucano). And also mobile fish-traps such as the *matapi* or the larger *jegui* (*kasaga* and *haruga* among the Tucanos), conical traps that are placed in the current. Then there are portable traps such as the *imiró* of the Tucanos *(p. 33)*. A trap that is today manufactured worldwide, but out of wire, and which I used at Gulf Fish Farms in Florida, where I worked for two years. The indigenous peoples of Amazonia also knew a form of angling long before the white man came, in which they literally lifted out the fishes; for example using the *kasatuti*, a basket with a conical base, woven from bast, in which fruit was placed; this was then attached to a stick with bast and lowered into the water, then lifted out quickly as soon as the fish "bit", ie swam in. They also had hand-nets made of *tucúm* (a strong fiber obtained from the

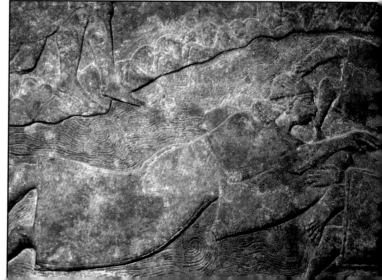

there was *S. geryi* in the Tocantins, long before the construction of the hydro-electric dam (the species was first described (by Jégu & Santos, 1988) after the completion of the dam and perhaps no longer survives there). In the Rio Negro drainage *S. gouldingi* has eaten discus fins and *S. humeralis* – one of the largest species – hunts discus in the lower Rio Xingú. A specimen of 43 cm SL has been caught on rod and line in the habitat, exceeding all the *S. rhombeus* (at 41.5 cm SL supposedly the largest of the piranhas) I have seen to date. Moreover I have come across the latter species in numerous discus regions, from the Xingú to the Igarapé Tacaná (Tabatinga region). In the upper Solimões drainage, where brown and blue discus occur, I have repeatedly found *Pygocentrus altus* Gill, 1870, frequently regarded as *P. nattereri* Kner, 1858 by recent authors, but *P. nattereri* was described from the Rio Guaporé, and is endemic to that river, and *P. altus* from where I have fished time and again, in the upper Solimões and Amazon region.

As well as those mentioned I have found additional piranha species in discus territory, but haven't always been able to identify them to species level. There is still much taxonomic confusion regarding the piranhas of the Amazon basin, despite heroic work by Jegú *et al.* in recent decades and Gery's work during four decades.

After the piranhas, it is undoubtedly the large predatory characins of the genus *Acestrorhynchus*, with their long, needle-sharp teeth, that represent the greatest danger to discus. On several occasions I have netted Heckel discus and Blues injured by them. In third place we have *Hydrolycus armatus* and *Rhaphiodon vulpinus (left-hand page)*, the largest of the predatory characins (a distinction they share with species of the genus *Salminus*, but the latter don't occur regularly in the discus habitat and as far as I know rarely visit it). But I doubt if any of these predators eats discus, although they regularly attack them. I have seen this.

I have often netted species of the predatory cichlid genus *Cichla* with discus and they are also almost always to be found near discus territory, but I have never yet seen them attack discus. And if they do, the discus are undoubtedly too quick for them

Other animals such as crocodiles, freshwater dolphins (below), and aquatic turtles visit discus habitats or swim past them, but I never saw any of them eat discus. Likewise water snakes, which I have repeatedly encountered. Fish otters may eat discus, but in my experience they prefer piranhas. For more than six months, when we were living among natives, a Brazilian fish otter, which I called Lontra, was my playmate and shared my hammock. Every day, at 12 noon on the dot, he would come waddling out of the Rio Guaporé (they have a unique, ungainly gait on land – not the environment for which they are designed) and lay a nice fat piranha at my feet for lunch. (It was then that I got my appetite for piranhas, and to the present day I still think them the most delicious food-fishes on Earth.) Unfortunately Lontra was poisoned because he bit an indian who was trying to rob some of my things, while I was away.

The Amazonian manatee eats only grass and a bird cannot catch a discus. The main threat to discus comes from Man, because of his relentless destruction of the environment.

Right-hand page: Rhaphiodon vulpinus (adult), dangerous for discus. *This page: Sorubimichthys planiceps* (left) do not eat discus, nor do freshwater dolphins such as *Ina geoffrensis* (above).

In this book I was able to show 37 discus communities and also most of their predators. In analysing what species actually live peacefully in company with discus, we are, of course, dealing with only a small part of the wealth of species in Amazonia, and predominantly with the larger peaceful cichlids. In first place we find species of the genera *Geophagus* and *Satanoperca* (17 times) followed by *Pterophyllum* (8) and *Mesonauta* (6). In addition *Uaru amphiacanthoides*, *Acarichthys heckelii*, and *Heros* species have been found sharing the habitat with discus frequently, were the later two are found, and Heckel discus in particular spend most of their time in the company of *Uaru*. *Acarichthys heckelii*, which is very widespread in almost the entire Amazon region, is often present. And *Heros* (of which there are more species than hitherto recognised) occur more frequently than I have caught them. But the latter occur often in shallow water. I have invariably found *Hoplarchus psittacus* only with Heckel discus in the rivers Negro, Urubu, and Nhamundá, as well as with Blues in the Trombetas. The two *Hypselecara* species are present now and then, but angelfishes frequently. All other species of the family Cichlidae are undoubtedly only "visitors" or occasional companions, for example *Biotodoma*, *Aequidens*, and *Acaronia* species, with the last of these, an omnivore and predator, rarely present. By contrast *Chaetobranchopsis* and *Chaetobranchus* are regularly found where they occur. *(Chaetobranchopsis orbicularis* is encountered with browns and blues at some places in the lower Amazon region; and I have found *Chaetobranchus semifasciatus*, whose distribution is restricted to the region between the Juruá and the Purus, south of the Solimões (all other records are suspect), only among blues there and in the region inhabited by greens.)

The other fish species recorded, perhaps with the exception of the larger members of the characin genus *Leporinus* (in particular species of the *Leporinus fasciatus* group, which I have recorded seven times) are undoubtedly only visitors. Also the groups of large *Crenicichla* species (with six records). The harmless cichlids of the genus *Astronotus* also occur in groups (large and small), but, just like species of the characin genera *Chalceus*, *Triportheus*, *Bryconops*, and *Boulengerella*, and species of the family Hemiodontidae, they are more visitors to the surface region in discus territory. In addition puffers of the genus *Colomesus* are also encountered mainly at the (surface) region.

Further – but rare – surface visitors are species of the characin family Gasteropelecidae, in particular the larger *Thoracocharax* and *Gasteropelecus* species. Rarely *Carnegiella* species, as they are almost exclusively inhabitants of shallow-water zones and rarely enter open (deep) water. (And of course, the first two are also primarily "at home" in the shallows.)

Likewise the piscivorous needlefishes of the genera *Potamorrhaphis* and *Pseudotylosurus* enter discus territory, but right at the surface. But they are harmless companions for discus (which don't fit into their mouths).

As already mentioned, I have now and then found loricariids in the discus habitat, mainly the genera *Liposarcus*, *Pterygophlichthys*, *Squaliforma*, and *Hypostomus*, rarely others. *Sturisoma*, *Ancistrus*, and perhaps *Farlowella* species only on the branches beneath which the discus group shelters. And species of the genera *Pimelodus*, *Calophysus* (there is undoubtedly more than one species of the latter in Amazonia), *Leptodoras*, and *Hassar*, inter alia. Large harmless catfishes that rummage around on the bottom or now and then through the "underworld" of the discus habitat. Rarely in groups.

Stingrays of the family Potamotrygonidae are often encountered beneath discus territory, right on the bottom, usually buried in the sand by day. They never interact with the discus. And stingrays are found in all discus distribution regions (and numerous others). Just like *Pterophyllum scalare*.

All the "harmless" fish species not listed here or mentioned on the preceding pages are truly rare visitors, which I won't discuss in further detail here. Likewise the shoals of large characins, such as *Curimata*, *Prochilodus*, *Brycon*, or *Psectrogaster* species, just passing by in their area.

But on to those that are dangerous for the Kings of Amazonia.

Naturally first place goes to the piranhas. I may have caught them only nine times (in the net or by rod and line) in the discus habitats discussed here, but there are piranhas in as good as every place where I have recorded discus. In the lower Amazon drainage region, with brown discus predominantly *Pristobrycon aureus*, *P. calmoni*, and *P. striolatus* as well as a *Pygopristis denticulatus*. As regards the genus *Serrasalmus* itself, I have repeatedly found *S. elongatus* Kner, 1858, among blue and brown discus in Purus tributaries, in the Manacapuru region, and elsewhere in central Amazonian discus habitats. Undoubtedly a different species to *S. pingke* Fernández-Yépez, 1951, described from the Orinoco system and regarded by many as a synonym. (And it is even possible that the so-called *S. elongatus* includes three species, as it again looks quite different in the Guaporé.) Then

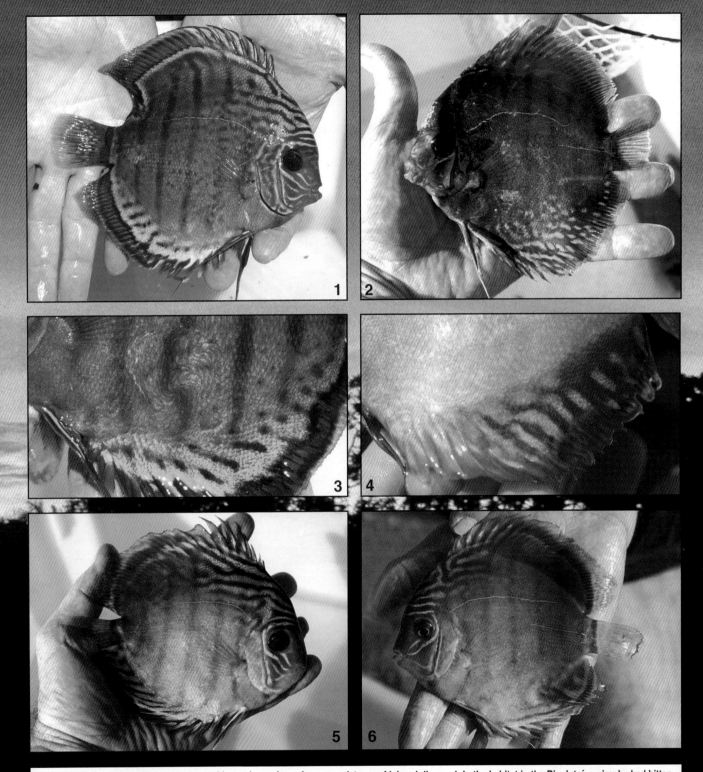

To conclude the photos of discus communities and enemies, a few more pictures of injured discus: 1. In the habitat in the Rio Jutaí: a piranha had bitten a whole piece out of the upper rear part of this green "leader", but it had grown back again. 2. At Terra Santa the head of this Blue, which was already debilitated at the end of the dry season and hadn't seen any food for a long time, had been damaged by a wimple piranha. 3. This Green has also lost scales to a predator, and they have grown back in the opposite direction to normal. 4. The anal fin of this blue from the Manacapuru region had been bitten by a piranha, but healed. 5. This green from the Mineruá region has had a piece bitten from the dorsal fin and tail region, but has healed. 6. A large piece had been bitten out of the anal of this brown from the Andirá, but it had healed up well.

Calderón

The most westerly (brown and blue) discus habitat:
1. Evening at a *lago* in the Igarapé Tacaná region. 2. Discus habitat with *Pistia stratiotes*. 3-4. Here there were very large numbers of *Mesonauta mirificus* ♂ (3) and ♀ (4). 5. A splendid, almost blue *Carnegiella strigata* beneath the floating plants. 6. This fish, also beneth *Pistia*, could be the true *Poecilocharax weitzmani* as it is from the type locality (p. 485) – or a *Crenuchus* species.

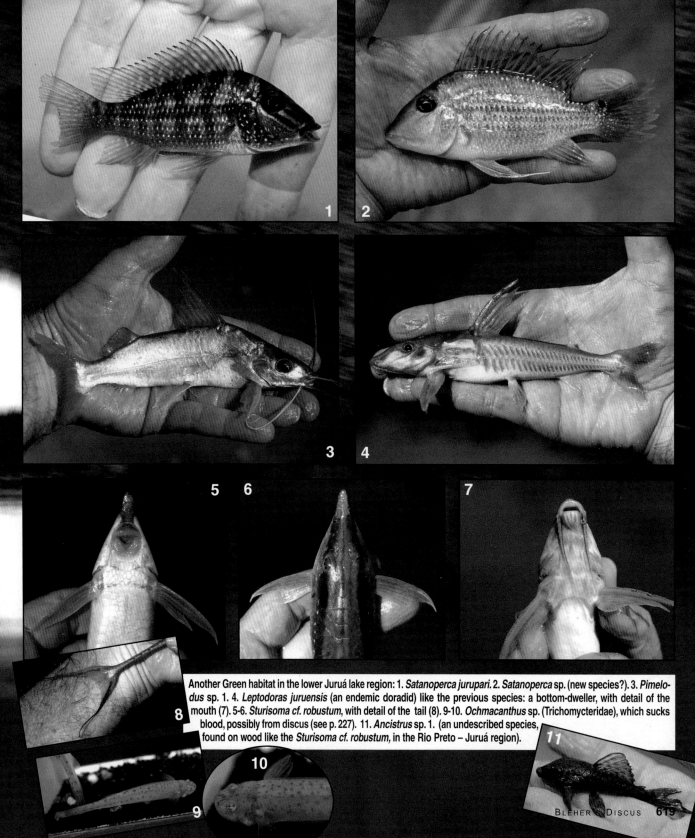

Another Green habitat in the lower Juruá lake region: 1. *Satanoperca jurupari*. 2. *Satanoperca* sp. (new species?). 3. *Pimelodus* sp. 1. 4. *Leptodoras juruensis* (an endemic doradid) like the previous species: a bottom-dweller, with detail of the mouth (7). 5-6. *Sturisoma cf. robustum*, with detail of the tail (8). 9-10. *Ochmacanthus* sp. (Trichomycteridae), which sucks blood, possibly from discus (see p. 227). 11. *Ancistrus* sp. 1. (an undescribed species, found on wood like the *Sturisoma cf. robustum*, in the Rio Preto – Juruá region).

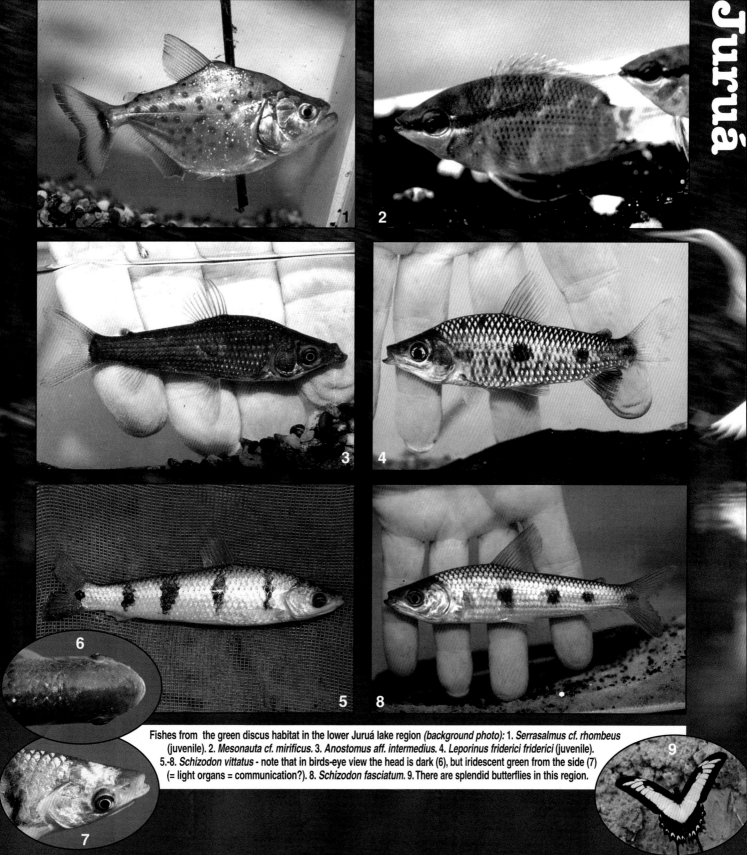

Fishes from the green discus habitat in the lower Juruá lake region *(background photo)*: 1. *Serrasalmus cf. rhombeus* (juvenile). 2. *Mesonauta cf. mirificus*. 3. *Anostomus aff. intermedius*. 4. *Leporinus friderici friderici* (juvenile). 5.-8. *Schizodon vittatus* - note that in birds-eye view the head is dark (6), but iridescent green from the side (7) (= light organs = communication?). 8. *Schizodon fasciatum*. 9. There are splendid butterflies in this region.

Lago Tefé

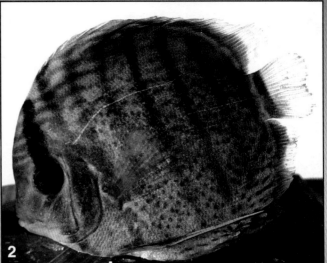

Left-hand page: Lago Jari *(background),* Blue habitat: 1. *Serrasalmus nigricans.* 2. *Serrasalmus rhombeus.* 3. *Uaru amphiacanthoides.* 4. *Colomesus asellus. This page:* Lago Tefé Green habitat (see also p. 85): 1. *Hypselecara temporalis.* 2. Green discus, debilitated "leader" (end of dry season, emaciated and attacked). 3. *Acestrorhynchus microlepis.* 4. *Acestrorhynchus falcatus.* 5. *Heros* cf. *severus.* 6. Boy with *pescada.* 7-8. Travelling in the Rio Tefé (green) habitat with Segderu (8).

Lago Jari

Lago Solitario

Left-hand page (and background): Lago 26 (Itaparaná (Purus) region, see also p. 419): 1. *Acestrorhynchus* sp. 2. *Boulengerella cuvieri.* 3. *Hypselecara temporalis.* 4. *Biotodoma cupido.* 5. *Heros* cf. *severus.* 6. *Satanoperca jurupari. This page:* Again blue discus habitat, in the Lago Solitario (Purus region): 1-2. *Osteoglossum bicirrhosum,* with detail of the two barbels used to find oxygen (1). 3. *Crenicichla* cf. *lugubris.* 4. *Uaru amphiacanthoides.* 5. *Aequidens* sp. 6. *Mesonauta* sp. 7. *Serrasalmus altus.* 8. *Astronotus* cf. *ocellatus.* 9. *Cichla monoculus.*

Lago 26

Lago Grande de Piracatuba

Fishes from the blue and brown discus habitat in the Lago Grande de Piracatuba (Purus region): 1. *Acaronia cf. vultuosa* (rare visitor). 2. *Acarichthys heckelii* (tail bitten by a piranha). 3. *Geophagus* sp. (new?). 4. *Mesonauta insignis*. 5. *Serrasalmus elongatus* (or a new species?). 6. *Leporinus* sp. (juvenile). 7. *Serrasalmus cf. eigenmanni* (juvenile), a "hit-and-run" fin-eater. 1, 2, 5, and 7 are "visitors".

There are other fish species in the blue discus habitat in the Lago Anamã, but they are almost entirely "visitors" (see also p. 398): 1. *Brachychalcinus orbicularis*. 2. *Thoracocharax securis*. 3. *Triportheus angulatus*. 4. *Chalceus aff. erythrurus*. 5. *Leporinus aff. friderici*. 6. *Hydrolycus wallacei* (juvenile). 7. *Pseudanos cf. trimaculatus*. Background: Lago Anamã.

Lago Anamã

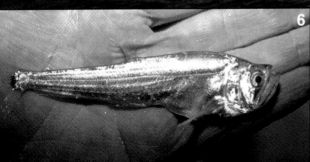

Manacapuru

I have already illustrated (pp. 352-352) the fishes from the blue and brown discus habitat in the Manacapuru region, but not these splendid *Pterophyllum scalare*, which I have often netted with discus and imported back in the 1960s. *Pt. scalare* with red backs also occur in the Tapajós, but those here have a particularly intense red. And of course, they don't show their glorious colours until they have been settled in the aquarium for a long time. My partner, Natasha, photographed them for me.

Rio Negro

More from the Rio Negro: 1. Heckel discus were once caught here at Manaus, but that is history. 2. *Peixe boi* (*Trichechus inunguis*) also sometimes turn up in the discus habitat in the Rio Negro (and elsewhere). These harmless manatees are officially protected. 3. Attempts are being made to breed them at INPA in Manaus. 4. They are totally vegetarian and eat only grass. 5. Their fat is sold in the market in Manaus, attractively packaged in boxes. 6. This large catfish, first discovered by Humboldt in the Rio Negro and in 1821 described as *Pimelodus zungaro* (= *Zungaro zungaro),* grows to 140 cm SL and sometimes turns up in discus territory, but it certainly doesn't eat discus, either here or elsewhere.

Rio Demini

Rio Negro region, Heckel discus habitat in the Demini: 1. Bank with yellow-flowered *ipé* tree. 2. *Hoplarchus psittacus* (juvenile). 3. *H. psittacus* (adult) that enjoyed being stroked! (blue colours as in the Nhamundá). 4-5. *Pterophyllum scalare* – sometimes called the "Rio Negro Altum". 6. *Cichla* sp. (a species from the Rio Negro that Schomburgk illustrated around 1840 (p. 381, Fig. 7) and called *C. flavo-maculata*). 7. *Dicrossus* sp. from the Demini. 8. *Serrasalmus* sp. "Demini".

Lago Bacururu

1. The northernmost distribution of the Heckel discus in in the Rio Negro region is the Lago Bacururu.
2. *Bryconops cf. giacopinii* are visitors here in the discus habitat, as well as the needlefish *Potamorrhaphis guianensis* (3) both near the surface.
4. *Hemiodus cf. semitaeniatus*. 5. *Hemiodus gracilis*.
6. Heckel discus habitat. 7. The incredible, hundreds of metres deep, layer of leaf litter everywhere in the Rio Negro (no wonder the latter is so black).

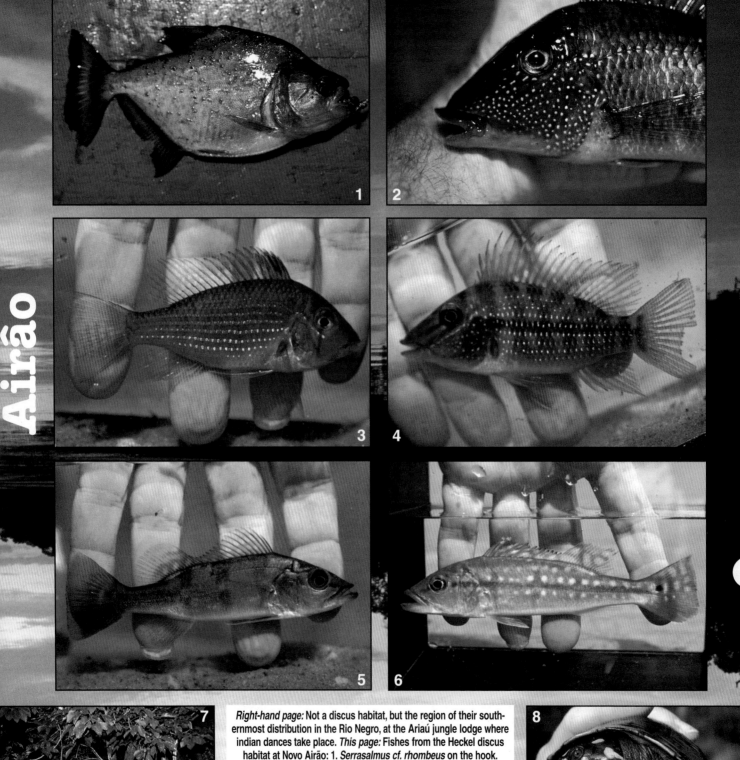

Right-hand page: Not a discus habitat, but the region of their southernmost distribution in the Rio Negro, at the Ariaú jungle lodge where indian dances take place. *This page:* Fishes from the Heckel discus habitat at Novo Airão: 1. *Serrasalmus* cf. *rhombeus* on the hook. 2. *Satanoperca* cf. *leucosticta*. 3. *Acarichthys heckelii*. 4. *Satanoperca* aff. *jurupari*. 5. *Cichla monoculus* juvenile. 6. *Cichla temensis* juvenile from the Rio Nhamundá habitat, for comparison. 7-8. Heckel discus habitat in the Novo Airão (7), where there are aquatic turtles (8).

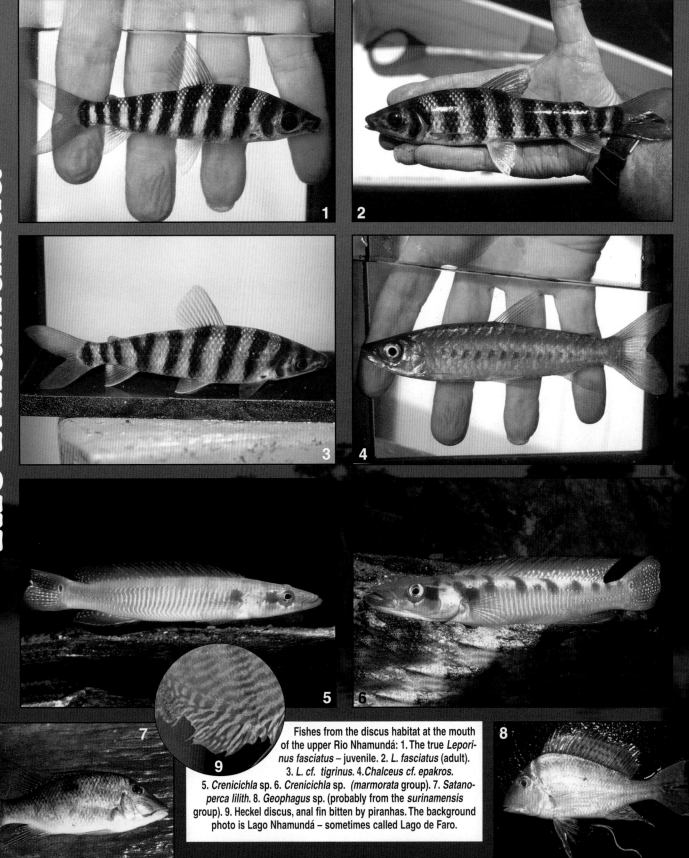

Fishes from the discus habitat at the mouth of the upper Rio Nhamundá: 1. The true *Leporinus fasciatus* – juvenile. 2. *L. fasciatus* (adult). 3. *L. cf. tigrinus*. 4. *Chalceus cf. epakros*. 5. *Crenicichla* sp. 6. *Crenicichla* sp. (*marmorata* group). 7. *Satanoperca lilith*. 8. *Geophagus* sp. (probably from the *surinamensis* group). 9. Heckel discus, anal fin bitten by piranhas. The background photo is Lago Nhamundá – sometimes called Lago de Faro.

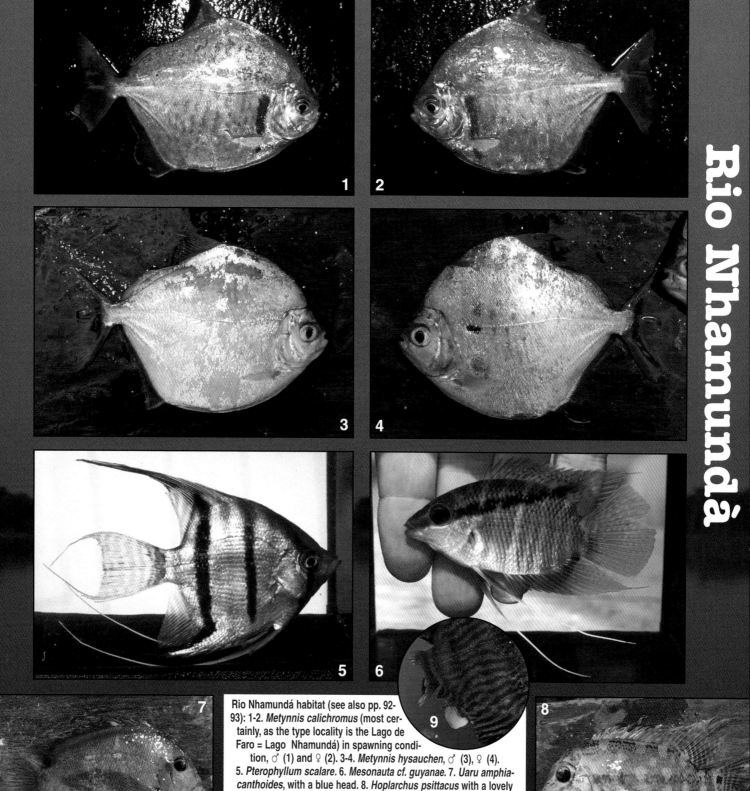

Rio Nhamundá habitat (see also pp. 92-93): 1-2. *Metynnis calichromus* (most certainly, as the type locality is the Lago de Faro = Lago Nhamundá) in spawning condition, ♂ (1) and ♀ (2). 3-4. *Metynnis hysauchen*, ♂ (3), ♀ (4). 5. *Pterophyllum scalare*. 6. *Mesonauta cf. guyanae*. 7. *Uaru amphiacanthoides*, with a blue head. 8. *Hoplarchus psittacus* with a lovely blue-red coloration, in the Nhamundá.

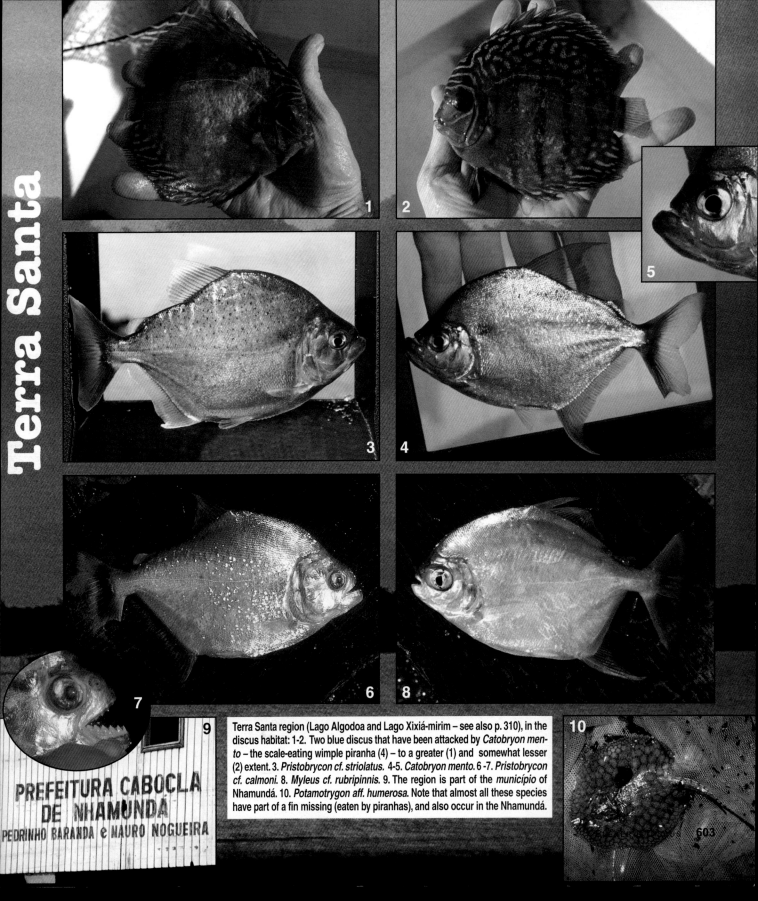

Terra Santa

Terra Santa region (Lago Algodoa and Lago Xixiá-mirim – see also p. 310), in the discus habitat: 1-2. Two blue discus that have been attacked by *Catobryon mento* – the scale-eating wimple piranha (4) – to a greater (1) and somewhat lesser (2) extent. 3. *Pristobrycon cf. striolatus.* 4-5. *Catobryon mento.* 6-7. *Pristobrycon cf. calmoni.* 8. *Myleus cf. rubripinnis.* 9. The region is part of the *município* of Nhamundá. 10. *Potamotrygon aff. humerosa.* Note that almost all these species have part of a fin missing (eaten by piranhas), and also occur in the Nhamundá.

Rio Tapajós

Fishes from discus habitats in the Tapajós region (see also pp. 265 and 269): 1. *Geophagus aff. surinamensis*. 2. *Pterophyllum scalare*. 3. *Potamotrygon* sp. (undoubtedly an undescribed species). 4. *Potamotrygon aff. signata*. 5. *P. cf. scobina*. *P. aff. signata* (caught at the end of the dry period – emaciated). 7. *Leporacanthicus* sp. (an as yet undescribed species). 8. Juvenile of an undescribed *Pseudacanthicus* species (L-273) which the *caboclos* call "Titanic". (The last two are rare in the discus habitat).

Rio Trombetas

Trombetas discus habitat (Lago Jacaré): 1. *Acestrorhynchus cf. falcatus*, a colourful enemy. 2. *Argonectes cf. longiceps*, a rare visitor that prefers flowing waters, as does 3. *Hemiodus huraulti*. 4. *Leporinus aff. granti* (upper) and *Leporinus* sp. (lower) (both new?). 5.-6. *Geophagus* spp. (possibly both *G. surinamensis*). 7. *Leporinus cf. affinis*. 8. *Heros* sp. – a black variant? (caught with bow and arrow). 9. *Liposarcus* sp. 10. *Colomesus asellus*.